HIERARCHIC ELECTRODYNAMICS AND FREE ELECTRON LASERS

CONCEPTS, CALCULATIONS, AND PRACTICAL APPLICATIONS

HIERARCHIC ELECTRODYNAMICS AND FREE ELECTRON LASERS

CONCEPTS, CALCULATIONS, AND PRACTICAL APPLICATIONS

VICTOR V. KULISH

CRC Press
Taylor & Francis Group
Boca Raton London New York

CRC Press is an imprint of the
Taylor & Francis Group, an **informa** business

CRC Press
Taylor & Francis Group
6000 Broken Sound Parkway NW, Suite 300
Boca Raton, FL 33487-2742

First issued in paperback 2017

ISBN: 978-1-138-11351-0 (pbk)
ISBN: 978-1-4398-4935-4 (hbk)

Visit the Taylor & Francis Web site at
http://www.taylorandfrancis.com

and the CRC Press Web site at
http://www.crcpress.com

To my Father

Vasyl' O. Kulish

(1911–1985)

Contents

Part II High-Current Free Electron Lasers

Preface

It is a well-known fact that development of some key military programs during the past fifty years systematically provoked critical situations in different parts of the applied physics. This happened mainly in cases when the appetite of the military exceeded the possibility of the existing, at that time, levels of science and technologies. The paradox of the situation is that every time, researchers, trying to solve arisen *applied* problems, made ultimately a next important step in *fundamental* sciences. As a result, we obtained systematically various new applied–fundamental achievements, which, naturally, were not foreseen initially by any military technical tasks.

We have a similar scenario in the case of the *hierarchic electrodynamics* and the *theory of hierarchic oscillations and waves.* It all began with ancient experiments with powerful plasma-beam instability, which were accomplished in 1950s through the 1960s (the famous Thermonuclear Plasma Fusion Program). Essentially multiharmonic wave processes were observed, which have been characterized by the generation of tens and even hundreds of plasma-wave harmonics with commensurable amplitudes (Chapters 13 and 14). At that time, however, the observed experimental physical picture did not get any satisfactory theoretical description. Especially difficult mathematical situations took place in cases when the amplitudes of higher harmonics exceeded the amplitudes of the lower ones.

The next share of difficulties has been generated by the (also famous) Star Wars program (the Strategic Defence Initiative—see Chapter 1 in detail) [1,2]. There the high-current free electron lasers (HFELs) [3–12] were one of the important inherent parts. The physics and the mathematics of some versions of the HFEL, as shown in Part II, turned out to be so complex that very often the traditional approaches were not able to give satisfactory results.

The discussed state of nonlinear theory of the mentioned class of objects held out for a long time. This situation took place until the appearance of more effective mathematical and physical tools [13,14]. They included the methods and ideology that were based, namely, on the principles of hierarchic electrodynamics (hierarchic methods). Very soon it was found also that the *fundamental* and especially *philosophical* aspects of this (initially purely applied) scientific direction can be even more interesting [14] than applied ones (Part I).

On the other hand, further development (after stopping the official stage of the Star Wars program) of the HFEL theory showed that a hidden potential of these kinds of technologies is actually much richer than what was considered. For instance, a possibility of attaining extremely high levels of amplification in the superheterodyne HFEL was found. Promising prospects were found also for the creation of a new class of superpower multiharmonic FELs for forming clusters of electromagnetic field, etc. It is important to note that the key point of these achievements is that all such bright possibilities and advantages have been realized practically owing only to the use of the hierarchic methods—*methods of the hierarchic electrodynamics.*

Thus, hierarchic electrodynamics, as a whole, and the theory of hierarchic oscillations and waves, in particular, have displayed their applied and fundamental effectiveness. For example, a number of results of the modern nonlinear theory of high-current FELs, which were obtained by the use of this theory, have opened very promising and sometimes fantastic prospects for practical applications. Summarizing the above, the author has decided that the time has arrived when the totality of the accumulated knowledge and new theoretical and applied ideas requires relevant generalization and comprehension. This book appears mainly as a result of the attempts of such generalization and comprehension.

Furthermore, let us say a few words about the main reasons of another nature. They have served historically also as a powerful push for writing this book. The first of them has purely pedagogical character.

It should be mentioned that the development of such highly effective calculation tools as the hierarchic methods [13,14] has given rise to a new, very specific problem. Summarizing his teaching experience, the author was astonished to discover that students (including PhD candidates) easily cope with the mathematical aspects of the hierarchic methods (basic concepts, calculation algorithms, etc.); whereas, many mature experts (even professors) have found serious difficulties in practical dealing with these methods. In the author's opinion, the key point of this paradoxical problem is in their (professors') former professional experience. The specific hierarchic philosophy, as is demonstrated in the book (Part I), is really so unusual and unexpected that its perception of value requires serious efforts for overcoming a number of traditional stereotypes. Students have another situation. Lacking great knowledge and experience (that is naturally), they were free from the problems related with overcoming the mentioned stereotypes. At the same time, their basic difficulties were concerned mainly with clarifying the physical meaning of the considered hierarchic oscillation-wave physical picture.

The reasons of the other type concern the fundamental and philosophical aspects of the discussed theory. Of course, material of such kind looks unnecessary in a book dedicated to the theory of FEL. Nevertheless, the author has decided to include it taking into consideration the two following important reasons.

1. As experience shows, this material assists in better understanding the key ideology of the hierarchic calculation methods as well as the general hierarchic theory of multiharmonic HFELs developed on such basis.
2. It is useful from a general education point of view because it enhances the reader's professional and fundamental level.

As was found, the theory of hierarchic oscillations and waves, being a very effective tool for a number of important practical calculations, proposes, at the same time a new, in principle, paradigm for modern natural sciences. First, it is because hierarchic oscillations and waves, as a universal physical phenomenon, can be regarded as a foundation of the surrounding world: all natural objects in the universe, as well as the universe as a whole, possess hierarchic oscillation wave nature (Chapters 2 and 8). Therein, this arrangement keeps strictly for each hierarchic level of the universe and all its sublevels. It is important also that it takes place independently of their physical nature (physical, biological, sociopolitical, economical, and so on).

But the most unexpected discoveries were found by trying to realize a true philosophy of the natural hierarchic oscillation-wave systems, such as the universe, the Earth, the solar system, atom, etc. It is a well-known fact that, as follows from the traditional general theory of the dynamic hierarchic systems [15,16], all such hierarchic objects must possess a cognitive capability. The following is, by far, not ordinary enigmatic fact of the same order: all natural dynamic hierarchic systems in reality are always open, whereas the hypothesis about the closed universe, as a dynamic hierarchic system, is the basis of the foundation of the modern physics. It is interesting that every time modern orthodox physicists came to realize the philosophical and physical sequences of these facts, they omitted discussion of inevitable eventual conclusions (see, for instance, references [15–25]). Moreover, the professional tradition of modern high physics considers any discussion of topics of such kind as indecent behavior. But why is it? The answer is very simple: in such situations, we come every time to the idea about existence in the world of a specific hierarchic physical-information-operation object, which can be called conditionally as the physical god (i.e., god of the universe as a complex hierarchic dynamic system—see relevant terminology in Chapter 2). Basically, it is not a true object: it is something that is out of the modern scientific classification. However, this great something can be successfully described in terms of modern physics and mathematics [14]. The main political and, at the same time, scientifically dramatic problem in this case is caused by that circumstance that the general properties of this physical god actually coincide with well-known properties of the Bible God.

In contrast to the existing tradition, the author does not avoid the discussion of this indecent problem. He understands basically that, in spite of the fact that he used a strictly scientific approach to the problem and gave a lot of convincing evidences of practical effectiveness of utilization of the developed new philosophy, this part of the book can be subjected to ostracism by many highly educated colleagues. Nevertheless, he realizes that the proposed new philosophy, which includes this concept as one of the important elements, might be interesting for many young people, who today form their scientific world outlook. The author believes that only the extended world outlook, which is based on the hierarchic-oscillation-wave representation of the surrounding in all its manifestations, will dominate science and everyday life of the decades of the twenty-first century. It should be stressed especially, however, that the problem of identity of both these concepts (physical god and Bible God, respectively) is not a subject of investigation of the proposed book.

Thus, the author's dream was to write a book on the theory of free electron lasers as hierarchic oscillation-wave electrodynamic systems, which would be accepted by students and their professors, as well as by the mentioned highly educated colleagues. Significant emphasis is on the specific philosophy and general ideology of the phenomenon of the oscillation and wave hierarchy, which were the basis of this theory. It is this book in which relevant hierarchic calculation technologies, physical, and applied results follow directly, clearly, and naturally from this incredible philosophy and ideology. However, the reader can appreciate the results of a practical realization of this author's dream.

Taking into account the above-mentioned peculiarities, the book's arrangement and the material balance are selected in accordance with the above-formulated author's dream, that is, it should be helpful equally for experienced experts as well as for students. Herewith, the author considers that the book (as a specific manual) should be more useful for students especially: it can assist in increasing their level of professional competitiveness in the future. However, the book principally is dedicated to everyone who expects to work in areas of modern relativistic physical electronics, radio physics, plasma electrodynamics, nonlinear optical and electronic engineering, accelerators, and various space (including special) technologies.

Writing the book, the author tried to convey to his readers the feeling of an immense astonishment by a new vision of the physical picture of the nature and oscillation-wave objects of modern electrodynamics (HFELs in our partial case). It was little enough to change the position for observation, and the habitual physical picture became unrecognizably changed. From the physical point of view, this little change in position for observation has been attained by introducing the principle of hierarchic resemblance into physics. As will be shown in the book, its acceptance as a new physical fundamental law has served as the main cause for the mentioned unrecognizable change.

The book consists of two parts. Part I contains eight chapters and it is dedicated to a short description of key points of the hierarchic electrodynamics. The second part has six chapters and its main subject of attention is the theory of HFELs as hierarchic electrodynamic systems. The main characteristic feature of the developed theory is that it is constructed on the basis of the methods of hierarchic electrodynamics (hierarchic methods).

References

1. Velikhov, E.P., Sagdeyev, R.Z., and Kokoshyna, A.A. 1986. *Space weapons: The security dilemma.* Moscow: Mir.
2. Duric, M., 2003. *The strategic defence initiative: US politicy and the Soviet Union*. Aldershot, UK: Ashgate Publishing.
3. Marshall, T.C. 1985. *Free electron laser.* New York: MacMillan.
4. Brau, C. 1990. *Free electron laser*. Boston: Academic Press.
5. Kulish, V.V. 2002. *Hierarchic methods. Undulative electrodynamic systems*. Vol. 2. Dordrecht: Kluwer Academic Publishers.
6. Freund, H.P., and Antonsen, T.M. 1996. *Principles of free electron lasers.* 2nd edition, Berlin: Springer.
7. Saldin, E.L., Scheidmiller, E.V., and Yurkov M.V. 2000. *The physics of free electron lasers.* Berlin: Springer.
8. Toshiyuki, S. 2004. *Classical relativistic electrodynamics: Theory of light emission and application to free electron lasers*. Berlin: Springer.
9. Schmuser, P., Ohlus, M., and Rossbach, J. 2008. *Ultraviolet and soft x-ray free electron lasers: Introduction to physical principles, experimental results, technological challenges.* Berlin: Springer.
10. Williams, R.E. 2004. *Naval electric weapons: The electromagnetic railgun and free electron laser.* London: Storm Media.
11. Allgaier, G.G. 2004. *Shipboard employment of a free electron laser weapon system.* London: Storm Media.
12. *Multi beam director for naval free electron laser weapons*. London: Storm Media, 2004.
13. Kulish, V.V. 1998. *Methods of averaging in non-linear problems of relativistic electrodynamics*. Atlanta: World Federation Publishers.
14. Kulish, V.V. 2002. *Hierarchic methods. Hierarchy and hierarchic asymptotic methods in electrodynamics*. Vol. 1. Dordrecht: Kluwer Academic Publishers.
15. Nicolis, J.S. 1986. *Dynamics of hierarchical systems. An evolutionary approach.* Berlin: Springer-Verlag.
16. Haken, H. 1983. *Advanced synergetic. Instability hierarchies of self-organizing systems and devices.* Berlin: Springer-Verlag.
17. Kaivarainen, A. 1997. *Hierarchic concept of matter and field.* Alaska: Earthpulse Press.
18. Ahl, V., and Allen, T.F.H. 1996. *Hierarchic theory*. New York: Columbia University Press.
19. Babb, L.A. 1975. *The divine hierarchy*. New York: Columbia University Press.
20. Pattee, H. 1976. *Hierarchic theory: The challenge of complex systems.* New York: George Braziler.
21. Boehm, C. 2001. *Hierarchy in the forest: The evolution of egalitarian behavior*. Cambridge, MA: Harvard University Press.
22. Graeber, D. 2007. *Possibilities: Essay on hierarchy, rebellion, and desire*. Oakland, CA: AK Press.
23. Harding, D.E. 1979. *The hierarchy of heaven and earth: A new diagram of man in the universe.* Gainesville, FL: University Press of Florida.
24. Bailey, A.A. 1983. *The externalization of the hierarchy*. New York: Lucis Pub.
25. Kaivarainen, A. 2008. *The hierarchic theory of liquids and solids: Computerized application for ice, water and biosystems*. Hauppauge, NY: Nova Science Publishers, Inc.

Acknowledgments

The author acknowledges the effort of all his colleagues, friends, followers, and present and former students who helped to make this book an accomplished reality.

"I have many things yet to say to YOU, but YOU are not able to bear them at present."

John 16:12

Part I

Hierarchic Electrodynamics

Key Concepts, Ideas, and Investigation Methods

As mentioned in the Preface, Part I consists of eight chapters. Chapter 1 has a purely ideological character and plays the role of a peculiar tuning fork for the book as a whole. Chapters 2, 3, and 4 are written as specific introductions into hierarchic electrodynamics. Chapters 5, 6, and 7 are dedicated to the general theory of hierarchic *oscillation–wave* systems, including free electron lasers (FELs). Chapter 8 occupies a special position in this book, considering some general and, at the same time, extraordinary philosophical problems of hierarchic theory.

Taking Part I as a whole, one can see that Chapters 5 and 6 are the most important. In Chapter 5 particularly, a general approach to the solution of the hierarchic oscillation–wave problems is described. The reader will also find a basic general approach to the analysis of electrodynamic *hierarchic* systems like FEL. The simplest two-level, nonresonant and resonant, scalar oscillation hierarchic systems with *total* derivatives, which can be treated as a general single-particle theory of the FEL-like electronic systems, are described in Chapter 6. A relevant generalization is also performed there for the multilevel hierarchic vector oscillation systems. Another key point of Chapter 6 is the description of the *method of averaged characteristic*, which could be treated as the most general hierarchic-wave version of the theory of electronic systems with long-time interaction. The method is used for the asymptotic integration of systems of differential equations with *partial* derivatives and nonlinear multifrequency and multiharmonic parts. Practical use of this method, historically, has allowed the solution of many difficult problems. Among such problems, we find the relevant multiharmonic, self-consistent, nonlinear models of the *plasma-beam* and *two-stream instabilities*, the parametrical and superheterodyne (including cluster) FEL, etc. (Part II).

Chapter 7 has dual objectives. The first is obviously illustrational: a discussion of various mathematical and physical peculiarities of the practical application of the hierarchic-oscillation theory developed in the preceding chapters and, generally, the hierarchic methods, in particular. The second objective is to demonstrate that the presented material could be treated as a *single-particle* FEL theory. The phenomena of the *cyclotron* and *parametric* single-particle resonances of different types (that play a key role in the FEL) are the main subjects of attention in this chapter. The three most characteristic types of these resonances are analyzed: *main*, *fractional*, and *coupled* (bounded) resonances.

Reviewing this abstract of Part I, one could think that it has—for a physical/mechanical book—traditional ideology and content. It may appear that the accomplished generalization of the new accumulated materials on hierarchic theory, methods, and their practical applications makes a serious argument for such an opinion. However, this is only partly true. Chapters 2 and 8 introduce a material of an essentially different type as a basis for serious discussion. We talk there about such topics as the concept of a physical god, the treatment of the universe (in all its appearances) as a gigantic information–operation hierarchic oscillation–wave system, a modern interpretation of the concept of the tree of life as an ancient forerunner of the most advanced modern physical theories, and so on. As mentioned in the Preface, the introduction of this kind of material into a professional scientific monograph could be useful in this case.

In Part I as a whole, it is necessary to note the fact that the new hierarchic mathematical calculation algorithms given here have been elaborated on by using the new physical hierarchic paradigm [1–3]. It is a widely accepted opinion that the basis of the modern physical theory of hierarchic dynamic systems is relatively new—it has been formed only within the past thirty years (synergetic, self-organization, etc.) [4–14]. Another point of this opinion is that this theory mainly has a philosophical cognitive significance. However, it is a little-known fact that many *mathematical* procedures characterized by evidently expressed *hierarchic* nature were applied in practice, at least in the *second decade of the twentieth century* [2,3,15–23]. Herewith, it is important to note especially that the specific philosophy, terminology, and concepts that are typical for the modern physical hierarchic theory [4–14] were not used at that time. Therefore, this fact long went unnoticed by the orthodox hierarchic experts.

In this regard, it is also important to note that researchers—van der Pol, Krylov, Bogolyubov, Leontovich [15–23], and others (Chapter 5)—were not interested in the *philosophical*–cognitive aspects. Their main interest has been the realization of the hierarchic-like (in accordance with the modern terminology), highly effective *calculation algorithms* for practical purposes (in electrical engineering, mechanics, radiophysics, etc.). Here the van der Pol method may be mentioned as well as the set of the averaging methods (Chapters 5 and 6). From the modern (i.e., new) point of view, these methods may be regarded as the simplest two-level versions of the hierarchic asymptotic analytical–numerical methods [1–3]. It can be substantiated that the specific basic transformations used there have an explicitly expressed hierarchic physical nature. We can say by this that these methods reflect, essentially, another physical approach to the treatment of the hierarchic dynamic systems, in this case, the oscillation–wave hierarchic systems. And, it is important to emphasize that this approach does not coincide with the traditionally accepted version [4–14].

The author is an advocate of this "old–new" alternative hierarchic approach, which serves as a basis for the main concepts of hierarchic electrodynamics. Because of this distinguishing conceptual basis, the new version of the hierarchic theory differs essentially from the well-known "traditional" one. This concerns the form of the representation as well as the inner content of both versions.

The key point of the new version is a specific set of the so-called hierarchic principles (Chapter 2). They provide the previously mentioned peculiarities for the arrangement of the theory as a whole. The most important among them is the principle of the hierarchic resemblance. It should be noted that never before was this principle formulated as the fundamental law of physics. In spite of that, it is well known in science as the holographic principle. This principle, for instance, has been used for interpretation of some unique physical objects, such as extraordinary physical theories of Stephen Hawking [24] and the no-less-extraordinary ancient philosophical doctrines (Chapter 8) [25–31].

The principle of the hierarchic resemblance could be formulated in the following form: each hierarchic level, in its general properties and arrangement, is similar to the system as a whole. This, in turn, leads to the conclusion that different reference frames should be used to describe a considered process at different hierarchic levels. Therefore, the hierarchic transactions between the levels should be organized in such a manner that each dynamic equation, written in the proper (for this level) variables, has the same general mathematical structure as the initial dynamic equation and the equations at all other hierarchic levels.

On the other hand, let us remember that the same reference frame is used for each level of the hierarchic system in traditional cases [4–14]. Moreover, the initial set of equations has a different form at different hierarchic levels. However, in fairness, it is assumed that different hierarchic levels could be characterized by their characteristic hierarchic parameters.

It should be mentioned that further application of the proposed new hierarchic physical/mathematical approach generated a number of nontrivial sequences. In particular, this stimulated a hierarchic grasping by the mind of some well-known electrodynamic problems [for instance, the physics of high-current FELs (HFELs); see Part II]. The results of this thought work turned out to be really unexpected. We were surprised that the hierarchic schemes of wave electronic devices like FEL and the universe and social and biological systems possess the same formal hierarchic description in principle [3]. Unfortunately, the fundamental reasons for this similarity are not yet understood, although some hypotheses for their explanations exist and are discussed in Chapters 2 and 8. In any case, this indicates that the proposed new hierarchic ideology possesses a more universal fundamental nature than it seems at first sight.

The key words of the book generally, and Part I particularly, are hierarchic electrodynamics and hierarchic oscillations and waves. The applied meaning of these words will be discussed in detail in Part II using various theoretical models of the HFELs.

References

1. Kulish, V.V. 2002. *Hierarchic methods. Undulative electrodynamic systems*, Vol. 2. Dordrecht/Boston/London: Kluwer Academic Publishers.
2. Kulish, V.V. 1998. *Methods of averaging in non-linear problems of relativistic electrodynamics*. Atlanta: World Federation Publishers.
3. Kulish, V.V. 2002. *Hierarchic methods. Hierarchy and hierarchic asymptotic methods in electrodynamics*, Vol. 1. Dordrecht/Boston/London: Kluwer Academic Publishers.
4. Nicolis, J.S. 1986. *Dynamics of hierarchical systems. An evolutionary approach*. Heidelberg, Germany: Springer-Verlag.
5. Haken, H. 1983. *Advanced synergetic. Instability hierarchies of self-organizing systems and devices*. Berlin-Heidelberg-New York-Tokyo: Springer-Verlag.
6. Kaivarainen, A. 1997. *Hierarchic concept of matter and field*. Anchorage, AL: Earthpulse Press.
7. Ahl, V., and Allen, T.F.H. 1996. *Hierarchic theory*. New York: Columbia University Press.

8. Babb, L.A. 1975. *The Divine hierarchy.* New York: Columbia University Press.
9. Pattee, H. 1976. *Hierachic theory: The challenge of complex systems.* New York: George Braziler.
10. Boehm, C. 2001. *Hierarchy in the forest: The evolution of egalitarian behavior.* Cambridge: Harvard University Press.
11. Graeber, D. 2007. *Possibilities: Essay on hierarchy, rebellion, and desire.* Oakland, CA: AK Press.
12. Harding, D.E. 1979. *The hierarchy of heaven and earth: A new diagram of man in the universe.* Gainesville, FL: University Press of Florida.
13. Bailey, A.A. 1983. *The externalization of the hierarchy.* New York: Lucis Pub.
14. Kaivarainen, A. 2008. *The hierarchic theory of liquids and solids: Computerized application for ice, water and biosystems.* New York: Vova Science Publishers, Inc.
15. van der Pol, B. 1935. *Nonlinear theory of electric oscillations.* Russian translation. Moscow: Swiazizdat.
16. Leontovich, M.A. 1944. To the problem about propagation of electromagnetic waves in the Earth atmosphere. *Izv. Akad. Nauk SSSR,* ser. Fiz., [Bull. Acad. Sci. USSR, Phys. Ser.], 8:6–20.
17. Moiseev, N.N. 1981. *Asymptotic methods of nonlinear mechanics.* Moscow: Nauka.
18. Krylov, N.M., and Bogolyubov, N.N. 1934. *Application of methods of nonlinear mechanics to the theory of stationary oscillations.* Kiev: Ukrainian Academy of Science Publishers.
19. Krylov, N.M., and Bogolyubov, N.N. 1947. *Introduction to nonlinear mechanics.* Princeton, NJ: Princeton University Press.
20. Grebennikov, E.A. 1986. *Averaging method in applied problems.* Moscow: Nauka.
21. Rabinovich, M.I. 1971. One asymptotic method in the theory of distributed system oscillations. *Dok. Akad. Nauk. USSR,* ser. Fiz. [*Sov. Phys.-Doklady*] 191:1248.
22. Sanders, J.A., Verhulst, F., Murdock, J. 2007. *Averaging methods in nonlinear dynamical systems.* Germany: Springer-Verlag.
23. Burd, V. 2007. *Method of averaging for differential equations on an infinite interval: Theory and applications.* Boca Raton, FL: Chapman & Hall/CRC.
24. Hawking, S. 2001. *The universe in a nutshell.* New York: Bantam.
25. Kapra, F. 1994. *Dao of physics.* St. Petersburg: ORIS.
26. Flood, G.D. 1996. *Introduction to Hinduism.* Cambridge, UK: Cambridge University Press.
27. Laitman, M. 1984. *Kabbalah.* Printed in Israel.
28. Fortune, D. 1991. *The mystical Qabalah.* New York: Alta Gaia Books.
29. Garbo, N. 1978. *Cabal.* New York-London: W. W. Norton & Company.
30. Ponce, C. 1995. *Kabbalah.* Wheaton, IL: Quest books and Adyar, Madras, India: The Thesophical Publishing House.
31. Parfitt, W. 1995. *The new living Qabalah.* Queensland: Element.

1

High-Current Free Electron Lasers as a Historical Relic of the Star Wars Epoch

For in the abundance of wisdom there is an abundance of vexation, so that he that increases knowledge increases pain.

Ecclesiastes 1:18

1.1 Star Wars Program from Today's Point of View

Almost thirty years have passed since March 23, 1983, when President Ronald Reagan announced the famous Star Wars program [the Strategic Defense Initiative (SDI)]. One does not have to be a historian to observe that our lives changed unrecognizably at that point. The old competitor of the United States for the world leadership—the Soviet Union—ceased to exist and the world became monopolar. The United States, which found itself in a position dreamed of by its financial and political elite, appears perplexed. Apparently, competing for the absolute leadership of the world was easier by far than carrying its heavy burden. The two wars with Iraq, the war in Afghanistan, the unprecedented activation of terrorism, and—as a consequence of all this—the enormous national debt are only a part of the cost of power over the life and death of the present civilization. Military science analysts are increasingly disturbed by a series of global trends that developed rapidly since the world did not simply change: it became much more dangerous and, even worse, unpredictable. The cost of overcoming the next share of difficulties is not yet known and might be much higher than that of the position of main world power.

The most striking changes that have developed in the time after President Reagan's speech were caused by the nature of the evolution of the global military, political, and financial/economic systems. We observed a significant geographic expansion of the centers of key decision-making and an increase in their influence on the present world system. Actually, it is the most pronounced effect of these changes. This is demonstrated by the appearance and the fortification of the new global financial/economic powers that appeared at the periphery of the so-called civilized world within this time. During the past decades, some of these powers acquired the form of a new potential military/political factor of influence. As long as these factors are very young and not yet very strong, the world is unconcerned. Nevertheless, they demonstrate exceptional ambition and endurance in the struggle for a position under the sun. This is most visible today during the global financial/economic and political crisis, when the more mature and potentially stronger global players are sluggish and indecisive. Our analysis reveals that the known leader among the new powers does not hide his principal strategic goal: to renew (at least) the lost bipolarity of the world, but with different participants.

There emerges a question: On account of which key military/political factors do they plan to accomplish this? The majority of the experts and politicians presumptuously answer: "This is unrealistic, since for them to reach the present-day technological level of the military of the United States, it would take at least 20 years and within this time we will develop further." These experts then start to enumerate the number of units and technical data of the present nuclear weapons, the means of their delivery and counteraction. One considers that just to name all the high-technology, trillion-dollar equipment should calm the so-called broad public. However, the most qualified among the experts know that, by far, this is not so.

First, in the present time, 20 or 30 years is too long a time period for predicting a reliable outcome. During a time of technological revolution, the situation changes very fast, sometimes rapidly. For example, consider computer technology; historically, there are even more instructive precedents. When a new principle weapon rapidly appeared, it practically leveled the fighting potential of the old one. It became obvious that, in such situations, nobody has to catch up with anybody, neither for the quantity nor for the quality of the old traditional weapons.

Let us think what would have happened to Europe after World War II if the Soviet Union had possessed nuclear weapons and the United States did not. In such a case, what would stop the well-mobilized and highly efficient 6-million-strong fighting force of the Soviet army in May 1945 from occupying not only Eastern Europe but Western Europe as well? It is not funny. At that time, the United States had not yet finished the war with Japan and the united military potential of Great Britain and France was incomparably weaker.

What conclusion can we derive from this hypothetical example? Quite simply, the sudden appearance of a new, in principle, type of weapon might drastically and fundamentally change the overall situation in the world. Moreover, it might change the direction of the course of history. This new weapon will render 6- or 10-million-strong armies obsolete, independent of how many aircraft or aircraft carriers those armies have. The very fact that the adversary has this new weapon, as it was with the nuclear weapon, would render the traditional weapons strategically ineffective, independent of its quantity or quality.

Let us continue the example using the nuclear weapon. Fortunately, the previously mentioned scenario did not occur, since the Soviet Union and the United States built their nuclear missile complexes practically simultaneously. As a result, a unique situation occurred, the so-called military strategic balance. The military theoreticians call such strategic situations mutually assured destruction, which still exists today in a weaker form. Nevertheless, let us imagine that suddenly one of the so-called partners of the balance comes out with a new wonder weapon that can reliably defend against a nuclear-rocket attack. Let us strain our imagination further and imagine that this wonder weapon was built rapidly by a third side—for example, the new military/political entities previously discussed. What will happen to the world? The answers to these rhetorical questions are so obvious that they do not require any comments.

It is obvious also that the reasoning ". . . for them to reach the present technical level of the military of the United States . . ." is based on a doubtful hypothesis that achieving some type of wonder weapon some place on the side is principally impossible today and for the next 40 to 50 years. But specifically why it is practically impossible? Maybe the scientific and technological experience obtained in the process of achieving the Star Wars program convinced us so. But, as is known, a circle of professionals proved the absolute opposite: the very revolutionary systems of the new weapon are, in principle, possible. For approximately 25 years, there existed a principal concept of how to do it. It was ascertained that it can be built by using the existing level of science and technology. Why, until today, was it not made? It is because the competitor (the Soviet Union) ceased to exist and worthy new competitors had not yet matured.

The name of this wonder weapon is the cluster electromagnetic system, the system that can generate super-powered clusters (clusters, not ordinary radiophysics pulses) of electromagnetic fields. The key characteristic of the clusters is their ability to penetrate, with practically no weakening of their power, into the critical areas of nuclear warheads and their carriers (missiles, aircrafts), control systems of the nuclear missiles, aircraft carriers or submarines, underground command points, nuclear reactors of modern power stations, etc. The clusters not only penetrate but also destroy their most important electronic and electrotechnical systems and their most delicate mechanical components.

Some contemporary experts will say that this is a simple electromagnetic impulse (EMI) weapon and we know how to efficiently protect against it. However, although it resembles an EMI weapon, it is not a standard EMI weapon and contemporary knowledge analysis reveals that protection from the cluster weapon does not exist. Clusters can pass through thick protective walls and implement themselves (by evolving electromagnetic, thermal, and mechanical energy) within the object without causing any outside mechanical destruction (all destructions are internal). Hence, considering its principal action, the cluster system is not the well-known EMI weapon. One can classify it as a specific version of a mixed EMI–laser weapon since it involves simultaneously a laser as well as EMI mechanisms of interaction with an object. In this very case, the laser component of the cluster with the EMI component passes freely and without attenuation through protective walls after passing through a layer of air, sea water, or rock and exhibits its absorbing properties within the very critical zone of an object of attack.

Hearing this, other experts mention the femtosecond laser weapon, the idea of which died quietly together with the Star Wars program. These experts will be only partially right. First, the femtosecond lasers generate super-short radio pulses, which are not clusters. Second, the ordinary femtosecond lasers cannot generate femtosecond electromagnetic signals of the required average power of tens or hundreds of kilowatts, while the femtosecond cluster FEL (CFEL) systems can achieve it. In contrast to the ordinary femtosecond lasers, this very fact makes the FEL-cluster version of the femtosecond idea much more viable.

To a standard potential accusation of "this is a fantasy," there is the following response. The author's experience is that during the past 20 years, the leading defense experts groundlessly ignored this scientific technological trend. That does not mean that the trend itself lost its right to exist anywhere in the world—it takes place now in the United States—without any visible reasons. This trend still exists today, and it develops, as this book shows, feebly perhaps, but it develops. Moreover, the accumulated scientific materials prove the correctness of the idea of a real possibility of the creation of a next generation of especially efficient new Star Wars systems. It is symptomatic that, during the past 20 years, we did not hear about the existence of any alternative scientifically substantiated opinions.

It appears that the other part of the world also sits quietly. But actually, it is not so. They now prepare the required general theoretical and technological bases. As an analysis shows, we can expect the first laboratory samples of combat cluster systems in the next 10 years. Within the next 20 years, their professionals will be able to create a first super-reliable, large-scale strategic cluster system of defense for the destruction of long-range ballistic missiles (LRBMs) and submarine-launched ballistic missiles (SLBMs) on any missile trajectory part. And, if they are successful, they will not have to catch up with anyone.

Before describing the complex nonlinear mathematics and physics of the theory of cluster devices, let us briefly consider a more simple matter, such as what is the old Star Wars program, what are its key ideas and elements, and what are the elementary physical principles of electromagnetic clusters.

1.2 Key Ideas and Potential Design Elements of the Star Wars Program

1.2.1 What Is the Star Wars Program?

The essence of the SDI is reduced to the formation of a strategic anti-LRBM and anti-SLBM defense system. The strike of the target, which is expected within all the sections of the trajectories of LRBM, is the peculiarity of the SDI. This is the difference between the SDI and other (tactical) types of antiballistic systems discussed during the past decades [1–3].

Let us discuss the basic elements of the trajectory of an LRBM, which attacks from the very moment of its start. Simultaneously we will emphasize the basic characteristics that are especially important from the point of view of the dependable defense [2].

As a rule, a missile trajectory is divided into four peculiar sections (Figure 1.1)

1. The booster stage, in which, owing to the performance of the missile sustainer engine, the missile reaches the burn-out velocity in 6–7 km/h
2. The post-booster stage, in which separation occurs between the warheads of the individual guidance systems and the false targets
3. The midcourse, where all the launched objects are moving along the trajectories of free flight
4. The terminal stage, where the warheads reenter and are directed to the impact zones

It was considered that the most efficient version of the SDI system should include, as one of its key elements, the means for the destruction of the attacking missiles at the booster stage. This conclusion is reached for the following reasons [1,2]:

1. The quantity of the objects that should be destroyed at this point in the trajectory is minimal because at this stage the warheads have not separated and the false targets have not yet been released.
2. At this point, it is easiest to detect the attacking missile by tracking facilities and its powerful torch formed by fuel spray.

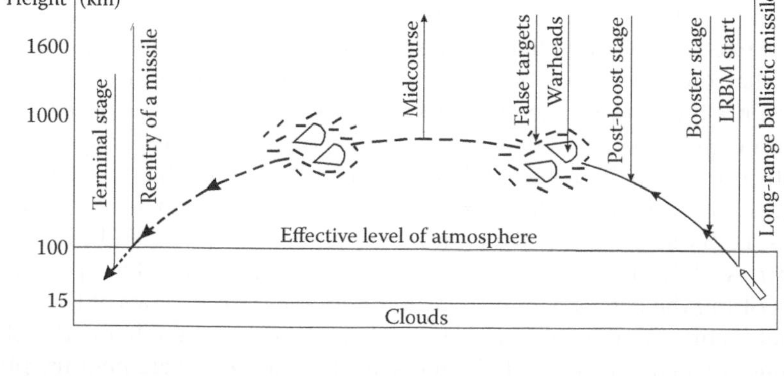

FIGURE 1.1
Elements of a long-range ballistic missile trajectory.

3. The missile as a whole is a much larger object than the warhead and, therefore, easier to observe.
4. The missile is more vulnerable because the protection of fuel tanks from thermal and impact effects is technologically much more complex than the protection of warheads.

Consequently, from the point of view of the side defending itself, the booster stage is the most convenient for beginning an attack on missiles, especially in a case of mass start. However, some specific characteristics of this part of the trajectory make it far from a simple project.

Two dimensions are the key characteristics of the booster stage: the time of the achievement of final velocity and the height at which this velocity is achieved. The time formulates the requirements of the tempo for the preparation of the appropriate layer of the system of antimissile defense for action. Also, the time determines the lower limit of the rate of fire of the means of defense in a case of a missile mass attack. The height at which the final velocity is achieved defines the technical means that might be used for striking the attacking missiles. In the 1980s, the typical time was defined by the number ~200 s and the height by ~200–350 km. One should notice that for most LRBMs, which today stand in operational readiness, the above-mentioned numerical dimensions remain approximately the same. One stresses that there were attempts to markedly lower the parameters to a time of 50 s and a height of 80–90 km. The eventual fate of these projects was not openly published.

The mentioned numerical dimensions reveal that a reliable observation of a missile at the booster stage can be achieved only from space. Indeed, because of the curvature of the earth, a missile reaching a height of 200 km can be observed from a distance of not farther than 1600 km. This distance can be increased to 2000 km in a case where the observer was elevated to 15 km. Remember that the dimensions of some countries and of the potentially interesting regions of the world are significantly larger. The means of the beyond-horizon radars, which use reflection of the direct and reverse radar beams of the ionosphere, cannot reach the needed level of reliability of observation.

Consequently, a reliable observation of the booster stage and striking of the LRBM requires a significantly complex combat space station. The key characteristics of such a system should be rapid firing and battle reserve.

One should stress that the satellite-basing of the combat space station makes the solution of both of the problems far from simple technologically. First, it depends on the specificity of the war operation; that is, it carries a global character. Not only each point of the adversary's territory but also the whole area of water of the world's oceans (where the missile-carrying submarines could hide) should be approachable for continuous observation. Taking into consideration the peculiarity of this task within the near-polar regions, the optimal number of the combat space stations in combat alert is not yet economically substantially estimated. Therefore, to solve this problem, parallel composite variants are investigated where the basic station is positioned on the ground and a system of distributing mirrors and targeting mirrors are placed in space. The utilization of an optimal combination of both types of systems appears the most reasonable. Nevertheless, as in the case of the satellite-basing, owing to exclusively objective reasons, the experts do not agree about the optimal combination of such a system.

The post-booster stage has different intrinsic peculiarities. Let us stress here that the platform with not-yet separated heads and the individual separated heads are significantly smaller objects than the very missile. The separation of the warheads from the platform is accompanied by the transitory work of small-thrust engines. This somehow simplifies the situation: the antimissile defense (AMD) systems can locate the platform and determine its position in space. Nevertheless, the destruction of the warheads at this stage of the trajectory is a much more complex task than at the booster stage because, in contrast to the missile, warheads do not have fuel containers. This makes a warhead better protected than a missile alone. One takes this under consideration when choosing a system for striking them. The above-mentioned combination system of a purely satellite-basing and a combined ground- and satellite-basing is considered the most expedient system for destruction of the warheads.

From the antimissile defense point of view, the most important characteristics of the midcourse stage are the maximal duration of the flight stage of the missile and the maximal number of targets, including false targets. For an LRBM, having a 10,000-km flight distance, the midcourse duration is 20–25 min. The apogee of the optimal (considering the energy expenditure) trajectory for such distance is 1000–1500 km. There are also other forms of the trajectory possible, but they are not energy-efficient, although they are practically effective.

At this flight stage, the number of targets exceeds multifold the number of starting missiles. This represents the basic complication of missile defense. It is typically estimated that a missile can carry 10 or more warheads and the same number of false targets. Each missile can also carry 100 or more simplified targets. This is done to attain maximal saturation of the antimissile defense within the midcourse. Metallized or thin-metal-wall inflatable balloons are used as false targets. Sometimes, real warheads are placed within such balloons. These factors increasingly complicate the situation. The defending side has a dilemma: destroy all targets, or select the true target from the false targets. Experience reveals that, considering the complexity, both of these actions are equivalent. One considers that in such cases, the defense systems should be satellite-based. At the final phase of the midcourse stage, the ground-based systems might be used.

At the terminal stage, the number of attacking targets decreases drastically. Entering the dense layers of the atmosphere, the false targets slow down and this allows the observing systems to perform a selection. However, this does not simplify the situation because the real warheads pass the terminal stage within approximately 1.5 min. At this stage, ground- and air-based are the most acceptable for defense. A generalization of the results of this analysis is given in Table 1.1 [1,2].

The described peculiarity of different sections of an attacking-missile trajectory allows explicating the needs of an antimissile defense system. Simultaneously, it allows establishing the requirements for the properties of the distraction means used for construction of such systems.

TABLE 1.1

Characteristics of Key Elements of the Long-Range Ballistic Missile Trajectory

No.	Section of Trajectory	Duration	Time of Motion in the Atmosphere	Number of Targets from Point of View of the ABM Defense	Main Objects of Destruction	Optimal Basing Means of the ABM Defense
1	Booster and postbooster stages	~200 s (or less)	100–200 s	Minimal	Fuel tanks	Satellite or satellite + ground
2	Midcourse stage	1000 s (or less)	None	Maximal	Warheads or all targets	Satellite
3	Terminal stage	100 s	All	Close to maximal	Warheads	Ground and air

Primary superficial analysis reveals that achieving the formulated aim by using the traditional weapons of the 1970s is technologically impossible. The same analysis shows that the formation of an effective system for striking LRBM and SLBM might be obtained only in one case, when the new systems will be constructed based on new physical principles. Consider the electromagnetic system of weapons (ESWs) with its key variety, the so-called directed energy beam weapon (DEBW).

1.2.2 Electromagnetic Systems of Weapons

Physical and technological experiments performed prior to the announcement of the Star Wars program allow reviewing the ESWs, which are the most promising for the solution of the above-mentioned problems. They may be classified as follows [1,2]:

1. A weapon of directed energy, in which energy is emitted into the thin surface layer of the target (DEBW-external). The electron-beam weapons and all types of 'ordinary' lasers belong here.
2. A weapon of directed energy with a deeper energy penetration into the material of the target (DEBW-internal). These are the ion-beam weapons, the electron-beam weapons based on the emission by electrons of transient X-rays within the thin surface layer of the target material, and the femtosecond quantum laser systems.
3. A kinetic weapon (KW) is a ballistic shell having a guidance system. It is accelerated to a very high velocity by a special electromagnetic system. A strike of a target occurs owing to purely mechanical effects.
4. EMI weapons, which in principle are similar to the action of an electromagnetic pulse (EMP) occurring during a nuclear explosion. It might be beams of millimeter emission or beams of high-current relativistic electrons that are emitted within a wide range of frequencies.

Further developments in the production of the specific variants of ESWs brought the concept of the directed-energy cluster electromagnetic systems weapon (CESW): the cluster-DEBW, which, in principle, can be classified as a specific variant of DEBW-internal. In contrast to the previously discussed types of ESW weapons, the experimental models of CEBW were not built because of the collapse of the Soviet Union.

The basic advantages of the DEBW over the traditional systems are related to its unique characteristics

1. It can instantly strike a target with the energy transfer either at the velocity of light or, as in the beam weapons, near the velocity of light. For example, at a distance of 3000 km, the target is hit at approximately 0.01 s. This means that, even in a case where the target is moving very fast, its displacement during the flight of the beam might consist of less than a few meters. Such displacement can be easily estimated since, during such small time intervals, a significant change of trajectory of the target motion is impossible.
2. The gravitational field of the earth has practically no effect on the trajectory of the energetic beam motion.
3. The real performance limitations regarding the large distance to the target are related to a transverse spreading of the beams, which at large distances leads to a significant decrease in their energy density at the target. However, in spite of this,

the distance ~10,000 km, which can be attained theoretically in such a situation, still retains the DEBW practical value.

As an illustration we present a short description of the two typical (for the Star Wars Program) representatives of DEBW.

1.2.3 Laser Weapons

As previously mentioned, laser beams act on the surface layer of materials. Their action involves two physical mechanisms: the thermal mechanism and the dynamic-percussive mechanism. The first is typical for the quasicontinuous and pulse lasers with a large duration of pulses. The dynamic-percussive mechanism outweighs the thermal mechanism in the case of short-pulse lasers. Such pulses are suitable for ruining different types of thin-wall casings, including the walls of fuel containers, sheathing of airplanes and helicopters, walls of oil reservoirs, and gas storage tanks. Therefore, the antiballistic missile (ABM) can be used for targeting missiles at the booster stage or from space for ground targets.

Owing to the damping of electromagnetic waves, the earth's atmosphere might represent a significant obstacle for the majority of laser beams. A transport of the intense laser beams in the atmosphere might be achieved only within the so-called transparency windows, which in the visible light range are within 0.3 to 1.0 μm. There are also some transparency windows within the infrared range, at which all kinds of atmospheric impurities do not absorb.

As a result, the presence of transparency windows determines one of the basic requirements for the laser's working frequency: it cannot be just any frequency. It should be mentioned that in spite of the transparencies of the windows, an appreciable absorption exists in reality. It becomes one of the factors for a case of large distances, which defines the possibility of the application of the system on the whole. Such a situation concerns the so-called ordinary lasers, those having the working pulse duration τ_{im} significantly longer than the critical duration τ_{cr}. The critical pulse duration τ_{cr} is considered the time at which a substance (in this case, air) stops absorbing the pulse energy. According to this parameter, all lasers are classified as precritical (pulses that pass through a substance without damping) and beyond critical (those absorbed by a substance). The value of τ_{cr} is usually within single to tens of femtoseconds. Because of this, the precritical lasers are also called femtosecond lasers. Theoretical and experimental studies reveal that the quantum precritical lasers are nonsuitable for weaponry. In spite of high pulsed power, the averaged power turns out to be too low. This, in turn, is caused by the fact that the rate of the subsequent pulses is also too low.

What concerns the nonquantum variety of lasers, the CFELs are somehow different. First of all, it should be noted that these systems were not developed within the American Star Wars program. Their concept appeared at the end of the 1980s in the USSR and it was not realized because of the collapse of the Soviet Union. We will discuss the basic concepts and potential characteristics of this type of system later in this chapter. Here, let us continue to discuss the typical properties of the ordinary, beyond critical quantum lasers.

To burn a hole in an aluminum sheet of 1 g/cm^2 specific thickness with a laser requires approximately 1000 J/cm^2 energy density of the laser beam. The same thickness of magnesium alloy requires approximately the same energy requirement; however, it requires 1.5 times more energy to burn a hole in a titanium sheet of the same thickness. Further thermal analysis might reveal that it takes pulse energy > 20 kJ/cm^2 to make a hole in a fuel tank wall with a laser weapon. Summarizing these data, we state that these numbers are the key data for calculating the optimal project solutions for design characteristics of the ordinary quantum lasers as the basis of the combat laser systems for striking a missile at the booster stage. Energy-wise, one should recognize that, in principle, these numbers can be practically realized. However, a complex project analysis of a system of satellite-basing reveals that there is a series of technological problems in the use of these systems in space conditions. One should mention that within the framework of the Star Wars program, such key problems were not solved [1,2].

For warheads with a high critical thermal protection that can survive the 100 MJ/cm^2, the use of a laser weapon appears almost unrealistic. Even when such energy pulses can be achieved in laboratory conditions, in reality, such lasers can be used only in a single-action system. The efficiency coefficient of this type of system is only a few percent. Therefore, a working laser cannot survive this thermal load during the pulse generation and is destroyed. Automatically, this disqualifies the use of these lasers as weapons on satellite bases. Apart from that, an analysis shows that the fighting efficiency of a satellite station should be higher than 1000 discharges at the discharge rate of a few hundred per second. Hence, the use of such lasers under these conditions on a satellite station is also unrealistic [1,2].

Another parameter that defines the range of action of any class of lasers or EMI weapons is the physical limitation of the beam parallelism. This, as is widely known, is based on the Huygens's principle. The maximal beam divergence is described by the so-called diffraction limit

$$\theta \sim \lambda/D, \tag{1.1}$$

where θ is the divergence angle, λ is the laser wave wavelength, and D is the laser mirror (antenna) diameter. We can write the dependency of the laser spot diameter d on the distance to target (range of action) R:

$$d = \sqrt{(\lambda R/D)^2 + (D\Delta R/R)^2}, \tag{1.2}$$

where ΔR is the so-called depth of focus, and the ratio $\Delta R/R$ is the range of defocusing. If we determine the limit beam diameter as $d_{lim} = D$, it is not difficult to derive from Expression 1.2:

$$R_{lim} = D^2/\lambda, \tag{1.3}$$

i.e., the larger the mirror diameter, the larger the range of the action R_{lim}, and vice versa: the larger the wavelength λ, the smaller the range of action R_{lim}. For the mirror diameter, for example, $D = 3m$ and the wavelength $\lambda = 0.3\mu$, the range of the laser weapon action $R_{lim} = 10^4$ km.

The last circumstance remains as the key argument in favor of using lasers as the key means for space defense at the booster stage. The only problem is to build such lasers, which would maximally fit their application in practical conditions of the discussed strategic defense system. One should mention that, to the very end of the program, this problem was not solved, especially for the case of satellite-basing. There was a lot of hope for four types of lasers that had the most success experimentally:

- Hydrogen fluoride chemical lasers
- Excimer lasers

- X-ray lasers with nuclear explosion pumping
- FELs

Once again, we want to stress that this work was not finished and that the final conclusions related to their applications were not derived. Nevertheless, it appears that there were enough results for deriving expert opinions for lasers promising potential applications [1,2]. Some of the results of the analysis are given in a condensed form in Table 1.2.

Table 1.2 reveals that, among the investigated lasers, the FEL received the most extensive attention. As discussed in Chapter 9, the first two FELs were built experimentally in 1976 by two research groups in the Naval Research Laboratory (NRL) (V. Granatstein and P. Sprangle) and Stanford University (L. Elias) [4–6]. Of the other variants—the competitors of FEL in those days—history retains only the chemical lasers. Today, they are successfully used in a series of tactical systems of laser weapons. One can consider the airborne lasers as the most demonstrative example of the important element of the present European tactical system of antimissile defense.

In this regard, it is worth mentioning that, in the Soviet Union, an identical program was initiated in the early 1970s as one of the fragments of the strategic ABM defense. An application of the chemical lasers in the satellite-basing systems—what was foreseen in the Star Wars program—appears very doubtful despite the obvious partial successes. To convince ourselves of this, it is enough just to calculate the combat reservoir of a station, considering approximately 1000 shots. By using the data in Table 1.2, it is easy to estimate the minimal necessary combat reservoir at approximately 2000 tons of fuel for a laser on one station. By considering a few tens (or hundreds, this question is still not solved today) it is obvious that, considering the cost of delivering fuel into orbit, supplying and maintaining chemical lasers for 20 years looks quite complex.

TABLE 1.2

Comparative Characteristics of Different Lasers Systems

No.	Laser Type	Working Wavelength	Passing through the Atmosphere	Method of Basing	Character of Action	Mass Characteristics	Energy Source
1	Chemical HF	2.8 μm	No pass	Satellite	Space–Space	~2 t fuel/shot	Internal (chemical)
2	Chemical DF	3.8 μm	Pass	Satellite	Space–Space Space–Earth	—	Internal (chemical)
3	Excimer	0.2–0.3 μm	Pass (under condition of wavelength shifting)	Ground with mirrors in space	Earth–Space Earth–Earth	—	External
4	X-ray	10 Å	Non pass	Satellite	Space–Space	Smallest mass among all types of lasers	Internal (Nuclear)
5	FEL	Any	Could be provided	Satellite or ground with mirrors in space	Space–Space Space–Earth Earth–Earth Earth–Space	—	External

For the sake of justice, we have to mention that, in the case of FELs, the problem of supplying energy to satellite-based lasers is also acute. However, it is more a diplomatic than a military–economical matter. Today, powerful radioactive sources with secondary electron emission or compact nuclear reactors are the most accepted types of onboard electric energy sources for satellite-based FELs. According to international agreements, launching such types of nuclear objects into space is forbidden.

Therefore, if this prohibition continues to exist, the future developers will have to choose one of two variants of practical applications for FELs, as was expected in the past. The first variant depends on the formation of a special ground-to-board electromagnetic energy transport system with its subsequent accumulation in board energy storages. Thermal accumulators based on the phase transition properties of substances as well as the electromagnetic (magnetic and electric) accumulators were considered for this purpose. An analysis showed that the best effect can be obtained by using a combination of various methods of accumulation. It should be mentioned that the most extraordinary project of an additional energy transport on board involved the ground-based FEL as the source of beams of electromagnetic energy. The second variant is more well known. It involves the ground-based FEL design scheme, the system arrangement of satellite-based distribution and the guidance mirrors.

In summarizing this topic, one should admit that within the Star Wars program [1,2], the most problematic questions regarding the technology of practical application of the ordinary quantum and X-ray lasers as the basic weapon for satellite-basing were not solved. Considering the FEL as a laser-based weapon, one should keep in mind the following.

In those days, Americans considered only their own variants of FEL, which were built within the framework of the primary, most primitive experiments of this class in 1976–1979 [4–6]. Soon, they realized that these were, by far, not the optimal versions of a practical FEL, considering the specificity of Star Wars' requirements. At the same time, the improved, alternative types of FELs [the multistage (1978), superheterodyne (1979), etc.] were proposed in the Soviet Union. It should be noted that some of them were more adequate technologically for the Star Wars' program (Chapters 9–14). During this time—the climax of Star Wars—the idea of the CFEL was also formulated.

Historically, a very specific situation occurred at this time in the United States. The program was formed, the finances were distributed, and a large amount of funds had been successfully spent. Naturally, no one wanted to explain anything to anybody, especially the fact that during the formulation of the application of the FEL to the program, an awkward strategic mistake was made and also—considering the final goal—that the officially approved and financed FEL project was wrong and did not have any chance for success, etc. As a result of such politics, the well-known and grandiose experimental realization of the FEL for the program was revealed to be, in principle, a loss. At the same time, all the alternate FEL variants were forgotten for many years and the fact of their existence was energetically hushed up as a side effect of this epopee [4]. In the Soviet Union, these alternate versions were included into the renewed military space program. In July 1991, the Soviet government approved this program; however, in a few months, this super-state ceased to exist.

The alternate direction, including the CFEL, did not die. The existence of this book and others [4–6] and many articles prove that it exists even today. The time of Star Wars, as well as the time of peak influence of its creators, passed long ago. Now one can—painlessly and unpunished (the author naively hopes)—discuss the historic reasons behind past failures as well as the existence of some alternate variants in the FEL biography. It is also reasonable to discuss some of the most interesting related physical/mathematical and philosophical aspects that originate from general FEL theory.

1.2.4 EMI Systems

It is well known that a nuclear explosion is accompanied by powerful pulses of electromagnetic emission. The peculiarity of the emission is that it strikes the electronic and weak current electrotechnical devices within a 5000-km radius. Approximately 10^{11} J is transferred into EMI energy at a megaton explosion. Within an approximately 1000-km distance, one can easily calculate the emitted electric field strength to be E ~180 V/cm. At this strength, a difference of the potential on surface of a 25-m space station might be ~450 kV. At frequency ~1 MHz, the resistance of the station frame is ~10 Ω. It is easy to estimate that, in such cases, ~45 kA currents flow on the station surface. Owing to various inductive and capacitance correlations, this current may penetrate into the station's depth and there can create electromagnetic-induced pulses of up to 100 V. When recalculated to the intensity, it is approximately equivalent to 100 W/cm^2. When such an induced electromagnetic signal reaches the most sensitive elements of computers and weak current systems (power supply chains, computer chips, and others), it has enough potential to break up communication, guidance, orientation, and other systems.

The main idea of the EMI weapon is an artificial imitation of the action of the same effect. It could be realized practically by irradiating the surface of an attack object with a narrow beam of electromagnetic waves of some intensity, for instance, ~100 W/cm^2, but in the millimeter range. Basically, this intensity can even be essentially smaller because, as is shown experimentally, the effectiveness of the EMI destruction increases with the growth of the electromagnetic wave frequency.

One might ask, "Can those-days emission sources create such conditions on the surface of an object of attack?" The answer to this question is "yes." Let us perform some numerical estimation. With a 10-m diameter antenna, the divergence of the mm wave beams is approximately 10^{-4} (see Expression 1.1). At a distance of 1000 km, such beam forms an ~100-m spot. So one needs a wave beam of approximately a few gigawatt millimeter-wave beam to create an intensity of ~100 W/cm^2 on the surface of a station. Let us recall that in the United States in the 1980s, a high-current induction advanced test accelerator (ATA) was built with a current beam strength of 10 kA and ~50 MeV of energy [8]. Hence, the pulse power of the electron beam was $\sim 5 \cdot 10^{11}$ W. Let us recall that, also in the 1980s, an FEL was built having an electronic efficiency of ~40% [9]. Theoretically (see Part II), one can obtain a higher value of efficiency. Let us take 50% as an example. It means that, even when using the obsolete 30-year-old basis, one could build powerful generators of tens or hundreds of gigawatt emissions in the millimeter range.

1.2.5 What Was the Star Wars Program?

A short discussion was given of the technological basis of the two key types of strategic weapons that constitute the basis of the Star Wars program. It is revealed that by the end of the program in the early 1990s, the practical realization of the first laser weapons, even from today's point of view, appears very doubtful. An analogous conclusion was derived considering the perspective of the creation of the strategic versions of the DEBW-internal and the KW. In this situation, some success with the EMI weapon did not solve the problem. First and foremost, this is because the strategic ABM systems might have a true practical value only when all, without exception, of their key subsystems are fully operational. Of course, significant parts of the technological achievements of the program were used during the past 20 years to make various types of weapons with tactical applications [10–12]. However, basically, the Star Wars program was a complete failure in the attempt to create a global system of strategic space defense.

But is this really so? Was the huge army of American scientists, military personnel, and politicians who participated in the project—including President Reagan and his administration—so naive and irresponsible as to inflict on the country the enormous financial losses related to the realization of such a doubtful project?

Our advantage over the Star Wars' contemporaries is our ability to analyze the situation retrospectively, knowing everything that happened after the program's launch. Such retrospective analysis reveals that the creation of the strategic space defense system was not the true aim of the Star Wars program. The objective point was to involve the Soviet Union in a new and extremely expensive leg of the weapons race. The real internal political and economic state of the country of workers and peasants had been analyzed in detail and the conclusion obtained indicated that such an opportunity should not be lost. The Star Wars program became the weighty means to create the very straw that would finally break the back of the Soviet camel.

From today's point of view, one should recognize that the idea of Star Wars was brilliant. Simultaneously, one can state that its realization was at a high level. Actually, the key ideas of the program were highly plausible, to the point that nobody among the experts could consider it unrealistic. There were some doubts [1]; however, the asking price was so serious that, in spite of the presence of doubts, fears to make a mistake did the risk unbearably high. Due to the armaments race, the greed of the Soviet military–industrial complex ("VPK") literally transformed it into an awful monster. In essence, it became a state within a state that was poorly controlled and, therefore, extremely dangerous for the country. Baiting this monster with such plausible ideas and conceptions, the creators of the Star Wars program could be sure that the fish would grab the hook enthusiastically.

As a result, the VPK obtained a very effective instrument with which to squeeze out the last juices from the weakened economy of its native country. One should also consider the psychology of the ruling class of the day, the age of which was between 70 and 80 years, and the corruption of the upper echelons of the ruling class that had also corroded the foundation of the state. The plan worked magnificently. The weapons system, which was never realized, without one shot fired, helped to win a decisive victory in the Cold War: the Soviet Union, the only competitor of the United States for global dominance, ceased to exist as a state.

Hence, the described paradoxical behavior of the subprogram managers in making the FEL-based laser weapon appears absolutely logical and cannot be judged. In reality, no one intended to bring this type of weapon to working capacity. Therefore, it was absolutely unimportant what type of FEL (an antiquated or modern one) was used as the basis of this program because the aim of the program was different than the one declared.

Nevertheless, as in any grandiose fraud, not everything went as planned. The attempt to make the conceptual principles of the program seem plausible could also play tricks on the program's initiators. Essentially, in many aspects this plausibility was brought to a level of truth, theoretically threatening to transform this great show into a reality that no one expected. The peculiarity of talent of Soviet scientists and engineers is known worldwide. It brought a missile-nuclear shield comparably equivalent to the United States' at a significantly more modest cost, the launches of Sputnik and the first astronaut into an orbit around the earth, etc. The awareness of the dubiousness of the scientific and technological basis of the American version of the program [1] forced the curators of its Soviet analogue to search for their own, alternate ways. As is now known, the Soviet scientists and engineers found many such ways.

One should stress that, from the standpoint of the Star Wars program, there were real scientific and technical breakthroughs that had unforeseen consequences for the adversary. Finally, luck appeared to be on the side of the American organizers of the whole epopee of Star Wars. If the Soviet Union had been able to stop its collapse and survive the crisis—as had happened many times in its history—the strategic situation of the world would have changed to its benefit. Historically, the American initiative provoked the USSR to accept the very idea of the possibility of building an absolute space defense system. As a result, in the time immediately before its collapse, the Soviet Union was ready to build not a mere show, but a real strategic ABM system of defense. Moreover, and especially important, the system would have been built using a completely different physical and technological conceptual basis. As with Hitler and his atomic bomb, the Soviet Union was simply short of time. Thus, the brilliant yet risky program of Star Wars is revealed to be more dangerous than it was first considered. In principle, it could have ended badly.

One should mention that the epopee of Star Wars was not exhausted with the collapse of the Soviet Union. By far, not all once sown seeds of the potential dangers sprouted and not all appeared sprouts were destroyed during these times. The ruinous ideas and knowledge are stored for future civilizations and, sooner or later, they will emerge. The question now becomes, "With whom?"

Let us pose a rhetorical question: What are the chances that, in the near future, someone different will finish this unfinished Soviet project? Taking into consideration the realities of the surrounding world and the knowledge gained by the Soviet Star Wars program, this idea does not seem as fantastic as one would like to believe. Taking into account the current global financial/economic crisis, consider the following scenario: Western civilization finds itself in a situation where the leaders (the United States and the European Union) are unable, within a short time, to raise significant financial and economic resources. Remember that the present national debt of the United States represents 80% of its GNP. Such a crisis might interfere with a sudden need to create a new trillion dollar strategic system like Star Wars. Simultaneously, this means that other world players, having an excess of resources and ambition, will be free to act.

There is no guarantee that one of them will not risk taking a chance, as the United States did once in relation to the Soviet Union. The initiators of the Star Wars program were not aware at the time that they had opened a Pandora's box. They did not imagine the potential remote consequences of their actions. This author believes that now is the time to seriously and analytically consider the past and our future. Let us review the key basis of the cluster electromagnetic systems of defense in the context of the science/technical illustration of the above-mentioned thoughts.

1.3 Femtosecond Laser Systems: Basis, Concepts, and Ideas

1.3.1 Femtosecond EMPs: Physical Mechanisms of Propagation

As previously mentioned, the main idea behind the femtosecond laser systems is an attempt to overcome the physical and technological difficulties of ordinary quantum laser weapons. The ideal femtosecond laser weapon was formulated as a scientific/military dream that should provide:

1. The possibility of propagation of electromagnetic signals in the earth's atmosphere with practically no damping
2. The capability of the electromagnetic signals to pass (also without damping) through various solid walls (metallic, ceramic, rock, concrete, etc.) and through thick layers of sea water
3. The capability of transformation of the electromagnetic signal energy into heat or mechanical energy only within crucial zones of the object of the attack (warhead, missile operation systems, underground command posts, operation systems, nuclear reactors, etc.)

In spite of that, the formulated scientific/military dream actually looks absolutely fantastic, the suitable physical mechanisms for its practical realization were found and studied in the framework of the Star Wars program.

The most unexpected and mysterious property of the femtosecond EMPs is their capability to pass through various material objects with practically no damping. Let us examine this phenomenon. Assume that we have the simplest quantum system shown in Figure 1.2. A photon with energy $\hbar\omega$ hits an electron that occupies the quantum energy level n. The situation is well known to us from a university general physics course: as a result of the disturbance of the electron quantum state, it can transit up the quantum system to one of the higher levels, for example, the level $n + 2$. As follows from the energy conservation law, the process of photon absorption (the effect of stimulation absorption) should accompany this transaction. In accordance with quantum electrodynamics, the probability (for the time unit) of this process is characterized by the value $w \sim 1/\tau_{cr}$, where τ_{cr} is the characteristic time of the considered quantum transaction.

Then, we assume that instead of the falling photon $\hbar\omega$ (see Figure 1.2), a very short (femtosecond) laser pulse of duration τ_{p2} hits the electron. This pulse, in accordance with quantum electrodynamics, can be treated as a quantum ensemble of photons. It is considered that this ensemble hits not only one electron but a totality of electrons in a given small volume. We also regard this totality as the quantum ensemble. But this time, it is the quantum ensemble of electrons (see Figure 1.3). The novelty of the principle of this model is that the result of the interaction of two of these quantum ensembles depends, essentially, on the ratio of two time intervals: the characteristic time τ_{cr} and the pulse duration

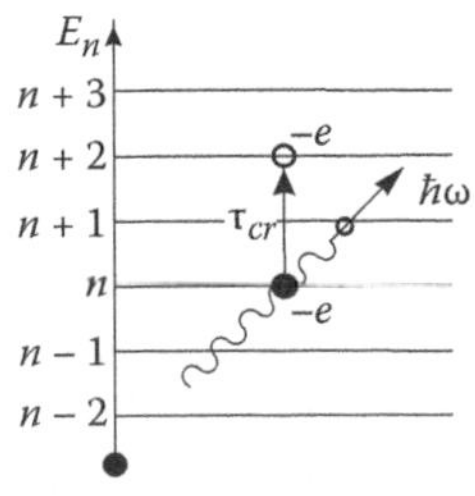

FIGURE 1.2
Illustration of the electron transaction in a quantum system (stimulated absorption). Here, E_n is the electron energy for level n, $\hbar\omega$ is the energy of photon with frequency ω, and τ_{cr} is the characteristic time of the quantum transaction.

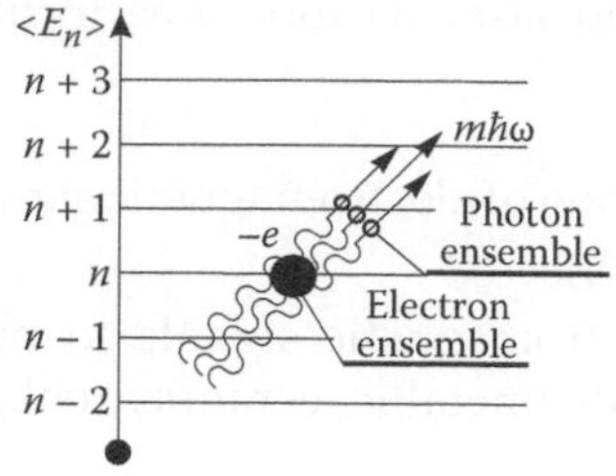

FIGURE 1.3
Illustration of the electron interaction with the multiharmonic quantum photon ensemble in the case of Condition 1.4 ($\tau_{cr} >> \tau_{p2}$, interaction without any absorption). Here, $m\hbar\omega$ is the photon energy with the harmonic number, and $m = 1,2,3,\ldots$, $\langle E_n \rangle$ is the averaged (on the ensemble) electron energy. Other designations are explained in Figure 1.2.

τ_{p2}, respectively. Quantitatively, it is obvious that in the case where the duration τ_{p2} of the photon ensemble (pulse) is much smaller than the characteristic time τ_{cr}:

$$\tau_{cr} >> \tau_{p2}, \tag{1.4}$$

the probability of the absorption of the femtosecond pulse should be very small. The reason for this is that, in this case, the electrons do not have time for a transition up in the quantum ensemble during the interaction with the pulse. As a result, the femtosecond pulse passes through the electron quantum system practically without absorption. In other words, the process of the propagation of the femtosecond EMP, as a whole within the medium under satisfying Condition 1.4, occurs practically without damping.

It should be noted that the simplest model, like the one shown in Figure 1.3, should be characteristic for gases, including the earth's atmosphere. As the relevant theory shows, the quantum representation of the condensed mediums (solid metallic or dielectric materials, liquids with admixtures, etc.), is more complex. However, the main result (a possibility of propagation without damping) does not change, even in these specific cases.

Corresponding experiments with the femtosecond pulses were performed in 1980s in the Soviet Union and the United States. Basically, they confirmed the described qualitative physical scenario of the propagation of the femtosecond pulses through different materials. In these experiments it was also found that the practical absence of the absorption is not the only unique property of the process. The second key physical effect is the appearance of a dispersion of the spectral harmonics of the femtosecond pulse. The applied significance of this discovery is very important. It opens a physical possibility of the destruction of targets at a given distance by the action of the femtosecond pulses. Let us discuss this problem in detail.

1.3.2 Femtosecond EMPs: Role of the Wave Dispersion

Let us begin with the question: How can the femtosecond pulses be used practically in the situation, when their energy is not absorbed by a substance (see Figure 1.3)? The answer is that it occurs only because of the dispersion of the pulse's spectral components. The point is that, in this case, the pulse duration τ_{p2} should increase in the process of its propagation. It is because the phase velocities of different spectral components are different. Or, in other words, the pulse duration becomes a function of the passed distance (see Figure 1.4). Consequently, the time should eventually come when Condition 1.4 begins to break.

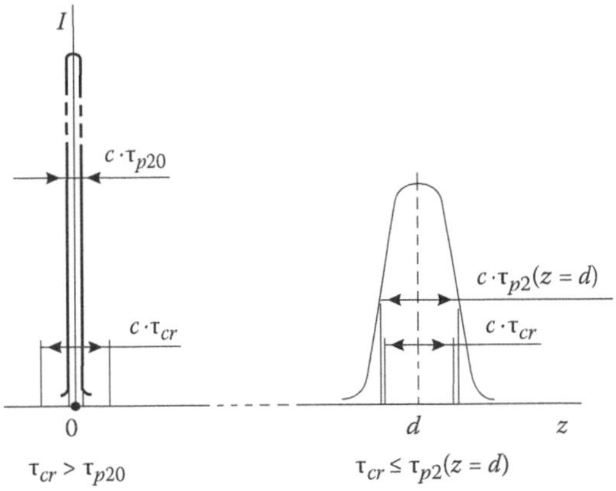

FIGURE 1.4
Effect of the pulse spreading caused by the wave dispersion. Here, $c \cdot \tau_{p20}$ is the initial pulse width, $c \cdot \tau_{p2}(z = d)$ is the terminal pulse width (within the target critical zone), $c \cdot \tau_{cr}$ is the critical pulse width, c is the light velocity, d is the hitting distance, and I is the pulse intensity.

This means that in such a situation we should observe the appearance of the pulse absorption. The process of the transformation of a noninteracting pulse into an interacting pulse begins just at the time of the break of Condition 1.4 (see Figure 1.4). The energy of the initial femtosecond pulse begins to transform into other forms (heat, mechanical, etc.). We treat this situation as the creation of necessary physical conditions for the destruction of the object of the attack. The hitting distance d of such a femtosecond weapon in practice could be changed by initially changing the pulse duration $\tau_{p20} = \tau_{p2}(d = 0)$ (see Figure 1.4).

Hence, based on the accomplished qualitative analysis, we can draw two important conclusions:

- The femtosecond pulse can propagate without absorption in a medium during some limited time (or, equivalent, at some limited distance) only.
- The time (and, consequently, the distance d) of the propagation can be changed by changing the initial pulse duration τ_{p20}.

The practical meaning of these conclusions is illustrated in Figures 1.5–1.9. Let us discuss a few possible practical situations, which can take place when the femtosecond beam of pulses attains a target.

The first situation, from when Condition 1.4 is satisfied until the femtosecond pulse attains the target wall, is illustrated in Figure 1.5. The specificity of this situation is that the initially nondamped beam 1 penetrating into the target wall at depth l_{wd} begins to transform into the damped (absorbed) electromagnetic signal. As a result, the wall substance at this depth, l_{wd} (in zone 3), begins to heat. However, the crucial zone of the target, in such a case, turns out to be protected by the remaining part of wall 2. Hence, the femtosecond beam does not attain, in this case, its true target (the crucial zone of the attacked object 4).

The second illustration example concerns the opposite physical situation: the length of the propagation of beam 1 within the wall material without damping l_{wd} is chosen a little longer than the wall thicknesses l_w: $l_{wd} = l_w + l'_w$ (here, $l'_w << l_w$ is the effective penetration depth within the material of critical zone 3; see Figure 1.6). In contrast to the situation

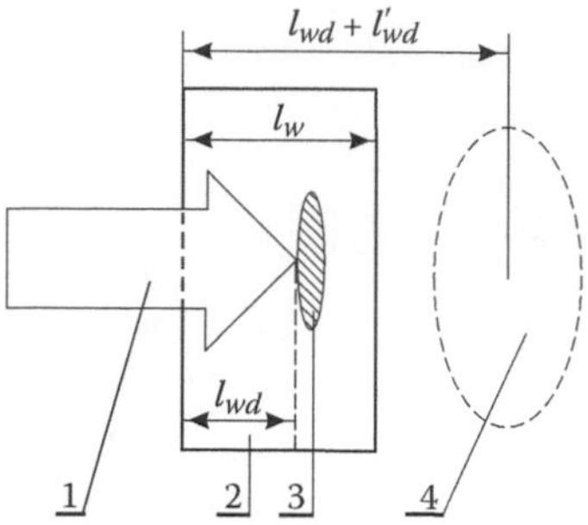

FIGURE 1.5
Illustration for the case $l_{wd} < l_w$, when a femtosecond attack does not attain, in this case, the true target (destruction of the crucial zone within the attacked object). Here, 1 is the beam of femtosecond pulses, 2 is the solid wall, 3 is the zone of wall 2 heated by beam 1, 4 is the crucial zone of the target; l_w is the thickness of wall 2, l_{wd} is the real penetration depth for beam 1 into the target without damping, l'_{wd} is the required penetration depth into the target, l'_w is the effective penetration depth into a part of the material of critical zone 4.

shown in Figure 1.5, the femtosecond beam attains, in this case, the true target (crucial zone of the attacked object 3).

The third possible situation is shown in Figure 1.7. Here the femtosecond pulse passes through the target without any interaction because the total penetration depth of the femtosecond pulse l_{wd} is larger than the sum of thicknesses $l_w + l'_w + l''_w$. Similar to the case illustrated in Figure 1.5, the target in this case also cannot be destroyed. On the other hand, this example, together with the example illustrated in Figure 1.6, demonstrates two rather unusual possibilities of femtosecond weapons:

1. The possibility of overcoming a multiscreened defense
2. The possibility of destroying targets hidden in the geometrical shadow of another object

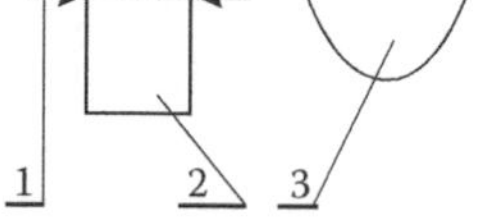

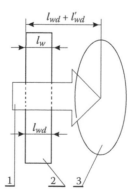

FIGURE 1.6
Illustration for the case $l_{wd} > l_w$, when a femtosecond attack attains the true target (destruction of the crucial zone of the attacked object). Here, 1 is the beam of a femtosecond pulse, 2 is the solid wall (screen), 3 is the crucial zone of the target, l_w is the thickness of wall 2, l'_w is the effective penetration depth within the material of critical zone 3, and l_{wd} is the total penetration depth of the femtosecond beam 1.

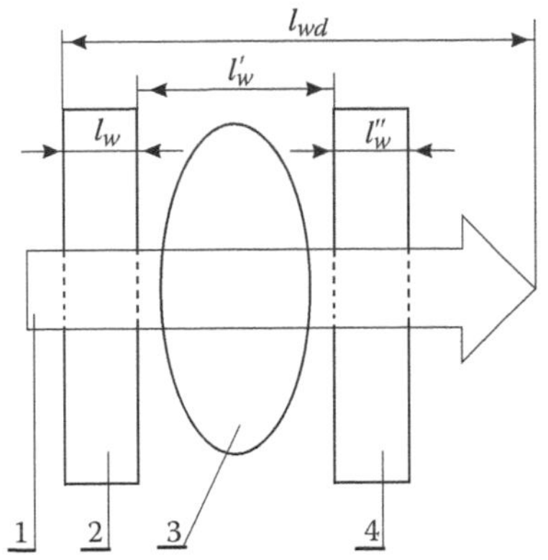

FIGURE 1.7
Illustration for the case $l_{wd} > l_w + l'_w + l''_w$, when a femtosecond attack does not attain the target successfully. The beam passes through the crucial zone of the attacked object without an interaction. Here, 1 is the beam of a femto second pulse, 2 is the first solid wall, 3 is the crucial zone of the target, 4 is the second wall, l_w is the thickness of the first wall, l'_w is the effective thickness of the crucial zone of the target, l''_w is the thickness of the second wall, and l_{wd} is the total penetration depth of the femtosecond pulse.

The main idea of the first possibility is illustrated in Figure 1.8. As is well known, the use of the multiscreening method is one of the most effective means of defense against the action of traditional EMI weapons. Figure 1.8 shows that this method is ineffective in the case of femtosecond weapons. It is obvious, as follows from Figure 1.8, that we basically have two possible tactics of attack in this case. The first consists of the adjustment of the pulse duration for attaining some chosen zone, which is considered most important. The second tactic consists of overcoming the levels of the system screens, step-by-step, by systematically changing the depth l_{wd}.

The second of the possibilities (see Figure 1.9) looks very unusual and illustrates the possibility of destroying a target hidden in the geometrical shadow of another target. Thus, changing the distance l_{wd}, we can create many different practical situations. However, in

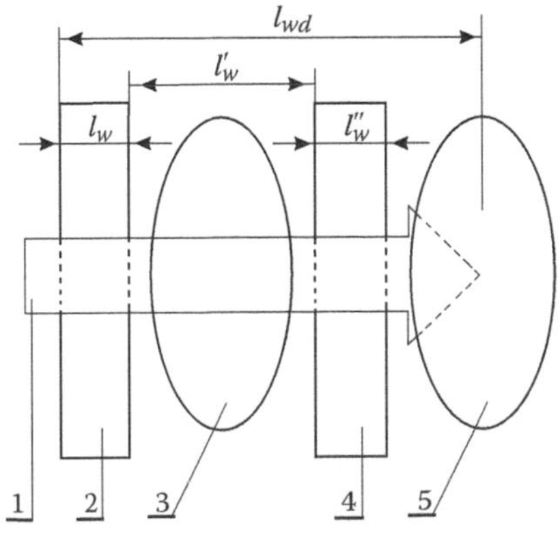

FIGURE 1.8
Illustration for the case of overcoming a multiscreened (multilevel) defense, ($l_{wd} > l_w + l'_w + l''_w$). Here, 1 is the beam of a femtosecond pulse, 2 is the first solid wall, 3 is the first crucial zone of the target, 4 is the second wall, 5 is the second (attacked) crucial zone of the target, l_w is the thickness of the first wall, l'_w is the effective thickness of the first crucial zone, l''_w is the thickness of the second wall, and l_{wd} is the total penetration depth of the femtosecond pulse.

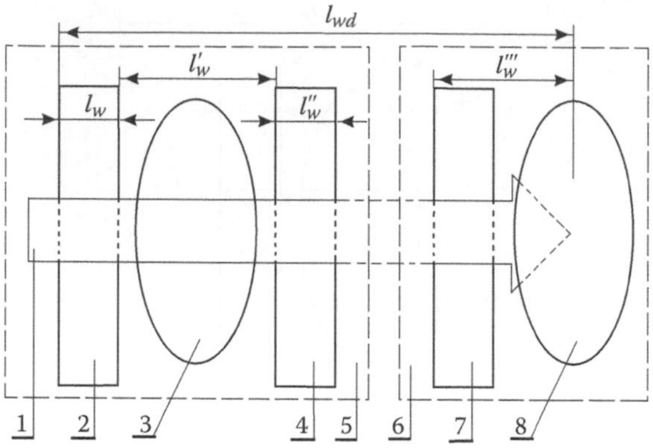

FIGURE 1.9
Illustration for the case of the destruction of a target hidden in the geometrical shadow of another target ($l_{wd} = l_w + l'_w + l''_w + l'''_w$). Here, 1 is the beam of a femtosecond pulse, 2 is the first solid wall, 3 is the crucial zone of the first target, 4 is the second wall, 5 is the first target, 6 is the second (true) target that is situated in the geometrical shadow of the first target (5), 7 is the wall thickness of the second target (6), 8 is the crucial zone of the second target (6), l_w is the thickness of the first wall, l'_w is the effective thickness of the first crucial zone, l''_w is the thickness of the second wall, l'''_w is the effective penetration depth of the second target, and l_{wd} is the total penetration depth of the femtosecond pulse.

spite of this difference, the common point of all these situations is that the desired transformation of the energy of femtosecond pulses can always be organized just within that part of a target's crucial zone, which is most interesting for us.

1.3.3 Main Technological Problems of the Classic Femtosecond Weapon Systems

Numerous real experiments performed in the 1980s in the Soviet Union and the United States basically confirmed the main physical properties of the femtosecond EMPs. However, until today, none of the real femtosecond systems of weapons was created. The main reason for this was that the system developers have used the femtosecond quantum lasers that generate the femtosecond radio pulses.

As is well known from standard university courses, a radio pulse is not really a true pulse like that we can observe in ordinary electrical circuits. It represents, in itself, a pulse of the amplitude envelope filled by some oscillations of carrier frequency. In the case of the femtosecond quantum lasers, we have, as a rule, a few oscillations only (see Figure 1.10).

The mean power of a beam of femtosecond pulses, which can be transformed into heat and mechanical energy, depends on the instant power of the pulses and the period of their repetition T. Unfortunately, the femtosecond quantum lasers used at that time were characterized by too low levels of mean power (not higher than tens of milliwatts). Because of that, the frequency of the pulse repetition $1/T$ (Figure 1.10), practically, did not exceed 10^4Hz. On the other hand the femtosecond weapon systems require a laser beam of mean power from a few kilowatts to hundreds of kilowatts for a successful operation. Such levels of magnitudes can be attained by the repetitious frequency of the pulses $1/T \sim 10^9 - 10^{11}$Hz only. But quantum femtosecond lasers with such repetitions of pulse levels are not yet constructed.

Apart from that, the radio pulses generated by the quantum femtosecond lasers correspond to the optical range carrier frequency. This makes the construction of the

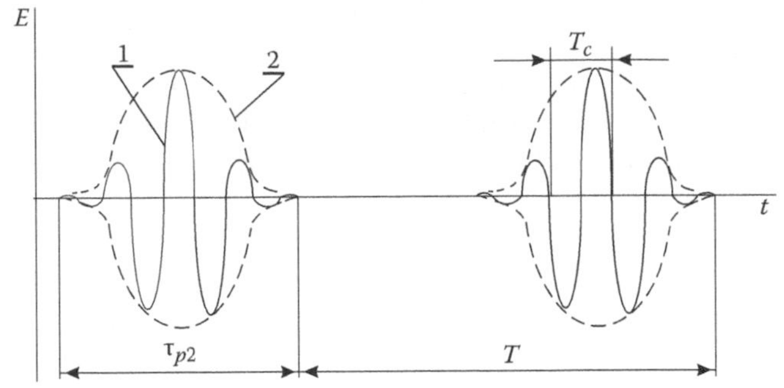

FIGURE 1.10
Illustration of the concept of femtosecond radio pulses. Here, 1 is the instant dependency of the pulse electric field on the current time t, 2 is the envelope line, E is the electric strength of the pulse oscillations, τ_{p2} is the pulse duration (see definitions of Condition 1), T is the period following the pulses, and T_c is the conditional period of the carrier oscillations.

femtosecond EMI weapons impossible since the optimal working frequencies of the EMI systems are within the millimeter range.

Thus, the practical realization of the scientific/military dream was impossible at the technological basis existing at that time. Considering this, the managers of the Soviet Star Wars program formulated the problem of the elaboration of the more effective physical mechanisms and technologies. As an answer, the required set of new conceptual scientific and technological ideas, which promised great success in the solution of the formulated problem, was proposed in the mid-1980s. One of the novelties in this set is the CFEL. It includes:

1. The method of forming the femtosecond clusters of electromagnetic energy
2. The physical mechanisms of the experimental realization of the proposed method for their formation
3. The technological means for practical realization of these mechanisms and method

It should be noted that the first item on this list does not look too obvious and banal because the concept of the electromagnetic energy clusters was not as well known in electrodynamics as was the concept of the radio pulses.

Unfortunately, at the time of the Star Wars program, the proposed conceptual ideas had only an essentially semiqualitative substantiation. The crux of the matter was that, at that time, fundamental science could not propose the value theories for their mathematical modeling. This was the main reason that the first quantitative high-level nonlinear theoretical models appeared first in the beginning of the 1990s. The existing technologies also could not answer all the formulated questions.

First of all, we did not have the needed mathematical methods for the solution of the just appearing new class of nonlinear multiharmonic wave-resonant problems (Part I). The difficulties increased, especially in situations where a higher wave harmonic possessed higher amplitude. However, just such situations were typical for many proposed versions of the femtosecond CFEL (Part II). We also did not have the necessary general physical approach to the theoretical description of the multiresonant and multiharmonic nonlinear wave models, especially in cases where the periods of different interacting waves differed by a few orders (Part I). At that time we did not have technologies for constructing the super-powerful, high-current compact accelerators. Such accelerators are designed for the formation of many dense and sufficiently stable parallel relativistic electron beams (see Part II). We did not know how to operate the generated signal of the femtosecond beams in situations where the substance of the relevant antennas (mirrors) was transparent to the beam radiation, and so on.

The required methods and the theoretical approaches were elaborated only at the beginning of the 1990s and (some of them) later [4–6]. The generalization of the new mathematics and physics was formulated as the theory of hierarchic oscillations and waves (Part I). Some of the most interesting results of the application of these new mathematics and physics to the theory of high-current multiharmonic FELs (including CFELs) are described in Part II of this book.

1.4 CFEL Systems: Methods for Formation and Application of Electromagnetic Clusters

The first key point of the three initiatives about the CFEL was the invention of a method for forming the femtosecond EMPs in the form of clusters of electromagnetic energy (Figure 1.11). As is readily seen in Figure 1.11, the clusters—in contrast to the radio pulses—are the real pulses. They are similar to the well-known pulses of current and voltage found in electrical circuits. The only difference is that the cluster can exist in a vacuum as well as in a substance, propagating there at light velocity. A periodically reversed sequence of the clusters (Figure 1.11) forms the so-called cluster waves. Let us discuss this in more detail.

1.4.1 Clusters of Electromagnetic Energy as a New Subject of Electrodynamics

It should be mentioned that the super-powerful clusters (tens of TW of instant power), as well as the methods of their generation, were absolutely new for electrodynamics at the time of Star Wars. One should realize that such situations commonly exist today. A suitable nonlinear theory that could adequately describe the physics of the clusters' interaction with a substance did not exist at that time or, in many ways, until now. Today, we have

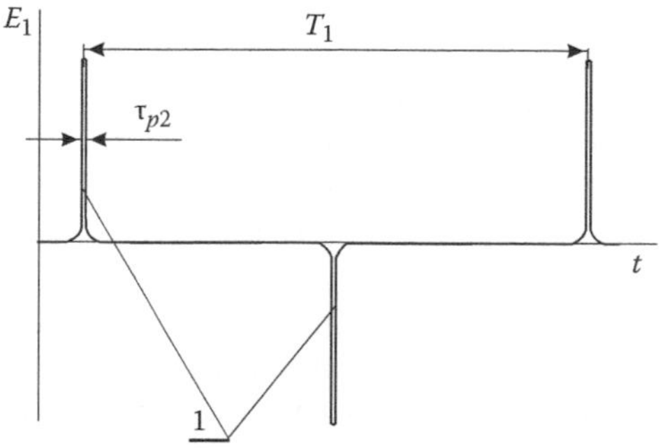

FIGURE 1.11
Concepts of cluster waves and clusters of electromagnetic energy. Here, 1 represents clusters of electromagnetic energy, τ_{p2} and E_1 are the duration and strength of the electric field, and T_1 is the period of cluster waves.

only the simplest version of the cubic nonlinear theory, which allows studying the interactions of the electromagnetic clusters with plasmas of relativistic electron beams (Part II). Concerning the theory of interaction with an ordinary substance, which consists of atoms and molecules, we do not presently have any serious theories. Actually, in this situation, we can talk about the new part of applied physics—cluster electrodynamics. Therefore, in the relevant project analysis, we were forced to use only semiphenomenological approaches based on results of the theory of femtosecond radio pulses.

The proposed method of formation of the femtosecond electromagnetic clusters (which is also called the *method of the compression* of the wave packages) is illustrated in Figure 1.12.

Let us assume that we have an initial harmonic (sine-like) electromagnetic wave, which is shown in Figure 1.12 by curve 1. It is obvious that this wave, in turn, can be imaginary, represented as a sequence (with period T) of the peculiar positive and negative half-sine pulses. The main idea of the method is to strongly compress these half-sine pulses into a sequence (with the same period T) of the delta-like femtosecond clusters (2). The compression factor f_{com}, which characterizes this process, can be defined by the following expression:

$$f_{com} = \tau_{p1}/\tau_{p2} \approx T/4\tau_{p2}, \tag{1.5}$$

where all the definitions are illustrated in Figure 1.12. Herewith, each compressed pulse (2) can be regarded as a solitary wave package (cluster) if the specific condition (4) $T/2 >> \tau_r$ is satisfied (the value τ_r is the medium's characteristic time, which characterizes the time of its relaxation after the passing of a preceding cluster).

At first sight, the described method looks very promising for the practical application, but a question arises: How to achieve such compression technologically? For this purpose, we should have a unique (from a technological point of view) multiharmonic source of electromagnetic waves. This uniqueness requires that such a source be able to generate at least tens or even hundreds of signal harmonics with commensurate amplitudes (see Figure 1.13). Let us discuss this topic in detail.

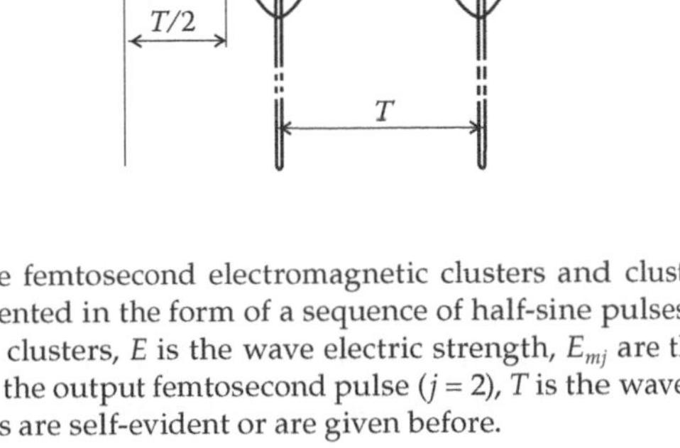

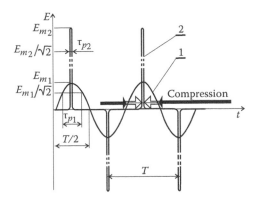

FIGURE 1.12
Method of compression of the femtosecond electromagnetic clusters and cluster waves. Here, 1 is an input harmonic signal that is represented in the form of a sequence of half-sine pulses, 2 is the sequence of the compressed (output) femtosecond clusters, E is the wave electric strength, E_{mj} are the maximal amplitudes of the initial (input) wave ($j = 1$) and the output femtosecond pulse ($j = 2$), T is the wave's one period, and t is the time coordinate. Other designations are self-evident or are given before.

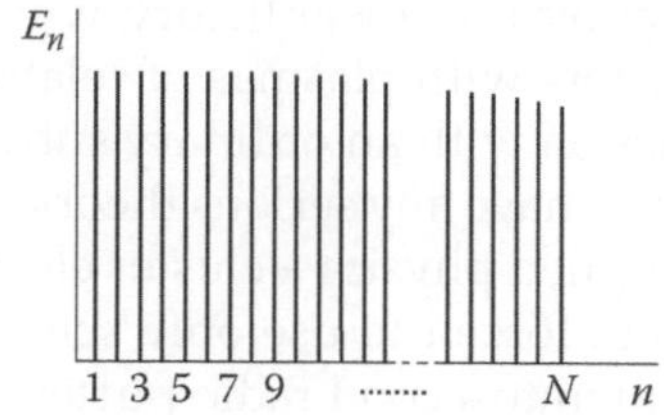

FIGURE 1.13
Fourier spectrum of the femtosecond cluster wave composed of the so-called spread Dirac delta functions (see Figure 1.11). Here, E_n are the amplitudes of harmonics of the Fourier spectrum, n is the current number of harmonics, N is the largest harmonic number taken into account.

Why do we need to have many wave harmonics in the case of the femtosecond system? The explanation is connected with the real multiharmonic nature of the cluster signal wave illustrated in Figure 1.12 (curve 2). This becomes obvious when we develop a function that represents this sequence of clusters in the Fourier series. For simplicity, it is considered that the form of each femtosecond cluster is close to the Dirac delta function. In this case, the Fourier spectrum of the cluster wave—composed of a periodical reversed sequence of the delta function (Figure 1.11)—can be represented as it is shown in Figure 1.13. It should be noted that, after accomplishing the Fourier development, we find that the discussed compression procedure (Figure 1.12) can be treated mathematically as a generation of many harmonics in the Fourier spectrum. From a physical point of view, the same procedure (i.e., the formation of the clusters) could be regarded as a synthesis of the required femtosecond clusters of many generated electromagnetic wave harmonics. Therefore, the proposed method for formation of the clusters is also called the method of the synthesis of the electromagnetic cluster waves. Let us summarize: the form of each femtosecond cluster should resemble the form of the Dirac delta function. This means, automatically, that the cluster Fourier spectrum must contain many harmonics with the commensurate spectral amplitudes.

The above-described method for cluster synthesis is not the only possible variant; its only merit is that it is the simplest one. As an example of a variant of the synthesis of the higher-order clusters we give the following imaginary experiment, illustrated by Figure 1.14. We consider that, instead of each of the first-order spectral component of the simple spectrum shown in Figure 1.13, we can generate a spectrum of "harmonic of harmonic," as shown in Figure 1.14 for the harmonic numbers $m = 1,3,5$. By composing such frequency "spectra of spectra," we gain the possibility of forming a second-order spectrum with the amplitudes E_{nm} presented in Figure 1.14. Accomplishing the reverse Fourier transformation, we obtain in the real space an extremely interesting cluster picture. It is easy here to be convinced that we also obtain the first-order clusters, which in such situation are the primary clusters. However, at this time, they are described by a much higher degree of compression (Figure 1.12) and amplitude. We also obtain a system of smaller clusters that have smaller amplitudes and wider widths. They are distributed among the primary clusters.

It is obvious that, by a similar scheme, one can build the third-, fourth-, and higher-order clusters. One should emphasize that when the methods of formation of higher-order cluster waves have been proposed, the physics of their interaction with matter is not yet studied enough, neither in the time of the Star Wars nor since.

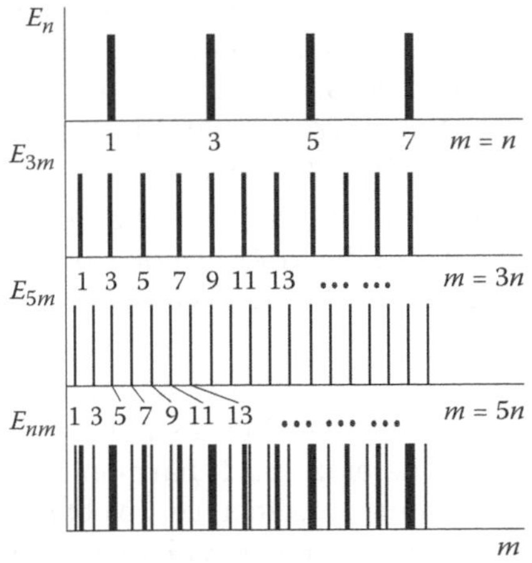

FIGURE 1.14
Illustration of the method of synthesis of second-order cluster waves. Here, E_n are the amplitudes of spectral harmonics of the simple (first-order) cluster wave shown in Figure 1.13, E_{3m} are the amplitudes of spectral harmonics—each of them, in turn, is the third harmonics ($m = 3$) of the cluster wave in Figure 1.13, E_{5m} are similar amplitudes, but for the case of fifth harmonics ($m = 5$), and amplitudes E_{nm} represent the total spectrum of the cluster of the second order.

1.4.2 Harmonic Coherence Problem and Some Key Principles of Construction of the Cluster Weapons

How does one realize technologically the idea of simultaneous generation of many harmonics for the synthesis of the femtosecond clusters? It should be mentioned that a few such technological solutions were proposed during the time of the Star Wars program. Now, the number of the known design variants is larger (Part II). However, before we begin to discuss them, let us describe some physical specificities of some of the previously described methods of the electromagnetic cluster synthesis.

How does one interpret the very process of compression of the wide semisinusoid pulses (1) into narrow clusters (2) as illustrated in Figure 1.12? Indeed, it is not clear how it happened that a part of the wave electric field (1) looks as if it disappears during the compression procedure within the interval between 0 and $\Delta = T/4 - \tau_{p2}/2$. Simultaneously, it strongly increases within the limit of the duration τ_{p2} of the cluster (2) (Figure 1.12). The harmonics of the wave (1) sum up the first interval, Δ, by taking into account their initial phases. The signs of some are positive, while the signs of others are negative. As a result of such algebraic summing up, the resulting amplitude of the wave's electric fields is equal to 0. In contrast, within the second interval τ_{p2} the signs of the harmonic amplitudes are positive; in this case, the resulting amplitude strongly increases. This means that the process of formation exhibits a specific obviously expressed interferential nature. Therefore, the discussed phenomenon of cluster synthesis has also been called the effect of multiharmonic interference. It is easy to conclude that all harmonics that participate in the formation of clusters should exhibit a high degree of coherence.

By comprehending the technological aspects that arise from the given physical observations, we can derive the key principles for the most rational ways to make cluster system weapons. The following are some examples:

1. In real cluster systems, the different wave harmonics cannot be generated by different electromagnetically mutually unconnected oscillators with self-excitation. This is primarily due to the stochastic nature of the initial phase of the vibrations that always occur at the initial process of harmonics generation and achievement of a high-level of mutual coherence is unreal.
2. The use of some nonlinear system (an element or a system of elements) that is attached to a master oscillator of the first harmonic of the cluster wave is the most obvious method for generation of the necessary collection of the mutually coherent harmonics.
3. It is obvious that by a simple duplication of harmonics in nonlinearity, one can obtain neither the necessary high level of instantaneous power nor the average power of the cluster wave. Therefore, each harmonic should undergo a stage of substantial amplification in some amplifier, i.e., to obtain a radiation source in the form of a multiharmonic generation block with an external excitation.
4. An analysis made it clear that only the FELs can serve as the key technological basis in the formation of this class of the amplifiers (Part II). All other known amplifiers of the electromagnetic waves (quantum as well as classical) are unsuitable to work within the needed wavelength range (from ~10 mm to hundreds of angstroms).
5. Only the high-current FEL amplifiers meet the key demands, which are formulated regarding the source of cluster system emission. Only they are able to secure the maximum power at a minimal overall size (Part II).
6. The harmonic (one frequency) as well as the multiharmonic types of FEL amplifiers may be used as the key elements in the formation of various cluster system weapons. In this, klystron design schemes and their arrangement can secure the best technological results (Part II).
7. The parametric one-beam and two-beam FELs as well as the superheterodyne two-beam and plasma-beam FELs are most promising for applications in the cluster systems (Part II).
8. The formation of the powerful cluster electromagnetic waves of many harmonics should take place in some synthesis system, built on the basis of different optical and quasioptical elements.
9. The stochastic component of each of the harmonics (electromagnetic noise) of a cluster wave should be minimal because only the coherent components of harmonics participate finally in the process of cluster formation.
10. The use of cluster sources of powerful electromagnetic emissions as cluster weapons foresees the presence of special antenna systems. In contrast to traditional antenna systems, these special antennas should secure efficient control and operation of the parameters of each cluster harmonic or groups of harmonics simultaneously. These are the key characteristics of the system. Such systems can be built, including on antenna mirrors that are especially arranged in space.

It is obvious that, in addition to these problems, the task of creating an effective cluster system weapon foresees the solution of many associated problems. In practice, many of them appear to be no simpler than the key problems. But it is important that, in contrast to the general science/technological situation considered during the Star Wars program (see Sections 1.1 and 1.2), this time the totality of the existing key and associated problems does not look so hopeless. In contrast, even a superficial analysis of the project reveals that the formation of a cluster-based strategic ABM system of defense is absolutely realizable. Moreover, this conclusion involves pure fundamental physics, as well as the overall state of temporary technologies in this sphere. The most critical problems related with this project are of a financial/political nature.

1.4.3 A New Star Wars System Based on CFEL: Is It Possible?

How did the initiators of the construction of cluster system weapons imagine their realization within the scope of the Star Wars program? In those days, a score of construction methods was proposed and each of them solved specific problems of antimissile defense (Sections 1.1 and 1.2). By considering the specificity of their construction, the totality of them can be classified as:

1. Systems based on FELs with an external block of harmonic generation and also external forming (synthesis) of cluster waves (see the example in Figure 1.15)
2. Systems based on FELs with an external block of harmonic generation and internal forming of cluster waves (see the example in Figure 1.16)
3. Systems based on FELs with an internal block of harmonic generation and also internal forming of cluster waves (see the example in Figure 1.17)
4. Combined systems

Here it is appropriate to comment on some key characteristics of cluster systems. We will start the investigation with a discussion of a system with an external harmonic generation block and external formation (synthesis) of cluster waves, as shown in Figure 1.15. The system performs in the following way: The master oscillator (1) generates the initial signal on the first harmonic of the cluster wave (ω_1, k_1). An analysis of the project reveals that one can get the most rational variant of the system as a whole when one chooses the frequency of the first harmonic of the cluster wave within the millimeter to sub-millimeter range. However, the final decision about the choice of the first harmonic range can be made only after obtaining enough data about the value τ_r, which describes the relaxation time of a substance after a cluster passes through. It depends, essentially, on the density and the

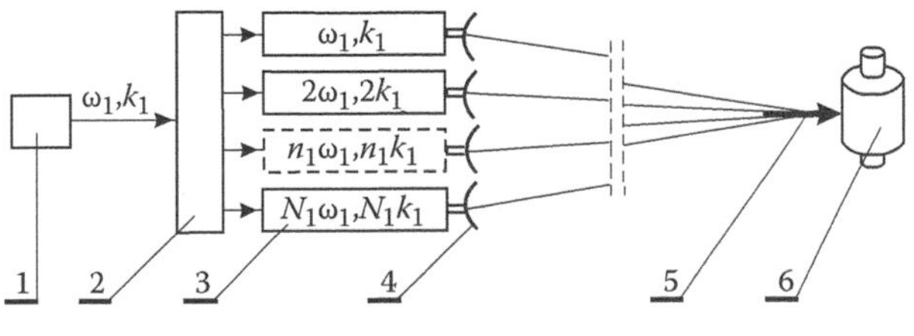

FIGURE 1.15
Design scheme of a cluster system based on FELs with an external block of harmonic generation and external formation of cluster waves. Here, 1 is the master oscillator, 2 is the block of nonlinear elements, 3 represents amplifiers of electromagnetic harmonics, 4 represents the antenna systems for electromagnetic harmonics or groups of harmonics, and 5 is the cluster electromagnetic signal that acts on the target (6). The values ω_1, k_1 are the cyclic frequency and wave number of the first harmonic of the cluster wave (5), and, at the same time, the cyclic frequency and wave number of the master oscillator (1). The values n_1 and N_1 are the current and highest harmonic numbers.

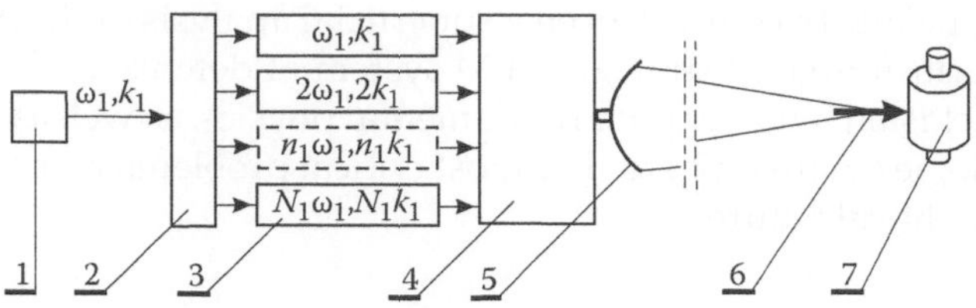

FIGURE 1.16
Design scheme of an FEL-based cluster system with an external block of harmonic generation and internal formation of cluster waves. Here, 1 is the master oscillator, 2 is the block of nonlinear elements, 3 represents the amplifiers of electromagnetic harmonics, 4 is the synthesis system, 5 represents the multiharmonic antenna systems, and 6 is the cluster electromagnetic signal that acts on the target, which is 7. The values ω_1, k_1 represent the cyclic frequency and wave number of the first harmonic of the cluster wave (5) and, at the same time, the cyclic frequency and wave number of the master oscillator (1). The values n_1 and N_1 are the current and highest harmonic numbers.

type of material that a cluster has to overcome in the process of penetration toward the target. One should not rule out the possibility that, in the course of the investigation of such interrelations, it will be discovered that the constructive variants in which the first cluster harmonic is within the optical range might have a practical application.

The harmonic signal (ω_1, k_1) is then directed into the block of nonlinear elements (2). A previous analysis revealed that the formation of such blocks does not involve complex technology; therefore, we may skip it for simplicity's sake with only a mention that, on exit from the block (2), the calibrated multiharmonic signal moves off and enters the block of harmonic amplifiers (3).

The block of amplifiers (3) is the key object of the entire cluster system. An analysis of the project shows that a few tens of equivalent variants are possible in a technological realization of the system presented by Figure 1.15. All of them are based on the high-current FEL, as a rule, the klystron-type FEL. This includes the simple and two-stream parametric FELs, the superheterodyne two-stream and plasma-beam FELs, and others. The final choice of the specific construction scheme of an FEL depend on its purpose and the specific characteristics of the working and task conditions in which it performs. Other physical and technological factors should also affect the choice; however, the majority of them have not been adequately studied until today. In Part II, a detailed analysis is given for the high-current FEL. Here we give only some general remarks.

In the first case, two types of possible arrangements of a system as a whole were analyzed. Their common characteristic was that the whole block of harmonic amplifiers, including their acceleration systems, was expected to perform like a single multichannel monoblock with common vacuum, power, control, and guidance systems. In the first case,

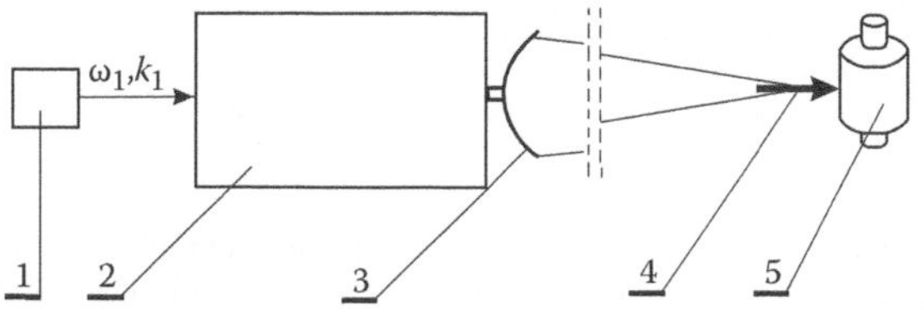

FIGURE 1.17
Design scheme of an FEL-based cluster system with internal harmonic generation and formation of cluster waves. Here, 1 is the master oscillator, 2 is the multiharmonic amplifier, 3 represents the multiharmonic antenna systems, and 4 is the cluster electromagnetic signal that acts on the target, which is 5.

it was proposed that the block be made in the form of a cylinder with the external part consisting of amplifiers and electron accelerators (including the multichannel and multibeam ones) only. The internal part would be used for arranging the accessory systems.

In the second case, the block could be flat, looking like a bookcase or honeycomb. The project analysis revealed that both cases had their strong and weak sides. Therefore the final choice can be made only in the process of concrete scientific research and project work.

The antenna systems for the electromagnetic harmonic waves or groups of harmonic waves (Figure 1.15) deserve special attention. As it follows from the above-formulated concept, the system is expected to perform as an aggregate of well-coordinated and phased separate (partial) antennas or antenna groups. The main purpose of such antenna systems is the formation of N_1 electromagnetic spots at each harmonic. This provides a common spot cluster on a target. Since the radiation at each of the harmonics differs essentially by wavelength, this automatically indicates the essentially different cross-section dimensions of partial antennas. Also, when the antenna for the first harmonic should be adjusted to put the target within the focus region, the higher harmonic antennas should be refocused with respect to it; the higher the number of harmonics, the higher the degree of its refocusing.

The basic characteristic of the described method of cluster wave synthesis is that, according to its electrodynamic properties, the emitted beam might be conditionally divided into two parts. The first part, which adjoins the antenna directly, has its partial emission beams on harmonics (or on groups of harmonics) emitted practically mutually independent of each other (at least in the nearest antenna region; see Figure 1.15). In the second part, the cluster waves (6) are formed and their main destination is to strike the target (7). Remember that in the interval between the system and the target, the emission beam might partially (or totally, depending on the system's destination) move through the atmosphere. It is obvious that the beam might pass it without attenuation only in the second part of its move, when the cluster wave is formed. This suggests that satellite-basing is the most rational method when using design versions of the system illustrated in Figure 1.15. In such cases, even when using the system for striking an object on earth, underground, or underwater, the cluster formation happens in a vacuum, without attenuation, and enters the atmosphere in the form of a cluster wave. The most rational design to use is the one illustrated in Figure 1.15 as a space–space, space–air, or space–ground type system.

An example of a cluster system FEL with an external block of harmonic generation and an internal formation (synthesis) of cluster waves is presented in Figure 1.16. It is obvious when one compares these two schemes, Figure 1.15 and Figure 1.16, that the basic difference between these two is the separate functions of cluster synthesis and the cluster wave direction to the target. The fulfillment of the first function is the synthesis system (4), and the fulfillment of the second function is the antenna system (5) (Figure 1.16). This takes place within the system (4), and as a result, a cluster wave (6) appears that is then aimed toward the target (7) by the antenna system (5).

A shortcoming of the design presented in Figure 1.16 is the impossibility of forming super-short clusters, which are able to pass freely through condensed matter without attenuation. However, this system is able to form more durable clusters that can pass through the atmosphere without attenuation. This paradox is explained by the transparency of the antenna (5) for a super-short cluster radiation. Therefore, using this type of construction scheme for the composite EMI laser weapon is the most rational method for a ground-based application, including ground–air, ground–space, or ground–space–ground (or air) classes. The low-frequency component of the cluster signal ensures an EMI effect mechanism, while the high-frequency component is responsible for the purely laser mechanism of the strike process. Overall, in the presented case, the cluster characteristics of the signal are used to secure the passage of the radiation through the atmosphere. Owing to the multichannel characteristics of the amplifier block (3), one can incidentally solve a series of complicated technological problems. This includes achieving very high levels of signal intensity (owing to the composition of various power generators), to easing thermal problems, and others.

Considering the previous discussion, the optimal variant of the combined system includes a combination of both of the design schemes presented in Figures 1.15 and 1.16. The peculiarity of such a combination is that at least a part of the partial antenna systems in Figure 1.15 are accomplished in the multiharmonic variant (like that shown in Figure 1.16). In general cases, we can have a few antenna subsystems where each of them emits a partial cluster wave. Each of theses waves is composed of partial clusters of durations sufficient for passing through the atmosphere without damping. But this duration is insufficient to pass through the material of the antenna. Then, the final step of the formation of the super-short clusters takes place in the process of propagation of a few partial cluster waves to a target. As a result, at the moment of reaching the target, the resulting clusters obtain a duration needed for overcoming the external obstacles for penetrating into the critical zone of the target.

Ground-basing is anticipated for such systems, including the satellite-based variants of their antenna systems. The systems of ground–air, ground–space, and ground–space–ground (or air) classes represent the most rational application. The most interesting and exotic applications of the combined systems might be their utilization for destroying underground command posts, moderate-depth submarine missile launch complexes, nuclear power stations, and other such targets.

In Figure 1.17, a design scheme is given in which the processes of multiplication, amplification of harmonics, and synthesis of cluster waves take place within the volume of one multiharmonic amplification block (2). A formed cluster wave is released from the exit of this block (2) and propagates toward the antenna system (3). As in the previous case, the shortcoming of this system is the impossibility of forming super-short clusters. The application field as well as the basis methods are the same as those discussed for the system presented in Figure 1.16. It should be mentioned that, within the scope of present knowledge in the field of multiharmonic FELs (Part II), it is impossible to give preference to one scheme over another similar one.

The solution of the formation of the super-short clusters might involve a combination of the systems presented in Figures 1.15 and 1.17, respectively. The system presented in Figure 1.17 forms a partial cluster wave that is able to propagate in the atmosphere without damping. The remaining multiharmonic sub-blocks generate harmonic signals or a few partial clusters that are able to propagate through the atmosphere without attenuation. Using the scheme presented in Figure 1.15, assembling these partial clusters into one allows the formation on the target of short clusters of the needed duration.

One more version of this type of complex system exists. It depends on a change of all the simple (harmonic) amplifiers presented in Figure 1.15 to the cluster blocks presented in Figure 1.17. The first harmonics of each of these cluster blocks should be chosen in a way that they will coincide with the simple harmonics of the master oscillator (1), which are generated by the nonlinear element blocks (2) in Figure 1.15. Consequently, a second-order cluster wave forms on the target. This method of cluster synthesis was previously discussed and illustrated in Figure 1.14. So, from a technological point of view, in principle, there exist practical possibilities for the formation of higher orders of cluster waves. However, today, the physics of the interactions of these electrodynamic objects with ordinary substances is much less studied than the physics of simple cluster waves. A brief qualitative analysis reveals a series of absolutely fantastic possibilities that open a series of extraordinary practical realizations of higher-order cluster systems.

Finally, let us turn to a specific type of cluster electromagnetic weapons: mixed (combined) EMI laser weapons. Generally, both physical mechanisms of action in the EMI as well as the laser mechanisms occur simultaneously. However, in some specific situations, when the EMI mechanism is prevailing, the system as a whole can be regarded conditionally as a cluster EMI weapon. In an opposite situation, it operates as a cluster laser weapon. This peculiarity is connected with the extremely wide spectrum of narrow clusters that penetrate into the crucial parts of the object of attack. The total spectrum can be conditionally separated into three parts: low frequency, high frequency, and intermediate frequency. The low-frequency part of the spectrum (the millimeter range waves, as a rule) acts as an EMI weapon, whereas the high-frequency part acts as a laser weapon. The action of the intermediate part has a mixed character. The EMI and laser systems have different thresholds of action with respect to pulse energy (the laser threshold is much higher than the EMI's; see Section 1.2). This criterion determines the correct classification of the CESW as an EMI weapon or a laser weapon.

1.4.4 Example of Multiharmonic FELs

Owing to the accumulated traditional experience of contemporary physical electronics and radiophysics, the physical and technological essence of the structural schemes presented in Figures 1.15 and 1.16 are understandable. This is not the case with the structural schemes of the multiharmonic CFEL amplifiers presented in Figure 1.17. Electronic devices that are able to generate and simultaneously amplify wide harmonic spectra, with approximately the same spectral amplitude, are still exotic today. The physics of such systems is discussed in sufficient detail in Chapter 14. Let us qualitatively analyze the basic principles of performance of this type of electronic devices through discussion of the simplest variant of the one-section cluster two-stream superheterodyne FEL (TSFEL) amplifier, the design of which is presented in Figure 1.18.

The system discussed in Figure 1.18 can be regarded as a transformer of the sine-like electromagnetic signal (curve 1 in Figure 1.12) into a periodical sequence of short clusters (curve 2 in Figure 1.12). The device works in the following manner. One-harmonic electromagnetic signal (1) enters the input system (2) and is directed into the work bulk of the TSFEL. High-current relativistic electron beams (3 and 6), which have different initial energies, are formed

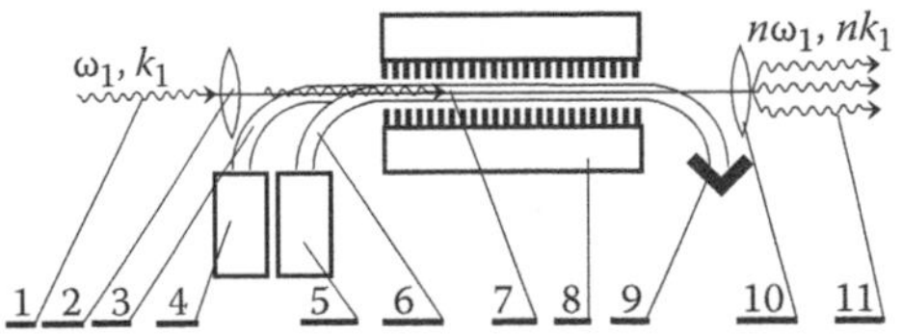

FIGURE 1.18
Simplest variant of the design scheme of the one-sectional CFEL amplifier. Here, 1 is the input harmonic wave signal from the master oscillator (where ω_1, k_1 are its frequency and wave vector), 2 is the input of the system, 3 is the first one-velocity high-current relativistic electron beam, 4 is the first high-current electron accelerator, 5 is the second high-current electron accelerator, 6 is the second high-current electron beam, 7 is the two-velocity (doubled) high-current electron beam, 8 is the multiharmonic pumping system, 9 is the electron collector [system for the recuperation of energy of the electron beam (7)], 10 is the output signal system, and 11 is the output multiharmonic signal in the form of a sequence of clusters (where $n_1\omega_1, n_1k_1$ are the harmonic frequencies and the wave vectors, and n_1 is the number of the Fourier-harmonic signals).

by the high-current accelerators (4 and 5). Both the beams (3 and 6) are merged together into one two-velocity (two-energy) high-current electron beam (7). This beam, in turn, is directed toward the operational part of the multiharmonic pumping system (8).

The characteristic feature illustrated in Figure 1.18 is the use of a multiharmonic pumping system (8). Its main destination is its dual function. The first function is assisting the multiharmonic modulation of the initial two-stream beam (7) with respect to electron density. The second purpose is assisting the generation and amplification of the highest harmonics of the initial electromagnetic signal (1).

The excitation (generation) and amplification of many of the space charge wave (SCW) harmonics occur via the superheterodyne amplification mechanism. It proceeds in the following stages. The first is the generation and the amplification of the first harmonic of the SCW, involving the effect of the tree-wave parametric resonance that is basic for all FELs. The second stage is related to the effect of the two-stream instability in the two-velocity electron beam (7). The superimposition of these two physical mechanisms leads to the appearance of an additional amplification of the first harmonic of the electromagnetic signal (1) and the excitation of many SCW harmonics. Then, these many SCW harmonics, interacting in a parametric-resonant manner with the harmonics of multiharmonic pumping (8), generate a spectrum of high signal harmonics. This physical mechanism is called the multiharmonic three-wave superheterodyne resonant interaction.

The most important physical feature of this mechanism is its explicitly expressed degenerated nature. Namely, each of the three harmonic waves (the harmonics of signal, pumping, and electron-beam waves) with the same harmonic number interacts with each other in a superheterodyne-resonant way. If we have exclusively, for example, N_1 harmonics for each of the three waves, then N_1 parallel three-wave superheterodyne resonances are realized simultaneously. As a result, the longitudinal multiharmonic electron-beam wave transforms (with amplification) into the transverse multiharmonic output electromagnetic signal (11) (Figure 1.18). In other words, in this case we have a unique resonant interaction of three multiharmonic clusters of different originations: the longitudinal electron-beam wave, the transverse pumping wave, and the transverse output signal clusters, respectively.

The Fourier spectrum of the output signal (11) should resemble the delta function Fourier spectrum (Figure 1.13). For an adjustment of the considered TSFEL for attaining such an optimal form of the spectrum, a number of various design versions were elaborated (Part II). However, as a rule, most of them are characterized by the same property: the output multiharmonic pumping systems have the regressive form of the Fourier spectrums when the amplitude of each following harmonic is smaller than the preceding one. At the same time, the spectrums of the electron-beam wave clusters have a progressive character (the amplitude of each following harmonic is higher than the preceding one). In such situations, the form of the spectrum of the output electromagnetic signal cluster can resemble the spectrum of the delta function (see the relevant results of numerical modeling in Part II).

1.5 Other Exotic Methods of Formation of Super-Powerful EMPs

1.5.1 History of the Problem

The problem of the generation, amplification, and transformation of electromagnetic waves became a burning question from the very moment in 1896 when Russian engineer Aleksandr Popov demonstrated (somewhat earlier than Marconi) the first radiotelegraphic apparatus. More than 100 years from this event passed in difficult and prolific work of many generations of scientists, engineers, and technologists. Our whole civilization changed as a consequence of the discoveries, transforming because of radio and television into something completely different, a unified social organism. The change touched everybody without exception and affected all facets of our lives from everyday life to military affairs.

However, despite the successes and varieties of existing classic methods, the generation, amplification, and transformation of electromagnetic signals is based on only one process: the transformation of the potential or kinetic energy of the motion of electrons into radiation energy. This means that traditional electronic instruments require the formation of either electron streams (in vacuum or solid material as in classic electronics) or special active inversion (laser) media (quantum electronics).

Simultaneously, during the time of the Star Wars program, new concepts were created and consolidated for electromagnetic energy generation, but by no means could they be considered classic electronics concepts. Within the framework of the program, concepts were studied as potential bases for the formation of pulses and clusters for new types of weapons. KW technology was considered as a possible technological basis. The basic idea of this concept depends on a direct transformation of the mechanic (kinetic) energy of motion of a macroscopic body into electromagnetic wave energy. The so-called explosion-magnetic generator (EMG) can be used as a convenient illustration of this type of system's basic principle (tactical weapon).

The principle of its action may be described as follows. Let us assume that we have an electric circuit with a significantly strong electric current, i (Figure 1.19a). Let us also assume that for some reason, e.g., an explosion, it undergoes a sudden compression (Figure 1.19b). According to Biot–Savart's law, the current (3) in the circuit (1) forms magnetic field $\vec{B}$ within the volume that the circuit encompasses. On compression, the magnetic field caused by the compression of the circuit (1, 3, and 4 in Figure 1.19) changes rapidly as a function of time. As a consequence of the electromagnetic induction, an electric field, $\vec{E}$, is generated and, subsequently, a change in the electric field results in the generation of a

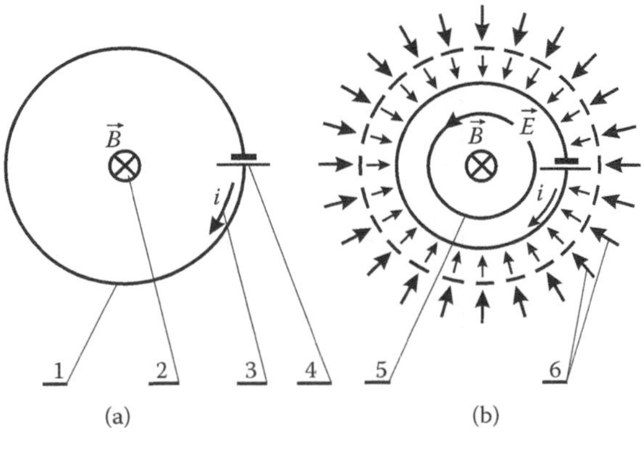

FIGURE 1.19
Operation principle of the EMG. Here, 1 is the closed electrical counter, 2 is the vector of magnetic field $\vec{B}$ generated by the electrical current, which is 3 (with strength i), 4 is the source of an electromotive force in the counter (1), 5 is the force line $\vec{E}$ of the electrical field generated by the changing magnetic field $\vec{B}$, and 6 is the external mechanical force.

magnetic field, and so on. Hence, a circuit becomes the source of an exceptionally powerful EMP, which is formed without the application of any electron streams or laser media.

Of course, in their traditional form, the EMGs cannot be used in any strategic systems, including those used in Star Wars. First, they generate spherical electromagnetic waves, the intensity of which diminishes strongly with the distance. The strategic system requires the formation of narrow emission beams, the intensity of which decreases only weakly with distance (Section 1.2). The aim was to find corresponding structural configurations for realization of the above-described ideas. Such conceptual approaches were proposed within the framework of the Stars Wars program.

Considering this, it is worth mentioning that the idea of a direct transformation of the mechanical energy of the motion of macroscopic bodies into radiation energy is historically not very new. In his time, Albert Einstein proposed this type of idea, which was named Einstein's mirror. Essentially, the Doppler effect occurs in light reflection from a moving mirror (Figure 1.20). Let us analyze this scheme in detail.

Assume that a flat mirror is moving in parallel with respect to itself and that its surface is perpendicular to its motion gradient (Figure 1.20). The electromagnetic radiation pulses with carrier frequency ω_{str} strike the mirror surface at the same right angle. In agreement with the known equation, the frequency of the reflected electromagnetic wave ω_{ref} will exhibit Doppler displacement.

$$\omega_{ref} = \omega_{str}\frac{1+v/c}{1-v/c}, \tag{1.6}$$

where v is the velocity of the mirror and c is the speed of light in vacuum. At the present time, mankind has learned how to better generate low-frequency electromagnetic radiation, e.g., in the radiomicrowave range; however, the generation of the millimeter and more shortwave signals is critical in many aspects (Sections 1.1 and 1.2).

Formally, this problem could be solved quite simply by using Einstein's mirror at relativistic velocities (at $c - v \ll c$; $\omega_{ref} \gg \omega_{str}$). Unfortunately, this idea cannot be practically realized, at least not in this way. In this case, even when using a very small mass mirror, it would require fantastic energy to accelerate it to the speed of light. The means for such acceleration do not exist at the present time. In principle, the maximal velocity to which

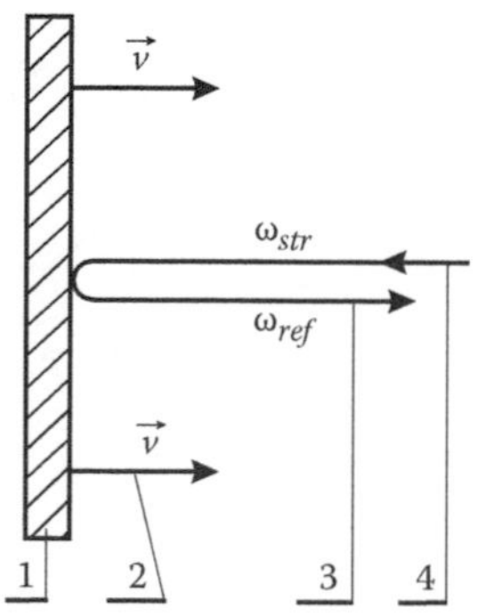

FIGURE 1.20
Model of Einstein's mirror. Here, 1 is the moving mirror, 2 is the vector of velocity of the mirror's translational motion, 3 is the reflecting electromagnetic wave with frequency ω_{ref}, and 4 is the striking wave with the frequency ω_{str}.

one can accelerate a low-mass macroscopic body is ≤3 · 10⁶ cm/s. In this case, the Doppler displacement of frequency,

$$\delta = \frac{\omega_{ref} - \omega_{str}}{\omega_{ref}} = \frac{2v/c}{1 - v/c}, \quad (1.7)$$

revealed by the insulting value $\delta \leq 2 \cdot 10^{-4}$ does not represent any practical interest. However, even in a case of mirror mass ~10 *g*, it requires the energy

$$\Delta E \approx \frac{mv^2}{2} \sim 4{,}5 \cdot 10^6 \text{ J}, \quad (1.8)$$

which is equivalent to the explosion of a powerful ordinary aviation bomb.

In such a way, in all its superficial attraction, the idea of Einstein's mirror in its original form will always remain in physics as an original intellectual exercise. In this context, it is worth mentioning that there were attempts to reconstruct this effect by using a plasma mirror, which forms by a plasma flux or high-current relativistic electron beam. Some aspects of this will be discussed further later in this chapter.

Simultaneously, as we will show, it is too early to consider the idea of Einstein's mirror as a lost cause. The general situation is not as hopeless as it appears. Further development of Einstein's mirror, although in a completely different context, allows this conclusion. In a very concrete case, it was used as an example of the effect of multistage increasing of frequency of an electromagnetic signal in a resonator with a moving wall. Here, EMG and a modified Einstein's mirror are combined. The modification involves multiple uses of small Doppler displacements of frequency during a large number of reflections.

In summation, we come to a conclusion that, at our present level of fundamental knowledge and technology as well as from a Star Wars point of view, the practical perspective of this class of systems is not yet finally clarified. Because of this, the scientific/technological trend must be controlled, despite its exotic basic idea.

1.5.2 Example of the Experimental Realization of a Resonator with a Moving Mirror Wall

Let us analyze an example for a possible experimental realization unit in the framework of which one would be able to generate powerful EMPs. Basically, this device is a further direct development of the EMGs (Figure 1.19). It is based on the model of EMP formation as a result of multiple reflections from a moving and static mirror (see the model in Figure 1.24 for detail). The simplest variant of this type is illustrated in Figures 1.21 through 1.23.

The operation principle consists of the following: The powerful pulse microwave oscillator (5) (e.g., a few tens of megawatt magnetron or klystron) is connected to a high-quality resonator (11) (Figure 1.21). In the resonator, it excites an intense standing wave (9). As it is illustrated in Figure 1.21, this standing electromagnetic wave can be separated into two traveling electromagnetic waves (8 and 10), which propagate in opposite directions. We consider the wave moving from left to right [(10) in Figure 1.21] the sticking wave, and the wave (8) (Figure 1.21) moving in the opposite direction as the reflected wave.

Simultaneously, with the beginning of the wave (8 and 10) excitation, the undermining apparatus (2) (Figure 1.21) generates an explosive pulse. This results in detonation (1). Under the pressure of the gun-powder gas that are formed in the explosion, the missile (3)

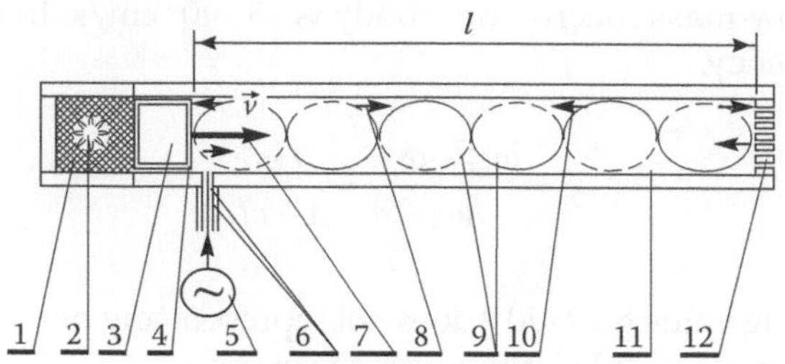

FIGURE 1.21
Variant of the design scheme of the electromagnetic gun with a traveling mechanical mirror. Here, 1 is the explosive, 2 is the undermining apparatus, 3 is the missile, the front part of which (4) serves as the mirror or mobile wall of the resonator (11), 5 is the microwave oscillator, 6 is the system for excitation of the electromagnetic waves within the resonator (11), 7 is the velocity vector of the mirror (4), 8 is the direction of the traveling electromagnetic wave propagated from left to right, 9 is the standing wave in the resonator (11), 10 is the direction of the traveling electromagnetic wave propagated from right to left [both traveling waves form the standing wave (9)], 11 is the resonator with the mobile wall, 12 is the immobile wall of the resonator (11), and l is the length of the resonator (11).

(Figure 1.21) begins to move along the longitudinal axis of the resonator (11) with acceleration a. As a result of this motion, the missile (3) blocks the system (6) for excitation of the electromagnetic waves within the resonator (11). It results in an electromagnetic isolation of the volume of the resonator (11) from the microwave oscillator (5): The resonator (11) remains electrodynamically closed. In such situations, it has neither an input nor an output.

The rear stationary (immobile) wall of the resonator (11) is accomplished in the form of a wall having a system of small transverse-size channels (Figure 1.21). Each of the channels can be regarded as a beyond-critical waveguide (microwaveguide). Therefore, in respect to the beyond-critical microwave standing wave (9) (Figure 1.21), the rear wall of the resonator (11) appears as a continuous conducting surface.

The front part of the missile (3) is made in the form of a resonator mirror. Its motion results in a compression of the resonator's electromagnetic field with a simultaneous increase of its frequency (see (2) in Figure 1.22). At the end of the terminal stage of the electromagnetic wave compression, when the missile mirror approaches the immobile wall of the resonator, the wavelength of the standing wave (9) (Figure 1.21) becomes near critical and, at the final stage of its motion, becomes higher than critical. Consequently, the microwaveguides, made in the immobile wall, transform from the beyond critical into

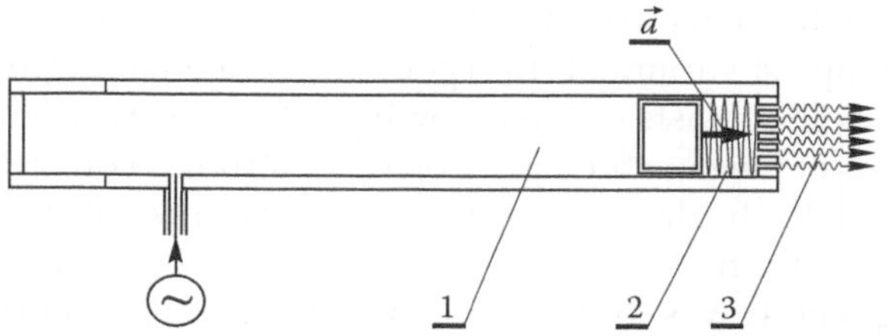

FIGURE 1.22
Terminal stage of compression of the electromagnetic standing wave in the electromagnetic gun represented in Figure 1.21. Here, 1 is the powder gas, 2 is the standing wave at the terminal stage of its compression, and 3 is the output electromagnetic radiation that begins to percolate through the apertures of the immobile wall of the resonator. All other elements of the drawing are explained in Figure 1.21.

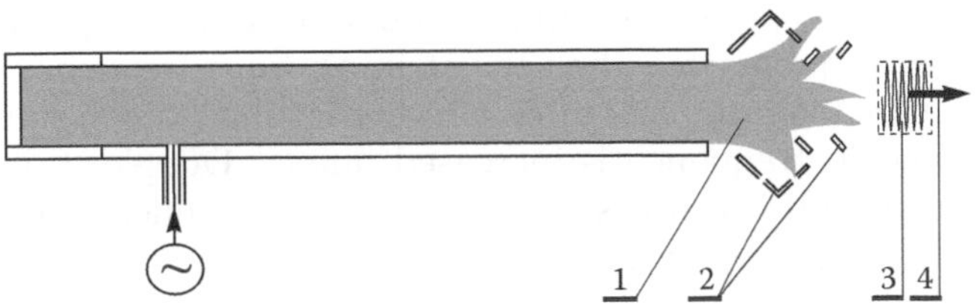

FIGURE 1.23
Final stage of the process of emission of EMPs by the electromagnetic gun represented in Figure 1.21. Here, 1 is the powder gas, 2 represents the fragments of the mirror and immobile resonator wall, 3 is the electromagnetic missile—EMP (cluster), and 4 is the direction of its propagation. All other elements of the drawing are explained in Figure 1.21.

subcritical. At this stage of the process, the EMP forms and is emitted as the output electromagnetic radiation (3) (Figure 1.22) through the subcritical microwaveguides. The very process of the formation of the pulse ends by the mechanical ruin of the immobile wall of the resonator (Figure 1.23).

1.5.3 Evolution of the Electromagnetic Wave Frequency in the Process of Multiple Reflections

In such a way, the model situation described above is realized in the system we investigated. Further, by using the scheme presented in Figure 1.21 as an imaginary experimental basis, we will discuss the evolution of the EMP parameters in the process of its multiple reflections [7].

We modernize Einstein's system presented in Figure 1.20 by adding one more parallel mirror. The mirror can move toward the first mirror or it can be immobile, as the one presented in Figure 1.21. Since this is not essential, we will use the model with a second immobile (stationary) mirror.

As in the case presented in Figure 1.21, we assume that, on the immobile mirror (5) (Figure 1.24), an electromagnetic signal (4) having the frequency ω_{str} is emitted. Then, in

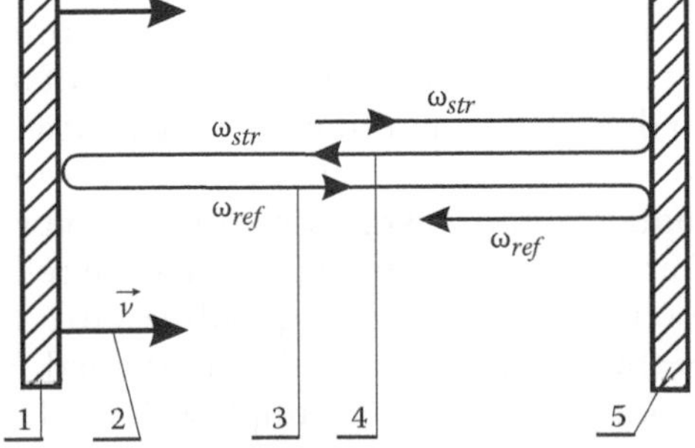

FIGURE 1.24
Model of a resonator with a moving mirror where the effect of the multistage frequency increase can be observed. Here, 1 is the moving mirror, 2 is the vector of velocity of the translation motion of the mirror 1, 3 is the part of electromagnetic wave where the frequency (ω_{ref}) does not change in the process of reflection from the immobile (stationary) mirror (5), 4 is the part of the wave whose frequency ω_{str} changes in the process of reflection from the moving mirror (1), and 5 is the (stationary) mirror.

agreement with Equation 1.5, the reflected electromagnetic wave (3) will have a $(1 + v/c)/(1 - v/c)$ times higher frequency ω_{ref} than the striking wave frequency ω_{str}. The striking wave, by moving toward the immobile mirror and reflecting from it, starts again to repeat its route. But this time, its frequency is increased due to Doppler transformation. As a translation development of this process, the frequency of the reflected from a moving mirror wave will become always higher and higher:

$$\omega'_{ref} = \omega_{str}\left(\frac{1+v/c}{1-v/c}\right)^N, \tag{1.9}$$

where N is the number of reflections from the moving mirror (1) (Figure 1.21), and ω'_{ref} is the frequency of the reflected wave after the N reflections. When the second mirror is also moving toward the first mirror with velocity v, then the equation for the frequency of the reflected wave coincides with Equation 1.9 with the exception that the N changes to $2N$. The described phenomenon has been named the effect of multistage increase of frequency. Another version of this phenomenon, which can take place in the multistage FEL, will be discussed in Part II.

Now, we will perform some numerical estimations. First, let us clarify if there is a possibility to solve the problem with an increase in the magnitude of the Doppler displacement of the frequency δ (see Formula 1.7). By taking into account Equation 1.9 in this case, we obtain

$$\delta = \left(\frac{1+v/c}{1-v/c}\right)^N - 1. \tag{1.10}$$

Similar to the preceding numerical example for Einstein's mirror, we take $v/c \sim 10^{-4}$ and assume $N = 3 \cdot 10^4$. Then, we directly from the equation and obtain $\delta \sim 4 \cdot 10^2 >> 1$. It is obvious that this value differs drastically from the case of Einstein's mirror (see comments regarding Formula 1.7). It must be taken into account that it is incorrect to expand in the power series in our example (Equation 1.10) because $N \cdot v/c = 3 > 1$.

In such a way, an elementary kinematical analysis allows us to ascertain that the proposed devices—considering those presented in Figure 1.21—might be effective in practice. For example, having a source of electromagnetic waves in the microwave range ($\omega_{str} \sim 10^{11}$ s^{-1}, which corresponds to the $\lambda_{ref} \sim$ 2cm wavelength), theoretically, we have a possibility of obtaining electromagnetic radiation in the infrared range $\omega_{ref} = \omega_{str}(\delta + 1) \approx \approx \omega_{str}\delta \sim 4 \cdot 10^{13}$ s^{-1}, which corresponds to the $\lambda_{ref} \sim 47$ μm wavelength. The presented numerical estimations illustrate quite clearly the sense and the principal idea of the given proposal.

Also, we should mention that a resonator in which the density of the wave energy is not zero can be considered as a complex electric circuit with currents. The currents, in agreement with the electrodynamics of the waveguide and resonator systems, flow on the surface of its inner walls and are responsible for a partial wave reflection process. Therefore, the system based on a resonator with a moving wall (Figure 1.21) can be considered as a peculiar modification of the EMG. More so, the traditional EMG as well as Einstein's mirror, from the model's point of view, is the simplest limited variant of the scheme of a resonator with a movable wall.

1.5.4 Process of Pulse Energy Transformation

Let us determine the change of the electromagnetic wave energy density in the process of its reflection from the moving mirror [7]. We will perform our analysis by using the energy conservation law. At first, we will perform this analysis without taking into account the electromagnetic field energy losses and then we will estimate its effect on the process of the pulse formation in this system.

Let us investigate the dynamic parameters of an electromagnetic wave at a single reflection from a mirror having surface S within a short time interval, from t to $t + \mathrm{d}t$. Figure 1.25 reveals that, within this time interval, the volume that the wave occupies decreases from

$$\mathrm{d}V_{str} = S(c + v)\mathrm{d}t \tag{1.11}$$

to

$$\mathrm{d}V_{ref} = S(c - v)\mathrm{d}t. \tag{1.12}$$

In Figure 1.25, the volume of the striking and reflecting waves is dashed and the position of the mirror at time $t + \mathrm{d}t$ is marked by a dashed line. The reflection of the electromagnetic wave involved the work of the mirror on the field.

$$\mathrm{d}A = F \cdot v \cdot \mathrm{d}t. \tag{1.13}$$

We define the interaction force of the field and the mirror (the force of the light pressure) as

$$F = \frac{w_{str}\mathrm{d}V_{str}}{c\mathrm{d}t} + \frac{w_{ref}\mathrm{d}V_{ref}}{c\mathrm{d}t}. \tag{1.14}$$

In Equation 1.14, w_{str} and w_{ref} are the densities of the striking and reflecting waves, respectively. The first and second terms in Equation 1.14 are derivatives of the striking

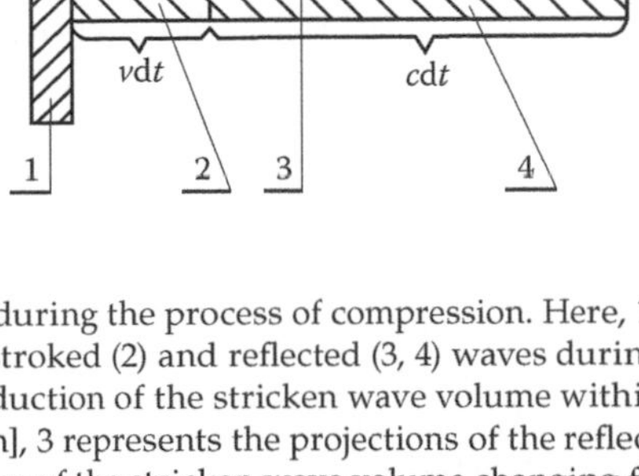

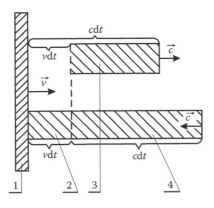

FIGURE 1.25
Evolution of the EMP parameters during the process of compression. Here, 1 is the moving mirror, 2–4 are the projections of the volumes of the stroked (2) and reflected (3, 4) waves during the time interval from t to $t + \mathrm{d}t$ [including (2) characterizes the reduction of the stricken wave volume within the time interval from t to $t + \mathrm{d}t$ on account of the mirror (1) motion], 3 represents the projections of the reflected wave volume changing from t to $t + \mathrm{d}t$, 4 represents the projections of the stricken wave volume changing from t to $t + \mathrm{d}t$, and v is the velocity vector of the mirror (1).

and reflected pulse of the electromagnetic field of the wave. Considering the energy conservation law we have

$$dA + w_{str\kappa} dV_{str} = w_{ref} dV_{ref}. \tag{1.15}$$

Then, by taking Equations 1.11 and 1.12 into consideration, we obtain

$$\frac{dw_{ref}}{dw_{str}} = \frac{(1+v/c)^2}{(1-v/c)^2}. \tag{1.16}$$

In such a way, the density of the reflecting wave increases on account of a decrease of the volume, which remained free for its spreading, as well as on account of the transfer of a part of the mechanical energy of the mirror to the field. One can easily obtain the relation in Expression 1.16 by using the equations for the transformation of the electric and magnetic field intensities at transition from one reference frame to another.

In our model, we consider that the striking electromagnetic wave represents a rectangular pulse of the duration τ_{str}. Also, we point out that the known Manley–Row relationship emerges from Expressions 1.6 and 1.16. The relationship describes only the conservation law for the number of electromagnetic field photons:

$$dN_{str} = dN_{ref}$$

where

$$dN_{ref,str} = \frac{w_{ref,str} dV_{ref,str}}{\hbar\omega_{ref,str}}, \tag{1.17}$$

and $\hbar$ is Planck's constant and dN_{ref}, dN_{str} are the numbers of the striking and reflecting photons in the laboratory reference frame.

Now, it is possible to determine how the energy of an EMP changes during a reflection

$$\frac{W_{ref}}{W_{str}} = \frac{w_{ref} dV_{ref}}{w_{str} dV_{str}} = \frac{1+v/c}{1-v/c}. \tag{1.18}$$

In this equation, W_{str}, W_{ref} are the energies of the striking and the reflecting pulse, respectively.

The duration of a pulse is directly related with its spatial length. Therefore, the ratio of the duration of the striking to the reflecting pulses is

$$\frac{\tau_{ref}}{\tau_{str}} = \frac{dV_{ref}}{dV_{str}} = \frac{1-v/c}{1+v/c}. \tag{1.19}$$

Then, we consider that the real mirrors absorb a part of the striking energy, i.e., the reflection coefficient $R < 1$. The reflection coefficient R is calculated by the ratio of the reflected energy to the striking energy, in the reference frame related to the immobile mirror. Therefore, it could be determined as

$$R = \frac{dN'_{ref}}{dN'_{str}}, \tag{1.20}$$

where dN'_{ref}, dN'_{str} is the number of the striking and reflecting photons. For the laboratory reference frame, analogous to Equation 1.20, we can write

$$R = \frac{dN_{ref}}{dN_{str}}. \tag{1.21}$$

It follows that the number of the striking or reflecting photons is an invariant value. In the reference frame related to an immobile mirror, the amount of the photons reaching it equals

$$dN'_{str} = \frac{w'_{str} c dt' S}{\hbar \omega'_{str}}. \tag{1.22}$$

By using Equations 1.17 and 1.11 in the laboratory reference frame, we have

$$dN_{ref} = \frac{w_{ref}(c+v)dtS}{\hbar \omega_{ref}},$$

then

$$dN_{str}/dN'_{str} = 1. \tag{1.23}$$

We used the well-known relations of the special theory of relativity to derive Equation 1.23.

$$\frac{w_{str}}{w'_{str}} = \frac{1-(v/c)^2}{(1+v/c)^2}; \quad \frac{dt}{dt'} = \frac{1}{(1-v^2/c^2)^{1/2}}; \quad \frac{\omega'_{str}}{\omega_{str}} = \frac{1+v/c}{(1-v^2/c^2)^{1/2}}.$$

An equation analogous to Equation 1.23 can be also obtained for the reflected photons. Then, by using Equations 1.6, 1.17, and 1.21, we have

$$\frac{W_{ref}}{W_{str}} = R \cdot \frac{1+v/c}{1-v/c}. \tag{1.24}$$

By generalizing the results of this subsection, we can stress that during a reflection from a moving mirror, the energy of the EMP increases under the condition $R(1 + v/c)/(1 - v/c) > 1$ (Equation 1.24), its frequency increases (Equation 1.6), and the pulse duration decreases (Equation 1.19). As was illustrated, after one reflection at nonrelativistic velocities, these values are of insignificant dimensions, whereas, in the multistage model, the amount of the reflections might be significantly large ($N >> 1$). Accordingly, the increase of the frequency and energy and the decrease of the pulse duration might be significant and experimentally weighty. Further, we move to an analysis of the dynamics of the EMP parameters at such multiple reflections.

1.5.5 Evolution of the EMP during Its Compression

Assume that the mirror moves along the axis z [(4) in Figure 1.21] at a constant acceleration of $a_z = a$. The velocity of the mirror is nonrelativistic. This case is the most interesting for an experiment. Here, we disregard the technical problems related to the presence and absence of electromagnetic radiation since they are immaterial in this case. We determine the dynamics of the momentum parameters in the process of decreasing the distance between the mirrors (Figure 1.24) [7].

We introduce the values v_n, l_n, which represent the mirror velocity and the distance between the mirrors, respectively, in the moment of n number of reflections from a moving mirror, t_n is the time between $n = 1$ and the nth number of the striking pulse on the moving mirror, and ω_n, W_n, and t_n are the frequency, energy, and duration of the pulse, respectively, after the nth reflection. A short pulse of the electromagnetic field forms near the surface of the movable mirror at the moment $t = 0$. Then, $v_1 = v_0$, $l_1 = l_0$, ω_0, W_0, and t_0 are the initial values of the frequency, energy, and pulse duration of the electromagnetic wave.

By using the introduced designations, we will write Equations 1.6, 1.18, and 1.24 as

$$\omega_n = \omega_{n-1}\frac{1+v_n/c}{1-v_n/c};$$

$$\tau_n = \tau_{n-1}\frac{1-v_n/c}{1+v_n/c};$$

$$W_n = W_{n-1}R\cdot\frac{1+v_n/c}{1-v_n/c}.$$

After N reflections, we have

$$\begin{aligned}\omega_N &= \omega_0\cdot X,\\ \tau_N &= \tau_0\cdot X^{-1},\\ W_N &= W_0\cdot R^{2N}\cdot X,\end{aligned} \tag{1.25}$$

where $X = \prod_{i=1}^{N}\left(\frac{1-v_i/c}{1+v_i/c}\right)$.

In the equation for energy W_N, the coefficient 2 in R^{2N} is stipulated by the reflection of the pulse (momentum) from the moving and immobile mirrors.

For the determination of the value X, we write the equations that correlate the velocity and distance for two subsequent reflections from a moving mirror as

$$\begin{aligned}l_n + l_{n-1} &= ct_n,\\ v_{n-1}t_n + at_n^2/2 &= l_{n-1} - l_n.\end{aligned} \tag{1.26}$$

By excluding t_n from Equation 1.26, we have

$$l_n = l_{n-1}\frac{1-v_n/c}{1+v_n/c}\cdot\frac{1-\alpha(1+v_{n-1}/c)/(1-v_n/c)}{1+\alpha}, \tag{1.27}$$

where

$$\alpha = \frac{2l_{n-1}a}{c^2(1+v_{n-1}/c)^2\left(1+\sqrt{1+4l_{n-1}a/(c^2(1+v_{n-1}/c)^2)}\right)} \approx \frac{l_{n-1}a}{c^2}. \tag{1.28}$$

By taking into account $v_{n-1}/c \ll 1$, we evaluate the multiplier in Equation 1.27 and get

$$l_n = l_{n-1}(1-2(v_{n-1}/c)-2\alpha) \approx l_{n-1}(1-2(v_{n-1}/c)-(v^2-v_0^2)/c^2). \tag{1.29}$$

Here v is the mirror velocity at $l = 0$. As we see, the third item in Equation 1.29 is significantly smaller than the second. Therefore, one can disregard the items that contain a. Then,

$$l_N = l_0\prod_{i=1}^{N-1}\left(\frac{1-v_i/c}{1+v_i/c}\right); \qquad X = \frac{l_0}{l_N}\cdot\frac{1+v_N/c}{1-v_N/c} \approx \frac{l_0}{l_N}, \tag{1.30}$$

where v_N, l_N are the velocity and distance between the mirrors at the moment of the Nth reflection. From the equations for the uniformly accelerated motion we get

$$v_N = v_1 + \frac{2a(l_1-l_N)}{v_1+\sqrt{v_1^2+2a(l_1-l_N)}}. \tag{1.31}$$

By considering Equation 1.30, we estimate the number of reflections

$$N \approx \ln(l_0/l_N)\cdot c/(v_1+v_N). \tag{1.32}$$

By taking into account Equations 1.25, 1.31, and 1.32, we get

$$\omega_N = \omega_0\frac{l_0}{l_N};$$
$$\tau_N = \tau_0\frac{l_N}{l_0}; \tag{1.33}$$
$$W_N = W_0(l_0/l_N)^{1-2(1-R)c/(v_1+v_N)}.$$

We considered that $1 - R \ll 1$ when we obtained Equations 1.33.

1.5.6 Discussion of Obtained Results

Equations 1.33 determine the nature and magnitude of changes in the frequency, energy, and duration of the EMP during the mirror motion from l_0 to l_N in the system presented in Figure 1.21 [7]. It appears that the evolution of the pulse duration depends only on the correlation l_0/l_N, which obviously means that these values are independent of the number of reflections. If we assume, for instance, that $l_0 = 1$ m, and $l_N = 1$ cm (i.e., $l_0/l_N = 100$), then the pulse momentum increases 100 times when the pulse duration decreases for the same 100 times. In a purely theoretical model in which one can neglect the wave energy absorption by mirrors, i.e., $R = 1$, we obtain that the energy of the final pulse also increases 100 times and the density of its energy 10^4 times.

Theoretically, the formation of the femtosecond radio pulses is possible in this way. Unfortunately, we do not today have suitable technological solutions for this. Almost always the real mirrors have a reflection coefficient smaller than 1, but their value might significantly approach 1. For example, within a limited range of optical frequencies, the dielectric mirrors might have $R \approx (1 - 10^{-4})$. Within the decimeter frequency range, when using superconductor resonators, $R = 1$ can be realized. However, the superconductivity is not realized starting from the centimeter frequency range, i.e., here the value of $R < 1$. Notice that even in this range, the value of R does not decrease very fast if frequencies are increased to include the millimeter waves. Here, in some chosen frequency intervals, we can count the R value from $(1 - 10^{-5})$ to $(1 - 10^{-4})$, but formation of the femtosecond pulses requires much wider frequency intervals. Within these intervals, there should be extremely high values of the reflection coefficient R, but the technology required to make this possible does not exist. In this case, the intervals we can provide based on existing technologies are much more narrow than required. In the model, by assuming that the increase of the pulse energy is compensated by losses on mirrors, i.e., the pulse energy is constant, we can obtain the limiting value R_{lim} by using Equation 1.33

$$R_{\text{lim}} = 1 - (v_1 + v_N)/(2c). \tag{1.34}$$

This means that at $v_1 \sim v_N \sim 3 \cdot 10^6$ cm/s we have

$$R_{\text{lim}} \approx 1 - 10^{-4}.$$

By taking into account the experimental values of R, we might discover that the described effect of the multistage frequency increase could be realized experimentally. Within the optical frequency range, the laser radiation can increase only a few times, whereas, within the microwave range, the situation might be better. Namely, when a superconducting resonator is used, the frequency ω_{str} might be in the decimeter range, whereas ω_{ref} will be in the millimeter range, which means that $\omega_{ref}/\omega_{str} \geq 10^2$. The discussed design variant can be used in practice only in cases where one does not have any specific demands for the degree of coherency of the transformed signal. This is, therefore, quite a complex field structure of higher modes and a corresponding proper frequency spectrum might exist in the microwave resonators. The latter is stipulated by the dispersion of electromagnetic waves, a phenomenon present in this system. Consequently, it is necessary to take into consideration two important conditions that essentially determine the overall physical picture of the case.

The first of them is related to the dependence of the phase velocity v_{ph} of the electromagnetic wave on the frequency, which can be presented as

$$v_{ph} = c/n(\omega), \tag{1.35}$$

where $n(\omega)$ is the retarding coefficient. For the resonators having a smooth waveguide in their construction, $n(\omega) < 1$ and it can change from 0 to 1. If a retarding system is used instead of the smooth waveguide, the value $n(\omega)$ can attain a few units. In a case where the resonator is filled by the magnetized plasmas, we can obtain for the unordinary wave mode the retarding coefficient $n(\omega) \sim 10^2$ (see in Part II in more detail).

The point $n(\omega_{cr}) = 0$ is called the critical point, and the frequency corresponding to it, ω_{cr}, is called the critical frequency. By substituting the definition in Equation 1.34 in Equation 1.9 (instead of the light velocity in vacuum), we may come to the conclusion that the obtained value of the frequency transformation coefficient [in the case where $n(\omega) < 1$] should, in practice, be diminished 10^2 times. On the other hand, some other possibilities present themselves. For instance, when a segment of the microwave retarding system or waveguides filled by the magnetized plasmas [$n(\omega) > 1$] are used, one can obtain a significantly higher value of the frequency transformation coefficient.

However, some additional problems appear in this case. They are related to the surface nature of the microwave retarded electromagnetic waves in such systems. Some even more impressive results can be obtained by filling the resonator's volume with magnetized plasma. A specific characteristic of such systems is the possibility to use the unordinary types of waves as the striking waves. The retarding coefficient of these waves can reach a few tens or even more. Evaluations reveal that the transformation coefficient in such systems can reach $\sim 10^4$ and higher, i.e., in systems where the microwave range waves can be transformed into optical range waves. However, the totality of the existing technological problems does not promise an easy life for the potential experimentalists.

A different problem is related to the existence of the critical frequencies. Within the process of an increase of the electromagnetic wave frequencies, the last passes consecutively through the critical points $\omega_{cr,j}$ ($j = 1,2,3,\ldots$), which can be considered as the bifurcation points in this case. Passing through each of these points in the system means a simultaneous excitation of a few electromagnetic waves of this frequency, but with different phase velocities (Equation 1.34) instead of an excitation of only one electromagnetic wave. Accordingly, for each higher-type wave, we obtain significantly different values for the transformation coefficient. In the presence of a high number for these points, which is realistic within the decimeter to millimeter wave range, we will obtain—instead of a monochromatic frequency signal ω'_{ref}—a broad and complex frequency spectrum that can practically be considered as an almost continuous spectrum. Also, the signal loses its coherency. Analysis reveals that these systems could have a certain application as EMI weapons of small- or intermediate-radius action.

1.5.7 Other Examples of Electromagnetic Guns Based on the Resonator with a Moving Wall

The given analysis reveals that the idea of an EMP formation in a resonator with a moving wall apparently has more value theoretically than practically. However, we reached this conclusion only because we put the traditional artillery system in the basic construction variant (Figure 1.21). This technological realization means automatically that the velocity of mirrors can be achieved in the compression process of the electromagnetic wave. Essentially, it cannot exceed the value $\sim 3 \cdot 10^5$ cm/s, which is the average thermal velocity of molecules of a powder gas. In light of the analysis in Sections 1.5.3 to 1.5.6, this value appears too small. First, at such velocities, the problem of excessive electromagnetic energy losses is strongly accentuated due to the repeated reflections from the mobile and immobile mirrors. In addition to this, there is a problem with the mechanical friction of the missile (the front wall of which is the mobile mirror) during its motion in the acceleration channel. This, in turn, causes overheating of the mirror and, consequently, a deterioration of its reflective properties.

As a consequence of the previous discussion, the researchers looked at the electromagnetic KW as a more acceptable technological basis, primarily because it can assure an order higher increase of the missile velocity to ~$3 \cdot 10^6$ cm/s [1]. (Notice that the same numerical values for the velocity of the mobile mirror were mentioned in the previous corresponding numerical evaluations.) Also, the problem of friction can be solved radically by using the so-called magnetic hanger of the missile in the acceleration channel (the effect of magnetic levitation).

Notice also that within the framework of the Star Wars program, a few types of KW were investigated. Many variants of their construction remain. It became clear that the best perspectives have systems built using the so-called magnetic gradient force. The operation principle of one of these types of KW is illustrated in Figure 1.26. In this case, the main idea depends on the formation of a steep wall of force, F_{gr}, which moves with the velocity, $\vec{v}_w$, along the acceleration channel (the upper part of Figure 1.26).

Technologically, such a wall can be formed at the rear of the accelerated mass (3) by using a system of special induction elements (2) (Figure 1.26). The effect of the motion of the magnetic wall in the channel (4) at a given velocity, $\vec{v}_w$, is achieved by using a consecutive switching of the elements (2), synchronized with the motion of the accelerated mass (3), at velocity $\vec{v}$. The given synchronism condition can be written as $\vec{v}_w = \vec{v}$ (Figure 1.26).

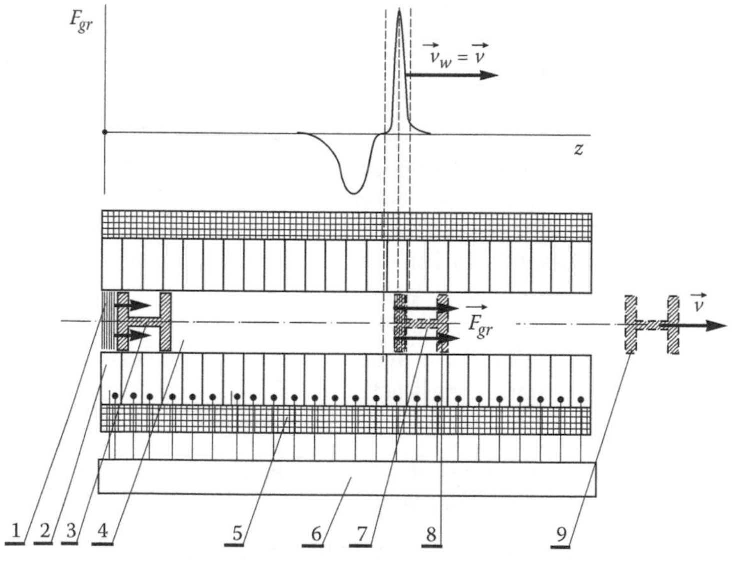

FIGURE 1.26
Variant of the design scheme of a KW, based on the idea of the moving gradient magnetic force. Here, 1 is the gradient magnetic force wall at the initial state, 2 represents the design elements of the system for the formation of a moving gradient magnetic wall (1), 3 is the accelerated ferro-magnetic mass, 4 is the acceleration channel, 5 is the magnetic system for creating the effect of magnetic levitation for the mass (3), 6 represents the driving and control systems, 7 is the position of the mass (3) in the process of acceleration, 8 represents the vectors of the gradient force $\vec{F}_{gr}$ acted on the mass (3), 9 is the position of the mass (3) after the acceleration, and $\vec{v}$ is the vector of the mass (3) velocity. The upper part of the drawing illustrates the synchronous motion of the gradient force (magnetic) wall (1) during the process of mass (3) acceleration. There, the value $\vec{v}_w$ is the velocity vector of the traveling magnetic wall (gradient force $\vec{F}_{gr}$).

In this case, the mass (3) is constantly under the effect of the gradient force, F_{gr}, that is responsible for the acceleration during motion in the channel (4).

Figure 1.27 illustrates variants of the practical realization of each individual element (2) in detail. The long line (coaxial or artificial) that is attached to the short coil (4), which might have only a few loops, is the basis of the element. The system is charged by the driving system (1) energy until it reaches a certain level of tension. This state of the system is described as the expectation mode. In the moment the system is started, a special starting pulse passes the driving system (1). The starting pulse reaches the commutation system (3) via the delay line (2). The key technological characteristic of the elements (2) (Figure 1.26) is that each has its individual tuning on the depreciation time delay.

Owing to this, the depreciation of each of the elements (2) subsequently moves from left to right. By this switching, a short tension pulse (of a few megavolts or more) forms in the first element of the forming line and generates a super-high pulse of current in the wings (5) (Figure 1.27). In turn, the passing of the megaampere electric current through the wings (5) results in the formation of a rapid strong pulse of the magnetic field (6). After a certain time interval, the following element depreciates and, as a result of this, the pulse (6) moves from left to right and then the third element depreciates, etc. As a result of this activity, the previously mentioned effect, the moving magnetic wall, is realized in the acceleration channel. The velocity of the magnetic wall is determined by the choice of the delay time of the depreciation of each element (2) in Figure 1.26.

The specific characteristic of the thus-formed magnetic wall is the steepness of its front side. This causes the formation of a pulse of the gradient magnetic force $\vec{F}_{gr}$ (8) (Figure 1.27). Since the generation of each of the force pulses by subsequent special elements (2) (Figure 1.27) takes place with a delay, the motion of the magnetic wall is accompanied by the effect of the moving gradient force $\vec{F}_{gr}$. Like the magnetic wall, the wall of the magnetic force propagates along the accelerator channel of the system with the velocity $\vec{v}_w$ (Figure 1.26). The mass (3) moves in the channel with velocity $\vec{v}$ under the action of the magnetic force gradient $\vec{F}_{gr}$ [(7) to (9) in Figure 1.26).

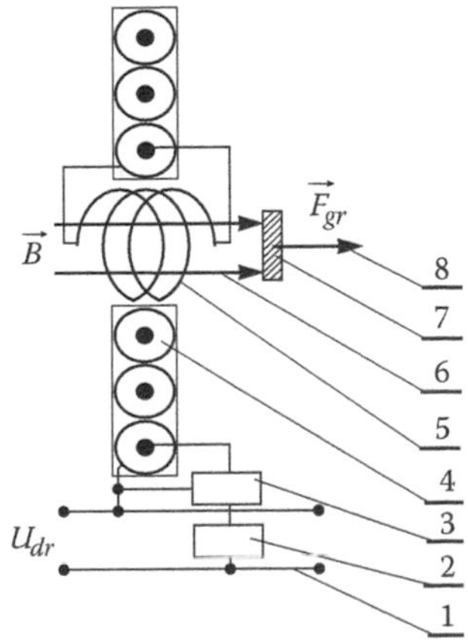

FIGURE 1.27
Elementary induction element of a system for the creation of a moving gradient magnetic force wall. Here, 1 is the driving system, 2 is the delay line, 3 is the commutation system, 4 is the forming line (coaxial or artificial line), 5 represents the wings with super-high pulsed current, 6 represents the vectors of induction $\vec{B}$ of the longitudinal magnetic field, 7 is the accelerated object, and 8 is the vector of the gradient force $\vec{F}_{gr}$.

In principle, the wall velocity $\vec{v}_w$, which has the physical meaning of a phase velocity, might be any value, including the velocity of light. Simultaneously the velocity of the accelerated mass (7) in Figure 1.26 is defined by the second law of dynamics and, consequently, cannot be arbitrary. It means that the acceleration process can have a continuous character only under the synchronism condition when $\vec{v}_w = \vec{v}$ is organized at each stage of the mass acceleration. This condition could be satisfied by the optimization of time delay tuning for each section (2) in Figure 1.26.

The shortcoming of the system presented in Figure 1.26 as a generator of EMPs involves the following matters. Theoretically, it is possible to form the traveling gradient force $\vec{F}_{gr}$ within a wide range of velocities, including the relativistic velocities. However, we cannot completely use this possibility in practice because the energy required for the acceleration of a mirror becomes unacceptably high. To verify this, let us use Equations 1.7 and 1.8: With a mirror weight of ~10 g, to obtain the velocity of $\sim 3 \cdot 10^6$ cm/s, the mirror has to use $\sim 4{,}5 \cdot 10^6$ J of energy, and according to the previously discussed theory, we need higher mirror velocities.

So, the system presented in Figure 1.26, which appeared to be an interesting KW, does not have any practical value as the theoretical basis for a generator of EMPs. At this stage of research, it becomes obvious that one should look for other ways to solve these formulated problems; for example, try to decrease cardinally the mass of the mirrors while simultaneously increasing its velocity significantly. It should be mentioned that such technical solutions were found. One of these solutions is demonstrated in Figure 1.28. There is a proposed design version of the so-called plasma gun, the work of which is based on the principle of the moving magnetic gradient force (Figures 1.26 and 1.27). As is obvious, the key design difference from the KW discussed earlier is the substitution of the solid-state mass (3), [(7) in Figure 1.26 by the plasma mirror, (11) in Figure 1.28].

The operation principle of the system presented in Figure 1.28 is as follows. The magnetic trap, a standard for plasma fusion, is the source of the plasma, which is generated and maintained in the trap (2). Plasma maintenance is realized by the magnetic trap (2)

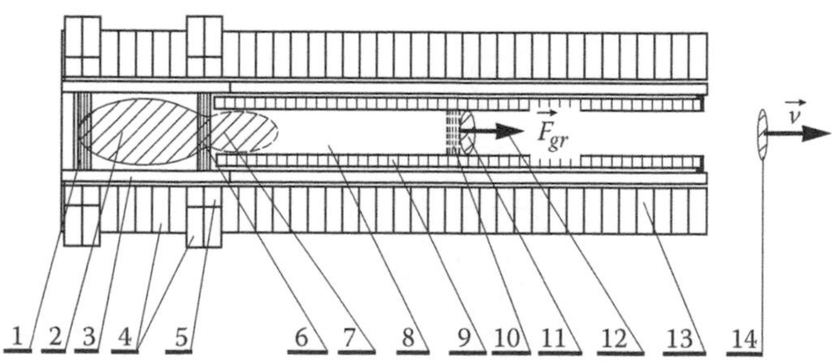

FIGURE 1.28
Variant of the design scheme of the plasma gun based on the effect of a moving gradient magnetic wall. Here, 1 is the magnetic wall in its initial state, 2 is the magnetized plasmas within the magnetic trap, 3 represents the magnetic systems for the creation and control of the magnetic walls for the plasmas (2) (magnetic trap), 4 represents the design element of the system for the creation of the moving gradient of magnetic wall in the volume of the magnetic trap (3), 5 is the magnetic system for the plasmas' (2) confinement, 6 is the magnetic wall in the input of the acceleration channel (8), 7 is the plasma bunch formed by the action of the magnetic wall (1, 6), 8 is the acceleration channel, 9 is the magnetic system for conveying the plasma bunches (7) in the acceleration channel (8), 10 is the magnetic wall (1, 6) in the process of the bunch (7, 11) acceleration, 11 is the bunch (7) in the process of acceleration, 12 is the vector of gradient magnetic force $\vec{F}_{gr}$, 13 represents the design element of the system for the creation of the moving gradient magnetic wall in the volume of the acceleration channel (8), and 14 is the accelerated plasma bunch formed by the plasma gun.

in which the so-called magnetic stoppers stop it from the left and right (Figure 1.28). It is important in this case that the formation of the traveling gradient magnetic force $\vec{F}_{gr}$ involves the region of the magnetic trap as well as the region of the acceleration channel [(4) and (13) in Figure 1.28].

The process of the magnetic bunch (7) formation in the acceleration channel begins by motion of the force gradient of the wall, starting in a position (1) in the left region of the magnetic trap (2). The motion of the force wall through the region of the magnetic trap (2) results in a drastic increase in the plasma pressure within its volume. This pressure cannot be stopped by the left trap magnetic stopper. As a result, the wall (1) squeezes out the plasma bunch (7) into the acceleration channel (8). Then, under the effect of this force wall [(6) in Figure 1.28], the plasma bunch accelerates to the given velocity, $\vec{v}$.

By comparing the systems presented in Figures 1.26 and 1.28, we can draw the basic conclusion that the plasma gun presented in Figure 1.28 is much more promising as the basis for the creation of an EMP generator than the system in Figure 1.26 because

1. The plasma mirror has a much lower mass than the equivalent solid-body analogue, i.e., an equivalent acceleration of the plasma mirror requires much less energy.
2. A plasma mirror can acquire a much higher velocity, up to ~10^8 cm/s, than the solid-matter analogue. This means that the same frequency value of the transformation coefficient can be obtained at a significantly lower number of reflections from the mirror (see the analysis in Sections 1.5.3–1.5.6). Hence, at the same values for reflection coefficients, we have significantly lower wave energy losses during a reflection from mirrors.

Within the scope of the information presented in this section, the principle of the electromagnetic gun based on the plasma mirror motion presented in Figure 1.29 is self-evident. Therefore, we will skip its description and move to a discussion of the more interesting, nontrivial problems of physics. For example, one should think that—owing to the multiple reflections—at an increase of the electromagnetic wave frequency, the reflection coefficient of a mirror should sharply decrease. This is explained by the well-known reflection characteristics of magnetized plasma. A question then arises: In such systems, can the transformation coefficient not be, in principle, a large value? However, detailed analysis reveals that, by far, it is not always so.

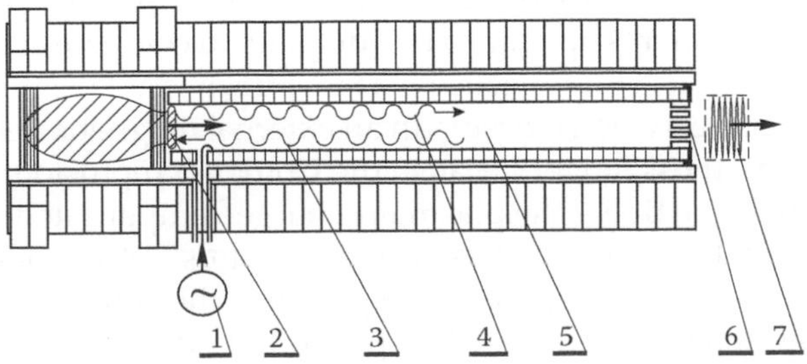

FIGURE 1.29
Variant of the design scheme of the electromagnetic gun based on a moving plasma mirror. Here, 1 is a microwave oscillator, 2 is a plasma mirror, 3 and 4 are the striking and reflected electromagnetic waves, respectively, within the resonator, which is 5, 6 is the system output, and 7 is the generated EMP (cluster). All other elements of the drawing are explained in the caption of Figure 1.28.

The point is that, in this case, one has to take into consideration the significantly high acceleration of a plasma mirror in the acceleration channel. The very plasma bunch as a whole exists in a strong force gradient field, where a stronger force acts on plasma particles near the force wall than on the more removed particles. In a case where plasma is sufficiently cold and the acceleration is high, an interesting situation arises where the density of the mirror plasma increases with an increase of its velocity. Accordingly, the frequency range moves up when the reflection coefficient is sufficiently high. When the velocity of a plasma mirror as a whole begins to exceed the average thermal velocity of the plasma ions, then the effect appears most strongly. Another no less interesting situation appears when the velocity of the plasma mirror exceeds the average thermal velocity of electrons. A practical realization of these effects allows a cardinal increase of the frequency of the transformation coefficient and, simultaneously, an increase of the especially promising rating of this type of system.

Historically, the described concept of moving gradient magnetic force was formulated more than 20 years ago. Regrettably (or perhaps happily, it is hard to know), after the collapse of the Soviet Union, this scientific trend is practically dead. During the time of the Star Wars program, the trend was considered a significantly serious physical basis for the creation of an alternative to the beam weapons of those days, like the plasma gun, laser, or EMI weapons. The theory showed that generators of powerful EMPs, on the basis of a resonator with a movable wall in the form of a plasma mirror, could have quite appealing qualities for practical utilizations.

When talking about beam weapons, it is necessary to mention the following. As is well known, the specificity of the so-called near space (near cosmos) as a potential theater of the war activity is the presence of the earth's magnetic field (the magnetosphere). The effect of this field on moving purely electronic or ionic beams could be essential, especially in the case of long-distance systems because of the curved trajectory of the charged particles in the process of their motion in the magnetic field. This effect makes an aimed strike of a target, from long distances problematic, to say the least. Replacing the purely electronic or purely ionic beams with beams of dense quasineutral plasmas opens an appealing possibility of a solution for the problem of the curved trajectory because, in this case, the magnetic Lorentz forces that affect the plasma electrons and ions are directed in mutually opposite directions. In contrast to the case of the purely charged beams, the effect of the magnetosphere on plasma bunches is confined to their partial traversal polarization. It is important that in the case of sufficiently dense plasmas, this polarization occurs without ruining the plasma bunch as a whole. In more detail, when the plasma bunch is dense (as is typical for the given systems) then the electric mutual attraction forces of the plasma electronic and ionic components appear to be adequate to compensate for the force of the magnetic field effect. Consequently, a dense plasma bunch is able to move along a rectilinear trajectory to a target even when it occurs in an external magnetic field. Also, the plasma bunch retains its compactness longer because the Coulomb repulsion forced between the same-sign charges are not acting – the plasma is charge-compensated as a whole.

What concerns the EMP generators, based on resonators with a mobile plasma mirror, the situation appears not completely clear. Despite that, the concept of such systems was developed during the time of the Star Wars program and did not provoke any serious objections by professionals, it didn't excite the curiosity of the developers of ESWs or theoreticians. With time, these ideas were forgotten, as was the very rich legacy of the ideas of those times. Only a few professionals probably remember today this important fragment of the history of the Star Wars program.

References

1. Velikhov, E.P., Sagdeyev, R.Z., and Kokoshyna, A.A. 1986. *Space weapons: The security dilemma.* Moscow: Mir.
2. U.S. Department of Defense (DOD). 1984. *The Strategic Defense Initiative. Defense Technology Study*. Washington, DC: DOD, April 1984.
3. Duric, M. 2003. *The strategic defense initiative: U.S. policy and the Soviet Union.* Farnham, UK: Ashgate Publishing.
4. Kulish, V.V. 1998. *Methods of averaging in non-linear problems of relativistic electrodynamics.* Atlanta: World Scientific Publishers.
5. Kulish, V.V. 2002. *Hierarchic methods. Hierarchy and hierarchic asymptotic methods in electrodynamics*, Vol.1. Dordrecht: Kluwer Academic Publishers.
6. Kulish, V.V. 2002. *Hierarchic methods. Undulative electrodynamic systems*. Vol. 2. Dordrecht: Kluwer Academic Publishers.
7. Kulish, V.V., and Lysenko, A.V. 1997. Effect of multistage increasing of frequency of an electromagnetic wave in the resonator with moving wall. *Gerald of Sumy State University* (Ukraine), 7:27–34.
8. Scarpetti, R.D., Boyd, J.K., Earley, G.G., et al. 1998. Upgrades to the LLNL flash X-ray induction linear accelerator (FXR). In *Digest of technical papers of the 11th IEEE International Pulsed Power Conference*, Vol. 2, 597–602. Baltimore, June 29–July 2, 1997.
9. Marshall, T.C. 1985. *Free electron laser*. New York: MacMillan.
10. Williams, R.E. 2004. *Naval electric weapons: The electromagnetic railgun and free electron laser.* Master's thesis, Naval Postgraduate School, Monterey, CA. http://www.dtic.mil/cgi-bin/GetTRDoc?AD=ADA424845&Location=U2&doc=GetTRDoc.pdf.
11. Allgaier, G.G. 2004. *Shipboard employment of a free electron laser weapon system.* Master's thesis, Naval Postgraduate School, Monterey, CA. http://www.dtic.mil/cgi-bin/GetTRDoc?AD=ADA420299&Location=U2&doc=GetTRDoc.pdf.
12. Mitchell, E.D. 2004. *Multiple beam directors for naval free electron laser weapons.* Master's thesis, Naval Postgraduate School, Monterey, CA. http://www.dtic.mil/cgi-bin/GetTRDoc?AD=ADA422357&Location=U2&doc=GetTRDoc.pdf.

2

Elements of the Theory of Hierarchic Dynamic Systems

The reader should resist the temptation to skip the first three sections of this chapter. They may appear boring and complicated, but there is no way around them. These sections provide a rigorous set of basic definitions, concepts, fundamental laws, and postulates. They serve as the basis for the general hierarchic theory of relativistic plasma-like systems developed later in the book, including the high-current FELs. Therefore, summon up the patience and read ahead.

2.1 Hierarchy and Hierarchic Dynamic Systems

What is hierarchy? Why does nature need hierarchy? What are the specific features of hierarchic systems? In this chapter, the author will try to give relevant answers to these and other similar questions.

2.1.1 What Is Hierarchy?

The term hierarchy evokes many different associations. We should recognize that it is the most unusual and, at the same time, enigmatic subject of our intellectual and spiritual life. It is also a trivial phenomenon of everyone's life. Indeed, there is hardly any doubt that we encounter hierarchy all around us. Each person lives in a completely hierarchic world. He or she occupies a fixed place in a social pyramid and the results of all events and situations depend essentially on that person's rank in a relevant hierarchic system (as the Latin proverb says, "Quod licet Jovi non licet bovi"). The hierarchic order of regional, state, national, and family levels reminds us of this continuously. In places of worship, each recalls that the Supreme Being has created our world with hierarchy and that the arrangements of the universe, human nature, and social order follow similar patterns. It seems incredible, but we can find some similarities in modern physics and engineering. In particular, there is a hierarchy in the condensed matter physics, cybernetics, coding and systems theories, economics, historiography, etc. [1–13]. Lately the concepts of hierarchy have been studied in oscillation-wave systems [1,3,14,15].

The essence of intuitive comprehension of the hierarchy phenomenon does not facilitate giving an adequate short and comprehensive definition of it. Therefore, let us confine ourselves to a not too precise or clear definition, but one that can, all the same, be accepted for our purposes. We will understand hierarchy as some ensemble of interacting parts that consists of a sequence of subunits, enclosed one within the other. But what is the ensemble of interacting parts? The answer is not too difficult: it is the so-called hierarchic system. However, another question arises: What are a system in general, and a hierarchic system in particular? Let us briefly consider this question.

2.1.2 Systems and Hierarchic Systems

We find a similar situation concerning the concept of a system. No traditional reference [16,17] offers a sufficiently simple and clear definition. Let us look at a typical traditional definition as an illustrative example: "We call the system an object of any nature (or an aggregate of interacting objects, including the objects of different nature) that has explicit 'system property' (properties). This system property does not belong to any individual part of the system for any method of its decomposition, and that does not follow directly from properties of the system parts" [17]. The systems that are arranged in accordance with hierarchic principles [1–15] (i.e., the systems consist of a sequence of subunits enclosed one within the other) are called hierarchic systems.

As is readily seen, the concept of special system property is the key point for understanding the concept of a system. In accordance with the given definition, this is such a new property—in principle—that it cannot be derived from the properties of any of its parts or elements. For example, air, a system of many molecules, has specific properties like the property to create wind, tornadoes, etc., which its individual molecules do not have. Another example: two young people decide to create their own family. Quite naturally, there emerges a problem: which of them shall be the head of the family? It is obvious that, before the wedding, this problem (i.e., the problem of the future family hierarchy) cannot exist in principle. Only the appearance of a special arrangement of social connections between the previously independent subjects (the potential wife, mother-in-law, family pets, etc.) makes up the system from the simple totality of these subjects. Or, in other words, it is this arrangement that generates a special system property in the form of various family relations.

We can also draw another conclusion from this life example: the fact (and the character) of interactions between the system subjects is key points of the concepts of a system and a hierarchic system (see Figure 2.1).

All systems can be classified as either dynamic or static systems. Dynamic systems are systems with motion. This means that, among the parameters that characterize such systems, we should always have the velocities of change of some of its characteristics.

A very important feature of dynamic systems is their ability to evolve. The evolution of a system can be observed only when its motion is unstable. Let us generalize the concept of

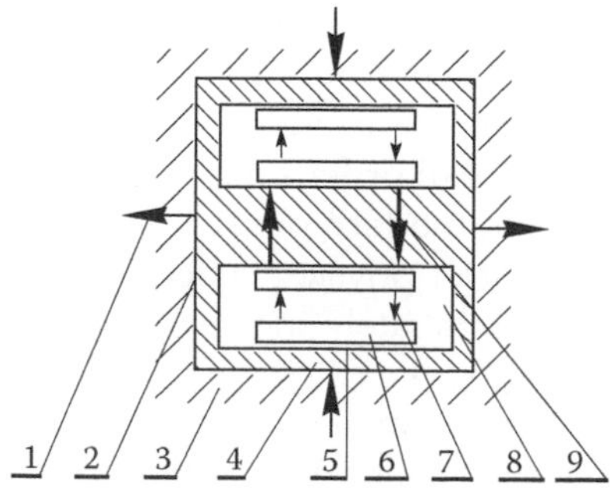

FIGURE 2.1
Illustration of the system concept. Here, 1 represents the connections (through relevant system inputs and outputs) with external systems, 2 is the considered system as a whole, 3 is the external system medium, 4 is the inner medium of the system (2), 5 represents the subsystems of the system (2), 6 represents the subsubsystems of the subsystems (5), 7 represents the connections between the subsubsystems (6), 8 represents the inner mediums of the subsystems (5), and 9 represents the connections between subsystems (5).

instability and introduce the concept of positive (increasing amplitudes of oscillations, for instance) and negative [decreasing (damped) amplitudes] instabilities. Taking these into account, we can reformulate the above to: All dynamic systems evolve due to their instability. We will exclude stationary dynamic systems (i.e., systems with stationary motion without instabilities and dissipations) from the discussion as these systems can be regarded as a special partial limit case of evolving unstable systems.

Static systems are systems consisting of immobile objects (i.e., they are systems without motion). Similarly to stationary systems, they can also be considered as a characteristic limiting case of dynamic systems when we neglect the motion of the system's subjects. An example of the system arrangement, which allows us to illustrate the general system's terminology, is given in Figure 2.1. Let us discuss it in more detail. First, we remember that separate parts of the system [see (5) in Figure 2.1] could have different properties from the properties of the system as a whole [see (2) in Figure 2.1]. In general, these parts can be treated as subsystems of the system, but the subsystems can also contain smaller parts. In turn, we treat these smaller parts as subsubsystems of subsystems [see (6) in Figure 2.1], and so on. On the other hand, the association of systems [(2) in Figure 2.1] forms the supersystem.

Hierarchic systems are a specific variety of system where each hierarchic level plays the specially arranged (in accordance with hierarchic principles) role of a subsystem, subsubsystem, and so on. Any system, supersystem, or subsystem has inputs and outputs for connections [see (1), (7), and (9) in Figure 2.1] with other systems, supersystems, or subsystems. The totality of these connections, outputs, and inputs forms the surrounding medium [see (3), (4), and (8) in Figure 2.1]. The exchanges of energy, matter, information, etc., are the main forms of communication (connection) between the systems, subsystems, subsubsystems, etc.

The concept of the system is rather relative. Each subsystem can be regarded as a system (with respect to a subsubsystem) or, equivalently, each supersystem can be considered a subsystem of some super-supersystem. As will be shown in this and the following chapters, the property of such relativity is one of the most peculiar characteristics of hierarchic systems.

The ordered aggregate of system parameters that determines the evolving system is called the state of the system. For instance, the state of a perfect gas in thermodynamic equilibrium is completely determined by three parameters: gas pressure p, temperature T, and volume V.

We distinguish open and closed systems. An open system has an exchange with a surrounding medium, whereas a closed system does not undergo any exchange. It is obvious that the system illustrated in Figure 2.1 should be classified as an open system. It should be emphasized that, strictly speaking, all natural and artificial systems are open. The concept of a closed system should actually be considered an idealization, like a material point in mechanics or a point charge in electrodynamics. Such a model (closed) could be obtained in a partial case provided we neglect the effect of all external surrounding systems and objects.

Let us turn to the hypothesis of our universe as a closed system that is accepted in traditional physics without any rigorous substantiation. This is done in spite of the presence of a number of well-known astrophysical observations, including those considered as the main experimental proofs of the big bang theory [47–49,50]. Therefore, the problem of openness of the universe is one of the most topical subjects in modern science [50]. This and other similarly intriguing issues will be discussed in this and the next chapters.

2.1.3 Complex Hierarchic Systems

In all its manifestations, our surrounding world can be regarded as a huge complex system. (Complex systems are systems that contain a large number of elements and connections between them.) According to the system theory [16,17], complex systems should be essentially unstable subjects. This seems to be a contradiction of both the theory and the real world, but let us discuss this topic in more detail.

The inclination to instability, according to the general theory of systems, increases with system complexity. Complexity is defined as the minimal length of an algorithm that describes the system structure and dynamics [1–3,16,17]. In turn, complexity increases with the number of elements and connections between them. Therefore, complex systems are believed, in principle, not to exist in nature. Nevertheless, a lot of complex systems obviously exist around us: our universe, the earth biosphere, human society, etc. What does it mean? Is the general system theory wrong? No, it is not. The point is that, among many types of possibly unstable complex systems, only one exception exists where the rule does not apply. This exception is hierarchic complex systems. Everyday practice shows that such systems are much more stable. All other natural complex systems formed with the birth of the universe seem to have ceased to exist long ago in the process of the universe's evolution. Thus, our world exists as a reality only because it is arranged in a hierarchic manner.

The most important and, at the same time, mysterious properties of natural complex systems are:

1. Uniqueness. Each complex system exists as an original or very rare subject. For instance, we know only one universe, only one human society, etc. Any doctor can say that every human organism is unique and exists in nature as an original. Each husband can attest that his wife and his mother-in-law, as complex systems, are also unique objects and each of them exists in nature as an original.
2. Weak predictability. One can obtain desired detailed information about a system's elements or its behavior at some interval $(-T, 0]$. However, this does not enable one to predict exactly the behavior of a complex system, as a whole, at further time interval $(0, \tau]$. The examples of the universe, human society, human civilization, and the wife and mother-in-law successfully illustrate this property.
3. Negentropyness or the purposefulness. Natural and quasinatural complex systems can control (to some degree) their own entropy to any random influences by external medium. The entropy characterizes the chaos within a system. The negentropy is the degree of order within the same system. The dynamics of negentropy characterizes the system's attempt to achieve some purpose. The purposefulness expresses a similar concept. This is the ability of a complex system to conserve and to amplify some basic dynamic process leading to the purpose. Therefore, both these concepts have a very similar physical meaning.

It should be recognized that the first two properties are intuitively comprehensible and, therefore, we will not pay much attention to them. However, the situation is diametrically opposite in the case of a third property. Moreover, it looks somehow mystical. Complex systems can demonstrate the ability to cognize and, because all existing complex systems are hierarchic, we can say that complex hierarchic systems can be treated as cognitive subjects. However, let us give more detailed consideration to this point.

2.1.4 Purposefulness and Self-Organization in Cognitive Hierarchic Systems

Thus, as previously mentioned, all complex natural systems possess the property of purposefulness. Hence, theoretically, some special mechanisms providing purposefulness should be found in nature. Experience shows that, indeed, such mechanisms do exist. For example, astrophysical analysis of cosmic catastrophes on the earth during the past 3 billion years [18] reveals that our biosphere has the obvious property of purposefulness. After every subsequent global catastrophe, the earth responds in a specific manner so that the negative effects on the biosphere are reduced to a minimum [18]. As a result, owing to purposefulness, the return to the main evolutionary line occurs every time in the shortest way possible.

So, the concept of purposefulness is closely connected with self-organization [1–17]. The self-organization is the appearance of far spatial and/or temporal coherence between the system parameters. It should be mentioned that only self-organizing hierarchic complex systems can possess the ability to cognize. This means that such systems have the ability to replicate another external system, each of its hierarchic levels (subsystem), and itself as a whole. This also means that the system knows how to model the surrounding medium and its hierarchic levels.

When we state that one system models another system, what we mean is that the first system is creating a compressed description of the second system. Hence, the modeling of any hierarchic level (system, subsystem, etc.) is always accompanied by information compression. We will give this property more consideration in the following sections.

2.1.5 Structural Hierarchy

A hierarchic system can be complex with respect to its structural and functional (dynamic) arrangements. The former characterizes the structural hierarchy, whereas the latter is related to the dynamic hierarchy.

Let us begin our discussion with the simplest example of a hierarchic system, referred to as the Russian matryoshka. The traditional version is shown in Figure 2.2. A typical matryoshka is commonly known to consist of a sequence of figures enclosed one within the other (see the definition of a hierarchic system in Section 2.1.2). With the exception of the smallest matryoshka, each larger matryoshka contains a smaller matryoshka.

The peculiar feature of matryoshkas is that one of its hierarchic levels differs from the other level by a characteristic amount (matryoshka height h_κ, in our case) only (see Figure 2.3). This is so because all matryoshkas have the same spatial proportion and arrangement.

FIGURE 2.2
Ordinary Russian matryoshka in an assembled state.

FIGURE 2.3
Illustration of the structural hierarchy concept. Here, the height of each matryoshka can be characterized by relevant hierarchic scale parameters h_κ. We can see that each of the following matryoshkas (or, in other words, the matryoshka of each higher hierarchic level) is of smaller height h_κ. Also, each matryoshka should contain a smaller number of structural elements (atoms and molecules in this particular case), i.e., is less complex.

Thus, the height of matryoshka h_κ could be regarded as a univalent (and rather convenient) structural hierarchic scale parameter (see Figure 2.3). All these parameters could be arranged in the form of the following structural hierarchic series:

$$h_1^{-1} < h_2^{-1} < \ldots < h_\kappa^{-1} < \ldots < h_m^{-1} < h_{m+1}^{-1}, \tag{2.1}$$

where m is the number of matryoshkas (i.e., the number of hierarchic levels), h_{m+1} is the structural comparison parameter (how tall a reader is, for example). Expressions like Equation 2.1 are called hierarchic series.

It is obvious that the provided example concerns a static hierarchic system only, which according to the above-mentioned terminology, is a system without motion. So, we can regard this system (i.e., the matryoshka) as the one possessing the structural hierarchy only.

Then, we can discover that every subsequent matryoshka contains a smaller number of structural elements (atoms and molecules of the matryoshka's material). It means that every higher hierarchic level of the matryoshka is characterized by a lower complexity than the previous one. By generalizing this, one can be sure that the formulated property is really common for all known natural hierarchic systems. We especially draw the reader's attention to this feature as it will be widely used in further discussion.

2.1.6 Dynamic Hierarchy

Besides the structural hierarchy, all real natural complex hierarchic systems also have a dynamic (functional) hierarchy. As mentioned above, this type of hierarchy is characteristic of natural dynamic systems.

Earlier, we introduced the concepts of structural scale parameters and hierarchic series (see Equation 2.1). Similar concepts could also be formulated for the case of dynamic hierarchy. Dynamic hierarchic scale parameters (in the form of velocities of changes in relevant system parameters) v_κ and dynamic hierarchic series could be introduced similarly with the preceding case

$$v_1 < v_2 < \ldots < v_\kappa < \ldots < v_k < v_{k+1}, \tag{2.2}$$

where k is the number of the highest hierarchic level, κ is the number of a considered hierarchic level, and v_{k+1} is the dynamic comparison parameter. The real existence of such relationships in nature could be illustrated by our universe (see Table 2.1).

It should be mentioned that hierarchic systems with purely structural hierarchies are known to exist. However, as a rule, all such systems have an artificial nature: various code systems (widely known in information theory), theoretical models like the spin glasses (in the physics of condensed matter), etc. Similarly, the systems with the purely dynamic hierarchy are also used in different theories (the models of oscillation hierarchic systems, for example, see Chapters 3 through 6). But, the human experience shows that all natural hierarchic systems are always combined, i.e., they have structural and dynamic hierarchies simultaneously (see the structural and dynamic parameters of the universe in Table 2.1, for example). Basically, hierarchic systems with different numbers of structural and dynamic hierarchic levels can be imaged theoretically, but these numbers are actually equal as a rule (again, see Table 2.1 for an example).

Thus, from a quantitative point of view, only one of two hierarchic series—Equations 2.1 and 2.2—is enough for characterization of this type of hierarchic systems. Normalizing Equations 2.1 and 2.2 with respect to the scale parameters h_{m+1}^{-1} and v_{m+1}, respectively, and, accepting $m = k$, we obtain the generalized hierarchic series in the normalized dimensionless form

$$\varepsilon_1 \ll \varepsilon_2 \ll \ldots \ll \varepsilon_\kappa \ll \ldots \ll \varepsilon_m \ll 1, \tag{2.3}$$

where $\varepsilon_\kappa = h_\kappa^{-1}/h_{m+1}^{-1} \sim v_\kappa / v_{k+1}$ is the normalized scale parameter of the κth hierarchic level. In the framework of the hierarchic methods described below, these parameters simultaneously play the role of relevant expansion parameters for the corresponding dynamic functions (see Chapters 3 through 7).

TABLE 2.1

Characteristics of Today's Universe as a Mixed Complex Hierarchic System

		Structural Hierarchy	**Dynamic Hierarchy**
		Scale of the Characteristic Sizes	**Scale of the Characteristic Angular Velocities (cyclic frequencies)**
Number of Hierarchic Level	**Physical Nature of Hierarchic Level**	**cm**	**s^{-1}**
1	Elementary particles	10^{-17}	10^{25}
2	Nuclei and electrons	10^{-10}	10^{19}
3	Atoms and molecules	10^{-6}	10^{15}
4	Substance (planet systems)	10^{8}	10^{-5}
5	Star systems	10^{15}	10^{-8}
6	Galaxy systems	10^{23}	10^{-15}
7	Metagalaxy	$>10^{30}$	$<10^{-37}$

Source: Kulish, V.V., *Hierarchic Methods. Hierarchy and Hierarchical Asymptotic Methods in Electrodynamics*, Vol. 1, Kluwer Academic Publishers, Dordrecht, the Netherlands, 2002.

2.2 Fundamental Principles in Natural Hierarchic Systems

As it is widely known, the method of principles (fundamental laws) is the basis of all modern natural sciences. The principles are formulated as a generalization of a great number of experiments and observations. The theory of the hierarchic dynamic systems is not an exclusion from this rule. Let us shortly discuss the fundamental principles that are the basis of the considered version of the theory of the hierarchic dynamic systems.

2.2.1 General Hierarchic Principle

Everything in the universe is hierarchic in nature. Indeed, it is impossible to find in nature or society any object that does not belong to a relevant level of some hierarchic system and does not have an intrinsic hierarchic structure. This means that hierarchy is really the most fundamental and widespread phenomenon of our world and, as it will be demonstrated in this book, the phenomenon of oscillations, in turn, is the most fundamental form of manifestation of this hierarchy.

2.2.2 Principle of Hierarchic Resemblance (Holographic Principle)

All principles discussed in this section are widely known and discussed in the professional scientific community [1–17]. However, there is one unexpected exception in this area and it is the so-called principle of hierarchic resemblance (holographic principle). This principle, despite being well known in modern physics, was not accepted as a fundamental law until now.

According to the principle of hierarchic resemblance, each hierarchic level in its general properties and arrangement represents the system as a whole. Or, mathematically speaking, the same type of dynamic equations describes every hierarchic level of the system, as well as the system as a whole. Each hierarchic level should be described in terms (variables) of its proper reference system. The search for a specific set of hierarchic transformations (from level to level) to provide the realization of the property of hierarchic resemblance is the main mathematical problem of the version of hierarchic theory under consideration [1,14,15].

We regard this principle also as a peculiar hierarchic invariance. The Boltzmann kinetic equation or the quasihydrodynamic equation can serve as evident illustration of practical application of the hierarchic resemblance principle [1,14,15] (see Part II in detail). From the considered theory principle, these equations have the same mathematical structure at every hierarchic level. However, every time this similarity occurs with respect to another (proper) set of dynamic variables, it is important to point out that this set is different for different hierarchic levels. Algorithms for finding such new variables and special hierarchic transformations form the basis of the theory of hierarchic asymptotic methods [1,14,15,19–29]. The key point of these algorithms is that they are constructed on the basis of the principle of hierarchic resemblance.

It looks mystical, but in spite of its obviousness, the principle of hierarchic resemblance is underappreciated for its true value in modern physics. However, its basic idea is very old and well known in ancient Indian, Middle East, Greek, Roman, Egyptian, and Mesoamerican doctrines and mythologies [30–32]. However, it is illustrated most clearly in the general cosmogony of the Old Testament and Kabbalah [33–37]. In accordance with its main doctrine, our surrounding world is organized within a peculiar self-resembling scheme. Every subject of any hierarchic world level, the level in itself, and the world as a whole have the same general arrangement. Some universal geometric pattern (a glif—a combined symbol—see Chapter 8) serves as a most elementary structural/dynamic element (module) for representation of a common basis of all possible varieties of this arrangement. This pattern is referred to as the tree of life (see Chapter 8 for more detail).

The tree of life as a peculiar universal structural module is the fundamental basis of everything existing in the world. Our universe as a whole, people, atoms, electronic devices (high-current FELs, for instance), and many other objects should have the same, in principle, general arrangement (see Chapters 3 through 8). Using the modern scientific language, we reformulate this ancient thought as follows: All natural hierarchic systems in the surrounding world meet the principle of hierarchic resemblance.

It means that each lower hierarchic level should model each higher level and, at the same time, the system as a whole. In such a situation, we can intuitively expect that the higher a hierarchic level is, the simpler its arrangement should be. This imparts sense to the hierarchic principle of information compression, which we will discuss later. Here, let us continue our discussion of the principle of hierarchic resemblance.

2.2.3 Sequences of the Principle of Hierarchic Resemblance

It should be noted that the principle of hierarchic resemblance has a few very interesting and useful sequences. The most important of them can be formulated as follows: The number of characteristic schemes and regularities in the universe is limited and the set of these schemes and regularities is universal for all manifestations of the world.

The formulated sequence has double significance. The first is methodological. It allows a far better understanding of the essence of the principle of hierarchic resemblance as a fundamental law. The second is of applied character. A new methodical tool can be developed on the basis of the discussed sequence. In this book, it is referred to as the method of the hierarchic resemblance.

Let us discuss the first of the two significances. In the process of living, many of us are amazed by the diversity and extreme complexity of the surrounding world. It gives an unusual feeling of endlessness to the variety of forms and regularities of the world's manifestations. Indeed, the beauty of living things and the frightening deepness of space, the charm of poetry and the cruelty of war—all these represent different sides of the same surroundings.

However, nature, possessing extremely rich diversity, is not endless. The illusion of infinite diversity, as well as its poetic component, disappears if we look at the world in the light of the principle of hierarchic resemblance. Then we find that a limited set of possible schemes and regularities allows us to describe the manifestations of the endless diversity. It is important to note that this set in itself is not too rich. The tree of life represents the key item of the set of universal schemes and regularities discussed in the chapter.

Just the properties of limitedness and universality reflect the key points of the suggested sequence of the hierarchic resemblance principle. The fundamental laws of physics provide sufficient evidence of it. For example, the phrase "The action equals the contraction" can be regarded as the well-known third law of dynamics in mechanics. On the other hand, the same statement can be used for a characteristic of an ordinary psychological life situation. At the same time, similar regularity is applicable for description of various political, economic, biological, and other dynamic processes. Or, in other words, this regularity does have a universal character, and its framework is not limited to mechanics. Likewise, examples could be proposed to illustrate the idea of the existence of universal schemes in the arrangement of natural hierarchic systems.

2.2.4 Method of Hierarchic Resemblance

The discussed sequence of the principle of the hierarchic resemblance mentioned in Section 1.2.3 forms the basis of the method of the hierarchic resemblance. Let us illustrate the essence of this method by performing a mental experiment. Assume that we know the arrangement and the main regularities that determine the physics of Bohr's atom, and we wish to construct a theoretical model, for example, a model of the solar system, which conditionally is considered an unknown. We also know that both systems belong to different levels of the same hierarchic system—our universe—and, consequently, both of them are governed by the principle of hierarchic resemblance.

The arrangement of Bohr's atom consists of a dense nucleus and electrons, which move (rotate) around this nucleus. This arrangement is known in physics as the planetary scheme (model) [48]. In this case, the main regularities are determined by the fact that electrons move in a potential field of the nucleus. In the considered partial case, the role of the potential field plays the electrostatic field.

Now, let us construct the scheme of the solar system as a hierarchic semblance of Bohr's atom. We consider the atom scheme and the regularities that we discussed as the characteristics. It means that the arrangement of the solar system should be also planetary: instead of the nucleus, we have the sun and instead of electrons, we have planets. Similar to the atom scheme, the planets should move in the potential field of the sun. In the discussed case, the gravitational field plays this role. Thus, the model of the solar system is constructed. It is not difficult to see that this model, in its most general properties, coincides with the well-known astronomical model.

Following the same methodical scheme, we can obtain the models of the planetary systems (like the systems of Jupiter or Saturn) or the galaxy systems. However, the given illustrative examples could appear too obvious because all of them belong to the astrophysical objects. Can we apply this method for constructing similar models of essentially different physical natures? Probably a reader would be surprised by the fact that we can obtain analogous schemes (arrangements) and regularities when trying to construct models of social/humanitarian and sociopolitical systems of different kinds (government, corporation, human society, etc.). Similar to the examples discussed here, we again have a dense nucleus (the highest authority of a different kind) and the electron–planet–satellites (groups of citizens, employees, etc.) that move within a potential field of the nucleus/authorities.

Generalizing, we can eventually construct a universal (for the discussed group of objects) model that contains concepts of a generalized nucleus, generalized satellite, generalized potential field, and generalized regularities of interaction.

Of course, the physical meaning of all partial models could be very different. However, in the situation discussed here, it is only important that they all be governed by the same, in principle, characteristic regularities and basic schemes of arrangement. The obvious external differences in this case characterize their different particular features, which are determined by individual physical properties of the modeled partial systems. They could be the number of the considered hierarchic level, the physical nature of the potential field, the number of structural elements and their characteristic sizes, and so on. Indeed, atoms, the solar system, galaxies, and social/humanitarian systems belong to different hierarchic levels and branches of the same universe as a dynamic hierarchic system. All of them are essentially different with respect to characteristic sizes (from ~10^{-6} cm to ~10^{23} cm), the number of structural elements (from ~10 to ~10^{12}), and so on. Nevertheless, in spite of this rich diversity of forms of existence, the number of characteristic schemes and regularities in the universe is limited and the set of these schemes and regularities is universal for all manifestations of the world.

2.2.5 Principle of Information Compression

As stated in Section 2.2, each higher hierarchic level is always simpler than the preceding one [the concept of system complexity is defined above (see Section 2.1.3)]. Our universe, human society, and other systems are the evident examples of this principle realization. We find that the smaller the number of structural elements (and corresponding connections between them), the higher the number of the hierarchic level. Indeed, we always have only one president or other head of state, a few heads of other power branches, a few tens of ministers, and so on. Simultaneously, millions of ordinary citizens are situated on the opposite end of the political or social hierarchic ladder.

Let us consider in detail the key points of the discussed principle as given in Figure 2.4. First, let us recall that any level of a hierarchic system can model other levels. But, what does it mean to model? This means creating a reflection (including the information reflection) of other levels. However, any model (reflection) should always be simpler than its original (Turing's theorem) [16,17]. Hence, the information about the state of the lower level (see Figure 2.4) should be compressed during the process of transference to the higher level. On the other hand, the upper levels could be treated as a model of the closer lower level. This means that it is simpler than the preceding one. Hence, any time that information is transferred from a lower level up, it is compressed by some information compressors. This compressor works similarly to the well-known computer programs like WinZip or Rar (see information compression in Figure 2.4).

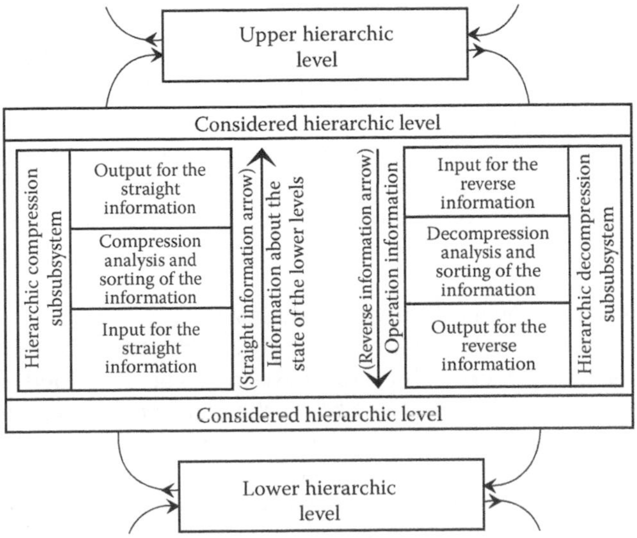

FIGURE 2.4
Treatment of the system hierarchic level (subsystem) as a compression operation subsystem (see also Figure 2.1 and corresponding comments).

But a question arises: Is any lower level superior in respect to an upper one? Indeed, if the lower level models the higher level, we can consider this as a manifestation of a certain superiority of the lower levels. However, it is not completely so.

We can illustrate this by the following obvious example. We will consider the hierarchic system of university staff and students–university rector (president). Of course, the rector, as the highest hierarchic level, exists only because of the existing university staff and students (lower hierarchic levels). This means that the rector position cannot exist, in principle, without the university professors and students. Similarly, professors (as positions) cannot exist without students. This can be considered as a peculiar superiority of students with respect to the professors and the rector. However, the students perceive that such superiority (which is considered as the democracy or ochlocracy in political systems) has a purely theoretical character in everyday reality. The point is that any one of us, being an element of some hierarchic system, always gets the reverse flow of information (see Figure 2.4), which moves top down. If the first flow of information (see Figure 2.4) corresponds to information about the state of levels as subsystems, the second flow concerns the algorithms of action for these levels, i.e., carrying the operation information (see Figure 2.4). Part of the information intended for the student level is generated at the rector level and contains a set of instructions or rules that completely governs real student behavior. As a result, practical and political student superiority realistically turns out to be rather doubtful.

Thus, taking that into consideration, the answer to the question of who is superior—students or rector—is obvious: authority always enjoys superiority. Why? Because the rector has the power; the rector makes the operational decisions (orders, instructions, etc.) that affect students' (as well as professors') lives. This observation is true and can be applied to any natural hierarchic system. Therefore, the numbering of hierarchic levels in the theory of hierarchic dynamic systems is opposite to the direction of operation information flow—from lower to upper levels (see Figure 2.4), i.e., bottom up.

2.2.6 Hierarchic Analog of the Second and Third Laws of Thermodynamics

So far we have not touched upon the question: Is the system considered determined or chaotic? As everyday experience shows, all natural hierarchic systems are a combination of order and chaos. A purely determined system is only a convenient idealization, like a perfectly rigid body in mechanics or a point charge in electrodynamics. The concept of entropy has been introduced in physics as the measure of chaos in a system. The inverse value (negentropy, see Sections 2.1.3 through 2.1.4) is, consequently, the measure of order in a system. The first and second laws of thermodynamics stipulate the behavior of thermodynamic entropy in physics. In particular, the second law says that in a closed system, entropy can increase or be constant but never decrease.

However, the concept of thermodynamic entropy can be generalized if we apply the idea that any physical (including thermodynamic) system can be described in terms of the theory of information [1–17]. As a result, we obtain the concepts of information entropy and information negentropy. In such an interpretation, they are also measures of chaos and order, but in this case we discuss information chaos and information order, respectively.

Let us assume that the discussed hierarchic system evolves with time. The evolution of our universe (see Chapter 8) could be illustrated in the following way. It is well known that this process proceeded from the lowest hierarchic level (elementary particles) to the highest level (the metagalaxy, which consists of galaxies, star systems, planet systems, etc.). It is interesting that the range of chaos in nature decreased during the process of evolution, i.e., during the creation of new and new hierarchic levels. Indeed, hadrons have formed nuclei, the nuclei and leptons (electrons) have formed atoms and molecules, and so on. Each new subsystem (hierarchic level) has been characterized by a higher order of negentropy than the preceding level.

Generalizing this observation in an arbitrary natural hierarchic system, we can formulate the hierarchic analogy of the second law of thermodynamics. In accordance with this law, each higher hierarchic level of a natural dynamic system has lower information entropy than the preceding one.

The third law of thermodynamics can also be generalized for the case of hierarchic dynamic systems by using the same scheme of reasoning. Indeed, the limit of system (subsystem) thermodynamic entropy approaches zero when its temperature approaches zero. But, the zero information entropy in the case of hierarchic subsystems (levels) corresponds to a subsystem consisting of one structural element only. The smaller the number of structural elements (and mutual connections between them), the lower the information entropy. Thus, the hierarchic analogy of the third law of thermodynamics can be formulated as follows: the highest level of any natural hierarchic system is characterized by vanishing information entropy.

Everyday life provides abundant examples or evidence of this. For instance, only one person (i.e., only one structural element of a system) occupies the top position on any social pyramid. As experience shows, this hierarchic principle holds true especially in everyday social life.

2.2.7 Physicalness Principle

According to this principle, all processes in natural systems are governed by principles of physics only. This means that any observed effect is always due to some physical cause that could be unknown to the observer. Moreover, these physical causes can always be, in principle, described by equations of physics. Hence, application of the physical principles is quite sufficient to explain any phenomena in the surrounding world. In other words, the idea of existence of some unknown nonphysical mystical power in nature is not accepted in modern natural sciences.

It should be mentioned, however, that the latter affirmation does not deny the existence of God, as is often asserted. It says only that God might be treated as a real physical subject, whose activity does not contradict the basic physical laws (because He created these laws). Hence, the concept of God might be formulated, in principle, in terms of physics. The example of this definition has been given, for instance, in monograph [1].

The physicalness principle is closely related to the so-called causality principle, but these two principles are not completely identical.

2.2.8 Principle of the Ability to Model

We previously considered before that one of the key peculiarities of the hierarchic systems is their ability to model one hierarchic level by another hierarchic level (see Figures 2.1 and 2.4 and relevant discussion). It should be especially pointed out that each such model level can reflect a limited set of peculiarities of the modeling level as well as the system as a whole. Owing to that and following from Turing's theorem, the model should always be simpler than the modeled system. Each complex system can be modeled by various methods because different reduced sets of system properties are used in different methods of modeling. In the general theory of systems, this statement is called the principle of ability to model [1,17]. It states that each particular group of properties of any complex system including the hierarchic systems and their subsystems (levels) can be modeled by models.

Thus, two types of such specific models can exist theoretically, particular and total models. The particular models can describe only a part of system properties, i.e., in this case we omit other parts of the information. On the other hand, it is clear that only the particular models could be of real interest in practice. The above-mentioned Turing's theorem explains this paradox: any total model cannot be simpler than the modeled system itself. This means that this case of modeling has no practical value. That is why we can say that any real model is simpler than the modeled subject.

However, the modeling process cannot always be arbitrary. In the case of a hierarchic dynamic system, we have relevant restrictions for the methods of obtaining partial (simpler) models (levels). These restrictions originate from the necessity to satisfy the considered hierarchic principles (hierarchic resemblance, information compression, analogs of the second and third thermodynamic laws, respectively).

2.2.9 Purposefulness Principle

As was already noted, purposefulness in the general system theory [16,17] is a functional trend to achieve a corresponding state of a complex system or to conserve (or to amplify) processes in the system. The essence of the purposefulness principle is that complex systems (including the hierarchic systems) have a specific purposefulness. The well-known Fermat's principle and its versions (principle of Maupertuis, variation principle, etc.) [48] are closely related to the purposefulness principle. It is the most incomprehensible and intriguing feature of the real natural hierarchic systems (indeed, why is it so?). In practice, purposefulness appears always to result from the evolution of the properties of the entire system. Obvious examples could be found in the life of the earth biosphere, human society, the universe, etc. (see examples in Chapter 8).

For instance, as mentioned above, the astrophysical analysis of cosmic catastrophes on earth over the past 3 billion years [18] shows that our biosphere possesses an obvious property of purposefulness. After every global catastrophe, the earth responds in a specific way, making the negative results of the catastrophic effect on the biosphere minimal [18]. In other words, owing to purposefulness each time, the preceding quasistable state is reached in the shortest time possible.

2.2.10 Two Approaches to the Theory of Hierarchic Dynamic Systems

The traditional version of hierarchic theory [2–13] is now generally accepted. The essential feature of this version is that all hierarchic levels of a considered dynamic system are described in the framework of the same reference system, i.e., the variables of the reference system are common for each hierarchic level, as well as for the system as a whole. As a result, mathematical structures of the relevant dynamic equations for each hierarchic level are found to be, as a rule, essentially different.

Another (new) version of the hierarchic approach differs from the traditional version because a proper reference system (and relevant set of dynamical variables) is introduced for each hierarchic level [1,14–23]. Therein, the proper variables are chosen in such a way that the mathematical structures of the dynamic equations for every hierarchic level, as well as for the system as a whole, are the same. We called this feature the principle of hierarchic resemblance [1,14,15,20]. The search for a specific set of hierarchic transformations that provide realization of the property of hierarchic resemblance creates the main mathematical problem of this new version of hierarchic theory [1,14,15,38–45].

All this brings about sufficient differences in the general arrangement of both versions of the hierarchic theory. These differences also determine the feasibility of their practical application. The traditional version is known to have mostly a philosophical/cognitive significance. This version was developed, predominantly, to extend our knowledge about the surrounding world. In contrast, the founders of the new version (van der Pol, Krylov, Bogolyubov, Leontovich, and others; see Chapter 5) [1,19,21–29] were not interested in the philosophical/cognitive aspects. Their main interest was in realization of the hierarchic-like (in accordance with the modern terminology) highly effective calculation algorithms for practical purposes (in electrical engineering, radiophysics, etc.). It is important to note that the specific philosophy, terminology, and basic concepts that are characteristic of modern hierarchic theory were not used in those times (1920s–1930s). As a result, this observation has been ignored by the orthodox hierarchic experts for a very long time. An interest in the traditional philosophical/cognitive aspects emerged only after relevant experience of the practical application of such calculation algorithms was accumulated [1,14,15,38–45]. The efforts to generalize and comprehend this experience led the author to write this and other books [1,19,20].

2.3 Postulates of the Theory of Hierarchic Systems

2.3.1 Integrity Postulate

Any natural complex system should be regarded as a single (or, at least, very rare) object of our world (the uniqueness principle). In the case of complex systems, this assertion is connected closely with the definition of the system as an object that possesses a specific system's property (Section 2.1.2). In turn, it is clear intuitively that the concept of the system's property reflects the property of system integrity. Summarizing, we can formulate this intuitive sense as the integrity postulate: Any decomposition of a system violates the system integrity inasmuch as such decomposition leads to vanishing of the system's property.

Let us illustrate this thought with the following mental experiment. Consider a volume of gas with the number of molecules $N >> 1$, so this object can be classified as a complex system. Divide this volume into N and also the separate parts. Then, assume that there is only one molecule within each partial volume. It is obvious that the integrity of the system as a whole is violated as a result of the decomposition. Indeed, any separate molecule does not have the collective properties of the gas.

A very peculiar situation can be observed for the hierarchic systems, which could be considered as resulting from a specific kind of decomposition. According to the above-defined principle, it may seem contradictory to the formulated general rule. In the case of hierarchic systems, we have the hierarchic resemblance principle (see Sections 2.2.1 through 2.2.3) that requires a relevant resemblance between the system levels (as subsystems) and the whole system. This means that all system levels reflect the same system property as the entire system. In other words, the hierarchic arrangement of the considered system should, at the first sight, violate the system integrity.

However, in reality it is not so. The hierarchic resemblance is not literal and it concerns only the similarity of the general arrangement of the system as a whole and its levels, in particular. In other words, this similarity is not word for word and, consequently, each level as a subsystem is not identical to the whole system. Therefore, in this case we actually have no contradictions to the integrity postulate. Thus, we can reformulate the integrity postulate as follows: The system as a whole can never be considered an exact equivalent of any of its isolated parts.

2.3.2 Autonomy Postulate

Every type of physical phenomena corresponds, as a rule, to a certain transformation group. Such a group generates proper geometry for the space problem. Therefore, the general theory of systems (including hierarchic ones) can also have a geometrical interpretation [18]. It is postulated that a type of geometry for a given complex system possesses autonomy. This geometry should be an invariant for any system decomposition, i.e., each part of the system is described by the same geometry as the entire system. This definition expresses the essence of the autonomy postulate.

Chapter 8 will illustrate that the autonomy postulate holds true for our universe as well as for the superuniverse. (The superuniverse contains our universe as a structural element of the fourth hierarchic level.) Both hierarchic systems exist in the same seven-dimensional space, where one part of the dimension manifests evidently and where the other part is always hidden.

2.3.3 Complementarity Postulate

The complementarity postulate was first formulated by Niels Bohr for quantum systems. However, further developments in physics reveal that it could be applied to macroscopic subjects of classical physics, too. In our case, the essence of the complementary postulate is as follows: A complex system reacts differently when subject to different external effects revealing its different set of properties. Of course, at first sight it may not seem clear and comprehensible; therefore, some commentary on this postulate is provided.

Any knowledge we possess about any natural complex system is always incomplete. This means that we never know the true arrangement of the systems or the true set of connections between its different parts. That is, why different parts exposed to the same action produce different responses in the system as a whole. Generally these responses can reveal the system properties that may seem to exclude each other. For instance, an electron has well-known corpuscular properties that appear in the photo-effect, the Compton effect, etc. But its behavior in other situations (diffraction of electrons on a crystal lattice, etc.) reveals the evidently expressed wave properties. Hence, the electron is a particle and a wave simultaneously. Ordinary common sense considers such a situation to be impossible. However, physical experiments prove that this really exists in nature.

2.3.4 Action Postulate

All known complex systems in different areas of science demonstrate one interesting peculiarity that could be referred to as the action postulate. In accordance with this postulate, any complex system reacts to external action as a threshold. This means that the system is stable when subjected to any rather small external actions. However, this situation remains until this action attains some threshold value. The system responds with remarkable changes only when the action is stronger than this threshold.

A number of examples are known in physics and engineering that illustrate this system property. For instance, electron transitions within the atom occur only when the energy of the perturbing photon exceeds some threshold value (Frank and Hertz experiments, the short wavelength limit of braking radiation, the red limit of photo-effect, etc.). Other types of threshold effects can be observed in such complex technological systems as FELs, EH accelerators, etc. (see Part II) [19,20].

2.3.5 Uncertainty Postulate

Similar to the complementarity postulate, the uncertainty postulate has also been introduced first in quantum mechanics (Heisenberg uncertainty principle). Scientific practice reveals that this principle holds for macroscopic complex systems, too. Its physical nature can be clarified by the known device problem. We have a possibility of obtaining any information about a system through appropriate measurement. However, any measuring process changes the true state of the investigated system. Therefore, the error of measuring is always finite. This means that a measured state of a system always differs from its true state, i.e., strictly true system behavior is uncertain, in principle. When there is a strong external effect on a device, its true dynamics can be essentially different from the measured one.

2.4 Hierarchic Trees and the Concept of the System God

2.4.1 Hierarchic Trees

The method of Feynman's diagrams [48–50] is well known in theoretical physics. It was introduced by R. Feynman for graphical uniform illustration of different types of physical interactions. The main advantage of this method is that, based on a universal qualitative approach, it can show a number of qualitative properties of different physical processes. This gives a good possibility of obtaining important qualitative information about the studied object without accomplishing a special complex quantitative analysis.

The method of the hierarchic tree plays the same role in the theory of hierarchic dynamic systems [1,14,15,19]. Similar to Feynman's diagrams, it also allows representing uniformly different hierarchic systems independently of their physical nature. It gives a very powerful tool for their preliminary qualitative analysis. The latter is obviously demonstrated in Chapters 3 through 7 for different partial cases of the oscillation-wave hierarchy. In this section, only the most general concepts and ideas of the method are covered.

Thus, the hierarchic tree is a specific graph that represents general properties of different hierarchic systems of different physical natures using a uniform (and universal) method of representation. The above-mentioned set of hierarchic principles and postulates made the basis of this method.

Let us illustrate the main idea of the method going back to the example of a university whose hierarchic tree is shown in Figure 2.5. For simplicity, we accept the following abstract idealized model. We assume that each lecturer teaches only two students, each department (with its head) is made up of two professors only, and so on. The rector (president) holds the top position of the university hierarchic pyramid. It is not difficult to see that the proposed graph (hierarchic tree) meets all hierarchic principles. But, let us discuss this topic in detail.

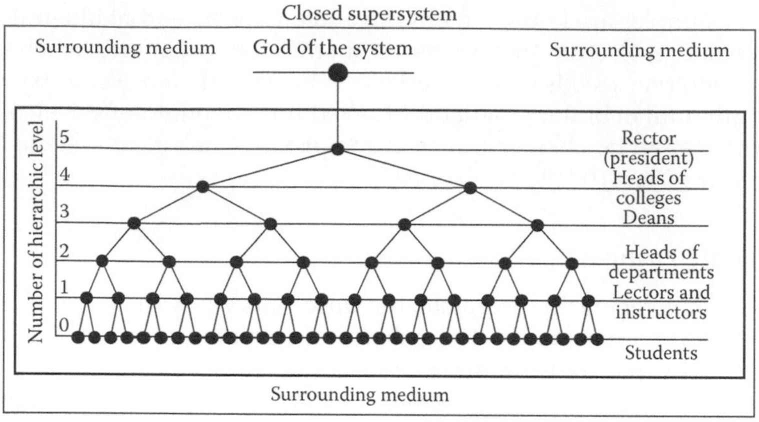

FIGURE 2.5
Hierarchic tree of a theoretical model of a university.

General hierarchic principle. Everything in the universe is hierarchic in nature. As it is readily seen, any university, as a complex dynamic system, indeed possesses a hierarchic arrangement, as demonstrated in Figure 2.5.

Principle of hierarchic resemblance. Each hierarchic level in its general properties and arrangement represents the system as a whole. Each hierarchic level, as well as the university as a whole (see Figure 2.5), is represented only by a totality of people. None of the levels represents structural elements of another biological species (dogs, cats, mice, or cockroaches, which are usually also present on a campus). Therefore, at any hierarchic level, all laws of interactions and communication between the structural elements are determined, generally, by the human nature of the objects of their interactions and communication. The latter, in turn, means that the relevant imaginary systems of equations (Section 2.5) that describe dynamics and behavior of the hierarchic levels should have the same (i.e., human) mathematical structure. The laws of collective psychology of human groups coincide, in general principles, with the laws of individual psychology [46].

In this case, possible mathematical differences in the level descriptions concern the sets of proper or appropriate variables used. These sets are specific for each hierarchic level and because of that, each of the totalities of students, teachers, heads of departments, deans, and so on (see Figure 2.5) is characterized by different social statuses, functions, and stereotypes of behavior. These very patterns of behavior perform the role of variables specific to different hierarchic levels.

Principle of information compression. Each higher hierarchic level is always simpler than the preceding one. The interpretation of this principle in this case is rather obvious. Indeed, the number of students (and their numerous connections, activities, and interactions) is always much larger than the number of professors, instructors, and so on. Consequently, relevant algorithms, which describe these connections, activities, and interactions, are always longer than a similar algorithm for the totality of lecturers or instructors. The complexity is the minimal length of such an algorithm.

Hence, all direct information (see Figure 2.4), from the student level to the level of lecturers and instructors, should be compressed at this higher level. It appears in the form of grades, credits, and other statistical data that describe the student level from the perspective of professors and instructors.

By contrast, the operation information (reverse information; see Figure 2.4) moves in the opposite direction. It determines student destiny and the course of their activities. It is clear that the operation information should be decompressed during this transition. Because of this, the general information from the level of professors and instructors should be personalized and delivered to each student at the student level. Obviously, the operation information concerns only student academic work; it does not concern private relationships with significant others, friends, or family. The same situation is typical of information interactions between the levels of lecturers and instructors, heads of departments, and so forth (see Figure 2.5). These areas of their lives belong to other hierarchic trees, which, in turn, exist parallel to the university in nature as well as in social/humanitarian, political/economic, and other hierarchic systems.

Hierarchic analog of the second law of thermodynamics. Each higher hierarchic level of the natural dynamic system has smaller information entropy than the preceding one. It is well known that the behavior of any person (student, lecturer, or dean; see Figure 2.5) in interactions with other persons is a specific mixture of chaos and order. It is also known that the individual peculiarities of a person determine the proportions of chaos and order in this mixture. Intuitively, it is clear that the smaller the number of structural elements is of a hierarchic level, the smaller should be the total intensity of such interactions. Consequently, the level of chaos (informative entropy) should also be smaller. However, the smaller the number of structural elements and connections is, the higher the hierarchic level will be. Hence, information entropy should decrease as the hierarchic-level number increases.

Hierarchic analog of the third law of thermodynamics. The highest level of any hierarchic system is characterized by vanishing information entropy. We can understand why the highest hierarchic level should always have zero entropy. The chaos at any hierarchic level is generated, first, by the interactions between its structural elements, i.e., the structural elements that belong to the same level. They include, for instance, student interactions with other students (zero hierarchic level), reciprocal interactions of lecturers and instructors (first hierarchic level), and so on. But the rector, as a structural element in our example (see Figure 2.5), is singular because he occupies the top position of the university hierarchic pyramid. This automatically means that he cannot interact with himself. That is why the information entropy at the top of any hierarchic system (including the university hierarchic pyramid) should be equal to zero.

Physicalness principle. Only physical laws and principles are sufficient explanations of any observed effect. In other words, any such effect results only from some physical cause. Any grades or credits obtained by a student are always the result of real physical (i.e., material) causes. And the student should not look for explanations involving the action of some nonmaterial mystical forces.

Principle of the ability to model. Each particular group of properties of any complex system, including the hierarchic systems and their subsystems (levels), can be modeled. The functioning of each higher hierarchic level of a university, as a hierarchic system, can be modeled by the lower preceding level. Or, in other words, any higher level can be considered as a specific (partial) model of the preceding lower level. For example, the hierarchic level of lecturers and instructors can be treated as a partial model that reflects student education (only) activity. Consequently (Sections 2.1 and 2.2), this model is always simpler than the original. It is attained in two main ways. The first is the compression of information (see the principle of hierarchic compression). The second is selection of information. Only information related to education processes is taken into consideration. The student's family relations, his or her problems with friends, and so on are left out in this modeling process. They are the subjects of other similar partial models. Hence, the current real education practice of universities confirms the correctness of the principle of ability to model.

Purposefulness principle. Complex systems (including the hierarchic systems) have specific purposefulness of behavior. The validity of this principle can be illustrated by an example from the history of the Civil War in the Ukraine in 1917–1920. The famous city of Odessa was occupied many times by different armies (the Red Army, the White Army, the Green Army, and so on). In spite of this, the reaction of the Odessa university was to continue its work, i.e., the education process as the main purpose of its activity was not broken and it involved the same students and professors, independent of the color of the occupant's political orientation. A similar illustrative analysis can be accomplished with respect to the main postulates of the theory of hierarchic dynamic systems (see Section 2.3. for more detail).

2.4.2 Concept of the System God

Eventually, we come to admit that the highest hierarchic level for a university in the form of a rector or president (see Figure 2.5) is not absolutely the highest level because a university, strictly speaking, is not a closed complex system. A real rector always has numerous higher authorities. They are relevant regional departments, government education administrations, etc. Different owner structures such as regional and state education institutions play the same role with private universities. Apart from that, universities always collaborate with external international education systems: union governments (as in Europe), relevant departments of large international corporations, etc. All this authority has an obviously higher level of hierarchy. Thus, in any event, the university rector is not absolutely free in his professional activities.

On the other hand, it should be stressed that all the above-mentioned structures do not belong to any university. They are placed beyond it (i.e., in the medium surrounding the universities; see Figures 2.1, 2.5, and relevant comments). Moreover, the rector never has absolutely complete information about the arrangement, methods of function, and hierarchic properties of this upper structure. Any direct information on the state and arrangement of the university levels (see Figure 2.4) can move only upward, in accordance with the hierarchic principles. By contrast, the reverse flow of information, i.e., the operation information (orders, instructions, etc.), for the levels (Figure 2.4) moves in the opposite direction. Therefore, the rector can have only partial information about the state and arrangement of the upper hierarchic structures and this information is always incomplete. These pieces of information are treated in the general theory of systems as the higher moments of information. Only direct information that moves upward is complete.

Modern physics is traditionally oriented mostly toward the analysis of closed systems. Most of the known fundamental laws are formulated only for closed systems and the traditional methodologies and approaches to analysis are developed mainly for such systems. On the other hand, the properties and behavior of any open system, strictly speaking, are determined (completely or partially) by external impact. Generally, this means that we cannot formulate universal fundamental laws as well as relevant methodologies and approaches if the information about this impact is not known. As a result, the same open systems exposed to the action of different external subjects can be characterized by different dynamic properties. However, we quite often have information about these externally acting subjects (for instance, due to the experimental observations). But, how can we construct a dynamic theory of such an open system using this information and the accumulated experience of the theory of closed systems? The method of the system god has been proposed for such practical situations [1,40].

The main idea of this method consists of the following. We introduce the point-like model (the god) into the open system, as shown in Figure 2.5. Herewith, we assign to this point the resulting action properties of all upper hierarchic levels that are unknown to us. As a result, we obtain a possibility of formally describing an open system by a quasiclosed system (see the closed supersystem shown in Figure 2.5).

Thus, the formal closeness of the system considered open could be graphically provided by the god of the system. It is placed over the highest known hierarchic level, but out of the system. In the case of a university, the highest level is the rector (as illustrated in Figure 2.5). By introducing the system god, we obtain the possibility of using all the above-mentioned traditional methodologies and approaches for studying such open systems.

Analyzing various natural and quasinatural artificial hierarchic systems, we find that the point of the system god at any time has the meaning of a source of some power. This is a source of political power in political systems, a source of financial power in financial systems, a source of military power in military systems, a source of energy power in technological systems, and a source of political, financial, military, energy, and other powers in the hierarchic system of human civilization. A logical question arises: How can we treat the point of system god in the case of our universe?

2.4.3 Hierarchic Tree of Our Universe

Using the obtained knowledge and conclusions, let us continue the consideration of the universe as a hierarchic dynamic system. First, we construct the hierarchic tree for the universe utilizing the data of Table 2.1 and the set of discussed hierarchic principles. We will also apply the reasoning that was used for constructing a similar graph of the hierarchic tree in Figure 2.5. After the relevant (not complex) work, the desired graph can be obtained in the form represented in Figure 2.6.

As can be easily seen, we accepted here (similar to Figure 2.5) a simplified ideal model of the universe. Namely, we assumed that each two elementary particles (nucleons) form one nucleus. Then, we assume that each nucleus and electron have the same hierarchic rank (see comments on Figure 2.5 and Table 2.1), form one atom, and so on. The metagalaxy is placed on the opposite side of the universe hierarchic pyramid (i.e., on its top). But, what is placed over the metagalaxy? In accordance with the discussion in Section 2.4.2, it should be the system god for the universe. For convenience, we call it the physical god (see Figure 2.6).

However, what is the physical god from the point of view of modern physics? Unfortunately, there is not yet a substantiated answer for this intriguing question. On the other hand, this author is sure that a believer of any faith in such a situation will be convinced that this is God, the Highest Supreme Being, Allah, Yahweh, and the Creator of the World. This very conclusion can be drawn by accomplishing a detailed comparison analysis of the physical properties of the point shown in Figure 2.6 on one hand and the appropriate materials of the sacred books of the main world religions on the other. Unfortunately, such detailed analysis is far beyond the interest and purpose of this book. Here, this author must emphasize once again that he pretends only to the formulation of the concept of the physical god as a system god of the universe as a hierarchic dynamic system—a physical object with some set of physical properties only. This author considers that physics today has a number of rather weighty arguments for acceptance of this concept as a reality.

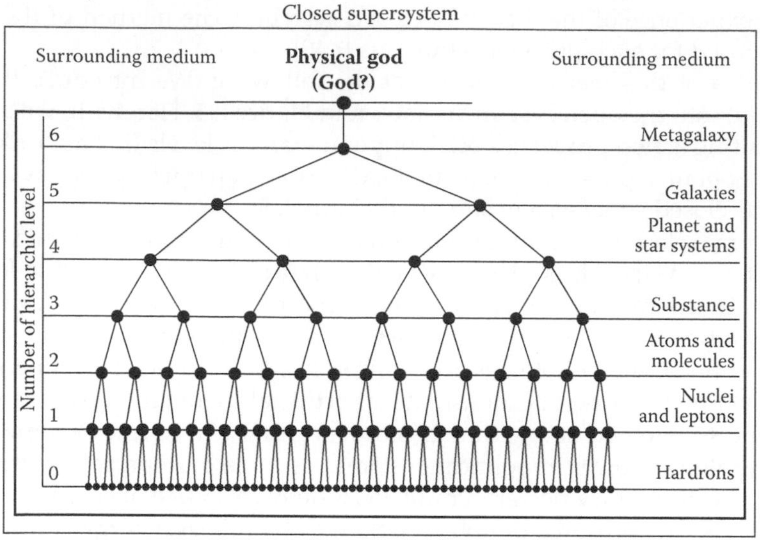

FIGURE 2.6
Simplest illustration for the concept of a hierarchic tree for our universe.

The first group of these physical arguments originates from known astrophysical observations, such as the phenomena of relic radiation, the expansion of the universe, etc. [47–50]. These observations, as is commonly accepted, confirm the correctness of the big bang theory (further discussion of this topic is provided in Chapter 8). In turn, this theory forms the basis of our ideas about the physical god of the universe [1]. This thought is derived, in particular, from one of the main conclusions of the big bang theory: our universe is an open complex dynamic system [50] because its birth was caused by the action of some external something. But, if so, the corresponding system god for the universe could be rigorously introduced into physics. We can use the totality of all known experimental facts like relic radiation and other facts for description of this point on the hierarchic tree (see Figure 2.6).

The second group of physical reasons comes from the experimental observation that all other known natural complex systems are open, too. This means that by rejecting the proposed idea of the god of the universe, we must recognize that our universe, as a natural complex dynamic system, is an exception from the general rule that any natural complex system has its system god. However, if the universe is indeed the exception, we should substantiate this fact rigorously. Because of this, such an approach to scientific research is the most important methodological rule of modern physics as a science. However, today we do not have any basis for such substantiation; moreover, there are a lot of experimental facts supporting the idea of the god of the universe.

2.4.4 Concept of the Hierarchic Besom

Let us turn again to the example of the university hierarchic tree (see Figure 2.5). This time we will assume that the considered hierarchic system is represented by a university department only, i.e., the university department (subsystem) as a separate hierarchic system. It means that each head of department in such a situation is at the top of the specific hierarchic pyramid (a subsubpyramid within the university structure). All upper hierarchic levels (including the dean, the rector, and the system god) appear to form a new pettier system god. We can regard it as a god of the department of the conventionally hierarchic subsubsystem. Thus, the totality of university departments forms a peculiar totality of the departments' gods.

A similar situation can be obtained by studying the totalities of faculties (the university subsystems), universities in a city (or a region), and so on. A common property of all these examples is that the highest elements of the hierarchic tree of any considered totality are characterized by the same rank (number) of hierarchic levels. Hence, the gods of the systems of all these hierarchic trees (for the departments, faculties, universities, etc.) of the same totality should also be of the same hierarchic rank.

Naturally, a question of how to treat these peculiar totalities of system gods could arise. The idea of such a graphical presentation is provided in Figure 2.7. We refer to this graph as the hierarchic besom. It describes the totality of many hierarchic trees; the elements resemble the arrangement of a traditional home besom. Obviously, the main difference between the ordinary hierarchic trees and the hierarchic besom (see Figures 2.5–2.7) is that, in the second case, we have a new subject—the common god of the systems. The physical meaning of this concept is self-evident in view of the above; it is nothing else but the totality of system gods.

It should be noted that situations that can be described as hierarchic besoms are quite common in engineering and science. Let us illustrate it with a complex engineering (artificial) system such as the FEL. It contains (Figure 2.8) the FEL section (i.e., the FEL itself), the electron accelerator, the driving systems (energy sources), the operation and control systems of the FEL as a whole, the vacuum system, and the operation systems for the generated laser beam. All these subsystems have the same hierarchic rank and their totalities make up the technological FEL system as an experimental unit. Each can be described by its personal hierarchic structure, which could be represented by the corresponding hierarchic tree. In the course of such subsystem elaboration, there is a separate project manager. All managers are mostly responsible for their partial or particular projects. Each of them can be regarded as the highest hierarchic level of corresponding subsystem/project.

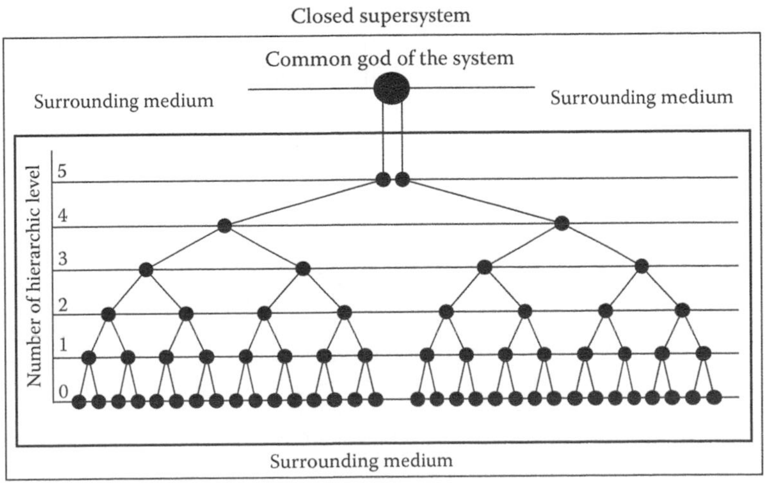

FIGURE 2.7
Illustration of the concept of the hierarchic besom.

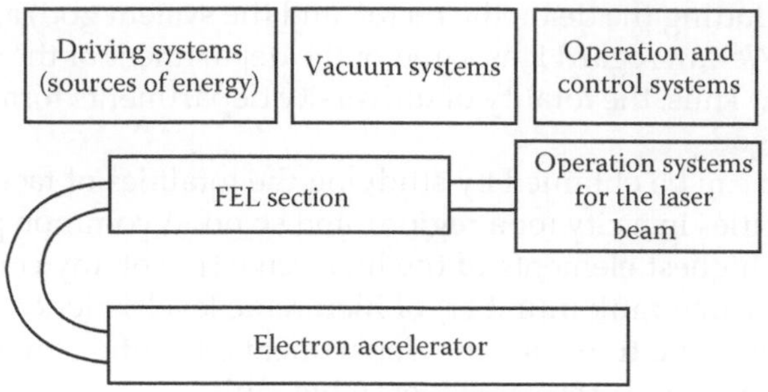

FIGURE 2.8
Block scheme of the technological units of a FEL.

However, the FEL project as a whole should have a top manager. This manager (and staff) is likely to play the role of a personified system god with respect to the ordinary project manager. This god provides finance, materials, documentation, instructions, etc., and also provides the required technological control, as well as checks and verification of the particular project programs (subprograms).

The peculiarity of the discussed situation is that the hierarchic ranks of all project managers are the same. In view of this, they have the same common god of the system: a general project manager. However, actually it is only a part of the complete common god of the system. The point is that the top project manager also has a personal higher authority. This authority, in turn, has a higher authority, and so on. Hence, from the point of view of the project managers of separate projects, it is not important because they do not have complete information about the real structure of the common system god (information about the state of their systems goes upward only).

Then, let us continue with the hierarchic besom. This time we take the superuniverse as a conventional example of the natural hierarchic complex system. We will use for this purpose the well-known model in theoretical physics—the bubble model of the superuniverse (see Figure 2.9) [50].

Is our universe single or are there also other universes somewhere in the surrounding world? This is one of the most topical and intriguing problems in modern theoretical physics. There is not a deficiency of various hypotheses and theories. The bubble theory of the superuniverse looks most interesting and likely among them [50].

In the big bang theory, our space and time and spatial coordinates were borne with the birth of the observed universe [47–50]. Or, in other words, the ordinary three-dimensional

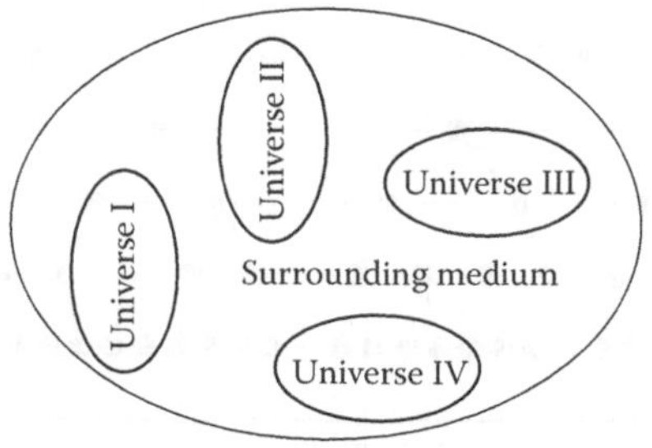

FIGURE 2.9
Illustration for Hawking's bubble theory of the superuniverse. (From Hawking, S., *The Universe in a Nutshell*, Bantam Books, New York, 2001. With permission.)

space and time that form the resulting four-dimensional space did not exist before the birth of the universe. It is interesting to note that in accordance with the M-theory, these four-dimensional coordinates describe actually no volume but a four-dimensional surface that envelops a five-dimensional volume [48,50]. This means automatically that the evolution of the universe can be imagined as blowing out a bubble, which is similar to the bubbles that we can observe in boiling water [50] (see Figure 2.9).

This extraordinary model has been proposed by modern theorists as an explanation of the birth and continued expansion (inflation) of our universe [50]. This explanation makes it possible for us to imagine our universe as a closed four-dimensional oscillated surface bubble (the so-called brane). It is important to note that these oscillations of the surface brane generate all observed matter (from the elementary particles to the galaxies) in the surrounding world. Hence, it—being the hierarchic dynamic system—is also a hierarchic-oscillation system. This very interesting topic will be discussed in detail in Chapter 8.

Our bubble universe is assumed to be the only bubble. All other possible bubbles (see Figure 2.9) are reciprocally isolated with respect to electromagnetic radiation (light), but can interact due to the gravitation field [50] (information about the first experimental observations of such phenomenon was recently published on the Internet). It is possible that all these partial universe bubbles (including ours) have been born simultaneously (unfortunately, it is not clear what the word "simultaneously" means in the situation because the time coordinate does not exist before the births of the bubbles) and developed in parallel. In other words, instead of the big bang theory, we can actually talk about the theory of plural big bangs.

It should be mentioned that Giordano Bruno is known for having been burned for the similar idea of a plurality of worlds, but within our universe only. Fortunately over the past few hundred years, times changed essentially and modern scientists (including Stephen Hawking, one of the main authors of this crazy multibubble idea [50]) enjoy the opportunity to freely generate and develop even more eccentric ideas (see Chapter 8).

How can we depict a situation with many bubble universes placed in the same seven-dimensional surrounding medium? Exactly, it can be done in the form of a hierarchic besom like the one shown in Figure 2.7. However, in this case we should extend the given definition of the concept of a physical god of our universe (see Figure 2.6) [1], accepting that he is also the common physical god of this multiuniverse (multibubble) world. The physics of this world is obviously very intriguing and we will touch upon it again in Chapter 8. However, it deviates too far from the main subject of this book, the theory of hierarchic waves and oscillations. Therefore, let us proceed to a discussion of the chapter's basic subject, the concepts of dynamic hierarchy and hierarchic systems.

2.5 Hierarchic Description: Basic Ideas and Approaches

All the material in this chapter is mainly of qualitative (verbal and graphical) character. However, the central part of the book (see Chapters 3 through 7) is dedicated to factorization (i.e., mathematical description) of the hierarchic laws and their corresponding sequences. Therefore, let us take the first step toward factorization by formulating a relevant set of general quantitative ideas and approaches. First, we should summarize some key items of the qualitative description, which will be helpful in accomplishing the factorization process.

2.5.1 Hierarchic Level as an Information–Diagnostic–Operation System

First, we will try to define the processes of transition and treatment of information flows in hierarchic systems. The general idea of these processes is illustrated in Figure 2.4. Using this scheme as a basis, we will modernize it by introducing a few new details. The result of such modernization is shown in Figure 2.10.

According to the illustrations in Figures 2.4 and 2.10 (see also Section 2.5), we have two main information flows in the considered hierarchic system: the straight one and the reverse one. The straight flow moves upward, whereas the reverse flow moves downward. The former is responsible for information about the state of the lower-lying hierarchic levels; the latter is connected with the operation information generated by the higher-lying levels.

But, what forms do the appearance of information about the state and operation information take in our everyday practice? The answer is unexpected: Any higher hierarchic level (as an authority with respect to the considered lower levels; see Figure 2.5) is interested mainly in the existing problems. The main function of any authority is to solve various problems at its level of competence only. In view of this statement, let us introduce further the concept of the main problem of the system. This is the problem formulated at the lowest (zero) hierarchic level and transferred up for solution.

A no-less-important function of the level authority is to put into practice the instructions and orders obtained from its higher authority. These instructions and orders, in turn, originate from the solution of higher-level problems, which are represented here by the reverse information flow (see Figure 2.10).

In summary, the information flow circulation within any hierarchic level could be presented in the following way. Straight information, i.e., the information about the state of lower levels, comes in the form of a formulated problem in the input of the block of hierarchic compression (see Figure 2.10). Then it goes to the information compressor, where it is compressed (see again Figure 2.10). The compressed problem comes further into the block of problem selection, which is managed by a level authority. Here we have a peculiar point

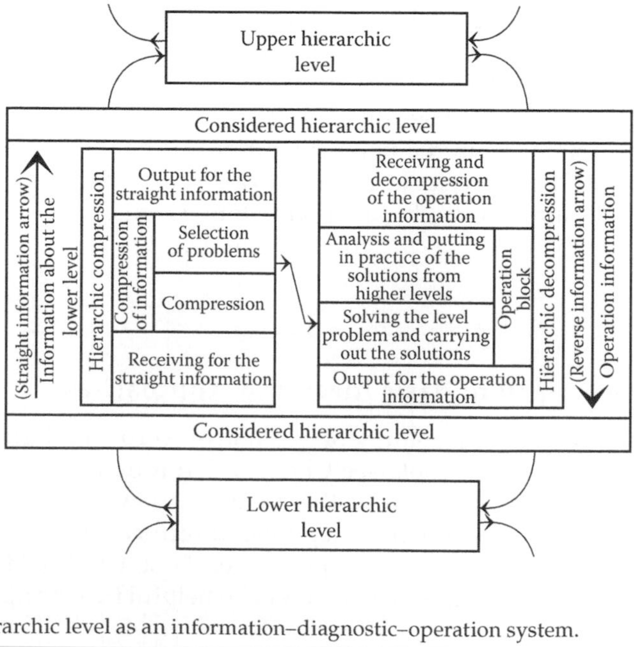

FIGURE 2.10
Block scheme of a hierarchic level as an information–diagnostic–operation system.

of branching. Namely, the problems that correspond to the competence of the considered hierarchic level (its authority) are transmitted to the operation block for realization. This block is placed on the neighboring (reverse) path for information flow (see Figure 2.10). The problems of the higher competence go to a higher hierarchic level through the level output (see Figure 2.10).

The flow of reverse (operation) information, which is presented in the form of solutions of higher-lying problems, has a different history. It comes from the higher hierarchic level through the reception and decompression of operation information in the operation block (see Figure 2.10). There, a decision about its further destiny is accepted after the required information analysis and solution of the level problem. Namely, if operation solutions correspond to the competence of the considered hierarchic level, they are put in practice immediately. If information belongs to the competence of the following lower-lying hierarchic level, it comes in the output to this lower-lying level (see Figure 2.10).

Thus, the main operation principles of a hierarchic level, as an information–diagnostic–operation system, are clarified qualitatively. Now let us discuss an approach to the description of these principles mathematically.

2.5.2 Hierarchic Principles and the Mathematical Formulation of the Hierarchic Problem

Let us assume that the main problem of the considered hierarchic system, as a whole, can be described by the following exact vector differential equation:

$$\frac{dz}{dt} = Z(z,t)\; z(t=0) = z_0, \tag{2.4}$$

where $z = \{z_1, z_2, \ldots, z_n\}$ is some vector and $Z = \{Z_1, Z_2, \ldots, Z_n\}$ is the relevant vector function in the n-dimensional Euclidean space $\mathbf{R}_n$, $t \in [0,\infty]$ (the laboratory time, for example). Using the hierarchic principles (see Sections 2.2 and 2.4), we will construct a mathematical description of the operation principles of the considered system.

General hierarchic principle. Everything in the universe is hierarchic in nature. This means that if we assume that our system is natural, we automatically accept the fact that it possesses a hierarchic structure. We consider that this hierarchy is strongly expressed and that the structural and dynamic hierarchic levels coincide (see Section 2.1 for details). We also accept that the set of hierarchic scale parameters ε_κ can be expressed in the form of a hierarchic series like Equation 2.3

$$\varepsilon_1 \ll \varepsilon_2 \ll \ldots \varepsilon_\kappa \ll \ldots \ll \varepsilon_m \ll 1, \tag{2.5}$$

where κ is the current number of hierarchic levels; m is the total number of system levels, and if we take into account the zero level, it is equal to $m + 1$.

Taking Equation 2.5 into consideration, we also regard each scale parameter ε_κ as a small parameter of the κth dynamic hierarchic problem. Given the strongly expressed hierarchy, the general definition for the scale (and, at the same time, small) parameter ε_κ could be described as follows:

$$\varepsilon_\kappa \sim \left|\frac{dz_j^{(\kappa)}}{dt}\right| \Bigg/ \left|\frac{dz_j^{(\kappa-1)}}{dt}\right| \ll 1, \tag{2.6}$$

where $z_j^{(\kappa)}$ are the components of the vector of proper (for the κth hierarchic level) variables $z^{(\kappa)} = \left\{ z_1^{(\kappa)}, \ldots, z_j^{(\kappa)}, \ldots, z_n^{(\kappa)} \right\}$. As a result, we can classify the initial set of the variables (components of the vector z; see Equation 2.4) according to the hierarchic levels κ. The parameters ε_κ also play a role as peculiar level markers.

Principle of hierarchic resemblance. In its general properties and arrangement, each hierarchic level represents the system as a whole. By translating this assertion into mathematical language, we can reformulate it. Each hierarchic level of the system can be described by Equation 2.4, which is the same, but with a general mathematical structure. Each of these equations should be written with the proper (i.e., very specific for this level) set of dynamical variables $z^{(\kappa)}$.

The initial main problem of the considered hierarchic system Equation 2.4 can be reformulated by taking a hierarchic series (Equation 2.5) and definition (Equation 2.6)

$$\frac{dz}{dt} = Z\left(z, t, \varepsilon_1, \ldots, \varepsilon_\kappa, \ldots, \varepsilon_m\right), \tag{2.7}$$

where each scale (small) parameter ε_κ marks the κth hierarchic level. Taking this into account and using the principle of hierarchic resemblance, we can write the dynamic equation for the zero level in the following way:

$$\frac{dz^{(0)}}{dt} = Z^{(0)}\left(z^{(0)}, z^{(1)}, \ldots, t, \varepsilon_1, \ldots\right), \tag{2.8}$$

where $z^{(0)} \equiv z$ and $Z^{(0)} \equiv Z$ are the proper vector variable and the vector function for the zero hierarchic level ($\varepsilon_0 = 1$), respectively.

Then, let us assume that we have some algorithm for transforming Problem 2.7 to the first, the second, or, in a general case, the κth hierarchic level. In this case, the equation that describes the problem of κth hierarchic level could be written as follows:

$$\frac{dz^{(\kappa)}}{dt} = Z^{(\kappa)}\left(z^{(\kappa)}, z^{(\kappa+1)}, \ldots, t, \varepsilon_\kappa, \varepsilon_{\kappa+1}, \ldots\right), \tag{2.9}$$

where $z^{(\kappa)}$ and $Z^{(\kappa)}$ are the vector variable and the vector function for the κth level, respectively. It is important that the vector function $Z^{(\kappa)}$ contains indirectly relevant vector variables and vector functions of all higher-lying hierarchic levels. However, it does not contain the analogous values that belong to the lower-lying levels. Thus, Equation 2.9 is of the same mathematical structure as Equation 2.4. We consider this as a manifestation of the principle of hierarchic resemblance. The difference only is that it is written in the proper variables for the chosen κth hierarchic level.

Principle of information compression. Each higher hierarchic level is always simpler than the preceding one. It means that each equation (like Equation 2.9), written for any higher hierarchic level, is simpler than the analogous equation for each lower-laying level. This simplification is attained without violation of the resemblance of the general mathematical structure of the equation.

Hierarchic analog of the second law of thermodynamics. Each higher hierarchic level of the natural dynamic system has smaller information entropy than a lower one. The analysis shows that the entropy is closely connected through the information temperature with the characteristic velocities. They are the velocities of changing in time of the proper (for this level) variables. We can formulate this as the following velocity theorem: The higher the entropy is, the higher the velocities are of changing in time of the proper variables. On the other hand, the higher the number of hierarchic levels is, the smaller the velocities should be in the left side of Equation 2.9. This is obvious in Equations 2.5, 2.6, and the velocity data for our universe given in Table 2.1. Thus, the higher the hierarchic level, the smaller the level of entropy—this hierarchic principle is satisfied by the discussed mathematical description.

Hierarchic analog of the third law of thermodynamics. The highest level of any hierarchic system is characterized by vanishing information entropy. In view of this, characteristic velocities for the highest hierarchic level should approach zero. Consequently, the entropy of this level should also approach zero.

2.5.3 Hierarchic Analog of the Concept of Short-Range Action

Now we should clarify the question of what regularities determine the methods of connection of the hierarchic levels. The answer is not too clear because these connections could generally be expressed by several figures. The illustration of such a hypothetical situation, where the same zero level can be connected simultaneously with a few higher levels, is given in Figure 2.11.

The general system theory shows that the larger the number of connections is between the levels, the more likely the system is to develop a global instability. On the other hand, human experience reveals that all natural hierarchic systems are obviously stable. A relevant analysis also shows that it can be practically observed only in a case where all connections in a natural hierarchic system are realized mainly between the neighboring levels. This observation has the rank of a fundamental law and is referred to as the principle of hierarchic short-range action.

In accordance with this principle, any effective interaction between two arbitrary hierarchic levels can only be realized when the levels have some immediate mutual information contact. In other words, it takes place when the levels are adjacent (see the item main flow

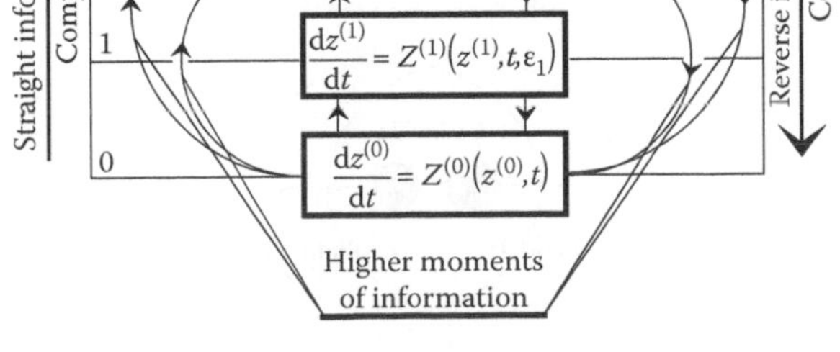

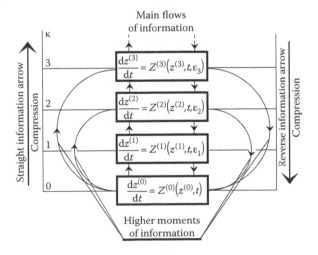

FIGURE 2.11
Scheme of interactions between the zero and other hierarchic levels of a dynamic system.

of information in Figure 2.11). The realization possibility of other types of interactions is characterized by vanishing probability. In the general theory of dynamic hierarchic systems, such other types of interactions are regarded as the highest moments of information exchanges (see Figure 2.11).

The formulated concept of hierarchic short-range action is similar to the short-range action, which is well known in general physics. It first appeared with attempts to interpret Coulomb's experiments. He studied the interaction of two electrically charged bodies that were separated spatially (see Figure 2.12). It was the first case where physical objects interact with each other at a distance, i.e., without any immediate contact. All physical experience accumulated before and obtained mainly within the framework of mechanics was destroyed. As a result, a fraction of these physicists reject the idea of the necessity of immediate contact for the objects of interaction. Instead, they propose the concept of long-range action (action at a distance), allowing interactions without immediate contact.

Another fraction of physicists, in turn, reject this idea. Despite the paradoxical character of Coulomb's experiments, they continue to keep the traditional concept of short-range action. Further developments in physics have confirmed the correctness of these conservative physicists. The concept discussed was recognized as a fundamental law—the principle of short-range action. The discovery of the electric (and then the electromagnetic) field has allowed the explanation of the observed paradoxical character of Coulomb's experiments.

The explanation is rather simple. Assume that we have two spatially-separated charged bodies (see Figure 2.12). The first charge q_1 is generating the electric field (Field 1 in Figure 2.12). This field, in turn, acts on the second charge q_2. The result of this action is the appearance of the force $\vec{F}_2$ (see Figure 2.12). The second charge q_2 generates Field 2, which, in turn, acts on the first charge with force $\vec{F}_1$. Hence, in spite of the spatial separation of charges, they interact rigorously in accordance with the short-range action principle. In this case, the electric field plays the role of mediator, providing an immediate contact.

First, the principle of short-range action is of a fundamental character and its action can be disseminated to other cases of interaction including interlevel interactions in hierarchic dynamic systems. In this case, the principle obtains the form of a hierarchic principle of short-range action, which means that only the neighboring information contacts between the hierarchic levels should be taken into account. We can travel through the hierarchic system only step by step, i.e., from one level to a neighboring level. Any other jumps are considered as highest moments of information (see Figure 2.11) and omitted as improbable.

Second, the universality of the practical applicability of the short-range action principle has been attained historically only due to the introduction of the concept of the field. It means that by further extending this concept to the hierarchic dynamic systems we should automatically introduce the equivalent concept of the information field. A preliminary study of this concept shows that it can serve as a very interesting and promising idea for

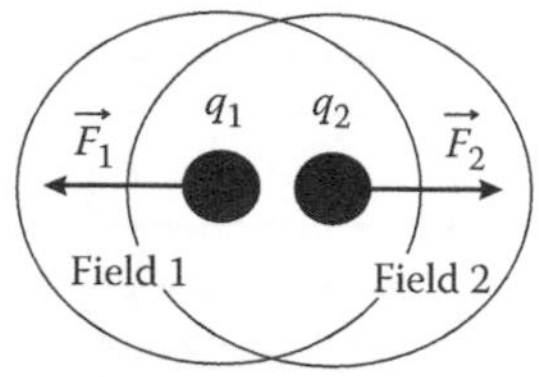

FIGURE 2.12
Illustration of the essence of the short-range action concept.

the explanation of many intriguing information aspects of the problem of life in nature. Also, we obtain a possibility of formulating a basic system of relevant equations for the information field. The role of such equations could be similar to the role of Maxwell's equations in electrodynamics. Unfortunately, because of the limitations of this book, we have to give up this interesting topic without further discussion.

2.5.4 Compression and Decompression Operators

Let us continue the factorization of the qualitative hierarchic theory. The concepts of the compression and decompression operators play key roles in this factorization. The compression operator accomplishes the compression of straight (forward) information, whereas the decompression operator fulfills the decompression of the reverse (backward) information flow (see Figures 2.10 and 2.11).

The compression operator could be introduced in the following general form:

$$Z^{(\kappa+1)}(z^{(\kappa+1)},t) = M^{(\kappa)}Z^{(\kappa)}(z^{(\kappa)},t), \tag{2.10}$$

where $Z^{(\kappa)}$ represents the vector functions for the κth hierarchic level (see Figure 2.11 and Equation 2.9). The action of the operator $M^{(\kappa)}$ (Equation 2.10) is illustrated in Figure 2.13. It can be constructed on the basis of some well-known integral transformations. They can be the averaging transformations, the Fourier or Laplace transformations, conformal mapping, convolution-type transformations, and others. In this book, we will use some types of the averaging transformations only [Krylov–Bogolyubov and Bogolyubov–Zubarev (Volosov) transformations; see Chapters 5 through 7].

The decompression operator $\hat{U}^{(\kappa)}$ is intended for decompression of the reverse flow of information (see Figures 2.10 and 2.11). It is, as a rule, in the form of a special algorithm of asymptotic (as a rule) transformation. The general definition for the operator $\hat{U}^{(\kappa)}$ could be given in the following way:

$$z^{(\kappa)} = \hat{U}^{(\kappa+1)}\left(z^{(\kappa+1)},t\right). \tag{2.11}$$

The action of the operator $\hat{U}^{(\kappa)}$ is illustrated graphically in Figure 2.14. Its real arrangement is determined by the choice of the form of the operator $M^{(\kappa)}$. Relevant variants of such algorithms are described below in this book (see Chapters 5 through 7).

From the point of view of the general theory, we may have two principally different mathematical problems. The first is a direct problem: to construct the operator $\hat{U}^{(\kappa)}$ for the given operator $M^{(\kappa)}$. The second problem is the reverse: to construct the operator $M^{(\kappa)}$ for the given operator $\hat{U}^{(\kappa)}$.

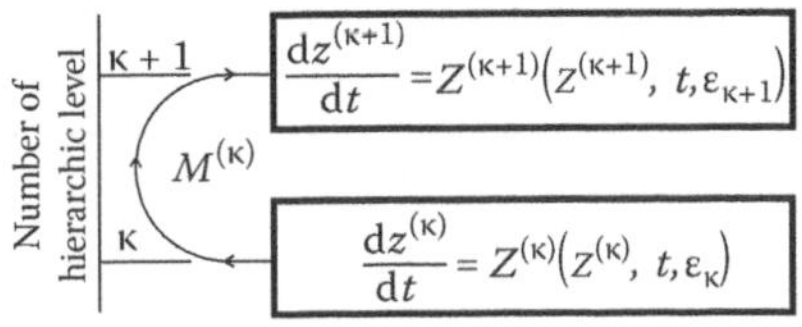

FIGURE 2.13
Illustration of the functioning of compression operator $M^{(\kappa)}$.

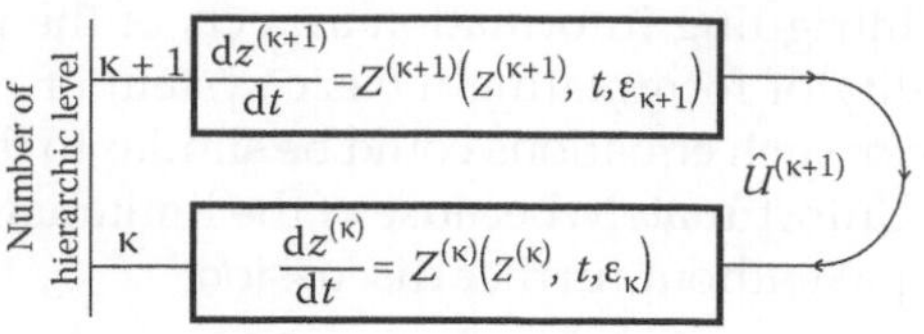

FIGURE 2.14
Illustration of the decompression operator $\hat{U}^{(\kappa)}$ action.

2.5.5 Main and Partial Competence Levels

Let us construct a general mathematical scheme for the functioning of hierarchic dynamic systems based on the previously introduced concepts and definitions.

For the given operators $M^{(\kappa)}$ and a formulated problem like Equation 2.4, we can construct a chain of hierarchic transformations like those given in Figure 2.11 (see the central column *Main flow of information*). It is clear that, ultimately, we should stop this transformation from running upward at some chosen hierarchic level m (see Figure 2.15). But how can we determine at what level to stop the transformation? This is the main level of competence (see Figure 2.15), where the initial main problem (Equation 2.4) can be solved completely. Herewith, relevant partial solutions can happen at all lower-lying hierarchic levels (Sections 2.5.2 and 2.5.3), which we call partial.

Analyzing, we can conclude that the main level of competence m is the highest level of the considered hierarchic system. The partial levels can exist, as a rule, due to the use of approximate calculation methods at a lower hierarchic level. Yet, let us continue the mathematical description of the hierarchic process running upward.

We have used the concept of compression operator $M^{(\kappa)}$ only for a mathematical description of the hierarchic passages between two neighboring levels (see Equation 2.10 for definition and also Figure 2.11). For the description of a total jump up from the zero level to the m level, we introduce the concept of complete compression operator $M^{(0,m)}$

$$M^{(0,m)} = M^{(0)}M^{(1)}...M^{(m-1)} = \prod_{j=0}^{m-1} M^{(j)}. \tag{2.12}$$

Such a jump is illustrated in Figure 2.15 by the left branch of the transformation scheme.

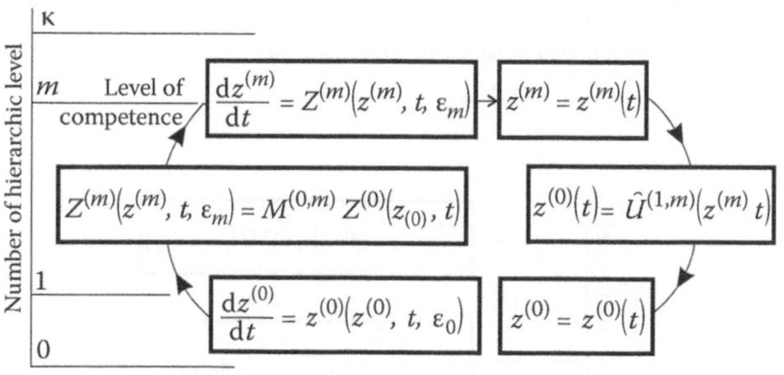

FIGURE 2.15
Illustration of the operation principles of the hierarchic system as an information–diagnostics–operation system (see Figure 2.10 for comparison).

Apart from a straight (forward) information flow, we also have the reverse flow. This flow begins at the level of competence m [see the item $z^{(m)} = z^{(m)}(t)$ in Figure 2.15 that is the solution of the mth hierarchic problem]. The following (symmetrical, with respect to the mathematical structure) complete decompression operator $\hat{U}^{(0,m)}$ is introduced for the reverse flow:

$$\hat{U}^{(1,m)} = \hat{U}^{(1)}\hat{U}^{(2)}...\hat{U}^{(m)} = \prod_{j=1}^{m} \hat{U}^{(j)}. \tag{2.13}$$

The right part of the transformation scheme in Figure 2.15 represents graphically the action of this operator. A complete solution for the problem $z^{(m)} = z^{(m)}(t)$ is transformed down to the zero hierarchic level [i.e., to the form $z^{(0)} = z^{(0)}(t) = z(t)$]. Herewith, going down the hierarchic level, the partial reverse information can be decompressed at other hierarchic levels. Each of these levels plays the role of the above-mentioned partial level of competence for this partial reverse information (i.e., for the information that concerns this level only).

2.5.6 Some Philosophical Aspects of the Theory of Hierarchic Dynamic Systems

Thus, the introduced operators $M^{(0,m)}$ (see Equation 2.12) and $\hat{U}^{(1,m)}$ (see Equation 2.13) allow us to describe mathematically the processes of jump up at the main level of competence and jump down to the initial (zero) level. But the question arises: How can we practically determine the main level of competence? Indeed, how can we know which level of competence is the main level, and which has the rank of a partial level only? What factors generally determine the hierarchic rank of the competence level? It is not difficult to be convinced that the search for relevant answers is not a simple task at all and touches upon rather complex aspects of the general theory of knowledge. Let us briefly discuss some of them.

First, we should note that, as analysis shows, the main competence level in any hierarchic system is determined by the following key factors:

1. The essence and the form of the formulated (initial) zero-level problem (see Figure 2.15)
2. The general arrangement and properties of the considered hierarchic system as a whole

The formulation of any initial problem always possesses an obviously expressed subjective character. This is a subject in practice. We designate it conditionally as an investigator. Formulating the initial problem for some given hierarchic system, the investigator actually plans to find a complete solution at a concrete main competence level. However, if the same investigator formulates another problem for the same hierarchic system, he obtains (in general) a complete solution of the other main competence level. Thus, the main level of competence depends on the investigator's purposes.

Let us focus on the second of the two formulated factors that concerns the general arrangement and properties of the considered hierarchic system. Let us assume that an investigator formulates an initial problem for a hierarchic system and, as a result, obtains a solution for the main level of competence. If he or she formulates the same initial problem for another hierarchic system, the investigator obtains the solution for another main level of competence. This means that the location of the main level of competence indeed depends on the general arrangement and properties of the considered hierarchic system.

Unfortunately, this illustration can explain only a part of the problem discussed because that somebody should make this system. This very somebody is responsible for the general arrangement and properties of the considered hierarchic system as a whole. Or, in other words, the other part of an explanation depends on the answer to the question, "Who is the designer of the considered hierarchic system?"

At first sight, the answer is obvious. It is the god of the system (Sections 2.4.2 through 2.4.4). This means that a more exact answer for the formulated question depends essentially on the personality of this god.

Let us point out that all systems known in the world can be conditionally divided into natural and artificial ones. The designer of all natural systems is God, the great creator of our world. A human being is trying to copy the activity of God by creating artificial systems (large computer networks, for example). We separate from the totality of the human-made systems the so-called quasinatural complex hierarchic systems. Their characteristic feature is that they are designed in accordance with the hierarchic principles (see Sections 2.2 through 2.4 for more details). In the case of the quasinatural system, a person—being the god of the system—plays also the role of the so-called designer who determines the general arrangement and properties of the system including the location of the main level of competence.

A much more complicated situation is found in the case of a natural system. Theoretically, God—being the creator—is also the formulator of the initial problems at the zero levels of all natural hierarchic systems. Humanity can only deal with the models of natural systems, not with the complete originals. This is because only God, as mentioned above, is the true Formulator and, at the same time, the creator, the operator, the investigator, and so on for all natural systems. Only He has complete information about the true arrangement and the true operation principles of natural systems.

Therefore, any time we talk about natural systems, in reality, we have in mind the imperfect models of natural systems. They can be regarded as the artificial simplified copies of true natural systems. In other words, they are artificial systems constructed to partially explain the operation principles of true natural systems. In accordance with the principle ability to model, these artificial simplified copies are incomplete. A lack of true information about natural systems leads us to obtain very nontrivial and complex (from a philosophical as well as mathematical point of view) sequences.

For instance, let us formulate the question: What happens if the formulated initial problem corresponds to the main competence level that is placed higher than the highest hierarchic system level? The answer is clear. This problem has no appropriate solution for human beings, in principle. This means that we will have two types of unsolved problems in the general theory of hierarchic systems. Problems of the first type are unsolvable, in principle, because the described situation corresponds to the level of competence that is localized higher than the top of the hierarchic pyramid. Problems of the second type are temporarily unsolvable; they can be solved, but we do not yet know how to solve them.

Let us bear in mind that the problem under consideration is well known in modern and ancient philosophy; the only novelty is in the unsolved problems' classifications in terms of the general theory of hierarchic dynamic systems.

2.5.7 Main Idea of the Hierarchic Methods

A variant of the approach to factorization (mathematical description) of the hierarchic system is described here. In all fairness, we should recognize that this achievement in itself cannot be considered too revolutionary. The traditional [2–12] versions of the theory of the dynamic hierarchic system were also an attempt to give an appropriate variant of the quantitative description.

Of course, both of these variants (ours and the well-known) are in essence ideologically different and our proposed new version seems much more progressive. But this advantage requires more evident and convincing arguments of illustration. The main argument, in our case, uses the proposed factorization scheme as a highly effective algorithm for the solution of the relevant class of differential equations. The traditional version is of cognitive and philosophical significance only.

What are the key points of the algorithm for the function of hierarchic systems? They are characterized by two main peculiarities (see Figure 2.15).

1. The existence of the hierarchic operation cycle: The formulation of the initial problem at the initial (zero) hierarchic level → the transformation of the problem at the level of competence → the solution of the problem at the competence level → the transformation of the obtained solutions again on the initial level.
2. The fact that the transformed equation at the main competence level is much simpler than the same equation at the initial level is due to the action of the principle of hierarchic compression (each higher hierarchic level is always simpler than the preceding one; see Sections 2.2 and 2.3). This means, automatically, that the mathematical procedure of solving the hierarchically transformed equation at the *m*th level (see Figure 2.15) should be much simpler than the direct solution of the initial equation (Equation 2.4) at the zero hierarchic level.

These two features represent the main idea of the hierarchic mathematical methods. Indeed, let us assume that we would like to solve some differential equation like Equation 2.4 and we know the evident form for the compression operator $M^{(\kappa)}$ (see Equation 2.10 and related comments). Regarding this equation as the initial problem of some hierarchic system, we accomplish mathematical procedures for all stages of the described hierarchic cycle: transformation of the initial differential equation at the main level of competence → solution of this problem at this level → transformation of the obtained solutions again at the initial (zero) level. Real algorithms of such a type are described and discussed in Chapters 5 through 7. A reader with enough patience and persistence has a chance to discover that such types of algorithms could indeed be very effective for many important practical applications. Examples of such practical applications are given in Chapter 6 (Part I) and in Chapters 9 through 13 (Part II).

Thus, contrary to the traditional version of hierarchic dynamic systems theory [2–12], our (new) version can be of obvious practical use. Although this fact alone is the main advantage of the proposed new version, this does not mean that the merits of the proposed version of the hierarchic approach are confined only to its purely applied value. In Chapter 8, the reader will learn that it also possesses a unique general philosophical side that is no less interesting than the practical applications.

According to the Bible, God created man in His own image. Being the Creator Himself, He gave man the ability to create. By mimicking God, man creates his human-made world—including artificial systems and models—by utilizing the same divine hierarchic principles. Observing the basic principles of the functioning of God's world, the man tries to apply them to his human world. As a result, all the described hierarchic principles in this human world acquire, eventually, the form of the hierarchic methods, i.e., the methods that are constructed on the basis of general divine algorithms (see Chapters 5 through 7). But let us take an in-depth look into the physical and philosophical nature of these divine [6] aspects of the proposed version of the theory of hierarchic oscillations and waves.

References

1. Kulish, V.V. 2002. *Hierarchic methods. Hierarchy and hierarchic asymptotic methods in electrodynamics*, Vol. 1. Dordrecht, the Netherlands: Kluwer Academic Publishers.
2. Nicolis, J.S. 1986. *Dynamics of hierarchical systems. An evolutionary approach.* Heidelberg, Germany: Springer-Verlag.
3. Haken, H. 1983. *Advanced synergetic. Instability hierarchies of self-organizing systems and devices.* Heidelberg, Germany: Springer-Verlag.
4. Kaivarainen, A. 1997. *Hierarchic concept of matter and field.* Anchorage, AL: Earthpulse Press.
5. Ahl, V., and Allen, T.F.H. 1996. *Hierarchic theory.* New York: Columbia University Press.
6. Babb, L.A. 1975. *The divine hierarchy.* New York: Columbia University Press.
7. Pattee, H. 1976. *Hierarchic theory: The challenge of complex systems.* New York: George Braziler.
8. Boehm, C. 2001. *Hierarchy in the forest: The evolution of egalitarian behavior.* Cambridge, MA: Harvard University Press.
9. Graeber, D. 2007. *Possibilities: Essay on hierarchy, rebellion, and desire.* Oakland, CA: AK Press.
10. Harding, D.E. 1979. *The hierarchy of heaven and earth: A new diagram of man in the universe.* Gainesville, FL: University Press of Florida.
11. Bailey, A.A. 1983. *The externalization of the hierarchy.* New York: Lucis Pub.
12. Kaivarainen, A. 2008. *The hierarchic theory of liquids and solids: Computerized application for ice, water and biosystems.* New York: Nova Science Publishers, Inc.
13. Rammal, R., Toulouse, G., and Virasoro, M.A. 1986. Ultrametricity for physicists. *Reviews of Modern Physics* 58(3):765–88.
14. Kulish, V.V. 1998. Hierarchic theory of oscillations and waves and its application to nonlinear problems of relativistic electrodynamics. In *Causality and locality in modern physics.* Dordrecht, the Netherlands: Kluwer Academic Publishers.
15. Kulish, V.V. 1997. Hierarchic oscillations and averaging methods in nonlinear problems of relativistic electronics. *The International Journal of Infrared and Millimeter Waves* 18(5):1048–1117.
16. Von Betralanffy, L. 1976. *General system theory: Foundations, development, applications.* New York: George Braziller.
17. Druzshynin, V.V., and Kontorov, D.S. 1985. *System techniques.* Moscow: Radio i Sviaz.
18. Vladimirskij, B.M., and Kislovskij, L.D. 1986. *The outer space influences and the biosphere evolution,* Vol. 1 of *Astronautics, Astronomy.* Moscow: Znanije.
19. Kulish, V.V. 1998. *Methods of averaging in non-linear problems of relativistic electrodynamics.* Atlanta, GA: World Scientific Publishers.
20. Kulish, V.V. 2002. *Hierarchic methods. Undulative electrodynamic systems,* Vol. 2. Dordrecht, the Netherlands: Kluwer Academic Publishers.
21. van der Pol, B. 1935. *Nonlinear theory of electric oscillations. Russian translation.* Moscow: Swiazizdat.
22. Leontovich, M.A. 1944. To the problem about propagation of electromagnetic waves in the Earth atmosphere. *Izv. Akad. Nauk SSSR, ser. Fiz.* [*Bull. Acad. Sci. USSR, Phys. Ser.*] 8:6–20.
23. Moiseev, N.N. 1981. *Asymptotic methods of nonlinear mechanics.* Moscow: Nauka.
24. Krylov, N.M., and Bogolyubov, N.N. 1934. *Application of methods of nonlinear mechanics to the theory of stationary oscillations.* Kiev: Ukrainian Academy of Science Publishers.
25. Krylov, N.M., and Bogolyubov, N.N. 1947. *Introduction to nonlinear mechanics.* Princeton, NJ: Princeton University Press.

26. Grebennikov, E.A. 1986. *Averaging method in applied problems*. Moscow: Nauka.
27. Rabinovich, M.I. 1971. One asymptotic method in the theory of distributed system oscillations. *Dok. Akad. Nauk. USSR, ser. Fiz.* [*Sov. Phys.-Doklady*] 191:1248.
28. Sanders, J.A., Verhulst, F., and Murdock, J. 2007. *Averaging methods in nonlinear dynamical systems*. Heidelberg, Germany: Springer-Verlag.
29. Burd, V. 2007. *Method of averaging for differential equations on an infinite interval: Theory and applications*. Boca Raton, FL: Chapman & Hall/CRC.
30. Flood, G.D. 1996. *Introduction to Hinduism*. Cambridge, UK: Cambridge University Press.
31. Andre, N. 1988. *Dictionary of the occult*. Herttordshire, UK: Wordworth Editions Ltd.
32. Kapra, F. 1994. *Dao of physics*. St. Petersburg: ORIS.
33. Laitman, M. *Kabbalah*. 1984. Printed in Israel.
34. Fortune, D. 1995. *The mystical Qabalah*. New York: Alta Gaia Books.
35. Garbo, N. *Cabal*. 1978. New York: W. W. Norton & Company.
36. Ponce, C. *Kabbalah*. 1995. Wheaton, IL: Quest Books and Adyar, Madras, India: The Thesophical Publishing House.
37. Parfitt, W. 1995. *The new living Qabalah*. Queensland: Element.
38. Kulish, V.V., and Lysenko, A.V. 1993. Method of averaged kinetic equation and its use in the nonlinear problems of plasma electrodynamics. *Fizika Plazmy* (*Sov. Plasma Physics*) 19(2):216–27.
39. Kulish, V.V. 1991. Nonlinear self-consistent theory of free electron lasers. Method of investigation. *Ukrainian Physical Journal* 36(9):1318–25.
40. Kulish, V.V. 1997. Hierarchic approach to nonlinear problems of electrodynamics. *Visnyk Sumskogo Derzshavnogo Universytetu* 1(7):3–11.
41. Kulish, V.V., Kosel P.B., and Kailyuk, A.G. 1998. New acceleration principle of charged particles for electronic applications. Hierarchic description. *The International Journal of Infrared and Millimeter Waves* 19(1):3–93.
42. Kulish, V.V. 1998. Hierarchic method and its application peculiarities in nonlinear problems of relativistic electrodynamics. General theory. *Ukrainian Physical Journal* 43:483–99.
43. Kulish, V.V., Kosel, P.B., Krutko, O.B., and Gubanov, I.V. 1998. Hierarchic method and its application peculiarities in nonlinear problems of relativistic electrodynamics. Theory of EH-Ubitron accelerator of charged particles. *Ukrainian Physical Journal* 43(2):33–138.
44. Kulish, V.V., and Kayliuk, A.G. 1998. Hierarchic method and its application peculiarities in nonlinear problems of relativistic electrodynamics. Single-particle model of cyclotron-resonant maser. *Ukrainian Physical Journal* 43(4):98–402.
45. Kulish, V.V., Kuleshov, S.A., and Lysenko, A.V. 1993. Nonlinear self-consistent theory of superheterodyne and free electron lasers. *The International Journal of Infrared and Millimeter Waves* 14(3):451–568.
46. Kahanetz, I. 1997. *Psychological aspects in Management: The Young's typology, the socionics, and the psycho-informatics*. Kiev: A.S.K.
47. Green, M.B., Schwartz, J.H., and Witten, E. *Superstring theory*. Cambridge, UK: Cambridge University Press.
48. Kuzmichov, V.E. 1989. *Laws and formulas of the physics*. Kiev: Naukova Dumka.
49. Sivukhin, D.V. 1989. *General course of physics. Atomic and nuclear physics*, Vol. V-2. Moscow: Nauka.
50. Hawking, S. 2001. *The universe in a nutshell*. New York: Bantam Books.

3

Hierarchic Oscillations

In the preceding chapters, we had a chance to assure ourselves that the discussed set of hierarchic principles represents one of the most important groups of fundamental laws. We also became acquainted with the hierarchic dynamic systems of the essentially different physical nature, origination, and hierarchic rank. And, in spite of this, they all obey the same general regularities.

The accumulated scientific experience reveals that such unanimity is connected, as a rule, with causes, which are common in all the mentioned systems. As is illustrated in the preceding chapters, the main cause of this is the similarity of their elementary basis. For example, the subjects of our world are characterized by impressive diversity of shapes, colors, physical states, scales, properties, etc. However, all this infinite variety can be reduced to rather limited types of atoms. The same atoms (carbon, hydrogen, and oxygen, for example), interconnected with each other in various ways, give a great number of variants of substances in nature. However, further development of natural science uncovered that atoms are not the most fundamental bricks of our universe. For a long time it is considered that the so-called elementary particles play this role (Chapter 8). But it has been disclosed that the elementary particles, having a double corpuscular-wave nature, are not really elementary because their physical arrangement is far from simple [1–5]. First of all, it is because of their complex inner oscillation-wave structure of a lower hierarchic level. Generalizing the material of Chapters 2 and 8, we can conclude that the oscillations and waves exist in nature in the form of the oscillation-resonant hierarchic dynamic systems. In this situation, the most incredible thing is that we met basically the same regularities in the theory of the oscillation-wave electrodynamic systems, in general, and in the FEL, in particular.

The following question arises: how can we describe mathematically these hierarchic oscillations and waves? Even more elementary questions arise: what are the oscillations and waves as a class of physical objects? What are their main physical properties and characteristics? The corresponding answers for these and other similar questions are given in this and in the following chapters.

Thus, generalizing, we had a chance to become convinced, without any exaggeration, that the hierarchic oscillations and waves are the most fundamental things of the surrounding world. And, correspondingly, the theory of the hierarchic oscillations and waves is the most universal fundamental basis for many branches of modern science. Therefore, let us continue by remembering the most general definitions and concepts of the theory of the hierarchic oscillations and waves. Because the author's field of professional expertise is *electrodynamics,* it is natural that all illustration examples below are chosen just from the field of electrodynamics, especially from the theory of FELs.

3.1 Oscillations as a Universal Physical Phenomenon

3.1.1 General Definitions and Classification: Free Linear Oscillations

Oscillations (vibrations) are a type of motion in which the considered moving system returns systematically to the initial state. All oscillations are classified into periodic and nonperiodic ones. The periodic oscillations take place in the case when the returning process occurs periodically.

Traditionally, periodic oscillations are the main subject of the study of the theory of oscillations. Herewith, harmonic oscillations occupy a special place. The most characteristic property of the harmonic oscillations is that they can always be described mathematically by the sine or cosine trigonometric functions

$$x = A \sin(\omega t + p_0) = A \sin p \text{ or } x = A \cos(\omega t + p_0) = A \cos p, \tag{3.1}$$

where A is the amplitude of oscillations, $\omega = 2\pi/T$ and $p = \omega t + p_0$ are their cyclic frequency and phase, respectively, T is the period, t is the laboratory time, and $p_0 = p(t = 0)$ is the initial phase.

Let us illustrate the harmonic oscillations by using an idealized model of the spring pendulum represented in Figure 3.1. We assume that all dissipation forces (the friction forces of different physical nature, for example) do not act on the oscillating mass m. This subject is referred to as the system with free oscillations.

In this case only two forces act on the mass m. They are the elasticity force $F_{el} = -kx$ and the force of inertia $F_{in} = ma = m\mathrm{d}^2x/\mathrm{d}t^2$ (where $a = \mathrm{d}^2x/\mathrm{d}t^2$ is the acceleration of the mass m during the oscillation process). In accordance with the second law of dynamics, we can write the condition of the force balance

$$F_{in} = F_{el}. \tag{3.2}$$

After corresponding substitutions and transformation, the second law of dynamics (Equation 3.2) can be rewritten in the form

$$\frac{\mathrm{d}^2x}{\mathrm{d}t^2} + \omega_0^2 x = 0 \tag{3.3}$$

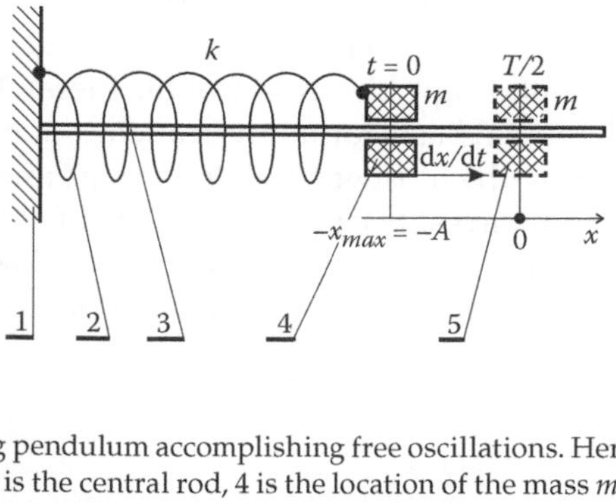

FIGURE 3.1
The simplest model of the spring pendulum accomplishing free oscillations. Here 1 is a hard wall, 2 is the spring with the stiffness coefficient k, 3 is the central rod, 4 is the location of the mass m for the initial time $t = 0$, 5 is the location of this mass for the time $t = T/2$, x is the spatial coordinate, $x_{\max}$ is the maximal deflection of the mass m, $\mathrm{d}x/\mathrm{d}t$ is the mass velocity, and A is the amplitude of oscillations.

where $\omega_0 = \sqrt{k/m}$ is the frequency of free oscillations. Expressions like Equation 3.3 are treated in mathematics as the linear homogeneous differential equations of the second order. Then we illustrate a standard algorithm of solution of such class of equations considering Equation 3.3 as a convenient example. We look for general solutions of the system (Equation 3.3) in the form of the so-called Euler substitution

$$x = C_1 e^{\alpha_1 t} + C_2 e^{\alpha_2 t}, \tag{3.4}$$

where $\alpha_{1,2}$ and $C_{1,2}$ are unknown constants, which should be determined further.

The constants $\alpha_{1,2}$ can be found if we substitute solution 3.4 in Equation 3.3 in the form of $x = Ce^{\alpha t}$

$$C\alpha^2 e^{\alpha t} + \omega^2 C e^{\alpha t} = 0. \tag{3.5}$$

After reducing the terms $Ce^{\alpha t}$ in both parts of Equation 3.5 we obtain the characteristic equation

$$\alpha^2 + \omega_0^2 = 0, \tag{3.6}$$

the solutions of which are evident

$$\alpha_1 = i\omega_0, \; \alpha_2 = -i\omega_0. \tag{3.7}$$

The constants $C_{1,2}$ in Equation 3.4 can be derived when the relevant initial conditions are given. We accept the conditions for the model, shown in Figure 3.1, in the following simplified form:

$$t = 0, \; x = -A, \; dx/dt = 0. \tag{3.8}$$

Then, by taking into account the general solution 3.4, expression 3.7, and conditions 3.8, it is easy to derive for the constants $C_{1,2}$ that

$$C_1 = C_2 = A/2. \tag{3.9}$$

Then, by substituting Equations 3.7 and 3.9 into Equation 3.4, we will find the solutions for the free oscillations of the spring pendulum represented in Figure 3.1

$$x = A \cos \omega_0 t. \tag{3.10}$$

It is not difficult to see that solution 3.10 coincides with the given general definition 3.1 in the particular case of $p_0 = 0$.

3.1.2 Conditionally Periodic Oscillations: Concept of Slowly Varying Amplitudes

Let us continue the discussion of the spring pendulum shown in Figure 3.1. Now we take into account the action of the friction force F_{fr} [it can be the friction within the spring material, between the mass m and rode 3 (see Figure 3.1), etc.]. Herewith, we assume that the dependency of the friction force F_{fr} on the mass velocity $v = dx/dt$ is linear

$$F_{fr} = -r\frac{dx}{dt}, \tag{3.11}$$

where the proportionality factor r is the friction coefficient. By replacing the earlier used expression 3.2 by the new equation for the force balance

$$F_{in} = F_{fr} + F_{el} \tag{3.12}$$

and by utilizing the algorithm of solution, we obtain, similarly with the preceding case, a linear homogeneous differential equation of the second order. However, the latter, at that time, contains an additional term that is proportional to the velocity dx/dt

$$\frac{d^2x}{dt^2} + 2\beta\frac{dx}{dt} + \omega_0^2 x = 0, \tag{3.13}$$

where $\beta = r/2m$ is the damping factor. The term $2\beta\, dx/dt$ in Equation 3.13 accounts for the action of the dissipation forces that, as will be shown later, cause the appearance of the damping of oscillations. Therefore, equations like Equation 3.13 are called the differential equation for the damping oscillations. The solutions of Equation 3.13 can be found (if $\beta \leq \omega_0$) in the form

$$x = A_0 e^{-\beta t} \cos \omega t. \tag{3.14}$$

Here the damping amplitude $A = A_0 \exp\{-\beta t\}$ is referred to also as the conditional amplitude because it is not a true oscillation amplitude. However, let us discuss this circumstance in detail.

First of all, we recall the above given condition for the periodic (including harmonic) oscillations. According to this definition, the system should return periodically to its initial state. This automatically means that the amplitude of these oscillations should be always constant and be equal to the initial amplitude. The above-mentioned free oscillations (see solution 3.10) satisfy this condition, whereas the damping oscillations do not satisfy it. The latter is evidently illustrated in Figure 3.2.

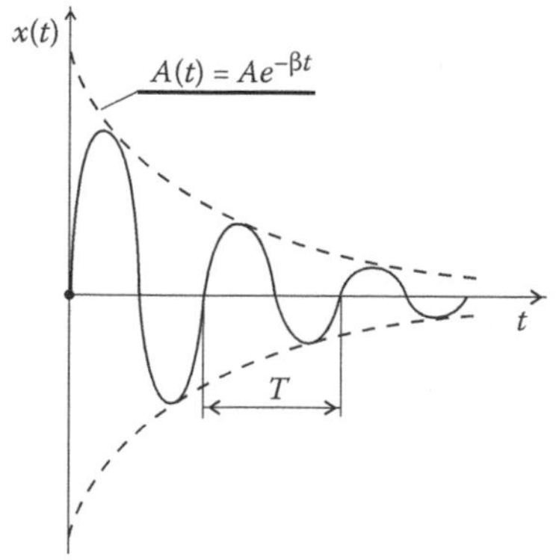

FIGURE 3.2
Graphic illustration of damping oscillations. Here T is the condition period, and $A(t)$ is the condition amplitude of oscillations.

As is readily seen in Figure 3.2, the amplitude of each oscillation is smaller than the amplitudes of the preceding oscillation:

$$Ae^{-\beta t} > A^{-(\beta t+kT)}, \tag{3.15}$$

where k is the number of the oscillation, $T = 2\pi/\omega$ is the conditional period, and

$$\omega = \sqrt{\omega_0^2 - \beta^2} = 2\pi/T, \tag{3.16}$$

is the conditional cyclic frequency of damping oscillations. The oscillations of this kind are called the conditionally periodic oscillations. We will see below in this book the conditionally periodic oscillations of different nature (including weakly nonlinear systems with resonant amplification or damping). But, in any case, all of them will be characterized by the same key property: the conditional frequency (period) of each oscillation process will always be a constant or very close to a constant (slowly varying in time value). The same behavior is characteristic for the conditional amplitudes. In this regard, let us discuss the concept of slowly varying amplitudes.

Let us assume that the damping process appears weak at the background of the oscillation process

$$\beta^2 << \omega_0^2. \tag{3.17}$$

It is obvious that the condition amplitude $A(t) = A\exp\{-\beta t\}$ in this case changes very feebly during one (or even a few) oscillation. We can observe a remarkable result only during a large time of observation in comparison with the period T. This means that, in this situation, the function $A(t) = A\exp\{-\beta t\}$ plays a peculiar role of the slowly varying amplitude. This observation will be used further in many similar situations.

The above-mentioned case of weakly damping oscillations in reality is the simplest illustration of the much more general class of oscillations. Furthermore, very often, we will have to do with the models where the oscillation amplitudes, as well as the cyclic frequency, are slowly varying functions of time. Herewith, in different cases, the causes of this change have a different physical nature. However, the main idea is always the same: we can neglect the change of amplitude during one or a few oscillations, and a remarkable result can only be attained with many oscillations.

3.1.3 Stimulated (Forced) Oscillations: Resonance

Until today we consider the processes in the closed oscillation systems, i.e., in the systems without any external acting forces. Let us then discuss some characteristic features of oscillation processes in the open systems. For simplicity we confine ourselves only to the case where the external force is represented by a harmonic stimulated force F_{stim}. We chose again the above-mentioned model of the spring pendulum (see Figure 3.3) as a convenient illustration example. But, contrary to the preceding cases, we assume that the external harmonic force

$$F_{stim} = F_0 \cos \Omega t \tag{3.18}$$

acts also on the mass m in the direction, which is parallel to the axis of rode 3 (see Figure 3.3).

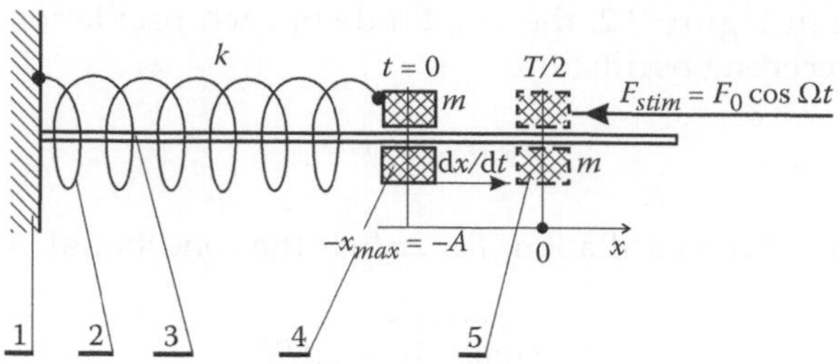

FIGURE 3.3
Model of the spring pendulum, which accomplishes the stimulated (forced) oscillations. Here 1 is a hard wall, 2 is the spring with the stiffness coefficient k, 3 is the central rod, 4 is the location of the mass m for the initial time $t = 0$, 5 is the location of this mass in the time $t = T/2$, x is the spatial coordinate, $x_{\max}$ is the maximal deflection of the mass m, dx/dt is the mass velocity, A is the amplitude of oscillations, F_{stim} is the acted external harmonic force, and F_0 and Ω are the amplitude and cyclic frequency of the force F_{stim}, respectively.

Let us use again the previously applied scheme of analysis. In the considered case, the equation for the force balance (like Equations 3.2 and 3.12) can be written in the form

$$F_{in} = F_{fr} + F_{el} + F_{stim}, \tag{3.19}$$

where all the values have been identified previously.

Substituting these definitions into Equation 3.19 and accomplishing the relevant similar transformations, we obtain the differential equation of stimulating oscillations

$$\frac{d^2x}{dt^2} + 2\beta\frac{dx}{dt} + \omega_0^2 x = f_0 \cos\Omega t, \tag{3.20}$$

where $f_0 = F_0/m$; other definitions are given above. This is (in accordance with the well-known mathematical classification) the linear nonhomogeneous differential equation of the second order, whose general solution can be found in the form

$$x = x^{(gen)} + x^{(part)}, \tag{3.21}$$

where $x^{(gen)}$ is the general solution of the corresponding homogeneous equation (see Equation 3.4)

$$x^{(gen)} = A_0 e^{-\beta t} \sin(\omega t + p_0), \tag{3.22}$$

where A_0 and p_0 are the first pair of unknown (for the time being) constants, $\beta = r/2m$, and $\omega = \sqrt{\omega_0^2 - \beta^2}$. Using the partial solutions of inhomogeneous equation 3.20, $x^{(part)}$ can be chosen, for instance, as

$$x^{(part)} = A \cos(\omega t + \Psi), \tag{3.23}$$

where A and Ψ are the second pair of unknown constants. Other values have already been determined above.

By putting the corresponding initial conditions and by using some well-known properties of the harmonic functions, we eventually obtain the solution 3.21 in an evident form.

Unfortunately, this solution turns out to be too cumbersome. Therefore, for simplicity let us limit ourselves to the analysis of a case of large observation time $t >> T$ [here $T = 2\pi/\omega$ is the conditional period (see the preceding section in more detail]. In this case we can neglect the general part of solution 3.22 and the eventual approximate solution can be written in the following simplified form:

$$x \approx A\cos(\omega t + \Psi), \tag{3.24}$$

where

$$tg\Psi = -2\beta\frac{\Omega}{\omega_0^2 - \Omega^2}, \tag{3.25}$$

$$A = \frac{f_0}{\sqrt{\left(\omega_0^2 - \Omega^2\right)^2 + (2\beta\Omega)^2}}. \tag{3.26}$$

The dependency of the oscillation amplitude A on Ω, which is the frequency of the stimulated force (Figure 3.4), illustrates the most characteristic properties of the considered model. First, we can see that this dependency has an explicitly expressed maximum. Herewith, the maximum location is determined by the so-called resonant condition. In the case $\beta^2 << \omega_0^2$ we can represent it as

$$\Omega = \Omega_{\max} \approx \omega_0, \tag{3.27}$$

where frequency Ω characterizes the stimulated (forced) oscillations, whereas frequency ω_0 is related with the proper oscillations. In summary, we can say that the resonance in the

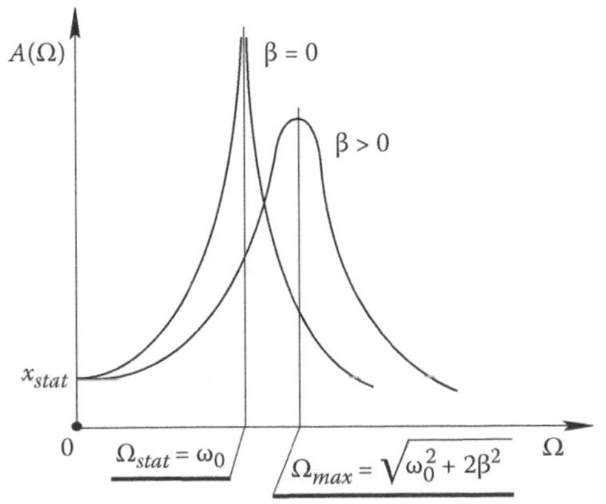

FIGURE 3.4
Dependencies of the oscillation amplitude A on Ω, which is the frequency of the stimulated force. Here Ω_{max} is the frequency when the oscillation amplitude attains a maximum, ω_0 is the oscillation frequency of free oscillations (when $\beta = 0$), β is the damping factor, and x_{stat} is the stationary amplitude of oscillations.

considered model can be determined as the closeness of the frequencies of the stimulated and proper oscillations.

If condition 3.27 is satisfied strictly ($\Omega = \omega_0$) we have the idealized oscillation system without damping ($\beta = 0$). In this case, the resonant amplitude approaches infinity (see Figure 3.4). In turn, the resonant amplitudes are limited in the more realistic case of a system with damping ($\beta > 0$). Herewith, the real resonant frequency is shifted (with respect to the ideal case $\beta = 0$)

$$\Omega = \Omega_{max} = \sqrt{\omega_0^2 + 2\beta^2}. \tag{3.28}$$

At last, we note that the discussed phenomenon of the resonance of the oscillations plays a very important role in many fields of the applied and fundamental physics. Therefore, in the future we will refer to it as the problem of resonance.

3.1.4 Nonlinear Oscillations

Now we discuss the linear oscillations only. Their characteristic feature is that they can be described by linear differential equations like Equations 3.3, 3.13, and 3.20. However, the practical value of such theoretical models is rather limited. In spite of the great role, which linear oscillations play in the education process, they are not a subject of main attention in professional science. Much more popular are the models with so-called nonlinear oscillations. Therefore, let us discuss some general concepts and ideas of the theory of nonlinear oscillations. But, taking into account the main trend of this book, we will limit ourselves to a setting forth of only those topics that are important for further discussion.

Conventionally, the mathematical pendulum is the simplest and the most evident example of nonlinear oscillation systems. The scheme of the mathematical pendulum is illustrated in Figure 3.5. The material point of the mass m oscillates around the vertical axis y. It is not difficult to obtain the differential equation for the description of the material point motion. As is known in the standard course of general physics [2,4], the generalization of the second law of dynamics for the rotational motion can be given in the form

$$M_z = I_z \varepsilon_z, \tag{3.29}$$

where M_z is the z-component of the moment of the force $\vec{F} = m\vec{g}$ (see Figure 3.5)

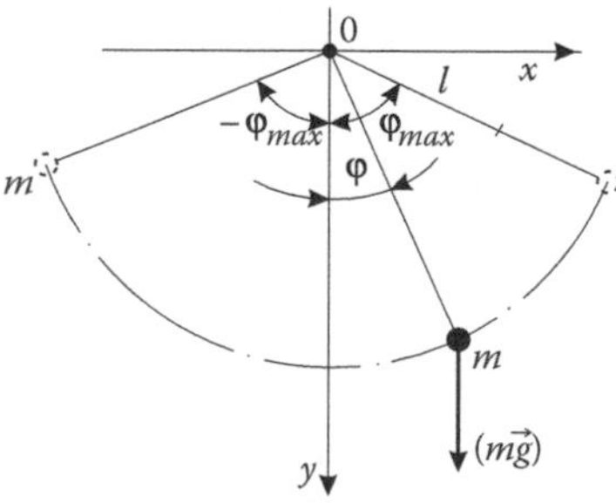

FIGURE 3.5
Simplest scheme of the nonlinear mathematical pendulum. Here m is the mass of material point, l is the length of the pendulum, φ is the angle of deviation, $\pm\varphi_{max}$ are the left and right maximal deflections of the pendulum, $\vec{g}$ is the vector of free fall acceleration, and x, y are the spatial coordinate axes.

$$\vec{M} = \left[\vec{r}\vec{F}\right] = \left\{M_x, M_y, M_z\right\}, \tag{3.30}$$

$\vec{g}$ is the free-fall acceleration vector, m is the mass of the oscillating material point, $\vec{r} = \left\{x, y, z\right\}$ is its radius vector, ε_z is the z-component of the angular acceleration vector

$$\vec{\varepsilon} = \frac{d^2\vec{\varphi}}{dt^2},$$

I_z is the moment of inertia of the material point with respect to the z-axis (which, in turn, is perpendicular to the plane in Figure 3.5), $\vec{\varphi} = \vec{n}\varphi$ is the vector of the angle turning, $\vec{n}$ is the unit vector along the rotation axis ($\vec{n}$ is collinear to the z-axis, i.e., it is normal to the drawing plane), φ is the magnitude of the angle turning, and t is the laboratory time.

The set of expressions 3.29 and 3.31 can be easily transformed into the form

$$I_z = ml^2;\ \varepsilon_z = \frac{d^2\varphi}{dt^2};\ M_z = -mgl\sin\varphi, \tag{3.31}$$

where l is the pendulum length (see Figure 3.5). After relevant elementary transformations the system of equations 3.31 can be reduced to the equation of nonlinear pendulum [2,4]

$$\frac{d^2\varphi}{dt^2} + \omega_0^2 \sin\varphi = 0, \tag{3.32}$$

where the physical meaning of the constant

$$\omega_0 = \sqrt{g/l} \tag{3.33}$$

is the cyclic frequency of the free oscillations for an equivalent linear system (Section 3.1.1).

Expression 3.32 represents a wide class of mathematical subjects that undergo nonlinear oscillations. But the following elementary question arises: why are these oscillations classified as nonlinear and the oscillations described in the preceding sections as linear?

The difference is determined by the mathematical structure of the functions, which are next to ω_0^2 in Equations 3.3, 3.20, and 3.32, respectively. In Equations 3.3 and 3.20, they are represented mathematically by the linear function $f(x) = \omega_0 x$, which presents geometrically the straight line. Therefore, these equations, as well as other similar oscillations, are classified as linear. In contrast, the analogous function in the case of Equation 3.32 has explicitly expressed a nonlinear (trigonometric in this particular case) mathematical structure. Therefore, equations of such kind (and the corresponding oscillations) are classified as nonlinear.

We should bear in mind that the considered example of the mathematical pendulum represents rather a narrow subclass of the nonlinear equations only. In general, the nonlinear equations form a very wide class of mathematical subjects [2,3]. However, in any case, the indicator is the same: if the right part contains a nonlinear function, we determine this equation as nonlinear. Correspondingly, all physical systems whose dynamics is described by nonlinear equations of any type are called nonlinear dynamic systems.

Let us illustrate a correlation between the concepts of linear and nonlinear systems (equations, oscillations). Besides that, we introduce the concepts of weak-nonlinear equations, weak-nonlinear oscillations, and weak-nonlinear systems, respectively.

Let us consider the fact that the pendulum shown in Figure 3.5 accomplishes small oscillations ($\varphi << 1$). We expand the sine in Equation 3.32 in the power series in small φ

$$\sin\varphi \approx \varphi - \frac{1}{3!}\varphi^3 + \cdots \tag{3.34}$$

By accounting only the first term in Equation 3.34, we can easily reduce Equation 3.32 to the above-mentioned (see Equation 3.3) form of the linear differential equation

$$\frac{d^2\varphi}{dt^2} + \omega_0^2\varphi = 0. \tag{3.35}$$

The solution algorithm expounded in Section 3.1.1 can be used for Equation 3.35. As a result, we obtain

$$\varphi = \varphi_0 \sin(\omega_0 t + p_0) = \varphi_0 \sin p, \tag{3.36}$$

where φ_0 is the angular amplitude, $p = (\omega_0 t + p_0)$ is the phase, and p_0 is the initial phase of oscillations.

Then we take into account the second term in expansion 3.34. Herewith, as before, we hold the supposition about the smallness of the angle φ

$$\varphi >> \varphi^3/6.$$

After routine calculations, it is not difficult to obtain the following nonlinear differential equation:

$$\frac{d^2\varphi}{dt^2} + \omega_0^2\varphi - \mu\varphi^3 = 0, \tag{3.37}$$

where $\mu = \omega_0^2/6$. Equations with such a mathematical structure are known in references as the *Duffing equations* [5–8].

The Duffing equation 3.37 is popular in the theory of nonlinear oscillations. Its analytical solutions are also well known (see, for instance, [5,8]). They are interesting first because they describe the simplest case of the so-called weak-nonlinear oscillations. Correspondingly, systems with such oscillations are referred to as weak-nonlinear dynamic systems.

Let us point out also the following important circumstance. We have in mind the phenomenon of the slowly changing oscillation amplitudes. The phenomenon is not new to us because we have already met a similar situation in the case of conditionally periodic linear oscillations (Section 3.1.2). Let us talk once more about the discussed concept of slowly varying amplitudes, but, before that, we will discuss first its weak-nonlinear version.

Strictly speaking, the concept of the amplitude has been introduced for the linear free oscillations only (see Equation 3.1). Therefore, a simple dissemination of such linear terminology under the weak-nonlinear case is a not too rigorous procedure. However, in this case, we do not have any contradiction if some special condition is satisfied. The point is that we have here a very close analogy in the behavior of the weak-nonlinear systems and the above-mentioned conditionally periodic weak-damped linear oscillation system (Section 3.1.2). Indeed, the oscillation amplitudes can be considered constant in both these cases if the observation interval is relatively small, for instance, when this interval is equal to one or a few oscillations only, which, in principle, is the same situation we have with respect to other oscillation parameters. In the weak-nonlinear case, the period T, the frequency ω, and the initial phase p_0 also can be considered as slowly varying values.

Thus, we can treat the weak-nonlinear system as a slowly evolving linear system. In this case the oscillation amplitude and other similar values can play the role of the slowly varying parameters (variables). In contrast, the phase of oscillations p in such situations should be treated as a fast-changing variable. It is important that we consider it as a fast value only at the background of small change in the slowly varying values. This means that, at another background, the same phase can demonstrate properties of the slowly varying value. We will widely use this methodological peculiarity for constructing various asymptotic hierarchic calculation algorithms with the fast and slow rotating phases (see Chapters 6 and 7, in particular).

3.1.5 Multiharmonic Nonlinear Oscillations

One of the most characteristic properties of nonlinear oscillation systems is a possibility to generate many oscillation harmonics. Let us illustrate this process on the chosen example of the nonlinear mathematical pendulum (see Figure 3.5). Its oscillation dynamics can be described, as shown above, by a nonlinear equation like Equation 3.32

$$\frac{d^2\varphi}{dt^2} + \omega_0^2 \cos(\varphi + p_0) = 0, \tag{3.38}$$

where p_0 is the initial phase. Let us find solutions of this equation.

We begin with the so-called first integral of motion that can be obtained from the equation of nonlinear pendulum (Equation 3.38). We multiply this equation by the value $d\varphi/dt$. After obvious simple transformations, we get

$$\frac{d}{dt}\left[\frac{1}{2}\left(\frac{d\varphi}{dt}\right)^2 + \omega_0^2 \sin(\varphi + p_0)\right] = 0. \tag{3.39}$$

Furthermore, the sought first motion integral after the integration can be derived in the form

$$\frac{1}{2}\left(\frac{d\varphi}{dt}\right)^2 + \omega_0^2 \sin(\varphi + p_0) = H = const. \tag{3.40}$$

Let us clarify the physical meaning of this integral, in general, and the constancy of H, in particular. For this we use the classic relationships that are well known in the mechanics of the rotating material point

$$v = R d\varphi/dt \tag{3.41}$$

where v is the linear velocity, and the value $d\varphi/dt = \omega$ (the cyclic frequency) plays also a role of the angular velocity of the material point. Multiplying both parts of Equation 3.40 by the value mR [where $R \equiv l$ (see Figure 3.5) is the rotation radius], we can reduce Equation 3.40 to the form

$$\frac{mv^2}{2} + U(y) = mRH = E = const \tag{3.42}$$

where the first term $mv^2/2$ characterizes the kinetic energy, the second term $U(y)$ is the potential function, and E is the total mechanical energy of the material point. Other definitions are given above (see also Figure 3.5). Thus, the motion integral 3.40 expresses nothing but the energy conservation law. Hence, the physical meaning of the constant H in Equation 3.40 is the normalized total mechanical energy.

Then, let us find the solutions of Equation 3.40. We introduce the energy parameter by means of the expression [9]

$$\kappa = \left[\frac{1}{2}\left(1 + \frac{H}{\omega_0^2}\right)\right]^{1/2}. \tag{3.43}$$

The solutions of the equation of the nonlinear pendulum, which has the same mathematical structure with Equation 3.39, are well known (see, for example, the work of Sagdeyev and Zaslavskiy [9]). In our designations they may be written as

$$\frac{d\varphi}{dt} = \pm 2\kappa \cdot \omega_0 \begin{cases} cn\left[\omega_0(t-t_0), \kappa\right], & (\kappa < 1); \\ dn\left[\omega_0(t-t_0), 1/\kappa\right], & (\kappa \geq 1); \end{cases} \tag{3.44}$$

where t_0 is the initial time. Using the Fourier expansions for the elliptic functions $cn[\ldots]$ and $dn[\ldots]$, we rewrite solutions 3.44 in the form

$$\frac{d\varphi}{dt} = \pm 8\omega \begin{cases} \sum_{n=1}^{\infty} \frac{(-1)^{n+1} a^{n-1/2}}{1 + a^{2n-1}} \cos\left[(2n-1)\omega(t-t_0)\right], & (\kappa \leq 1); \\ \frac{1}{4} + \sum_{n=1}^{\infty} \frac{a^n}{1 + a^{2n}} \cos\left[2n\omega(t-t_0)\right], & (\kappa \geq 1). \end{cases} \tag{3.45}$$

$$\varphi = \pm 8 \begin{cases} \sum_{n=1}^{\infty} \frac{(-1)^{n+1} a^{n-1/2}}{(2n-1)\left(1 + a^{2n-1}\right)} \cos\left[(2n-1)\omega(t-t_0)\right], & (\kappa \leq 1); \\ \sum_{n=1}^{\infty} \left\{\frac{1}{8n} + \frac{a^n}{2n\left(1 + a^{2n}\right)}\right\} \cos\left[2n\omega(t-t_0)\right], & (\kappa \geq 1). \end{cases} \tag{3.46}$$

which determines the dependence of the functions $d\varphi/dt$ and $\varphi(t)$. Here, the following designations are introduced: n is the number of oscillation harmonic,

$$a = \exp(-\pi F'/F), \tag{3.47}$$

$$F' = F\left(\frac{\pi}{2}, \sqrt{1-\kappa^2}\right), \quad F = F\left(\frac{\pi}{2}, \kappa\right)$$

are the full elliptic integrals, and ω is the nonlinear cyclic frequency of the nonlinear proper oscillations

$$\omega = \omega(H) = \frac{\pi}{2}\omega_0 \begin{cases} \dfrac{1}{F(\pi/2;\kappa)} & (\kappa < 1); \\ \dfrac{k}{F(\pi/2;1/\kappa)} & (\kappa \geq 1). \end{cases} \tag{3.48}$$

As is readily seen, the nonlinear frequency ω depends on the normalized energy H (or on the amplitude of oscillations). Therefore, the nonlinear pendulum can be treated as a nonisochronous multiharmonic oscillator, in contrast to the above-mentioned variants of the linear oscillators.

Thus, by analyzing solution 3.45, we can conclude that, in contrast to the case of the linear oscillation (Section 3.1.1), the oscillation process of the nonlinear pendulum has an explicitly expressed multiharmonic nature. As it follows from references [5–9] this is an inherent property of most of the nonlinear oscillation systems.

3.1.6 Rotating and Oscillation Phases

Analyzing solutions 3.45 and 3.46, it can be seen that they describe two different types of pendulum motion. The first characterizes a limited motion [with respect to the angle φ ($\kappa < 1$); see Figure 3.5] and the second represents the case of unlimited motion ($\kappa \geq 1$; see Figure 3.6). Hence, we can introduce the concepts of oscillating and rotating phases $\varphi \equiv p$.

One can think that both these phases express the same, in principle, physical categories. Indeed, we can represent any oscillation mathematically as a sum of two rotations, which occur in the reciprocally opposite directions, and vice versa—any rotation can be represented as a sum of two oscillations, which are encountered on two reciprocally perpendicular axes. But this is not so. The theoretical physics says that the rotational motion is more elementary than the oscillation one. Therefore it is considered, for instance, that photon has circular polarization. On the other hand, the rotational motion is much more widespread in nature than the purely oscillatory motion. For example, electrons in an atom rotate around the nucleus, the moon rotates around the earth, the earth rotates around the sun, the sun, in turn, rotates around the galaxy center, and so on. Therefore in Chapters 5 through 7, we will use the term "rotating phase" for characterization of any oscillation/rotational motion, considering that the oscillating phases are a specific partial case of the rotating one.

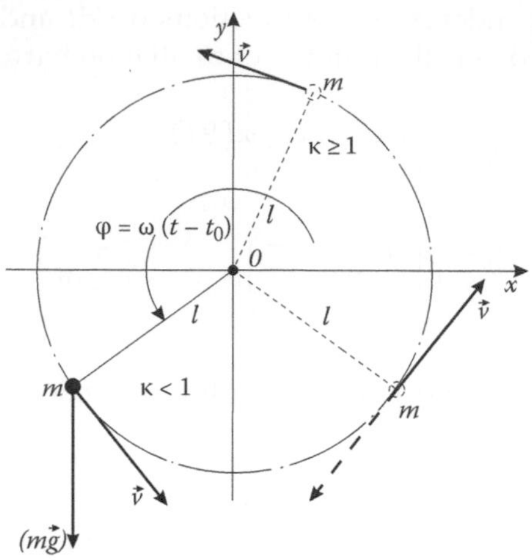

FIGURE 3.6
Rotation type of pendulum motion. Here $\vec{v}$ is the velocity vector of the mass m. Other designations are the same with Figure 3.5.

3.1.7 Hidden and Explicit Phases

We divide all oscillation phases into hidden and explicit ones. Let us begin with the concept of the hidden oscillation phase. For this we recur to the discussion of stimulated oscillations. As is obviously seen in Figure 3.3, in this case, we have to deal with the phases of two principally different kinds (see also solutions 3.21 through 3.23). The first is the hidden phase, which is represented here by the phase of free (or damping) oscillations

$$p_{gen} = \omega_0 t + p_0. \tag{3.49}$$

The key property of this phase is that we can find it only within the course of solving the oscillation problem. In the considered case it is determined by the hardness k (see Figure 3.3) of the spring and the mass m of the material point ($\beta = 0$)

$$p_{gen} = \left(\sqrt{k/m}\right)t + p_0. \tag{3.50}$$

In the case of the mathematical pendulum (see Figure 3.5), the hidden phase of oscillation also depends on the system parameters (the free fall acceleration g and the pendulum length l)

$$p = p_{gen} = \omega_0 t + p_0 = \left(\sqrt{g/l}\right)t + p_0. \tag{3.51}$$

Thus, an evident form of any hidden oscillation phase is unknown until one undergoes an oscillation problem. Figuratively speaking, it is hidden to us at the initial stage of the solution procedure.

Then, let us again consider the second-type phases. In contrast to the hidden phases, the explicit oscillation phases are always given initially. The phase of the stimulated force F_{stim}

(see Figure 3.3 and solutions 3.21 through 3.23) can serve as a conventional example of the explicit phase

$$p_{part} = \Omega t + \Psi. \tag{3.52}$$

In contrast to the hidden phases, the stimulated phase p_{part} is explicitly given initially. That is why it is classified as the explicit phase.

In general, the explicit phases can characterize the stimulated as well as the proper oscillations, whereas the hidden phases can only be proper ones.

3.1.8 Main Resonance

Let us discuss once again the concept of resonance. For this we revert to the subject of stimulated oscillations (Section 3.1.3). In this regard, let us recall the resonant condition 3.27

$$\Omega \approx \omega_0. \tag{3.53}$$

By taking into consideration the definitions for phases 3.50 through 3.52, we can rewrite condition 3.54 in the following more general form:

$$\frac{dp_{stim}}{dt} \approx \frac{dp_{prop}}{dt} \tag{3.54}$$

where $p_{stim} \equiv p_{part}$ is the phase of stimulated oscillations, and $p_{prop} \equiv p_{gen}$ is the phase of proper oscillations in solutions 3.21 through 3.23. The resonant conditions like Equation 3.54 are called the condition for the main resonance.

As experience shows, the resonant condition in the form Equation 3.54 is much more convenient in calculation and physical analysis than it seems at first sight. First, this is because the phases p_{stim} and p_{prop}, depending on time t, additionally depend very often on the spatial coordinates x,y,z, too. Let us discuss, as an evident illustration example, the simplest model of the FEL with the Dopplertron pumping [10–15]. It is the model with the resonant motion of an electron in the field of two collinearly propagated (in reciprocally opposite directions) plane harmonic electromagnetic waves (pumping and signal). In this case the explicit electron oscillation phases can be written as [10,14]

$$p_1 = \omega_1 t - k_1 z + p_{01},\ p_2 = \omega_2 t + k_2 z + p_{02}, \tag{3.55}$$

where $k_j = \omega_j/c$ are the wave numbers, $j = 1,2$ are the numbers of the waves, c is the light velocity in vacuum, and z is the longitudinal spatial coordinate of the electron. Differentiating Equation 3.55 on time, resonant condition 3.54 can be rewritten in the form

$$\omega_1 - k_1 v \approx \omega_2 + k_2 v \tag{3.56}$$

or, after noncomplex transformations ($k_j = \omega_j/c$), as

$$\omega_1 \approx \frac{1+v/c}{1-v/c}\omega_2 \tag{3.57}$$

were $v = dz/dt$ is the longitudinal electron velocity. For convenience, we accepted $p_{stim} \equiv p_1, p_{prop} \equiv p_2$.

Expression 3.57 is well known in literature (see, for instance, [10–22]) as the resonant condition for FELs with pumping by an electromagnetic wave. The physical meaning of the coefficient $(1 + v/c)/(1 - v/c)$ in Equation 3.57 is that it is the ratio of the mutual Doppler's shifts for frequencies of both electromagnetic waves in the reference system bounded with the longitudinal motion of electron. In this reference system (i.e., in the case $v' = 0$) we have the standard (see Equation 3.54) resonant condition $\omega_1' \approx \omega_2'$.

It is easily evident that the procedure of obtaining the same resonant condition 3.57 from definition 3.54 immediately turns out to be much more complicated. First, because of this condition, as mentioned above, we should preliminarily accomplish a passage into the moving (with the longitudinal velocity v) reference system. It follows immediately from the physical meaning of the discussed phenomenon.

Then, taking into account the introduced new definitions, let us continue the discussion of the concept of resonance. First, in spite of the self-obviousness of this concept, some confusion exists in references with the precise definition of resonance. For instance, sometimes any sharp increase in the oscillation amplitude is called resonance (see, for instance, [22]). However, there are a few rather strange phenomena in electrodynamics, such as the plasma-beam instability or the two-stream instability [10,23–25]. They are also characterized by a rather sharp increase in the oscillation amplitude but do not have truly resonant nature. Similar phenomena are also known in hydrodynamics (the Helmholtz instability [26,27]) and other areas of science. The point is that in this type of phenomena, all acting forces have explicitly expressed proper nature. In electrodynamics such processes, as a rule, are called instabilities and not resonances. Therefore, following the tradition, we consider the resonance as a specific kind of instability, when conditions such as Equation 3.54 are satisfied. This means also that, by far, each instability cannot be treated as resonance. As shown above, the distinguishing feature of resonance is that here, one of the acting stimulated forces should always be external.

Let us also say a few words about another limit case, when both oscillation phases are evident. Such a model, basically, can also be treated as the resonant system (see Equations 3.55 through 3.57 and corresponding discussion). However, it is not the above-mentioned ordinary classic resonance, when at least one of the interacted phases is hidden. In this case we have to do with a specific case of resonance, which is known in literature as the parametric resonance. We will discuss this type of resonance in detail in Chapters 5 through 7.

Thus, we accept in the book the following definition of resonance: this is such a sharp increase in amplitude, when the frequencies of oscillations are sufficiently close (see Equation 3.54). Herewith, at least one of the phases of oscillations is evident. In other words, we can also formulate resonance as the closeness of velocities of changing of phases of two or more oscillations.

3.1.9 Resonances at Harmonics

Apart from the main resonance, resonances at harmonics are also known in the theory of oscillations (see Chapter 6 for more detail). Let us discuss this class of resonant interactions in detail.

As is shown above (see expression 3.46 and corresponding comments), the characteristic feature of nonlinear oscillation systems is a possibility to generate oscillation harmonics. The theory of nonlinear oscillations states [1,8,10,14,28] that resonant conditions for oscillation harmonics can also be realized.

As before, for further explanation, we use the following model: an electron in the field of two opposite electromagnetic waves. Analogous to Equation 3.46, we can write the phases of harmonics in the considered case as

$$p_{1n} = n_1 p_1, \; p_{2n} = n_2 p_2, \tag{3.58}$$

where n_j are the harmonic numbers. Besides that, we take into account that each of the electromagnetic waves can be multiharmonic initially. We designate this kind of harmonics by m_j. This means that the definitions for phases of electron oscillations (Equation 3.58) in such a multiharmonic electromagnetic field can be generalized as

$$p_{1nm} = n_1 m_1 p_1, \; p_{2nm} = n_2 m_2 p_2. \tag{3.59}$$

Then, we reformulate the resonant condition like Equation 3.54 for the discussed case of resonance on harmonics

$$n_1 m_1 \frac{dp_1}{dt} \approx n_2 m_2 \frac{dp_2}{dt}, \tag{3.60}$$

which is the same as

$$\frac{dp_1}{dt} \approx \frac{n_2 m_2}{n_1 m_1} \frac{dp_2}{dt}. \tag{3.61}$$

Expressions such as Equations 3.60 and 3.61 are the necessary conditions for the realization of resonances at harmonics.

All resonances at harmonics can be divided into integer and fractional resonances. The coefficient $n_1 m_1 / n_2 m_2$ is the criterion for such classification. In particular, if it is an integer number, we have ordinary (integer) resonance at harmonics. When this coefficient is a fraction, we have fractional resonances [10].

It is clear that expressions 3.60 and 3.61 can also be treated as general definitions for the pair (two-multiply) main and harmonic resonances. Indeed, accepting, for example, that $n_j = m_j = 1$, we obtain the main resonance such as Equation 3.54 as a partial case. Assuming that $n_j, m_j > 1$, we have a typical resonance at harmonics.

Then, we accept additionally the assumption about the existence of the negative as well as positive harmonics, i.e., $n_j, m_j = \pm 1, \pm 2, \ldots$ (we can obtain in such harmonics developing periodic functions in a Fourier series in a complex form). Furthermore, in the book, we will use the mathematical definition of any pair of (two-multiply) resonances just in the form Equation 3.60 or 3.61, where the accepted assumption is taken into account.

We will discuss later the concept of nonpair resonances (three-multiply, four-multiply, and so on), which corresponds to the case when more than two phases take part in one resonance simultaneously [10].

3.1.10 Slow and Fast Combination Phases: Resonances

Let us rewrite expression 3.60 in the form

$$\frac{d}{dt}\left(n_1 m_1 p_1 - \sigma_{12} n_2 m_2 p_2\right) = \frac{d\theta}{dt} \approx 0 \tag{3.62}$$

where

$$\theta = n_1 m_1 p_1 - \sigma_{12} n_2 m_2 p_2 \tag{3.63}$$

is the so-called slow combination phase (because its velocity of change is very small); $\sigma_{12} = \pm 1$ is the sign function, which we introduce for the convenience of representation or, more exactly, for compensation of the inconvenience connected with the possibility of existing negative numbers of harmonics. For the sake of convenience, we introduce also the symmetrical (sign-conjugated) phase

$$\psi = n_1 m_1 p_1 + \sigma_{12} n_2 m_2 p_2. \tag{3.64}$$

Comparing definitions 3.63 and 3.64, it is not difficult to see that the phase ψ should be classified as the *fast combination phase* because $|d\psi/dt| >> |d\theta/dt|$.

Furthermore, once more we extend in Equations 3.63 and 3.64 the definitions for the numbers of oscillation harmonics. Namely, we suppose that apart from the positive and negative harmonics, the zero harmonics can also exist

$$n_j, m_j = 0 \pm 1, \pm 2, \ldots \tag{3.65}$$

This means that not only combination phases θ can have small velocity of changing but some from the ordinary phases and their harmonics ($n_1, m_1 = 0$ or $n_2, m_2 = 0$) can be slowly changing, too.

Using the above introduced values θ and ψ, we can rewrite the generalized resonant conditions 3.60 and 3.61 in the following nondimensional forms:

$$\begin{aligned} \left|\frac{d\theta/dt}{d\psi/dt}\right| &= \left|\frac{d\left(n_1 m_1 p_1 - \sigma_{12} n_2 m_2 p_2\right)/dt}{d\left(n_1 m_1 p_1 + \sigma_{12} n_2 m_2 p_2\right)/dt}\right| \\ &= \left|\frac{d\left(p_1 - \sigma_{12}\left(n_2 m_2/n_1 m_1\right) p_2\right)/dt}{d\left(p_1 + \sigma_{12}\left(n_2 m_2/n_1 m_1\right) p_2\right)/dt}\right| \sim \varepsilon_1 << 1 \end{aligned} \tag{3.66}$$

where ε_1 is a small parameter of the problem for the considered pair (two-multiply) resonance. It characterizes the closeness of frequencies (velocities of the phase change). Small parameters like Equation 3.66 also play the role of hierarchic small parameters (see, for instance, expression 2.5).

Resonances like Equations 3.62 and 3.65 are classified, in turn, into parametric and quasilinear resonances. The parametric resonances, as mentioned above, take place if both phases p_j are classified as explicit. The other possible variant, when one of the phases is always represented by a hidden phase, is called quasilinear resonance.

3.1.11 Hierarchy of Resonances

We discuss the concept of n-multiply (where $n \geq 2$) resonances. Let us assume that the combination phase θ (Equation 3.63) can, in turn, take part in a resonance with another changing phase p_3

$$\frac{d\theta}{dt} \approx \sigma_{13} n_3 m_3 \frac{dp_3}{dt}, \tag{3.67}$$

where $n_3, m_3 = \pm 1, \pm 2, \ldots$ are the numbers of the relevant harmonics. This situation illustrates clearly the relative character of the terms "slow" and "fast." Indeed, we classified the combination phase θ as a slowly changing value at the background of fast-oscillation phases p_j and ψ (see definitions 3.66). On the other hand, by applying additional condition 3.67 we can formulate the new superslow

$$\Theta = \theta - \sigma_{13} n_3 m_3 p_3 = n_1 m_1 p_1 - \sigma_{12} n_2 m_2 p_2 - \sigma_{13} n_3 m_3 p_3 \tag{3.68}$$

and the new fast

$$\Psi = \theta + \sigma_{13} n_3 m_3 p_3 = n_1 m_1 p_1 - \sigma_{12} n_2 m_2 p_2 + \sigma_{13} n_3 m_3 p_3 \tag{3.69}$$

combination phases. Here $\sigma_{13} = \pm 1$ is another sign function. As a result, we can rewrite Equation 3.67 in the form

$$\frac{d\Theta/dt}{d\Psi/dt} = \varepsilon_2 << 1. \tag{3.70}$$

As a result, we obtain the obviously expressed hierarchy of the small parameters ε_1 and ε_2

$$\varepsilon_1 << \varepsilon_2 << 1. \tag{3.71}$$

In Chapter 1 (see, for instance, definition 2.5), we treated expressions such as Equation 3.71 as the *hierarchic series*. In the framework of the general theory of dynamic hierarchic systems, it characterizes (see Chapter 2 for more detail) the hierarchic arrangement of a considered model. Hence, in the discussed case of the two connected resonances (Equation 3.71), we can regard the considered model as the simplest (three-level, if we take into account the zero-level) hierarchic resonant system. In the general case of multi-level resonance hierarchic series, hierarchic series 3.71 can be represented in the following extended form:

$$\varepsilon_1 << \varepsilon_2 << \ldots << \varepsilon_\kappa << << \ldots << \varepsilon_m << 1, \tag{3.72}$$

where each term in Equation 3.72 characterizes a separate hierarchic oscillation level. Here κ is the current number of the hierarchic level, and m is the highest hierarchic level of the considered multiresonant system. Let us discuss this interesting topic in detail.

3.1.12 Slowly Varying Amplitudes and Initial Phases: Complex Amplitudes

In Sections 3.1.2 and 3.1.4 we have mentioned the concept of slowly varying amplitudes. Let us now discuss it in detail.

The classic form of linear solutions like Equation 3.36

$$z_k = A_k \sin\left[\omega t + \varphi_{0k}\right]$$

is, strictly speaking, illegal in the case of nonlinear oscillation systems. Nonetheless, such a form of representation can be disseminated in the case of weak-nonlinear models under the condition that some additional specific suppositions are accepted. These additional specific suppositions concern with new extended definitions for the oscillation amplitude and the initial phase. The difference with the linear case like that in Equation 3.36 is that, at that time, the oscillation amplitude and the initial phase are found to be slowly varying functions on time t. So, the corresponding weak-nonlinear solutions can be written in the following modernized general form:

$$z_k = A_k(t) \sin [\omega t + \varphi_{0k}(t)], \tag{3.73}$$

where z_k is the kth component of some n-dimensional vector z, $A_k(t)$ is the kth component of the slowly varying amplitude vector $A(t)$, $\varphi_0(t)$ is the kth component of the slowly varying initial oscillation phases vector $\varphi(t)$, and ω is the (permanent) cyclic frequency. It is easy to be convinced that some other version of introducing the slowly varying values is also possible

$$z_k = A_k(t) \sin [\omega(t)t + \varphi_{0k}], \tag{3.74}$$

where $\omega(t)$ is the slowly varying cyclic frequency, and φ_{0k} is the permanent initial phase. Experience shows that both discussed representations 3.73 and 3.74 are equivalent. However, inasmuch as it is more convenient, the first version is more widespread in practice.

Thus, the basic weak-nonlinear resonant mathematical problem can be formulated as determining the slow dependencies $A(t)$ and $\varphi_0(t)$ [or $A(t)$ and $\omega(t)$], respectively. One of the main interests in this book is the description of the methods for constructing such dependencies.

Apart from the real amplitude $A(t)$ and phase $\varphi_0(t)$, the so-called complex slowly varying amplitudes are used for practical calculations. The definition of the complex slowly varying amplitude can easily be obtained from Equation 3.73 by utilizing the well-known de Moivre's formula

$$\sin\alpha = \frac{1}{2i}\left(e^{i\alpha} - e^{-i\alpha}\right) = \left(\frac{1}{2i}e^{i\alpha} + c.c.\right). \tag{3.75}$$

Using Equation 3.75 in Equation 3.73, we obtain the representation for Equation 3.73 via the slowly varying complex amplitudes

$$z_k = \left[\frac{1}{2i}A_k(t)\exp\left\{i\varphi_{0k}\right\}\right]e^{ip} + \left[\frac{-1}{2i}A_k(t)\exp\left\{-i\varphi_{0k}\right\}\right]e^{-ip} = \left[\frac{1}{2i}A_{ck}(t)e^{ip} + c.c.\right], \tag{3.76}$$

where the clear oscillation phase p is defined as

$$p = \omega t, \tag{3.77}$$

$A_{ck}(t) = A_k(t)\exp\{i\varphi_{0k}\}$ is the slowly varying complex amplitude, and the designation $c.c. = A_k^*(t) = A_k(t)\exp\left\{-i\varphi_{0k}\right\}$, i.e., it is the complex-conjugated term.

Thus, the weak-nonlinear oscillations formally can be described by means of the concepts of slowly varying amplitudes and initial phases. Herewith, oscillations like Equations 3.73, 3.74, and 3.76 are called quasiharmonic oscillations because, as mentioned above, they are not the true harmonic oscillations.

3.2 Hierarchic Oscillations and Hierarchic Trees

3.2.1 Hierarchic Series and the Hierarchic Trees

In general, the classification of resonances is not a simple mathematical as well as physical procedure. The point is that practically we can have a great variety of possible interaction variants, schemes, and versions for any of the real considered multiresonant systems. Herewith, the same oscillation phase in the same system can be classified as slow and fast at the same time. The result of such classification depends on the velocities of change of other phases also existing in the system. Some of these phases can form with this phase different variants, schemes, and versions of the resonances. On the other hand, the characteristic peculiarities of these variants, schemes, and versions determine completely the mathematical arrangement of an algorithm, which we apply to the investigation of the considered (multiresonant, in general) system.

In such a situation it is obvious that we should have a qualitative method for preliminary classification and analysis of complex multiresonant systems. It is also obvious that the most preferable variant of such a method is a graphical representation like that used, for instance, in the famous Feynman's diagrams [2,3].

The proposed method plays a similar role in the theory of hierarchic oscillations and waves. This is the method of hierarchic trees. It is described in detail in Sections 2.1.4 through 2.1.6 for the general case of the dynamic hierarchic system. In this chapter and further in this book, we will discuss and use a specific oscillation-wave variant of this method. This variant is destined, predominantly, for studying the oscillation-wave hierarchic systems. In other words, we will use the so-called method of oscillation-wave hierarchic trees.

As is mentioned already, the key point of the quantitative description of any hierarchic dynamic system is the use of hierarchic series like Equations 2.3 and 3.72

$$\varepsilon_1 << \varepsilon_2 << \ldots << \varepsilon_\kappa << \ldots << \varepsilon_m << 1. \tag{3.78}$$

This series is different for different concrete systems and can be regarded as a peculiar system passport. Herewith, in accordance with the above, each term of series 3.78 corresponds to a separate hierarchic level of the considered oscillation-wave system.

Thus, we can state that the smallness of the parameter ε_j in Equation 3.78 characterizes the order (rank) of the hierarchy of a group of oscillating phases.

Another important feature of the discussed version of the method is that the hierarchic trees can illustrate evidently the relationships of causes and effects in the considered oscillation-wave system. For instance, hierarchic trees allow clearing up the physical nature of the oscillation phases, which take part simultaneously in different resonant and nonresonant interactions. In other words, the method allows us to determine, for instance, the origination of the phases: is this oscillation phase initially given or does it appear as a result of the nonlinear (including the resonant) interaction of two (or more) oscillations in the system?

However, the main destination of the series such as Equations 3.72 and 3.78 is the classification of oscillation phases into slow and fast phases for each hierarchic level. In general (i.e., for the system as a whole), such classification can be rather conditional because the phases that are treated as slow phases for one hierarchic level can be classified as fast phases for the next level. Series such as Equations 3.72 and 3.78 allow us to know the status of each of such phase at any stage of the hierarchic transformation.

Let us illustrate the above by the simplest variants of the hierarchic trees of different types.

3.2.2 Pairwise Main Resonances: Simplest Version

We begin our discussion with the simplest example of an oscillation hierarchic tree, which corresponds to the case of the main pair resonances (Section 3.1.8). Let us assume that we have a system with three oscillation phases p_1,p_2,p_3 only, which are characterized by the velocities of changing approximately the same order. Besides that, we suppose that two of these phases (for example, p_1 and p_2) can interact with one another in a resonance manner. Hence, the third phase p_3 can be classified in this case as nonresonant.

In accordance with Equation 3.66, the resonant condition for the pair of phases p_1,p_2 in the general case can be written as

$$\left|\frac{d\theta/dt}{d\psi/dt}\right| = \left|\frac{d\left(n_1 m_1 p_1 - \sigma_{12} n_2 m_2 p_2\right)/dt}{d\left(n_1 m_1 p_1 + \sigma_{12} n_2 m_2 p_2\right)/dt}\right| \sim \varepsilon_1 << 1. \tag{3.79}$$

On the other hand, according to the definition of the main resonances (Section 3.1.8), we should accept in Equation (3.79) $n_1 m_1 = +1$, $n_2 m_2 \sigma_{12} = +1$ because the main resonance, according to the definition, takes place for the first harmonics only. After a noncomplex transformation we can rewrite the general resonant condition 3.79 in the form

$$\frac{d\theta_{12}}{dt} << \frac{d\psi}{dt}, \frac{dp_1}{dt}, \frac{dp_2}{dt}, \frac{dp_3}{dt} \quad \text{or} \quad \frac{dp_1}{dt} \approx \frac{dp_2}{dt}, \tag{3.80}$$

where, for convenience, we have changed the designation $\theta \equiv \theta_{12} = p_1 - p_2$.

By taking into consideration the above set of suppositions, we can classify phases p_j and ψ (see definitions 3.64 and 3.79) as fast phases, whereas $\theta \equiv \theta_{12}$ (see definitions 3.63 and 3.79) should be considered as the *slow combination phase*. This means that phases p_j and ψ belong to the zero hierarchic level ($\kappa = 0$), whereas the phase θ_{12} is related to the first level ($\kappa = 1$). Graphically this situation can be pictured in the form of the hierarchic tree shown in Figure 3.7.

The discussed type of the pair resonance is typical for most electronic devices with the protracted interaction: magnetoresonant devices, including the gyrotrons, the FELs, etc. [5,10–16,19–21,23–25,30–34].

3.2.3 Pairwise Main Resonances: Bounded Resonances

The above-mentioned case of the oscillation system with three phases p_j (see Figure 3.7) looks, at first sight, rather trivial. However, its simplicity is really deceptive. A variety of

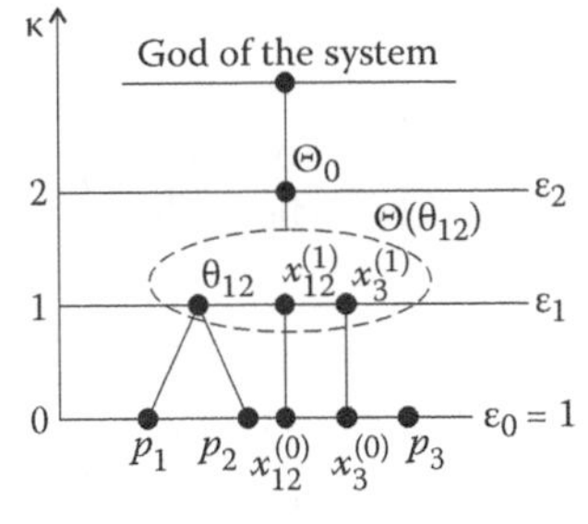

FIGURE 3.7
Hierarchic tree for the pair (two-multiply) resonance. Here the fast phases p_1, p_2, p_3 and other variables $x_{12}^{(0)}$, $x_3^{(0)}$ belong to the zero hierarchic level ($\kappa = 0$). Two phases (p_1, p_2) form the resonance. The slow combination phase θ_{12} (first hierarchic level $\kappa = 1$) is generated as a result of nonlinear interactions in the system. The values $x_{12}^{(1)}(\theta_{12})$ and $x_3^{(1)}(\theta_{12})$ represent the other slow variables of the first hierarchic level (in Figures 3.8 through 3.18 we will omit them for simplicity). Totality of the variables of the first hierarchic level is represented by the vector $\Theta(\theta_{12})$. The value Θ_0 characterizes the highest (second) hierarchic level. The values ε_j ($j = 0,1,2$) are the hierarchic small parameters (see hierarchic series 2.3, 3.72, and 3.78 and corresponding comments). The so-called god of the system is beyond the system. In the considered case, it characterizes an external energy source for the system as a whole (see Chapter 2 for more details).

possible interaction schemes, even in the simplest three-phase case, turn out to be richer than they seem. Let us continue studying the discussed three-phase system, illustrating its general physical nature by means of the hierarchic trees shown in Figure 3.8.

First we discuss the hierarchic tree represented in Figure 3.8. This is the so-called case of bounded (including coupled) resonances [10]. The FEL with bichromatic pumping [10,29,30] can be given as a practical example. The peculiarity of this interaction scheme is that one of the three phases (for instance, p_2) is situated in the resonance simultaneously with the two other phases

$$\frac{dp_1}{dt} \approx \frac{dp_2}{dt}, \quad \frac{dp_3}{dt} \approx \frac{dp_2}{dt}. \tag{3.81}$$

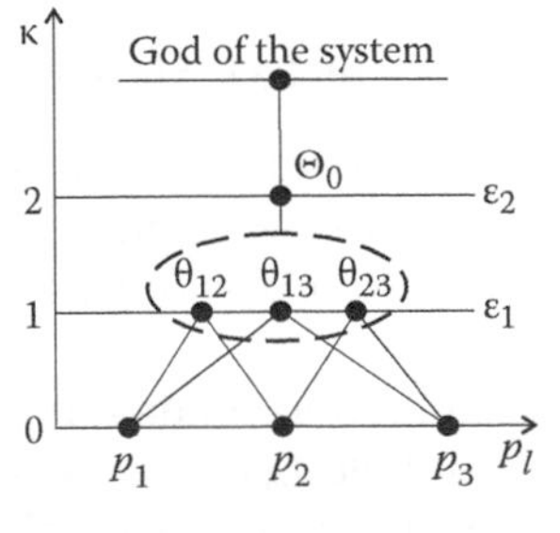

FIGURE 3.8
Hierarchic tree for the system of phases with two of the bounded resonances [10,29,30]. Here the phase p_2 takes part in two resonances (with phases p_1 and p_3) simultaneously. Herewith, the resonant condition for phases $p_{1,3}$ is satisfied automatically. It is important that the formed slow combination phases θ_{12}, θ_{13}, θ_{23} cannot form the resonance of the next rank of the hierarchy. Here and further in Figures 3.9 through 3.18, we confine ourselves to discussing only the behavior of the hierarchy of phases (the ordinary and the combined). Other designations are the same as those given in comments to Figure 3.7. (From Kulish, V.V., *Hierarchic Methods. Undulative Electrodynamic Systems*, Kluwer Academic Publishers, Dordrecht, 2002. With permission. From Bolonin, O.I. et al., *Acta Phys. Pol.*, A76, 455–473, 1989. With permission. From Kohmanski, S.S., and Kulish, V.V., *Acta Phys. Pol.* A68, 41–748, 1985. With permission.)

That is accompanied by the appearance of the two slow combination phases θ_{12} and θ_{23}

$$\frac{d(p_1 - p_2)/dt}{d(p_1 + p_2)/dt} = \frac{d\theta_{12}/dt}{d\psi_{12}/dt} \sim \varepsilon_1 << 1,$$

$$\frac{d(p_2 - p_3)/dt}{d(p_2 + p_3)/dt} = \frac{d\theta_{23}/dt}{d\psi_{23}/dt} \sim \varepsilon_1 << 1. \quad (3.82)$$

It is interesting to note that satisfying conditions 3.81 and 3.82 leads automatically to satisfying the resonant conditions for the resonance between phases p_1 and p_3

$$\frac{dp_1}{dt} \approx \frac{dp_3}{dt}; \quad \frac{d(p_1 - p_3)/dt}{d(p_1 + p_3)/dt} = \frac{d\theta_{13}/dt}{d\psi_{13}/dt} \sim \varepsilon_1 << 1. \quad (3.83)$$

At first sight (by analogy with the preceding case), it seems that the slow combination phases θ_{12}, θ_{13}, and θ_{23} can, in turn, also form the resonances of the second ($\kappa = 2$) hierarchy if

$$\frac{d\theta_{12}}{dt} \approx \frac{d\theta_{23}}{dt}; \quad \frac{d\theta_{12}}{dt} \approx \frac{d\theta_{13}}{dt}; \quad \frac{d\theta_{13}}{dt} \approx \frac{d\theta_{23}}{dt}. \quad (3.84)$$

But it is not so. A simple analysis of relationships 3.80 through 3.84 shows that resonant conditions 3.84 cannot be satisfied because neither the combinations $\theta_{12} - \theta_{23} = \Theta_{123}$, $\theta_{12} - \theta_{13} = \Theta'_{123}$, nor the combination $\theta_{13} - \theta_{23} = \Theta''_{123}$ is slow (or, more exactly, superslow), i.e., they are not slowly varying functions of the second hierarchic order ($\kappa = 2$). This peculiarity is clearly illustrated in Figure 3.8.

3.2.4 Pairwise Main Resonances: Case of Two Slow Phases

Earlier we considered interaction schemes with resonance of the fast phases that belong to the zero hierarchic level $\kappa = 0$. However, the slow phases can also interact with each other in a resonant-like manner. Illustration examples of this type are given in Figures 3.9 and 3.10.

The characteristic feature of the interaction scheme represented in Figure 3.9 is that the third phase p_3 is the slow phase, i.e., it belongs initially to the first hierarchic level ($\kappa = 1$). Moreover, it has resonance with respect to the slow combination phase $\theta_{12} = p_1 - p_2$

$$\frac{d\theta_{12}}{dt} \approx \frac{dp_3}{dt}; \quad \frac{d(p_1 - p_2 - p_3)/dt}{d(p_1 + p_2 + p_3)/dt} = \frac{d\Theta_{123}/dt}{d\Psi_{123}/dt} \sim \varepsilon_2 << 1. \quad (3.85)$$

It can appear unexpected, but this interaction scheme is basic for the main physical mechanisms of FELs (see Chapter 4) [5,10–16,19,20,29–33]. In this case, phases $p_{1,2}$ are caused by the electron oscillations in the electromagnetic fields of the signal and pumping waves.

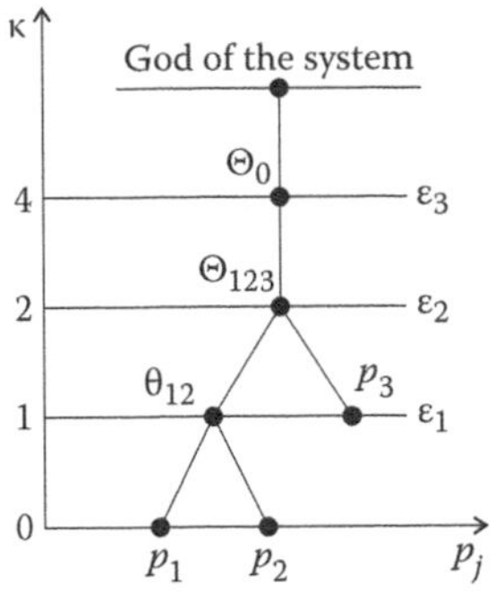

FIGURE 3.9
Hierarchic tree for the system of phases with the fast and slow resonances. It explains the operation principles of the FELs [5,10,14,19,20,29–33]. The phases p_1 (signal) and p_2 (pumping) form the single-particle parametric (fast) resonance. The slow combination phase θ_{12} in this case has the physical meaning of the phases of stimulated electron oscillations. That and the initial slow phase p_3 (the phase of electron oscillations in the proper electron-beam wave) form, in turn, the wave parametric (slow) resonance. Other designations are the same as those given in comments to Figure 3.7.

The combination phase θ_{12} is the phase of stimulated oscillations, whereas the phase p_3 characterizes the electron oscillation in the field of the proper electron-beam waves (the so-called SCWs) [5,10–16,19,20,29–33].

Thus, as is shown, a specific hierarchic combination of two parametric resonances (the fast and the slow) forms the physical basis of the FEL operation principles. The first (fast) resonance is the single-particle parametric resonance [5,10–16,19,20,29–33]. Herewith, in the ordinary terms of the theory of oscillations, the slow combination phase θ_{12} can be treated also as a phase of relevant mismatch (detuning) of the resonant condition. Waves with this phase are called the beat of waves.

The second (slow) resonance (see condition 3.84) is treated as the three-wave parametric resonance (see Chapter 4). It takes place between the stimulated wave (beat of waves) θ_{12} and the proper electron-beam wave (SCW) p_3 [5,10,14,19,20,29–33].

As is readily seen in Figure 3.9, the single-particle parametric resonance is realized at the lowest (zero: $\kappa = 0$) hierarchic level, whereas the wave parametric resonance takes place at the next (first: $\kappa = 1$) level [5,10,14,19,20,29–33]. The new superslow combination phase Θ_{123}

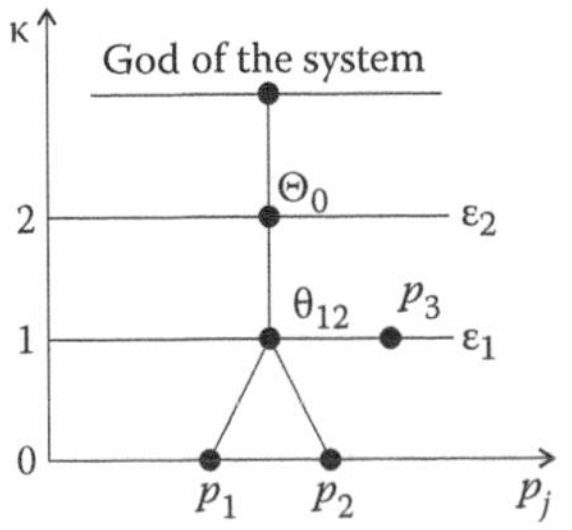

FIGURE 3.10
Hierarchic tree for the system of phases with two fast resonant oscillation phases $p_{1,2}$ and one slow nonresonant phase p_3. Here the fast resonant phases $p_{1,2}$ form the slow combination phase θ_{12}. In this case (in contrast to the situation shown in Figure 3.9) the resonant condition 3.84 is not satisfied. As a result, the initial slow phase p_3 is not resonant. Other designations are the same as those in the comments to Figure 3.7.

is the last oscillation level (top of the hierarchic pyramid) of the discussed FEL model as a hierarchic dynamic system. Beyond the top we have only the highest (nonoscillation) level point $\Theta_0 = const$. In the case of the FEL, it is represented by the static (more precisely the quasistatic) electric and magnetic fields, constant component of the electron longitudinal motion, etc. This point can be treated mathematically as a zero harmonic of an oscillation. The god of the system, in accordance with the definition (see Chapter 2), is situated outside the system. It can be represented by blocks of driving systems, operation systems, and so on, and by the serving stuff.

It should be noted that the above given physical treatment can also be obtained from a relevant analytical numerical analysis of basic FEL physical mechanisms. However, the physical picture, constructed in the framework of the traditional FEL theory [11–13,15], is not too obvious and clear for an average researcher. In this case, clarifying this picture requires essential additional efforts that make the general situation even more complicated than it seems at first sight [5,10–16,19,20,29–33]. As a result, we observe a number of relevant confusions in recent literature dedicated to the traditional (nonhierarchic) theory of FELs. On the other hand, the discussed physical picture, described in terms of the method of the hierarchic tree (see, for instance, Figure 3.9 and some other similar below), strongly allows us to simplify this part of the physical analysis. And, as is shown, for example, in a study by Kulish [10], owing to this, researchers obtain a possibility to avoid the mentioned methodological difficulties in a rather simple way.

In the case when resonant condition 3.84 for the slow resonance is not satisfied, we obtain the situation illustrated in Figure 3.10. Basically, it is not characteristic for the FELs because both resonant conditions 3.83 and 3.84 always coincide in this case. A characteristic kinematical feature of the basic FEL mechanism is such that satisfying the first (fast) condition (Equation 3.83) leads automatically to the satisfaction of the condition for slow resonance (Equation 3.84) [5,10,14].

The next characteristic variant of the hierarchic tree for the system with two slow oscillation phases is shown in Figure 3.11. It takes place if the velocity of changing one of the three initial phases (phase p_3, for instance) is much higher than the velocities of the remaining two phases (phases $p_{1,2}$). A similar situation can be realized practically in the single-particle FEL models with the super-strong longitudinal magnetic field [10,14,34].

As is readily seen, the two initial slow phases form the slow resonance of the first hierarchy ($\kappa = 1$). In this case, the corresponding superslow combination phase Θ_{12} belongs to the third hierarchic level ($\kappa = 2$).

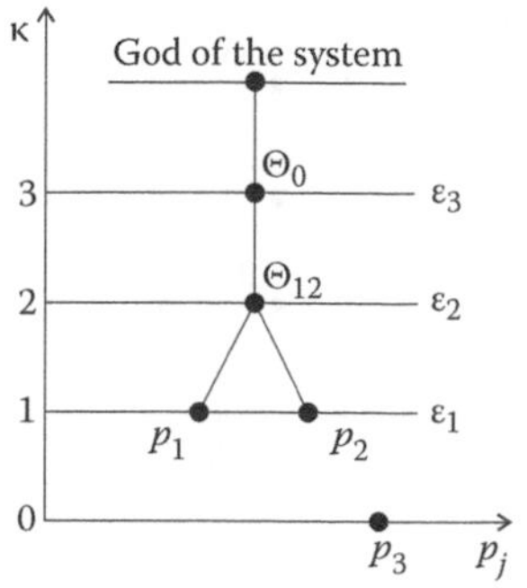

FIGURE 3.11
Hierarchic tree for the system of phases with the two initial slow resonant phases $p_{1,2}$ and one initial fast nonresonant phase p_3. All designations are the same as those in the comments to Figure 3.7.

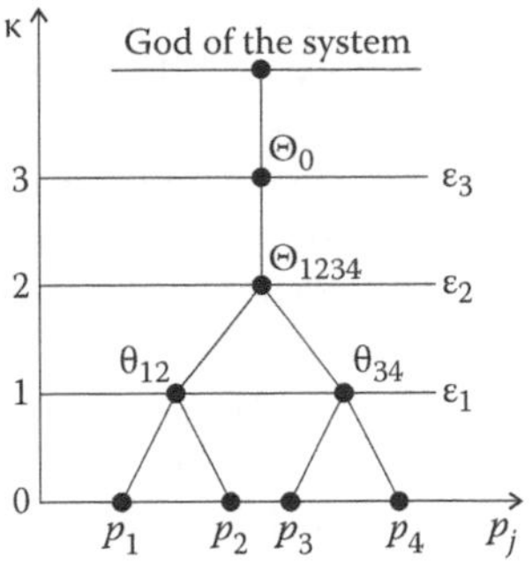

FIGURE 3.12
Hierarchic tree for the system of phases with two pairs of the initial fast resonant phases $p_{1,2,3,4}$. It is readily seen that the two fast resonances are formed by the two pairs of initial fast phases $p_{1,2}$ and $p_{3,4}$. In contrast, the third (slow) resonant phase appears as a result of the resonant interaction of the two slow combination phases θ_{12} and θ_{34}, which do not exist initially. Other designations are the same as those in comments to Figure 3.7.

The schemes of the oscillation-resonant interactions given in Figures 3.7 through 3.11 represent the most typical set of elementary schemes with the pair resonances. In general, by combining these schemes, we can obtain a great variety of hierarchic arrangements in real oscillation-resonant systems. Two examples of such kind, which are shown in Figures 3.12 and 3.13, illustrate this thought evidently.

The hierarchic tree in Figure 3.12 represents the case of two fast resonances and one slow resonance. A characteristic feature of this case is that the slow resonance is realized between the two slow combination phases θ_{12} and θ_{34}. This means that initially we have only two pairs of the fast ($p_{1,2}$ and $p_{3,4}$) phases at the zero hierarchic level ($\kappa = 0$). The two other resonant slow phases θ_{12} and θ_{34} appear as a result of the nonlinear-resonant interactions of the fast phases $p_{1,2}$ and $p_{3,4}$. In other words, in this case, the slow resonance is realized between the combination phases, which do not exist initially. This is the simplest case of the so-called cascade resonances.

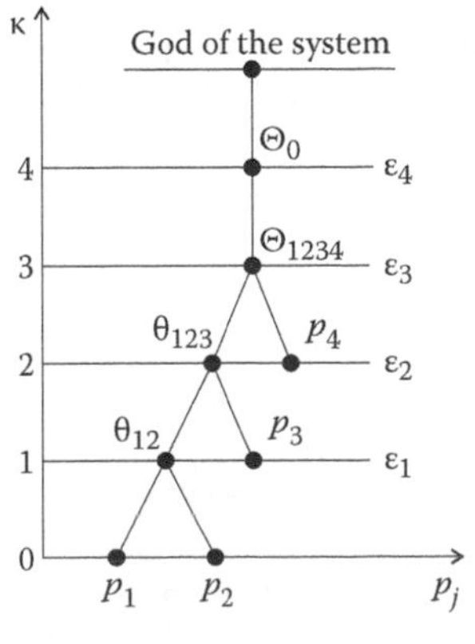

FIGURE 3.13
Hierarchic tree for the system of phases with the two initial fast resonant phases $p_{1,2}$ and the pair of slow resonant phases $p_{3,4}$, which belong to different hierarchic levels. The scheme illustrates the peculiar sequent chain of the resonances of different hierarchy. Other designations are the same as those given in comments to Figure 3.7.

The peculiarity of the second illustration example shown in Figure 3.13 is that both the initial phases $p_{3,4}$ of the second pair are slow and they belong to different hierarchic levels. Phase p_3 is situated at the first level, whereas p_4 is at the second level. These initial phases interact in a nonlinear-resonant manner with the combination phases θ_{12} and θ_{123}, correspondingly. It is important to note that both new slow resonances have a different rank of hierarchy. If the first among them belongs to the first level ($\kappa = 1$), then the second one is realized at the second hierarchic level ($\kappa = 2$).

Thus, by comparing both discussed interaction schemes shown in Figures 3.12 and 3.13, we can see that for the same number of initial phases (four), they form hierarchic trees of different rank of hierarchy. If in the first of them the highest point of the hierarchic pyramid is localized at the second level ($\kappa = 2$; see Figure 3.12), then in the second case (see Figure 3.13), the same point is in the third hierarchic level ($\kappa = 3$).

Analogously we can construct more and more complex interaction schemes with the participation of many pair resonances and oscillation phases. But, in each case, all this diversity can always be represented as various combinations of the above-mentioned elementary schemes. Herewith, the use of the method of the hierarchic trees, as is shown above, allows us to simplify essentially the analysis of the complex oscillation-resonant systems of such kind.

3.2.5 Multiple Resonances

As is mentioned above, multiple resonances can take place if the resonant condition is satisfied for more than two oscillation phases of the same hierarchic rank. The graphical illustration of the multiple (three in this particular case) resonance is given in Figure 3.14. The resonant condition herewith can be written in the form

$$\frac{d\left(p_1 \pm p_2 \mp p_3\right)/dt}{d\left(p_1 \mp p_2 \pm p_3\right)/dt} = \frac{d\theta_{123}/dt}{d\psi_{123}/dt} \sim \varepsilon_1 << 1, \tag{3.86}$$

where all used designations are self-evident in view of the above discussion and Figure 3.14.

In general, the characteristic feature of multiple resonances is that the slow and fast combination phases contain more than two fast initial phases. In the accepted three-phase

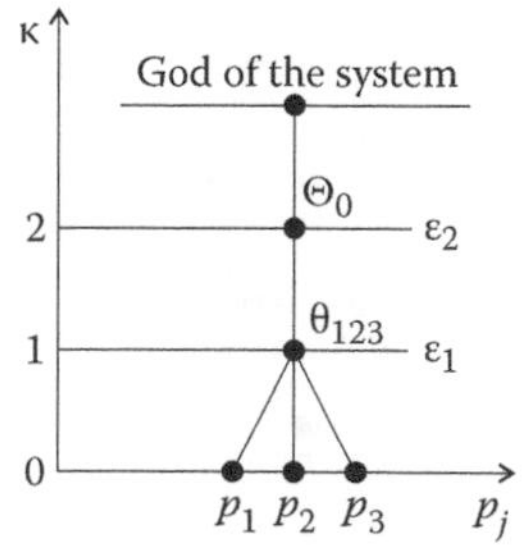

FIGURE 3.14
Hierarchic tree for the system of phases with the three-multiple resonance. Here all fast phases $p_{1,2,3}$ (the zero hierarchic level $\kappa = 0$) are resonant simultaneously. They form the slow combination phase θ_{123} that belongs to the first hierarchic level ($\kappa = 1$). Other designations are the same with those given in comments to Figure 3.7.

illustration example, which is shown in Figure 3.14, these combination phases (see Equation 3.86) have the form

$$\theta_{123} = p_1 \pm p_2 \mp p_3; \quad \psi = p_1 \mp p_2 \pm p_3. \tag{3.87}$$

By comparing condition 3.86 with similar condition 3.85, we can find that the mathematical structures of the combination phases Θ_{123} and θ_{123} in both cases look the same. However, the hierarchic trees, given in Figures 3.9 and 3.14, respectively, show that their physical meaning is essentially different. As is readily seen in Figure 3.14, the slow combination phase θ_{123} in the case of three-multiple resonance consists of only the initial *fast* phases $p_{1,2,3}$. In contrast, in the case represented in Figure 3.9, the similar (with respect to the mathematical structure) combination phase Θ_{123} is formed as a result of nonlinear superposition of two-pair (two-multiply) resonances. Herewith, one of them is fast, whereas the second is slow.

Apart from them, in both compared cases, the interaction schemes have essentially different hierarchic arrangement. In the three-multiple case, it is represented by the two-level hierarchic tree (see Figure 3.14), whereas a similar graph in the case of two superimposed pair resonances (see Figure 3.9) has the explicitly expressed three-level hierarchic arrangement.

Resonances of this kind are characteristic, for example, for the so-called four-wave resonances in nonlinear optics [35,36] and plasma physics [37].

Hence, in spite of the above-mentioned similarity of resonant conditions, both compared oscillation systems represented in Figures 3.9 and 3.14 have essentially different physical nature.

The combined variant of the pair and three-multiple resonances is illustrated in Figure 3.15. Its characteristic methodological feature is that, in this case, the superslow combination phase Θ_{1234} contains four different initial phases. Herewith, such a mathematical structure appears as a result of realization of the ordinary pair (two-multiple) slow phases of the first hierarchy (θ_{123} and p_4, respectively), i.e., the presence of many oscillation initial phases in the definition for a slow combination phase does not mean, in general, that we have to do with a multiple resonance in this case. It is because the construction of

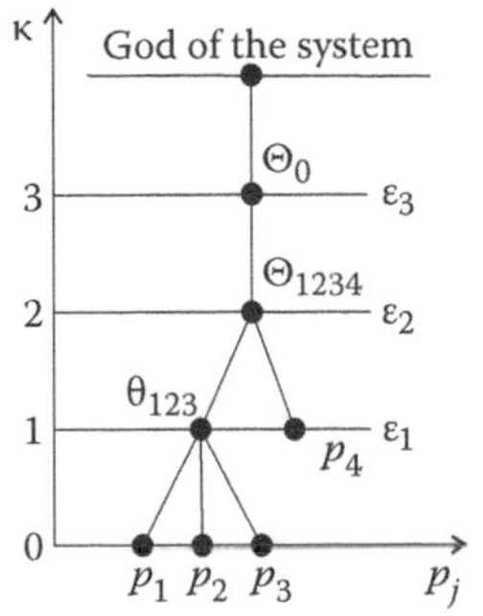

FIGURE 3.15
Hierarchic tree for the system of phases with the combined pair and three-multiply resonances. Here the fast initial phases $p_{1,2,3}$ form the three-multiple (see also Figure 3.12) resonance of the zero hierarchy ($\kappa = 0$). It is characterized by the slow combination phase θ_{123}. This phase, in turn, and the initial slow phase p_4 form the four-multiple resonance of the first hierarchy $\kappa = 1$. Other designations are the same as those given in comments to Figure 3.7.

the multiple resonant condition is not only a way for forming the multiphase combination phases of such kind.

3.2.6 Resonances at Harmonics

We now discuss multiharmonic resonant oscillation systems (Section 3.1.6). Let us confine ourselves by considering the pair harmonic resonances only. In general, the resonances at harmonics can be characterized by the resonant condition like (see Equation 3.61)

$$\frac{dp_1}{dt} \approx \frac{n_2 m_2}{n_1 m_1} \frac{dp_2}{dt}, \tag{3.88}$$

where n_j are the numbers of harmonics stipulated by the nonlinear properties of the considered system, and m_j are the numbers of harmonics of the given initial phases p_j. In the theory of resonances at harmonics, the ratio n_1m_1/n_2m_2 is the criterion of the classification of the types of resonances in which if the ratio is an integer, we call such resonances as integer resonances, and if the ratio is a fraction, we call these resonances as fractional resonances.

For the sake of simplicity, in Figures 3.16 through 3.18, we accept the resulting harmonic numbers $n_jm_j = 1,2,3$.

The case of integer resonance at the second harmonics is shown in Figure 3.16. It illustrates the situation when the second harmonics of the initial fast phase p_1 interacts resonantly also with the second harmonic of phase p_2 (i.e., $n_1m_1 = n_2m_2 = 2$). In accordance with the above accepted classification, this is the integer harmonic resonance of type 2/2. The case of a fractional resonance is illustrated in Figure 3.17. The second harmonics of phase p_2 is there in the resonance with the third harmonic of phase p_2 ($n_1m_1 = 2, n_2m_2 = 3$).

The nonlinear theory of the electron-beam devices with the protracted interaction shows that the harmonic resonances are not unique physical phenomena. The multiharmonic model of the maser at the cyclotron resonance [38] can be used as a convenient illustration for this assertion (see Chapter 7 for more details). Most often they use the resonances on the cyclotron harmonic. The working modes with the participation of harmonics of the signal electromagnetic waves also, basically, are possible (see, for instance, Section 7.2.3). However, up to now, any evidently useful practical applications of the devices for this type of the fractional resonance are not known. The multiharmonic two-stream and plasma-

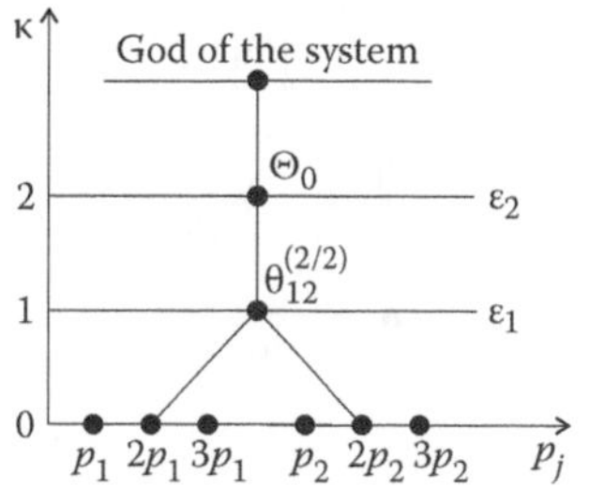

FIGURE 3.16
Hierarchic tree for the system of phases with the integer resonance of the type 2/2. Here we have the initial fast phases $p_{1,2}$ with the resonant second $2p_{1,2}$ resulting harmonics. Other designations are the same with those given in comments to Figure 3.7.

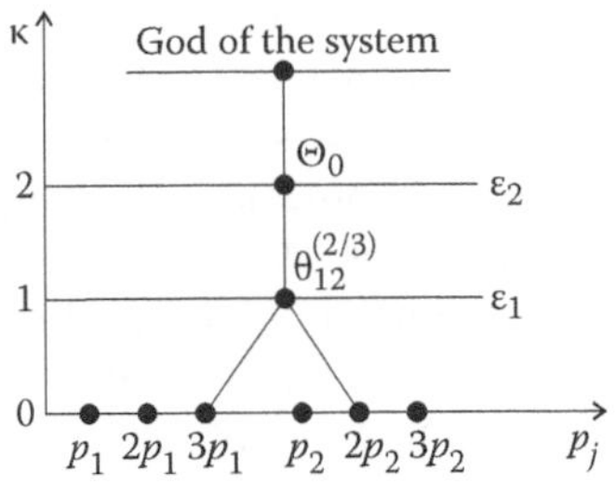

FIGURE 3.17
Hierarchic tree for the system of phases with the fraction resonance of the type 2/3. Here we have the initial fast phases $p_{1,2}$ with the resonant third $3p_1$ and second $2p_2$ resulting harmonics. Other designations are the same with those given in comments to Figure 3.7.

beam superheterodyne FELs (SFELs) [10,14,38–41] can be considered as an exclusion from this rule (see Part II). But let us discuss this topic in detail.

The essence of the principle of the superheterodyne amplification [10,14,42,43] involves the following: the initial electromagnetic signal comes into the device in the form of a transverse electromagnetic wave. It is transformed into the form of the longitudinal electron-beam (SCW) wave. It occurs within the volume of the relativistic electron beam (REB), which moves in the field of a transverse wave-like electromagnetic pumping. Then it is amplified due to the plasma-beam or two-beam instabilities. The amplified SCW is then transformed into the form of the amplified electromagnetic signal wave. The mechanism of three-wave parametric resonance is used for the direct and reverse transformations of the signal wave.

The characteristic feature of the plasma-beam and two-beam versions of the effect of superheterodyne amplification is a possibility to work at essentially multiharmonic modes.

A number of different versions of SFELs [10,14,39–44] are proposed. Among them, multiharmonic systems look the most promising for practical applications. The unique property of such systems is that, in contrast to traditional microwave devices, the maximum of the amplitude in relevant Fourier SCW spectrums is for the higher (tenth or fiftieth, for example) harmonic. This means that multiharmonic SFELs, apart from the amplification, can also be used as highly effective transformers upward of the frequency of the electromagnetic signals.

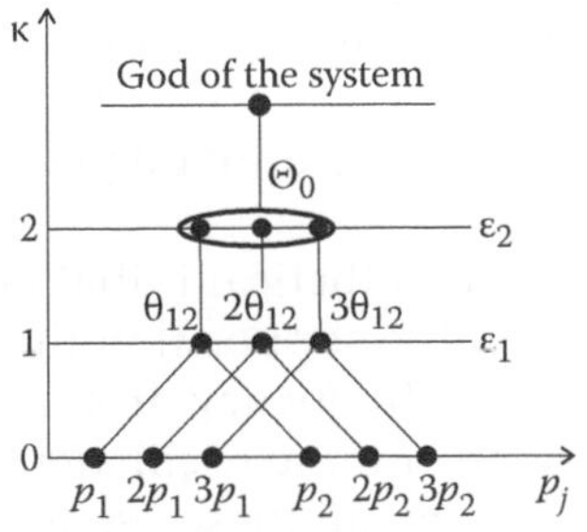

FIGURE 3.18
Hierarchic besom for the multiharmonic system of phases with three parallel resonances at harmonics. Here the same resonant condition is satisfied simultaneously for the three pairs of the harmonics. Other designations are the same as those given in comments to Figure 3.7.

Figures 3.16 and 3.18 illustrate two types of such multiharmonic SFEL transformers. The first (see Figure 3.16) represents the SFEL, with a special multiharmonic pumping [10,14,39,40,44]. The transformation process was carried out in two stages. The superheterodyne-resonant interaction of the monochromatic initial signal and pumping waves takes place at the first stage. The multiharmonic SCW is generated as a result of the realization of such interaction. The reverse transformation occurs at the second stage: one of the higher harmonics (tenth or fiftieth, for example) of the multiharmonic SCW interacts with the same harmonic of the pumping wave. As a result of such interaction, the same harmonic of the transformed signal wave is generated (see Figure 3.16). A possibility of generation of other higher signal harmonics is suppressed by means of the use of some technological means. Multiharmonic FELs of this type allow us to generate rather high-frequency powerful electromagnetic signals (in the UV to soft X-ray range) by utilizing relatively low-energetic high-current electron beams.

The interaction scheme represented in Figure 3.18 illustrates the main idea of the multiharmonic SFELs of the second type [40,41]. There all interaction wave fields (signal, pumping, and SCW, respectively) are multiharmonic. Herewith, the same resonant condition satisfies all the single-particle resonances on harmonics (see Figure 3.18). In this case, we obtain two new applied possibilities. The first is an amplification of the essentially multiharmonic complex signal spectrums [40]. The second is the formation of the femtosecond clusters of the superintensive electromagnetic field [40,41]. The main physical peculiarity of these clusters is a possibility to pass through the air, water (including sea water), or solid walls practically without damping [40,41]. Relevant numerical estimations for such femtosecond SFELs give the following possible project parameters: the mean power ≤300 kW, the pulsed power ~10^{12}W (and more), and the pulse duration, which is a few femtoseconds [41].

The graph in Figure 3.18 also has a methodological meaning. It illustrates the concept of hierarchic besom (Section 2.4.4) in the specific case of the oscillation-resonant hierarchy.

Thus, as is illustrated above, the method of the hierarchic tree can be very promising for analysis of complex multioscillation hierarchic systems.

3.2.7 Our Universe as an Oscillation-Resonant Hierarchic System

We ask once more whether the oscillations and waves are actually the foundation of our world. Contrary to similar discussion in Section 8.2.2, we will analyze this problem here from the point of view of the above-mentioned theory of hierarchic oscillations.

In Section 8.3.2 we will discuss the hypothesis about the cascade oscillation-resonant character of the universe evolution. The simplest version of such a process is illustrated in Figure 3.19. Basically, in view of the above set forth in this chapter, the represented graphical illustration is self-evident. Therefore we limit ourselves to the discussion of some key points only.

As will be shown in Chapter 8, a hypothetical initial influence has caused the appearance of initial singularity. We accept the supposition that this influence has also played the role of a source of the initial multifrequency and multiharmonic oscillations of our brane universe. We assume that some of these oscillations can interact with each other in a hierarchic-resonant manner. Herewith, all these oscillation-resonant interactions go on the background of the extended brane surface. This means that effects that are similar to the effects of the relict photons also exert an influence at periods of the mentioned resonant oscillations. We will call in Chapter 8 such a version of the universe evolution as the cascade oscillation-resonant scenario.

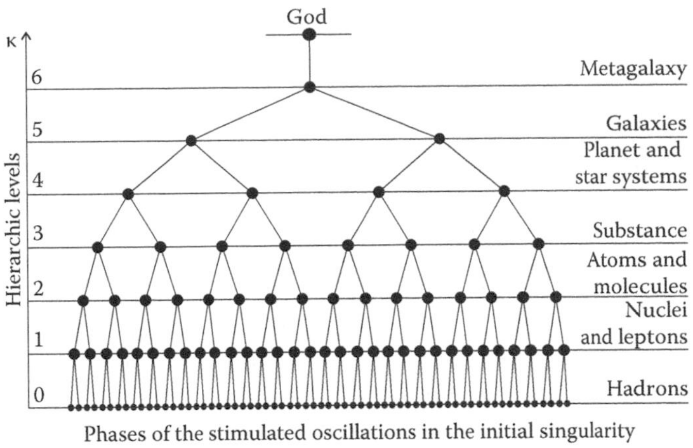

FIGURE 3.19
Illustration for the parametric resonant cascade scenario of the universe evolution process. Here the small dots at the lower hierarchic level characterize the oscillation phases p_j of the four-dimensional initial singularity. The combination resonant phases θ_{vg} at the higher levels are designated by the large dots. It is a further generalization of the hierarchic oscillation-resonant system represented in Figure 3.12. See also, for comparison, Figure 2.5 and the relevant discussion.

What type of elementary resonances plays a key role in this scenario? An analysis shows that it is the physical mechanism of the hierarchic parametric resonance. The general concept of this phenomenon is discussed in detail in this chapter at a semiqualitative level. Furthermore, in the following chapters, its relevant quantitative theory is constructed. As is shown, the parametric resonance, in general, can be realized within a group of many oscillations and their harmonics: the so-called multifold (multiple) multiharmonic resonance. Its general hierarchic arrangement, basically, is far from simple (see, for instance, discussion in this chapter and in Chapters 4 through 7). Therefore, furthermore in this section, let us confine ourselves to a discussion of a simplified version of the pairwise main parametric resonance. In this regard, we recall that the pairwise main resonance takes place in the case when only first harmonics of two oscillations takes part in the resonant interaction. Hence, for the sake of simplicity, we omit the various multiharmonic and multiple resonances and other similar interactions.

The main key points of the discussed cascade oscillation-resonant scenario are the following. We accept that parametric resonant conditions are satisfied for some pairs of many initial oscillations p_j (see Section 3.1 for more details)

$$\frac{dp_\nu}{dt} \approx \frac{dp_g}{dt}, \tag{3.89}$$

where $p_{\nu,g}$ are the resonant phases, and $j = \{\nu,g,\ldots\}$, $\nu,g = 1,2,\ldots$, $\nu \neq g$. It is important to remember that any nonlinear resonance in physics is always characterized by the presence of resonant detuning (mismatch). In the framework of the theory of hierarchic oscillations, we interpret it as a velocity of the slow change in the combination phase

$$\frac{d\theta_{\nu g}}{dt} = \frac{d\left(p_\nu - p_g\right)}{dt} << \frac{dp_\nu}{dt}, \frac{dp_g}{dt}. \tag{3.90}$$

Expression 3.90 illustrates two important facts.

1. The resonant oscillations with phases p_{vg} generate a new, in principle, type of oscillations, i.e., the combination oscillations with phases θ_{vg}.
2. These new oscillations are slow at the fast background of changing the initial phases p_j.

In other words, the totality of the combination phases θ_{vg} forms the next hierarchic level in the considered oscillation-resonant hierarchic system (see Figure 3.19). Then the described forming process repeats cyclically, i.e., some from the combination phases θ_{vg} interact with each other in a parametric resonant manner. As a result, we have the formation of a following (a new) hierarchic level, which consists of the superslow combination phases like $\Theta_{vgkl} = \theta_{vg} - \theta_{kl}$, and so on. The hierarchic chain of this formation process is completed at the highest (sixth in our case; see Figure 3.19) hierarchic level.

Let us note that the formation of each oscillation-resonant level requires some limited time. This means that we can interpret it also as a peculiar transient process. Herewith, the higher the formed hierarchic level is, the larger the characteristic transient time should be. It is important to note the continuous character of the discussed chain of transformations: the completion of any transient process is accompanied always by the beginning of the formation of a next hierarchic level. We can imagine this chain of transient processes as a cascade of the hierarchic resonances (see Figure 3.19). That is why we regard the discussed scenario as the cascade-oscillation-resonant interpretation of evolution of the universe.

Thus, the hierarchic oscillations and waves should actually be regarded as the foundations of our world. Each hierarchic level of the universe has oscillation-wave-resonant nature with some characteristic frequencies and wavelengths. Herewith, all these frequencies and wavelengths evolve on time because of the universe extension. They determine the specific characteristic time and spatial interval of each hierarchic level (see, for instance, Table 2.1). The only source of all oscillation-resonant hierarchic levels is the initial multifrequency push whose action caused the beginning of our world.

3.3 Relativistic Electron Beam without the Proper Magnetic Field as a Hierarchic Oscillation System

Any attempt to add the methods of the theory of hierarchic oscillations and waves to the set of the traditional formal procedures, employed in theoretical studies, proves very soon to be, by far, a more difficult task than it seems. The first difficulties arise when a researcher who is formulating the problem, in particular, tries to give an adequate description of a physical model in terms of these methods. Following the author's experience, both terminology and concepts of the methods are fairly unusual within the framework of the traditional electrodynamics. Thus, translating the hierarchic description of an object under consideration from traditional into foreign (hierarchic) language requires additional training.

The main methodological purpose of this section (and other similar illustration materials given below) is to help the reader in mastering the technology using the above concept.

The reader is advised to use analogies. We consider some typical physical situations of relativistic electrodynamics in order to introduce the above-mentioned concepts of the methods.

3.3.1 Model Description

An electron beam, as well as any other flow of charged particles, constitutes an electric current in space. Therefore, within the context of the Biot–Savart–Laplace law, it can produce a magnetic field both within and outside the volume occupied by the beam. We shall call it the proper field to be distinct from the external field produced by the external (with respect to the beam) sources. Both degree and nature of the proper magnetic field influence on the electron motion depend, to a considerable extent, on the beam density, geometry, energy, and other parameters. First, we discuss a case when the effect of the proper magnetic field influence on the motion of beam electrons can be disregarded.

We consider an azimuthally symmetric monoenergetic (for each cross-section S) electron beam which drifts as a whole along the external magnetic field induction vector $\vec{B}_0$ with the velocity $\vec{v}_e$ (see Figure 3.20).

We assume that both the magnitudes and directions of vectors $\vec{v}_e$ and $\vec{B}_0$ can change slowly with time, i.e., $\vec{v}_e$ and $\vec{B}_0$ are *slowly varying* functions of time (Section 3.1.2). The following question arises: what is the mathematical meaning of the term "slowly varying" in this case? Let us formulate the relevant definition.

The assumption of beam azimuthal symmetry implies that electrons move in the transverse plane along the circular trajectories with angular velocities $\omega_e = d\varphi_e/dt$ (see Figure 3.20) and drift along the axis of symmetry of the beam. In other words, the electron trajectory is a helix with the lead

$$h_s = \frac{2\pi}{\omega_e} v_{ez}, \tag{3.91}$$

where v_{ez} is the z-component of the vector $\vec{v}_e$ (see Figure 3.20). If both $\vec{B}_0$ and $\vec{v}_{ez}$ change insignificantly with distances $\sim h_s$, then $\vec{B}_0(\vec{r},t)$ and $\vec{v}_{ez}(\vec{r},t)$ are slowly varying functions. Their variations are appreciable only for many electron revolutions in the transverse plane $N_e >> 1$, i.e., at large distances $\Delta l \sim N_e h_s$ and long times $\Delta t \sim N_e(2\pi/\omega_e)$.

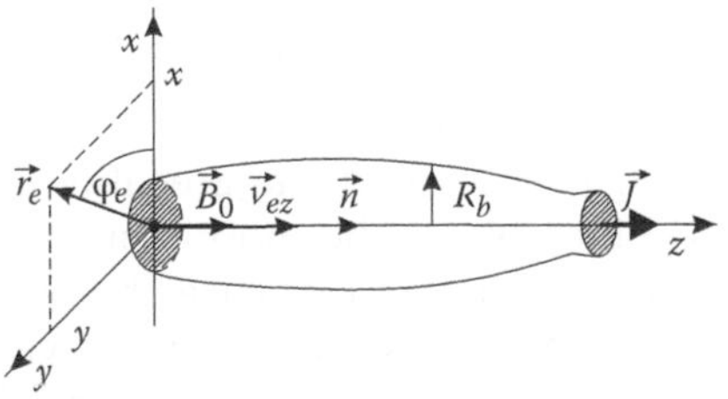

FIGURE 3.20
Model of the electron beam with the slowly varying (along the z-axis) equilibrium radius R_b and the electrons that rotate fast within the transverse plane XY. Here: $\vec{B}_0 = B_0\vec{n}$ is the induction vector of the external longitudinal magnetic field, which changes slowly on the coordinate z, $\vec{v}_{ez}$ is the axial component of the electron beam velocity (it is also the slowly varying value), $\vec{j}$ is the vector of the current density of the electron beam, and $\vec{n}$ is the unit vector along the z-axis.

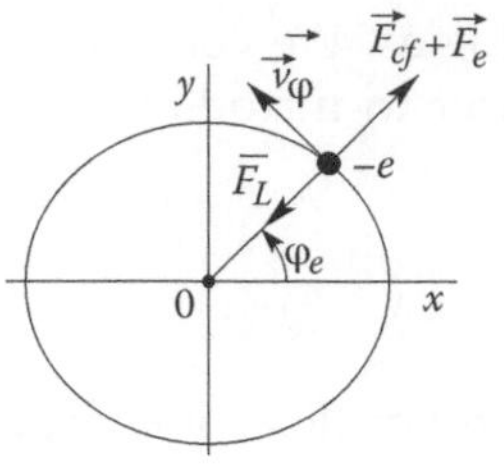

FIGURE 3.21
Balance of the radial forces $\vec{F}_L$ (magnetic Lorentz's force), $\vec{F}_{cf}$ (centrifugal), and $\vec{F}_e$ (electrical Lorentz's force) acting on an electron. Here $\vec{v}_\varphi$ is the azimuthal linear electron velocity, and φ_e is the electron rotating phase. The equilibrium state of the beam takes place if the condition $\vec{F}_L + \vec{F}_{cf} + \vec{F}_e = 0$ is satisfied in every beam cross-section S.

The beam electrons are affected by competing forces, which violate the beam shape stability and counteract these violations. In the case when all such forces are mutually compensated, the beam is referred to as in the state of equilibrium (see Figure 3.21).

In our model, the equilibrium beam is described for each cross-section S by the equilibrium radius R_b (see Figure 3.20) which, in the general case, is a slowly varying function of the space coordinates and time. It is obvious that the possibility of slow variations concerns as well the direction of the vector $\vec{B}_0$. To be rigorous, this means that the model no longer possesses an azimuthal symmetry, especially when one considers its dynamics for large distances and long times—the axial line of the beam can slowly bend. For moderate intervals, however, we can assume that the azimuthal symmetry is preserved with respect to some axis which is treated as a straight line.

Another question arises: which quantity plays the role of a fast rotating phase that is one of the most important concepts in the calculation approach taken as the basis in our consideration (see Section 3.1.6 for more details)? In our case, electron azimuthal coordinate φ_e (see Figures 3.20 and 3.21) can perform this role. The rate of change of the fast rotating phase (fast phase) is then nothing but the angular velocity of the electron motion along the circular trajectory in a transverse plane, i.e.,

$$\frac{d\varphi_e}{dt} \equiv \dot{\varphi}_e = \omega_e. \tag{3.92}$$

Electron beams are usually described in terms of hydrodynamics or aerodynamics [23,24]. Electron streams are treated as flows of charged fluids or gases [23,24] and described in terms of the Lagrange or Euler formalisms. In the first approach, the researcher considers an individual particle (electron or compact aggregation of electrons—large particle) and studies its dynamics and kinematics. The motion of the whole beam is treated as a superposition of all one-particle motions. In this case, the one-particle characteristics are referred to as Lagrange—the Lagrange phase, the Lagrange velocity, and so on. The reader can easily see that the above-mentioned linear and angular velocities, $\vec{v}_e$ and ω_e, as well as the phase φ_e, are Lagrange variables.

In the second approach, the researcher considers a spatial point with the radius-vector $\vec{r}$ and registers the velocities $\vec{v}$ of electrons occurring at this point at each time instant t. Therefore, each point of the beam volume is put in correspondence to a certain point of the field of velocities $\vec{v}(\vec{r},t)$. The velocity function $\vec{v}(\vec{r},t)$, the phase $\varphi(\vec{r},t)$, and the angular velocity $\omega(\vec{r},t)$ are Euler variables.

Thus, Lagrange formalism provides the one-particle description of the system, whereas Euler formalism is associated with a many-particle description. This distinction will help us further to avoid misunderstanding in the analysis of the nature of quantities to be considered in what follows. Below, we will employ both the Lagrange and Euler approaches.

3.3.2 Analysis

To simplify the model, we assume that $\vec{B}_0 = const$ and that the electron density distribution is homogeneous ($n_e = const$). As a result, the axial beam velocity must be constant. We pass to the coordinate system K', which moves progressively with the velocity $v_{ez} = (\vec{v}_e \vec{n})$ ($\vec{n}$ is a unit vector along the z-axis). For symmetry reasons, electron trajectories may be regarded in this model as circles $z_e = const$ whose centers lie on the axis of the beam.

Without ions, the charge of the electron plasma is not compensated and beam electrons are affected by the Coulomb repulsion force $\vec{F}_e$ (see Figure 3.21). The field of these forces is described by the scalar electric field strength

$$\vec{E} = -\frac{\vec{F}_e}{e}. \tag{3.93}$$

In the cylindrical coordinate system $\{r,\varphi,z\}$, we have

$$\vec{E} = \vec{E}_r = E_r \vec{e}_r; \quad \vec{E}_z = E_z \vec{n} = 0,$$

where $\vec{e}_r$ is the unit vector of the transverse coordinate $\vec{r}$. Along with the Coulomb (i.e., electric Lorentz) force $\vec{F}_e$, the electron is affected by the magnetic Lorentz force

$$\vec{F}_L = -\frac{e}{c}\left[\vec{v}_\varphi \vec{B}_0\right], \tag{3.94}$$

and the mechanical centrifugal force

$$\vec{F}_{cf} = \left(\frac{mv_\varphi}{r^2}\right)\vec{e}_r. \tag{3.95}$$

Here $\vec{v}_\varphi$ is the azimuthal component of the linear velocity $\vec{v}_e$, $-e$ is the electron charge, c is the light velocity in vacuum, m is the electron mass at rest, $r = |\vec{r}|$, and $v_\varphi = |\vec{v}_\varphi|$. The beam can only be in equilibrium state provided that the force balance condition is satisfied (see Figure 3.21), i.e., for

$$\vec{F}_e + \vec{F}_L + \vec{F}_{cf} = 0. \tag{3.96}$$

Having projected Equation 3.96 at the transverse radius $\vec{r}$, we obtain

$$\left(\frac{mv_\varphi^2}{r}\right) - eE_r - e\left(\frac{v_\varphi}{c}\right)B_0 = 0\cdot \tag{3.97}$$

Then we make use of the Poisson equation which relates the electron density n_e with the electric field E_r

$$\frac{1}{r}\frac{\partial}{\partial r}(rE_r) = -4\pi e n_e. \tag{3.98}$$

Having integrated over r from 0 to R_b, we find the solution for E_r to be given by [24]

$$E_r = -\left(\frac{m\omega_p^2}{2e}\right)r; \quad 0 \le r \le R_b, \tag{3.99}$$

where

$$\omega_p = \sqrt{\frac{4\pi e^2 n_e}{m}} \tag{3.100}$$

is the plasma (Langmuir) frequency of the electron beam plasma. We employ the relation between electron linear and angular velocities

$$v_\varphi = \omega_e r, \tag{3.101}$$

and substitute Equations 3.99 and 3.101 in Equation 3.97. As a result, we obtain a quadratic equation with respect to the (Lagrange) angular velocity

$$\omega_e^2 - \Omega_e\omega_e + \frac{\omega_p^2}{2e} = 0, \tag{3.102}$$

where $\Omega_e = eB_0/mc$ is the cyclotron frequency. Solving Equation 3.12 is straightforward; the result is given by [24]

$$\omega_e = \omega^{\pm} = \frac{\Omega_e}{2}\left[1 \pm \left(1 - \frac{2\omega_p^2}{\Omega_e^2}\right)^{1/2}\right] = const\cdot \tag{3.103}$$

According to Equation 3.92, the law of time evolution of the phase φ_e is linear, i.e.,

$$\varphi_e = \omega_e t + \varphi_0; \quad (\varphi_0 = \varphi_e(t = 0)). \tag{3.104}$$

From Equation 3.103, the rotation velocity of the phase φ_e is given by an ambiguous relation. To eliminate the ambiguity, we have to introduce relevant initial conditions.

In reality, the electron beam moves in the ion background rather than in vacuum. The ion background can be formed either via ionization of residual gases by the beam (nonrelativistic and slightly relativistic systems) or by means of some special methods (essentially relativistic systems). The ratio of the ion density n_i to the electron density n_e

$$f = \frac{n_i}{n_e} \tag{3.105}$$

is referred to as the coefficient of charge neutralization. The ion background influence on the processes, treated by the model under consideration, may be described in a manner similar to the above analysis [24], i.e., in terms of

$$\omega_e = \omega^{\pm} = \frac{\Omega_e}{2}\left\{1 \pm \left[1 - \frac{2\omega_p^2}{\Omega_e^2}(1-f)\right]^{1/2}\right\}. \tag{3.106}$$

For $f = 0$, Equation 3.106 reproduces Equation 3.103. The case $f = 1$ ($n_i = n_e$) is referred to as complete charge compensation, and the relevant beams are called charge-compensated beams. In this case, $\omega^- = 0$; $\omega^+ = \Omega_e$, i.e., electrons move with the cyclic frequency Ω_e and the beam space charge field influence is fully compensated by the ion background. The processes occurring in such (wide) beams are considered in the illustration example described in Chapter 9. For $f = 0$ but $2\omega_p^2 = \Omega_e^2$, we have $\omega^+ = \omega^- = \Omega_e/2$ Brillouin beams. The latter are employed in the traditional nonrelativistic microwave electronics.

In all above models, however, $h_s << L$, where L is the working region length. This means that an electron that moves through the interaction region of length L makes several revolutions about the beam axis. Let us estimate the actual scale of values of the quantities under consideration. In practice, the typical values are the following: $\Omega_e \sim 10^{10}$–$10^{11} s^{-1}$, $\omega_p \sim 10^9$–$10^{11} s^{-1}$ (moderate-intensive beams). We specify $\Omega_e \sim 10^{11} s^{-1}$, $f = 1$, $L \sim 1m$, $(v_{ez}/c) \sim 0.1$. Then we find the number of revolutions (in the laboratory coordinate system) to be $N_e \cong (\Omega_e L/2\pi v_{ez}) \cong 5 \cdot 10^2 >> 1$. Therefore, the assumption that the rotation phase is fast is quite reasonable for the given realistic situation.

3.3.3 Hierarchic Standard System

Let us turn to Section 2.5, i.e., to the problem of hierarchic description of the considered model. In accordance with Section 2.5, we should reduce the initial equation to the so-called hierarchic standard form

$$\frac{dz}{dt} = Z\left(z, t, \varepsilon_1, \ldots, \varepsilon_\kappa, \ldots, \varepsilon_m\right), \tag{3.107}$$

where $z = \{z_1, z_2, \ldots, z_n\}$ is some vector, $Z = \{Z_1, Z_2, \ldots, Z_n\}$ is the relevant vector-function in the n-dimensional Euclidean space $\mathbf{R}_n$, $t \in [0,\infty]$ (the laboratory time, for example), ε_κ are the hierarchic scale parameters, and κ is the number of the hierarchic level.

We generalize Equation 3.92 bearing in mind that, in our model, z_e, v_φ, v_{ze}, etc., are constants. The physical situation corresponding to this model may be described by means of

$$\frac{dz_e}{dt} = 0; \quad \frac{dv_\varphi}{dt} = 0; \quad \ldots; \quad \frac{d\varphi}{dt} = \omega_e.$$

We transform this set of equations to another form; its meaning will become clear from the analysis below. We write

$$\frac{dx}{dt} = \varepsilon_1 X\left(x, \psi, t, \varepsilon_1, \ldots \varepsilon_\kappa, \ldots \varepsilon_m\right);$$
$$\frac{d\psi}{dt} = \omega\left(x\right) + \varepsilon_1 Y\left(x, \psi, t, \varepsilon_1, \ldots, \varepsilon_\kappa, \ldots, \varepsilon_m\right), \tag{3.108}$$

where x is the vector whose components are the constants z_e, v_φ, R_b, and so on; X and Y are periodic functions with respect to ψ which vanish for $\varepsilon_1 = 0$; $\psi \equiv \varphi_e$ is the fast rotating phase. Thus, within the context of Equation 3.108, the hierarchic small parameter ε_1 is the measure of accuracy within which the condition $x = const$ is satisfied. The latter, in turn, owes to the assumption $\vec{B}_0 = const$. However, absolutely constant fields cannot exist in nature. Bearing this in mind, we shall regard the quantity $\vec{B}_0$ as a given slowly varying function of time and the coordinate z. In our case, the slowly varying function is the cyclotron frequency Ω_e (the notation was introduced in Equation 3.102) and hence the angular velocity ω_e (see Equations 3.103 and 3.106). If the initial velocity azimuthal component is given, then, according to Equations 3.101 and 3.103, the equilibrium radius R_b, the electron density n_e, the plasma frequency ω_p (see Equation 3.100), etc., are also slowly varying functions. We carry out calculations in the above manner by taking into account slow variations of the function $\vec{B}_0\left(z, t\right)$ and thus obtain a more accurate set of equations of the same form as Equation 3.108 (we do not write the functions X and Y since these are insignificant now). However, the hierarchic small parameter ε_1 is no longer equal to zero, in contrast to the above considered case $\vec{B}_0 = const$. In the general case $\varepsilon_1 \geq 0$, the systems of the type Equation 3.108 are referred to as the hierarchic standard systems with fast rotating scalar phases. The mathematical structure of the set 3.108 yields the formal definition for the hierarchic small parameter ε_1 to be the relative rate of change of slow quantities, i.e.,

$$\varepsilon_1 \sim \frac{dx_k/dt}{d\psi/dt} << 1,$$

where x_k is the kth component of the vector x.

In view of the above discussion, we can conclude that systems 3.107 can be regarded as a partial case of system 3.108. In this case, the variables x and ψ in Equation 3.108 can be treated as components of the vector z in Equation 3.107, and the vector-function X and function $\omega(x) + \varepsilon_1 Y$ as the components of the vector-function $Z(\ldots)$ in Equation 3.107.

But the following question arises: what is the meaning of other values (including the small parameters $\varepsilon_2, \ldots, \varepsilon_\kappa, \ldots, \varepsilon_m$) in the general equation 3.107 in terms of the considered model of the electron beam? The answer is in this model, we do not have hierarchic levels higher than the first level. Therefore, we can put zero for all ε_κ for $\kappa > 1$. The two-level hierarchic tree, which corresponds to the discussed model, is represented in Figure 3.22.

In some cases (see, e.g., Chapters 7 and 9), it is more convenient to employ the large parameter

$$\xi_\kappa = \frac{1}{\varepsilon_\kappa} >> 1 \tag{3.109}$$

rather than the small one ε_κ. The relevant standard system 3.108 can also be modified. Then we have

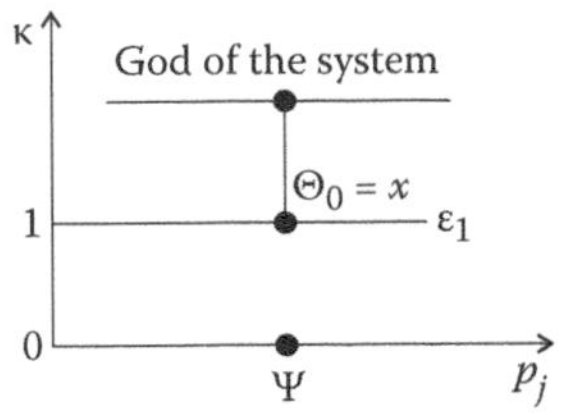

FIGURE 3.22
Hierarchic tree for the cylindrical model of the electron beam without proper magnetic field. Here we have at the zero level only one initial fast phase ψ. The first hierarchic level is formed by the slowly varying components of the vector x. Other designations are the same as those given in comments to Figure 3.7.

$$\frac{dx}{dt} = X\left(x, \psi, t, \xi_2, \ldots, \xi_\kappa, \ldots, \xi_m\right);$$
$$\frac{d\psi}{dt} = \xi_1 \omega(x) + Y\left(x, \psi, t, \xi_2, \ldots, \xi_\kappa, \ldots, \xi_m\right). \tag{3.110}$$

The asymptotic integration methods for the hierarchic standard systems of the form Equations 3.108 and 3.110 are described in Chapters 5 and 6.

3.3.4 Model with Two Rotation Phases

In the above examples, the initial problem was reduced to the standard systems 3.108 and 3.110 with one ($l = 1$) fast rotating scalar phase. Now let us consider an example that illustrates the situation with the number of phases greater than one, namely, $l = 2$.

Again, let the basic model be a homogeneous cylindrical beam ($\vec{B}_0 = const$). However, we no longer assume that the centers of individual electron circles lie on the beam axis. Let us consider the character of motion and find the form of the standard system of the type Equation 3.108.

The equation of electron motion in the Cartesian coordinate system is given by

$$\frac{d^2\vec{r}}{dt^2} = -e\vec{E} - \frac{e}{c}\left[\frac{d\vec{r}}{dt}\vec{B}_0\right]. \tag{3.111}$$

We assume that there is no ion background ($f = 0$). Then, according to the above scheme, we find that the Coulomb electric field strength vector (see Equation 3.99) can be described by the expression

$$\vec{E} = -\frac{m\omega_p^2}{2e}\left(x\vec{e}_1 + \vec{e}_2 y\right), \tag{3.112}$$

where $\vec{e}_{1,2}$ are unit vectors along the axes x and y, respectively. Within the context of Equation 3.112, Equation 3.111 may be reduced to

$$\frac{d^2x}{dt^2} = \left(\frac{\omega_p^2}{2}\right)x - \Omega_e\frac{dy}{dt};$$
$$\frac{d^2y}{dt^2} = \left(\frac{\omega_p^2}{2}\right)y + \Omega_e\frac{dx}{dt}. \tag{3.113}$$

We pass to the coordinate system K'' which rotates with the velocity $\omega_e = \omega^+$ or $\omega_e = \omega^-$, i.e.,

$$x' = x\cos\omega_e t + y\sin\omega_e t;$$
$$y' = y\cos\omega_e t + x\sin\omega_e t. \tag{3.114}$$

In the system K'', Equations 3.113 are transformed to

$$\frac{d^2x'}{dt^2} = (\Omega_e - 2\omega_e)\frac{dy'}{dt} + \left(\omega_e^2 - \Omega_e\omega_e + \frac{\omega_p^2}{2}\right)x';$$
$$\frac{d^2y'}{dt^2} = (\Omega_e - 2\omega_e)\frac{dx'}{dt} + \left(\omega_e^2 - \Omega_e\omega_e + \frac{\omega_p^2}{2}\right)y'. \tag{3.115}$$

From Equation 3.102, the expression within the parentheses before x' and y' is equal to zero. Therefore [24],

$$\frac{d^2x'}{dt^2} = (\Omega_e - 2\omega_e)\frac{dy'}{dt};$$
$$\frac{d^2y'}{dt^2} = (\Omega_e - 2\omega_2)\frac{dx'}{dt}. \tag{3.116}$$

The system 3.116 can be easily integrated. It is not difficult to verify that, in the system K'' (which rotates with the velocity ω_e), the electron trajectory is circular. The rotation phase is then determined by the angular velocity $\omega_{ev} = \Omega_e - 2\omega_e$, i.e.,

$$\varphi_{ev} = \omega_{ev}t + \varphi_{v0}; \quad \varphi_{v0} = \varphi_{v0}(t = 0). \tag{3.117}$$

The quantity ω_{ev} is often referred to as the vortex frequency [24]. Returning back to the system K', we find that electron motion in the transverse plane is described by two phases (see Figure 3.23) which are determined by Equations 3.104 and 3.117; the proportion of the angular velocities of the two rotations can vary in a wide range. Indeed, the definition of the vortex frequency ω_{ev} yields

$$|\omega_{ev}| = \omega^+ - \omega^- = \Omega_e\left(1 - \frac{2\omega_p^2}{\Omega_e^2}\right)^{1/2}.$$

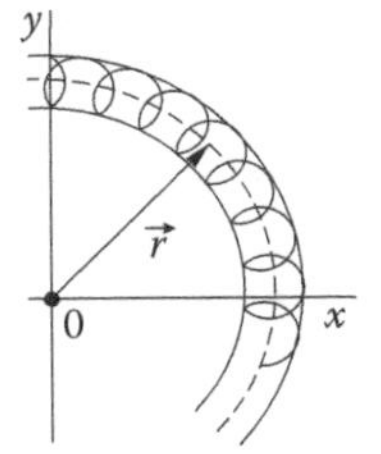

FIGURE 3.23
Electron trajectory in the transverse plane in the case of the disturbed electron-beam model.

For $\left|2\omega_p^2 - \Omega_e^2\right| = \Delta^2 << 2\omega_p^2$ (a beam close to the Brillouin case), we find that (see Figure 3.24)

$$|\omega_e| >> |\omega_{ev}| = |\Delta|. \tag{3.118}$$

On the other hand, for $\Omega_e^2 >> \omega_p^2$ (the case of strong magnetic field; see Figure 3.25), we have

$$|\omega_e| \cong |\omega_{ev}| \cong \Omega_e. \tag{3.119}$$

Thus, it is reasonable to assume that the system under consideration can possess a hierarchy of the rates of change of the fast phases φ_e and φ_{ev}, respectively.

Now, let us consider the model with the slowly varying magnetic field induction $\vec{B}_0(\tau, \chi)$, where $\tau = \varepsilon_1 t = t/\xi_1$ is the slow time, and $\vec{\chi} = \varepsilon_1 \vec{r} = \vec{r}/\xi_1$ is the slow spatial coordinate. Having reduced the initial equations to the standard form, we find the latter to be given by the formal expression which reproduces, e.g., Equation 3.110, i.e.,

$$\begin{aligned} \frac{dx}{dt} &= X\left(x, \psi, t, \hat{\xi}\right); \\ \frac{d\psi}{dt} &= \hat{\xi}\omega(x) + Y\left(x, \psi, t, \hat{\xi}\right). \end{aligned} \tag{3.120}$$

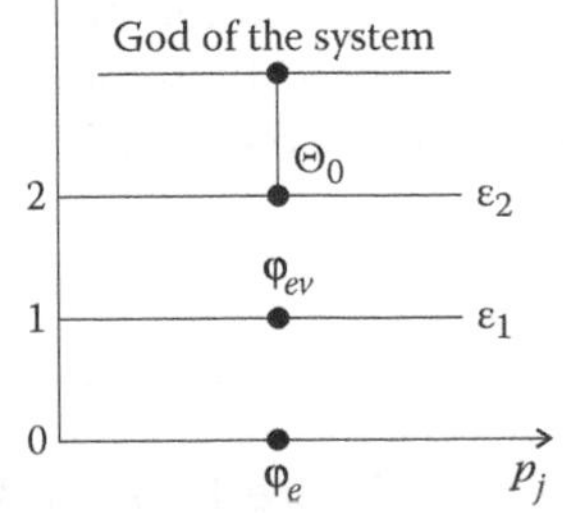

FIGURE 3.24
Hierarchic tree for the disturbed model of the electron beam without proper magnetic field (the beam close to the Brillouin's model). Here we have only one initial fast phase φ_e of electron rotations at the zero level. The slow components of the vector $x(\varphi_{ev})$ (which is not shown in the figure) and the slow vortex phase φ_{ev} belong to the first hierarchic level. The value Θ_0 corresponds to nonoscillated adiabatically slowly changing motion in the system. See also Figure 3.22 for comparison. Other designations are the same as those given in comments to Figure 3.7.

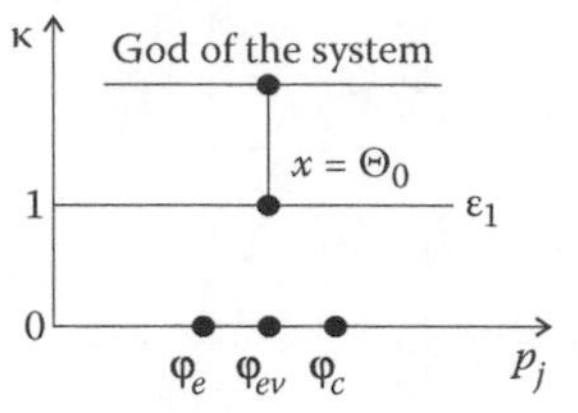

FIGURE 3.25
Hierarchic tree for the disturbed model of the electron beam without proper magnetic field (the case of strong external magnetic field). Here we have three initial fast phases φ_e, φ_{ev}, φ_c, and the slow components of the vector x (see also Figures 3.22 and 3.24 for comparison). φ_c is the phase of cyclotron oscillations of electrons ($d\varphi_c/dt = \Omega_e$). In spite of the closeness of the fast phases, they do not form a resonance. Other designations are the same as those given in Figure 3.7.

The difference is that now the fast rotating phase ψ is a vector, and the large parameter $\hat{\xi}$ is a tensor:

$$\psi = \begin{pmatrix} \varphi_e \\ \varphi_{ev} \end{pmatrix}; \quad \omega = \begin{pmatrix} \omega_e \\ \omega_{ev} \end{pmatrix}; \quad \hat{\xi} = \begin{pmatrix} \xi_1 & 0 \\ 0 & \xi_2 \end{pmatrix}, \tag{3.121}$$

where

$$\xi_1 \sim \frac{|d\varphi_e/dt|}{|dx_k/dt|} >> 1; \quad \xi_2 \sim \frac{|d\varphi_{ev}/dt|}{|dx_k/dt|} >> 1 \tag{3.122}$$

are large scalar parameters of the system, and x_k is the kth component of the slow variables vector x. The rates of change of all the components x_k in the scale of rates of change of the phases φ_e, φ_{ev} are assumed to be commensurable. In the special case $\xi_1 \sim \xi_2 >> 1$, the system reduces to the form with one large scalar parameter and vector fast phase. In the case $\xi_1 >> \xi_2 >> 1$, we have a hierarchy of large parameters, with the relevant inequalities being referred to as the hierarchy series of large parameters. This series has the same meaning as the hierarchic series of small hierarchic parameters such as Equations 3.72 and 3.78. The asymptotic integration algorithms for the systems of the type Equation 3.120, as well as Equations 3.108 and 3.110, are discussed in Chapters 5 and 6.

3.4 Relativistic Electron Beam with the Proper Magnetic Field as a Hierarchic Oscillation System

Among the intense beams, the high-current REBs (HCREBs) are of special interest [23,24,45–51]. Usually, the currents in such beams are greater than hundreds of amperes and energies are higher than hundreds of kiloelectron volts. From the point of view of physics, HCREBs are important for understanding the nature of processes occurring in thermodynamically nonequilibrium media. The electrodynamics of the latter is rather complicated and in many cases has been studied insufficiently [23,24,45,50,51]. Sometimes, combinations of parameters are such that the physical situation seems to be nothing but exotic [23,24,45–51]. A possibility to generate a strong proper magnetic field is one of the exotic peculiarities of HCREBs.

The physical peculiarities of HCREBs, as subjects of study, are stipulated for their unique set of energetic parameters: the pulsed power can attain values ~10^{13}W. That is why they are used in practice mainly as super–high-power energy sources in relativistic electronic devices [10–15,45,48]. Other much more extraordinary application possibilities are connected with their utilizations, for instance, as sources of superpowerful electromagnetic radiation in free space [46,47]. Such systems can be realized due to the use of the strong proper magnetic field of such beams.

It should be mentioned that the electronic devices of the discussed class are very complicated and expensive in practice. Therefore, their calculation requires a clear understanding of the physical processes occurring in the system—the penalty for a mistake is too high. Thus, for the time being, studies of HCREB electrodynamics still remain important and topical.

However, our interest to these unique objects is connected with other aspects of the theory of high-current electron beams. The point is that the discussed hierarchic methods in this book, as practice shows, can be very effective for investigation of such beams. However, the examples considered in the previous section may mislead the reader to the apparent conclusion that their mathematical procedures, for instance, the separation of fast phases, are a straightforward routine job. However, the actual situation is far from it. For example, let us consider a relatively simple model called the Hammer–Rostoker model [24,51] (see Figure 3.26).

3.4.1 Description of the Model

Suppose that the electron motion in the HCREB is associated with the conservation of the integrals of motion for the total energy $\mathcal{H}$ and the z-component of the canonical momentum $\mathcal{P}_z$, i.e.,

$$\frac{d\mathcal{H}}{dt} = \frac{\partial \mathcal{H}}{\partial t} = 0; \quad \mathcal{H} = \gamma_0 mc^2 = const; \tag{3.123}$$

$$\frac{d\mathcal{P}_z}{dt} = -\frac{\partial \mathcal{H}}{\partial z} = 0; \quad \mathcal{P}_z = \gamma_0 mc\beta_0; \tag{3.124}$$

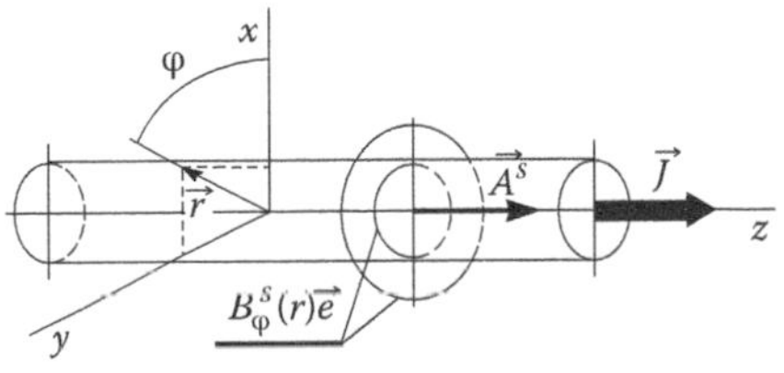

FIGURE 3.26
Model of the high-current electron beam (the Hammer–Rostoker model [51]). Here $B_\varphi^s(r)\vec{e}$ is the magnitude of the azimuthal symmetrical proper magnetic field, $\vec{J}$ is the vector of current density, $\vec{A}^s = A^s\vec{n}$ is the axial vector-potential component of the proper magnetic field, $\vec{n}$ is the unit vector along the z-axis, and $\vec{r}$ is the radius vector of a spatial point, $r = |\vec{r}|$. (From Hammer D.A., and Rostoker, N., *Phys. Fluids*, 13, 1831–1843, 1970. With permission.)

In these equations, $\mathcal{H}$ is the Hamiltonian (total energy expressed in terms of the canonical momentum and coordinates),

$$\mathcal{H} = \sqrt{m^2c^4 + c^2\left(\vec{\mathcal{P}} + \frac{e}{c}\vec{A}\right)^2} + e\Phi, \tag{3.125}$$

$\vec{\mathcal{P}}$ is the electron canonical momentum,

$$\vec{\mathcal{P}} = \vec{p} - \frac{e}{c}\vec{A} = \{\mathcal{P}_r, \mathcal{P}_\varphi, \mathcal{P}_z\}, \tag{3.126}$$

$\vec{A}$ and Φ are vector and scalar potentials of the fields acting on the beam electrons, $\vec{p}$ is the electron mechanical momentum, $\beta_0 = \beta_z(r = 0)$, $c\beta_z = v_z$ is the electron velocity z-component, and $\gamma_0 = \mathcal{H}/mc^2 = const.$

Let us consider the cylindrical system of coordinates and the cylindrical beam model, i.e., the beam is assumed to be stationary, azimuthally symmetric, and axially homogeneous. The effect of possible external electric and magnetic fields will be disregarded. Then the electron is affected only by the proper fields,

$$\vec{A} = \vec{A}^s; \quad \Phi = \Phi^s, \tag{3.127}$$

and, of the three components of the vector potential $\vec{A} = \{A_r, A_\varphi, A_z\}$, only the axial one, A_z, does not vanish ($A_r = A_\varphi = 0$; $A_z \neq 0$). Such a model can be attributed to an electron beam moving in the free space [46,47].

We describe the motion of the whole beam in terms of the one-particle distribution function $f^0(\vec{r}, \vec{\mathcal{P}}, t)$. In view of the above assumptions (stationarity, azimuthal symmetry, and axial homogeneity), the distribution function explicitly depends only on the momentum $f = f(\vec{\mathcal{P}})$. Moreover, by virtue of the azimuthal symmetry of the system, we have

$$\mathcal{P}_\varphi = \mathcal{P}_{\varphi 0} = const. \tag{3.128}$$

For the sake of simplicity, we assume that the beam is monoenergetic and that $\mathcal{P}_{\varphi 0} = 0$ (i.e., electrons move only in the planes r,z; see Figure 3.26). We make use of Equation 3.125 to obtain the canonical momentum r-component,

$$\mathcal{P}_r = \sqrt{(\mathcal{H} - e\Phi)^2 - c^2\left(\mathcal{P}_z + \frac{e}{c}A^s\right)^2}, \tag{3.129}$$

and thus find the solution of the Vlasov kinetic equation [25] to be given by [51]

$$f^0(\mathcal{H}, \mathcal{P}_z) = \frac{\overline{n}_e}{2\pi\gamma_0 m}\delta\left[\mathcal{P}_z - \gamma_0 m\beta_0 c\right]\delta\left[\mathcal{H} - \gamma_0 mc^2\right], \tag{3.130}$$

where $\delta[\ldots]$ is the Dirac delta function, and $\overline{n}_e$ is the cross-section-average electron density.

3.4.2 Solutions for the Fields and the Velocities

The knowledge of the distribution function 3.130 enables us to find the current density $\vec{J}$ and spatial charge ρ defined by the expressions [25]

$$\vec{J} = -e\int_{-\infty}^{+\infty} \vec{v} f^0\left(\vec{r}, \vec{\mathcal{P}}, t\right) d^3\mathcal{P}; \quad \rho = -e\int_{-\infty}^{+\infty} f^0\left(\vec{r}, \vec{\mathcal{P}}, t\right) d^3\mathcal{P}, \tag{3.131}$$

where $\vec{v} = \{v_r, v_\varphi, v_z\}$ is the electron linear velocity vector. Substituting the expressions for $\vec{J}$ and ρ, calculated with regard to the above assumptions, into Maxwell equations [25], we obtain the solutions for the beam proper fields given by [24,51]

$$A_z^s(r) = -\frac{\gamma_0 mc^2\beta_0}{e}\left[1 - I_0\left(\frac{r}{\delta}\right)\right];$$
$$\Phi^s(r) = -\frac{\gamma_0 mc^2\beta_0}{e}\left\{1 - I_0\left[(1-f)^{1/2}\frac{r}{\delta}\right]\right\}, \tag{3.132}$$

where $I_0(x)$ is the zero-order modified Bessel function,

$$\delta = \sqrt{\gamma_0 c^2/\overline{\omega}_p^2} \tag{3.133}$$

is the collisionless skin layer thickness,

$$\overline{\omega}_p = \sqrt{4\pi e^2 \overline{n}_e/m} \tag{3.134}$$

is the cross-section-average plasma frequency of the REB electron plasma, and $f = \overline{n}_i/\overline{n}_e$ is the mean coefficient of charge neutralization ($\overline{n}_i$ is the cross-section-average ion density). We assume the ions to be immovable; the ion mass is much greater than the electron mass: $m_i >> m_e$. To simplify the problem, we put $f = 1$, i.e., the beam is assumed to be fully charge-compensated. Then Equation 3.129 yields

$$\Phi^s(f = 1) = 0. \tag{3.135}$$

Therefore, only Lorentz magnetic force and centrifugal force act on the beam electrons. The manner of motion can be easily obtained from the integrals of motion 3.125 and solutions 3.129 making use of Equations 3.124, 3.129, and 3.135. Thus we find that

$$\frac{dz}{dt} = v_z = c\beta_z = c\beta_0 I_0\left(\frac{r}{\delta}\right); \tag{3.136}$$

$$\frac{dr}{dt} = v_r = c\beta_r = \pm c\beta_0\left[I_0^2\left(\frac{R_b}{\delta}\right) - I_0^2\left(\frac{r}{\delta}\right)\right]^{1/2}, \tag{3.137}$$

where R_b is the equilibrium beam radius that is determined [24,51] by the equation

$$\gamma_0^{-2} - 1 + \beta_0 I_0^2 \left(\frac{R_b}{\delta} \right) = 0. \tag{3.138}$$

As in the previous example, the ambiguity of Equation 3.136 can be removed by an appropriate choice of the initial direction of motion (in this case—up or down; see Figure 3.27). It should be noted that for $f = 1$, the constant γ_0 acquires the physical meaning of the relativistic factor:

$$\gamma_0 = \left(1 - \beta_{r0}^2 - \beta_{z0}^2\right)^{-1/2}, \tag{3.139}$$

where $\beta_{r0} = \beta_r(r = 0)$, and the electron density turns out to be homogeneous over the cross-section, $n_e = \bar{n}_e$.

3.4.3 Separation of the Fast Phase

From the mathematical structure of solutions 3.136 and 3.137, in this case, the fast variable is the transverse radius r (see Figure 3.27). At the same time, it is easy to see that the set of Equations 3.136 and 3.137 is not standard as compared with Equations 3.108 and 3.110. First, the variables are not divided into slow and fast ones. Hence, the small (large) parameter of the problem, which was defined as the rates of change ratio of the velocities of slow and fast variables (see Equation 3.109), is not separated out. Second, the structure of the equation for the fast variable is not reduced to the hierarchic standard form for which the first term on the right-hand side of the equation for the phase (corresponding to the rotation velocity of the fast phase) should be separated out as a slowly varying quantity that depends only on slow variables. In the general case, reducing the set of Equations 3.104 and 3.105 to the hierarchic standard form is a fairly involved problem. In order to simplify it, we restrict the consideration to the special case

$$\frac{R_b}{\delta} = \sqrt{\tilde{\varepsilon}} << 1, \tag{3.140}$$

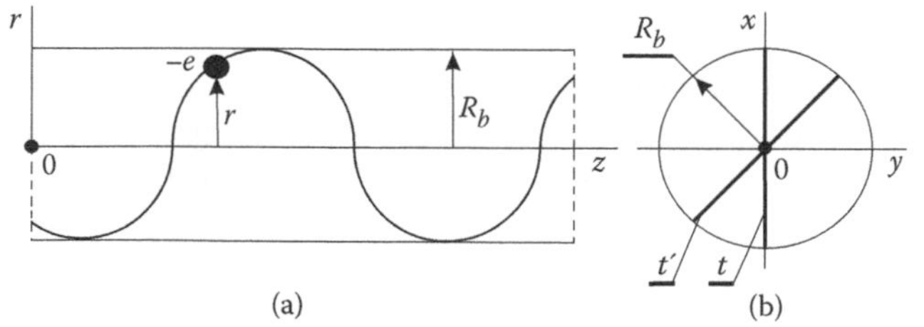

FIGURE 3.27
The electron trajectory in the considered model of the electron beam. Here (a) corresponds to the plane rx, whereas (b) represents the situation in the transverse plane xy (in the case $r = x$). R_b is the equilibrium beam radius, t is the trajectory in the plane xz, and t' is the trajectory in the other plane rz. Thus, all trajectories transverse the z-axis.

where $\tilde{\varepsilon}$ is the expansion parameter that should not be mistaken for the small parameter of the problem, ε_1, defined by Equation 3.109.

We rewrite the transverse coordinate r in the form

$$r = R_b \sin\psi, \tag{3.141}$$

which is suggested by the character of the electron trajectory in the model (see Figure 3.27). Here ψ is some function. Let us show that ψ may be taken for the fast rotating phase. We employ the approximate expression for the modified Bessel functions,

$$I_0(z) + \sum_{k=0}^{\infty} \frac{(z/2)^{2k}}{(k!)^2}, \tag{3.142}$$

and retain only two expansion terms. Then we differentiate the definition 3.141 with respect to t and make use of Equation 3.136. Thus we obtain

$$\frac{d\psi}{dt} = \pm \frac{v_0\left[I_0^2(R_b/\delta) - I_0^2(r/\delta)\right]^{1/2}}{R_b \cos\psi} \cong \pm \frac{v_0}{\sqrt{2\delta}} = \omega = const. \tag{3.143}$$

Within the context of the definition 3.141, the set of Equations 3.123, 3.124, and 3.143 may be rewritten in the required (two-level) hierarchic standard form, i.e.,

$$\frac{dx}{dt} = \varepsilon_1 X(x, \psi, t, \varepsilon_1);$$
$$\frac{d\psi}{dt} = \omega(x) + \varepsilon_1 Y(x, \psi, t, \varepsilon_1), \tag{3.144}$$

where the slow variable vector x contains the total energy $\mathcal{H}$ and the longitudinal canonical momentum $\mathcal{P}_z$ as its components, $[x = (\mathcal{H}, \mathcal{P}_z)]$, and the small parameter (Equation 3.109),

$$\varepsilon_1 \sim \frac{|d\mathcal{H}/dt|}{|d\psi/dt|} \sim \frac{|d\mathcal{P}_z/dt|}{|d\psi/dt|} << 1, \tag{3.145}$$

is assumed to be equal to zero like in Equation 3.108, $\varepsilon_1 = 0$.

3.4.4 Disturbed Model

A disturbed model is one in which the system under consideration is influenced by some small regular disturbance whose nature is not specified. In this case, the integrals of motion $\mathcal{H}$ and $\mathcal{P}_z$ and the fast phase rotation velocity ω can become slowly varying functions. Then, according to the definition 3.109, the small parameter ε_1, given by Equation 3.145, serves as the measure of their slowness. Now the set of Equations 3.144 takes the full-fledged hierarchic standard form with $\varepsilon_1 > 0$ which reproduces Equations 3.108 and 3.110 (if we pass from the small parameter ε_1 to the large parameter ξ_1).

Thus, the example of two models that concern different physical processes shows that it is possible, in principle, to reduce the initial systems of physical equations to hierarchic standard forms containing fast rotating phases. Moreover, we see that at least three such forms occur: Equations 3.108, 3.110, and 3.120. The common feature of these forms is that they describe the so-called nonresonance states. Relevant resonant illustration example will be discussed in Chapter 7.

References

1. Green, M.B., Schwartz, J.H., Witten, E. 1988. *Superstring theory*. New York: Cambridge University Press.
2. Kuzmichov, V.E. 1989. *Laws and formulas of the physics*. Kiev: Naukova Dumka.
3. Sivukhin, D.V. 1989. *General course of the physics. Atomic and nuclear physics*. Vol. V-3. Moscow: Nauka.
4. Menzel, D.H. 1960. *Fundamental formulas of physics*. Two volumes. Dover Publications, Inc.: New York.
5. Kulish, V.V. 2002. *Hierarchic methods. Hierarchy and hierarchic asymptotic methods in electrodynamics*. Vol. 1. Dordrecht: Kluwer Academic Publishers.
6. Hildebrand, F.B. 1987. *Introduction to numerical analysis. Second edition*. New York: Dover Publishers, Inc.
7. Dodd, R.K., Eilbeck, J.C., Gibbon, J.D., and Morris, H.C. 1982. *Solutions and nonlinear wave equations*. London: Academic Press.
8. Grebennikov, E.A. 1986. *Averaging method in applied problems*. Moscow: Nauka.
9. Sagdeyev, R.Z., and Zaslavskiy, G.M. 1988. *Introduction in nonlinear physics*. Moscow: Nauka.
10. Kulish, V.V. 2002. *Hierarchic methods. Undulative electrodynamic systems*. Vol. 2. Dordrecht: Kluwer Academic Publishers.
11. Marshall, T.C. 1985. *Free electron laser*. New York: MacMillan.
12. Brau, C. 1990. *Free electron laser*. Boston: Academic Press.
13. Luchini, P., and Motz, U. 1990. *Undulators and free electron lasers*. Oxford: Clarendon Press.
14. Kulish, V. V. 1998. *Methods of averaging in non-linear problems of relativistic electrodynamics*. Atlanta: World Scientific Publishers.
15. Saldin, E.L., Schneidmiller, E.V., and Yurkov, M.V. 2000. *The physics of free electron lasers (advanced texts in physics)*. New York: Springer.
16. Silin, R.A., Kulish, V.V., and Klymenko, Ju.I. 1991. Soviet Inventors Certificate, SU No. 705914, Priority 18.05.72, 15.05.91. *Inventions Bulletin* 26.
17. Nicolis, J.S. 1986. *Dynamics of hierarchical systems. An evolutionary approach*. Berlin: Springer-Verlag.
18. Kondratenko, A.N., and Kuklin, V.M. 1988. *Principles of plasma electronics*. Moscow: Energoatomizdat.
19. Kulish, V.V., and Lysenko, A.V. 1993. Method of averaged kinetic equation and its use in the nonlinear problems of plasma electrodynamics. *Fizika Plazmy (Sov. Plasma Physics)* 19(2):216–217.
20. Kulish, V.V., Kuleshov, S.A., and Lysenko, A.V. 1993. Nonlinear self-consistent theory of superheterodyne and parametrical free electron lasers. *The International Journal of Infrared and Millimeter Waves* 14(3):451–568.
21. Kulish, V.V. 1997. Hierarchic oscillations and averaging methods in nonlinear problems of relativistic electronics. *The International Journal of Infrared and Millimeter Waves* 18(5):1053–1117.
22. Khapaev, M.M. 1988. *Asymptotic methods and equilibrium in theory of nonlinear oscillations*. Moscow: Vysshaja shkola.

23. Ruhadze, A.A., Bogdankevich, L.S., Rosinkii, S.E., and Ruhlin, V.G. 1980. *Physics of high-current relativistic beams*. Moscow: Atomizdat.
24. Davidson, R.C. 1974. *Theory of nonlinear plasmas*. Reading: Benjamin.
25. Sitenko, A.G., and Malnev, V.M. 1994. *Principles of plasma theory*. Kiev: Naukova Dumka.
26. Landau, L.D., Lifshitz, E.M. 1986. *Theoretical physics*, vol. 6. Hydrodynamics. Moscow: Nauka.
27. Rabinovich, M.I., and Trubetskov, D.I. 1984. *Introduction in theory of oscillations and waves*. Moscow: Nauka.
28. Moiseev, N.N. 1981. *Asymptotic methods of nonlinear mechanics*. Moscow: Nauka.
29. Bolonin, O.I., Kochmanski, S.S., and Kulish, V.V. 1989. Coupled paramagnetic resonances in the FELs. *Acta Physica Polonica* A76(3):455–473.
30. Kohmanski, S.S., and Kulish, V.V. 1985. To the nonlinear theory of free electron lasers with multi-frequency pumping. *Acta Phys. Polonica* A68(5):741–748.
31. Butuzov, V.V., Zakharov, V.P., and Kulish, V.V. 1983. Parametric instability of a flux of high current relativistic electron in the field of dispersed electromagnetic waves. *Deposited manuscript. Deposited in Ukrainian Scientific Research Institute of Technical Information*, Kiev, 297 Uk-83:67.
32. Zakharov, V.P., and Kulish, V.V. 1985. Explosive instability of electron flux in the field of dispersing electromagnetic waves. *Ukrainian Physical Journal* 30(6):878–881.
33. Kalmykov, A.M., Kotsarenko, N.Ja., and Kulish, V.V. 1978. Possibility of transformation of the frequency of the laser radiation in electron flux. *Pisma v Zhurnal Technicheskoj Fiziki* (Soviet: Letters in the Journ. of Technical Physics) 4(14):820–822.
34. Kohmanski, S.S., and Kulish, V.V. 1985. Parametric resonance interaction of electron in the field of electromagnetic waves and longitudinal magnetic field. *Acta Phys. Polonica* A68:525–736.
35. Sukhorukov, A.P. 1988. *Nonlinear wave-interactions in optics and radiophysics*. Moscow: Nauka.
36. Bloembergen, N. 1965. *Nonlinear optics*. New York: Benjamin.
37. Weiland, J., and Wilhelmsson, H. 1977. *Coherent nonlinear interactions of waves in plasmas*. Oxford: Pergamon Press.
38. Kulish, V.V., and Kaylyuk, G.A. 1998. The hierarchic method and peculiarities of its practical application in the nonlinear electrodynamic problems. Single-particle model of the maser on the cyclotron resonance. *Ukrainian Physical Journal* 43(4):398–402.
39. Kulish, V.V., Lysenko, A.V., and Savchenko, V.I. 2003. Two-stream free electron lasers. General properties. *International Journal on Infrared and Millimeter Waves* 24(2):129–172.
40. Kulish, V.V., Lysenko, A.V., and Savchenko, V.I. 2003. Two-stream free electron lasers. Physical and project analysis of the multi-harmonic models. *International Journal on Infrared and Millimeter Waves* 24(4):501–524.
41. Kulish, V.V., Lysenko, A.V., Savchenko, V.I., and Majornikov, I.G. 2005. The two-stream free electron laser as a source of electromagnetic femto-second wave packages. *Laser Physics* (12):1629–1633.
42. Kotsarenko, N.Ya., and Kulish, V.V. 1980. On the possibility of superheterodyne amplification of electromagnetic waves in electron beams. *Zhurnal Tekhnicheskoy Fiziki* 50:220–222.
43. Kotsarenko, N.Ya., and Kulish, V.V. 1980. On the effect of superheterodyne amplification of electromagnetic waves in a plasma-beam system. *Radiotekhnika I Electronika* 35(11):2470–2471.
44. Kulish, V.V., Lysenko, A.V., and Savchenko, V.I. 2003. Two-stream free electron lasers. Analysis of the system with monochromatic pumping. *International Journal on Infrared and Millimeter Waves* 24(3):285–309.
45. Buts, V.A., Lebedev, A.N., and Kurilko, V.I. 2006. *The theory of coherent radiation by intensive electron beams*. New York: Springer.
46. Ablyekov, V.K., Babajev, Yu.N., and Pugachiov, V.P. 1980. Radiation of a self-focusing relativistic electron beam in the mode of phase correlation. *Doklady Adademii Nauk SSSR* (Soviet: Reports of Academy of Sciences of USSR) 225(4):848–849.
47. Bolonin, O.N., Kulish, V.V., and Pugachiov, V.P. 1989. Stimulated radiation of high-current electron beam in the proper magnetic field. *Deposited in Ukrainian Scientific Research Institute of Technical Information*, Kiev, 431 Uk-89.

48. Tsimring, S.E. 2006. *Electron beams in microwave vacuum electronics*. Wiley Series in Microwave and Optical Engineering. Hooken, New Jersey: John & Sons, Inc.
49. Davidson, R.C., and Qin, H. 2001. *Physics of Intensive Charged Particle Beams in High Energy Accelerators*. Singapore: World Scientific.
50. Abramyan, E.A. 1988. *Industrial electron accelerators and applications*. New York: Hemisphere Publishing Corporation.
51. Hammer, D.A., and Rostoker, N. 1970. Propagation high current relativistic electron beams. *Phys. Fluids*. 13:1831–1843.

4

Hierarchic Waves

4.1 Waves

4.1.1 Concept of Waves

The definition of a wave is based on the concept of oscillations discussed in Section 3.1. A wave consists of oscillations occurring in time and space simultaneously. Let us discuss this concept and definition using the simplest example of a wave in an elastic solid medium.

We begin our discussion by modeling the elastic properties of the considered solid medium. A system of mathematical pendulums interconnected by elastic springs (see Figure 4.1) is chosen as a model. It is assumed also that the initial system state ($t = t_2$) is unperturbed ($x_1 = x_2 = \ldots = x_5$).

Then, we assume that one of the pendulums (the first, for example) is deflected from the initial unperturbed state (Figure 4.2). This leads to deflection (with a time lag) of the second neighboring pendulum (Figure 4.2). In this case, the appearance of the time lag is caused by the fact that all pendulums are connected by the elastic springs. Furthermore, a deflection of the second pendulum disturbs (also with a time lag) the initial equilibrium state of the third pendulum, and so on. The most important fact in this situation is that the deflection of each subsequent pendulum is delayed by the preceding deflection. Or, in the other words, the deflection of each subsequent pendulum begins with a time lag Δt (Figure 4.2).

However, the deflection of any pendulum leads to the appearance of oscillations with some proper frequency and an initial phase (Section 3.1.4). Inasmuch as the pendulums are interconnected by the springs, we have the above-mentioned time lag. This means that the oscillations of each subsequent pendulum occur with a shifted (with respect to the preceding pendulum) initial phase (see Figure 4.3). As a result, the spatial distribution of amplitudes of different pendulums on the initial phases obtains a wave-like form (see Figure 4.3).

Thus, the collective oscillations of many elastically interconnected pendulums can be regarded as a peculiar spatial oscillation process (Figure 4.4). For us, it is a new kind of oscillation because in Chapter 3 we dealt with temporal oscillations only (i.e., oscillations in time; see Figure 4.5). Here we will introduce an analogous set of parameters for the spatial oscillations. Some are illustrated in Figure 4.4.

As is evident in Figure 4.4, the spatial oscillations (as well as the temporal ones in Figure 4.5) are characterized by the same amplitude of oscillations A. The comparison of materials in Figures 4.4 and 4.5 shows a functional analogy between other characteristic parameters of the spatial and temporal oscillations: the value λ (wavelength; Figure 4.4) plays the same role as the conventional period T (temporal oscillations; Figure 4.5). The value Δx

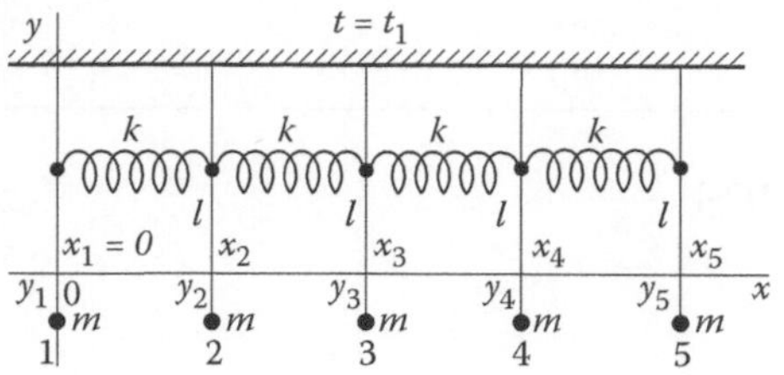

FIGURE 4.1
Model of a one-dimensional elastic medium as a sequence of mathematical pendulums interconnected by springs (the unperturbed system state for the time $t = t_1$). Here, m and l are the masses of material points and the lengths of the pendulums, k is the stiffness coefficient of the springs, x_j are the coordinates of the points m for the time $t = t_1$, and their y-coordinates are the same for this time instance: $y_1 = y_2 = \ldots = y_5$.

(spatial lag; Figure 4.4) has its temporal equivalent in the form of the time lag Δt. The wave number k can be treated as a spatial analogy of the cyclic frequency ω, and so on. However, this analogy is not really as clear as it seems in the one-dimensional model. Unlike time, space is a three-dimensional subject. Because of this, in the system of equivalency, we should take the three-dimensional nature of the interval (spatial lag) $\Delta\vec{r}$ instead of the one-dimensional Δx; in addition to the wave number $k = 2\pi/\lambda$ we should introduce its vector analogy, the wave vector $\vec{k}$ ($k = |\vec{k}|$).

4.1.2 Phase and Group Wave Velocities

Thus, between oscillations in two spatial points x_1 and x_2, the time lag is $\Delta t = t_1 - t_2$ (Figure 4.5). Consequently, the propagation velocity of the perturbation (pendulum deflection; see Figures 4.2 and 4.3) between these material points can be determined by

$$v = \Delta x/\Delta t \tag{4.1}$$

(all notations are given in Section 4.1.1). However, it should be mentioned that, in general, the notation v determined by such a method describes two different physical values. The first is the motion velocity of a fixed phase point on the wave front. The second is the velocity of the energy transfer by the wave (the wave front is a surface in space that is formed by points with equal oscillation phases). Here, the simplest cases of both of these values coincide. However, these velocities can be different: they are called the phase and group velocities, respectively.

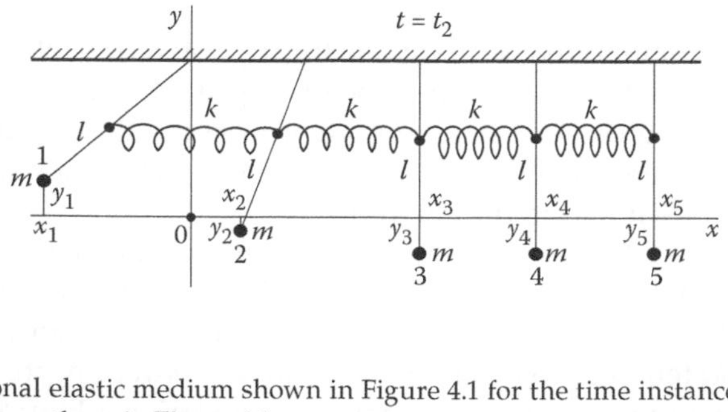

FIGURE 4.2
Model of a one-dimensional elastic medium shown in Figure 4.1 for the time instance $t = t_2 = t_1 + \Delta t$. Here, all designations are the same as those in Figure 4.1.

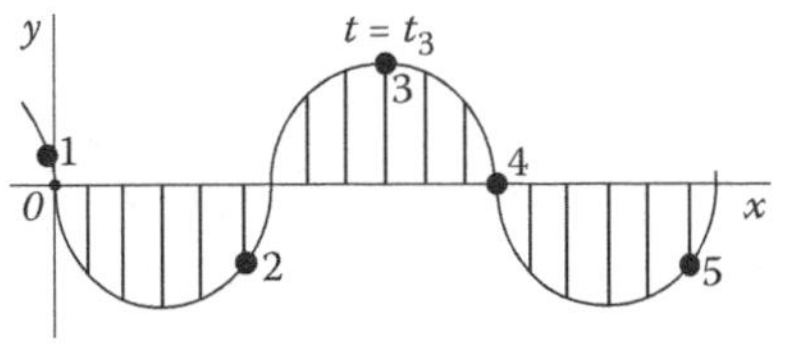

FIGURE 4.3
Illustration of the process of a wave forming in the modeled elastic medium (see also Figures 4.1 and 4.2, and relevant explanations). Here, points 1 through 5 give the y-coordinates of the modeling material points 1 through 5 in Figures 4.1 and 4.2 for an arbitrary time $t = t_3$. As it is easily seen, all pendulums in the considered model oscillate along the y-axis with different initial phases. The spatial distribution of the y-deflection of the pendulums has a wave-like form.

The following rigorous definition for the phase velocity is accepted:

$$\vec{v}_{ph} = \omega\left(\frac{1}{k_x}\vec{e}_x + \frac{1}{k_y}\vec{e}_y + \frac{1}{k_z}\vec{e}_z\right), \tag{4.2}$$

where $\vec{k} = (k_x, k_y, k_z) = \vec{k}_0 k = k_x\vec{e}_x + k_y\vec{e}_y + k_z\vec{e}_z$ is the wave vector, and $\vec{e}_{x,y,z}$ represents the unit vector along the axes x, y, and z, respectively; $|\vec{k}_0| = 1$, $k = \sqrt{k_x^2 + k_y^2 + k_z^2}$ is the wave number, and $k_{x,y,z}$ are the corresponding modules of the vector $\vec{k}$ components. The group velocity $\vec{v}_{gr}$ could be defined as:

$$\vec{v}_{gr} = \frac{d\omega}{d\vec{k}}. \tag{4.3}$$

Let us note that the directions of vectors of the phase and group velocities coincide only in the simplest case of isotropic medium. Generally, in anisotropic mediums, these vectors can be oriented in different directions. FELs provide a convenient illustrative example [1–6]. For instance, as a rule, the phase and group velocities of the signal electromagnetic

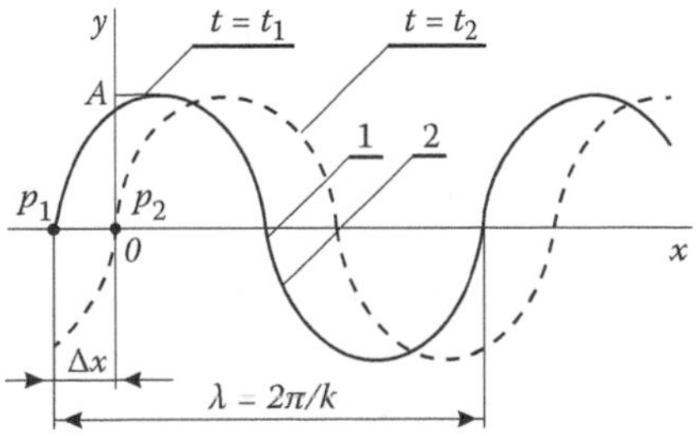

FIGURE 4.4
Illustration of the process of spatial oscillations in the model represented in Figure 4.1 and 4.2. Here, $p_{1,2}$ are the material points of two neighboring pendulums (Figures. 4.1 and 4.2), curve 1 corresponds to the time $t = t_1$, whereas curve 2 illustrates the phase of the spatial oscillation process for the time $t = t_2$, Δx is the distance (spatial lag) between the considered pendulums, λ is the spatial period of oscillations (wavelength), $k = 2\pi/\lambda$ is the wave number, and A is the amplitude of oscillations. (From Kulish, V.V., *Hierarchic Methods. Hierarchy and Hierarchical Asymptotic Methods in Electrodynamics*, Vol. 1, Kluwer Academic Publishers, 2002. With permission.)

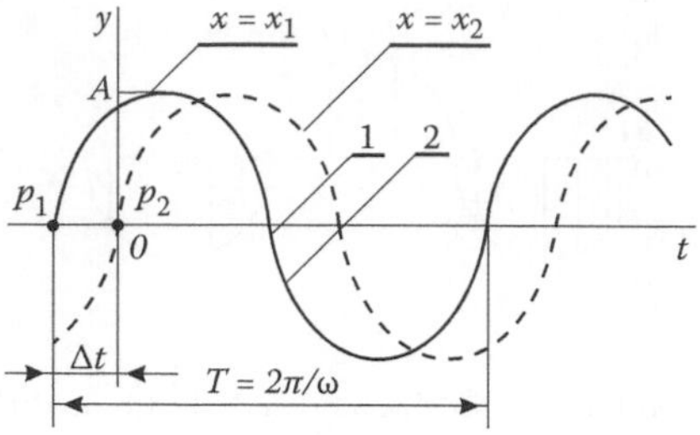

FIGURE 4.5
Illustration of the process of temporal oscillations in the model represented in Figures 4.1 and 4.2. Here, $p_{1,2}$ represents the material points of the two neighboring pendulums, curve 1 corresponds to oscillations of the pendulum in the initial point $x = x_1$, and curve 2 illustrates the oscillations of the neighboring pendulum in the initial point $x = x_2$. Δt represents the time lag in the beginning of oscillations of the pendulums (Figures 4.1 and 4.2), T is the temporal period of oscillations, and $\omega = 2\pi/T$ is the cyclic frequency. (From Kulish, V.V., *Hierarchic Methods. Hierarchy and Hierarchical Asymptotic Methods in Electrodynamics*, Vol. 1, Kluwer Academic Publishers, 2002. With permission.)

wave in traditional FEL-H-Ubitron arrangements coincide [1–6]. However, some other rather exotic physical situations can be realized in some Dopplertron-type FELs. Their characteristic design feature is the use of intense electromagnetic waves (including microwaves) as FEL-pumping waves [7]. Therein, the situations with different directions of the phase and group velocities can occur [1,5]. This includes the cases of opposite or perpendicular reciprocal directions of both vectors.

4.1.3 Phase of a Wave

Thus, each jth oscillation point in the considered model medium can be characterized by a proper oscillation phase (Figures 4.3 and 4.4)

$$p_j = \omega t + \varphi_0. \tag{4.4}$$

Therein, each next point p_{j+1}, owing to the above-mentioned lag effect, oscillates with the shifted phase

$$p_{j+1} = \omega(t - \Delta x/v) + \varphi_0 \tag{4.5}$$

(all designations are given above). However, we should remember that the choice of the jth observation point is arbitrary. It could be written for any oscillating point of the medium

$$p = \omega(t - x/v) + \varphi_0 = \omega t - kx + \varphi_0, \tag{4.6}$$

where, as before, $k = \omega/\nu = 2\pi/\lambda$ is the wave number, and λ is the wavelength (see Figure 4.4).

Taking into consideration the definitions for phase velocity (Definition 4.2), we can generalize the definition for the wave phase (Equation 4.6) for the case of an arbitrary wave process

$$p = \omega t - \vec{k}\vec{r} + \varphi_0, \tag{4.7}$$

where $\vec{r} = (x, y, z)$ is the radius vector of an observation point in the three-dimensional space.

It should be noted that the definitions (Equations 4.6 and 4.7) are formulated within the framework of the Euler description. The difference between the Lagrange and Euler descriptions is discussed in Section 3.3.1.

4.1.4 Transverse and Longitudinal Waves

Earlier we discussed examples (Figures 4.3 and 4.4) in which the plane of oscillations of the material points in a medium is normal to the direction of the wave propagation. These waves are called transversal waves. In the opposite situation (i.e., when the material points oscillate along the direction of wave propagation), we have the longitudinal waves. It should be mentioned that rather exotic wave types with mixed (longitudinal–transversal) structures can exist in some complex electrodynamic systems. They cannot be classified as either purely transversal or longitudinal waves. For example, in some type of retarding electrodynamic systems, the L-waves can be excited (the TM- or TE-waves in wave guides, etc.). As a rule, the electromagnetic signal waves in FELs can be classified as transverse-type waves, while the pumping waves in FEL-Dopplertron are purely transversal (laser radiation, for instance) as well as complex mixed-type waves [1,5]. The space charged (plasma) waves could be excited within the working bulk of relativistic electron beams. Among these waves, both transversal and longitudinal waves can be found.

4.1.5 Surface and Volumetric Waves Dispersion

Surface and volumetric waves are distinguished by their physical nature. Surface waves are characterized by physical mechanisms of excitation. The total wave energy is localized within the layer of medium nearest to the surface. The characteristic thickness of such a layer roughly equals the length of the wave. Waves on water are evidently surface waves. Volumetric waves propagate through volumes whose characteristic size can exceed the wavelength. The light and sound waves serve as other obvious examples of the volumetric waves.

Dispersion is characterized by the presence of a dependency of the phase wave velocity on frequency

$$\vec{v}_{ph} = \vec{v}_{ph}(\omega). \tag{4.8}$$

Taking this into account, we can rewrite Definition 4.2 in a more general form

$$k_{x,y,z} = \frac{\omega}{v_{x,y,z}(\omega)}, \tag{4.9}$$

called the dispersion relation or the dispersion law. Solving the corresponding dispersion equation can obtain

$$D\left(\omega, \vec{k}\right) = 0, \tag{4.10}$$

where $D(\omega,\vec{k})$ is the dispersion function. Waves for which the dispersion law satisfies dispersion equations like Equation 4.9 are defined as proper waves of the considered system. For example, light waves are proper waves for vacuum as an electrodynamic system. The case where Condition 4.10 is not satisfied

$$D(\omega,\vec{k}) \neq 0, \tag{4.11}$$

corresponds to the improper waves. Condition 4.11 is called the condition for existence of improper waves. The stimulated electron waves in plasmas of an electronic beam or FELs are examples of improper waves.

4.1.6 Waves with Negative, Zero, and Positive Energy

By performing corresponding calculations for the considered model of a mechanical wave in an elastic medium, we obtain the well-known expression for wave energy density:

$$u = \frac{d\mathcal{E}}{dV} = \frac{\rho A^2 \omega^2}{2}, \tag{4.12}$$

where $d\mathcal{E}$ is the wave energy in an elementary volume dV, ρ is the medium density, and A and ω are the wave amplitude and the frequency, respectively. At first sight, it seems that the wave energy u should be a quite positively determined quantity because ρ, A^2, $\omega^2 > 0$. As we know, energy describes the ability of bodies to work. Therefore, the existing negative energy or zero energy seems to be impossible. However, the concepts of a wave with negative energy and a wave with zero energy are often used in modern physics.

Why is this a problem? As simple analysis shows, the apparent contradiction is of a purely terminological nature. Talking about wave energy, we always consider the energy difference. It is the difference between a medium energy with a wave and the medium energy without a wave. Therefore, strictly speaking, for the wave density we should write the following definition, which is more precise than Equation 4.12:

$$u = \frac{d\left(\mathcal{E}_{with} - \mathcal{E}_{without}\right)}{dV}. \tag{4.13}$$

It can easily be seen that the sign of the energy density u (and the wave energy, too) depends on the correlation of the magnitudes of the two different energies, $\mathcal{E}_{with}$ and $\mathcal{E}_{without}$, in Equation 4.13. In particular, we have the ordinary case of the positive wave energy

$$\mathcal{E}_{with} > \mathcal{E}_{without}. \tag{4.14}$$

The opposite case describes a system with negative wave energy. If both energies are equal, we have zero wave energy. All three types of waves are known in electrodynamics generally and in FEL theory [1,5] in particular. For instance, the electromagnetic signal wave in FELs is characterized always by positive wave energy. In electron beam plasmas, the slow space charge wave (SCW) is a wave with negative energy. An example of a wave with zero wave energy is shown by the increased and dissipated waves in the beam models with two-stream instability [1,5].

4.2 Electron Beam as a Hierarchic Wave System

In Section 3.3, we considered models that employ the Lagrange (i.e., one-particle) description. Since the initial one-particle equations for the electron motion in external fields are total differential equations, the relevant standard systems of the forms similar to those in Equations 3.108, 3.120, and 3.144 are systems of ordinary differential equations.

The situation with an analysis of the distributed wave models is quite different. The latter are usually described by systems of partial differential equations—Maxwell equations and quasihydrodynamic equations for charged particle beam motion, and Vlasov and Boltzmann kinetic equations, for example (see Chapters 9 and 10). Therefore, relevant asymptotic methods of hierarchic description based on the hierarchic principles we have discussed (see Chapter 2) must take these distinctions into account.

All mathematical problems associated with the description of such systems may be divided into two groups [5]: problems that can be reduced to the standard systems of ordinary differential equations such as Equations 3.108, 3.120, and 3.144, and problems that cannot be reduced to such systems. The examples are considered in Chapters 5 and 6 in the discussion of the methods of averaged kinetic and quasihydrodynamic equations [5]. Problems of the second type often arise when one considers Maxwell equations. Generally, these equations are very complicated, with difficulties outnumbering achievements. In special cases, when both the nonlinearity and the dispersion of the medium are weak, the analysis is simplified. The methods that proved to be effective—based on hierarchic ideas (see Chapter 2) and described in Chapters 5 and 6—employ a somewhat different technique.

This section should be treated as an adapted introduction to the theory and the applications of the hierarchic methods for the case of spatially distributed system. We will describe several typical physical cases for which these methods may be efficient. Moreover, we introduce the concepts of slowly varying amplitudes (once more), proper and induced (stimulated) waves, parametric and quasilinear wave resonances, degenerated and nondegenerated interaction modes, etc. Inasmuch as the hierarchic methods described in Chapters 5 and 6 essentially employ the properties of linear proper waves of the system under consideration [1], the latter are a subject for special discussion.

4.2.1 Some Criteria

Any charged particle beam may be treated as a flow of drifting plasma. However, in many cases, collective plasma properties are weak compared to the background of numerous processes occurring in the system. That is why it is important to introduce criteria that would enable us to estimate when and to what extent plasma properties of an electron beam should be taken into account. Such criteria are well known in electrodynamics [16].

The Debye radius is associated with the geometry of a system. The scale is determined by the quantity

$$r_D = \sqrt{k_B T_e / 4\pi e^2 n_e}, \tag{4.15}$$

which is referred to as the Debye screening radius. Here k_B is the Boltzmann constant, and T_e and n_e are the beam electron temperature and density, respectively. For the collective properties of the beam plasma to be manifested appreciably, the inequality

$$r_D << d \tag{4.16}$$

must be satisfied, where d is the characteristic beam dimension. Most often it is the transverse dimension, that is, the radius (if the beam is cylindrical), thickness (for a strip beam), etc. For typical relativistic electron beam parameters, common sense suggests that the collective effects must be fairly pronounced in high-current beams. This is not so clear as far as moderate-current beams in amperes (A) are concerned. Therefore, it seems useful to estimate Criterion 4.16 from below in view of characteristic conditions of the first FEL experiments, as an example. To specialize the problem, we assume the parameters to be those given by Elias [43–45], i.e., we take the beam current to be equal to 2A, and the dimensionless velocity thermal spread $\beta_T \cong 5.6 \cdot 10^{-3}$ ($\beta_T = v_T/c$, v_T is the electron thermal velocity). We assume that, within the FEL interaction range, the beam is compressed to diameter $d \cong 2$ mm. Then, it is not difficult to obtain $r_D \cong 0.3$ mm $<< d \cong 2$ mm. Thus, Criterion 4.16 is satisfied, though rather weakly. That this infers a possibility that the collective properties of electron plasma can be manifested should be borne in mind, even for A-level beams.

Criterion 4.16 suggests rather than provides manifestations of the collective plasma properties in a relativistic electron beam (REB). The necessary condition is that electrons must travel within the interaction region for a time longer than the plasma oscillation period, i.e.,

$$t > T_p = \frac{2\pi}{\omega'_p}, \tag{4.17}$$

where ω'_p is the plasma (Langmuir) frequency given by

$$\omega'_p = \sqrt{4\pi e^2 n_e / m\gamma}, \tag{4.18}$$

$\gamma = \mathcal{E}/mc^2$ is the relativistic factor, and $\mathcal{E}$ is the total energy of an electron. It is not difficult to find that Criterion 4.17 is satisfied for $L > 2.5$ m (L is the interaction region length). For comparison, the interaction region length was about ~6m in Elias's experiments and Criterion 4.17 was satisfied.

As follows from these estimates, situations in which electron beams manifest collective plasma properties are fairly typical in relativistic electrodynamics. We can conclude that in cases where both Criteria 4.16 and 4.17 are satisfied, the so-called transversally unbounded models of electron beams can be used for the simplification of calculations. We will demonstrate this in the illustrative examples.

4.2.2 Proper and Stimulated (Induced or Forced) Electron Waves

It is well known in the theory of FELs [1–7] that pumping and signal wave fields cause the formation of electron bunches (clusters). Since electrons are charged particles, relativistic electron beam bunching is always accompanied by an excitement of the longitudinal wave of the electrostatic field—a satellite of the electron density wave. The latter is the wave of induced (stimulated, forced) oscillations because it is excited by external, with respect to the beam, forces. At the same time, the Coulomb forces induce electron plasma oscillations with the frequency ω'_p (Equation 4.18). Thus, such a REB is actually a system of coupled oscillators. This infers that any disturbances of the system produce proper wave oscillations. In the case under consideration, these are the SCWs. The wave resonance occurs in the system when some proper wave has the same frequency and the wave number as the wave of the stimulated (induced) oscillations.

Let us illustrate the above reasoning with the simplest example—when both stimulated and proper oscillations are excited in the beam by a superimposition of two collinear electromagnetic waves. Treating this configuration as the simplest Dopplertron FEL model [5], we can attribute these waves as pumping and signal waves. We consider the processes occurring far from the transverse boundaries of the beam and assume that Criteria 4.16 and 4.17 are satisfied. Then the model may be regarded as charge-compensated homogeneous in the transverse direction and unbounded in the plane XY and along the z-axis. Inasmuch as relativistic effects are irrelevant in this problem, we employ the moving coordinate system K' connected with the longitudinal velocity of the beam as a whole. Apart from that, we assume that all electron motions in this system are nonrelativistic.

In our model, electrons move in the field of two opposite direction plane electromagnetic waves of the form

$$\vec{E} = \frac{1}{2}\sum_{j=1}^{2} \vec{E}_j \exp\left\{i\left[\omega_j t - \left(\vec{k}_j \vec{n}\right) z\right]\right\} + c.c., \tag{4.19}$$

where $\vec{E}$ is the resultant electric field strength vector, $\vec{E}_j$, ω_j, and $\vec{k}_j$ are the vector complex amplitudes, the frequencies, and the wave vectors of electromagnetic waves ($\left(\vec{k}_1\vec{n}\right) = -k_1$, $\left(\vec{k}_2\vec{n}\right) = +k_2$, respectively, and $\vec{n}$ is a unit vector along the z-axis. We assume the waves to be linearly polarized.

We make use of the quasihydrodynamic beam model [1,5] (see also Chapter 6). The equation of motion may be written as [5]

$$\frac{\partial \vec{v}}{\partial t} + \vec{v}\left(\vec{\nabla}\vec{v}\right) = -\eta\left(\vec{E} + \left[\vec{v}\vec{H}\right]\right) - \nu_e \vec{v} - \frac{v_T^2}{\rho_0}\vec{\nabla}\rho, \tag{4.20}$$

where $\eta = e/m$ is the specific electron charge, $\vec{v}$ is the Euler velocity, ρ is the beam space charge density ($\rho = \rho_0 + \tilde{\rho}$, $\vec{v} = \vec{v}_0 + \tilde{\vec{v}}$), $\rho_0, \vec{v}_0$ and $\tilde{\rho}, \tilde{\vec{v}}$ are the undisturbed and disturbed components of the relevant quantities, respectively, ν_e is the collision frequency, v_T is the mean square-root electrons thermal velocity, and $\vec{\nabla}$ is the nabla operator.

We linearize Equation 4.20 with respect to the electron wave amplitudes. It should be noted that—unlike the examples of Sections 3.3 and 3.4—electrons in this case are affected by the superposition of the space charge Coulomb forces and the wave fields (Equation 4.19). When calculating the latter, we disregard the space charge field effect on the transverse electron motion. We neglect the terms of the order of and higher than E_j^2 and, within the context of the equation of motion (Equation 4.20), find that propagating waves (Equation 4.19) stimulate high-frequency transverse electron oscillations with the velocities

$$\tilde{\vec{v}}^{(1)} = i\eta\sum_{j=1}^{2}\left(\frac{\vec{E}_j}{2\omega_j}\right)\exp\left\{i\left[\omega_j t - \left(\vec{k}_j\vec{n}\right)z\right]\right\}.$$

In the next (second) approximation with respect to the amplitudes $\vec{E}_j$, we find the force with which the fields (Equation 4.19) act on electrons—the stimulated force

$$\vec{F}=\sum_{s=1}^{4}\vec{F}_s\exp\left\{i\left[\omega_s t-\left(\vec{k}_s\vec{n}\right)z\right]\right\}, \tag{4.21}$$

where $\omega_s = \omega_j$, $2\omega_j$, $\omega_1 \pm \omega_2$, $\vec{k}_s=\vec{k}_j, 2\vec{k}, \vec{k}\pm\vec{k}_2$, and $j = 1,2$. Now we consider only the longitudinal component of the difference frequency force $\vec{F}$, i.e.,

$$\vec{F}_z = F_0\vec{n}\exp\left\{i\left(\omega t-kz\right)\right\}, \tag{4.22}$$

where $F_0=e\eta\left(\vec{E}_1\vec{E}_2\right)k/2\omega_1\omega_2$, $k=\left(\vec{k}\vec{n}\right)=k_1+k_2$, $\vec{k}=\vec{k}_1-\vec{k}_2$, and $\omega = \omega_1 - \omega_2$. We linearize Equation 4.20 with respect to the SCW amplitudes described by the function $\tilde{\rho}$. Using Equation 4.22, we obtain a linear inhomogeneous differential equation and solve it by means of the standard procedure.

Using Equation 4.22 and substituting the relevant exponential functions, ~exp$\{i\omega t\}$ for all time-dependent quantities, we reduce Equation 4.20 to the form

$$v_0\frac{\partial v_z}{\partial t}+(\nu_e+i\omega)v_z+\eta E_z+\frac{v_T^2}{\rho_0}\frac{\partial\tilde{\rho}}{\partial z}=\frac{F_z}{m}. \tag{4.23}$$

We supplement Equation 4.23 with the field equations (Poisson and continuity equations, respectively)

$$\frac{\partial E_z}{\partial z}=-4\pi\tilde{\rho};\quad \frac{\partial J_z}{\partial z}-i\omega\tilde{\rho}. \tag{4.24}$$

Here $J_z=\rho_0 v_z+\tilde{\rho}v_0$ is the difference frequency component of the current density and $\tilde{\rho}, v_z$ are variable components of the relevant quantities.

Equations 4.23 and 4.24 may be refined by introducing the kinetic potential [5]

$$V_k = V_1 + V_T + V_c, \tag{4.25}$$

where $V_1=4\pi\rho_0 v_0 v_z/\omega_p^2$, $V_T=4\pi\hat{v}_T^2\tilde{\rho}/\omega_p^2$, $V_c=i4\pi\nu' v_0 J_z/2\omega_p^2$, $\hat{v}_T=v_T/v_0$, and $\omega_p=\sqrt{4\pi e^2 n_e/m}=\sqrt{4\pi e\rho_0/m}$. Then, Equations 4.21 through 4.24 formally reduce to the equations of long line with an external source that moves with the velocity v_0 [5], i.e.,

$$\begin{aligned}&\left(\frac{\partial}{\partial z}+i\beta_{eT}\right)V_k+Z_{eT}\beta_{pT}J_z=\frac{F_e}{e}\\&\left(\frac{\partial}{\partial z}+i\beta_{eT}\right)J_z+\frac{\beta_{pT}}{Z_{eT}}V_k=0,\end{aligned} \tag{4.26}$$

where $\beta_{pT}=\beta_p\sqrt{\hat{D}}/\left(1-\hat{v}_T^2\right)$, $Z_{eT}=4\pi\sqrt{\hat{D}}\omega$, $\beta_{eT}=\beta_e\left(1-i\nu'/2\right)/\left(1-\hat{v}_T^2\right)$ and $\hat{D}=1-\hat{v}_T^2+\omega^2\omega_p^{-2}\left[\left(1-i\nu'\right)v_T^2-\left(\nu'/2\right)^2\right]$, and $\beta_p=\omega_p/v_0$, $\beta_e=\omega/v_0$.

We employ the standard procedures of the theory of linear differential equations and find the solutions of Equation 4.26 to be given by

$$V_k = i\frac{k-\beta_{eT}}{eD}F_0\exp\{-ikz\} + \sum_{j=1}^{2}\frac{Z_{eT}\beta_{pT}}{\beta_{eT}q_j}C_j\exp\{-iq_jz\} \tag{4.27}$$

$$J_z = i\frac{\beta_{pT}}{eZ_{eT}D}F_0\exp\{-ikz\} + \sum_{j=1}^{2}C_j\exp\{-iq_jz\}, \tag{4.28}$$

where C_j is the integration constant, q_j is the wave number of the proper wave of the beam, and $D(\omega,k) = (k-\beta_{eT})^2 - \beta_{pT}^2$ is the dispersion function. The wave number, q_j, is determined by the dispersion equation

$$D(\omega,k) = (k-\beta_{eT})^2 - \beta_{pT}^2 = 0. \tag{4.29}$$

It is not difficult to find by solving Equation 4.29 that

$$q_1 = \beta_{eT} + \beta_{pT}, \tag{4.30}$$

$$q_1 = \beta_{eT} - \beta_{pT}. \tag{4.31}$$

Thus, the wave field (Equation 4.19) excites three electron waves in the beam plasma

1. A wave of stimulated oscillations [$\sim\exp\{-kz\}$]
2. A slow SCW [$\sim\exp\{-i(\beta_{eT} + \beta_{pT})z\}$]
3. A fast SCW [$\sim\exp\{-i(\beta_{eT} - \beta_{pT})z\}$]

It is not difficult to verify that the phase velocity of the slow SCW is lower than v_0, while the phase velocity of the fast SCW is higher than v_0; these properties are reflected in their names. The group velocities of the two waves are similar and equal to $v_0' = v_0\left(1-\hat{v}_T^2\right)/\left(1-\mathrm{v}'^2/4\right)$ (with neither collisions nor diffusion being allowed for, we have $v_0' = v_0$).

In order to find the constant C_j in Equations 4.27 and 4.28, we employ the boundary condition

$$V_k|_{z=0} = J|_{z=0} = 0. \tag{4.32}$$

After some algebra, the solutions to Equations 4.27 and 4.28 may be written as

$$\begin{aligned} V_k = {} & i\frac{k-\beta_{eT}}{eD}F_0\exp\{-ikz\} + i\frac{k}{2eD}F_0\exp\{-i(\beta_{eT}+\beta_{pT})z\} + \\ & + i\frac{k-2\beta_{eT}}{2eD}F_0\exp\{-i(\beta_{eT}-\beta_{pT})z\}; \end{aligned} \tag{4.33}$$

$$J_k = i\frac{\beta_{eT}}{eZ_{eT}D}F_0 \exp\{-ikz\} + i\frac{k\beta_{pT}}{2eD\beta_{eT}Z_{eT}}F_0 \exp\{-i(\beta_{eT}+\beta_{pT})z\}$$
$$-i\frac{\beta_{pT}(k-2\beta_{eT})}{2eD\beta_{eT}Z_{eT}}F_0 \exp(-i(\beta_{eT}-\beta_{pT})z). \quad (4.34)$$

The numerical analysis of the thermal diffusion and collision corrections to V_k, J_z, q_j shows that their values are small in realizable beams and may be neglected in most cases of practical interest. However, it is useful to consider them for methodological reasons since this approach provides a fairly simple way to illustrate the physical nature of both proper and stimulated waves. In particular, as follows from Expressions 4.33 and 4.34, the SCW are damped waves, i.e., their amplitudes are maximum for $z = 0$. It is clear that damping is caused by diffusion and collisions. Though the damping rate is small, we can see that both slow and fast SCWs are excited near the plane $z = 0$. This implies that they are of a surface nature. The wave of induced oscillations is not damped, which indicates that its excitation mechanism is of a bulk nature (the waves are produced by the external electromagnetic fields; see Equation 4.19).

4.2.3 Parametric Wave Resonance

Excited electron wave amplitudes are proportional to the stimulating force amplitude F_0. That is why the excitation efficiency depends considerably on the mutual orientation of the vector $\vec{E}_{1,2}$ and the relations between the frequency $\omega_{1,2}$ and the wave number $k_{1,2}$. For example, according to Equation 4.22, $F_0 \sim (\vec{E}_1\vec{E}_2)$. Therefore, the efficiency is maximum for coinciding planes of electromagnetic wave polarizations and minimum for their perpendicular mutual orientation.

As follows from Equations 4.29 and 4.31, the dispersion function approaches zero, $D \to 0$, as $k \to q_j$, and Expressions 4.33 and 4.34 become divergent. This implies that the system is in a parametric wave resonance state and must be treated in terms of the theory based on another version of solutions. We mentioned that the wave parametric resonance conditions ($q_j = k$) may be written as

$$\omega_1 - \omega_2 = \omega; \quad k_1 + k_2 = q_{1,2}; \quad (D(\omega,k) \cong 0), \quad (4.35)$$

or $D(\omega,k) = D(\omega,q_j)\big|_{q_j \to k}$. We remind the reader that Condition 4.35 is written for the moving coordinate system K'. The conditions may be modified considerably in a laboratory system. So, the signal frequency ω_1 is much higher here than the pumping frequency ω_2. The resonance can occur for waves with sum frequencies along with difference frequency (retarded) waves [1,5]. The main physical property, however, is the same: the wave parametric resonance occurs in a system when the wave of the stimulated oscillations has the same frequency and wave number as some SCWs of the electron beam.

Now we put $\nu_e = 0$, $v_T = 0$ and find that $k_{1,2} \cong \omega_{1,2}/c$, $q_j = \omega/v_0 \pm \pm\omega_p/v_0$. Then, Condition 4.35 yields

$$\omega_1 = \frac{1+\beta}{1-\beta}\omega_2 \pm \frac{\omega_p}{1-\beta}. \quad (4.36)$$

We compare the result to Condition 3.57 and see that when the collective properties of the beam plasma are taken into account, the single-particle resonance Condition 3.57 is split into two conditions: lower and upper. The values of the down- and up-shifts are similar and proportional to the plasma oscillation frequency ω_p. Therefore, the allowance for the Coulomb interaction of beam electrons essentially modifies the qualitative interaction picture in the case of FELs.

Thus, we see that for $\nu_e = v_T = 0$ and without resonance, the wave occurring in the beam is a superposition of three constant amplitude waves (see Expressions 4.33 and 4.34). With the resonance condition (Condition 4.35) being satisfied, the solutions (Expressions 4.33 and 4.34) are, generally speaking, invalid. The first reason for this is that the wave of the stimulated oscillations coincides with one of the proper waves (simultaneously) and becomes physically indistinguishable. The second reason is that, with Condition 4.35 satisfied, the proper wave amplitudes approach infinity. The latter observation should be interpreted as follows. The observation time in the analyzed nonresonant system is formally infinite [the waves (Equation 4.19) are switched on for $t \to -\infty$]. Therefore, though the amplitude increase rate is finite, the amplitude can grow arbitrarily large. Hence, we have to reformulate the problem to consider the evolution of the system in the regime of resonance beginning from some switch-on time instant $t = 0$. The formulation and methods of solving this problem are described later. However, let us discuss qualitative aspects of this quantitative description.

4.2.4 Slowly Varying Complex Amplitudes: Raman and Compton Interaction Modes

If the system is in a resonant state (parametric resonance as well), then the wave amplitudes evolve in the course of interaction. Having restricted consideration to the analysis of weakly nonlinear systems with weak dispersion, one may expect the evolution to be slow. Generally, the initial wave oscillation phases are also slowly varying. Hence, it is convenient to consider linear combinations, i.e., the slowly varying complex vector amplitudes. The main essence of the problem is finding the laws for amplitude variation in time and space.

The slow evolution of amplitudes makes the situation quite different from the case considered above. First of all, the waves of Equations 4.19 and 4.27, and Expression 4.34 are now wave packets rather than harmonic waves. The consequences of the model in question are fairly peculiar.

Since the waves of stimulated oscillations are produced by the interaction of electrons with pumping and signal fields—represented by wave packets—they must also be wave packets with half-widths Δk_s (see Figure 4.6). On the other hand, according to Equations 4.30 and 4.31, the difference of the wave numbers of the slow SCW $q_1 = k_s = \beta_{eT} + \beta_{pT}$ and the fast SCW $q_2 = k_f = \beta_{eT} - \beta_{pT}$ is given by

$$\Delta k_{sf} = k_s - k_f = 2\beta_{pT} >> \Delta k_s, \Delta k_f \tag{4.37}$$

where Δk_s, Δk_f are the half-widths of the slow and fast SCW wave packets (spectral lines), respectively. Two limiting cases of the hierarchy of the quantities Δk_s and Δk_f occur

$$\Delta k_{stim} << \Delta k_{sf} \text{ (see Figure 4.7)} \tag{4.38}$$

$$\Delta k_{stim} >> \Delta k_{sf} \text{ (see Figure 4.8).} \tag{4.39}$$

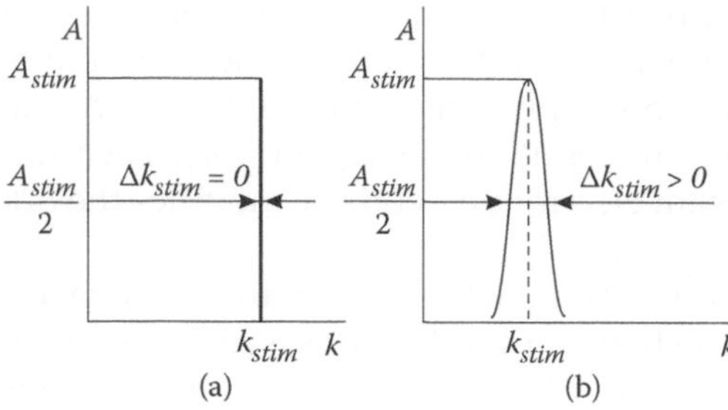

FIGURE 4.6
Fourier spectrums (for wave number k) for a stimulated electron wave. Here, (a) corresponds to the case of a constant amplitude, (b) describes a spectral line of a slowly varying amplitude, k_{stim} is the wave number, and Δk_{stim} is the half-width of the spectral line. As is readily seen, the appearance of the slowly changing amplitude A_{stim} leads to a spread of the half-width of the spectral line from $\Delta k_{stim} = 0$ (a) to the value $\Delta k_{stim} > 0$ (b).

In the first case, only one SCW can take part in the parametric resonance (the second one is asynchronous). This case is referred to as the Raman interaction mode (see Figure 4.7). This effect is of a pure collective nature associated with electron plasma oscillations. By virtue of Conditions 4.35 and 4.36, the Raman mode occurs mainly in high-current FEL systems that operate in the millimeter and submillimeter ranges [1,5].

The case where the inequality is satisfied (Criterion 4.17) corresponds to the Compton interaction mode (see Figure 4.8) [1,5], where both fast and slow SCWs can be involved in the parametric resonance. This indicates that the physical mechanism under consideration is of a one-particle nature [1,5]. Indeed, the system does not distinguish the slow wave from the fast one. We remind the reader that an occurrence of the two types of waves is a manifestation of the collective properties of REB plasma. Therefore, the intensity of the interaction of each electron with the external fields obviously dominates the intensities of collective processes. The Compton mode may also be obtained for Δk_s, $\Delta k_f > \Delta k_{sf}$ [1,5].

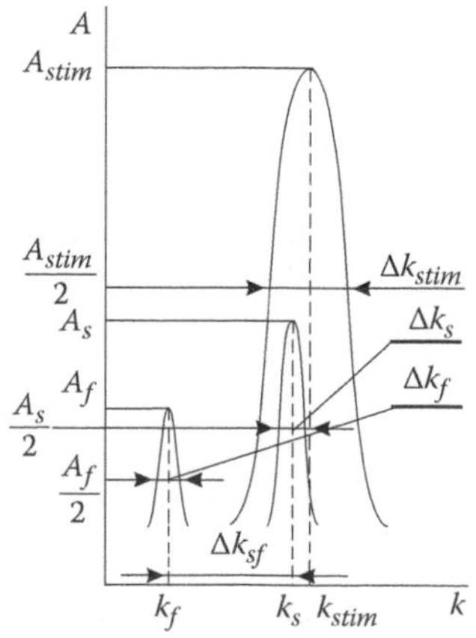

FIGURE 4.7
Illustration of the Raman interaction mode. Here, k_f and k_s are the wave numbers of the fast and slowly varying SCWs, respectively, Δk_f and Δk_s are half-widths of their spectral lines, Δk_{sf} is the difference of the wave numbers k_f and k_s, and A_f and A_s are the amplitudes of the fast and slowly varying SCWs. Other values are the same as those in Figure 4.6.

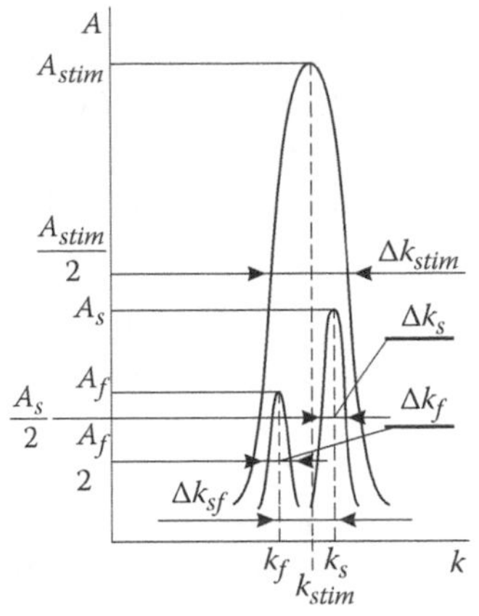

FIGURE 4.8
Illustration of the Compton interaction mode. Here, all designations are the same as those in Figures 4.6 and 4.7.

In this case, both SCWs merge from the very beginning and are indistinguishable at later stages of interaction. This can occur for moderate-density strongly thermalized beams.

It is not difficult to verify that the intermediate states of the system can also occur in the FELs. They are referred to as the combination Compton–Raman mode. It corresponds to the case when the resonance line half-width for the stimulated oscillations wave is close to the wave number difference Δk_{sf}. Though the individual nature of interaction of each SCW is clearly manifested here, both waves are simultaneously involved in the resonance. It is peculiar to this mode that the Chirikov criterion can be satisfied and, as a result, the FEL can operate in the stochastic regimes.

The peculiarities of the interaction mechanisms discussed here have a great influence on the ideology of the approaches that are employed in FEL theory [1,5]. In particular, the one-particle nature of the interaction is widely exploited in the study of Compton lasers, and the methods of plasma electrodynamics are fairly popular in the theory of Raman modes.

4.2.5 SCWs in a Limited Electron Beam

In previous sections, we considered the infinite relativistic electron beam model. The fact that the beam is bounded in the transverse direction introduces modifications of the wave processes that are fairly important in some cases. Qualitatively, two effects are of the greatest interest. The first one is the Coulomb field sagging outside the transverse boundary of the beam. This weakens the SCW field in the bulk of the beam—the phenomenon is referred to as the beam wave depression. It is convenient to describe this effect in terms of the reduced plasma frequency ω_q. The second effect is the violation of the longitudinal character of the SCW and, as a result, the multimode structure of the latter. Let us consider briefly the first of these effects.

As before, we deal here with the system K', in which the beam motion is nonrelativistic. Let us show that the SCW field in a transversely bounded cylindrical beam is $1/\Gamma_d$-times weaker than in an unbounded beam (Γ_d is the depression coefficient). For the sake of simplicity, we will ignore for a while the violation of the SCW longitudinal character. The rigorous theory of Ruhadze et al. [14] shows that this assumption holds even for not very narrow beams. The depression coefficient Γ_d depends on the product of the equilibrium beam radius a and the SCW wave number β_e [5]. We expand the function $\Gamma_e(\beta_e a)$ in a power series of its argument in the vicinity of the point $(\beta_e a)$ (where β_e is equal to the wave number of the unbounded-beam SCW) and retain only the first and second terms of the expansion. Then, making use of Equation 4.24, we find the SCW field amplitude to be given by [5]

$$E_{SCW} = -\frac{1}{i\omega}\left[\Gamma'_e + i\frac{d\Gamma_d}{d\beta}\left(\frac{\partial}{\partial z} + i\beta_e\right)\right]J_z, \tag{4.40}$$

where $\Gamma_e = \beta_e a$; $d\Gamma_e/d\beta\big|_{\beta=\beta_e} = \Gamma'_e$, Γ_d is the depression coefficient. Then, we employ the standard procedure used to obtain a corrected expression for SCW wave numbers with regard to the depression, i.e.,

$$\beta = \beta_{ed} \pm \beta_{pd},$$

where

$$\beta_{ed} = \frac{\beta_e}{1 - v_T'^2}\left(1 - i\frac{v'_e}{2} + \frac{\beta_p^2}{2\beta_e}\Gamma'_e\right)$$

$$\beta_{pd} = \frac{\beta_p}{1 - v_T'^2}$$

$$\times\left\{\frac{\omega^2}{\omega_p^2}\left[v'_T\left(1 - i v'_e\right) - \frac{v'^2}{4}\right] + \Gamma'_e\beta_e\left[v_T'^2 - i\frac{v'_e}{2} + \frac{\omega'_p}{4\omega^2}\beta_e\Gamma'_e\right] + \Gamma'_e\left(1 - v_T^2\right)\right\}^{\frac{1}{2}}$$

Equation 4.40 is written by taking into account the plasma wave reduction, i.e.,

$$\beta_{2,1} + \beta_e \pm R_{2,1}\beta_p, \tag{4.41}$$

where $R_{2,1}$ represents the reduction factors of the fast and slow waves, respectively,

$$R_{2,1} = \sqrt{\Gamma_e + \left(\Gamma'_e\frac{\beta_p}{2}\right)^2} \pm \Gamma'_e\frac{\beta_p}{2}. \tag{4.42}$$

The rough numerical estimates may be done using a simpler formula [5], i.e.,

$$R_1 \cong R_2 \cong R \cong 1 - \exp\{-0.7\omega a/v_0\}. \tag{4.43}$$

Both the diffusion and the collisions are disregarded here. This formula clearly reveals the dependence $R(\omega)$, which infers that the reduced plasma frequency is a function of ω_q, too, i.e.,

$$\omega_q = \omega_q(\omega) \cong R(a)\omega_p. \tag{4.44}$$

The existence of the dependence $\omega_q(\omega)$ implies interesting physical consequences. A possibility arises that parametric resonance conditions (Condition 4.35) can be satisfied for the case where all three waves are SCW of the same monoenergetic beam. The relevant parametric resonance mechanism has been known in microwave electronics since 1950s and is employed in the parametric electron wave amplifiers. A physically similar picture occurs in the two-stream beams (see Chapter 9). It should be noted that such physical mechanisms are associated with the nonlinear properties of beam plasmas and are essentially nonlinear (in this case quadratic). The relevant quasilinear mechanisms of inherent REB instabilities occur as well (see Chapter 8). For qualitative analysis of this chapter, these types of interactions are of interest for the following reasons. When considering the FEL operation mechanism in terms of one-particle and wave approaches, we found that the one-particle (see Equation 3.57) and wave (Condition 4.35) resonances can occur in the system simultaneously. This might lead to a false conclusion that these phenomena are identical. Moreover, each wave resonance must be accompanied by a relevant one-particle resonance. It is, however, peculiar to the above-mentioned quasilinear (two-stream) instability that the one-particle mechanism is not observed in this process. In such a sense, this instability is nonresonant [1,5]. This point is rather specific for the traditional interpretations and requires deep comprehension. For example, it is of special interest in analyzing the correlation between the instability and resonance concepts in electrodynamic systems (Section 3.1).

Let us turn, however, to other physical aspects of this problem. First, we note that the transverse boundedness of the beam distorts the REB field structure. As a result, the SCWs are no longer pure longitudinal waves because the transverse component of the wave vector $\vec{\beta}$ arises, for which the multimode picture occurs. These aspects are analyzed in detail in literature on the REB electrodynamics [14,15,40,41].

4.2.6 Beam Waves of Other Types

The external (focusing) magnetic field can considerably influence the general picture of the wave processes occurring in the REB plasma. In particular, transverse electron waves can be excited along with the longitudinal (Langmuir) waves. On the other hand, it is well known that an intense REB can be transferred in the FEL interaction region by neutralizing the beam charge with the plasmas [1–7]. The electron waves are the proper waves for such magnetized plasmas and possess some interesting distinctive features. They can be of great use when employed for FEL pumping [34]. Therefore, it is of interest to consider the physical aspects of wave propagation in magnetized plasma.

We assume that a transverse unbounded beam drifts in a uniform magnetic field of finite magnitude. As before, we employ the moving coordinate system K'. The analysis will be restricted to the linear approximation with respect to the wave amplitudes. We start from the set of Equations 4.20 and 4.24. Suppose the beam waves are plane, transverse, and propagate along the z-axis (see Equation 4.20). We take the focusing magnetic field induction vector to be directed along the z-axis $\vec{B}_0 = B_0\vec{n}$. Then, after having been linearized, Equation 4.20 may be rewritten in the form [5]

$$\left[\frac{\partial}{\partial z} + i\beta_e(1 - i\nu')\right]v_x = \frac{\eta}{v_0}E_x + \beta_c v_y - \eta B_x;$$
$$\left[\frac{\partial}{\partial z} + i\beta_e(1 - i\nu')\right]v_y = \frac{\eta}{v_0}E_y - \beta_c v_x + \eta B_y, \quad (4.45)$$

where $\beta_c = \Omega_c/v_0$, and $\Omega_c = \eta B_0/c$ is the electron cyclotron frequency. The other quantities have been defined in this section. We proceed to the polarization variables of the form $v_\pm = v_x + iv_y$ and $E_\pm = E_x + iE_y$ and make use of the Maxwell equations. After some straightforward algebra, we obtain

$$\left[\frac{\partial}{\partial z} + i\beta_e(1 - i\nu') \pm \beta_c\right] v_\pm = -i\frac{\eta}{\omega}\left(\frac{\partial}{\partial z} + i\beta_e\right) E_\pm;$$
$$\eta\left(\frac{1}{k^2}\frac{\partial^2}{\partial z^2} + 1\right) E_\pm = i\frac{\omega_p^2}{\omega} v_\pm, \tag{4.46}$$

where $k = \omega/c$.

Solving Equation 4.46 yields the dispersion equation in the form

$$\left(\beta^2 - k^2\right)\left[\beta_e(1 - i\nu') - \beta - \beta_c\right] = \frac{\omega_p^2}{\omega^2} k^2 (\beta - \beta_c). \tag{4.47}$$

For low-density beams ($\omega_p \sim 0$), Equation 4.47 is reduced to two equations. The first equation describes the electromagnetic waves that propagate in vacuum, i.e.,

$$\beta = \pm k.$$

The second equation describes the fast and slow cyclotron waves

$$\beta_\pm = \beta_e(1 - i\nu') \pm \beta_c,$$

which are proper waves of the beam. The upper and lower signs in the subscripts "±" correspond to the clockwise- and counterclockwise-polarized waves, respectively. Generally, electromagnetic and cyclotron waves are coupled and take part in the same wave processes.

Let us consider the case $v_0 = 0$, i.e., the immovable plasma model. In this case, Equation 4.47 yields

$$\beta^2 - k^2 = -k^2 \frac{\omega_p^2}{\omega(\omega \pm \Omega_c - i\nu_e)}. \tag{4.48}$$

For $\nu_e >> \omega$ (low-frequency electromagnetic waves), $\Omega_c >> \nu_e$, $\omega_p^2 >> \omega\Omega_c$ (high cyclotron and plasma frequencies), the waves propagating in the magnetized plasma may be retarded. This case is peculiar to the neutralizing background plasma rather than REB since conditions $\nu_e >> \omega$ can hardly be satisfied for the latter. Of the two helicon waves given by

$$\beta_+ = k\frac{\omega_p^2}{\omega\Omega_c}\left(1 - i\frac{\nu_e}{\Omega_c}\right), \tag{4.49}$$

$$\beta_- = k\frac{\omega_p}{\sqrt{\omega\Omega_c}}\left(\frac{\nu_e}{2\Omega_c} - 1\right), \tag{4.50}$$

only the clockwise-polarized (extraordinary) wave can propagate with small damping. The retardation factor $N = c\beta_+/\omega$ may attain values ~10^2, which suggests that if such waves are employed for pumping, the signals generated in the FEL may attain high frequencies for moderate REB energies [1,5,34].

It is of special interest to analyze the system under the cyclotron resonance conditions, $\omega \cong \Omega_c$. This phenomenon is widely employed both for research and technological purposes. The most popular applications are fusion plasma heating in TOKAMAKs, electronic microwave generators, and amplifiers [gyrotrons, cyclotron resonance masers (CRM), etc.]. The hierarchic methods are highly efficient when applied to the analysis of such systems (e.g., see Chapter 7) [1,5].

4.2.7 Again: Parametric Wave Resonance

In the preceding sections, we introduced the concepts of slowly varying amplitudes and wave resonance. Here, we give a quantitative illustration of the methodology of the problem for the simplest example of the three-wave parametric resonance in FELs.

Let us continue to analyze the model of Section 3.3. We simplify the model, assuming that the beam is a cold collisionless plasma. This means that the collision frequency ν_e and the electron thermal velocity v_T vanish. We assume all processes occurring in the system to be steady, and then the model is stationary—the amplitudes can vary only along the z-axis. Moreover, we put the interaction mode to be Raman and the Dopplertron [1,5] pumping.

Distinct from the passive model considered above, we assume that the wave parametric resonance condition (Condition 4.35) can be satisfied. We start with the Maxwell equations

$$\left[\vec{\nabla}\vec{H}\right] = \frac{1}{c}\frac{\partial \vec{E}}{\partial t} + \frac{4\pi}{c}\vec{j}; \quad div\vec{E} = 4\pi\rho;$$
$$\left[\vec{\nabla}\vec{E}\right] = -\frac{1}{c}\frac{\partial \vec{H}}{\partial t}; \quad div\vec{H} = 0, \tag{4.51}$$

and the quasihydrodynamic equation of motion (Equation 4.20), for $\nu_e = v_T = 0$. Here,

$$\rho = \rho_i + \rho_e; \quad \rho_i = 4\pi e n_0/c = const; \quad \rho_e = -4\pi e n_e/c \tag{4.52}$$

are space charge densities, and n_0 and n_e are the ion and electron concentrations, respectively;

$$\vec{j} = \vec{j}_i + \vec{j}_e; \quad \vec{j}_i = 0; \quad \vec{j}_e = -e n_e \vec{v}, \tag{4.53}$$

where $\vec{j}_i$ and $\vec{j}_e$ are the ion and electron current densities, respectively.

The calculation scheme is semiempirical. First, we linearize the set of Equations 4.20 and 4.50, with respect to the wave amplitudes and assume that all nonlinear terms vanish. The solutions of the resultant linear set of equations are given by

$$\hat{E}_{1,2}(z,t) = \frac{1}{2}\left(E_{1,2}\exp\{ip_{1,2}\} + c.c.\right);$$

$$\hat{E}_3(z,t)=\frac{1}{2}\left(E_3\exp\{ip_3\}+c.c.\right);$$
$$v_x=v_{x1}+v_{x2};$$
$$v_{x1,2}=\frac{1}{2}\left(\frac{ie}{m\omega_{1,2}}E_{1,2}\exp\{ip_{1,2}\}+c.c.\right); \tag{4.54}$$
$$v_z=v_0+\tilde{v}_z;$$
$$\tilde{v}_z=\frac{1}{2}\left[\frac{e(\omega_3-v_0k_3)}{m\omega_p^2}E_3\exp[ip_3]+c.c.\right],$$

and therefore the linear dispersion laws may be written in the form

$$k_{1,2}=\frac{\omega_{1,2}}{c}\sqrt{1-\frac{\omega_p^2}{\omega_{1,2}^2}};$$
$$k_3\frac{\omega_3}{v_0}+r\frac{\omega_p}{v_0}. \tag{4.55}$$

Here, $r=\pm1$ is the sign function ($r=+1$ and $r=-1$ correspond to the slow and fast SCWs, respectively). At this stage, we assume the complex wave amplitudes $E_{1,2,3}$ to be constants. Now we again employ the initial set of equations. This time we retain the quadratic terms on the right-hand side of the equations. We remind the reader that in this model both nonlinearity and dispersion are assumed to be small. This forms the reason for taking the solution of Equation 4.54 in the quadratic approximation in the same form as in the linear case, but assuming that the wave amplitudes are slowly varying functions. We separate out the small dimensionless parameter of the problem in the form

$$\varepsilon\sim\frac{e|E_j|}{mc\omega_j}=\frac{|v_{xj}|}{c}<<1 \tag{4.56}$$

(the motion in the transverse plane is nonrelativistic). Then, the derivatives of the slowly varying amplitudes with respect to z may be estimated as

$$\left|\frac{dE_k}{dz}\right|\le\varepsilon^2,\quad k=1,2,3. \tag{4.57}$$

We substitute Equation 4.54 in the initial system by bearing in mind all the above assumptions and employing a successive approximations method. We equate the coefficients of similar exponential functions on the left- and right-hand sides of relevant equations with the use of the parametric resonance conditions (Condition 4.35). With the notation as given above, these conditions reduce to

$$\omega_1-\omega_2=\omega_3;\quad k_1+k_2=k_3. \tag{4.58}$$

After some calculations, we obtain a set of truncated equations for the slowly varying complex wave amplitudes, i.e.,

$$\frac{dE_1}{dz} = -A_1 E_2 E_3; \quad \frac{dE_2}{dz} = -A_2 E_1 E_3^*; \quad \frac{dE_3}{dz} = -rA_3 E_1 E_2^*, \tag{4.59}$$

with the nonlinear coefficients (matrix elements) A_k being given by (for $\omega_p/\omega_k << 1$)

$$A_1 \cong \frac{e\omega_1 k_3}{mc^2 \omega_2 k_1}; \quad A_2 \cong \frac{e\omega_2 k_3}{mc^2 \omega_1 k_2}; \quad A_3 = \frac{e\omega_p \left(c^2 k_3 - v_0 \omega_3\right)}{mc^2 \omega_1 \omega_2}.$$

The systems of equations whose mathematical structure is similar to that of Equation 4.59 have analytic solutions in terms of the Jacobi elliptic functions [1,5,26–28]. They were employed for the first time in nonlinear optics [26,27] and then in the study of nonlinear wave interactions in plasmas [28]. It should be mentioned that the method applied for deriving Equation 4.59 is insufficiently substantiated mathematically (see [5] for details). In the quadratic approximation, however, it does not lead to appreciable errors. For the cubic and higher approximations, one can use more rigorous methods (e.g., see Chapter 6).

Let us discuss the treatment of the considered FEL model as an oscillation-wave hierarchic system (see Figure 4.9). In accordance with the discussions in this chapter, we should

- Classify all variables in accordance with the hierarchic levels
- Clear up the scheme of phase resonances

In order to accomplish these tasks, we have to deal with two sets of variables that have different physical/mathematical natures. Each of theses sets is determined by the choice of method of description.

Let us recall that the first case was treated as a single particle because the motion of a separate particle is observed in the working bulk during the interaction process. The

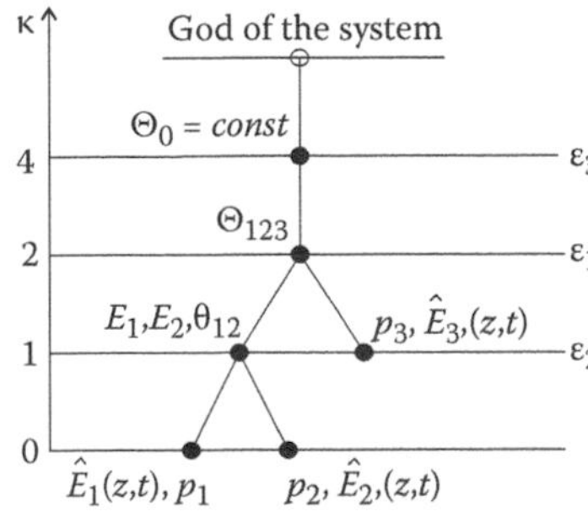

FIGURE 4.9
Illustration of the FEL as an oscillation-wave three-level hierarchic system represented in Lagrange variables. Here, the Lagrange phases p_1 (signal) and p_2 (pumping) form the single-particle parametric (fast) resonance (see Condition 3.57). Electron oscillations with the combination phase θ_{12} form the wave of stimulated oscillations. Its slow phase (also Lagrange) p_3 (the phase of electron oscillations in the proper electron-beam waves) forms the wave parametric (slow) resonance. The electric strength of the electromagnetic waves $\hat{E}_{1,2}(z,t)$ are fast values, whereas the analogous value for the proper electron-beam waves is initially slow (as the Lagrange variable). All complex amplitudes $E_{1,2,3}$ are slowly varying functions of z, $\Theta_{1,2,3}$ is the super-slow combination phase, and Θ_0 is the god of the system.

second method carries an explicitly expressed collective (multiparticle) nature. This is because, in this case, we fix a spatial point and observe the velocities of particles that pass through this point. The Lagrange description was used in Sections 3.1–3.4, whereas the Euler description is applied in this and the preceding section.

The method of the oscillation (oscillation-wave) hierarchic tree (Section 3.2) was elaborated predominantly for the case of single-particle (Lagrange) description. Therefore, to analyze the corresponding hierarchy of waves in the FEL, we must use the Lagrange description. The results of this analysis are represented in Figure 4.9.

Here the Lagrange phases p_1 (signal) and p_2 (pumping) form the single-particle parametric (fast) resonance (see Condition 3.57). Electron oscillations with the slow combination phase θ_{12} form the wave of stimulated oscillations. Apart from the stimulated wave, the proper electron-beam waves are also excited by the above-mentioned mechanisms. The physical peculiarity of the FEL operation mechanism is that the Lagrange phase of this wave p_3 is slow initially. The phase p_3 and the phase θ_{12} slow phase form, in turn, the wave parametric (slow) resonance. The first (single-particle) resonance belongs to the zero hierarchic level and the second (wave resonance), as shown in Figure 4.9, belongs to the second level. The electric strength of the electromagnetic waves $\hat{E}_{1,2}(z,t)$ is initially the fast values, whereas the analogous value for the proper electron-beam waves $\hat{E}_3(z,t)$ is slow initially (within the framework of the Lagrange description). All complex amplitudes $E_{1,2,3}$ are slowly varying functions on z. The value $\Theta_{1,2,3}$ is the super-slow combination phase of the third hierarchic level, whereas Θ_0 is the god of the system. As previously mentioned, this characterizes the external energy sources and relevant operation systems in this case.

4.2.8 Superheterodyne Amplification Effect in FEL

The general concept of the superheterodyne amplification of a wave was formulated by Yu.V. Guliaev and P.E. Zilberman [46]. It was done for the case of parametric resonance of three electromagnetic (pumping, signal, and idle) waves in a nonlinear active (laser) medium. The parameters of the medium were chosen in order to provide that the idle wave frequency would get into the amplification band associated with some auxiliary external mechanism. By virtue of nonlinear coupling, this amplification is translated to a signal wave. This effect was interpreted as the superheterodyne amplification. In practice, this implies the following.

The pumping and the signal frequencies, ω_1 and ω_2, are fitted to be close so that the idle wave frequency would satisfy the requirement $\omega_1 - \omega_2 = \omega_3 << \omega_{1,2}$. Therefore, the superheterodyne amplification makes it possible to shift the amplification up the frequency range, namely, from the low-frequency range (where an efficient amplification mechanism exists and can be employed as the auxiliary mechanism) toward the range of higher frequencies of the signal wave (where such mechanisms were unavailable).

An analogous idea for the FEL was formulated by Kotsarenko and Kulish [33,34]. Here the practical reason was somewhat different. In FEL, the frequencies of the signal and idle (SCW) waves, ω_1 and ω_3, are close or equal. So, shifting the amplification along the frequency range cannot be accomplished. On the other hand, highly efficient amplification mechanisms for the longitudinal beam waves have been known in vacuum and plasma electronics for a long time: for example, the plasma beam, the two-stream, and the parametric two-stream mechanisms of amplification [1,5], which provide very high amplification levels. For many years, however, no efficient ways have been known to transform longitudinal (beam) wave amplification into transverse (electromagnetic) waves. Kotsarenko and Kulish [33,34] solved this problem by means of superheterodyne amplification. In other words, the wave parametric resonance was employed as a transformer of the amplification from a longitudinal electron-beam wave into an electromagnetic (transverse) signal wave.

The distinctive features of the superheterodyne amplification mechanism in the FEL will be considered for the stationary Raman model with some arbitrary quasilinear mechanism of additional amplification [33,34]. The normalized wave amplitudes are described by the equations whose structure is analogous to Equation 4.58 (see other works by Kulish [1,5] for details), i.e.,

$$\frac{da_1}{dz} = -a_2 a_3; \quad \frac{da_2}{dz} = -a_1 a_3; \quad \frac{da_3}{dz} - \hat{\Gamma} a_3 = -a_1 a_2. \tag{4.60}$$

The linear term in the third equation (Equation 4.59) describes the negative damping of the SCW (i.e., amplification) due to the external auxiliary mechanism. Here a_k ($k = 1,2,3$) are the normalized amplitudes of the pumping wave ($k = 2$), the signal wave ($k = 1$), and the SCW ($k = 3$), and $\hat{\Gamma}$ is the normalized growth rate that describes the auxiliary amplification mechanism. Though physical forms of the latter are various, all models of that kind possess mutual peculiarities. We shall discuss them in terms of their solutions of the set of Equation 4.59.

Suppose the pumping is strong ($|a_2| >> |a_{1,3}|$), so that its variations in the course of interaction can be ignored ($a_2 \cong a_{20}$). We assume that the signal and the SCWs propagate toward the positive direction of the z-axis and write the initial conditions as

$$a_1(z = 0) = a_{10}; \quad a_3(z = 0) = 0. \tag{4.61}$$

We take the solution of the set of Equation 4.60 in the form $\sim\exp\{\alpha z\}$. It is not difficult to show that the growth rates $\alpha_{1,2}$ are given by [33,34]

$$\alpha_{1,2} = \frac{\hat{\Gamma}}{2} \pm \sqrt{\frac{\hat{\Gamma}^2}{4} + a_{20}^2}\ . \tag{4.62}$$

Within the context of the boundary conditions, we find the electromagnetic wave gain factor

$$K_1 = \frac{a_1(L)}{a_{10}} = \frac{\alpha_2 \exp\{\alpha_1 L\} - \alpha_1 \exp\{\alpha_2 L\}}{\alpha_2 - \alpha_1}, \tag{4.63}$$

where L is the interaction region length, $a_1(L) = a_1(z = L)$. For $4a_{20}^2 << \hat{\Gamma}^2$ (this assumption is quite natural), Equation 4.62 can be reduced to a simpler one [33,34],

$$K_1 \cong 1 + \frac{a_{20}^2}{\hat{\Gamma}^2} \exp\{\hat{\Gamma} L\}, \tag{4.64}$$

which is suitable for interpreting the essence of the phenomenon under consideration. We see that the electromagnetic wave is amplified mainly due to the auxiliary amplification mechanism (with the growth rate $\hat{\Gamma}$) rather than the parametric (with the growth rate α_{20} similar to the traditional FEL). We introduce the value

$$S = \frac{a_{20}^2}{\hat{\Gamma}^2} << 1 \tag{4.65}$$

and find that the parametric resonance cannot be too weak. Namely, it must be sufficient to provide that $S\exp\{\hat{\Gamma}L\} >> 1$. The numerical analysis of Equations 4.63 and 4.64 shows that the superheterodyne FEL can provide amplification of the same order as other FELs for considerably lower (by an order of magnitude and more) values of the pumping wave amplitudes [1,5].

4.3 Elementary Mechanisms of Wave Amplification in FELs

Thus, a wave is a specific kind of oscillation that occurs simultaneously in time and spatial coordinates. The temporal oscillations are closely connected with the spatial ones by elementary physical mechanisms. The characteristic properties of these oscillations are illustrated in this book by examples of two principally different types: mechanical systems and plasma systems. The first example is illustrated by the wave properties of a solid-state elastic medium (Section 4.1), while the second represents electron waves in an electron beam (Section 4.2). These examples reflect the essential differences in the elementary mechanisms of the formation of waves as totalities of many temporal and spatial oscillations. Although such an elementary mechanism is rather obvious in the case of mechanical waves (Section 4.1.1), it is not so obvious in the case of plasma systems. Therefore, let us fill in the knowledge gaps by using the FEL as a convenient illustrative example. Following the tradition of physical electronics [1,5,47], we will treat the forming (excitation) of the electron-beam waves as a grouping (bunching) process.

One of the most interesting physical features of the FEL is that the excitation of the longitudinal stimulated waves occurs as a result of nonlinear interactions of electrons with the field of two transverse electromagnetic waves: signal and pumping. This means that two principally different elementary grouping mechanisms can be separated in the FEL: the longitudinal and the transverse. Let us begin with the first.

4.3.1 Longitudinal Grouping Mechanism in TWT

As shown before, the basic operation principles of FELs allow the hierarchic treatment. In this chapter, we will discuss hierarchic treatment in view of the basic concepts and definitions of classical microwave electronics [47]. For the sake of convenience, we will use the comparison method of analysis. For this, we will study successively relevant elementary processes occurring in both traveling wave tubes (TWTs) [47] and FELs.

It should be mentioned that an equivalent concept to the concept of hierarchy (see Chapter 1) has existed in microwave physical electronics for a long time. It is the so-called quasistationary interaction principle. It is the basis for the theories of all microwave devices with long time interactions [47]. According to this principle, the energy transfer in a beam wave system is efficient only in a case where the electrons related with the beam—in a proper coordinate system—see the decelerating (accelerating) component of the wave electric field as almost static (quasistationary). Regarding FELs, this means that moving electrons are in approximately the same phase of the signal wave field. Therefore, when such a quasistationary condition holds, the phases of electron oscillations in the wave field are slowly varying functions, i.e., they could be treated as quantities of the next hierarchic level. Thus, we can say that the quasistationary interaction principle is actually a sequence of the basic concepts of the theory of hierarchic waves and oscillations.

For illustration of the FEL operation principles, let us choose the well-known classical microwave device, the TWT [47]. The beam of electrons in this device moves along the straight lines of the electric fields of amplified waves. These lines, in turn, are directed along the direction of the electromagnetic wave propagation. This means that the amplified electromagnetic wave in TWTs, in contrast to FELs, is not purely a transverse wave but contains the longitudinal component of the electric field. The distinctive feature of the wave in TWTs is that the phase velocity v_{ph} is smaller than c, the light velocity in vacuum.

We pass to the coordinate system K', moving in the longitudinal direction with mean beam velocity v_0 (the proper beam reference system). The quasistationarity condition (which is a sequence of the quasistationary interaction principle) can be satisfied if the condition $v'_{ph} \approx v'_0 \approx 0$ is held. This condition in the laboratory coordinate system can be rewritten in the following form:

$$v_{ph} \approx v_0. \tag{4.66}$$

Thus, only in this case can the electrons see approximately the same phase of the amplified field of the signal wave in the beam proper system K'. In microwave electronics, conditions such as Equation 4.65 are referred to as synchronism conditions. It is clear from Equation 4.65 that the Lagrange phase of electron oscillations in the field of this type of wave is a slowly varying function.

Since the input beam is continuous in the course of interaction, some electrons must match the accelerating phase of the Lorentz electric force F'_e, while others will match the decelerating phase (see Figure 4.10). Therefore, electrons of the first group are slowly catching up with the electrons of the second group. Electrons of the third group lag behind electrons of the first group and, thus, electron bunches are formed in the vicinity of the points

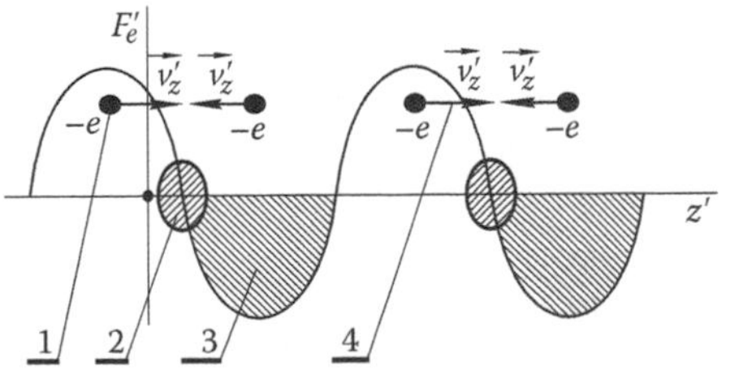

FIGURE 4.10
Illustration of the grouping process in a TWT. Here, 1 is the electron charge $-e$, 2 represents the electron bunches (clusters), 3 represents the accelerative (along the positive direction of the z') phases of the force $F'_e = |\vec{F}'| = -eE'_1$, $\vec{F}'_e$ is the magnitude of the vector of the electric Lorentz force, E'_1 is the intensity of the longitudinal component of the electric field of the amplified electromagnetic wave, and $\vec{v}'_z$ (4) represents the z'-components of the vector of the electron velocity $\vec{v}'$ for different phases of the force $\vec{F}'_e$. The illustrated grouping process can also be treated as an elementary mechanism of amplification of the longitudinal electron wave. The bunches (2) in this case correspond to the wave maximums, and the middle points between the maximums correspond to the wave minimums.

of meeting (see Figure 4.10). If the condition of Equation 4.65 is satisfied exactly (i.e., $v_{ph} = v_0$), then these bunches are formed in the vicinity of the point at which the field strength vanishes. The mean energy transfer in the beam wave system equals zero. If we assume $v_0 > v_{ph}$ (but $|v_{ph} - v_0| << v_0$) then, sooner or later, bunches formed near zero-field points slip to the decelerating phase of the electric Lorentz force F_e'. In this case, each bunch constitutes a large charged particle. The continuous beam is transformed into a periodic chain of bunches, each of them being decelerated by the longitudinal field of the electromagnetic wave. We treat this chain as a longitudinal stimulated electron wave in the TWT and the described elementary mechanism is nothing but the single-particle quasi-Cherenkov's [47] resonance.

The appearance of longitudinal periodical inhomogeneity in the beam electron concentration leads to the generation of the longitudinal electric wave of stimulated oscillations. Simultaneously, the slow and fast proper electron waves are excited within the plasma of the beam. In the traditional versions of TWTs with moderate electron beams [47], these waves merge together with the stimulated wave, forming a common electron wave of complex structure, an SCW, in the TWT. However, the role of the proper electron waves becomes essential in the case of intensive (including relativistic) beams. In this case we can observe the interaction modes that are similar to the Raman or Compton modes in FELs [47].

The beam kinetic energy in TWT is transferred to an electromagnetic (signal) wave and the latter is amplified. This is the operation principle of TWTs. For $v_0 < v_{ph}$ (but $|v_{ph} - v_0| << v_0$), the picture is inverse: bunches drift in the accelerating phase of the field of the force F_e' and the wave energy is spent on the beam acceleration. This is the elementary operation principle of the linear radio frequency accelerator that is the system inverse to the TWT [47].

It is obvious that the discussed simple mechanism can be classified as a one-level hierarchic oscillation system. However, we obtain such results within the framework of the simplest single-particle TWT model only. For more perfect models, other types of oscillations can appear (cyclotron, nonsynchronous electron waves, etc.) [47]. Therefore, oscillation models of such types could be classified as two-level, three-level, and so on, hierarchic systems.

4.3.2 Longitudinal Grouping Mechanism in FEL

Now, let us apply this knowledge to our main object, the FEL. At first sight, the difference between TWTs and FELs is so great that one hardly expects to find any similarity. Indeed, beams of electrons in FELs move along a slalom (in the case of linearly polarized waves; see Figure 4.11) or some helix-like (in the case of circular polarization) trajectories [1–7], distinctly different from the straight-line trajectories of TWTs [47]. The amplified electromagnetic wave $\{\omega_1, k_1\}$ in FELs is transverse (i.e., has no longitudinal components) and propagates with the phase velocity $v_{ph} \cong c > v_0$. Nonetheless, we see from the following analysis that a deep physical analogy exists between processes in both of these systems. This analogy appears to be especially evident in the case where we accomplish the transition at the first hierarchic level of the FEL as a hierarchic dynamic system (see Figure 4.9). Let us discuss this topic in more detail.

As follows from the formulation of the quasistationary interaction principle, to employ it for FEL analysis, one should turn to the proper beam reference system. However, a straightforward procedure is inconvenient since undisturbed electron trajectories in the working region of an FEL have transverse oscillatory characters [1–7]. Therefore, the proper beam

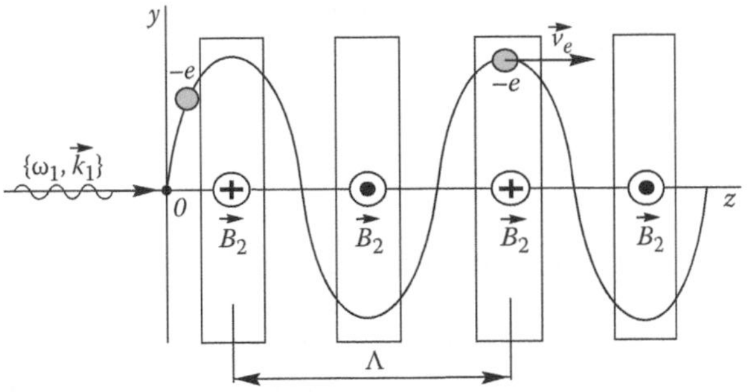

FIGURE 4.11
Electron trajectory in the work bulk of the H-Ubitron FEL with linearly polarized pumping. Here, $\{\omega_1, \vec{k}_1\}$ is the direction of the propagation of the signal wave (whose vector of electric field strength $\vec{E}_1'$ is directed collinearly to the axis y), ω_1 and $\vec{k}_1$ are their signal cyclic frequency and wave vector, respectively, $-e$ is the electron on its trajectory in the plane YZ, $\vec{v}_e$ is the vector of electron velocity, $\vec{B}_2$ represents the vector of induction of the H-pumping magnetic field that is perpendicular to the plane YZ, and Λ is the period of this field. In the K' reference system the H-magnetic field, according to the pumping theorem, is transformed into an electromagnetic pumping wave propagated opposite to the signal wave $\{\omega_1, \vec{k}_1\}$.

reference system is noninertial and, consequently, its treatment could be very complicated. We proceed in the following way. We represent the total beam motion as a sum of transverse (oscillatory) and longitudinal (progressive) motions. We pass to the reference system K', which moves at the beam velocity of progress v_0. In the reference system K', there are two types of electron transverse oscillation phases: electron oscillations in the signal wave field, p_1', and in the pumping field, p_2'. But a question arises: What are the expressions for pumping and signal fields in the new reference system K'? The answer is that, according to the pumping theorem in [1,5], any pumping field in an FEL (H-Ubitron, for example; see Figure 4.11) transforms in the system K' into an oppositely propagated electromagnetic wave. Respectively, there is a consequence from the pumping theorem: Any electromagnetic wave, transformed into a forward-direction electromagnetic wave field in the proper coordinate system, can be treated as a signal in an FEL [1,5].

Thus, within the context of the pumping field theorem and its sequence, the Lagrange phases of electron oscillations in the signal and pumping fields (in the system K') are

$$p_1' = \omega_1' t' - k_1' z' \cong \omega_1' \left(t' - \frac{z'}{c} \right); \tag{4.67}$$

$$p_2' = \omega_2' t' + k_2' z' \cong \omega_2' \left(t' + \frac{z'}{c} \right). \tag{4.68}$$

Here, for simplicity, we assumed $v_0 \sim c$; $v_{phj} \approx c$ [where v_{phj} are the signal ($j = 1$) and pumping ($j = 2$) phase velocities]. It can be verified that, unlike the case of TWTs, the phases illustrated in Equations 4.67 and 4.68 are fast phases of the zero hierarchic level. However, the slow (combination) phase of the electron oscillations θ_{12} (the value of the first hierarchic level; see Figure 4.9) can be separated in this case too. We will use the formulated quasistationary interaction principle for this purpose. According to this principle, an electron oscillating in the field of the pumping wave with the phase p_2' sees approximately the same signal wave phase p_1'. Hence the rotation velocities (rates of change – see Section 3.1.6) of these phases must be near, i.e.,

$$\frac{dp_1'}{dt} \approx \frac{dp_2'}{dt}, \tag{4.69}$$

where phase velocities dp_j'/dt could be calculated easily from Equations 4.67 and 4.68.

As defined in the preceding chapter (Sections 3.2 and 3.3), conditions like Equation 4.69 determine the resonant state of the considered system. When both phases p_j' are wavy (as in the case discussed here), the condition of Equation 4.68 could be treated as the one-particle main parametric resonance condition for FEL. It is clear that due to Equation 4.69, the quantity

$$\theta_{12}' = p_1' - p_2' \tag{4.70}$$

is slow and the sign-conjugated combination

$$\psi_{12}' = p_1' + p_2' \tag{4.71}$$

is a fast quantity. Corresponding to definitions given in Sections 3.2 and 3.3, the phase θ_{12}' is called the slow combination phase, and ψ_{12}' is the fast combination phase. The slow phase can be classified as a phase of the first hierarchic level of FELs, being a hierarchic oscillatory system. The phase θ_{12}' in FELs (the first hierarchic level) is a functional analogy of the electron oscillation phase in the signal electromagnetic wave in TWTs (the zero hierarchic level). Hence, in contrast to the simplest model of a TWT (defined as a one-level hierarchic system), the simplest FEL model can be considered at least as a two-level (zero + first levels) hierarchic system.

Inasmuch as we accept electron motion in the system K' as a nonrelativistic ($|dz'/dt| \ll c$) Condition 4.69, in view of Equation 4.67, Equation 4.68 may be rewritten as

$$\omega_1' \approx \omega_2'. \tag{4.72}$$

Expression 4.72 is another (specific for the chosen supposition set) representation of the one-particle parametric resonance (Equation 4.68). Let us proceed then to the laboratory coordinate system. As a result we get from Equation 4.71 the classical formulas [1–7] in the laboratory coordinate system for the FEL-Dopplertron (see also Equation 3.57)

$$\omega_1 \approx \omega_2 \frac{1+\beta_0}{1-\beta_0} \approx 4\gamma_0^2 \omega_2. \tag{4.73}$$

Here, we used, as before, the supposition $v_0 \approx c$; $\gamma_0 = 1/\sqrt{1 - v_0^2/c^2}$ is the relativistic factor, $\beta_0 = v_0/c$. By an analogous method, the equivalent formula can be obtained for the H-Ubitron FEL [1–7]

$$\omega_1 \approx \frac{v_0 k_2}{1-\beta_0} \approx 2\gamma_0^2 k_2 v_0 \approx 4\pi\gamma_0^2 \frac{c}{\Lambda}, \tag{4.74}$$

where $\Lambda = 2\pi/k_2$ is the period of magnetic undulator (H-Ubitron pumping system; see Figure 4.11). As follows from Equations 4.73 and 4.74, high operation frequencies ω_1 can be obtained in an FEL only in the essentially relativistic case $\beta_0 \approx 1$. As mentioned earlier, this is the main reason why FEL design employs relativistic electron beams.

Then, we study the grouping mechanisms in an FEL, which is treated here as a hierarchic oscillation system (see Figure 4.9 and related comments). For this, let us write the Lorentz equation for the electron motion in system K' as

$$\frac{d^2\vec{r}'}{dt^2} = \frac{-e}{m}\left\{\vec{E}_1' + \vec{E}_2' + \frac{1}{c}\left[v'\left(\vec{H}_1 + \vec{H}'\right)\right]\right\}, \tag{4.75}$$

where $\vec{r}'$ is the electron radius vector in the system K'. For simplicity we assume that both the pumping and the signal fields are circularly polarized (in contrast to the linear polarization, shown in Figure 4.11) and the electron's motion (in the chosen system K') is nonrelativistic. Thus

$$\vec{E}_1' = E_1'\left\{\vec{e}_1 \sin\omega_1'\left(t' - \frac{z'}{c}\right) + \vec{e}_2 \cos\omega_1'\left(t' - \frac{z'}{c}\right)\right\};$$

$$\vec{E}_2' = E_2'\left\{\vec{e}_1 \sin\left[\omega_2'\left(t' + \frac{z'}{c}\right) + \varphi_{20}'\right] + \vec{e}_2 \cos\left[\omega_2'\left(t' + \frac{z'}{c}\right) + \varphi_{20}'\right]\right\}, \tag{4.76}$$

where all notations are given earlier. Moreover, for simplicity, we assume $E_1' << E_2'$ (given pumping field approximation) and $\varphi_{20}' = 0$. Then, we find the solution of the equation in the first approximation with respect to the normalized amplitudes $\gamma_{1,2}' = eE_{1,2}'/mc\omega_{1,2}'$. Therein, it is considered that $\gamma_1' << \gamma_2' << 1$. The notations are

$$x^{(1)'} = \frac{c\gamma_2'}{\omega_2'} \sin\omega_2'\left(t' + \frac{z'^{(1)}}{c}\right);$$

$$y^{(1)'} = \frac{c\gamma_2'}{\omega_2'} \cos\omega_2'\left(t' + \frac{z'^{(1)}}{c}\right);$$

$$z'^{(1)} = z_{10}', \tag{4.77}$$

where z_{10}' is the constant of integration; the superscript indicates the number of approximations with respect to the chosen small parameter. The quantities $\gamma_{1,2}'$ are referred to as the acceleration parameters [1,5]. For $\gamma_2' > 1$, the pumping is called a wiggler; for $\gamma_2' < 1$, the pumping is referred to as an undulator [1–7]. In this case, similar to a uniform magnetic field, electrons move along circles [5]. The circle radius, according to Equation 4.77, is

$$R'^{(1)} = \sqrt{\left(x'^{(1)}\right)^2 + \left(y'^{(1)}\right)^2} = c\gamma_2'/\omega_2'. \tag{4.78}$$

In the second approximation, with respect to the small parameter $\gamma'_{1,2}$, we find an expression for the z'-component of the magnetic Lorentz force acting on the electron (force of the combination wave of the stimulated oscillations) to be given by

$$F'_L = -\gamma'_1\gamma'_2 mc\left(\omega'_1 + \omega'_2\right)\sin\theta'_{12}, \tag{4.79}$$

where θ'_{12} is the slow combination phase (Equation 4.70). It is interesting to note that Expression 4.79 corresponds to the analogous result that can be obtained by using the relevant hierarchic averaging procedure (see Chapter 4). This means that the force F'_L in Expression 4.79 belongs to the first hierarchic level. This situation is a characteristic feature of the chosen model with the circular polarizations only. In the case of a linear polarization of wave fields, after accomplishing corresponding averaging procedures, we can obtain a similar result for the first hierarchic level only.

Thus, as follows from Expression 4.79, the longitudinal magnetic Lorentz force in an FEL with a circularly polarized pumping is a slowly varying function. Its performance in FELs is completely analogous to the effect of the electric Lorentz force in the TWTs (see Figure 4.10 and corresponding comments). As in TWTs, electrons in the FELs are also accelerated or decelerated if they get in the positive or negative phases of the force (Expression 4.79).

Electron bunches are formed similarly. However, in the FELs, two distinguishing bunching (grouping) mechanisms are realized simultaneously. The first is referred to as the longitudinal (linear) mechanism. The difference between the physical picture under consideration and that of TWTs is that the magnetic Lorentz force (Expression 4.79) is solenoidal and cannot work on electrons (it can only change the directions of their motion). The electric Lorentz force does the work. However, in FELs, this force is transverse (not longitudinal) by virtue of Equation 4.75. Therefore, the elementary mechanism of energy transfer in FELs occurs in the transverse plane, while the longitudinal bunching only provides its coherence. We can treat it as an elementary mechanism of wave amplification.

Let us now consider further the mechanism of transverse energy transfer in the system beam + pumping + signal. It should be noted in this connection that, according to the basic principle of microwave electronics [47], some specific transverse grouping mechanism should accompany this type of energy transfer. It is because that, in such a case only, the energy exchange has a coherent character. Hence, the mentioned elementary mechanism of wave amplification should contain, in turn, two different submechanisms: the transverse grouping (bunching) mechanism and the mechanism of energy transfer itself.

Let us begin with the transverse grouping (bunching) mechanism—the first submechanism.

4.3.3 Elementary Mechanism of Longitudinal Electron Wave Amplification and Transverse Grouping

Let us recall that electron kinetic energy can transfer into the signal wave energy only in the case when a certain phasing of electron oscillations is provided. According to the quasistationarity principle, the electron oscillations in pumping and signal fields must be cophase. When the signal wave is absent ($E_1 = 0$), electrons rotate in the pumping field with equal angular velocities ω_2 and they are characterized by different initial oscillation phases in different cross-sections z'.

The effect of the signal wave results in the appearance of magnetic Lorentz force (Equation 4.79). In turn, electrons begin to group along the z-axis (the process of the longitudinal electron wave amplification). Therein, the grouping electrons have different initial phases. This means that during the formation of linear (longitudinal) bunches of different electrons, they could introduce different mismatches (detuning) of their initial transverse oscillation phases. In such a case, the bunch looks more like a cloud rather than a large particle. In this case, the coherent energy transfer should be impossible. However, inasmuch as the formation of coherent longitudinal-transverse bunches really occurs, there exists a mechanism of transverse self-phasing oscillations, i.e., the peculiar phase-bunching mechanism (see Figure 4.12).

Under the influence of the Lorentz force F'_L (Expression 4.79), an electron gains velocity along the z'-axis (see Figure 4.13)

$$\frac{dz'}{dt} = \frac{dz'}{dt}(z',t) = \frac{1}{m}\int F'_L dt'. \tag{4.80}$$

Then the electron rotation frequency in pumping field ($E'_2 >> E'_1$) is changed by the Doppler effect and, in view of Equations 4.67 through 4.69, we have

$$\omega'^{*}_e = \frac{dp'_2}{dt} = \omega'_2\left(1+\frac{\dot{z}'}{c}\right) = \omega'_2\left(1+\frac{1}{mc}\int F'_L\, dt'\right). \tag{4.81}$$

Hence, the electron rotation frequency ω'^{*}_e is a slowly varying function of coordinate z' and time t'. Moreover, electrons moving along the positive z'-axis ($dz'/dt > 0$) rotate faster than the electrons located at the points with $dz'/dt = 0$. At the same time, the electrons moving in the opposite direction rotate slower than electrons moving along the positive z'-axis (Figure 4.12). Therefore, phases overlap so that fast and slow rotating electrons approach each other along the z'-axis, meet in an electron bunch with the same oscillation phase, and form a phase bunch (see Figure 4.12). We can regard this mechanism as an elementary mechanism of the longitudinal SCW amplification.

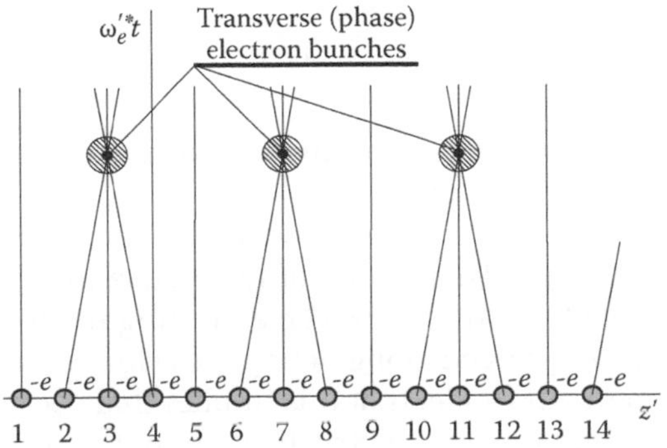

FIGURE 4.12
Transverse (phase) mechanism of electron grouping (bunching). Here, a series of dependencies of the rotation phases $(\omega'^{*}_e t)$ of 14 electrons is presented. It can be easily seen that the dependencies of the rotating phases on coordinate z' for different electrons ($-e$) can be different. These differences are stipulated by the differences of their initial oscillation phases. The transverse (phase) electron bunches are formed in the vicinity of the third, seventh, and eleventh electrons.

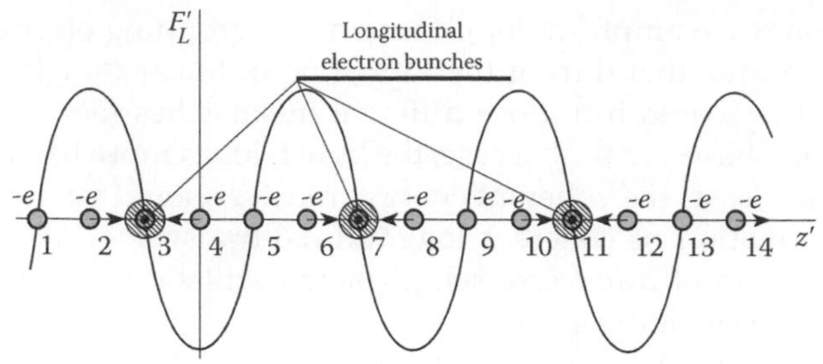

FIGURE 4.13
Illustration of the elementary mechanism of the longitudinal electron wave (SCW) amplification (compare with the analogous process in TWTs shown in Figure 4.10). Here, $F'_L(z')$ is the dependency of the magnetic Lorentz force on the coordinate z' for the same electrons that are shown in Figure 4.12. The longitudinal electron bunches are formed (similar to the transverse ones; see Figure 4.12) in the vicinity of the third, seventh, and eleventh electrons. These points correspond to the maximums of the SCW, while the points of the first, fifth, ninth, and thirteenth electrons correspond to the SCW minimums. All other designations are the same as in Figure 4.12. In the laboratory reference system, this picture as a whole moves along the z-axis with the velocity v_0.

Therefore, processes of linear (along the z'-axis; see Figure 4.13) and transverse phase-bunching [phases $p'_e = (\omega'_e * t')$; see Figure 4.12] are matched. This suggests the existence of a linear (linearly transverse) phase-bunching mechanism in FELs. It also means that the longitudinal elementary mechanism for the signal wave—the transverse mechanism—also should exist.

4.3.4 Elementary Mechanism of the Signal Wave Amplification and the Energy Transfer

Let us discuss the peculiarities of the energy exchange mechanism in FELs. As mentioned, it can be treated as an elementary mechanism of wave amplification. The transformation mechanism of electron kinetic energy in FELs is purely transverse because, in this model, the work cannot be done by any other force than the electric Lorentz force,

$$\vec{F}_e = -e\vec{E}, \tag{4.82}$$

which is transverse since $(\vec{E}\vec{n}) = 0$. The longitudinal force [magnetic Lorentz force (Equation 4.79)] is solenoidal and does not work.

By virtue of the law of conservation of energy, the kinetic energy of electron motion is transferred into the wave energy. This occurs only because the electric field provides the coherent deceleration of the electrons. Moreover, according to the quasistationary interaction principle, the decelerating force is long-acting, i.e., slowly varying from the electron point of view. We mention in our model that this implies that vectors of the whole electron bunch velocity and of the signal wave electric field strength vary in phase (see Equation 4.69). The positive effect of the interaction is because the bunch electron oscillations, caused by the pumping field, are decelerated by the electric field of the signal. In this case, we have decreasing electron-beam kinetic energy and increasing signal wave energy. We interpret such a process of energy transfer as the elementary mechanism of signal wave amplification. To verify this, let us calculate the average (over fast oscillations) power of the force in Equation 4.82. Using the laboratory coordinate system, we obtain

$$\bar{P} \cong \left\langle \left(\vec{F}_{\perp 1} + \vec{F}_{\perp 2} \right)\left(\vec{v}_{\perp 1} + \vec{v}_{\perp 2} \right) \right\rangle = \left\langle \left(\vec{F}_{\perp 1}\vec{v}_{\perp 2} \right) \right\rangle + \left\langle \left(\vec{F}_{\perp 2}\vec{v}_{\perp 1} \right) \right\rangle < 0;$$

$$\left\langle \left(\vec{F}_{\perp 1}\vec{v}_{\perp 1} \right) \right\rangle = \left\langle \left(\vec{F}_{\perp 2}\vec{v}_{\perp 2} \right) \right\rangle, \tag{4.83}$$

where $\vec{F}_{\perp 1,2}$ represents electric Lorentz forces of the signal and pumping waves in the laboratory coordinate system (Equation 4.75), and $\vec{v}_{\perp 1}$ and $\vec{v}_{\perp 2}$ are the electron velocity transverse components given by the individual effect of the forces $\vec{F}_{\perp 1}$ and $\vec{F}_{\perp 2}$. It should be noted that for the H-Ubitron pumping wave, we have $\vec{F}_{\perp 2} = 0$. By the system K', we find that the condition $\bar{P} > 0$ is satisfied only for $\omega_1' \neq \omega_2'$ and, provided the mismatch is sufficiently small, $|\omega_1' - \omega_2'| << \omega_{1,2}'$. Thus, we have

$$\bar{P}' \cong -e\left(R_2' E_1' \omega_2' + R_1' E_2' \omega_1' \right)\sin\left(\omega_2' - \omega_1' \right)t; \quad \omega_2' > \omega_1'. \tag{4.84}$$

In calculating Equation 4.84, we use the solutions of Equations 4.75 through 4.77 and analogous others for signal waves (switched-off pumping).

Here we can clear up the physical meaning of the synchronism condition (Equation 4.72). For the interaction to be quasistationary, the frequencies ω_1' and ω_2' in the proper coordinate system K' must be equal; however, the mean power $\bar{P}'$ (Equation 4.84) does not vanish as long as this equality is approximate. When the synchronism condition (Equation 4.71) is satisfied exactly, then an electron bunch is formed near the zero decelerating field (see Figures 4.12 and 4.13) and, hence,

$$\bar{P}'\left(\omega_1' = \omega_2' \right) = 0. \tag{4.85}$$

With a small mismatch, $\Delta = \omega_2' - \omega_1'$, the electron bunch slowly drifts through the region of the decelerating field amplitudes until it enters the accelerating phase region (details of the dynamics of this process are shown in Figure 4.14).

For simplicity, the pumping field is assumed here to be strong ($E_2' >> E_1'$), and wave polarizations are assumed to be circular. Initially, an electron bunch is formed near the point

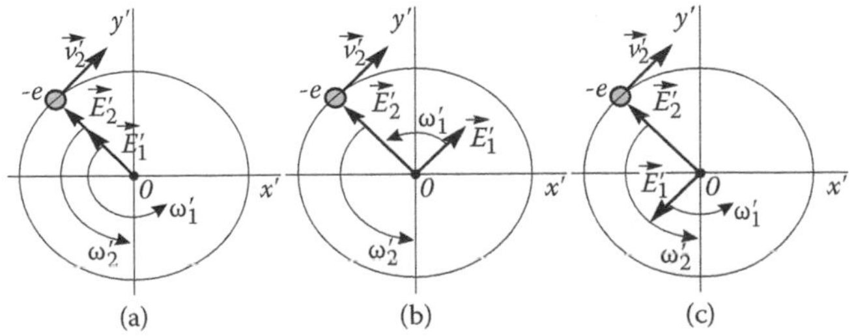

FIGURE 4.14
Reciprocal positions of the vector of electromagnetic wave strength $\vec{E}_{1,2}'$ and the transverse electron velocity vector $\vec{v}_2'$ for the different phases of the energy exchange process. (From Kulish, V.V., *Physical Process in Parametrical Electronic Lasers (Free Electron Lasers)*, Thesis for the Scientific Degree of Doctor of Physical-Mathematical Sciences, Institute of Physics of Academy of Sciences of Ukraine, 1985.)

where the difference in the oscillation phases vanishes (see Figure 4.12). This system state is shown in Figure 4.14a. Therein, we have a paradox: for the small frequency mismatch $\omega'_{1,2}(\omega'_2 > \omega'_1)$, we have an opposite situation in the laboratory system K ($\omega_1 >> \omega_2$) due to the reciprocal shifting of the frequencies because of the Doppler effect.

The slowly rotating vector E'_1 in the system K' lags behind vector E'_2. As a result, the force $F_e = -eE'_1 \sin\left[\left(\omega'_2 - \omega'_1\right)t'\right]$ is quasistationary and decelerating. Since $E'_2 >> E'_1$, the action of the signal wave on the electron motion in the pumping field is weak. As soon as the bunch passes through the region of the decelerating wave field phases (i.e., when the angle between vectors $\vec{E}'_1$ and $\vec{E}'_2$ becomes greater than π), it enters the accelerating phase of the signal wave field. Instead of energy transferring to the wave, energy absorption occurs. Thus, the useful length of the working region is determined by how the bunch escapes from the signal field decelerating phase. In FEL theory [1–7], this length is called the saturation length.

The longitudinal progressive motion of electrons is related to their transverse oscillations caused by the wave magnetic field. Therefore, electron deceleration in the transverse plane immediately decreases the longitudinal velocity of the whole electron beam. When the synchronism conditions of Equations 4.71 through 4.73 are violated, the process becomes nonresonant. In terms of the theory of hierarchic oscillations and waves, this infers that the velocity of the slow rotation of vector $\vec{E}'_1$ with respect to $\vec{E}'_2$ increases and that the classification of the combination phases θ_{12}, ψ_{12} as slow and fast phases, respectively, is not valid. The bunch goes through the decelerating phase region much quicker and preserves a greater portion of energy. The interaction becomes nonstationary and the electron efficiency of the interaction decreases.

If resonance condition (Equation 4.72) is satisfied and $\omega'_1 > \omega'_2$, then according to the discussed physical situation, the signal wave must be absorbed. The absorbed energy is converted into kinetic energy of the electron beam. This system, with a design similar to that of an FEL, is called the surfotron accelerator [1–7]. It is a total analog of the linear radio frequency accelerator. Its relation to the FELs resembles that of the relation of the linear radio frequency accelerator to the TWTs.

References

1. Kulish, V.V. 2002. *Hierarchic methods. Undulative electrodynamic systems*, Vol. 2. Dordrecht, the Netherlands: Kluwer Academic Publishers.
2. Marshall, T.C. 1985. *Free electron laser*. New York: MacMillan.
3. Brau, C. 1990. *Free electron laser.* Boston: Academic Press.
4. Luchini, P., and Motz, U. 1990. *Undulators and free electron lasers*. Oxford: Clarendon Press.
5. Kulish, V.V. 1998. *Methods of averaging in non-linear problems of relativistic electrodynamics*. Atlanta, GA: World Scientific Publishers.
6. Saldin, E.L., Schneidmiller, E.V., and Yurkov, M.V. 2000. *The physics of free electron lasers. Advanced texts in physics*. Heidelberg, Germany: Springer-Verlag.
7. Silin, R.A., Kulish, V.V., and Klymenko, Ju.I. 1991. Soviet Inventors Certificate, SU No 705914, Priority 18.05.72, 15.05.91. *Inventions Bulletin* 26.
8. Nicolis, J.S. 1986. *Dynamics of hierarchical systems. An evolutionary approach.* Heidelberg, Germany: Springer-Verlag.

9. Kondratenko, A.N., and Kuklin, V.M. 1988. *Principles of plasma electronics*. Moscow: Energoatomizdat.
10. Kulish, V.V., and Lysenko A.V. 1993. Method of averaged kinetic equation and its use in the nonlinear problems of plasma electrodynamics. *Fizika Plazmy* (*Sov. Plasma Physics*) 19(2):216–227.
11. Kulish, V.V., Kuleshov, S.A., and Lysenko, A.V. 1993. Nonlinear self-consistent theory of superheterodyne and free electron lasers. *The International Journal of Infrared and Millimeter Waves* 14(3):451–568.
12. Kulish, V.V. 1997. Hierarchic oscillations and averaging methods in nonlinear problems of relativistic electronics. *The International Journal of Infrared and Millimeter Waves* 18(5):1053–1117.
13. Khapaev, M.M. 1988. *Asymptotic methods and equilibrium in theory of nonlinear oscillations*. Moscow: Vysshaja Shkola.
14. Ruhadze, A.A., Bogdankevich, L.S., Rosinkii, S.E., and Ruhlin, V.G. 1980. *Physics of high-current relativistic beams*. Moscow: Atomizdat.
15. Davidson, R.C. 1974. *Theory of nonneutral plasmas*. Reading, MA: W. A. Benjamin.
16. Sitenko, A.G., and Malnev, V.M. 1994. *Principles of plasma theory*. Kiev: Naukova Dumka.
17. Landau, L.D., and Lifshitz, E.M. 1986 *Hydrodynamics*, Vol. 6 of *Theoretical physics*. Moscow: Nauka.
18. Rabinovich, M.I., and Trubetskov, D.I. 1984. *Introduction in theory of oscillations and waves*. Moscow: Nauka.
19. Moiseev, N.N. 1981. *Asymptotic methods of nonlinear mechanics*. Moscow: Nauka.
20. Bolonin, O.I., Kochmanski, S.S., and Kulish, V.V. 1989. Coupled paramagnetic resonances in the FELs. *Acta Physica Polonica*, A76(N3):455–473.
21. Kohmanski, S.S., and Kulish, V.V. 1985. To the nonlinear theory of free electron lasers with multi-frequency pumping. *Acta Phys. Polonica* A68(5):741–748.
22. Butuzov, V.V., Zakharov, V.P., and Kulish, V.V. 1983. *Parametric instability of a flux of high current relativistic electron in the field of dispersed electromagnetic waves*. Deposited manuscript. Deposited in Ukrainian Scientific Research Institute of Technical Information, Kiev. 297 Uk-83:67.
23. Zakharov, V.P., and Kulish V.V. 1985. Explosive instability of electron flux in the field of dispersing electromagnetic waves. *Ukrainian Physical Journal* 6:78–881.
24. Kalmykov, A.M., Kotsarenko, N.Ja., and Kulish, V.V. 1978. Possibility of transformation of the frequency of the laser radiation in electron flux. *Pisma v Zhurnal Technicheskoj Fiziki* (*Letters in the Journal of Technical Physics*) 4(14):820–822.
25. Kohmanski, S.S., and Kulish, V.V. 1985. Parametric resonance interaction of electron in the field of electromagnetic waves and longitudinal magnetic field. *Acta Phys. Polonica* A68:6525–6736.
26. Sukhorukov, A.P. 1988. *Nonlinear wave-interactions in optics and radiophysics*. Moscow: Nauka.
27. Bloembergen, N. 1965. *Nonlinear optics*. Singapore: World Scientific Publishing.
28. Weiland, J., and Wilhelmsson, H. 1977. *Coherent nonlinear interactions of waves in plasmas*. Oxford: Pergamon Press.
29. Kulish, V.V., and Kayliuk, A.G. 1998. Hierarchic method and its application peculiarities in nonlinear problems of relativistic electrodynamics. Single-particle model of cyclotron-resonant maser. *Ukrainian Physical Journal* 43(4):398–402.
30. Kulish, V.V., Lysenko, A.V., and Savchenko, V.I. 2003. Two-stream free electron lasers: General properties. *International Journal on Infrared and Millimeter Waves* 24(2):129–172.
31. Kulish, V.V., Lysenko, A.V., and Savchenko, V.I. 2003. Two-stream free electron lasers: Physical and project analysis of the multi-harmonic models. *International Journal on Infrared and Millimeter Waves* 24(4):501–524.
32. Kulish, V.V., Lysenko, A.V., Savchenko, V.I., and Majornikov, I.G. 2005. The two-stream free electron laser as a source of electromagnetic femto-second wave packages. *Laser Physics* 12:1629–1633.
33. Kotsarenko, N.Ya., and Kulish, V.V. 1980. On the possibility of superheterodyne amplification of electromagnetic waves in electron beams. *Zhurnal Tekhnicheskoy Fiziki* 50:220–222.

34. Kotsarenko, N.Ya., and Kulish V.V. 1980. On the effect of superheterodyne amplification of electromagnetic waves in a plasma-beam system. *Radiotekhnika I Electronika* 35(N11):2470–2471.
35. Kulish, V.V., Lysenko, A.V., and Savchenko, V.I. 2003. Two-stream free electron lasers: Analysis of the system with monochromatic pumping. *International Journal on Infrared and Millimeter Waves* 24(3):285–309.
36. Buts, V.A., Lebedev A.N., Kurilko, V.I. 2006. *The theory of coherent radiation by intensive electron beams*. Heidelberg, Germany: Springer-Verlag.
37. Ablyekov, V.K., Babajev, Yu. N., and Pugachiov, V.P. 1980. Radiation of a self-focusing relativistic electron beam in the mode of phase correlation. *Doklady Adademii Nauk SSSR (Reports of Academy of Sciences of USSR)*, 225(4):848–849.
38. Bolonin, O.N., Kulish, V.V., and Pugachiov, V.P. 1989. *Stimulated radiation of high-current electron beam in the proper magnetic field*. Deposited in Ukrainian Scientific Research Institute of Technical Information, Kiev, No. 431 Uk-89.
39. Tsimring, S.E. 2006. *Electron beams in microwave vacuum electronics. Microwave and Optical Engineering Series*. New York: Wiley.
40. Davidson, R.C., and Hong, Qin. 2001. *Physics of intensive charged particle beams in high energy accelerators*. Singapore: World Scientific Publishers.
41. Abramyan, E.A. 1988. *Industrial electron accelerators and applications*. Boca Raton, FL: CRC Press.
42. Hammer, D.A., and Rostoker, N. 1970. Propagation of high current relativistic electron beams. *Phys. Fluids* 13:1831–1843.
43. Ramian, G., and Elias, L. 1988. The new UCSB compact far-infrared FEL. *Nuclear Instrument and Methods in Physics Research* A272:81–88.
44. Elias, L.R. 1987. Electrostatic accelerator FEL. *Society of Photo-Optical. Instrumentation Engineers* 738:28–35.
45. Ramian, G., and Elias, L. 1988. The new UCSB compact far-infrared FEL. *Nuclear Instrument and Methods in Physics Research* A272:81–88.
46. Guliaev, Yu.V., and Zilberman, P.E. 1971. Superheterodyne amplification of electromagnetic waves. *Fizika Tvyordoho Tela (Soviet Solid State Physics)* 13(4):955–957.
47. Trubetskov, D.I., and Khramov, A.E. 2003. *Lectures on microwave electronics for physicists*, Vol. 1 and 2. Moscow: Fizmatlit.

5

Hierarchic Description

In Chapter 2 (Section 2.5), we discussed the main ideas and peculiarities of the hierarchic calculation method. As was shown, their specific calculation ideology and technologies can be developed directly from the formulated hierarchic laws. The latter completely determine the characteristic properties of the considered hierarchic dynamic systems as well as the elaborated methods of their treatment.

The key idea of the formulated hierarchic approach is the introduction of the specific compression and decompression operators (Figures 2.7 through 2.9 and relevant comments). We have discussed only the ideological side of the considered hierarchic problem; the corresponding calculation algorithms were not described there. This and the following chapters will fill this gap. This chapter, in particular, will discuss the most general calculations for constructing such possible algorithms. Their detailed partial realizations are described in other chapters.

All hierarchic problems can be divided into oscillations and waves (see Chapter 2). The first are characteristic for the so-called lumped systems and can be described by relevant exact differential equations. The problems of motion of a charged particle in the external electromagnetic fields can serve as an obvious example of such systems.

A characteristic feature of the second type of problem is that they can be described mathematically by some differential equations with partial derivatives. Such objects are usually classified as distributed systems. The problems of wave propagation in various physical mediums (including nonlinear) are as a rule described by the system with partial derivatives.

Thus, the main subjects of future discussion in this book are general descriptions of the mathematical procedures for constructing decompression and compression operators. This will be done both for the lumped and distributed oscillation-wave systems.

5.1 Decompression and Compression Operators: General Case of Lumped Systems*

We regard the material of this chapter as a natural continuation of Section 2.5. Therefore, we recommend the reader review that section once again.

* From Kulish, V.V. et al., *Int. J. Infrared Millimeter Waves*, 19, 33–94, 1998. With permission. Kulish, V.V., *Hierarchic Methods. Hierarchy and Hierarchic Asymptotic Methods in Electrodynamics*, Vol. 1, Kluwer Academic Publishers, Dordrecht/Boston/London. 2002. With permission.

5.1.1 Decompression Operators

Let us continue our general discussion with the study of a hierarchic system that could be described by an initial dynamic equation such as Equation 2.4. We use the definition of Equation 2.11 as a connection between variables of the zero and first hierarchic levels

$$z^{(0)} = \hat{U}^{(1)}(z^{(1)}, t, \varepsilon_1), \tag{5.1}$$

where the functional operator $\hat{U}^{(1)}$ in Chapter 2 is called the decompression operator. Let us discuss one of the possible algorithms for constructing its evident form. Differentiating Expression 5.1 in time t, we consider the dynamic equation for the zero hierarchic level (Equation 2.4). After relevant transformations, we yield the so-called generalized hierarchic equation of the first order for the unknown decompression operator $\hat{U}^{(1)}$

$$\left(\frac{\partial \hat{U}^{(1)}}{\partial z^{(1)}}, Z^{(1)}(z^{(1)}, t, \varepsilon_1) \right) + \frac{\partial \hat{U}^{(1)}}{\partial t} = Z^{(0)}(\hat{U}^{(1)}(z^{(1)}, t, \varepsilon_1), t, \varepsilon_1). \tag{5.2}$$

In the coordinate form, Equation 5.2 can be written as

$$\sum_{i=1}^{n} \frac{\partial \hat{U}_k^{(1)}}{\partial z_i} Z_i^{(1)}\left(z_1^{(1)}, \ldots, z_n^{(1)}, t, \varepsilon_1\right) + \frac{\partial \hat{U}_k^{(1)}}{\partial t} = Z_k^{(0)}\left(\hat{U}_1^{(1)}, \ldots, \hat{U}_n^{(1)}, t, \varepsilon_1\right). \tag{5.3}$$

The first term of Equation 5.2 is the product of Jacobi matrix $\partial D^{(1)}/\partial z^{(1)}$ of the order $n \times n$ and the column vector $Z^{(1)}$. It is assumed that the compression operator $M^{(o)}$ (see Definition 2.10 and corresponding discussion) is given initially. It means that, in view of Definition 2.10, the new dynamic function of the first hierarchic level $Z^{(1)}$ can be considered known.

According to the resemblance hierarchic principle, the relevant dynamical equation for the next hierarchic level should have a mathematical structure analogous to Equation 2.8. At this time, it takes place with respect to a new (proper) dynamic variable set of the first level only

$$\frac{dz^{(1)}}{dt} = Z^{(1)}(z^{(1)}, t, \varepsilon_1), \quad z^{(1)}(t_0) = z_0^{(1)}. \tag{5.4}$$

The new dynamic equation for the first hierarchic level (Equation 5.4), related with the above-mentioned principle of hierarchic compression, should be simpler than the initial equation for the zero hierarchic level (Equation 2.8). Therefore, these equations are usually referred to as truncated equations. Relevant solutions of Equation 5.4 allow one to find the decompression operator $\hat{U}^{(1)}$ (see Equations 5.2 and 5.3) and, as a result, to construct solutions for variables of the zero hierarchic level (see Definition 5.1). Owing to the application of the hierarchic method, this calculation algorithm is found to be simpler than the solutions of the initial problem directly (Equation 2.4).

When the discussed dynamic system, besides zero, has higher levels of hierarchy (the first and second, and so on), we can process them analogously. Namely, we consider Equation 5.4 as a new initial equation (analogous to Problem 2.7) that can be solved analogously:

$$z^{(1)} = \hat{U}^{(2)}(z^{(2)}, t, \varepsilon_1, \ldots, \varepsilon_\kappa, \ldots, \varepsilon_m), \tag{5.5}$$

$$\frac{dz^{(2)}}{dt} = Z^{(2)}(z^{(2)}, t, \varepsilon_2, \ldots, \varepsilon_m), \quad z^{(2)}(t_0) = z_0^{(2)}, \tag{5.6}$$

$$\left(\frac{\partial \hat{U}^{(2)}}{\partial z^{(2)}}, Z^{(2)}(z^{(2)}, t, \varepsilon_2, \ldots, \varepsilon_m)\right) + \frac{\partial \hat{U}^{(2)}}{\partial t} = Z^{(1)}(\hat{U}^{(2)}(z^{(2)}, t, \varepsilon_2, \ldots, \varepsilon_m), \tag{5.7}$$

$$z^{(m-1)} = \hat{U}^{(m)}\left(z^{(m)}, t, \varepsilon_{m-1}, \varepsilon_m\right), \tag{5.8}$$

$$\frac{dz^{(m)}}{dt} = Z^{(m)}\left(z^{(m)}, t, \varepsilon_m\right); \quad z^{(m)}\left(t_0\right) = z_0^{(0)}; \tag{5.9}$$

$$\left(\frac{\partial \hat{U}^{(m)}}{\partial z^{(m)}}, Z^{(m-1)}(z^{(m)}, t, \varepsilon_m)\right) + \frac{\partial \hat{U}^{(m)}}{\partial t} = Z^{(m-1)}(\hat{U}^{(m)}, z^{(m)}, t, \varepsilon_m). \tag{5.10}$$

By summarizing, we can formulate once again the general idea of the discussed hierarchic algorithm (see Section 2.5). This idea consists of the fact that each new truncated dynamic equation for a higher κth hierarchic level is simpler than the preceding one. At the same time, it possesses an analogous mathematical structure with the initial (and preceding) equation. It is a result of practical realization of the compression and resemblance hierarchic principles (see Chapter 2). This means automatically that the solution procedure for the dynamical equation of the last hierarchic level indeed should be essentially simpler than the analogous procedure of the initial equation of the zero level.

Furthermore, we propose an unexpected analogy with a similar (this author thinks) situation in biology (see Figure 5.1). In our illustrational example, the initial problem is formulated on the lowest (zero) hierarchic level (problem in Figure 5.1). We can treat this as an order for an evolution. In accordance with the description in Chapter 2, this order should be transformed on the highest hierarchic level (level of competence) to the form of truncated equations (Equation 5.9). We treat these compressed equations as a peculiar genome (set of genes) of the considered dynamical system, whereas we consider the solutions of this equation $z^{(m)} = z^{(m)}(t)$ to be an embryo. Then, we can treat the complete set of equations (Equation 5.8 through 5.10) for the decompression operator $\hat{U}^{(\kappa)}$ as a peculiar medium for our embryo. It is obvious that in such a case, the complete set of properties of the initial whole system (Equation 2.7) is enciphered in the hierarchic manner by genes (Equation 5.9).

Providing some culture medium in the form of the decompression operators $\hat{U}^{(\kappa)}$ and the hierarchic set of equations such as Equation 5.2, 5.7, and 5.10 for their determination, we open for our mathematical embryo, $z^{(m)} = z^{(m)}(t)$, a possibility for evolution (Figure 5.1). The embryo transforms (as a result of such evolution) into desired solutions $z = z(t)$ (organism; see Figure 5.1) for the initial system (Equation 2.7). It is not difficult to find that it happens similarly to, for instance, the development of a human embryo within a human organism.

This analogy clarifies the fact that, in the case of hierarchic-like algorithms, we can hope for a simplification of the general solution procedure. It is because any embryo in nature is always simpler than the whole organism.

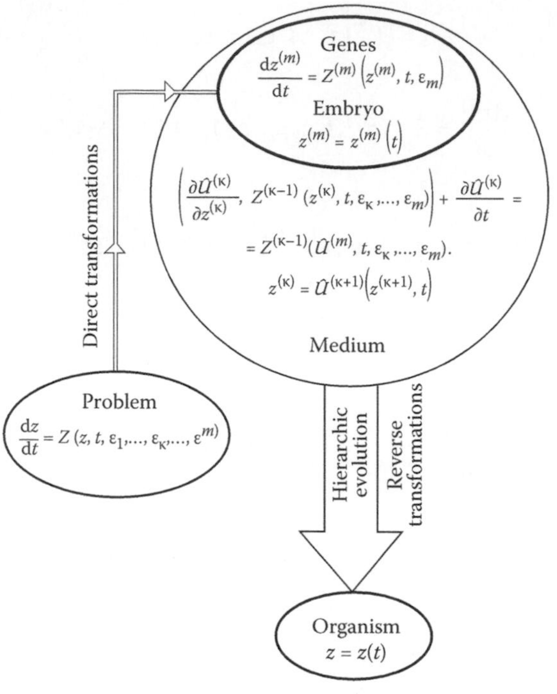

FIGURE 5.1
Illustration of the hierarchic solution algorithm (direct and reverse transformations) of an embryo developing into a whole biological organism.

Once again, this analogy illustrates that all natural hierarchic dynamic systems apparently have a common nature (see Chapter 2). Moreover, we might hope that this common nature could be described within the framework of the proposed new hierarchic approach. Finally, it should be mentioned that the discussed method can have some difficulties. Most originate from the characteristic features of the procedures of integration of generalized hierarchic equations (Equations 5.2, 5.3, 5.7, and 5.10). These difficulties are determined, mainly, by the special form of the chosen compression operator $M^{(\kappa)}$. It is obvious that such calculation schemes are of practical interest only in the one case where the total labor intensity of the joint solution procedure (for hierarchic dynamical equations such as Equations 5.4, 5.6, and 5.9, and corresponding generalized ones such as Equations 5.2, 5.3, 5.7, and 5.10) is shorter than the procedure for solving the initial equation (Equation 2.4) directly. This result can be attained when some analytical algorithms can be constructed for the solution of generalized hierarchic equations such as Equations 5.2 and 5.3. This holds in practice for special compression operators $M^{(\kappa)}$ only. The general situation is simplified in the case of periodic (or conditionally periodic) oscillation systems. A number of such structural operators can be constructed for similar situations. It should be mentioned, however, that most widely used operators are based on the well-known Riemann theorem of the integral averaged value [1]

$$\lim_{p\to\infty}\left[\int_a^b f(x)\sin(px)\,\mathrm{d}x\right]\to 0. \tag{5.11}$$

Let us discuss some compression (averaging) operators $M^{(\kappa)}$ that can be constructed on the basis of the Riemann theorem (Equation 5.11).

5.1.2 Compression Operators

The averaging (smoothing) operators satisfy the above-mentioned labor intensity test for the compression operators in an optimal manner. It is explained by the wide prevalence of these operators in electrodynamics and mechanical problems. Most often, the averaging operators are used for some evident function of time, *t*, a space coordinate, z_s, or of *t* and z_s simultaneously. Formally, these operators can be constructed in the form

$$M_z\left[Z(z,\varepsilon)\right]=\bar{Z}(\varepsilon)=\frac{1}{(2\pi)^n}\int_0^{2\pi}\ldots\int_0^{2\pi}Z(z,\varepsilon)dz_1\ldots dz_s\ldots dz_n, \tag{5.12}$$

the compression operator in the form of an averaging operator in spatial coordinates z_s, and

$$M_t\left[Z(z,t,\varepsilon)\right]=\bar{Z}(z,t_0,\varepsilon)=\lim_{T\to\infty}\frac{1}{T}\int_t^{t_0+T}Z(z,t,\varepsilon)dt \tag{5.13}$$

is the compression operator in the form of the averaging operator in time *t*. The condition of the function $Z(z,t,\varepsilon)$ periodicity, in the latter case, could be written as

$$Z(z,t+(2\pi),\varepsilon)=Z(z,t,\varepsilon).$$

It is assumed that this condition is satisfied. In the first case (Equation 5.12), the function $Z(z,\varepsilon)$ is the supposed 2π-periodical on all components of the *n*-dimensional vector $z = \{z_1,\ldots,z_s,\ldots,z_n\}$ within the domain of the definition $z \in G^n$, i.e.,

$$Z(z+(2\pi),\varepsilon)=Z(z,\varepsilon), \tag{5.14}$$

where the vector $(z_1 + 2\pi,\ldots,2_n + 2\pi)$ is denoted by $(z + (2\pi),\varepsilon)$. If the function $Z(z,t,\varepsilon)$ is periodical only on a part of coordinates $z_1,\ldots,z_n$, the operator M_z (Equation 5.12) can be determined as

$$M_{zs}\left[Z(z,\varepsilon)\right]=\bar{Z}\left(z_{s+1},\ldots,z_n,\varepsilon\right)dz_1\ldots dz_s\ldots dz_n, \tag{5.15}$$

if

$$Z\left(z_1+(2\pi),\ldots,z_s+(2\pi),z_{s+1},\ldots,z_n,\varepsilon\right)=Z\left(z_1,\ldots,z_s,z_{s+1},\ldots,z_n,\varepsilon\right)$$

$$0\le s\le n. \tag{5.16}$$

This means that the averaging procedure is performed only on part of the variables $z_1,\ldots,z_s$, and that the function Z is periodical in these variables only. While integrating in Equation 5.13, the variables $z_1,\ldots,z_s$ are considered relevant constants. Here, we will neglect the hierarchic index κ.

In spite of the formal resemblance of both versions of the averaging operator, their mathematical nature is essentially different. Namely, the averaging in Equation 5.13 is accomplished over the independent scalar variable t (laboratory time, for instance). We have to deal with similar procedures (Equation 5.12) with the components of some vector z. But, generally, the vector z is a relevant function of variable t. This functional dependence is determined by relevant dynamical equation. Nevertheless, the coordinates z_s are considered as constants in accomplishing the relevant averaging procedures in Equation 5.12.

It should be noted that the procedure for constructing a corresponding averaging operator, accounting for the dependencies of the type $z_s(t)$, is possible also (see, for example, [21]). This book will not have such calculation situations.

In practice, a mixed version of function $Z(z,t,\varepsilon)$ periodicity (i.e., in time t and vector z coordinates simultaneously) is used rather often. In view of the above-mentioned definitions, the construction of averaging operators of such type is self-obvious. However, let us note one special case of mixed averaging operators. It is characteristic for situations where the considered forces are of the wavy form (see examples in Chapters 3 and 4). In particular, the discussion in Chapter 4 of FELs is an evident case of a mixed problem. The characteristic peculiarity of its model is the presence of the linear combinations of time t and spatial coordinates z_s. Such special combinations have, as a rule, the form of scalar phases p_s (see Chapters 3 and 4 for examples).

$$p=\{p_1,\ldots,p_i,\ldots,p_q\}$$
$$p_i=\omega_i t-\sum_j\left(k_j^{(i)}z_j\right), \tag{5.17}$$

where ω_i are the components of the vector of cyclic frequencies ω. The values $k_j^{(i)}$ are the components of the wave vectors $k^{(i)}$ in n-dimensional space

$$\omega=\{\omega_1,\ldots,\omega_i,\ldots,\omega_q\};\quad k^{(i)}=\{k_1^{(i)},\ldots,k_j^{(i)},\ldots,k_n^{(i)}\}=k(\omega_i). \tag{5.18}$$

In problems of electrodynamics and mechanics, a case where function $Z(z,t,\varepsilon)$—expandable in a Fourier series—holds a special place. It is determined by the fact that the Dirichlet–Jordan theorem [2] holds. As is well known, this theorem concerns a possibility of developing a function in a Fourier series. In our case, the n-multiplied Fourier series is expressing the vector function $Z(z,t,p,\varepsilon)$ in the domain G_n

$$Z(p,\varepsilon)=\sum_{\|m\|\geq 0}Z(\varepsilon)\exp\{i(mp)\}, \tag{5.19}$$

where $\|m\|=|m_1|+\ldots+|m_q|$ is the norm of the q-dimensional harmonic vector $m=\{m_1,\ldots,m_q\}$, the components of which are integer numbers $m_i=0,\pm1,\pm2,\ldots;\ i=1,\ldots,q;\ (mp)$ is a scalar product of two q-dimensional vectors $(mp)=\sum_{i=1}^{q}m_ip_i$.

According to the given definitions, m_i plays the role of the harmonic numbers and the functions $Z_m(\varepsilon)$ are relevant Fourier amplitudes.

5.2 Distributed Hierarchic Systems*

5.2.1 Stochastic Hierarchic Distributed Systems

The general characteristics of the oscillation-like and wave problems have been discussed in Chapters 3 and 4. Our study has concerned the so-called determined systems. Let us dwell shortly on the hierarchic stochastic distributed systems.

As is known in physics, the stochastic properties considered in practice models can be determined by the following:

1. Random character of initial conditions for scalar components of vector z for corresponding Equations 2.4, 5.4, 5.6, and 5.9 (a spread in thermal particle velocities in the input of a plasma-like system, for example).
2. Stochastic dynamics in physical mechanisms described by functions $Z^{(\kappa)}$ (collision processes in kinetic systems).
3. Stochasticity of the connections between the hierarchic levels of dynamic systems; it could be expressed practically, for example, as the nonsynonymy of relevant solutions for the decompression operators $\tilde{U}^{(\kappa)}$. It is obvious that the stochasticity of such systems could be also caused by the stochastic nature of the compression operator $M^{(\kappa)}$.

Due to the stochastic mechanisms, we have the transformation of an initially determined hierarchic tree into a stochastic ensemble of hierarchic trees. Here, we discuss only the general ideas of possible stochastic versions of the presented hierarchic theory. Its more complete versions could be found in other works [4–12,36–38].

For simplicity, we consider that only the first two of the three above-mentioned processes determine the stochasticity of the considered system. To describe this system we construct the distribution function

$$f = f\left(z_1,\dots,z_j,\dots,z_n,t,\varepsilon_1,\dots,\varepsilon_\kappa,\dots,\varepsilon_m\right), \tag{5.20}$$

where z_j represents the components of z-vector, and $\varepsilon_\kappa = \varepsilon_1,\dots,\varepsilon_\kappa,\dots,\varepsilon_m << 1$, as before, are the hierarchic scale parameters. In this case, the dynamics of the phase volume is determined by the single-particle dynamics of the vector components z_j. Most often, there are space coordinates of particles and their velocities among these components.

The motion of each separate particle (structural element) of the considered stochastic system is described by the single-particle equation of the same type as Equations 5.4 and 5.6. In the stochastic model discussed here, the fact of the stochasticity of the particle motion is determined only by the first two causes. In this case, these causes are the random character of the initial conditions for scalar components of vector z and the stochastic character of the dynamics of the basic physical mechanisms reflected by the dynamic functions $Z^{(\kappa)}$, respectively. We assume that the determined part Z' can be separated from the total function $Z(z,t,\varepsilon)$ (see Equations 5.4 and 5.6) in the additive way

$$Z = Z' + Z'', \tag{5.21}$$

* From Kulish, V.V. et al., *Int. J. Infrared Millimeter Waves*, 19, 33–94, 1998. With permission. Kulish, V.V., *Hierarchic Methods. Hierarchy and Hierarchic Asymptotic Methods in Electrodynamics*, Vol. 1, Kluwer Academic Publishers, Dordrecht/Boston/London. 2002. With permission.

where the function Z'' describes the above-mentioned stochastic component of the general particle dynamics. It can be the stochastic addendum in the Langevin equation, and so on. Hence, the condition $Z' = 0$ corresponds to the case of the purely stochastic model. In contrast, in the case $Z'' = 0$, we have the system where stochasticity is determined by the initial (boundary) conditions only. For simplicity, we accept that the considered stochastic system has properties of a quasicontinuous medium.

We differentiate the definition of Equation 5.21 in time t and change the momentum differential into particular derivatives

$$\frac{\mathrm{d}}{\mathrm{d}t} \to \frac{\partial}{\partial t} + \sum_{j=1}^{n} \frac{\mathrm{d}z_j}{\mathrm{d}t}\frac{\partial}{\partial z_j}. \tag{5.22}$$

Subsequently, we consider single-particle Equation 2.7, with the assumption that Equation 5.21 is satisfied. By applying the procedure of Equation 5.22, we obtain as a result the so-called general form of Boltzmann's kinetic equation

$$\left\{\frac{\partial}{\partial t} + \sum_{j=1}^{n} Z_j' \frac{\partial}{\partial z_j}\right\} f\left(z_1,\ldots,z_n,t,\varepsilon_\kappa\right) = \left\{\sum_{j=1}^{n} Z'' \frac{\partial}{\partial z_j}\right\} f\left(z_1,\ldots,z_n,t,\varepsilon_\kappa\right). \tag{5.23}$$

Often, the function Z'' describes the collision processes in a stochastic system. In this case, the right-hand side of Equation 5.23 is referred to as the collision integral, which we designate ςI_{st}

$$\left\{\sum_{j=1}^{n} Z''\left(z_1,\ldots,z_n,t,\varepsilon_\kappa\right)\frac{\partial}{\partial z_j}\right\} f\left(z_1,\ldots,z_n,t,\varepsilon_\kappa\right) = \zeta I_{st}, \tag{5.24}$$

where ζ is the collision scale parameter, which can generally be arbitrary. In the case of the six-dimensional space (three space coordinates plus three components of canonical momentum), we formally can reduce Equations 5.23 and 5.24 to traditional forms of the Boltzmann equation. Equations similar to Equations 5.23 and 5.24 represent a rather wide class of distributed stochastic dynamic problems. It is not difficult to be convinced that the hierarchic asymptotic methods can also be applied to the investigation of such problems. Let us discuss this in detail.

We see that the stochastic dynamics of the system as a whole is determined by the single-particle dynamics. This circumstance allows us to use the above-mentioned version of the hierarchic approach to the investigation of kinetic systems. Indeed, by using formula 5.26 and Definition 5.24, we can reduce kinetic equation 5.23 to the form with a total differential

$$\frac{\mathrm{d}f}{\mathrm{d}t} = Z_f\left(f,t,\varepsilon_1,\ldots,\varepsilon_m\right), \tag{5.25}$$

where all notations are self-obvious in view of the above discussion. It is evident that the mathematical structures of Equations 5.25 and 2.7 are the same. Hence, the hierarchic calculative scheme set forth here can be applied to solving stochastic problems. However, it is clear that it is not as simple as it might seem at first glance. A simple analysis shows that, in this case, we should have a number of some methodical difficulties.

The point is that the single-particle (lumped) problems are formulated in the so-called Lagrange's variables, whereas the distributed ones always are formulated in Euler's variables (see Chapters 3 and 4 for further discussion of these variables). Formulating the initially evident distributed problem of Equation 5.23 in the purely lumped-like form of Equation 5.25, we automatically obtain the problem of adequateness of the obtained solution versus the desired solution. Mathematically this problem can be regarded as a problem of back transformations of solutions for a system with total derivatives like Equation 5.25 into the adequate solutions of a system with partial derivatives like Equation 5.23.

The essence of the idea for accomplishing such transformations and the technical ways of overcoming of mentioned difficulties will be discussed in Chapter 6. Here we point out only that generally this type of problem is a particular variety of the wider class of problems called the hierarchic wave problems. Let us discuss their most general characteristic properties.

5.2.2 Decompression Operator in the General Case of Hierarchic Wave Problems

In the averaging method [12–21], Krylov and Bogolyubov [12,13] have elaborated the first calculation approach for asymptotic solution of nonlinear oscillation problems that we now regard as the hierarchic one. Bogolyubov's algorithm in itself is the simplest version of the two-level hierarchic calculation scheme. Hence, the developed hierarchic technologies can be considered a relevant generalization of Bogolyubov's ideas [13–16,19,21,36–38].

The essential characteristic feature of this class of methods involves the exact differential equations that are written in some specific forms. These forms are called the standard systems. As is clear from the preceding discussion, the calculative situations using the exact differential equations are typical for lumped systems.

The use of partial derivatives is the most characteristic point for the methods of description of the distributed systems. Therefore, as the first step in constructing relevant hierarchic calculative procedures, we should consider the formulation of analogous concepts of the standard system. It should be mentioned that, depending on the type of distributed problem, we can have a few different variants of the standard form. Let us begin with the formulation of the general standard form.

Let us assume that we have a class of dynamical wave systems that can be described by the following differential matrix equation:

$$A\frac{\partial U}{\partial t}+(BP)U+CU=R(U,\partial U/\partial t,(PU),z,t), \tag{5.26}$$

which we will consider the general standard form. Here, A,B,C are the square matrices of the size $n \times n$, $U = U(z,t)$ is some vector function in Euclidean n-dimensional space R^n with coordinates $\{z_1,z_2,\ldots,z_n\}$, i.e., $\forall z \in R^n$ $z = (z_1,z_2,\ldots,z_n)^T$, $\forall z_I \in (-\infty, +\infty)$, $I \in (1,2,\ldots,n)$, P is some linear differential operator in the space R^n, $R(\ldots)$ is a given weak-nonlinear vector function, and t is a scalar variable such as laboratory time. Equation 5.26 can be written in the coordinate form as

$$\sum_{l=1}^{n}\left(a_{pl}\frac{\partial}{\partial t}+b_{pl}P_{pl}+c_{pl}\right)U_l=R_p\left(U,\partial U/\partial t,\left(PU\right),z,t\right), \tag{5.27}$$

where a_{pl}, b_{pl}, c_{pl} are the elements of the matrices A,B,C, P_{pl} are the elements of the operator P, and U_l, R_p are the components of the vectors U and R.

One may be convinced that Maxwell equations, the Boltzmann kinetic equation, the quasihydrodynamic equation, and many other equations can be reduced to the general standard form like Equations 5.26 and 5.27 [14–20]. Unfortunately, a general algorithm for the integration of equations such as Equations 5.26 and 5.27 is not currently known. Therefore, researchers must work with particular cases only, as we will do here.

First, we use the general hierarchic principle. In accordance with this principle, the considered dynamic system should have a hierarchic structure. Then, we apply the principle of information compression: the dynamic equation for each higher level should be simpler than that for a preceding level. Each hierarchic level of the system is described by the equations of the same mathematical structure (the principle of hierarchic resemblance). Therein, for each hierarchic level, we have a set of the proper (for that level only) dynamical variables (the principle of information compression), and so on. Then, we introduce corresponding compression and decompression operators.

The compression operator $M^{(\kappa)}$, in particular, can be introduced similarly to Definition 2.10

$$R^{(\kappa+1)} = R^{(\kappa+1)}\left(U^{(\kappa+1)},\ldots\right) = M^{(\kappa)}R^{(\kappa)}\left(U^{(\kappa)},\ldots\right), \tag{5.28}$$

where $R^{(\kappa)}$ is the given vector function defined before as the right-hand side of Equation 5.26. This function characterizes the nonlinear dynamical properties of the κth hierarchic level. It is obvious that the operator of Equation 5.28 satisfies all hierarchic principles, too.

The decompression operator (a function in this particular case) describes the dynamics of connections between the proper variables of two neighboring hierarchic levels. We can determine it analogously to Equation 2.11

$$U^{(\kappa)}(z^{(\kappa)},t) = \hat{U}^{(\kappa+1)}(U^{(\kappa+1)}(z^{(\kappa+1)},t)). \tag{5.29}$$

Both operators are not really independent mathematical objects. We consider that the compression operator $M^{(\kappa)}$ is given in the form of some integral operator. Then, we act on Equation 5.29 by the differential operator, defined by the left-hand side of Equation 5.26, and use the principle of hierarchic compression (with respect to Equation 5.26 and the definition of Equation 5.29). The general hierarchic transformation equation can be obtained as a result of the accomplished procedures

$$\begin{aligned}&\left\{A\frac{\partial}{\partial t}+\left[B\left(\frac{\partial z^{(\kappa+1)}\left(z^{(\kappa)}\right)}{\partial z^{(\kappa)}}\bigg|_{z^{(\kappa)}=z^{(\kappa)}\left(z^{(\kappa+1)}\right)}\frac{\partial}{\partial z^{(\kappa+1)}}\right)\right]+C\right\}\hat{U}^{(\kappa+1)}\left[U^{(\kappa+1)}\left(z^{(\kappa+1)},t\right)\right]\\&=\left[M^{(\kappa+1)}\right]^{-1}R^{(\kappa+1)}\left(U^{(\kappa+1)},\frac{\partial U^{(\kappa+1)}}{\partial t},\frac{\partial U^{(\kappa+1)}}{\partial z^{(\kappa+1)}},z^{(\kappa+1)},t\right)\end{aligned}, \tag{5.30}$$

where the required vector dependencies $z^{(\kappa+1)} = z^{(\kappa+1)}(z^{(\kappa)})$ and $z^{(\kappa)} = z^{(\kappa)}(z^{(\kappa+1)})$ can be found by using the definition of Equation 5.29; $[M^{(\kappa+1)}]^{-1}$ is the inverted operator $M^{(\kappa+1)} = M^{(\kappa)}|_{\kappa\to\kappa+1}$. These operators satisfy the condition $M^{(\kappa+1)}[M^{(\kappa+1)}]^{-1} = I$, where I is the unit vector. Here also the corresponding operator $P = P^{(\kappa)}$ is written in the variables of the (κ+1)th hierarchic level. In this case we choose it in the simplest differential form: $P^{(\kappa+1)} = \partial/\partial z^{(\kappa+1)}$. Therefore, with the given structural operator $M^{(\kappa+1)}$ (or the reversed operator $[M^{(\kappa+1)}]^{-1}$), Equation 5.30 gives us relevant hierarchic transformation function $U^{(\kappa+1)}$ (Equation 5.29). Furthermore, we assume the function $R^{(\kappa+1)}$ is weakly nonlinear and expandable in the Fourier series.

It should be mentioned that these assumptions hold for a wide class of problems in electrodynamics, hydrodynamics, etc. One type of such problems formally correlates with the problem of a charged particle motion (beam) in predetermined electromagnetic fields [7–11,29–34,36–38] (see also relevant examples in Chapter 4). The self-consistent dynamics of plasma-like systems is the second example (see Chapter 9) [3–6,36–38].

5.2.3 Some Particular Cases of the Hierarchic Problems

Hierarchic algorithms can be conditionally divided into two groups

- Algorithms reducible to the integration of the standard single-particle equation like Equation 2.7 (the single-particle case)
- Algorithms reducible to the so-called Rabinovich standard form (the Rabinovich case) [26–28,36–38]

Let us comment on the specific features of both of these cases.

Single-particle case. This case realizes that some calculative procedure for constructing some equivalent set of the exact differential equations can be accomplished. A number of these types of procedures are known and represented in the references of this book. They include the method of characteristics (see Chapter 6), the method of lines, the method of total differential, and others. Let us illustrate this idea with the simplest example of the method of total differential. For this, we suppose that the two first terms on the left-hand side can be rolled up to the total derivatives because

$$A = B = I;$$

$$\left(BP\right) = \left(\frac{dz^{(\kappa)}}{dt} \times \frac{\partial}{\partial z^{(\kappa)}}\right);$$

$$R = R^{(\kappa)}(U^{(\kappa)}, z^{(\kappa)}, t);$$

$$\frac{dz^{(\kappa)}}{dt} = Z^{(\kappa)}\left(z^{(\kappa)}, t\right), \tag{5.31}$$

where I is the unit matrix; the expression on the right-hand side of the second equation is the product of the diagonal Jacobian matrix of order $n \times n$ and the column operator $\partial/\partial z^{(\kappa)}$, and $Z^{(\kappa)}$ is a nonlinear dynamical vector function defining the single-particle motion in the system (see Equation 5.22). One may be confident that the above-mentioned stochastic hierarchic problem can be reduced to the form of Equations 5.26 and 5.31. Equation 5.31, after a relevant transformation equation (Equation 5.26), can be rewritten as

$$\frac{dU^{(\kappa)}}{dt} = R'^{(\kappa)}\left(U^{(\kappa)}, z^{(\kappa)}, t\right), \tag{5.32}$$

where $R'^{(\kappa)} = R^{(\kappa)} - CU^{(\kappa)}$ is a given nonlinear wave vector function of the κth hierarchic level. Furthermore, we introduce the new extended coordinate vector for the κth hierarchic level:

$$z'^{(\kappa)} = \{z^{(\kappa)}, U^{(\kappa)}\}, \tag{5.33}$$

and a new extended nonlinear dynamic vector function

$$Z'^{(\kappa)} = \{Z^{(\kappa)}, R'^{(\kappa)}\} = Z'^{(\kappa)}(z'^{(\kappa)}, t). \tag{5.34}$$

As a result of the accomplished transformation, the initial system (Equation 5.26) obtains the form of a relevant true single-particle equation like Equation 2.9

$$\frac{dz'^{(\kappa)}}{dt} = Z'^{(\kappa)}\left(z'^{(\kappa)}, t\right), \tag{5.35}$$

whose general algorithm of the asymptotic integration is described above. This algorithm should be completed by a relevant algorithm for the solution of the problem of back transformation (from the Lagrange description to the Euler description). In the particular case $R^{(\kappa)} = 0$, we obtain the pure single-particle version of the discussed hierarchic problem.

Thus, a spatially distributed wave-resonant dynamic problem described by Equation 5.26 in the case of Equation 5.31 can be reduced to a quasi-single-particle problem (Equation 5.35). This means that all calculative technologies developed for the asymptotic integration of the single-particle exact equations can be successfully used for the solution of this type of distributed problem. However, there are some essential differences between the calculative algorithms in the purely lumped and distributed cases, respectively. We talked about the problem of back transformation; let us review that once more.

It can be readily seen that Equation 5.35 allows us to find solutions for the function $U^{(\kappa)}$ in the form $U^{(\kappa)} = U^{(\kappa)}(t)$. The form of the initial distributed Equation 5.26 supposes the eventual solution's dependency on the form $U^{(\kappa)} = U^{(\kappa)}(z^{(\kappa)}, t)$. Specific calculative technology can be developed for the solution of the discussed problem of the reverse transformation [36–38]. The following main ideas are based on such calculative technology.

Constructing the truncated equation for the mth hierarchic level like Equation 2.9

$$\frac{dz'^{(m)}}{dt} = Z'^{(m)}\left(z'^{(m)}, t, \varepsilon_1, \ldots, \varepsilon_m\right), \tag{5.36}$$

and solving it according to the above-mentioned general hierarchic scheme, we can obtain formal solutions for the $(m-1)$th level in the form:

$$z'^{(m-1)} = \hat{U}'^{(m)}\left(z'^{(m)}, t, \varepsilon_1, \ldots, \varepsilon_m\right). \tag{5.37}$$

Then, taking into account the definitions of Equations 5.33 and 5.34, we rewrite the solution (Equation 5.37) as

$$U^{(m-1)}\left(z^{(m)}\right)=\hat{U}_U^{(m)}\left(z^{(m)},U^{(m)},t,\varepsilon_1,\ldots,\varepsilon_m\right),$$

$$z^{(m-1)}\left(z^{(m)}\right)=\hat{U}_z^{(m)}\left(z^{(m)},U^{(m)},t,\varepsilon_1,\ldots,\varepsilon_m\right), \tag{5.38}$$

where all notations are self-evident in view of the previous discussion. Furthermore, we expand the decompression operator $U^{(m-1)}(z^{(m)})$ in a series within close vicinity of the point $z(m-1)$

$$U^{(m-1)}\left(z^{(m-1)},t\right)\cong U_U^{(m)}\left(z^{(m-1)},U^{(m)}\left(z^{(m-1)}\right),t,\varepsilon_1,\ldots,\varepsilon_m\right)$$

$$+\frac{1}{2!}\frac{\partial\hat{U}_U^{(m)}}{\partial z^{(m)}}\Delta z^{(m-1)}\bigg|_{z^{(m)}=z^{(m-1)}}+\ldots, \tag{5.39}$$

where Δz_{m-1} is the difference between $z^{(m)}$ and $z^{(m-1)}$, which can be found from the corresponding expansion of the function $z^{(m)}$ within close vicinity of the point $z^{(m-1)}$.

Subsequently, we perform the analogous procedure with hierarchic levels $(m-1)$, $(m-2)$, and so on. Moving backward by this method, we eventually reach the zero level

$$U\left(z,t\right)\equiv U^{(0)}\left(z^{(0)},t\right)=\hat{U}^{(1)}\left(z^{(1)},U^{(1)},t,\varepsilon_1,\ldots,\varepsilon_m\right). \tag{5.40}$$

After performing the expansion in a series (Equation 5.40) within close vicinity of the point $z^{(1)}=z$, we obtain the dependency $U(z,t)$.

5.2.4 Decompression Operator in the Form of a Krylov–Bogolyubov Substitution

As will be shown here, the case of the function $Z^{(\kappa)}(z^{(\kappa)},t)$ on the right-hand side of Definition 2.10, when it is periodical (or conditionally periodical), is the most interesting for us. The experience shows that the most promising versions of the hierarchic method for studying the hierarchic oscillation systems can be constructed on the basis of the averaging methods. Let us say a few words about these types of calculation schemes, which are based on the so-called Krylov–Bogolyubov substitution. This is interesting for other reasons, too. Historically, Krylov and Bogolyubov (and many followers after them) [13–21] began studying averaging methods as the simplest case of hierarchic-like calculation technology. Therefore, the averaged methods can be regarded as the first practically effective realization of the hierarchic idea in practical calculative algorithm.

We assume that the function $Z'^{(\kappa)}$ in Equation 5.35 is periodical and slowly varying in time

$$Z'^{(\kappa)}\left(z^{(\kappa)},t,\varepsilon_\kappa\right)=\varepsilon_\kappa\hat{Z}^{(\kappa)}\left(z^{(\kappa)},t,\varepsilon_\kappa\right), \tag{5.41}$$

where the prime is ignored for simplicity. In this case, the decompression operator $\hat{U}^{(\kappa)}$ can be found by using the Krylov–Bogolyubov substitution

$$z^{(\kappa-1)} = \hat{U}^{(\kappa)}\left(z^{(\kappa)}, t, \varepsilon_{\kappa}\right) = z^{(\kappa)} + \varepsilon_{\kappa}\hat{U}'^{(\kappa)}\left(z^{(\kappa)}, t, \varepsilon_{\kappa}\right). \tag{5.42}$$

The relevant general hierarchic equation for the vector function $\hat{U}'^{(\kappa)}$ (Equation 5.27) can be written in another form

$$\varepsilon_{\kappa}\left(\frac{\partial \hat{U}'^{(\kappa)}}{\partial z^{(\kappa)}}, Z^{(\kappa-1)}\left(z^{(\kappa)}, t, \varepsilon_{\kappa}\right)\right) + \frac{\partial \hat{U}'^{(\kappa)}}{\partial t} = Z^{(\kappa-1)}\left(z^{(\kappa)} + \varepsilon_{\kappa}\hat{U}'^{(\kappa)}, t, \varepsilon_{\kappa}\right)$$

$$-Z^{(\kappa)}\left(z^{(\kappa)}, t, \varepsilon_{\kappa}\right). \tag{5.43}$$

Here, we would like to stress, especially, that Equation 5.43, as will be shown in the next chapter, can be solved analytically [12,13]. Due to this, the application of the hierarchic scheme based on substitution (Equation 5.42) is very effective.

5.2.5 Rabinovich's Case

Let us turn to the second partial case of the problem discussed in Section 5.2.3. Let us assume that the function R in standard form (Equation 5.26) can be developed in a Fourier series. Separate out the zero Fourier harmonics

$$R_0 = R - \tilde{R}, \tag{5.44}$$

where $\tilde{R}$ is the oscillatory part of the function R. Then we consider that the function R can be represented as a power series of the highest hierarchic small parameter ε_1

$$R = \sum_{n=1}^{\infty} \varepsilon_1^n R_U^{(n)}\left(U, \partial U/\partial t, (PU), z, t, \varepsilon_1, \ldots, \varepsilon_m\right). \tag{5.45}$$

With Equations 5.44 and 5.45, we can write the standard equation (Equation 5.26) as a set of two equations

$$A\frac{\partial U_0}{\partial t} + (BP)U_0 + CU_0 = \sum_{n=1}^{\infty} \varepsilon_1^n R_{U0}^{(n)}\left(U, \frac{\partial U}{\partial t}, (PU), z, t, \varepsilon_1, \ldots, \varepsilon_m\right);$$

$$A\frac{\partial \tilde{U}}{\partial t} + (BP)\tilde{U} + C\tilde{U} + \sum_{n=1}^{\infty} \varepsilon_1^n \tilde{R}_U^{(n)}\left(U, \frac{\partial U}{\partial t}, (PU), z, t, \varepsilon_1, \ldots, \varepsilon_m\right), \tag{5.46}$$

where $R_{U0}^{(n)}$, $\tilde{R}_U^{(n)}$ are relevant expansion functions, $U = U_0 + \tilde{U}$.

Each equation of Equation 5.46 describes an essentially different class of objects for study. In particular, the first governs a smooth (quasistationary) dynamics of the system. For example, it might concern a generation of some smooth quasistatic electric and magnetic fields within a plasma-like relativistic system [38]. Standard numerical methods or exact or approximate calculative methods can be used for solving this part of the problem.

The second part of Equation 5.46 describes nonlinear dynamics of waves in the system, including the wave–resonant interactions of various kinds. A characteristic feature of this problem is the presence of a number of fast oscillations and their harmonics. General calculations are complicated most by the nonlinear resonant nature of these wave interactions. Problems of such types can be solved using the hierarchic analytical and numerical methods discussed in this book [36–38].

5.3 Decompression Operator in the Case of the van der Pol Method*

5.3.1 A Few Introductory Words

Let us recall the fact that the new hierarchic mathematical calculation algorithms have been elaborated using the new physical hierarchic paradigm described in Chapter 2. In this regard, we turn the reader's attention to the following nontrivial circumstances. It is a widely accepted opinion that the basis of the modern theory of hierarchic dynamic systems is relatively new—formed during the past 50 years (synergetic, self-organization, etc.; see Chapter 2). This theory has a mainly philosophical/cognitive significance. However, it has been a little-known fact until today that many calculative procedures—characterized by evidently expressed hierarchic natures—were used at least into the second decade of the twentieth century [13–21].

It is important to note that the researchers—van der Pol, Krylov, Bogolyubov, Leontovich, and others [13–21,23–28,37,38]—did not have any interest in the philosophical or cognitive aspects. Their main interest was the realization of the hierarchic-like highly effective calculation algorithms for practical purposes (in electrical engineering, radio-physics, etc.). From the modern point of view, these methods can be regarded as the simplest (two-level) versions of the general hierarchic asymptotic analytical numerical methods [37,38]. This is because the specific bases of using their transformations have an explicitly expressed hierarchic nature. Therefore, we can say that these methods essentially represent another approach to the investigation of the oscillation-wave dynamic systems.

The key methodological point of these hierarchic analytical calculation technologies is the choice of the compression operator $M^{(\kappa)}$. As shown in preceding sections, the operator $M^{(\kappa)}$, which is constructed on the basis of the averaging procedure, is the most promising for our purposes. Let us say a few words about this procedure and its role in the applied problems.

First, the mathematical averaging procedure has been employed in the celestial mechanics as early as in Gauss's time [21]. With time, it became a very important element in algorithms for the construction of approximate solutions of some types of nonlinear differential equations. However, it happened only after Flemish radio engineer van der Pol

* From Kulish, V.V. et al., *Int. J. Infrared Millimeter Waves*, 19, 33–94, 1998. With permission. Kulish, V.V., *Hierarchic Methods. Hierarchy and Hierarchic Asymptotic Methods in Electrodynamics*, Vol. 1, Kluwer Academic Publishers, Dordrecht/Boston/London. 2002. With permission.

successfully applied the averaging idea to the solution of some typical nonlinear problems of electrical and radio engineering [24].

It is interesting that—not being sufficiently mathematically substantiated—the method has been widely employed by practical engineers, while mathematicians have considered it not worth their professional attention. The situation, unfortunately, is not new; they treated Heaviside's papers on operational calculus in the same manner. Only after his death did they change their opinions.

Mathematicians N.M. Krylov and N.N. Bogolyubov initiated the crucial changes in the 1930s. They employed van der Pol's ideas to develop a rigorous mathematical approach to the analysis of some specific types of nonlinear ordinary differential equations. Having separated the slow and fast variables, they presented the solution in the asymptotic series form where the first term has reproduced the relevant van der Pol result [13,14]. Later, the approach was developed and substantiated by Ukrainian mathematicians of Bogolyubov's school (Yu.A. Mitropolsky, B.I. Moiseenkov, A.M. Samoilenko, and others) [15,16,19].

5.3.2 van der Pol Variables

Let us suppose that some lumped oscillation system is described by the following second-order equation:

$$\ddot{z}+\omega^2 z=\varepsilon\varphi(z,\dot{z});\; z(t_0)=z_0, \tag{5.47}$$

where z is the spatial coordinate, ω is some constant (linear cyclic frequency, for example; see Chapter 3), $\varepsilon \ll 1$ is the small parameter of the problem, and $\varphi(z,\dot{z})$ is a given (generally nonlinear) function of coordinate z and velocity $\dot{z}$. Here the sign dot is a total (material) derivative in time t (i.e., $\dot{z} \equiv dz/dt$). It is important that the requirements for the function φ are not too strict, e.g., discontinuous. The condition $\varepsilon \ll 1$ implies that the system (Equation 5.47) must be weakly nonlinear (see Chapter 3). This means that it describes nearly linear oscillations. Then we introduce the generating equation

$$\ddot{z}+\omega^2 z=0, \tag{5.48}$$

which immediately follows from Equation 5.47 for $\varepsilon = 0$. As is readily seen, the generation equation 5.48 is an ordinary linear differential equation (compared with Equation 3.3) and its solution is well known to be (see Equations 3.4 and 3.10)

$$z = x\cos\psi, \tag{5.49}$$

where x is the amplitude, and

$$\psi = \omega(t - t_0) \tag{5.50}$$

is the linear oscillation phase (Section 3.1). The variables x and ψ are referred to as the van der Pol variables. The main idea of van der Pol is to assume that the solution of nonlinear equation 5.47 is formally described by the same expression as the solution of linear equation 5.48. The difference is that the amplitude x is a slowly varying function of t. The physical meaning of the term "slowly varying function," (discussed in detail in Chapter 3) means that the amplitude x weakly changes during one oscillation period $T = 2\pi/\omega$. The result of this change becomes appreciable only after a long time $t \gg T$ that includes many oscillation periods: $N = \psi/2\pi \gg 1$. If amplitude x is slow, then the phase ψ, for the above-mentioned reasons, is the fast variable (again, see Chapter 3). It is reasonable to assume that the weak nonlinearity does not modify the general behavior of the solutions, though it gives rise to small nonlinear corrections in the relevant weakly nonlinear equation. We derive such equations from the slow amplitude and the fast phase.

We require that the new functions $x(t)$ and $\psi(t)$ satisfy, along Equation 5.49, the condition

$$\dot{z} = -\omega x \sin\psi. \tag{5.51}$$

We differentiate Equation 5.51 with respect to t and substitute the result of Equation 5.47 to obtain

$$-\dot{x}\omega \sin\psi - \omega x\dot{\psi}\cos\psi + x\omega^2\cos\psi = \varepsilon\varphi\left(x\cos\psi, -\omega x\sin\psi\right). \tag{5.52}$$

Moreover, we again differentiate Equation 5.49 with respect to t and make use of the condition of Equation 5.51 to obtain

$$\dot{x}\cos\psi - x\dot{\psi}\sin\psi + \omega x\sin\psi = 0. \tag{5.53}$$

Thus, we have derived a set of two differential equations for $x(t)$ and $\psi(t)$. Solving the equations with respect to $\dot{x}$ and $\dot{\psi}$ reduces the system of Equations 5.52 through 5.53 to the form

$$\dot{x} = -\frac{\varepsilon}{\omega}\varphi(x\cos\psi, -x\omega\sin\psi)\sin\psi \equiv \frac{\varepsilon}{\omega}\varphi_1(x,\psi); \tag{5.54}$$

$$\dot{\psi} = \omega - \frac{\varepsilon}{\omega x}\varphi(x\cos\psi, -x\omega\sin\psi)\cos\psi \equiv \omega - \frac{\varepsilon}{\omega x}\varphi_2(x,\psi). \tag{5.55}$$

Equations 5.54 and 5.55 evidently illustrate the above-mentioned concepts of slowly varying and fast variables. Indeed, as is readily seen, the ratio of the velocities of the changing variables x and ψ is proportional to the small parameter ε

$$|\dot{x}|/|\dot{\psi}| \sim \varepsilon << 1. \tag{5.56}$$

Taking into consideration the definition for hierarchic small parameters (Equation 2.6) we can consequently conclude that these variables are related to different hierarchic levels of the system. We can also detect that van der Pol's description could be treated as the simplest and earliest version of the hierarchic approach to the theory of oscillations and waves.

5.3.3 Truncated Equations and Their Hierarchic Sense

Equations 5.54 and 5.55 are nothing but alternative versions of the initial equation (Equation 5.47) and do not have evident advantages over it. However, the form of Equations

5.54 and 5.55 suggests a way to find approximate solutions. Following van der Pol, we replace the slow varying amplitude x with its time-averaged value using the averaging operator equation 5.13, i.e.,

$$x \cong \bar{x};$$

$$\dot{x} \cong \dot{\bar{x}}. \tag{5.57}$$

For the phase, we substitute

$$\psi \cong \bar{\psi};$$

$$\dot{\psi} \cong \dot{\bar{\psi}}. \tag{5.58}$$

Here the line is the sign of averaging by using the mixed operator M_ψ, constructed on the basis of Equations 5.12 and 5.13. Then we carry out the averaging in Equations 5.54 and 5.55 with regard to Equation 5.57 and obtain an approximate system of equations

$$\dot{x} = \frac{\varepsilon}{\omega}\bar{\varphi}_1(x); \tag{5.59}$$

$$\dot{\psi} = \omega - \frac{\varepsilon}{\omega x}\bar{\varphi}_2(x), \tag{5.60}$$

where the used compression (averaging) operator M_ψ is determined in the following manner:

$$\bar{\varphi}_{1,2} = \frac{1}{2\pi}\int_0^{2\pi} \varphi_{1,2}(x,\psi)d\psi \equiv M_\psi \varphi_{1,2}(x,\psi). \tag{5.61}$$

Equations 5.59 and 5.60 are referred to as truncated. In terms of hierarchic ideology, they are equations of the first hierarchic level. The small parameter ε can be treated as the hierarchic scale parameter of the first hierarchic level.

Thus, we have a possibility to determine that the van der Pol method is indeed the simplest realization of hierarchic methods. Deriving the system of truncated equations (Equations 5.59 and 5.60) is the most important part of the van der Pol method. This system possesses a simpler mathematical structure (principle of information compression) than both the initial equation (Equation 5.47) and the system (Equations 5.54 and 5.55), since Equations 5.59 and 5.60 can be disconnected. Indeed, one can solve the first equation independently and write the solution for the phase ψ in quadratures. As a result, solving the truncated equations is simpler than any procedure of integrating Equation 5.47 directly. Moreover, in many problems of electrodynamics, the researcher is interested in obtaining the rate of change of the phase $\dot{\psi} = \omega^*$, rather than the phase ψ in itself. The quantity ω^* has a physical meaning of the nonlinear cyclic oscillation frequency of the system. In particular, the dependency $\omega^*(t)$ [to be more exact, $\omega^*(x(t))$] is the basis for the realization of the elementary physical mechanism that provides for the operation of many electronic devices, including the FELs, CRMs, etc. [37,38]. With $x(t)$ being given, the slow dependence $\omega^*(t)$ is immediately determined by the truncated Equation 5.60, i.e.,

$$\omega^*\left(x(t)\right) = \omega - \frac{\varepsilon}{\omega x(t)}\bar{\varphi}_2\left(x(t)\right), \tag{5.62}$$

which makes this approach even more attractive for the researcher.

The idyllic situation does have some complications. In particular, it has been unclear for a long time whether the solutions obtained by the van der Pol method are sufficiently accurate, i.e., to what extent the exact solutions of the initial equation (Equation 5.47) are equivalent to those of the truncated systems (Equations 5.59 and 5.60). This is the essence of the mathematicians' main objection against the van der Pol method [35]. This aspect was much less disturbing to the practical engineers. They have been quite satisfied with the observation that, in most cases, the results obtained by this method had been in fairly good agreement with experimental data. The few rare cases of obvious discrepancies did not influence the favorable attitude. However, the negative experiences accumulated as the range of application of the method expanded. The problems were solved by Krylov and Bogolyubov, and then by Mitropolsky, Zubarev, Andronov, Pontriagin, Tichonov, Volosov, Grebennikov, and others [13–21]. van der Pol's ideas obtained a new mathematically rigorous form and established a foundation for an extended mathematical theory—Bogolyubov's approach being its part. The latter made it possible to develop efficient asymptotic integration algorithms of hierarchic types for a wide spectrum of differential equations.

5.4 Decompression and Compression Operators in the Case of the Averaging Methods*

The main object of the application of hierarchic versions of the averaging methods is the mathematical model that can be described by the following vector equation:

$$\frac{dz}{dt} = Z(z, t, \varepsilon_1, \ldots, \varepsilon_m), \tag{5.63}$$

where $z = \{z_1, z_2, \ldots, z_n\}$, as before, is some vector, $Z = \{Z_1, Z_2, \ldots, Z_n\}$ is the relevant vector function in n-dimensional Euclidean space $\mathbf{R}^n$, $t \in [0,\infty]$, and ε_κ are the hierarchic small parameters. In this section and in Chapter 6, we will discuss particular varieties of this model, where only the influence of neighboring hierarchic levels is really important. In this case (see Chapter 2), the general hierarchic problem (Equation 5.63) can be reduced to successively catching hold of two-level hierarchic systems. As a result, we eventually determine that the basic element of any hierarchic problem of the same type as Equation 5.63 is utilization of algorithms of asymptotic integration of the simplest two-level hierarchic system. Therefore, here and in the next chapter, we will pay attention mainly to the investigation of two-level hierarchic systems of various types. We begin our study with the so-called Bogolyubov's standard system [21].

* From Kulish, V.V., *Hierarchic Methods. Hierarchy and Hierarchical Asymptotic Methods in Electrodynamics*, Vol. 1, Kluwer Academic Publishers, Dordrecht/Boston/London, 2002. With permission.

5.4.1 Bogolyubov's Standard System

Traditionally the particular variety of general systems (Equation 5.63)

$$\frac{dz}{dt} = \varepsilon Z(z,t,\varepsilon),\ z(0) = z_0 \tag{5.64}$$

is referred to as Bogolyubov's standard system. Here, for simplicity's sake, we accept $\varepsilon \equiv \varepsilon_1$. Let us construct the asymptotic algorithm for its integration based on averaging operator Equation 5.13. This means that the function $Z(z,t,\varepsilon)$ for the first hierarchic level can be written in the form

$$\bar{Z}(z,\varepsilon) = M_t Z(z,t,\varepsilon) = \lim_{T\to\infty}\frac{1}{T}\int_0^T Z(z,t,\varepsilon)dt. \tag{5.65}$$

We assume that the function $Z(z,t,\varepsilon)$ is 2π-periodical on t, i.e.,

$$Z(z,t + 2\pi) \equiv Z(z,t) \tag{5.66}$$

for all $z \in G_n$.

According to the general hierarchic scheme (see Section 5.1), the dynamic equation for the first hierarchic level in this case should have a mathematical structure similar to Equation 5.64

$$\frac{d\bar{z}}{dt} = \varepsilon\bar{Z}(\bar{z},t,\varepsilon),\ \ \bar{z}(0) = z_0. \tag{5.67}$$

We find the connection of the proper variables of the zero and first hierarchic level in the form of the Krylov–Bogolyubov substitution (see also Equation 5.42 and corresponding comments)

$$z = \bar{z} + \varepsilon u(\bar{z},t,\varepsilon). \tag{5.68}$$

The generalized hierarchic equation of the first order (Equation 5.2) for the unknown decompression operator u in this case can be obtained in the form

$$\varepsilon\left(\frac{\partial u}{\partial \bar{z}}, Z_0(\bar{z})\right) + \frac{\partial u}{\partial t} = Z_0(\bar{z} + \varepsilon u, t, \varepsilon) - \bar{Z}(\bar{z},\varepsilon). \tag{5.69}$$

Then we expand the function Z in a Fourier series (see Equation 5.19 and relevant comments)

$$Z(z,t) = \sum_{|k|\geq 0} Z_k(z)\exp\{ikt\}, \tag{5.70}$$

$$\bar{Z}(\bar{z}) = Z_0(\bar{z}). \tag{5.71}$$

Using the expansion Equations 5.70 and 5.71, we rewrite the general hierarchic equation 5.69 in the form

$$\varepsilon\left(\frac{\partial u}{\partial t}, Z_0(\bar{z})\right) + \frac{\partial u}{\partial t} = Z_0(\bar{z}+\varepsilon u) - Z_0(\bar{z}) + \sum_{|k|\geq 1} Z_k(\bar{z}+\varepsilon u)\exp\{ikt\}. \tag{5.72}$$

Furthermore, we accept the following supposition: the function $Z(z,t)$ is considered analytical with respect to z within domain G_n. In this case, we have a possibility to construct a formal solution for $u(\bar{z},t)$ [21].

We will look for the change of variables (Equation 5.68) in the form

$$z = \bar{z} + \sum_{n=1}^{\infty} \varepsilon^n u_n(\bar{z},t). \tag{5.73}$$

Then we substitute Equation 5.73 in the general hierarchic Equation 5.72, equalizing terms of equal order (with respect to the small parameter ε) on the left- and right-hand sides, respectively. As a result, after relevant calculations, we obtain the following system of equations

$$\frac{\partial u_1}{\partial t} = Z(\bar{z},t) - Z_0(\bar{z}) = \sum_{|k|\geq 1} Z_k(\bar{z})\exp\{ikt\};$$

$$\frac{\partial u_2}{\partial t} = \left(\frac{\partial Z(\bar{z},t)}{\partial \bar{z}}, u_1\right) - \left(\frac{\partial u_1}{\partial \bar{z}}, Z_0\right);$$

$$\frac{\partial u_3}{\partial t} = \left(\frac{\partial Z(\bar{z},t)}{\partial \bar{z}}, u_2\right) + \frac{1}{2!}\left(\left(\frac{\partial^2 Z(\bar{z},t)}{\partial \bar{z}^2}, u_1\right), u_1\right) - \left(\frac{\partial u_2}{\partial \bar{z}}, Z_0\right); \tag{5.74}$$

A very important feature of System 5.74 is the possibility of obtaining its approximate analytical solutions

$$u_1(\bar{z},t) = \sum_{|k|\geq 1} \frac{1}{ik} Z(\bar{z})\exp\{ikt\} + \varphi_1(\bar{z});$$

$$u_2(\bar{z},t) = \sum_{|k|\geq 1} \frac{1}{ik}\left(\frac{\partial Z_{-k}(\bar{z})}{\partial \bar{z}}, Z_k(\bar{z})\right)t + \sum_{|k|\geq 1}\sum_{|s|\geq 1} \frac{1}{i^2(k+s)}\left(\frac{\partial Z_k(\bar{z})}{\partial \bar{z}}, Z_s(\bar{z})\right)\exp\left\{i(k+s)t\right\}$$

$$-\left(\sum_{|k|\geq 1} \frac{1}{i^2 k^2}\frac{\partial Z_k(\bar{z})}{\partial \bar{z}}\exp\{ikt\}, Z_0(\bar{z}) + \varphi_1(\bar{z})t\right) + \varphi_2(\bar{z}). \tag{5.75}$$

For obtaining of total solution of Equation 5.64 (zero hierarchic level) we must

1. Substitute solutions for Equation 5.75 into Equation 5.73.
2. Find relevant solutions for the dynamic equation of the first hierarchic level (Equation 5.67) and substitute them into Equation 5.73.
3. Coordinate initial conditions.

The last item means that we should choose the arbitrary functions $\varphi_1(\bar{z}_0)$, $\varphi_2(\bar{z}_2)$, … in such a way that the equation is satisfied

$$\bar{z}_0 + \varepsilon u_1(\bar{z}_0, 0) + \varepsilon^2 u_2(\bar{z}_0, 0) + \ldots = z_0. \tag{5.76}$$

Thus, it is shown that in the view of the general hierarchic approach (Sections 5.1 through 5.3), the Bogolyubov method can indeed be regarded as a particular variety of hierarchic method. In principle, we might also have other algorithms for the averaging method [21]. Such an algorithm can be realized when the idea of the generating equation (see Equation 5.48 and comments) is used. Such equations are also referred to as equations of comparison [21]. In this case, the compression operator can be chosen in the following manner

$$\hat{M}_t Z(z,t,\varepsilon) = \bar{Z}(\bar{z},\varepsilon)\big|_{\varepsilon=0} = \bar{Z}(\bar{z},0). \tag{5.77}$$

Correspondingly, the dynamic equation of the first hierarchic level can be written as

$$\frac{d\bar{z}}{dt} = \varepsilon \bar{Z}(\bar{z},0), \quad \bar{z}(0) = \bar{z}_0. \tag{5.78}$$

The calculation scheme is the same. However, it should be mentioned that, in general, the form of the asymptotic solution here turns out to be somewhat different from that of the solutions of Equations 5.73 through 5.75 [21].

5.4.2 Compression Operator and the Problem of Secular Terms

An unpleasant peculiarity of the above-mentioned versions of the calculation scheme of the average method is the presence of so-called secular terms of the type $\varphi_s(\bar{z})t$ in asymptotic solutions (Equation 5.75). This problem can be solved (completely or particularly) by using the modernized version of the calculation scheme. Let us choose the following form of the compression operator

$$\hat{M}_t Z(z,t,\varepsilon) = Z_0(\bar{z}) + \varepsilon A_2(\bar{z}) + \varepsilon^2 A_3(\bar{z}) + \ldots, \tag{5.79}$$

where A_2, A_3, … are some unknown (at this stage of the calculation procedure) functions, and $Z_0(\bar{z})$, as before, is the zero term of a Fourier series of the type like Equation 5.70. Then, we follow the above-described calculation scheme. As a result, instead of the system of Equation 5.75, we obtain the modernized system

$$\frac{\partial u_1}{\partial t} = Z(\overline{z},t) - Z_0(\overline{z});$$

$$\frac{\partial u_2}{\partial t} = \left(\frac{\partial Z(\overline{z},t)}{\partial \overline{z}}, u_1\right) - \left(\frac{\partial u_1}{\partial \overline{z}}, Z_0\right) - A_2(\overline{z});$$

$$\frac{\partial u_3}{\partial t} = \left(\frac{\partial Z(\overline{z},t)}{\partial \overline{z}}, u_2\right) + \frac{1}{2!}\left(\left(\frac{\partial^2 Z(\overline{z},t)}{\partial \overline{z}^2}, u_1\right), u_1\right) - \left(\frac{\partial u_2}{\partial \overline{z}}, Z_0\right) - A_3(\overline{z}). \tag{5.80}$$

The next stage of the modernized version of the calculation scheme is the determination of the unknown function $A_s(\overline{z})$. This procedure follows from the comparison of the systems of Equations 5.74 and 5.80, respectively. It is understood that the function $A_s(\overline{z})$ can be derived, for example, from the condition of the elimination of the relevant secular terms by corresponding addend-functions $A_s(\overline{z})$. As a result, we obtain the following definitions for these functions

$$A_2(\overline{z}) = \lim_{T\to\infty}\frac{1}{T}\int_0^T\left[\left(\frac{\partial Z(\overline{z},t)}{\partial \overline{z}}, u_1\right) - \left(\frac{\partial u_1}{\partial \overline{z}}, Z_0\right)\right]dt;$$

$$A_3(\overline{z}) = \lim_{T\to\infty}\frac{1}{T}\int_0^T\left[\left(\frac{\partial Z(\overline{z},t)}{\partial \overline{z}}, u_2\right) + \frac{1}{2!}\left(\left(\frac{\partial^2 Z(\overline{z},t)}{\partial \overline{z}^2}, u_1\right), u_1\right) - \left(\frac{\partial u_2}{\partial \overline{z}}, Z_0\right)\right]dt. \tag{5.81}$$

We will also use the above-described method of elimination for the secular terms in other versions of the averaging method.

5.5 Decompression Operator in the Case of Systems with Slow and Fast Variables*

5.5.1 General Case of Systems with Slow and Fast Variables

The Bogolyubov's standard system like Equation 5.64 represents the simplest object for application of the averaging methods. Historically, such systems have been the first full-fledged realizations of hierarchic ideas [13,14]. Later, a few other more general versions of such realizations were developed [15–21]. Let us illustrate one of these more general versions with an example of the so-called system with slow and fast variables.

* From Kulish, V.V., *Hierarchic Methods. Hierarchy and Hierarchical Asymptotic Methods in Electrodynamics*, Vol. 1, Kluwer Academic Publishers, Dordrecht/Boston/London, 2002. With permission.

Bogolyubov's standard system (Equation 5.64) can be regarded as a particular case of the more general system

$$\frac{dz}{dt} = Z(z,t,\varepsilon). \tag{5.82}$$

Indeed, assuming

$$Z(z,t,\varepsilon) = \varepsilon Z'(z,t,\varepsilon), \tag{5.83}$$

we easily come to the standard form (Equation 5.64). Thus, as is readily seen, it is so if the function $Z(z,t,\varepsilon)$ is a slowly varying function in time t.

But that particular case is not the only possible scenario. Let us discuss another possible case, namely, when different velocities of change take place for the components of different groups of the vector z. In other words, we might say that its specific characteristic is the presence of an evidently expressed dynamic hierarchy of the initial variables with respect to the velocities of their change. For an illustration of this idea, we consider a two-level hierarchy.

Let us represent the vector z in the form:

$$z = \{x,y\}, \tag{5.84}$$

where x represents the slow part of the vector z and y represents the fast part. Such a standard system can be written in the form

$$\frac{dx}{dt} = \varepsilon X(x,y,t,\varepsilon);$$

$$\frac{dy}{dt} = \omega(x,y,t) + \varepsilon Y(x,y,t,\varepsilon), \tag{5.85}$$

where x, X are the m-dimension vectors, and y, Y, and ω are the n-dimension vectors, $t \in R_1$. The vectors y, Y, and ω are determined within some $m + n + 2$-dimensional space

$$G_{m+n+2} = \left\{(x,y,t,\varepsilon):\ x \in P_m,\ y = G_n,\ t \in R_1,\ \varepsilon \in [0,\hat{\varepsilon}]\right\}. \tag{5.86}$$

The vector ω comprises, as components, the scalar cyclic frequencies (angular velocities of rotation) $\omega_1,\ldots,\omega_j,\ldots,\omega_n$. Hence, we can say that Equation 5.85 describes a multifrequency two-level hierarchic system with slow and fast variables. In this case, the vector x is referred to as the vector of the slow variables, and the vector y is the vector of the fast variables.

It is not difficult to show that Bogolyubov's standard system (Equation 5.64), in itself, is the simplest particular case of the system with slow and fast rotating phases. Indeed, we rewrite Equation 5.64 in the parametric form (with respect to the parameter $y = t$ and for $z \equiv x$, $Z \equiv X$) to obtain

$$\frac{dx}{dt} = \varepsilon X(x,y,t,\varepsilon);$$

$$\frac{dy}{dt} = 1. \tag{5.87}$$

It is obvious that the standard system can be considered to be equivalent to the standard form with slow and fast phases (Equation 5.85) in the following particular case:

$$X(x,y,t,\varepsilon) \rightarrow X(x,t,\varepsilon);$$

$$\omega(x,y,t) = 1;$$

$$Y(x,y,t,\varepsilon) = 0. \tag{5.88}$$

The general algorithm for obtaining analytical solutions of this type of system is not known today [21]. Therefore, in practice, only particular cases are used as a rule. The systems with fast rotating phases represent one such convenient particular case. Owing to their specific features, only these systems are found to be most interesting practically.

5.5.2 Two-Level Systems with Fast Rotating Phases

The particular case of systems with slow variables and fast phases like Equation 5.85 is referred to as a system with fast rotating phases when

$$X(x,y,t,\varepsilon) = X(x,y,\varepsilon);$$

$$\omega(x,y,t) = \omega(x);$$

$$X(x,y,t,\varepsilon) = Y(x,y,\varepsilon). \tag{5.89}$$

In this case the fast variable y is called the fast rotating phase, $\psi \equiv y$.

Thus, the two-level multifrequency hierarchic standard system with a fast rotating vector phase can be written as

$$\frac{dx}{dt} = \varepsilon X(x,\psi,\varepsilon);$$

$$\frac{d\psi}{dt} = \omega(x) + \varepsilon Y(x,\psi,\varepsilon). \tag{5.90}$$

Another version of the form for writing the hierarchic standard system with fast rotating phases will also be used further

$$\frac{dx}{dt} = X(x,\psi,\xi);$$

$$\frac{d\psi}{dt} = \xi\omega(x) + Y(x,\psi,\xi), \tag{5.91}$$

where $\xi = 1/\varepsilon$ is the large-scale parameter of the problem. Various algorithms of asymptotic integration of systems similar to Equations 5.90 and 5.91 are set forth in Chapter 6.

References

1. Khapaev, M.M. 1988. *Asymptotic methods and equilibrium in theory of nonlinear oscillations*. Moscow: Vysshaja Shkola.
2. Korn, G.A., and Korn, T.W. 1961. Mathematical handbook for scientists and engineers. NY: McGraw-Hill.
3. Kulish, V.V. 1991. Nonlinear self-consistent theory of free electron lasers. Method of investigation. *Ukrainian Physical Journal* 36(9):1318–1325.
4. Kulish, V.V., and Lysenko, A.V. 1993. Method of averaged kinetic equation and its use in the nonlinear problems of plasma electrodynamics. *Fizika Plazmy* (*Sov. Plasma Physics*) 19(2):216–227.
5. Kulish, V.V., Kuleshov, S.A., and Lysenko, A.V. 1993. Nonlinear self-consistent theory of superheterodyne and free electron lasers. *The International Journal of Infrared and Millimeter Waves* 14(3):451–568.
6. Kulish, V.V. 1997. Hierarchic oscillations and averaging methods in nonlinear problems of relativistic electronics. *The International Journal of Infrared and Millimeter Waves* 18(5):1053–1117.
7. Kulish, V.V. 1997. Hierarchic approach to nonlinear problems of electrodynamics. *Visnyk Sumskogo Derzshavnogo Universytetu* 1(7):3–11.
8. Kulish, V.V., Kosel P.B., and Kailyuk, A.G. 1998. New acceleration principle of charged particles for electronic applications. Hierarchic description. *The International Journal of Infrared and Millimeter Waves* 19(1):3–93.
9. Kulish, V.V. 1998. Hierarchic method and its application peculiarities in nonlinear problems of relativistic electrodynamics. General theory. *Ukrainian Physical Journal* 43:483–499.
10. Kulish, V.V., Kosel, P.B., Krutko, O.B., and Gubanov, I.V. 1998. Hierarchic method and its application peculiarities in nonlinear problems of relativistic electrodynamics. Theory of EH-Ubitron accelerator of charged particles. *Ukrainian Physical Journal* 43(2):133–138.
11. Kulish, V.V., and Kayliuk, A.G. 1998. Hierarchic method and its application peculiarities in nonlinear problems of relativistic electrodynamics. Single-particle model of cyclotron-resonant maser. *Ukrainian Physical Journal* 43(4):398–402.
12. Kulish, V.V. 1998. Hierarchic theory of oscillations and waves and its application to nonlinear problems of relativistic electrodynamics. In *Causality and locality in modern physics*. Dordrecht, the Netherlands: Kluwer Academic Publishers.
13. Krylov, N.M., and Bogolyubov, N.N. 1934. *Application of methods of nonlinear mechanics to the theory of stationary oscillations*. Kiev: Ukrainian Academy of Science Publishers.
14. Krylov, N.M., and Bogolyubov, N.N. 1947. *Introduction to nonlinear mechanics*. Princeton, NJ: Princeton University Press.
15. Bogolubov, N.N., and Mitropolskii, Ju.A. 1963. *Methods of averaging in the theory of nonlinear oscillations*. Moscow: USSR Academy of Sciences.
16. Bogolubov, N.N., and Zubarev, D.N. 1955. Asymptotic approximation method for the system with rotating phases and its application to the motion of charged particles in magnetic fields. *Ukrain. Mathem. Zhurn.* (*Ukrainian Mathematics Journal*) 7:201–221.
17. Arnold, V.I. 1965. Applicability conditions and error estimates for the averaged method applied to the resonant systems. *Dok. Akad. Nauk. SSSR* [*Sov. Phys.-Doklady*] 161(1):9.
18. Andronov, A.A., Vitt, A.A., and Khaikin, S.E. 1959. *Theory of oscillations*. Moscow: Fizmatgiz.
19. Mitropolsky, Y.A., and Moseenkov, B.I. 1968. *Lectures on the asymptotic applications to the solution of partial differential equations*. Kiev: Mathematics Institute.
20. Moiseev, N.N. 1981. *Asymptotic methods of nonlinear mechanics*. Moscow: Nauka.

21. Grebennikov, E.A. 1986. *Averaging method in applied problems*. Moscow: Nauka.
22. Dodd, R.K., Eilbeck, J.C., Gibbon, J.D., and Morris, H.C. 1982. *Solutions and nonlinear wave equations*. London: Academic Press.
23. Leontovich, M.A. 1944. To the problem about propagation of electromagnetic waves in the Earth atmosphere. *Izv. Akad. Nauk SSSR, ser. Fiz.*, [*Bull. Acad. Sci. USSR, Phys. Ser.*] 8:6–20.
24. van der Pol, B. 1935. *Nonlinear theory of electric oscillations. Russian translation*. Moscow: Swiazizdat.
25. Kohmanski, S.S., and Kulish, V.V. 1985. To the nonlinear theory of free electron lasers with multi-frequency pumping. *Acta Phys. Polonica* A68(5):741–748.
26. Gaponov, A.V., Ostrovskii, L.A., and Rabinovich, M.I. 1970. One-dimensional waves in nonlinear disperse media. *Izv. Vysh. Uchebn.*, Ser. Radiofizika (*Sov. Radiophysics*) 13(2):169–213.
27. Rabinovich, M.I., and Talanov, V.I. 1972. *Four lectures on the theory of nonlinear waves and wave interactions*. Leningrad: Leningrad University Publishers.
28. Rabinovich M.I. 1971. On the asymptotic in the theory of distributed system oscillations. *Dok. Akad. Nauk. SSSR*, ser. Fiz. [Sov. Phys.-Doklady], 191:253–1268.
29. Kohmanski, S.S., and Kulish, V.V. 1985. To the nonlinear theory of free electron lasers. *Acta Physica Polonica* A68(5):749–756.
30. Kohmanski, S.S., and Kulish, V.V. 1985. To the nonlinear theory of free electron lasers with multi-frequency pumping. *Acta Phys. Polonica* A68(5):741–748.
31. Kohmanski, S.S., and Kulish, V.V. 1985. Parametric resonance interaction of electron in the field of electromagnetic waves and longitudinal magnetic field. *Acta Phys. Polonica* A68:6525–6736.
32. Kulish, V.V., Dzedolik, I.V., and Kudinov, M.A. 1985. *Motion of relativistic electrons in periodically reversed electromagnetic field, Part I*. Deposited in Ukrainian Scientific Research Institute of Technical Information on 23.07.85, Kiev Vol. 1490, Uk-85, 110 pages.
33. Kulish, V.V., and Dzedolik, I.V. 1985. *Motion of relativistic electrons in periodically reversed electromagnetic field, Part II*. Deposited in Ukrainian Scientific Research Institute of Technical Information on 20.09.85, Kiev. Vol. 2257, Uk-85, 54 pages.
34. Jakovlev, V.S., Kulish, V.V., Dzedolik, I.V., Motina, V.G., and Kohmanski, S.S. 1983. Generation of energy by relativistic electrons moving in the field of two electromagnetic waves in presence of longitudinal magnetic field. *PrePrint of Institute of Electrodynamics Academy of Sciences of Ukraine* 321, 41 pages.
35. Mandelshtam, L.I., and Papaleksi, N.D. 1934. On substantiation of an approximate method of solving differential equations. *Zh.Eksp.Teor.Fiz* [*Sov. Phys.-JETP*] 2:220–234.
36. Kulish, V.V. 1998. *Methods of averaging in non-linear problems of relativistic electrodynamics*. Atlanta: World Scientific Publishers.
37. Kulish, V.V. 2002. *Hierarchy and hierarchic asymptotic methods in electrodynamics*, Vol. 1 of *Hierarchic methods*. Dordrecht: Kluwer Academic Publishers.
38. Kulish, V.V. 2002. *Undulative Electrodynamic Systems*, Vol. 2 of *Hierarchic methods*. Dordrecht: Kluwer Academic Publishers.

6

Hierarchic Systems with Fast Rotating Phases

The hierarchic systems with fast rotating phases are the most popular among asymptotic hierarchic methods in different branches of applied physics for two reasons. First, it is because the physical models that can be described by such equations are in widespread use. Second, it is because—as is mentioned in Chapter 5—analytical algorithms of asymptotic integration of this type of system can be constructed without excessive difficulty. Taking this into consideration, we will pay special attention to this in later chapters when studying systems with fast rotating phases.

6.1 General Approach*

6.1.1 Formulation of the Hierarchic Oscillation Problem

We begin this chapter with a discussion of general peculiarities of the basic ideas of the theory of hierarchic oscillations in some typical practical situations. We will do this by using examples of real electrodynamic problems and taking an interest in the mathematical part of the problems. These problems can often be reduced to the asymptotic integration of just the hierarchic systems with fast rotating phases.

We start with a discussion of some general features of the hierarchic version of the classical single-particle electrodynamic problem. It is a problem of periodical (or quasiperiodical) motion of a charged particle in an electromagnetic field.

We accept that the acting electromagnetic field consists of wave and nonwave parts. The nonwave part represents slowly varying, nonperiodic quasistationary and quasihomogeneous fields. Thus, in the considered case, the arrangement of the electric and magnetic fields that act on an electron can be represented in the form

$$\vec{E} = \frac{1}{2}\sum_{l}\left[\vec{E}_l \exp\left\{im_l p_l\right\} + c.c.\right] + \vec{E}_0;$$

$$\vec{B} = \frac{1}{2}\sum_{j}\left[\vec{B}_j \exp\left\{im_j p_j\right\} + c.c.\right] + \vec{B}_0, \tag{6.1}$$

where $\vec{E}$ and $\vec{B}$ are the electric field strength and the magnetic field induction, respectively. The values $\vec{E}_l = \vec{E}_l\left(\vec{r},t\right)$ and $\vec{B}_j = \vec{B}_j\left(\vec{r},t\right)$ are slowly varying complex amplitudes of electric and magnetic induction fields of partial waves (proper and stimulated), $\vec{E}_0 = \vec{E}_0\left(\vec{r},t\right)$, $\vec{B}_0 = \vec{B}_0\left(\vec{r},t\right)$ are slowly varying quasistationary and quasihomogeneous electric and

* From Kulish, V.V., et al., *Int. J. Infrared Millimeter Waves*, 19, 49–71 and 74–83, 1998. With permission.

magnetic induction fields, and m_l, m_j are the wave harmonic numbers (m_j, m_l = ±1, ±2, …). The quantities

$$p_q = \omega_q t - \vec{k}_q \vec{r}, q = j, l \tag{6.2}$$

are phases, ω_q is the angular frequency, k_q is the wave vector, t is the laboratory time, and $\vec{r}$ is the position vector. In the case of the proper waves, we have dispersion relations between ω_q and k_q of $\left(\vec{k}_q = \vec{k}_q\left(\omega_q\right)\right)$. These concepts are discussed in detail in Chapters 3 and 4.

In terms of general hierarchic theory (see Chapters 2 and 4), assigning Form 6.2 to the fields determines the form of the compression operator $M^{(\kappa)}$ (see Equation 2.10). We will illustrate some practical features of the studied calculation procedures by using examples of various hot electrodynamic systems. As a rule, they are always related to objects such as charged particle beams or plasma fluxes that move within electromagnetic fields. Usually, the single particle and multiparticle (including collective) problems are distinguished.

In this book, we are interested in electron beams of two extreme types: low and high intensity (see Chapter 4) [1–3]. The characteristic feature of low-intensity beams is the relatively low intensity of the Coulomb interaction between the charged particles (rarified plasma beams). In fact, such beams are just streams of nearly independent drifting charged particles. Hence, we have the possibility of studying a separate particle interaction with the fields. The single-particle theory of motion of charged particles in electromagnetic fields appears as a result of such study. In turn, the single-particle problem is the main point of this theory. The particular case of the multiparticle problem can be realized here by simply summing up all the separate single-particle interactions.

In contrast, the particle–particle Coulomb interactions can be rather intense in the high-intensity beams (intensive, including high-current, beam). At sufficiently high densities, the beams show a collective behavior and can be treated as a quasicontinuous flow of drifting charged (or quasineutral) plasmas. Hence, in this case, the charged particle beam should be regarded as a whole electrodynamic object.

It is obvious that due to the above-mentioned differences, the mathematical descriptions of these objects should also be different. Furthermore, in the first four sections of this chapter, we give hierarchic algorithms for the treatment of various single-particle systems. Relevant methods of the asymptotic treatment of high-density beams will be set forth in the last two sections.

The first type of beam motion (rarified plasma beams) can be considered as the motion of an aggregate of individual particles (Lagrange formalism). In this case, the beam motion can be described by the single-particle equations in the Hamiltonian form

$$\frac{d\mathcal{H}_\alpha}{dt} = \frac{\partial \mathcal{H}_\alpha}{\partial t}$$

$$\frac{d\vec{\mathcal{P}}_\alpha}{dt} = -\frac{\partial \mathcal{H}_\alpha}{\partial t}$$

$$\frac{d\vec{r}}{dt} = \frac{\partial \mathcal{H}_\alpha}{\partial \vec{\mathcal{P}}_\alpha}, \tag{6.3}$$

or by the Lorenz equation

$$\frac{d\vec{\beta}_\alpha}{dt} = \frac{q_\alpha}{m_\alpha \gamma_\alpha c}\left\{\vec{E} + \left[\vec{\beta}_\alpha \vec{B}\right] - \vec{\beta}_\alpha\left(\vec{\beta}_\alpha \vec{E}\right)\right\}. \tag{6.4}$$

Here, $\mathcal{H}_\alpha = \sqrt{m_\alpha^2 c^4 + c^2\left(\vec{\mathcal{P}}_\alpha - \frac{q_\alpha}{c}\vec{A}\right)} + q_\alpha$ is the Hamiltonian of a charged particle of the sort α, q_α is its charge ($q_\alpha = -e$ corresponds to electrons, $q_\alpha = +Ze$ for ions, e is the electron charge, and Z is the charge number of the ion), m_α is the rest mass of a particle of sort α, $\vec{A}$ is the vector potential, φ is the scalar potential of the electromagnetic fields [4], $\vec{\beta}_\alpha = \vec{v}_\alpha/c$ is the nondimensional particle velocity (here, $\vec{v}_\alpha$ is the velocity, and c is the light velocity in vacuum), $\vec{\mathcal{P}}_\alpha$ is the canonical momentum, $\vec{r}_\alpha$ is the position vector of particle α, $\vec{E}$ and $\vec{B}$ have been defined already, and $\gamma_\alpha = \left(1-\beta_\alpha^2\right)^{-1/2}$ is the relativistic factor (here $\beta_\alpha = \left|\vec{\beta}_\alpha\right|$).

The hierarchic part of the problem consists of two main stages. The first is the reduction of Equations 6.3 or 6.4 to one of the standard forms. The second is the asymptotic integration of the standardized equations. In this chapter and later, we will study initial sets 6.3 and 6.4, which are reduced to the standard form hierarchic system with fast rotating phases [5–7]. In the general case of the m-level hierarchic oscillation systems, the standard system with fast rotating phases can be written in the following form [5–7]

$$\frac{dx}{dt} = X\left(x, \psi_1, \ldots, \psi_\kappa, \ldots, \psi_m, \xi_1, \ldots, \xi_m\right)$$

$$\frac{d\psi_1}{dt} = \xi_1 \omega_1(x) + Y_1\left(x, \psi_1, \ldots, \psi_\kappa, \ldots, \psi_m, \xi_1, \ldots, \xi_m\right)$$

$$\frac{d\psi_m}{dt} = \xi_m \omega_m(x) + Y_m\left(x, \psi_1, \ldots, \psi_\kappa, \ldots, \psi_m, \xi_1, \ldots, \xi_m\right), \tag{6.5}$$

where scale parameters $\xi_\kappa = 1/\varepsilon_\kappa$ are the relevant terms of the strong hierarchic series

$$\xi_1 >> \xi_2 >> \ldots >> \xi_\kappa >> \xi_m >> 1, \tag{6.6}$$

x is the vector whose components are slowly varying variables only, ψ_κ represents partial vectors of different hierarchies of the vector of fast rotation (revolving) phases ψ, ω_κ represents a slowly varying part of the velocity vector of fast rotating phases, Y_κ represents relevant vector functions, κ is the current number of the hierarchic level, and m is the number of the highest hierarchic level.

In view of the above, the reduction of initial Equations 6.3 and 6.4 to the standard form of System 6.5 means that some hierarchization law (compression operator $M^{(\kappa)}$) is determined. The large hierarchic parameters ξ_κ in Systems 6.5 and 6.6 can be determined by

$$\xi_\kappa \sim \left|\frac{d\psi_{\kappa l}}{dt}\right| \Big/ \left|\frac{dx_q}{dt}\right|, \tag{6.7}$$

where $\psi_{\kappa l}$ is the lth component of vector ψ_κ, and x_q is the qth component of vector x (see Equation 2.6 and related discussion). Here we consider the rate of variation of x_q as not exceeding the rates of variations of other slow variables, and the component $\psi_{\kappa l}$ is the slowest of other fast components of vector ψ_κ. Thus, with the general definition of hierarchic scale parameter (Equation 2.6), with Equation 6.7 we affirm that the variables $\psi_{\kappa l}$ (in the case $\kappa = 1$) and x_q represent two different neighboring hierarchic levels.

6.1.2 Oscillation Phases and Resonances

According to the general hierarchic principles, we should reduce the initial systems (Equations 6.3 or 6.4) to the standard form (System 6.5). The first step is the determination of all elements of the slow vector x and the fast phase vector ψ in System 6.5. In other words, the first step consists of the classification of the total set of variables of Equations 6.3 and 6.4 into the slow and fast ones, respectively.

As a rule, the Hamiltonian $\mathcal{H}$ and the variables $\vec{\mathcal{P}}$, $\vec{\beta}$ (or $\vec{v} = \vec{\beta}c$), and $\vec{r}$ might be considered as slow variables. In contrast, a classification of the Lagrange phases of particle oscillations is more complicated. In this case, we should separately find the set of phases by formation of the component basis of vectors ψ and x. In addition, we must divide the rotating phases of particle oscillations into those associated with the explicit and hidden periods of oscillation (see the definitions in Chapter 3).

The phases of the first group are related with the phases of the particle oscillations in the wave-type fields. For example, they are the particle oscillation phases of the undulated fields of various types, of the fields of electromagnetic waves, or of the fields of proper and stimulated beam waves, and others (see examples in Chapters 3 and 4). The periodicity of these fields is responsible for the periodic character of the electron motion with respect to the same wave phases (Equation 6.2)

$$p_q = \omega_q t - \vec{k}_q \vec{r}. \tag{6.8}$$

The second type of phases is much more complicated. It can be associated with the phases of the periodic particle motion in nonperiodical quasistationary intrinsic beam fields, external focusing, and beam-forming fields (see examples in Chapters 3 and 4). Procedures for separating the hidden periods of the arbitrary functions are known in mathematics as Lantzos's method [8,9] and Grebennikov's method [10], for example. It could be said that there are also some semiempirical approaches based on the qualitative peculiarities of a studied physical picture [6,7]. Generally, however, these procedures are not at all simple and require a relevant practical skill.

We assume that the procedure for finding all particle oscillation phases (hidden and explicit) has been completed. By analyzing the rates of varying particle phases, we find that their total set involves both types of phases: the slow components of the vector x and the fast components of the vector ψ. Some nonlinear combinations of two, three, or more Lagrange fast phases produce the slowly varying functions, too. Each slow phase resulting from this combination action (generally m-fold; see Chapter 3) corresponds to some physical mechanism of the particle resonance. We distinguish the quasilinear and parametric resonance (see relevant definitions in Chapter 3). The quasilinear resonances are characterized by the presence of two-phase (or m-phase) combinations, where only one of the phases is associated with an explicit period of the system. The magnitude of the stimulated wave force acting on the particle is always linearly dependent on the wave field amplitude in the lowest order in some small parameter (see examples in Chapters 3 and 4). Therefore, these resonances could be classified as quasilinear. Examples include cyclotron resonance and various types of synchrotron resonances (see Chapter 3).

The parametric one-particle resonances correspond to cases where all phases forming the slow combination have a wave nature (see Chapter 3). For instance, the parametric resonances of third-, fourth-, and higher orders [11–18,31–33] can be realized in FELs in parametric electronic devices (Adler's lamp, parametric electron-wave lamps) [17,18,31–33]. In general, slow θ_{vg} and fast ψ_{vg} combined phases of coupled-pair parametric or quasilinear resonances can be defined as

$$\theta_{vg} = \frac{m_v n_v}{m_g n_g} p_v - \sigma_{vg} p_g$$

$$\psi_{vg} = \frac{m_v n_v}{m_g n_g} p_v + \sigma_{vg} p_g, \qquad (6.9)$$

where $\sigma_{vg} = \pm 1$ represents the sign functions, m_v and m_g are the numbers of wave field harmonics, and n_v and n_g are the numbers of electron oscillation harmonics in these fields (see Chapter 3). Hence, the lowest order of the relevant amplitudes of the stimulated waves in the case of parametric resonances is quadratic. Therefore, we can classify the parametric resonances as nonlinear.

According to the hierarchic principles set forth in Chapter 2, we define components of the scale parameter tensor according to the reciprocal rate of variation of the fast to slow combined phases (Equation 6.9; compare with Equation 6.7)

$$\xi_{vgrk} \sim \left|\frac{d\psi_{vg}}{dt}\right| \Big/ \left|\frac{d\theta_{vg}}{dt}\right| >> 1. \qquad (6.10)$$

This means that the slow θ_{vg} and fast ψ_{vg} combined phases are variables of different hierarchic levels (see also Chapter 3). We assume the rates of change of the velocities of the phases θ_{vg} are of the same order as the rates of velocities of other slow variables (see Equation 6.7).

Then, by using a similar calculation scheme, we take into account the nonresonant phases of oscillations. The only difference is that, instead of the slow and fast combined phases, we classify the ordinary slow and fast phases. Furthermore, we construct a hierarchic series like Inequality 6.6, which consists of the components of both tensors of scale parameters.

Hierarchic series (Inequality 6.6) are the essential points of the considered theory. Each term of the series corresponds to a single oscillation or a resonance, or a group of oscillations and resonances of an equivalent hierarchy.

6.1.3 Hierarchic Transformations

Because of the accomplished transformations, the initial systems in Equations 6.3 and 6.4 were reduced to the hierarchic standard form of System 6.5. Let us briefly discuss the general algorithm of asymptotic integration of systems like System 6.5.

Within the definition of the κth scale parameter of Equation 6.10, the mathematical meaning of the hierarchic series of Inequality 6.6 is equivalent to the statement that a system has hierarchy over the velocities of the characteristic fast oscillation (rotation) phases of particles, i.e.,

$$\left|\frac{d\psi_{1j}}{dt}\right| >> \left|\frac{d\psi_{2i}}{dt}\right| >> \left|\frac{d\psi_{\kappa l}}{dt}\right| >> \ldots >> \left|\frac{dx_q}{dt}\right|. \tag{6.11}$$

Apart from the scale parameter of the first hierarchic level ξ_1 (the leading term in the hierarchic series of Inequality 6.6, all other fast phases ψ_κ $\kappa > 1$ are relatively slow quantities. This observation is useful for constructing a required hierarchic algorithm of asymptotic integration reducing complex total System 6.5 to an essentially simpler two-level hierarchic form of the same type as Equation 3.108

$$\frac{dx'}{dt} = X'(x', \psi_1, \xi_1)$$

$$\frac{d\psi_1}{dt} = \xi_1\omega_1(x') + Y_1(x', \psi_1, \xi_1), \tag{6.12}$$

where x' is the expanded vector of slow variables [here, all partial fast phases ψ_κ, except the first one ψ_1 ($\kappa = 1$) are considered as its components], and so on. Then we write the Krylov–Bogolyubov substitution for the system with rotating phases through the decompression operator $\hat{U}\{u, v\}$

$$x'_q = \bar{x}'_q + \sum_{n=1}^{\infty} \frac{1}{\xi_1^n} u_q^{(n)}(\bar{x}', \bar{\psi}_1)$$

$$\psi_{1l} = \bar{\psi}_{1l} + \sum_{n=1}^{\infty} \frac{1}{\xi_1^n} v_{1l}^{(n)}(\bar{x}', \bar{\psi}_1), \tag{6.13}$$

where the averaged variables $\bar{x}'_q$ and $\bar{\psi}_{1l}$ (components of relevant vectors $\bar{x}' = \{\bar{x}'_q\}$, $\bar{\psi}_1 = \{\bar{\psi}'_{1l}\}$) can be found from the truncated (shortened) system of equations for the next (first) hierarchic level

$$\frac{d\bar{x}'}{dt} = \sum_{n=1}^{\infty} \frac{1}{\xi_1^n} A^{(n)}(\bar{x}')$$

$$\frac{d\bar{\psi}'_1}{dt} = \xi_1\omega_1(\bar{x}') + \sum_{n=1}^{\infty} \frac{1}{\xi_1^n} B_1^{(n)}(\bar{x}'), \tag{6.14}$$

which are obtained by using the given compression operator $M^{(1)}$. Here $u_q^{(n)}$, $v_{1l}^{(n)}$, $A^{(n)} = \{A_q^{(n)}\}$, $B_1^{(n)} = \{B_{1l}^{(n)}\}$, and $\omega_1(\bar{x}')$ are functions calculated below. As we will show, Equations 6.12 and 6.14 are equivalent with respect to general mathematical structure. This means that the hierarchic resemblance principle holds. Analogously, the principle of information compression is also satisfied, because Equation 6.14 describes the simplest physical situation; nonaveraged fast vector phases ψ_1 are absent in Equation 6.14. By virtue of the same cause, the hierarchic analogy of the second thermodynamic principle also holds here.

At the next stage of hierarchic analysis, we use the hierarchic features of the system in Equation 6.14. Namely, we separate out the group of fast scalar phases from the temporary components of slow variables of the vector $\bar{x}'$. These phases define the dynamics of the vector of partial fast phases ψ_2 (see Inequality 6.11). The other components of vector $\bar{x}'$ and the components of averaged phase $\bar{\psi}_{1l}$ serve here as a basis for the formation of a vector of slow variables of the next (second) hierarchy

$$\frac{dx''}{dt} = X''(x'', \bar{\psi}_2, \xi_2)$$

$$\frac{d\bar{\psi}_2}{dt} = \xi_2\omega_2(x'') + Y''(x'', \bar{\psi}_2, \xi_2)$$

$$x''_{q'} = \bar{x}''_{q'} + \sum_{n=1}^{\infty} \frac{1}{\xi_2^n} \bar{\bar{u}}_{q'}^{(n)}(\bar{x}'', \bar{\psi}_2)$$

$$\bar{\psi}_{2l'} = \bar{\bar{\psi}}_{2l'} + \sum_{n=1}^{\infty} \frac{1}{\xi_2^n} \bar{\bar{v}}_{2l'}^{(n)}(\bar{x}'', \bar{\psi}_{2l'}), \tag{6.15}$$

where $x'' = \{x''_{q'}\}$ is the part of the vector $\bar{x}' = \{\bar{x}'_q\}$ where the vector of fast phase $\bar{\psi}_2 = \{\psi'_{2l}\}$ is separated. Therefore, the meaning of new functions X'', ω_2, Y''_2, and similar others are evident. The algorithm for their calculation will be described later.

We will treat the system of Equation 6.15 in terms of a hierarchic approach. We separate out a new group of variables from the vector $\bar{x}''$ (their dynamics is determined by the vector phase $\bar{\psi}_3$, which we consider as a new fast rotating phase). The procedure repeats cyclically until we attain the level of competence, which as a rule is determined by characteristic features of the considered system (for the concept of the competence level, see Figure 2.15 and related discussion). The limit case of the competence level is the last terms in the hierarchic series of Systems 6.6 and 6.11. In other words, we repeat the above-mentioned procedure as long as we take into account all the partial vector phases ψ_κ that form a complete vector of the total fast phases ψ. As an eventual result, we obtain a set of m-fold averaged equations that most often can be solved relatively easily. This property is the main advantage of this version of hierarchic method, making it very attractive for practice. Approximate nonaveraged solutions of initial System 6.5 are obtained by the successive use of an inverse transformation formula like Equation 6.13.

6.2 Decompression Operator in the Simplest Case of One Scalar Phase*

6.2.1 Formulation of the Problem

The procedure of the asymptotic integration of a general multilevel system (System 6.5) can be essentially simplified in some special cases. In particular, the integration of systems (System 6.5) can be reduced to a hierarchic chain of much more simple integration procedures of two-level systems like Equations 6.12 and 6.14. Let us begin the discussion of this type of algorithms with the simplest case of one scalar fast rotating phase $\psi \equiv \psi_1$.

We rewrite Equation 6.12 in a form with a small parameter $\varepsilon = 1/\xi \ll 1$ ($\xi \equiv \xi_1$; see Equation 3.107)

$$\frac{dx}{dt} = \varepsilon X(x, \psi, \varepsilon);$$

$$\frac{d\psi}{dt} = \omega(x) + \varepsilon Y(x, \psi, \varepsilon), \tag{6.16}$$

where $\omega \equiv \omega_1$, $Y \equiv Y_1$, and we neglect the sign of prime. It is obvious that for the scalar phase ψ, the function Y is scalar, too.

Without loss of generality, we put the period T to be equal to 2π. Moreover, we assume that functions x, Y, ω, are differentiable with respect to the phase ψ as many times as is the number of the terms retained in the asymptotic series expansion.

We follow with the calculative scheme discussed in Chapter 5. This means that the problem is to find the substitution for a form like Equation 5.42, which will make it possible to separate the slow and fast variables. Taking into consideration the above-mentioned problem of secular terms (see Chapter 5 for details), let us write the Krylov–Bogolyubov substitutions in the form of the asymptotic series

$$x = \bar{x} + \sum_{i=1}^{\infty} \varepsilon^i u_i(\bar{x}, \bar{\psi}) = \bar{x} + \varepsilon u(\bar{x}, \bar{\psi}, \varepsilon);$$

$$\psi = \bar{\psi} + \sum_{i=1}^{\infty} \varepsilon^i v_i(\bar{x}, \bar{\psi}) = \bar{\psi} + \varepsilon v(\bar{x}, \bar{\psi}, \varepsilon), \tag{6.17}$$

with the unknown components of the decompression operators u_i and v_i to be defined further. We also represent the equation for the first hierarchic level in the form of the asymptotic series

$$M^{(1)} \frac{dx}{dt} = \frac{d\bar{x}}{dt} = \sum_{i=1}^{\infty} \varepsilon^i A_i(\bar{x}) = \varepsilon A(\bar{x}, \varepsilon);$$

* From Kulish, V.V., *Hierarchic Methods. Hierarchy and Hierarchic Asymptotic Methods in Electrodynamics*. Kluwer Academic Publishers, Dordrecht/Boston/London. 2002. With permission.

$$M^{(1)}\frac{d\psi}{dt}=\frac{d\bar{\psi}}{dt}=\omega(\bar{x})+\sum_{i=1}^{\infty}\varepsilon^{i}B_i(\bar{x})=\omega^{*}(\bar{x},\varepsilon), \tag{6.18}$$

where $M^{(1)}$ is the compression operator.

The coefficients A_i and B_i are unknown; they will be found later along with the components of the decompression operators u_i and v_i. In order to specify the problem, we impose additional restrictions on the unknown functions u_i and v_i. Namely, we assume them to be bounded functions of $\bar{\psi}$ for $\bar{\psi}\to\infty$, which is the similar assumption we accepted before (with respect to time t only; see Chapter 5) for the solution of the secular term problem.

6.2.2 Compression and Decompression Operators

As shown here and in Chapter 5, the problem of determining the compression and decompression operators $M^{(\kappa)},\hat{U}^{(\kappa)}$ in our case is reduced to the problem of determining the functions A_i, B_i, u_i, and v_i. Let us describe the relevant versions of the calculation schemes in the simplest case.

In a manner described in Chapter 5, we differentiate the representation (Equation 6.17) with respect to t and use the system in Equations 6.16 and 6.18. Moreover, we note that

$$\frac{du_i}{dt}=\frac{\partial u_i}{\partial t}\frac{d\bar{x}}{dt}+\frac{\partial u_i}{\partial t}\frac{d\bar{\psi}}{dt}=\frac{\partial u_i}{\partial\bar{x}}\sum_{i=1}^{\infty}\varepsilon^{i}A_i(\bar{x})+\frac{\partial u_i}{dt}\sum_{i=1}^{\infty}\left(\omega(\bar{x})+\sum_{i=1}^{\infty}\varepsilon^{i}B_i(\bar{x})\right). \tag{6.19}$$

After some straightforward algebra, the relevant generalized hierarchic equation of the first order of the type given in Equation 5.72) reduces to the expansion

$$\begin{aligned}&\varepsilon A_1(\bar{x})+\varepsilon^2A_2(\bar{x})+\ldots+\varepsilon\frac{\partial u_1}{\partial\bar{x}}\left(\varepsilon A_1(\bar{x})+\ldots\right)+\varepsilon\frac{\partial u_1}{\partial\bar{\psi}}\left(\omega(\bar{x})+\varepsilon B_1(\bar{x})+\ldots\right)\\&\quad+\varepsilon^2\frac{\partial u_2}{\partial\bar{\psi}}\left(\varepsilon A_1(\bar{x})+\ldots\right)+\varepsilon^2\frac{\partial u_2}{\partial\bar{\psi}}\left(\omega(\bar{x})+\varepsilon B_1(\bar{x})+\ldots\right)\qquad;\\&\quad=\varepsilon X\left(\bar{x}+\varepsilon u_1+\ldots,\bar{\psi}+\varepsilon v_1+\ldots,\varepsilon\right)\end{aligned}$$

$$\begin{aligned}&\omega(\bar{x})+\varepsilon B_1(\bar{x})+\varepsilon^2B_2(\bar{x})\\&\quad+\ldots+\varepsilon\frac{\partial v_1}{\partial\bar{x}}\left(\varepsilon A_1+\ldots\right)+\varepsilon\frac{\partial v_1}{\partial\bar{\psi}}\left(\omega(\bar{x})+\varepsilon B_1(\bar{x})+\ldots\right)\\&\quad+\varepsilon^2\frac{\partial v_2}{\partial\bar{x}}\left(\varepsilon A_1+\ldots\right)+\varepsilon^2\frac{\partial v_2}{\partial\bar{\psi}}\left(\omega(\bar{x})+\varepsilon B_1(\bar{x})+\ldots\right)=\omega\left(\bar{x}+\varepsilon u_1(\bar{x},\bar{\psi})+\ldots\right).\\&\quad+\varepsilon Y\left(\bar{x}+\varepsilon u_1+\ldots,\bar{\psi}+\varepsilon v_1+\ldots\right)\end{aligned} \tag{6.20}$$

We equate the coefficients of the equal-order terms with respect to ε on the left- and right-hand sides of Equation 6.20 to obtain a system of interconnected equations that determine the required functions, i.e.,

$$\frac{\partial u_1}{\partial \bar{\psi}}\omega(\bar{x}) = X(\bar{x},\bar{\psi},0) - A_1(\bar{x}) \equiv g_1(\bar{x},\bar{\psi}) - A_1(\bar{x});$$

$$\frac{\partial v_1}{\partial \bar{\psi}}\omega(\bar{x}) = Y(\bar{x},\bar{\psi},0) + D_1(u_1) - B_1(\bar{x}) \equiv h_1(\bar{x},\bar{\psi}) - B_1(\bar{x});$$

$$\frac{\partial u_2}{\partial \bar{\psi}}\omega(\bar{x}) = \frac{\partial X}{\partial \bar{x}}u_1 + \frac{\partial X}{\partial \bar{\psi}}v_1 + \frac{\partial X}{\partial \varepsilon} - \frac{\partial u_1}{\partial \bar{x}}A_1 - \frac{\partial u_1}{\partial \bar{\psi}}B_1 - A_2(\bar{x})$$

$$\equiv g_2(\bar{x},\bar{\psi}) - A_2(\bar{x});$$

$$\frac{\partial v_2}{\partial \bar{\psi}}\omega(\bar{x}) = \frac{\partial Y}{\partial \bar{x}}u_1 + \frac{\partial Y}{\partial \bar{\psi}}v_1 + \frac{\partial Y}{\partial \varepsilon} - \frac{\partial v_1}{\partial \bar{\psi}}B_1 - \frac{\partial v_1}{\partial \bar{x}}A_1 +$$

$$+D_1(u_1) + D_2(u_1) - B_2(\bar{x}) \equiv h_2(\bar{x},\bar{\psi}) - B_2(\bar{x}). \tag{6.21}$$

Here,

$$D_1(u) = \sum_j \frac{\partial}{\partial \bar{x}^{(j)}}u^{(j)};\ D_2(u) = \frac{1}{2}\sum_{j,k}\frac{\partial^2 \omega}{\partial \bar{x}^{(k)}\partial \bar{x}^{(j)}}u^{(j)}u^{(k)},$$

the functions D_1 and D_2 are obtained by expanding $\omega(\bar{x})$ in a Taylor series with respect to ε; the components of the relevant vectors $\bar{x}^{(j)}$, $u_k^{(j)}$, $A_k^{(j)}$, and all derivatives are calculated for $\varepsilon = 0$. It may be easily verified that the system (Equation 6.21) generally may be rewritten as

$$\frac{d\bar{\psi}_1'}{dt} = \xi_1\omega_1(\bar{x}') + \sum_{n=1}^{\infty}\frac{1}{\xi_1^n}B_1^{(n)}(\bar{x}');$$

$$\omega(\bar{x})\frac{\partial v_k}{\partial \bar{\psi}} = h_k(\bar{x},\bar{\psi}) - B_k(\bar{x}), \tag{6.22}$$

where, as before, $\xi_1 = 1/\varepsilon$.

We shall find the unknown functions u_k and v_k by means of successive integration of the system of equation 6.21. Each next equation is solved by taking into account the solution of the previous one, but not vice versa. A complicating point is that Equations 6.21 and 6.22 contain the unknown functions $A_k(\bar{x})$ and B_k. This difficulty may be overcome because both functions u_k and v_k are bounded for $\bar{\psi} \to \infty$.

So, let us consider Equation 6.21. We integrate it over $\bar{\psi}$ (bearing in mind that the derivative with respect to $\bar{\psi}$ is partial) to find the formal solution in quadratures to be given by

$$u_1(\bar{x},\bar{\psi}) = \frac{1}{\omega(\bar{x})}\int_{\bar{\psi}_0}^{\bar{\psi}} \{g_1(\bar{x},\bar{\psi}) - A_1(\bar{x})\} d\bar{\psi} + \varphi_1^{\bullet}(\bar{x}), \tag{6.23}$$

where $\varphi_1^{\bullet}(\bar{x})$ is an unknown function (integration constant). When calculating the integrals of the form of Equation 6.23, we assume the variable $\bar{x}$ to be constant. We see that the Solution 6.23 still contains an unknown function $A_1(\bar{x})$. We note that the function g_1 is periodic with respect to $\bar{\psi}$ [by virtue of periodicity $X(\bar{x},\bar{\psi},0)$; see Equation 6.21]. Let us calculate the average over the period $T = 2\pi$ of the integrand

$$\{\bar{g}_1(\bar{x},\bar{\psi}) - \bar{A}_1(\bar{x})\} = \frac{1}{2\pi}\int_{\bar{\psi}}^{\bar{\psi}+2\pi} \{g_1(\bar{x},\bar{\psi}) - A_1(\bar{x})\} d\bar{\psi} = c(\bar{x}). \tag{6.24}$$

In order to find the value of the function $c(\bar{x})$, we employ the boundedness condition for the function u_k,

$$\lim_{\bar{\psi}\to\infty} u_k(\bar{x},\bar{\psi}) < \infty. \tag{6.25}$$

We consider the limit

$$\lim_{m\to\infty} u_1(\bar{x},\bar{\psi}_0 + 2\pi m) = \frac{1}{\omega(\bar{x})}\lim_{m\to\infty}\left[2\pi c(\bar{x})m\right] \tag{6.26}$$

to find that, for $m \to \infty$, the Condition 6.25 is satisfied provided that

$$c(\bar{x}) = 0. \tag{6.27}$$

Making use of Equations 6.24 and 6.27, we obtain the required expression for the function A_1, i.e.,

$$A_1(\bar{x}) = \bar{g}_1(\bar{x}) = \frac{1}{2\pi}\int_0^{2\pi} g(\bar{x},\bar{\psi}) d\bar{\psi} = \bar{X}(\bar{x}). \tag{6.28}$$

This, in turn, makes it possible to find

$$u_1(\bar{x},\bar{\psi}) = \frac{1}{\omega(\bar{x})}\left[\int_{\bar{\psi}_0}^{\bar{\psi}} X(\bar{x},\bar{\psi}) d\bar{\psi} - \bar{X}(\bar{x},\bar{\psi})\bar{\psi}\right] + \varphi_1^{\bullet}(\bar{x}). \tag{6.29}$$

The mathematical structure of the arbitrary function $\varphi_1^{\bullet}(\bar{x})$ suggests that the latter should be chosen in accordance with the initial (boundary) conditions of the problem. For example, if the conditions are

$$x(t_0) = \bar{x}(t_0);\ \psi(t_0) = \bar{\psi}(t_0), \tag{6.30}$$

then $\varphi_1^\bullet(\bar{x}) = 0$. In principle, the solutions can be normalized to the function $\varphi_1^\bullet(\bar{x})$ in many other ways. In particular, if the initial system is the Hamiltonian, then one can require that the system for the first hierarchic level (Equation 6.18) should be the Hamiltonian also [21].

Now, let us find other unknown functions contained in the transformations 6.17 and representations 6.18. In a manner similar to solving Equation 6.29, we obtain

$$B_1(\bar{x}) = \bar{h}_1 = \bar{Y} + D_1 u_1; \tag{6.31}$$

$$v_1(\bar{x},\bar{\psi}) = \frac{1}{\omega(\bar{x})}\left\{\int_{\bar{\psi}_0}^{\bar{\psi}} h_1(\bar{x},\bar{\psi})d\bar{\psi} - \bar{h}_1(\bar{x})\bar{\psi}\right\} + \psi_1^\bullet(\bar{x}), \tag{6.32}$$

and so forth. This result may be generalized. Thus, we have

$$A_1 = \bar{g}_1; B_1 = \bar{h}_1, \tag{6.33}$$

$$u_i(\bar{x},\bar{\psi}) = \frac{1}{\omega(\bar{x})}\left\{\int_{\bar{\psi}_0}^{\bar{\psi}} g_i(\bar{x},\bar{\psi})d\bar{\psi} - \bar{g}_i(\bar{x})\bar{\psi}\right\} + \varphi_i^\bullet(\bar{x}); \tag{6.34}$$

$$v_i(\bar{x},\bar{\psi}) = \frac{1}{\omega(\bar{x})}\left\{\int_{\bar{\psi}_0}^{\bar{\psi}} h_i(\bar{x},\bar{\psi})d\bar{\psi} - \bar{h}_i(\bar{x})\bar{\psi}\right\} + \psi_i^\bullet(x). \tag{6.35}$$

Substituting Equations 6.34 and 6.35 (with regard to Equation 6.33) in Equation 6.17 yields the required transformations and the solution of the problem—provided, of course, that the solutions of the system of the truncated equation (Equation 6.18) are known.

Thus, the essence of the Bogolyubov method is to integrate the simpler (truncated) equations for the first hierarchic level (Equation 6.18) rather than complicated equations (Equation 6.16). In other words, finding the solutions for the truncated equations is sufficient for the asymptotic integration of the initial system (Equation 6.16) to be completed. The accuracy of the solutions is determined by the number n of terms retained in the series (Equations 6.17 and 6.18). Let us consider this aspect in detail.

6.2.3 Accuracy of the Approximate Solutions

As we shall show, the accuracy problem is important for the application of the hierarchic asymptotic methods. We will use the averaging method to illustrate its peculiarities. Suppose the number n of terms retained in asymptotic representation similar to Equation 6.14 is fixed, i.e.,

$$\frac{d\bar{x}}{dt} = \sum_{i=1}^{n} \varepsilon^i A_i(\bar{x}_n). \tag{6.36}$$

Integrating Equation 6.36 over t yields components of vector $\bar{x}_n$. We substitute the latter for the phase $\bar{\psi}_n$ in Equation 6.17 to obtain

$$\bar{\psi}_n(t) = \bar{\psi}_0 + \int_0^t \{(\omega(\bar{x}_n) + \varepsilon B_1(\bar{x}_n) + \ldots)\} dt. \quad (6.37)$$

Now we treat the solutions $\bar{x}_n$, $\bar{\psi}_n$ as approximations of the true solutions $\bar{x}$ and $\bar{\psi}$. Then, solutions of the initial equation (Equation 6.16) can be written as

$$x(t) \cong \bar{x}_n(t) + \sum_{m=1}^{N_1} \varepsilon^m u_m(\bar{x}_n, \bar{\psi}_n);$$

$$\psi(t) \cong \bar{\psi}_n(t) + \sum_{m=1}^{N_1} \varepsilon^m v_m(\bar{x}_n, \bar{\psi}_n). \quad (6.38)$$

A reasonable question arises: How accurate are the approximations? We remind the reader that an uncertain degree of accuracy of the approximate expressions of Equations 5.57 and 5.58 has been the main point of the mathematicians' objections against the van der Pol method.

Thus, let us estimate the accuracy of approximation for Equation 6.38. Because $d\bar{x}/dt$, according to Equation 6.36, is on the order of

$$\frac{d\bar{x}}{dt} = \dot{\bar{x}}_n + 0(\varepsilon^{n+1}), \quad (6.39)$$

then, in view of the mathematical structure of Equation 6.36, the estimate for $d\bar{x}/dt$ can be written as

$$\bar{x} = \bar{x}_n + 0(\varepsilon_n). \quad (6.40)$$

The asymptotic integration procedure implies that $\omega(\bar{x})$ is a slowly varying function of $\bar{x}$. Therefore, its accuracy is given by

$$\omega(\bar{x}) = \omega(\bar{x}_n) + 0(\varepsilon^n). \quad (6.41)$$

We recall that calculating the derivative of a slowly varying function increases the order of the magnitude of the result by ε. Therefore, integrating a slowly varying function decreases its order of magnitude by ε as well. Hence, we estimate the accuracy of terms contained in the integrand of Equation 6.37 as

$$\int_0^t \omega(\bar{x})dt = \int_0^t \omega(\bar{x}_n)dt + 0(\varepsilon^{n-1});$$

$$\varepsilon^n \int_0^t B_n(\bar{x})dt = \varepsilon^n \int_0^t B_n(\bar{x}_n)dt + 0(\varepsilon^{n-1}).$$

Thus, for the quadrature (Equation 6.41) we have

$$\bar{\psi}_n(t) = \bar{\psi}(0) + \int_0^t \left\{\omega(\bar{x}_n) + \varepsilon B_1(\bar{x}_n) + \ldots + \varepsilon^{n-1} B_{n-1}(\bar{x}_n)\right\}dt. \tag{6.42}$$

This means that if, in the nth approximation, $\bar{x}_n$ is found within the accuracy of ε^n, then $\bar{\psi}_n$ is found within ε^{n-1}. This conclusion is important and must be kept in mind when estimating calculation accuracy. In the lowest approximation ($n = 1$), truncated equations are

$$\frac{d\bar{x}}{dt} = \varepsilon \bar{X}(\bar{x});\ x \cong \bar{x};$$

$$\frac{d\bar{\psi}}{dt} = \omega(\bar{x});\ \psi \cong \bar{\psi}, \tag{6.43}$$

(a so-called zeroth approximation; see examples in Chapter 7). In a general approximation, the solutions $x(t)$ and $\psi(t)$ are

$$x = \bar{x} + \varepsilon u_1(\bar{x}, \bar{\psi}) + \ldots + \varepsilon^{n-1} u_{n-1}(\bar{x}, \bar{\psi});$$

$$\psi = \bar{\psi} + \varepsilon v_1(\bar{x}, \bar{\psi}) + \ldots + \varepsilon^{n-2} v_{n-2}(\bar{x}, \bar{\psi}). \tag{6.44}$$

It is important to mention the correlation between the time interval t_n, for which the obtained solutions are valid, and the number of the approximation in the averaging method

$$t_n \le 1/\varepsilon^n = \xi^n. \tag{6.45}$$

This means that, if the solutions are obtained in the nth approximation, then the physical processes described by them can be adequately interpreted only for the interval of Equation 6.45. Ignoring this fact leads to mistakes. If some physical effect is predicted theoretically, one has to verify whether the criterion of Equation 6.45 is satisfied. Otherwise, the validity of the scientific results is doubtful.

Then we turn to the problem of accuracy of the van der Pol solutions (Equations 5.57 and 5.58). When Equations 6.44 and 6.43 (for $n = 1$) are compared to the relevant results obtained by the van der Pol method (Equations 5.57 through 5.60), it may seem that Equation 5.60 is written with an excessive accuracy with respect to Equation 6.13. However, this is a premature conclusion. When deriving Equations 5.57 through 5.60, we assume $\omega = const$, i.e., the integral $\int_0^t \omega \, dt$ can be calculated exactly, the error in the coefficient $B_i(\bar{x}_n)$ lowers to (ε^n). As a result we have

$$x = \bar{x} + \varepsilon u_1(\bar{x}, \bar{\psi}) + \ldots + \varepsilon^{n-1} u_{n-1}(\bar{x}, \bar{\psi});$$

$$\psi = \bar{\psi} + \varepsilon v_1(\bar{x}, \bar{\psi}) + \ldots + \varepsilon^{n-1} u_{n-1}(\bar{x}, \bar{\psi});$$

$$\frac{d\bar{x}}{dt} = \varepsilon A_1(\bar{x}) + \ldots + \varepsilon^n A_n(\bar{x});$$

$$\frac{d\bar{\psi}}{dt} = \omega + \varepsilon B_1(\bar{x}) + \ldots + \varepsilon^n B_n(\bar{x}). \tag{6.46}$$

For $n = 1$, Equation 6.46 reduces to van der Pol results (Equations 5.57 through 5.60), i.e.,

$$x = \bar{x};$$

$$\frac{d\bar{x}}{dt} = \varepsilon \bar{X}(\bar{x});$$

$$\psi = \bar{\psi};$$

$$\frac{d\bar{\psi}}{dt} = \omega + \varepsilon Y(\bar{x}). \tag{6.47}$$

Thus, the above-mentioned contradiction does not actually take place.

6.2.4 Case of Successive Approximations

In preceding sections, we discussed an asymptotic integration algorithm for the initial set (Equation 6.16) based on the hypothesis that both functions X and Y are n-times differentiable in ψ. Now, following works by Moiseev and Kulish [30,31], we show that this assumption is not fundamental and is employed for convenience only.

Again, we consider the set of Equation 6.16. We write the Krylov–Bogolyubov substitution in the standard form

$$x = \bar{x} + \varepsilon u(\bar{x}, \bar{\psi}, \varepsilon);$$

$$\psi = \bar{\psi} + \varepsilon v(\bar{x}, \bar{\psi}, \varepsilon). \tag{6.48}$$

The relevant system of the equations of the first hierarchic level (Equation 6.18) is given by

$$\frac{d\bar{x}}{dt} = \varepsilon A(\bar{x});$$

$$\frac{d\bar{\psi}}{dt} = \omega(\bar{x}) + \varepsilon B(\bar{x}). \tag{6.49}$$

Substituting Equations 6.48 and 6.49 in Equation 6.16 yields

$$A(\bar{x},\varepsilon) + \varepsilon\frac{\partial u}{\partial \bar{x}}A(\bar{x},\varepsilon) + \frac{\partial u}{\partial \bar{\psi}}\left(\omega(\bar{x}) + \varepsilon B(\bar{x},\varepsilon)\right) = X\left(\bar{x} + \varepsilon u; \bar{\psi} + \varepsilon v; \varepsilon\right);$$

$$\begin{gathered}\omega(\bar{x}) + \varepsilon B(\bar{x},\mu) + \varepsilon^2\frac{\partial v}{\partial \bar{x}}A(\bar{x},\varepsilon) + \mu\frac{\partial v}{\partial \bar{\psi}}\left(\omega(\bar{x}) + \varepsilon B(\bar{x},\varepsilon)\right) \\ = \omega(\bar{x} + \varepsilon u) + \varepsilon Y\left(\bar{x} + \varepsilon u, \bar{\psi} + \varepsilon v, \varepsilon\right)\end{gathered}. \tag{6.50}$$

We analyze Equation 6.50 by means of successive approximations. The standard (for this method) iteration procedure implies

$$\frac{\partial u^{(k)}}{\partial \bar{\psi}}\omega(\bar{x}) = g_k - A^{(k)};$$

$$\frac{\partial v^{(k)}}{\partial \bar{\psi}}\omega(\bar{x}) = h_k - B^{(k)}, \tag{6.51}$$

where k is the iteration number,

$$g_k = X\left(x + \varepsilon u^{(k-1)}, \bar{\psi} + \varepsilon v^{(k-1)}, \varepsilon\right) - \varepsilon\frac{\partial u^{(k-1)}}{\partial \bar{x}}A^{(k-1)} - \varepsilon\frac{\partial u^{(k-1)}}{\partial \bar{\psi}}B^{(k-1)};$$

$$\begin{aligned}h_k = {} & \frac{\omega\left(\bar{x} + \varepsilon u^{(k-1)}\right) - \omega(\bar{x})}{\varepsilon} + Y\left(\bar{x} + \varepsilon \bar{u}^{(k-1)}, \bar{\psi} + \varepsilon v^{(k-1)}, \mu\right) \\ & - \varepsilon\frac{\partial u^{(k-1)}}{\partial \bar{x}}A^{(k-1)} - \varepsilon\frac{\partial v^{(k-1)}}{\partial \bar{\psi}}B^{(k-1)}\end{aligned}. \tag{6.52}$$

We compare the systems of Equation 6.52 and 6.22 and find them to be of similar mathematical structure. Repeating the sequence of operations results in solutions similar to Equations 6.34 and 6.35.

6.2.5 Case of Fourier Transformation

We have shown that asymptotic integration of initial System 6.16 reduces to solving partial differential Equation 6.22. The latter can be written in generalized form as

$$\frac{\partial F(x,\psi)}{\partial \psi} = V(x,\psi) - D(x). \tag{6.53}$$

We remember that $V(x,\psi)$ is a periodic function of ψ with the period $T = 2\pi$. It is peculiar to the integration procedure applied to this type of equation that the function $D(x)$ is unknown and found from the condition that the solution $F(x,\psi)$ is bounded for $\psi \to \infty$. The Fourier version of the averaging method is the most convenient. Therefore, it is taken as the basic version later in this book. We substitute a Fourier series for $V(x,\psi)$, i.e.,

$$V(x,\psi) = \sum_{k=-\infty}^{\infty} a_k(x) e^{ik\psi}, \tag{6.54}$$

where a_k is the Fourier amplitude defined in the conventional way. Let us consider the zero harmonic of the Fourier expansion 6.54. Its amplitude is equal to the mean value of the function $V(x,\psi)$ for the period T, i.e.,

$$a_0 = \frac{1}{2\pi} \int_{-\pi}^{\pi} V(x,\psi) d\psi = \bar{V}(x). \tag{6.55}$$

Then Equation 6.53 can be written as

$$\frac{\partial F}{\partial \psi} = \sum_{k \neq 0} a_k(x) e^{ik\psi} + \bar{V}(x) - D(x). \tag{6.56}$$

Let us expand the function F in a Fourier series. Within the context of Equation 6.56, we have

$$F(x,\psi) = \sum_{k=-\infty}^{\infty} b_x(x) e^{ik\psi} + c(x)\psi + (x), \tag{6.57}$$

where $\varphi(x)$ is the integration constant,

$$b_k(x) = \frac{a_k(x)}{ik} (k = 0),$$

$$c(x) = \bar{V}(x) - D(x). \tag{6.58}$$

It follows from this result that $c(x) = 0$. Indeed, function F is bounded for $\psi \to \infty$ only provided that the secular term $c(x)\psi$ contained in Equation 6.57 vanishes, and that is possible for $c(x) = 0$ only. Thus, we obtain the result from the preceding analysis, i.e.,

$$D(x) = \bar{V}(x). \tag{6.59}$$

The described procedure is efficient when applied to electrodynamics.

6.3 Case of Two Fast Rotating Scalar Phases*

In Section 6.1, we formulated a general calculation of a hierarchic problem. This problem mathematically consists of the development of an algorithm of asymptotic integration of hierarchic standard systems of the type of System 6.5. It should be mentioned that, generally, the arbitrary number of components of the total vector of fast rotating phases is a characteristic feature of these systems. In Section 6.2, we discussed the simplest variant of a hierarchic system where the vector of fast rotation is represented by one scalar phase ($j = 1$) only.

In this connection, systems with a number of rotating phases where $j > 1$ possess some essentially new specific features (in comparison to the discussed case where $j = 1$). This concerns the possibility of realization of the resonances in such multiphase (multifrequency) objects. Apart from that, some specific features also characterize the nonresonant multiphase systems. Unfortunately, the general scheme of asymptotic integration of multiphase hierarchic systems similar to System 6.5, (given in Section 6.4) seems too cumbersome for a beginner. So, we begin the study of the multiphase (multifrequency) hierarchic systems with a discussion of the simplest system, where the phenomenon of resonance can be realized. As an analysis shows, this is a system with two rotating scalar phases of the same hierarchy.

6.3.1 Formulation of the Problem

Thus, we consider a system with two fast scalar phases $\psi_j (j = 1,2)$ and one large parameter scalar $\xi \equiv \xi_1 = 1/\varepsilon >> 1$ given by the following standard form

$$\frac{dx}{dt} = X(x, \psi_1, \psi_2, \xi);$$

$$\frac{d\psi_1}{dt} = \xi\omega_1(x) + Y_1(x, \psi_1, \psi_2, \xi);$$

$$\frac{d\psi_2}{dt} = \xi\omega_2(x) + Y_2(x, \psi_1, \psi_2, \xi), \tag{6.60}$$

where x and X are vectors, and ω_j, Y_j, ψ_j, ξ are scalars. Contrary to the situation studied in the previous section, X and Y_j are now two-period functions with respect to phase ψ_j with the periods

$$T_j = 2\pi/q_j. \tag{6.61}$$

* From Kulish, V.V., *Hierarchic Methods. Hierarchy and Hierarchic Asymptotic Methods in Electrodynamics*. Kluwer Academic Publishers, Dordrecht/Boston/London. 2002. With permission.

In applied problems, coefficient q_j, as a rule, is usually equal to 1. Hence, we can put $q_j = 1$ for simplicity. The generalization for the arbitrary case $q_j \neq 1$ is straightforward and does not involve any difficulties (see the general algorithm described in Section 6.4).

We write the system of truncated equations (the system of the first hierarchic level, according to our terminology) in a form analogous to Equation 6.18, i.e.,

$$\frac{d\bar{x}}{dt} = \sum_{n=1}^{\infty} A^{(n)}(\bar{x});$$

$$\frac{d\bar{\psi}_j}{dt} = \xi\omega_j(\bar{x}) + \sum_{n=1}^{\infty} \frac{1}{\xi^n} B_j^{(n)}(\bar{x}). \tag{6.62}$$

According to the general procedure of the averaging method, we find the change of variables when reducing Equation 6.60 to the form of Equation 6.62. For this, we use substitution of the standard form given by

$$x = \bar{x} + \sum_{n=1}^{\infty} \frac{1}{\xi} u^{(n)}(\bar{x});$$

$$\psi_j = \bar{\psi}_j + \sum_{n=1}^{\infty} \frac{1}{\xi^n} v_j^{(n)}(\bar{x}). \tag{6.63}$$

When the solutions of the equation of the first hierarchic level (truncated equations—Equation 6.62) are known, then the substitutions of Equation 6.63 play the role of formal solutions to the initial system Equation 6.60. Therefore, the main task in this case is to find the unknown components of the decompression operators $u^{(n)}(\bar{x})$ and $v_j^{(n)}(\bar{x})$, as well as the unknown functions $A^{(n)}(\bar{x})$ and $B_j^{(n)}(\bar{x})$ in the compression operator.

6.3.2 Nonresonant Case

We carry out a procedure similar to that set forth in Section 6.2 and find the first approximation ($n = 1$) equations for the transformation functions given by

$$\frac{\partial u^{(1)}}{\partial \bar{\psi}}\omega_1(\bar{x}) + \frac{\partial u^{(1)}}{\partial \bar{\psi}_2}\omega_2(\bar{x}) = X(\bar{x}, \bar{\psi}_1, \bar{\psi}_2, \infty) - A^{(1)}(\bar{x});$$

$$\frac{\partial v_1^{(1)}}{\partial \bar{\psi}_1}\omega_1(\bar{x}) + \frac{\partial v_1^{(1)}}{\partial \bar{\psi}_2}\omega_2(\bar{x}) = Y_1(\bar{x}, \bar{\psi}_1, \bar{\psi}_2, \infty)$$
$$+ \omega_1(\bar{x})u^{(1)}(\bar{x}, \bar{\psi}_1, \bar{\psi}_2) - B_1^{(1)}(x);$$

$$\frac{\partial v_2^{(1)}}{\partial \bar{\psi}_1}\omega_1(\bar{x})+\frac{\partial v_2^{(1)}}{\partial \bar{\psi}_2}\omega_2(\bar{x})=Y_2(\bar{x},\bar{\psi}_1,\bar{\psi}_2,\infty)$$
$$+\omega_2(\bar{x})u^{(1)}(\bar{x},\bar{\psi}_1,\bar{\psi}_2)-B_2^{(1)}(\bar{x}). \quad (6.64)$$

It is not difficult to show that the next approximation $n > 1$ equations of a similar mathematical structure can be obtained.

The Krylov–Bogolyubov system of equations can be written in a form similar to Equation 6.53, i.e.,

$$\frac{\partial F(\bar{x},\bar{\psi}_1,\bar{\psi}_2)}{\partial \bar{\psi}_1}\omega_1(\bar{x})+\frac{\partial F(\bar{x},\bar{\psi}_1,\bar{\psi}_2)}{\partial \bar{\psi}_2}\omega_2(\bar{x})=V(\bar{x},\bar{\psi}_1,\bar{\psi}_2)-D(\bar{x}). \quad (6.65)$$

In the discussed case, in contrast to Equation 6.53, the fast phase is represented by the two-component vector $(\bar{\psi}_1,\bar{\psi}_2)$, and, correspondingly, we have the two-periodic (with respect to the periods $T_1 = T_2 = 2\pi$) functions F and V. Let us employ the Fourier procedure described in Section 6.2.5. We expand the functions $u^{(1)}$ and X in a double Fourier series, i.e.,

$$u^{(1)}(\bar{x},\bar{\psi}_1,\bar{\psi}_2)=\sum_{k=-\infty}^{\infty}\sum_{s=-\infty}^{\infty}b_{ks}(\bar{x})\exp\{ik\bar{\psi}_1+is\bar{\psi}_2\}+c(\bar{x})\bar{\psi}_1+d(\bar{x})\bar{\psi}_2, \quad (6.66)$$

$$X(\bar{x},\bar{\psi}_1,\bar{\psi}_2)=\sum_{k=-\infty}^{\infty}\sum_{s=-\infty}^{\infty}a_{ks}^{(x)}\exp\{ik\bar{\psi}_1+is\bar{\psi}_2\}. \quad (6.67)$$

Then we substitute Equations 6.66 and 6.67 in the first equation of Equation 6.64 and assume that the coefficients of similar exponential functions are equal. Thus, we obtain

$$b_{ks}=\frac{a_{ks}^{(x)}}{i(k\omega_1+s\omega_2)}; \quad (6.68)$$

$$c(\bar{x})\omega_1(\bar{x})+d(\bar{x})\omega_2(\bar{x})=a_{00}^{(x)}-A^{(1)}(\bar{x}). \quad (6.69)$$

The condition that the function $u^{(1)}$ is bounded for $\bar{\psi}_j\to\infty$ is satisfied when the secular terms in Equation 6.66 vanish, i.e.,

$$c(\bar{x})=0;\, d(\bar{x})=0. \quad (6.70)$$

In view of Equation 6.70, relation 6.69 yields

$$A^{(1)}=a_{00}^{(x)}(\bar{x})=\bar{X}(\bar{x}), \quad (6.71)$$

where

$$\bar{X}(\bar{x}) = \frac{1}{T_1 T_2} \int_0^{T_1}\int_0^{T_2} X(\bar{x}, \bar{\psi}_1, \bar{\psi}_2, \infty) d\bar{\psi}_1 d\bar{\psi}_2. \quad (6.72)$$

The other expansion terms of the first approximation can be calculated in the same manner. Thus, we obtain

$$A^{(1)} = \bar{X}(\bar{x});$$

$$B_1^{(1)} = \bar{Y}_1 + \omega_1(\bar{x}) \langle u_1(\bar{x}, \bar{\psi}_1, \bar{\psi}_2) \rangle;$$

$$B_2^{(1)} = \bar{Y}_2 + \omega_2(\bar{x}) \langle u^{(1)}(\bar{x}, \bar{\psi}_1, \bar{\psi}_2) \rangle;$$

$$u^{(1)} = \sum_{s,k \neq 0} \frac{a_{ks}^{(x)}(\bar{x})}{i(k\omega_1 + s\omega_2)} \exp\{ik\bar{\psi}_1 + is\bar{\psi}_2\} + a_0^{(x)}(\bar{x});$$

$$v_j^{(1)} = \sum_{k,s \neq 0} \frac{a_{ks}^{(\psi_j)}(\bar{x})}{i(k\omega_1 + s\omega_2)} \exp\{ik\bar{\psi}_1 + is\bar{\psi}_2\} + a_0^{(\psi_j)}(\bar{x}), \quad (6.73)$$

where $a_{ks}^{(\psi_j)}$ are the coefficients of the Fourier expansions of the functions $Y_j - \omega_j(\bar{x}) u^{(1)}$; the arbitrary functions $a_0^{(x)}, a_0^{(\psi_j)}$ should be found from the initial (boundary) conditions, and $\langle u^{(1)}(\bar{x}, \bar{\psi}_1, \bar{\psi}_2) \rangle$ is the twice-averaged (on the phases $\bar{\psi}_1$ and $\bar{\psi}_2$) function $u^{(1)}$

$$\langle u^{(1)}(\bar{x}, \bar{\psi}_1, \bar{\psi}_2) \rangle = \frac{1}{T_1 T_2} \int_0^{T_1}\int_0^{T_2} u^{(1)}(\bar{x}, \bar{\psi}_1, \bar{\psi}_2) d\bar{\psi}_1 d\bar{\psi}_2.$$

We note that the choice of the functions $a_0^{(x)}, a_0^{(\psi_j)}$ does not influence the accuracy of the solutions obtained.

Let us analyze the mathematical structure of the obtained solutions for Equation 6.73. We see that they contain the specific resonance denominators in the form

$$\left(k\omega_1(\bar{x}) + s\omega_2(\bar{x})\right). \quad (6.74)$$

The problem in the discussed case of the two scalar phases has been formulated under the assumption that both phase rotation velocities are commensurable, i.e.,

$$\left|\frac{d\psi_1}{dt}\right| \sim \left|\frac{d\psi_2}{dt}\right| \text{ or } |\omega_1(\bar{x})| \sim |\omega_2(\bar{x})|. \quad (6.75)$$

This looks clearer when the following circumstance is taken into consideration. By virtue of the characteristic structure of Equation 6.60 in the discussed case, the difference in the scales of rates of change of the slow and fast variables is determined by the same large parameter of the problem

$$\xi \sim \left|\frac{d\psi_1}{dt}\right| \Big/ \left|\frac{dx_q}{dt}\right| \sim \left|\frac{d\psi_2}{dt}\right| \Big/ \left|\frac{dx_q}{dt}\right| >> 1, \tag{6.76}$$

where x_q is the qth component of the vector x. The harmonic number in Equation 6.74 is an algebraic quantity, i.e., $k,s = \pm 1, \pm 2, \ldots$. Also, the angular velocities of the fast phase rotation [frequencies $\omega_j(\bar{x})$] are algebraic quantities, too. Therefore, in principle, some denominator in Equation 6.74 at some points can equal zero, i.e.,

$$k\omega_1(\bar{x}) + s\omega_2(\bar{x}) = 0. \tag{6.77}$$

This is the well-known problem of small denominators [10]. Equation 6.77 determines the resonance condition (the physical essence of this concept is described in Chapter 3). It is clear that, if this condition is satisfied, then the solution of Equation 6.73 is divergent. This infers that this calculation procedure is inapplicable in the resonance case of Equation 6.77. To describe the resonance states of the system, the constructed asymptotic integration algorithm should be somewhat modified. Let us discuss one of the methods of such modification.

6.3.3 Resonant Case

Let us find asymptotic solutions that would adequately describe the dynamics of the system of Equation 6.60 in the neighborhood of the resonance point (Equation 6.77), including the point itself. In order to describe the resonance state, we introduce its mismatch in the form

$$h^{\bullet}(\bar{x}) = k\omega_1(\bar{x}) + s\omega_2(\bar{x}). \tag{6.78}$$

Solving Equation 6.78 for $h^{\bullet}(\bar{x}) = 0$, we obtain the resonant point $\bar{x}^{\bullet}$. Then, we expand the function $\omega_j(\bar{x})$ in the vicinity of the point $\bar{x}^{\bullet}$. The analysis of the expansion shows that $h^{\bullet} \sim 1/\xi$. Thus in the resonance case, the large parameter Equation 6.76 may also be interpreted as a measure of the approach to the resonance state of the system in the near-resonance region.

Two types of resonances are distinguished—the main resonance (see Chapter 3) and the resonances associated with harmonics. The main resonance occurs for $|k| = |s| = 1$. The harmonic resonances occur for $|k| > 1$, or $|s| > 1$, or when both inequalities are satisfied simultaneously: $|k|,|s| > 1$ (see also Chapter 3).

Let us begin with the case of the main resonance for the constant frequencies. It means that the supposition that the frequencies $\omega_j \neq \omega_j(\bar{x})$ are constants is satisfied. If the main resonance occurs in such a model, we may write

$$\omega_1 = \omega_2 + (1/\xi)h^{\bullet}. \tag{6.79}$$

Within the context of Equation 6.79, the initial system Equation 6.60 may be rewritten as

$$\frac{dx}{dt} = X(x, \psi_1, \psi_2, \xi);$$

$$\frac{d\psi_1}{dt} = \xi\omega_1 + Y_1(x, \psi_1, \psi_2, \xi);$$

$$\frac{d\psi_2}{dt} = \xi\omega_1 + \left[Y_2(x, \psi_1, \psi_2, \xi) + h^\bullet\right]. \tag{6.80}$$

We introduce a new variable

$$\theta = \psi_2 - \psi_1, \tag{6.81}$$

which is referred to as the combination phase (see Chapter 3 for more detail). We differentiate Equation 6.81 with respect to time t and make use of Equation 6.80. Thus we obtain

$$\begin{aligned} \frac{d\theta}{dt} = \frac{d\psi_1}{dt} - \frac{d\psi_2}{dt} &= Y_2(x, \psi_1, \psi_2, \xi) + h^\bullet - Y_1(x, \psi_1, \psi_2, \xi) \\ &= \Theta(x, \psi_1, \psi_2, \xi) \end{aligned}. \tag{6.82}$$

Inasmuch as the phase rates of the change have been assumed to be close (see Equation 6.79), their difference $(d\psi_1/dt) - (d\psi_2/dt)$ can be small. Therefore, both the new variable θ and the new function Θ entering Equation 6.82 are slowly varying quantities under the resonance conditions. So, we can rewrite the system Equation 6.80 by taking into account Equations 6.81 and 6.82. As a result we obtain

$$\frac{dx}{dt} = X(x, \theta, \psi_2, \xi);$$

$$\frac{d\theta}{dt} = \Theta(x, \theta, \psi_2, \xi);$$

$$\frac{d\psi_2}{dt} = \xi\omega_1 + \left[Y_2(x, \theta, \psi_2, \xi) + h^\bullet\right]. \tag{6.83}$$

It is readily seen from the latter system that, as a result of the accomplished transformations: 1) the variable θ can be regarded as a component of the new slow variables x', and 2) the functions X, Θ, and Y_2 are periodic with respect to one fast phase ψ_2 only (rather than two fast phases ψ_1 and ψ_2 of Equation 6.60).

In other words, the mathematical properties of the system Equation 6.83 are similar to those of the system Equation 6.16, in which one should use the form of the initial equations with large parameter ξ (instead of the form with small one ε). Hence, we can employ the relevant asymptotic integration algorithm described in Section 6.2.

Let us turn our attention to the more general case of slowly varying frequencies $\omega_j = \omega_j(x)$. At first, we consider the main resonance $|k|$, $|s| = 1$, for $\omega_j = \omega_j(x)$, bearing in mind that the function $\omega_j(x)$ is an algebraic quantity. Taking this into account, the main resonance condition similar to Equation 6.79 may be written in this case in a form like Equation 6.79

$$-\sigma\omega_1(x) = \omega_2(x) + (1/\xi)h^{\bullet}(x), \tag{6.84}$$

where $\sigma = ks = \pm 1$ is the sign function (because $k,s = \pm 1$, $sign\{\omega_j\} = +1$). Formally, there seems to be no difference between the Conditions 6.84 (for $\sigma = -1$) and 6.79. However, a rather important distinction exists. It is associated with the very possibility that variations of the frequencies $\omega(x)$ can be slow under nonlinear resonance interactions in the system. This means that the evolution of the system can be such that the system can slowly get in and out of the resonance state. Therefore, for each interval of the values of x, we have to check up whether the solutions obtained really contain the resonance denominators (6.74). The results of the checkup determine whether the nonresonance or resonance asymptotic integration procedure should be employed. In the general case, when solutions of both types occur (associated with different stages of interaction), we have to join the solutions together in terms of some special procedure [7].

We now turn to discussion of the algorithm of asymptotic integration in the discussed resonance case. At the first stage, the procedure of separating out the slowly varying combination phase θ could be modified in order to generalize it. Generally, this type of multiphase procedure is described in the monograph by Kulish [31] (see also Chapter 3). In this case of two scalar phases only, it reduces to the following simplified form. We introduce the trivial nondegenerate linear transformation, which relates the fast phase vector $\{\psi_1, \psi_2\}$ to the vector of the combination phases $\{\theta, \psi\}$, to be given by

$$\theta = \psi_2 + \sigma\psi_1; \ \psi = \psi_2 - \sigma\psi_1. \tag{6.85}$$

Since the combination phase θ has been shown to be slow, it is associated with a component of the vector of slow variables $x' = \{x, \theta\}$. In turn, the combination phase ψ, by virtue of the definition of Equation 6.85, must be considered a fast one. The relations between the old and new variables in such a situation are given by the following simple linear transformations:

$$\psi_1 = \frac{\sigma}{2}(\theta - \psi); \psi_2 = \frac{1}{2}(\theta + \psi). \tag{6.86}$$

Then we substitute Equation 6.86 into Equation 6.60, make use of Equation 6.85, and carry out some calculations. As a result we obtain

$$\frac{dx}{dt} = X(x, \theta, \psi, \xi);$$

$$\frac{d\theta}{dt} = \left[Y_2(x, \theta, \psi, \xi) + \sigma Y_1(x, \theta, \psi, \xi) - h^{\bullet}(x)\right];$$

$$\frac{d\psi}{dt} = \xi\left[2\omega_2(x)\right] + \left[Y_2(x, \theta, \psi, \xi) - \sigma Y_1(x, \theta, \psi, \xi) + h^{\bullet}(x)\right]. \tag{6.87}$$

At the next stage of the calculation procedure, we introduce the obvious notation

$$Y_2(x,\theta,\psi,\xi) + \sigma Y_1(x,\theta,\psi,\xi) - h^{\bullet}(x) = \Theta(x,\theta,\psi,\xi);$$

$$2\omega_2(x) = \omega(x);$$

$$Y_2(x,\theta,\psi,\xi) - \sigma Y_1(x,\theta,\psi,\xi) + h^{\bullet}(x) = Y(x,\theta,\psi,\xi);$$

$$x' = \{x,\theta\};\ X = X'(X,\Theta). \tag{6.88}$$

As a result of the performed transformations, we avoid problems concerning the small denominations: The system (Equation 6.87) reduces to the standard form with a single fast rotating phase similar to Equation 6.16, where we should accomplish the change $\varepsilon = 1/\xi$ with this relevant transformations, i.e.,

$$\frac{dx'}{dt} = X'\left(x',\psi,\xi\right);$$

$$\frac{d\psi}{dt} = \xi\omega\left(x'\right) + Y\left(x',\psi,\xi\right). \tag{6.89}$$

The algorithm for solving such systems is described above in Section 6.2. This means that the formulated resonant problem is solved.

We turn our attention to the case of harmonic resonance (see Chapter 3 for detail). Analogous to Equation 6.84, the relevant harmonic resonance condition (i.e., in the case $|k| > 1$ or $|s| > 1$ or $|k|,|s| > 1$ simultaneously) may be written as

$$\sigma\frac{k}{s}\omega_1\left(x\right) = \omega_2\left(x\right) + \frac{1}{\xi}h^{\bullet}\left(x\right). \tag{6.90}$$

We define the slow and fast combination phases by the analogous method

$$\theta = \sigma\frac{k}{s}\psi_2 + \psi_1;\ \psi = \sigma\frac{k}{s}\psi_2 - \psi_2, \tag{6.91}$$

where, as before, $\sigma = sign\{\omega_1\omega_2\}$, and $k,s > 0$. The relation between the old and new variables is now given by

$$\theta = \frac{1}{2}\left(\theta - \psi\right);\ \psi_2 = \sigma\frac{k}{2s}\left(\theta + \psi\right). \tag{6.92}$$

We carry out substitutions analogous to those considered in the previous case and again obtain a system in the form of Equation 6.89.

It should be noted that, in practice, the definition of harmonic concepts sometimes causes misunderstanding. In electrodynamic problems, for example, one has to distinguish the harmonics of wave fields, in which the electron moves, from the harmonics of its nonlinear oscillations in these fields. In the asymptotic integration procedure, the harmonics of the external wave fields are taken into account in the definitions of the fast phases (see Chapter 3), i.e.,

$$\psi_{m1} = m_1\psi_1;\ \psi_{m2} = m_2\psi_2, \tag{6.93}$$

where $m_j = 1,2,\ldots$ are the numbers of the external wave field harmonics. The above-mentioned quantities k and s are harmonics of electron oscillations in these fields. Then, the relevant combination phases (Equation 6.91) should be written in the form of Equation 6.9. The integration procedure for the initial resonance system Equation 6.60 remains, in principle, similar.

6.4 Case of Many Rotating Scalar Phases*

6.4.1 Formulation of the Problem

Let us generalize the described algorithms in a case of an arbitrary number of rotating phases (see system Equation 6.12). Introducing the vector of the fast phases in the form $\psi = \{\psi_1,\psi_2,\ldots,\psi_m\}$, we rewrite Equation 6.12 in the following manner:

$$\frac{dx}{dt} = X(x,\psi,\xi);$$

$$\frac{d\psi}{dt} = \xi\omega(x) + Y(x,\psi,\xi), \tag{6.94}$$

where all notations are self-evident. We look for a solution to system Equation 6.94 in the form of Bogolyubov substitutions (Equation 6.13)

$$x_q = \bar{x}_q + \sum_{n=1}^{\infty} \frac{1}{\xi^n} u_q^{(n)}(\bar{x},\bar{\psi}), \qquad (q = 1,2,3,\ldots,k);$$

$$\psi_j = \bar{\psi}_j + \sum_{n=1}^{\infty} \frac{1}{\xi^n} v_j^{(n)}(\bar{x},\bar{\psi}), \qquad (q = 1,2,3,\ldots,m), \tag{6.95}$$

where x_q, ψ_j are elements of the vectors x and ψ, and k and m are numbers of the components; other values have been determined above. In accepted notations, one can rewrite the equations of the first hierarchic level (Equation 6.14) as

$$\frac{d\bar{x}_q}{dt} = \sum_{n=1}^{\infty} \frac{1}{\xi^n} A_q^{(n)}(\bar{x});$$

* From Kulish, V.V., *Hierarchic Methods. Hierarchy and Hierarchic Asymptotic Methods in Electrodynamics*. Kluwer Academic Publishers, Dordrecht/Boston/London. 2002. With permission.

$$\frac{d\bar{\psi}_j}{dt} = \xi\omega_j(\bar{x}) + \sum_{n=1}^{\infty} \frac{1}{\xi^n} B_j^{(n)}(\bar{x}). \tag{6.96}$$

6.4.2 Compression and Decompression Operators

Strict determination of the unknown components of the compression and decompression operators in Equations 6.95 and 6.96 is ambiguous. This is due to arbitrariness in attributing different terms of the series. Following the discussed calculation schemes, we can eliminate this arbitrariness by assuming that all $u_q^{(n)}$ and $v_l^{(n)}$ are free of zero (in $\bar{\psi}_l$) Fourier harmonics. Thus, we postulate that the whole averaged motion is described by $\bar{x}_q$ and $\bar{\psi}_l$.

Let us resume the differentiation of Equation 6.95 and, taking into account Equation 6.96, substitute the obtained result into Equation 6.94, equating the coefficients of equal powers ξ^{-1}. Therefore, we obtain the infinite sequence of relations

$$\sum_{j=1}^{m} \frac{\partial u_q^{(1)}}{\partial \psi_j} \omega_j = X_q - A_q^{(0)}; \tag{6.97}$$

$$\sum_{j=1}^{m} \frac{\partial u_q^{(2)}}{\partial \psi_j} \omega_j = \sum_{i=1}^{k} \left(\frac{\partial X_q}{\partial x_i} u_i^{(1)} - \frac{\partial u_q^{(1)}}{\partial x_i} A_i^{(0)} \right) + \sum_{j=1}^{m} \left(\frac{\partial X_q}{\partial \psi_j} v_j^{(1)} - \frac{\partial u_q^{(1)}}{\partial \psi_j} B_j^{(0)} \right) - A_q^{(1)}, \tag{6.98}$$

$$\sum_{l=1}^{m} \frac{\partial v_j^{(1)}}{\partial \psi_l} \omega_j = \sum_{i=1}^{k} \frac{\partial \omega_j}{\partial x_i} u_i^{(1)} + Y_j - B_j^{(0)} \equiv \psi_j(x,\psi) - B_j^{(0)}; \tag{6.99}$$

$$\sum_{l=1}^{m} \frac{\partial v_l^{(2)}}{\partial \psi_l} \omega_j = \sum_{i=1}^{k} \frac{\partial \omega_j}{\partial x_i} u_i^{(2)} + \frac{1}{2} \sum_{i,l=1}^{k} \frac{\partial^2 \omega_j}{\partial x_i \partial x_l} u_i^{(1)} u_l^{(1)} + \sum_{i=1}^{k} \frac{\partial Y_j}{\partial x_i} u_i^{(1)} + \sum_{l=1}^{m} \frac{\partial Y_l}{\partial \psi_l} v_l^{(1)} - \sum_{i=1}^{k} \frac{\partial v_j^{(1)}}{\partial x_i} A_q^{(0)} - \sum_{l=1}^{m} \frac{\partial v_j^{(1)}}{\partial \psi_l} B_l^{(0)} - B_j^{(1)}. \tag{6.100}$$

Hereafter, the averaging sign will be omitted for simplicity. Taking into account Equation 6.97, it is easy to see that function x_q is periodic in ψ_l. Using this, we expand it into a Fourier series of multiplicity m

$$X_q = \sum_{sk\ldots p} a_q^{sk\ldots p}(x) \exp\left[2\pi i \left(\frac{s\psi_1}{T_1} + \ldots + \frac{p\psi_m}{T_m} \right) \right]. \tag{6.101}$$

Here T_j is the period associated with phase ψ_j ($j = 1,2,\ldots,m$), and $s,k,\ldots,p$ are the relevant Fourier harmonics. Proceeding from similar considerations as well as Equation 5.18, we represent unknown function $u_q^{(n)}$ as

$$u_q^{(1)} = \sum_{sk\ldots p} b_q^{sk\ldots p}(x)\exp\left[2\pi i\left(\frac{s\psi_1}{T_1}+\ldots+\frac{p\psi_m}{T_m}\right)\right]+\sum_{j=1}^{m} C_{qj}(x)\psi_j. \quad (6.102)$$

Upon substituting Equation 6.101 into Equation 6.97 and equating the coefficients with equal exponents, we obtain expressions for amplitude $b_q^{sk\ldots p}(x)$ and function $A_q^{(0)}$

$$b_q^{sk\ldots p}(x) = a_q^{sk\ldots p}(x)\left[2\pi i\left(\frac{s\psi_1}{T_1}+\ldots+\frac{p\psi_m}{T_m}\right)\right]^{-1}, \quad (6.103)$$

$$A_q^{(0)} = a_q^{00\ldots 0} - \sum_{j=1}^{m} C_{qj}(x)\omega_j, \quad (6.104)$$

where function $C_{qj}(x)$ in limits on $u_q^{(1)}$ satisfies the normalization condition

$$C_{qj} = 0. \quad (6.105)$$

Accordingly, the definitions for coefficient $A_q^{(0)}$ (Equation 6.104) can be rewritten as

$$A_q^{(0)} = a_q^{00\ldots 0}(x) = \frac{1}{T_1T_2\ldots T_m}\int_0^{T_1}\ldots\int_0^{T_m} X_q\, d\psi_1\ldots d\psi_m. \quad (6.106)$$

Taking into account the absence of zero harmonics on ψ_j and $u_q^{(1)}$, for the latter we formulate the definition differently from Equation 6.102 only by the exclusion of terms in $s, k, \ldots, p \neq 0$. Similar to Equation 6.104, the second approximation equation is solved. From the structure of Equation 6.103, in particular, it follows that

$$\frac{s\omega_1}{T_1}+\ldots+\frac{p\omega_m}{T_m} \neq 0, \quad (6.107)$$

i.e., there are no resonances among the components of the vector of fast phases. Failure of Equation 6.107 means that, at the stage of classification of phases (and their linear combinations), not all slow phases have been singled out. Following the procedure, we obtain expressions for other unknown functions, in particular

$$u_q^{(1)} = \sum_{ks\ldots p} \frac{1}{2\pi i}\frac{a_q^{sk\ldots p}(x)}{\dfrac{s\omega_1}{T_1}+\ldots+\dfrac{p\omega_m}{T_m}}\exp\left(2\pi i\left(\frac{s\omega_1}{T_1}+\ldots+\frac{p\omega_m}{T_m}\right)\right), \quad (6.108)$$

$$v_j^{(1)} = \sum_{ks\ldots p} \frac{1}{2\pi i} \frac{d_j^{sk\ldots p}(x)}{\frac{s\omega_1}{T_1} + \ldots + \frac{p\omega_m}{T_m}} \exp\left(2\pi i\left(\frac{s\omega_1}{T_1} + \ldots + \frac{p\omega_m}{T_m}\right)\right); \tag{6.109}$$

$$A_q^{(1)} = \left\langle \sum_{i=1}^{k} \frac{\partial X_q}{\partial x_i} u_i^{(1)} + \sum_{j=1}^{m} \frac{\partial X_q}{\partial \psi_j} v_j^{(1)} \right\rangle; \tag{6.110}$$

$$B_1^{(0)} = \langle Y_1 \rangle, \tag{6.111}$$

$$B_j^{(1)} = \left\langle \sum_{i,l=1}^{k} \left[\frac{1}{2} \frac{\partial^2 \omega_j}{\partial x_i \partial x_l} u_i^{(1)} u_l^{(1)} \frac{\partial Y_j}{\partial x_i} v_l^{(1)} \right] + \sum_{l=1}^{m} \frac{\partial Y_j}{\partial \psi_l} v_l^{(1)} \right\rangle, \tag{6.112}$$

where $\langle \ldots \rangle$ means the averaging in all fast phases (similar to Equation 6.106); $d_j^{sk\ldots p}(x)$ are factors of expansions in the Fourier series of the functions of this form.

6.5 Method of Averaged Characteristics

As shown in this and the preceding chapters, the single-particle methods that are destined for the study of the oscillation (lumped) systems are characterized by the standard systems of the exact differential equations. In contrast, the wave (distributed) systems are described by differential equations with partial derivatives. Hence, at the first sight, the single-particle methods cannot be used in this case for the construction of wave-like asymptotic solutions.

However, it is not so always. As shown, for example in *Undulative Electrodynamic Systems* by Kulish [33], the single-particle methods can be used rather often as a basis for the highly efficient calculation algorithms for asymptotic integration of systems with partial derivatives. Let us illustrate one such possibility by using the method of averaged characteristics [33] as a convenient illustrative example.

6.5.1 Some General Concepts and Definitions

Let us recall some general concepts and definitions that are usually studied in a standard course on differential equations with partial derivatives. The following general type of equation is usually replaced by a differential equation with partial derivatives of the nth order:

$$F\left(z_1, \ldots, z_n, u, \frac{\partial u}{\partial z_1}, \ldots, \frac{\partial u}{\partial z_n}, \frac{\partial^2 u}{\partial z_1 \partial z_1}, \frac{\partial^2 u}{\partial z_2 \partial z_2}, \ldots, \frac{\partial^2 u}{\partial z_n \partial z_n}, \ldots, \frac{\partial^n u}{\partial z_n \ldots \partial z_n}\right) = 0. \tag{6.113}$$

This equation contains at least one nth order partial derivative of an unknown function $u(z_1,\ldots,z_n)$; z_j $(j = 1,\ldots,n)$ are independent variables.

It is necessary to find such a function $u(z_1,\ldots,z_n)$ that the function F becomes equal to zero after a substitution of $u(z_1,\ldots,z_n)$ into Equation 6.113. Such a function is called the solution or the integral of a differential equation with partial derivatives. All partial solutions (excepting the additional solutions) can be obtained from the general integral.

It is well known that any general solution of an exact differential equation (equation with momentum derivatives) can be found with the accuracy of some arbitrary constants. The specific feature of the equation with partial derivatives is that instead of the arbitrary constants, we obtain some arbitrary functions. Use of some initial or boundary conditions for the function u and its derivatives can define the arbitrary functions. Generally, their number is equal to the order of the differential equation. If we have n independent variables, then the arbitrary functions are the functions of $n-1$ variables.

The systems of the equations for a few unknown functions $u_1(z_1,\ldots,z_n),\ldots,u_k(z_1,\ldots,z_n),\ldots$ can be written in the form

$$F_i\left(z_j,u_1,\ldots,u_k,\frac{\partial u_1}{\partial z_j},\ldots,\frac{\partial u_k}{\partial z_j},\ldots,\frac{\partial^n u}{\partial z_j\ldots\partial z_l}\right)=0, \tag{6.114}$$

where $i=1,\ldots,h$, $j,l=1,\ldots,n$, k is the current number of the unknown functions, and h is the total number of the system equations.

In the case $k=h$, the system in Equation 6.114 is called the determined one and, in the cases $k<h$ or $k>h$, we have to deal with the overdetermined or underdetermined systems.

Each differential equation of the type of Equation 6.113 or system of equations with higher derivatives like Equation 6.114 can be reduced to the system with only first derivatives (system of the first order).

The system of differential equations (Equation 6.114) allows obtaining the solutions $u_1(z_1,\ldots,z_n)$, $u_2(z_1,\ldots,z_n),\ldots$ only in the case when the given function F_i and its derivatives satisfy the conditions of compatibility. These conditions guarantee that the differentiation of two or more equations (Equation 6.114) leads to the coincidence of the highest derivatives of the sought function u_k. The conditions of compatibility can be obtained by excluding the function u_k and its derivatives from some sequences of the equations. In turn, these sequences could be obtained by the differentiation of equations (Equation 6.114).

6.5.2 Concept of the Standard Form

As analysis shows, the method of averaged characteristics is suitable for the asymptotic integration of the systems like Equation 6.113 that could be described by using the so-called standard form (also know as the standard system with partial derivatives; see Equation 5.26 and corresponding comments)

$$A'(U,z,t)\frac{\partial U}{\partial t}+\left(Z'(U,z,t)\times\frac{\partial}{\partial z}\right)U+C'(z,t)U=R'(U,z,t), \tag{6.115}$$

where A',Z',C',R',Z,R are square matrices of size $l\times l$, $U=U(z,t)$ is some vector function in Euclidean n-dimensional space R^n with coordinates $\{z_1,z_2,\ldots,z_n\}$, i.e., $\forall z\in R^n z=(z_1,z_2,\ldots,z_n)^T$, $\forall z_i\in(-\infty,+\infty)$, $i\in(1,2,\ldots,n)$, $R(\ldots)$ is a given weakly nonlinear periodical (m-fold, in general case) vector function, and t is some scalar variable such as laboratory time. It is assumed that some hierarchy of the dynamic values (in time or spatial coordinates) can be determined.

Let us turn to the terminology of the theory of method of characteristics [32]. It is not difficult to prove that the standard form (Equation 6.115) can be transformed into the so-called quasilinear (with respect to derivatives) homogeneous equation

$$\frac{\partial U}{\partial t}+Z(U,z,t)\frac{\partial U}{\partial z}=R(U,z,t). \tag{6.116}$$

Furthermore, in this section, we shall study the standard equations (and systems of equations) of the form of Equation 6.116.

6.5.3 General Scheme of the Method

The calculation procedure of the method of averaged characteristic includes three specific steps. The first is the straight transformation of an initial system of equations with partial derivatives to some κth higher level. The solution of this equation is the second step. Finally, the transformation of the obtained solution from this κth hierarchic level to the initial (zero) level (backward or reverse transformation) is accomplished during the third step. The first step consists of three smaller substeps (substages). All these steps and substeps are illustrated graphically in Figure 6.1.

We begin with the discussion of the first step of the considered calculative algorithm. As can be easily seen in Figure 6.1, it includes three more particular substeps. The first substep is based on the known properties of the characteristics of the initial differential equation (Equation 6.116). Because of these properties [32], we have a possibility of reducing the initial equation with partial derivatives (Equation 6.116) to some system with momentum derivatives (exact differential equation; see Figure 6.1)

$$\frac{dz}{dt}=Z(U,z,t);$$

$$\frac{dU}{dt}=R(U,z,t), \tag{6.117}$$

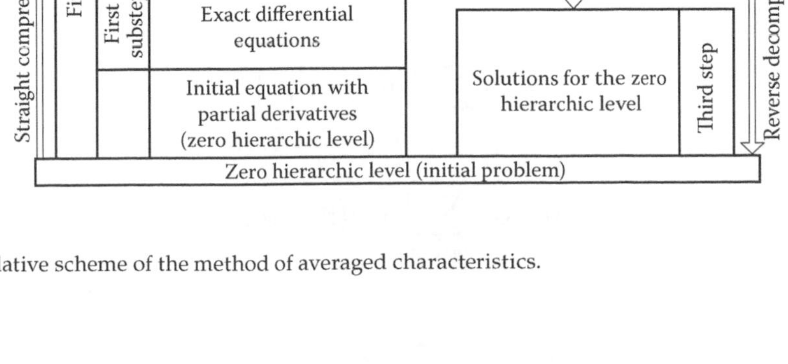

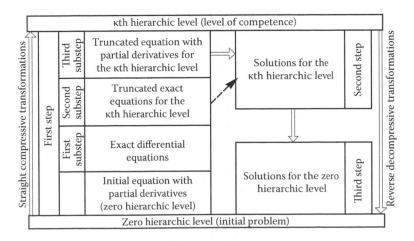

FIGURE 6.1
General calculative scheme of the method of averaged characteristics.

where all designations are self-evident. Corresponding essential simplification of Equation 6.116 can be attained as a result of relevant transformations by means of the compression operators. This simplification could be explained by the fact that the integration procedure for the equations with momentum derivatives like Equation 6.117 is usually simpler than the procedure for finding solutions of the corresponding equations with partial derivatives like Equation 6.116 (principle of information compression).

Unfortunately, the simplifying that is achieved is very often insufficient for the successive solution of Equation 6.116. The point is that Equation 6.117, especially in the case of multiperiodical nonlinear oscillations, could be far from simple from a calculative point of view. Therefore, the goal of the second substep (see Figure 6.1) is the accomplishment of the asymptotic integration of Equation 6.117 by the hierarchic methods (including versions with fast rotating phases). Thus, the hierarchic calculative technology developed in this and preceding chapters is proposed as a methodical basis for this part of the discussed general algorithm.

As before, equations such as Equation 6.117 can be regarded as another form of the initial equation of the zeroth hierarchic level (see Figure 6.1). Let us rewrite them by introducing the fast rotating phases as parameters

$$\frac{dz}{dt} = Z(U, z, \psi, t);$$

$$\frac{dU}{dt} = R(U, z, \psi, t);$$

$$\frac{d\psi}{dt} = \Omega(z, U) + Y(U, z, \psi, t), \tag{6.118}$$

where the essence of all designations is obvious from what was set forth in this chapter. Then, according to the general procedure, we separate out the complete set of the problem large parameters and construct the hierarchic series

$$\xi_1 >> \xi_2 >> \ldots >> \xi_\kappa >> \xi_m >> 1, \tag{6.119}$$

where all large parameters are determined by the described standard method. The asymptotic solutions of the system (Equation 6.118) can be represented in the form of a hierarchic sequence of the Krylov–Bogolyubov substitutions (see previous sections in this chapter). In particular, in the case of a two-level hierarchic system, the asymptotic solutions can be represented in the following form:

$$U = \bar{U} + \sum_{n=1}^{\infty} \frac{1}{\xi^n} u_u^{(n)}(\bar{z}, \bar{U}, \bar{\psi});$$

$$z = \bar{z} + \sum_{n=1}^{\infty} \frac{1}{\xi^n} u_z^{(n)}(\bar{z}, \bar{U}, \bar{\psi});$$

$$\psi = \bar{\psi} + \sum_{n=1}^{\infty} \frac{1}{\xi^n} v^{(n)}(\bar{z}, \bar{U}, \bar{\psi}), \tag{6.120}$$

where all averaged values can be found from the equations for the first hierarchic level

$$\frac{d\bar{U}}{dt} = \sum_{n=1}^{\infty} \frac{1}{\xi^n} A_u^{(n)}(\bar{z},\bar{U});$$

$$\frac{d\bar{z}}{dt} = \sum_{n=1}^{\infty} \frac{1}{\xi^n} A_z^{(n)}(\bar{z},\bar{U});$$

$$\frac{d\bar{\psi}}{dt} = \Omega(\bar{z},\bar{U}) + \sum_{n=1}^{\infty} \frac{1}{\xi^n} B^{(n)}(\bar{z},\bar{U}), \tag{6.121}$$

$\xi \equiv \xi_1$. All nonlinear components of the compression operator can be taken into account by using the algorithms described in this chapter. For instance, the result of such calculations for the first (by the small parameter $1/\xi$) approximation could be represented in the form

$$A_u^{(1)} = \frac{1}{2\pi} \int_0^{2\pi} R(\bar{z},\bar{U},\bar{\psi})\, d\bar{\psi}; \tag{6.122}$$

$$u_u^{(1)} = \frac{1}{\Omega} \int_0^{\psi} \left(R(\bar{z},\bar{U},\bar{\psi}) - A_u^{(1)} \right) d\bar{\psi}; \tag{6.123}$$

$$A_z^{(1)} = \frac{1}{2\pi} \int_0^{2\pi} Z(\bar{z},\bar{U},\bar{\psi})\, d\bar{\psi}; \tag{6.124}$$

$$u_z^{(1)} = \frac{1}{\Omega} \int \left(Z(\bar{z},\bar{U},\bar{\psi}) - A_z^{(1)} \right) d\bar{\psi}; \tag{6.125}$$

$$B^{(1)} = \frac{1}{2\pi} \int_0^{2\pi} \left(\frac{\partial \Omega}{\partial U} u_u^{(1)} + \frac{\partial \Omega}{\partial \bar{z}} u_z^{(1)} + Y(\bar{z},\bar{U},\bar{\psi}) \right) d\bar{\psi}; \tag{6.126}$$

$$v^{(1)} = \frac{1}{\Omega} \int_0^{2\pi} \left(\frac{\partial \Omega}{\partial U} u_u^{(1)} + \frac{\partial \Omega}{\partial \bar{z}} u_z^{(1)} + Y(\bar{z},\bar{U},\bar{\psi}) - B^{(1)} \right) d\bar{\psi}. \tag{6.127}$$

Taking into account the parametrical nature of the variable ψ, the equations for some κth hierarchic levels can be written as

$$\frac{dz^{(\kappa)}}{dt} = Z^{(\kappa)}\left(U^{(\kappa)}, z^{(\kappa)}\right);$$

$$\frac{dU^{(\kappa)}}{dt} = R^{(\kappa)}\left(U^{(\kappa)}, z^{(\kappa)}\right), \tag{6.128}$$

where $R^{(\kappa)}(U^{(\kappa)},z^{(\kappa)})$ is some nonlinear functional matrix, and the upper index κ denotes that the corresponding values belong to the κth hierarchic level (see Figure 6.1). The values $z^{(\kappa)}$ and t in Equation 6.128 are, in principle, dependent variables (because we passed here to the Lagrange formulation of the problem). However, and this is very important, they are considered here to be the parameters of the problem. The equations for any κth hierarchic level by themselves are the κ-fold averaged equations (Equation 6.117), that is, the equations for the κth hierarchic level.

Then, keeping the parametrical form of Equation 6.128, a special hypothesis is put in the basis of the considered step of the discussed algorithm. Namely, it is supposed that Equation 6.128 can be treated as characteristics of some other equation with partial derivatives (third substep; see Figure 6.1)

$$\frac{\partial U^{(\kappa)}}{\partial t} + Z^{(\kappa)}\left(U^{(\kappa)}, z^{(\kappa)}, t\right)\frac{\partial U^{(\kappa)}}{\partial z} = R^{(\kappa)}(U^{(\kappa)}, z^{(\kappa)}, t). \tag{6.129}$$

It should be specially pointed out that, in contrast to Equation 6.128, the variables $z^{(\kappa)}$ and t are independent (because we again come back to the Euler-form representation).

Comparing Equation 6.129 and the initial equation (Equation 6.116) we find that, in principle, both these equations have analogous mathematical structure, i.e., the hierarchic resemblance principle holds here. Because of the special form, we call Equation 6.129 the averaged (truncated) quasilinear equation of the κth hierarchic level. Also, the second hierarchic principle is satisfied (hierarchic compression principle; see Chapter 2 for details). According to this principle, the initial hierarchic system, Equation 6.116 (zero hierarchic level), reproduces itself on each hierarchic level in its general mathematic structure. It grows simpler as the hierarchy level becomes higher. Hence, the other two hierarchic principles are satisfied here, too.

The passage from the equation with partial derivatives (Equation 6.116) to the equations of characteristics (Equation 6.117) (exact differential equations; see Figure 6.1) is similar to the passage from the Euler form to the Lagrange form. Correspondingly, the passage from Equation 6.128 to Equation 6.129 can be described as a reverse passage from the Lagrange to the Euler form. However, it should be mentioned that we are talking about some resemblance between both procedures only in a general case. Exact equivalence takes place only in some particular cases. Here, a mutual resemblance of these passages is retained due to the use of the parametrical (with respect to the variables $z^{(\kappa)}$ and t) form of the representation in corresponding differential equations.

It can be seen easily that Equation 6.129 could be treated as the initial Equation 6.116 (zero hierarchic level), which is transformed to the κth hierarchic level by means of the straight compressive transformations (see Figure 6.1). So, taking this into consideration, we can affirm that the eventual result of the first step of the discussed calculative algorithm is obtaining a κ-fold averaged equation with the partial derivatives like Equation 6.129 (i.e., an equation of κth hierarchic level). We treat the totality of the calculative procedures of the first step as the straight compressive transformation of initial equation 6.116 into a hierarchically equivalent equation (Equation 6.129) for some κth level (see Figure 6.1).

The main task of the second step of the discussed algorithm is the solution of Equation 6.129. Analogous to the situation that takes place in the framework of the hierarchic methods for exact differential equations, the transformed equation for any κth hierarchic level of Equation 6.129 is also simpler than the initial equation 6.116. Due to this, the general procedure of solving a κ-fold averaged equation such as Equation 6.129 is found to be simpler than the analogous procedure for Equation 6.116. In principle, maximal simplification could be attained in the case where straight transformation of initial equation 6.116 is accomplished up to the highest hierarchic level $\kappa = m$. But, it should be mentioned especially that there is not often a strong need for performing such complete hierarchic transformations. The point is that sufficiently simplified solutions (because they are averaged) could be found on some of the lower hierarchic levels $\kappa < m$. In Chapter 2, we called them the levels of competence (see Figure 6.1).

In what follows, it should be noted that any known analytical or numerical calculative method can be applied for the solution of Equation 6.129. However, a very interesting situation occurs in the application for the purpose of the method of characteristics. The characteristics are exact equations (equations with momentum derivatives). This means that when the characteristics method is used, strictly speaking, we do not need to construct the κ-fold averaged equation 6.129 (see the dashed line in Figure 6.1). Their solutions can be constructed directly on the basis of the solutions of equations for characteristic equation 6.128. But, in any case, independent of the solution method, we eventually obtain the solutions of Equation 6.129 in the form

$$U^{(\kappa)} = U^{(\kappa)}(z^{(\kappa)},t). \tag{6.130}$$

Constructing a solution like Equation 6.130 completes the main task of the second step of the discussed calculative algorithm (see Figure 6.1).

Analyzing the form of Solution 6.130 and the possibilities of constructing corresponding solutions of the initial problem (Equation 6.116), we can observe a new peculiar calculative situation realized due to the use of the hierarchic procedure developed earlier for solving the exact equations. However, let us discuss this problem in detail.

First, we turn the reader's attention to the following circumstance. The problems discussed earlier in this and preceding chapters can be classified as one-dimensional because all dynamic variables change in this case on the time coordinate t only. On the other hand, we used the same time t for the description of functional dependencies on any κth hierarchic level, i.e., $z^{(\kappa)} = z^{(\kappa)}(t)$, $z^{(\kappa-1)} = z^{(\kappa-1)}(t)$, and so on. The spatial coordinates are different for different hierarchic levels. However, in the case of the exact equations, this does not lead to any difficulties because we have the temporal dependencies like $z^{(\kappa)} = z^{(\kappa)}(t)$ on any hierarchic level. Because of this, we get a chance to construct solutions for any lower hierarchic level, substituting the solutions for higher hierarchic levels into the Krylov–Bogolyubov substitutions

$$x^{(\kappa-1)}(t) = x^{(\kappa)}(t) + \sum_{n=1}^{\infty} \frac{1}{\xi_\kappa^n} u_{x\kappa}^{(n)}\left(x^{(\kappa)}(t), \psi^{(\kappa)}(t)\right)$$

$$\psi^{(\kappa-1)}(t) = \psi^{(\kappa)}(t) + \sum_{n=1}^{\infty} \frac{1}{\xi_\kappa^n} v_\kappa^{(n)}\left(x^{(\kappa)}(t), \psi^{(\kappa)}(t)\right), \tag{6.131}$$

where all designations were given earlier in this chapter. Thus, the relationships like Equation 6.131 play a double role in the discussed situation. Using them, we have accomplished the straight as well as the reverse transformations within the framework of the method described in this chapter.

A different, in principle, situation occurs in the case of differential equations with partial derivatives. Here, besides the temporal functional dependencies, we also have dependencies on the spatial coordinates (see the form of standard equations 6.115 and 6.116). In contrast to the preceding case, the variables of both these types are independent from each other. But, as mentioned before, the spatial coordinates are different for different hierarchic levels. This means that on each hierarchic level, the corresponding compression operator $U^{(\kappa)}$ depends on a proper set of the variables for this level. As before, the Krylov–Bogolyubov substitutions give us a connection between the different neighboring hierarchic levels

$$U^{(\kappa-1)}\left(z^{(\kappa-1)},t\right)=U^{(\kappa)}\left(z^{(\kappa)},t\right)+\sum_{n=1}^{\infty}\frac{1}{\xi_\kappa^n}u_{U\kappa}^{(n)}\left(U^{(\kappa)},z^{(\kappa)},t\right), \tag{6.132}$$

i.e., it could also be used for the straight transformations in this case. But, contrary to the discussed situation with the exact differential equations, here we cannot directly use the relationships like Equation 6.122 for constructing the solutions of the initial problem (Equation 6.116) (i.e., for performing the reverse transformations). For this, we should also have some special correlation between the proper coordinates of the neighboring hierarchic level like $z^{(\kappa)} = z^{(\kappa)}(z^{(\kappa-1)})$. This type of correlation can be constructed. Modernized in such a manner, the calculative procedures embody the main task of the third step of the considered algorithm (see Figure 6.1). We call the totality of all procedures that allow accomplishment of the transformation of the solutions from any κth hierarchic level to the zeroth hierarchic level the reverse (backward) transformations (see Figure 6.1).

The main idea of the reverse transformations is based on the following observation. If we look at the Krylov–Bogolyubov substitutions (Equation 6.132), we can see that the difference between any variables of the two neighboring hierarchic levels [(κ – 1)th and κth, for example] is proportional to $1/\xi_\kappa \ll 1$

$$U^{(\kappa-1)}\left(z^{(\kappa-1)},\psi^{(\kappa-1)},t\right)-U^{(\kappa)}\left(z^{(\kappa)},\psi^{(\kappa)},t\right)=\sum_{n=1}^{\infty}\frac{1}{\xi_\kappa^n}u_{U\kappa}^{(n)}\left(U^{(\kappa)},z^{(\kappa)},\psi^{(\kappa)},t\right). \tag{6.133}$$

Taking this into account, let us expound the function $U^{(\kappa)}(z^{(\kappa)},\psi^{(\kappa)},t)$ in the Taylor series in the vicinity of the equivalent spatial point of the (κ – 1)th hierarchic level. Substituting all differences like Equation 6.123 into this series, we obtain the required set of approximate correlation between any two neighboring hierarchic levels. For instance, we can do it with respect to the difference between the first (averaged) and the zero (initial) levels, respectively

$$U(z,t)=U(\bar{z},\bar{\psi})\Big|_{\substack{\bar{z}=z\\ \bar{\psi}=\psi}}+\left\{\frac{\partial\bar{U}(\bar{z},\bar{\psi})}{\partial\bar{z}}\left(\sum_{n=1}^{\infty}\frac{1}{\xi^n}\mathrm{u}_z^{(n)}\right)+\frac{\partial\bar{U}(\bar{z},\bar{\psi})}{\partial\bar{\psi}}\left(\sum_{n=1}^{\infty}\frac{1}{\xi^n}\mathrm{v}^{(n)}\right)\right\}\Bigg|_{\substack{\bar{z}=z\\ \bar{\psi}=\psi}}+\cdots \tag{6.134}$$

and so on. Then, we accomplish the chain of the successive analogous reverse transformations between each pair of the neighboring hierarchic levels. As a result, the reverse transformation procedure could be constructed eventually. It allows us to accomplish the reverse transformations from any higher hierarchic level (including the level of competence κ) to the initial (zero) hierarchic level.

6.6 One Example of the Application of the Method of Averaged Characteristics

As we can see, the arrangement and the calculative technology of the method of averaged characteristics are far from simple. Practice shows that their comprehension sometimes requires some additional training. This section is dedicated to discussion of an illustrative example that allows better understanding of the main calculative peculiarities of the method. This example has a purely abstract nature.

A simple system of two equations with oscillation in the right parts was chosen as a basic illustrative object. The goal of this example is to demonstrate the practical calculative technology of the method using the simplest means. The examples of other kinds that are given in Part II have another purpose. Namely, specific features of the application of the considered method in real electrodynamic problems are shown there. This includes the problem statement (i.e., reducing the initial electrodynamic equations to the standard form) and the solving of the obtained standard equations. The nonlinear multiharmonic theory of parametrical and superheterodyne FELs is also regarded as a convenient illustrative object.

6.6.1 Initial Equations

Let us write the considered system of equations in the following form:

$$\frac{\partial u_1(z,t)}{\partial t} + \frac{1}{u_0}\frac{\partial u_2(z,t)}{\partial z} = a_1 \cos(\omega t - kz);$$

$$\frac{\partial u_2(z,t)}{\partial t} + u_0 c^2 \frac{\partial u_1(z,t)}{\partial z} = a_2 \cos(\omega t - kz), \tag{6.135}$$

where u_1 and u_2 are some unknown functions, and u_0, c, a_1, a_2 are given parameters of the problem.

6.6.2 Characteristics

According to the general procedure of the method of averaged characteristics, we should construct an equation for the characteristics of Equation 6.135 in the first step (see Figure 6.1). Following the standard calculation scheme of the ordinary method of characteristics, we can construct the determinant

$$\det\begin{pmatrix} 1 & 0 & 0 & \dfrac{1}{u_0} \\ 0 & 1 & u_0 c^2 & 0 \\ dt & 0 & dz & 0 \\ 0 & dt & 0 & dz \end{pmatrix} = dz^2 - c^2 dt = 0. \tag{6.136}$$

This means that characteristics of the system can be written in the form

$$\frac{dz}{dt} = \pm c. \tag{6.137}$$

Using the well-known definition for momentum derivative

$$\frac{du_i}{dt} = \frac{\partial u_i}{\partial t} + \frac{\partial u_i}{\partial z}\frac{dz}{dt}, \tag{6.138}$$

we get the following two systems (from Equation 6.135):

$$\begin{cases} \dfrac{dz}{dt} = c; \\ \dfrac{du_1}{dt} = a_1 \cos(\omega t - kz), \end{cases} \tag{6.139}$$

and

$$\begin{cases} \dfrac{dz}{dt} = -c; \\ \dfrac{du_2}{dt} = a_2 \cos(\omega t - kz), \end{cases} \tag{6.140}$$

respectively.

6.6.3 Passage to the First Hierarchic Level

Taking Equations 6.139 and 6.140 as equations for the zeroth hierarchic level, we perform the passage to the first (i.e., the one-fold averaged) hierarchic level. For this, we transform them into the standard form (like Equation 6.116). The first stage of this transformation procedure is the introduction of the new dependent variable (oscillation phase ψ)

$$\psi = \omega t - kz. \tag{6.141}$$

The next stage is the subdivision of all variables into slow and fast ones. Taking into consideration the specific mathematical structure of systems 6.139 and 6.140, we can assume that the variables u_j and z can be treated as slow variables (variables with magnitudes that do not essentially change during the system period; see Chapter 3). The phase ψ plays the role of the fast variable because it changes greatly during the system period. Then, the large parameter of the system (hierarchic large parameter) could be determined by the standard method

$$\xi \sim \left|\frac{d\psi}{dt}\right| \Big/ \left|\frac{dx_i}{dt}\right| >> 1, \tag{6.142}$$

where $x = \{x_i\} = \{x_1, x_2\}$, $x_1 \equiv u_j$, $x_2 \equiv z$, and $j = 1,2$.

Differentiating the definition of Equation 6.141 by t, it is not difficult to obtain the differential equation for the phase ψ. As a result of the performed transformation, we can reconstruct Equations 6.139 and 6.140 into two parametrical standard forms like Equation 6.61)

$$\begin{cases} \dfrac{dz}{dt} = c; \\ \dfrac{du_1}{dt} = a_1 \cos(\psi); \\ \dfrac{d\psi}{dt} = \omega + kc, \end{cases}$$

and

$$\begin{cases} \dfrac{dz}{dt} = -c; \\ \dfrac{du_2}{dt} = a_2 \cos(\psi); \\ \dfrac{d\psi}{dt} = \omega - kc. \end{cases} \tag{6.143}$$

Then we transform Equation 6.143 into the first hierarchic level. The formal solutions of systems 6.143 can be represented in the form of Krylov–Bogolyubov substitutions

$$u_1 = \bar{u}_1 + \sum_{n=1}^{\infty} \frac{1}{\xi^n} U_{u1}^{(n)}(\bar{z}, \bar{\psi});$$

$$u_2 = \bar{u}_2 + \sum_{n=1}^{\infty} \frac{1}{\xi^n} U_{u2}^{(n)}(\bar{z}, \bar{\psi});$$

$$z = \bar{z} + \sum_{n=1}^{\infty} \frac{1}{\xi^n} U_{z}^{(n)}(\bar{z}, \bar{\psi});$$

$$\psi = \bar{\psi} + \sum_{n=1}^{\infty} \frac{1}{\xi^n} V^{(n)}(\bar{z}, \bar{\psi}), \tag{6.144}$$

where the unknown components of the decompression operators $U_{u,z}^{(n)}$ and $V^{(n)}$ can be found by using corresponding procedures of the method described in this chapter. The corresponding equations for the first hierarchic level could be obtained analogously. If we regard these equations as characteristics of the corresponding averaged equations in partial derivatives, we can construct them in the form

$$\left(\frac{\partial}{\partial t}+c\frac{\partial}{\partial \bar{z}}\right)\bar{u}_1(\bar{z},t)=0;$$

$$\left(\frac{\partial}{\partial t}-c\frac{\partial}{\partial \bar{z}}\right)\bar{u}_2(\bar{z},t)=0. \tag{6.145}$$

The solutions of truncated system 6.145 can easily be obtained

$$\begin{aligned}\bar{u}_1(\bar{z},t)&=\bar{u}_{01}(\bar{z}-ct);\\ \bar{u}_2(\bar{z},t)&=\bar{u}_{02}(\bar{z}+ct),\end{aligned} \tag{6.146}$$

where the initial (or boundary) conditions for the arbitrary functions $\bar{u}_{01}$, $\bar{u}_{02}$ (solutions) coincide with the analogous conditions for the equivalent functions of Equation 6.135. The reverse transformations should be performed (see Figure 6.1) to obtain the complete asymptotic solutions for the zeroth hierarchic level.

6.6.4 Reverse Transformations

It is not difficult to reconstruct the first two Krylov–Bogolyubov substitutions (Equation 6.143) into the form (first approximation of the small parameter $1/\xi$)

$$u_{1,2}(\bar{z},\bar{\psi},t)\cong\bar{u}_{1,2}(\bar{z},t)+\frac{a_{1,2}}{\omega\pm kc}\sin(\bar{\psi}). \tag{6.147}$$

Then, we use the calculative scheme of the reverse transformations described in the preceding section. We note that differences between any variables of the first and zeroth hierarchic levels are proportional to $1/\xi_\kappa \ll 1$

$$u_j-\bar{u}_j=\sum_{n=1}^{\infty}\frac{1}{\xi^n}\mathrm{u}_{ui}^{(n)}(\bar{z},\bar{u}_i,\bar{\psi});$$

$$z-\bar{z}=\sum_{n=1}^{\infty}\frac{1}{\xi^n}\mathrm{u}_z^{(n)}(\bar{z},\bar{u}_i,\bar{\psi});$$

$$\psi-\bar{\psi}=\sum_{n=1}^{\infty}\frac{1}{\xi^n}\mathrm{v}^{(n)}(\bar{z},\bar{u}_i,\bar{\psi}). \tag{6.148}$$

Using this observation we can accomplish corresponding expansions in the Taylor series in Equation 6.147. As a result, we obtain transformation formulas like Equation 6.134

$$u_j(z,t) = u_j(\bar{z},\bar{\psi})\Big|_{\substack{\bar{z}=z\\ \bar{\psi}=\psi}} + \left\{ \frac{\partial u_j(\bar{z},\bar{\psi})}{\partial \bar{z}} \left(\sum_{n=1}^{\infty} \frac{1}{\xi^n} \mathrm{u}_z^{(n)} \right) + \frac{\partial u_j(\bar{z},\bar{\psi})}{\partial \bar{\psi}} \left(\sum_{n=1}^{\infty} \frac{1}{\xi^n} \mathrm{v}^{(n)} \right) \right\}\Bigg|_{\substack{\bar{z}=z\\ \bar{\psi}=p_3}} + \cdots \text{ and so on.} \tag{6.149}$$

After not very difficult transformations and restricting ourselves to keeping only terms of order $1/\xi$ in Equation 6.149, we eventually obtain

$$u_1(z,t) = \bar{u}_{01}(z-ct) + \frac{a_1}{\omega + kc} \sin(\omega t - kz); \tag{6.150}$$

$$u_2(z,t) = \bar{u}_{02}(z+ct) + \frac{a_2}{\omega - kc} \sin(\omega t - kz). \tag{6.151}$$

Thus, the initial problem (Equation 6.135) is solved analytically by using the method of averaged characteristics.

References

1. Ruhadze, A.A., Bogdankevich, L.S., Rosinkii, S.E., and Ruhlin, V.G. 1980. *Physics of high-current relativistic beams*. Moscow: Atomizdat.
2. Davidson, R.C. 1974. *Theory of nonneutral plasmas.* Reading, MA: W. A. Benjamin.
3. Sitenko, A.G., and Malnev, V.M. 1994. *Principles of plasma theory.* Kiev: Naukova Dumka.
4. Landau, L.P., and Liftshitz, E.M. 1974. *Theory of field.* Moscow: Nauka.
5. Kulish, V.V. 1991. Nonlinear self-consistent theory of free electron lasers. Method of investigation. *Ukrainian Physical Journal* 36(9):1318–1325.
6. Kulish, V.V., and Lysenko, A.V. 1993. Method of averaged kinetic equation and its use in the nonlinear problems of plasma electrodynamics. *Fizika Plazmy* (*Sov. Plasma Physics*) 19(2):216–227.
7. Kulish, V.V., Kuleshov, S.A., and Lysenko, A.V. 1993. Nonlinear self-consistent theory of superheterodyne and free electron lasers. *The International Journal of Infrared and Millimeter Waves* 14(3):451–568.
8. Serebriannikov, M.G., and Pervozvansky, A.A. 1965. *Discovery of hidden periodicities.* Moscow: Nauka.
9. Lancos, L. 1961. *Practical methods of applied analysis.* Moscow: Fizmatgiz.
10. Grebennikov, E.A. 1987. *Introduction to the theory of the reasonable systems.* Moscow: Izdatelstvo MGU.
11. Sukhorukov, A.P. 1988. *Nonlinear wave interactions in optics and radiophysics.* Moscow: Nauka.
12. Bloembergen, N. 1965. *Nonlinear optics.* Singapore: World Scientific Publishing.
13. Weiland, J., and Wilhelmsson, H. 1977. *Coherent nonlinear interactions of waves in plasmas.* Oxford: Pergamon Press.
14. Vainstein, L.A., and Solnzev, V.A. 1973. *Lectures on microwave electronics.* Moscow: Soviet Radio.

15. Gaiduk, V.I., Palatov, K.I., and Petrov, D.M. 1971. *Principles of microwave physical electronics*. Moscow: Soviet Radio.
16. Alexandrov, A.F., Bogdankevich, L.S., and Ruhadze, A.A. 1978. *Principles of plasma electrodynamics*. Moscow: Vyschja Shkola.
17. Kondratenko, A.N., and Kuklin, V.M. 1988. *Principles of plasma electronics*. Moscow: Energoatomizdat.
18. Zshelezovskii, B.E. 1971. *Electron-beam parametrical microwave amplifiers*. Moscow: Nauka.
19. Olemskoi, A.I., and Flat, A.Ya. 1993. Application of the factual concept in the condensed-matter physics. *Physics-Uspekhy* 163(12):1–104.
20. Rammal, R., Toulouse, G., and Virasoro, M.A. 1986. Ultrametricity for physicists. *Reviews of Modern Physics* 58(3):765–788.
21. Kulish, V.V., Kosel P.B., and Kailyuk, A.G. 1998. New acceleration principle of charged particles for electronic applications: Hierarchic description. *The International Journal of Infrared and Millimeter waves* 19(1):3–93.
22. Marshall, T.C. 1985. *Free electron laser*. New York: MacMillan.
23. Brau, C. 1990. *Free Electron Laser*. Boston: Academic Press.
24. Luchini, P., and Motz, U. 1990. *Undulators and free electron lasers*. Oxford: Clarendon Press.
25. Landau, L.P., and Liftshitz, E.M. 1974. *Theory of field*. Moscow: Nauka.
26. Liftshitz, E.M., and Pitayevskiy, L.P. 1979. *Physical kinetics*. Moscow: Nauka.
27. Gaponov, A.V., Ostrovskii, L.A., and Rabinovich, M.I. 1970. One-dimensional waves in nonlinear disperse media. *Izv. Vysh. Uchebn.*, Ser. Radiofizika (*Sov. Radiophysics*) 13(2):169–213.
28. Rabinovich, M.I., and Talanov, V.I. 1972. *Four lectures on the theory of nonlinear waves and wave interactions*. Leningrad: Leningrad University Publishers.
29. Dzedolik, I.V., and Kulish, V.V. 1987. To the nonlinear theory of parametrical resonance of electromagnetic waves in plasmas of a high-current relativistic electron flux. *Ukrainian Physical Journal* 32(11):1672–1677.
30. Moiseev, N.N. 1981. *Asymptotic methods of nonlinear mechanics*. Moscow: Nauka.
31. Kulish, V. V. 1998. *Methods of averaging in non-linear problems of relativistic electrodynamics*. Atlanta: World Scientific Publishers.
32. Kulish, V.V. 2002. *Hierarchy and hierarchic asymptotic methods in electrodynamics*, Vol. 1 of *Hierarchic methods*. Dordrecht: Kluwer Academic Publishers.
33. Kulish, V.V. 2002. *Undulative electrodynamic systems*, Vol. 2 of *Hierarchic methods*. Dordrecht: Kluwer Academic Publishers.

7

Electron Oscillations in FEL-Like Electronic Systems

In previous chapters, we discussed the general ideology and calculation algorithms of the theory of hierarchic oscillations and waves and the methods based on that ideology and those algorithms. However, as experience shows, most of them are not easily utilized for practical purposes. Moreover, while many physicists have an idea of the hierarchic-type methods, only some know how to use them in real practice.

The main cause of this situation, as mentioned in the Preface, is the extraordinariness of the psychology, structure, basic ideology, and special language of the methods. This explains why the hierarchic methods (including the averaging methods) are not too popular in physics and engineering. On the other hand, they demonstrate an obviously high level of practical effectiveness that explains their true attractiveness (Part II).

The author's personal experience shows that the best teaching effect can be achieved in such situations only when all discussed calculation technologies are illustrated clearly through simple applied examples. This chapter presents such examples.

It should be stressed that the described calculation technologies have an explicitly expressed universal character. Consequently, they could be used practically in many areas of modern science. However, this author is predominantly an expert in electrodynamics generally, and FEL theory particularly. Therefore, the relevant illustration examples given here are in the area of FELs. All of them concern the oscillation resonant interactions of an electron that moves in the multifrequency field of electromagnetic signal and pumping waves and the longitudinal magnetic and electric fields. The models of this type are characteristic of single-particle FEL theory. We will show that, in spite of the deceptive simplicity of these models, the physics is far from simple in reality. Surprisingly, one electron in those fields can form various dynamic hierarchic structures.

7.1 Formulation of the Problem*

7.1.1 Electron Motion in the Field of Electromagnetic Waves and Magnetic and Electrostatic Fields

Let us consider the motion of an electron in a field of electromagnetic waves and longitudinal electric and magnetic fields. The general arrangement of the considered model is illustrated in Figure 7.1. We determine the strength of the longitudinal static electric and magnetic fields as

$$\vec{E}_0 = E_0\vec{n};\ \vec{H}_0 = H_0\vec{n},$$

* From Kulish, V.V. et al., *Int. J. Millimeter Waves*, 19, 33–94, 1998. With permission.

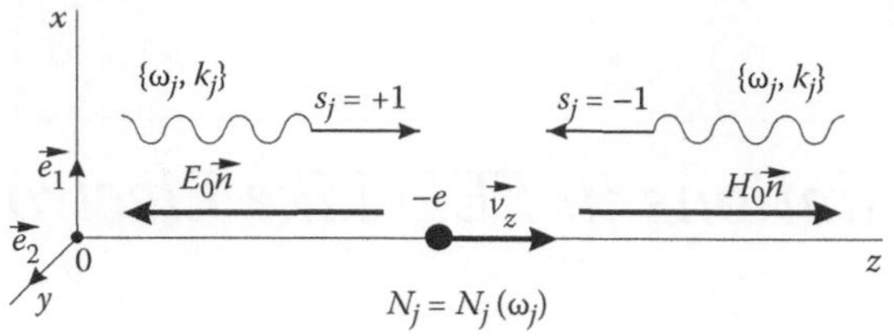

FIGURE 7.1
General arrangement of the motion of an electron in a field of electromagnetic waves and a longitudinal electric and magnetic fields. Here, $-e$ is the electron charge, $\vec{e}_1$ and $\vec{e}_2$ are the unit vectors along the x- and y-axes, $\vec{v}_z$ is the vector of longitudinal velocity, $\{\omega_j, k_j\}$ are electromagnetic waves with the frequency ω_j and wave number k_j ($j = 1,2,\ldots,n$), $s_j = \pm 1$ are the sign functions that describe the possible variants of the wave propagation, E_0 is the strength of the longitudinal electrostatic field (the field of support), H_0 is the strength of the longitudinal magnetic field (the focusing field), $\vec{n}$ is the unit vector along the z-axis, $N_j = ck_j/\omega_j$ are the wave retardation factors (refraction coefficients) in the model medium (simulated magnetodielectric), and c is the light velocity in vacuum.

where $\vec{n}$ is the unit vector along the z-axis.

We chose the vector potential of the total electromagnetic field acting on the electron in the form

$$\vec{A} = \vec{A}_0 + \sum_{j=1}^{n} \hat{\vec{A}}_j, \tag{7.1}$$

where $\vec{A}_0 = H_0\left[\vec{n}\vec{r}\right]/2$ is the vector potential of the longitudinal magnetic field, $\vec{r}$ is the radius vector in the three-dimensional space, and $\hat{\vec{A}}_j$ describes the wave component of the total field

$$\hat{\vec{A}}_j = -\frac{c\mathcal{E}_j}{\omega_j}\left[\vec{e}_1 \cos\hat{\psi}_j \sin m_j\left(p_j + \chi_{j1}\right) - \vec{e}_2 \sin\hat{\psi}_j \cos m_j\left(p_j + \chi_{j2}\right)\right], \tag{7.2}$$

$\hat{\psi}_j$ and χ_{ji} ($i = 1,2$) are the wave polarization parameters, $\mathcal{E}_j$ is the electric field amplitude of the jth transverse electromagnetic wave, p_j and ω_j are its oscillation phase and cyclic frequency, respectively, and c is the light velocity in vacuum.

We accept that the plane waves in Equation 7.2 propagate collinearly to the z-axis. It is also assumed that the interaction volume is filled with a dispersive modeling medium (simulated magnetodielectric) [1,2] with the refraction coefficient (retardation factor) $N_j = N_j(\omega_j)$. For this medium, we also consider that it is transparent for the electron and exerts an influence only on the electromagnetic waves. The medium parameters can be considered as slowly (adiabatically) varying functions along the z-axis [1,2].

By using this set of suppositions, the oscillation phases p_j can be represented in the form

$$p_j = \omega_j t - s_j k_j z = \omega_j\left(t - s_j\frac{N_j z}{c}\right) = \omega_j \zeta_j \tag{7.3}$$

where $\vec{k}_j = s_j k_j \vec{n}$ are the wave vectors, $k_j = \omega_j N_j / c$ are the wave numbers, $s_j = sign\{\vec{k}_j \vec{n}\} = \pm 1$ are the sign functions that determine the direction of the wave propagation along the z-axis (see Figure 7.1), $m_j = 1,2,3\ldots$ are the wave harmonic numbers, and $\zeta_j = t - s_j N_j z / c$. The relation between the strengths $\vec{E}$ and $\vec{H}$ and vector potential (Equation 7.1) is determined by the well-known relationship

$$\vec{E} = -\frac{1}{c}\frac{\partial \vec{A}}{\partial t} - \left(\vec{\nabla}\varphi\right), \vec{H} = \mu^{-1}\left[\vec{\nabla}\vec{A}\right], \tag{7.4}$$

where φ is the longitudinal scalar potential (i.e., $\left(\vec{\nabla}\varphi\right) = E_0 \vec{n}$), and μ is the magnetic permeability of the stimulated magnetodielectric.

Then, we find the possible integrals of motion for the electron in the field (Equations 7.1 and 7.2). For this, we use the method of the proper time [1–3]. In the framework of this method, the electron motion can be described by the following four-dimensional Lorentz equation [1–3]

$$m\frac{dv_\mu}{d\tau} = -\frac{e}{c}H_{\mu\nu}v_\nu , \tag{7.5}$$

where v_μ is the electron four-dimensional velocity

$$v_\mu = v_\mu\left\{\vec{u}, i'\frac{d(ct)}{d\tau}\right\} = v_\mu\left\{\frac{d\vec{r}}{dt}, \frac{i'c}{\sqrt{1-\beta^2}}\right\}; \tag{7.6}$$

$H_{\mu\nu}$ is the electromagnetic field tensor

$$H_{\mu\nu} = \begin{pmatrix} 0 & H_z & -H_y & -iE_x \\ -H_z & 0 & H_x & -iE_y \\ H_y & -H_x & 0 & -iE_z \\ iE_x & iE_y & iE_z & 0 \end{pmatrix}, \tag{7.7}$$

τ is the proper time, $\beta = v/c$; $E_{x,y,z}$ are the spatial components of the three-dimensional vector of the strength of the electric field $\vec{E}$, $H_{x,y,z}$, analogously, are the spatial components of the three-dimensional strength vector of the magnetic field $\vec{H}$, e is the electron's absolute charge, m is its rest mass, and i' is the primed imaginary unit for which $i'^* = i'$. The rest of the notations are standard.

We rewrite Equation 7.5 in the three-dimensional form containing derivatives with respect to the proper time τ

$$m\frac{d^2\vec{r}}{d\tau^2} = -\frac{e}{c}\left\{\vec{E}\frac{d(ct)}{d\tau} + \left[\vec{v}\vec{H}\right]\right\}; \tag{7.8}$$

$$m\frac{d^2(ct)}{d\tau^2} = -\frac{e}{c}\left(\vec{E}\vec{v}\right). \tag{7.9}$$

Now, let us introduce the electron mechanical momentum

$$\vec{p} = m\frac{d\vec{r}}{d\tau}, \tag{7.10}$$

and the total relativistic energy

$$\boldsymbol{\mathcal{E}} = \frac{mc^2}{\sqrt{1-\beta^2}}. \tag{7.11}$$

Within Equations 7.10 and 7.11, the set of Equations 7.29 and 7.9 can be written as

$$\dot{\vec{p}} = -\frac{e\boldsymbol{\mathcal{E}}}{mc^2}\vec{E} - \frac{e}{mc}\vec{n}\left(\vec{p}\vec{E}\right) + \frac{e}{mc}\vec{E}\left(\vec{p}\vec{n}\right) - \frac{eH_0}{mc}\left[\vec{p}\vec{n}\right], \tag{7.12}$$

$$\dot{\boldsymbol{\mathcal{E}}} = -\frac{e}{m}\left(\vec{E}\vec{p}\right). \tag{7.13}$$

Here, the dot denotes the total derivative with respect to the proper time τ, and $\vec{n}$ is the unit vector along the z-axis. The hidden phases p_j in Equations 7.5 through 7.13, unlike the phases in Equation 7.2, have the physical meaning of the Lagrange variables, whereas in Definition 7.2, they play the role of the Euler variables (for definitions of the Lagrange and Euler variables, see Chapter 3).

Furthermore, we find the motion integrals. For this, we pass in Equations 7.12 and 7.13 from the differentiation with respect to the proper time τ to that with respect to the parameter $\zeta_j = p_j/\omega_j$. As a result, we obtain

$$\dot{\vec{p}}_\perp = \frac{e}{mc^3}\sum_{j=1}^{n}\left\{\vec{A}_j'(\boldsymbol{\mathcal{E}} - s_jcN_jp_z)\right\} - \frac{eH_0}{mc}\left[\vec{p}\vec{n}\right]; \tag{7.14}$$

$$\dot{\vec{p}}_z = \vec{n}p_z = \vec{n}\frac{e}{mc^2}\sum_{j=1}^{n}s_jN_j\left(\vec{p}\vec{A}_j'\right); \tag{7.15}$$

$$\dot{\boldsymbol{\mathcal{E}}} = \frac{e}{mc}\sum_{j=1}^{n}\left(\vec{p}\vec{A}_j'\right), \tag{7.16}$$

where prime denotes the total derivative with respect to variable ζ_j, and $\vec{p}_\perp$ and $\vec{p}_z$ are the transverse and longitudinal components of the mechanical momentum $\vec{p} = \vec{p}_\perp + \vec{p}_z$. By using the obvious relation

$$\dot{\zeta}_j = \frac{d}{d\tau}\left(t - s_jN_j\frac{z}{c}\right) = \frac{1}{mc^2}(\boldsymbol{\mathcal{E}} - s_jN_jcp_z), \tag{7.17}$$

we obtain from Equation 7.14 an expression for the integral of the transverse motion (Redmond integral of motion) [3,4]

$$\vec{p}_{\perp} - \frac{e}{c}\sum_{j=1}^{n}\vec{A}_j - \frac{eH_0}{c}\left[\vec{n}\vec{r}\right] = \vec{f}_0 = const. \tag{7.18}$$

Within the context of the canonical momentum definition

$$\vec{\mathcal{P}} = \vec{p} + \frac{e}{c}\vec{A} = \vec{\mathcal{P}}_{\perp} + \vec{n}\mathcal{P}_z, \tag{7.19}$$

and the definition of the vector potential of the longitudinal magnetic field $\vec{A}_0$ (see Equation 7.1 and relevant comments) and integral of motion (Equation 7.18) can be rewritten as

$$\vec{\mathcal{P}}_{\perp} + \frac{e}{c}\vec{A}_0 = \vec{f}_0 = const. \tag{7.20}$$

Until now, we have not made special assumptions concerning the proportion of the rates of change of the Lagrange phases p_j of electron oscillation or the number of the partial waves and their dispersion properties. Therefore, the expression for the integral of motion (Equation 7.20), thus derived, is valid for a study of electron motions in both resonance and nonresonance cases. Moreover, one can be sure that the integral of motion (Equation 7.20) preserves its form in the presence of an arbitrary longitudinal electrostatic field (Equation 7.4). We will use these properties of the integral (Equation 7.20) for our future analysis.

In the case $H_0 = 0$, Equations 7.18 and 7.20 yield the known Volkov integral of motion [2,3]

$$\vec{p}_{\perp} + \frac{e}{c}\vec{A} = \vec{\mathcal{P}}_{\perp} = const. \tag{7.21}$$

Apart from Equations 7.20 and 7.21, other motion integrals can be found in various cases of the studied model. However, such integrals are not interesting to us because they concern only the nonresonant variants of the electron motion.

At this stage of the resonant problem, we exhaust all means for the exact solution of the problem. Further analysis requires using some approximate methods. For this, we will use the hierarchic methods described in Chapter 6.

7.1.2 Reducing the Initial Equations to the Hierarchic Standard Form

Let us illustrate the calculation and physical peculiarities of the electron resonant motion using the example of the discussed model. As before, we assume that the electron moves in the fields of Equations 7.1 through 7.4. According to the general theory of hierarchic oscillations, a complete set of the oscillation phases should be separated at the first stage of the calculation procedures. This set can contain evident as well as hidden phases (see Chapter 3 for more detail). The procedure of the separation of the evident phase in our problem is obvious: These phases simply reproduce the form of the Euler phases of electromagnetic waves p_j in Equations 7.2 and 7.3. The only peculiarity is the fact that, at this time, these phases should be treated as Lagrange variables, where the time and spatial coordinates are dependent [$z = z(t)$]. This is in contrast to the Euler phases, where the variables z and t are independent.

We then find that all sets of the hidden phases (i.e., those associated with the hidden periods) are represented by only one magnetic phase p_0

$$p_0 = \operatorname{arctg}(\mathcal{P}_y/\mathcal{P}_x), \tag{7.22}$$

where we represent the canonical momentum $\vec{\mathcal{P}}$ in the form of the sum of the transverse and longitudinal components: $\vec{\mathcal{P}} = \vec{\mathcal{P}}_\perp + \vec{n}\mathcal{P}_z$. Here, $\vec{\mathcal{P}}_\perp = \mathcal{P}_x, \mathcal{P}_y$, $\mathcal{P}_{x,y}$ are the x and y components of the transverse canonical momentum

$$\vec{\mathcal{P}}_\perp = (\mathcal{P}_\perp \cos p_0)\vec{e}_1 + (\mathcal{P}_\perp \sin p_0)\vec{e}_2, \mathcal{P}_\perp = \sqrt{\mathcal{P}_x^2 + \mathcal{P}_y^2}, \tag{7.23}$$

$\vec{e}_{1,2}$ are the unit vectors along the x- and y-axes.

We select the Hamilton equations as a basic set of the motion equations

$$\frac{d\mathcal{H}}{dt} = \frac{\partial \mathcal{H}}{\partial t};$$

$$\frac{d\vec{\mathcal{P}}}{dt} = -\frac{\partial \mathcal{H}}{\partial t};$$

$$\frac{d\vec{r}}{dt} = \frac{\partial \mathcal{H}}{\partial \vec{\mathcal{P}}}, \tag{7.24}$$

where $\mathcal{H} = \sqrt{m^2c^4 + c^2\left(\vec{\mathcal{P}} + \frac{e}{c}\vec{A}\right)^2} - e\varphi$ is the electron Hamiltonian in the fields of Equation 7.1. Let us perform the required procedures of reducing Equation 7.24 to the so-called hierarchic standard form (see Chapters 5 and 6 for more detail). As the first step, we represent System 7.24 in the parametric (with respect to the phases p_0,p_j) form

$$\frac{d\mathcal{H}}{dt} = \frac{\partial}{\partial t}\mathcal{H}\left(t, \vec{r}, \vec{\mathcal{P}}, p_j, p_0\right);$$

$$\frac{d\vec{\mathcal{P}}_\perp}{dt} = -\frac{\partial}{\partial \vec{r}_\perp}\mathcal{H}\left(t, \vec{r}, \vec{\mathcal{P}}, p_j, p_0\right);$$

$$\frac{d\mathcal{P}_z}{dt} = -\frac{\partial}{\partial z}\mathcal{H}\left(t, \vec{r}, \vec{\mathcal{P}}, p_j, p_0\right);$$

$$\frac{dp_j}{dt} = \omega_j - s_j\frac{\omega_j}{c}N_j(z)\frac{\partial}{\partial \mathcal{P}_z}\mathcal{H}\left(t, \vec{r}, \vec{\mathcal{P}}, p_j, p_0\right);$$

$$\frac{dp_0}{dt} = \frac{1}{\mathcal{P}_\perp^2}\left(\mathcal{P}_y \frac{\partial}{\partial x} - \mathcal{P}_x \frac{\partial}{\partial y}\right)\mathcal{H}\left(t, \vec{r}, \vec{\mathcal{P}}, p_j, p_0\right);$$

$$\frac{d\vec{r}_\perp}{dt} = \frac{\partial}{\partial \mathcal{P}_\perp}\mathcal{H}\left(t, \vec{r}, \vec{\mathcal{P}}, p_j, p_0\right);$$

$$\frac{dz}{dt} = \frac{\partial}{\partial \mathcal{P}_z}\mathcal{H}\left(t, \vec{r}, \vec{\mathcal{P}}, p_j, p_0\right). \tag{7.25}$$

We eliminate the transverse coordinate $\vec{r}_\perp = \{x, y, 0\}$ from System 7.25 by using the integral of motion (Equation 7.20) and the given definition for the vector potential $\vec{A}_0$ (see Equation 7.1).

Now we have to classify the slow and fast variables. For physical reasons, we take the energy $\mathcal{H}$, the transverse and longitudinal moments $\mathcal{P}_\perp$ and $\mathcal{P}_z$, and some phases as the slow variables. Then, we introduce the combination phases θ and ψ. The combination phases θ are the slowly varying quantities in accordance with the definition (see Chapters 3, 5, and 6)

$$\theta_{\nu g} = \frac{m_\nu n_\nu}{m_g n_g} p_\nu + \sigma_{\nu g} p_g, \tag{7.26}$$

where the numbers $\nu, g = j$, $\nu \neq g$ and $m_{\nu,g} = 1,2,3\ldots$ are the harmonic numbers of the wave fields in Equation 7.2, $n_{\nu g} = 1,2,3\ldots$ are the harmonic numbers of nonlinear electron oscillations in these fields, and $\sigma_{\nu g} = \pm 1$ are the sign functions. By definition, the fast variables are the sign-conjugated combination phases ψ

$$\psi_{\nu g} = \frac{m_\nu n_\nu}{m_g n_g} p_\nu - \sigma_{\nu g} p_g. \tag{7.27}$$

The rotating phases p_ν and p_g, $\nu, g = j$ given by Equation 7.4 are the components of the Lagrange phase vector p_μ. Pairs of these components form the pair-wise combination phases of Equations 7.26 and 7.27.

Basically, combination phases (Equation 7.27) can form the slow combination phases with other combination and ordinary fast phases of the vector $\{p_j, p_0\}$, which are not included in Equations 7.26 and 7.27. However, we do not consider such complex versions of the multifrequency resonances. Therefore, for simplicity, we can assume that the vector of slow combination phases includes only pair-wise combinations of the form of Equation 7.26.

To write the hierarchic large parameter ξ, we employ the procedure considered in Chapters 3, 5, and 6. Namely, we arrange the large parameters

$$\xi_{\nu g f k} \sim \left|\frac{d\psi_{fk}}{dt}\right| \Big/ \left|\frac{d\theta_{\nu g}}{dt}\right| \tag{7.28}$$

as a hierarchy series and determine the leading term ξ_1, and so on. As a result of the accomplished transformations, System 7.25 is reduced to the required standard form (see Chapters 5 and 6).

7.1.3 Classification of the Models

Then, analogous to the general theory in Chapter 3, let us discuss some of the resonances determined by Condition 7.28. In Chapter 3, we distinguished the so-called quasilinear and parametric resonances. Here, the quasilinear resonance of the zero hierarchic level is the cyclotron resonance only

$$n_0 \frac{dp_0}{dt} \approx m_j n_j \frac{dp_j}{dt}, \tag{7.29}$$

where $n_0 = \pm 1, \pm 2, \ldots$ are the cyclotron harmonic numbers; all other designations are given above. Generally, the definition of Equation 7.29 describes the multiharmonic cyclotron resonance. In the simplest partial case $n_0 = m_j n_j = 1$, we obtain the main cyclotron resonance (see Chapters 3, 5, and 6 for details). Both variants of the cyclotron resonance are illustrated in this chapter.

The situation with the parametric resonances is much more complicated. We must deal with the harmonics of the multifrequency field of the electromagnetic wave field m_j (Equations 7.1 and 7.2) and the harmonics of the nonlinear electron oscillations in the fields n_j at the same time. Apart from that, we should consider that these wave-like fields can be realized physically in many different technological ways. They can be presented in the form of electromagnetic waves in the microwave waveguides or in the retarding systems. They can be the laser radiation in optical or quasioptical systems. We can also use them for creation of such fields as the magnetic or the electrical undulation systems (H- and E-undulators), including the crossed magnetoelectrical undulation fields (EH-undulators) [1,2]. This means that in real situations, the parametric resonance can be in essentially different technological forms. However, each time, it demonstrates the same basic physical regularities. Some of these will be illustrated in the examples of various technological designs of the FEL models.

This means that in practice, we may have to deal simultaneously with a lot of various evident oscillation phases of different hierarchies. As a result, the general arrangement of the hierarchic oscillation resonant physical picture could be rather complicated. Some of the most typical oscillation resonant combinations were discussed in Chapter 3 (see Figures 3.7 through 3.19 and corresponding comments). Therefore, in this chapter, we confine ourselves to a discussion of only a few of the simplest variants. They are the pair-wise parametric resonances

$$m_v n_v \frac{dp_v}{dt} \approx m_g n_g \frac{dp_g}{dt}, \tag{7.30}$$

and the bounded (coupled) resonances, where one of the oscillation phases (p_g, for example) takes place in two resonances simultaneously

$$m_v n_v \frac{dp_v}{dt} \approx m_g n_g \frac{dp_g}{dt} \approx m_l n_l \frac{dp_l}{dt}. \tag{7.31}$$

Here, the numbers ν, g, and l are the indices of the relevant electromagnetic waves: $\nu, g, l = j$, but $\nu \neq g \neq l$. In the case $\nu = 0$, $m_\nu = m_0 = 1$, Condition 7.31 describes the coupled parametric cyclotron resonances.

Taking into account the physical meaning of the cyclotron phase and the cyclotron frequency, we obtain a possibility of classifying the discussed hierarchic models with respect to the magnitudes of the longitudinal magnetic field H_0 (let us recall that $H_0 \sim |dp_0/dt|$)

1. Weak magnetic field

$$\left|\frac{dp_0}{dt}\right| << \left|\frac{d\psi_{vg}}{dt}\right|. \tag{7.32}$$

As follows from the analysis of the motion integral (Equation 7.20), the strict inequality holds for a weak magnetic field if

$$\mathcal{P}_\perp^2 >> \left(\frac{e}{c}A_0\right)^2. \tag{7.33}$$

The subregion of the superweak field can be separated in the region of the interaction volume with the weak magnetic field. It is situated immediately in the vicinity of the point $H_0 = 0$, such that

$$\left|\frac{dp_0}{dt}\right| \leq \left|\frac{d\theta_{vg}}{dt}\right|, \mathcal{P}_\perp >> \frac{e}{c}A_0. \tag{7.34}$$

If the condition

$$\left|\frac{dp_0}{dt}\right| \approx \left|\frac{d\theta_{vg}}{dt}\right| \tag{7.35}$$

holds, the phenomenon of the quasiresonance can be realized in the system [2,6]. As analysis shows, it is not a true resonance because we do not observe the essential growth of the oscillation amplitude. It is accompanied only by a qualitative anomaly in the character of the transverse electron motion [5,6].

2. Strong magnetic field

The rates of change of the magnetic and the fast combination phases in this case have the same order,

$$\left|\frac{dp_0}{dt}\right| \sim \left|\frac{d\psi_{vg}}{dt}\right|. \tag{7.36}$$

The magnetic field affects the character of the transverse motion considerably if

$$\mathcal{P}_{\perp} \sim \frac{e}{c} A_0. \tag{7.37}$$

For

$$\left|\frac{dp_0}{dt}\right| \approx \left|\frac{d\psi_{vg}}{dt}\right|; \quad \left|\frac{d\theta_{vg}}{dt}\right|, \left|\frac{d\theta_{0j}}{dt}\right| << \left|\frac{d\psi_{vg}}{dt}\right|, \left|\frac{d\psi_{0j}}{dt}\right|, \tag{7.38}$$

we have the above-mentioned case of the doubled parametric cyclotron resonance [5,6].

3. Superstrong magnetic field

A superstrong magnetic field takes place when [5,6]

$$\left|\frac{dp_0}{dt}\right| >> \left|\frac{d\psi_{vg}}{dt}\right|; \tag{7.39}$$

$$\mathcal{P}_{\perp} << \frac{e}{c} A_0. \tag{7.40}$$

In this case, the influence of the magnetic field on the electron motion in the wave-like fields has a suppressive character [5,7]. Therefore, such magnetic fields are used for focusing electron beams in the microwave electronic devices with longitudinal basic waves [7].

7.2 Cyclotron Resonances*

7.2.1 Model and the General Hierarchic Tree

We assume that the electromagnetic wave defined by Expressions 7.1 and 7.2 contains only one spectral component ($j = n = 1, m_j = m_1 = 1$). Then, the resonant condition (Equation 7.29) can be rewritten in the following simplified form:

$$n_0 \frac{dp_0}{dt} \approx n_j \frac{dp_j}{dt}. \tag{7.41}$$

The hierarchic tree, which illustrates this simplified situation, is represented in Figure 7.2. As can be readily seen, its characteristic feature in this case has to do with the harmonics

* From Kulish, V.V. et al., *Int. J. Millimeter Waves*, 19, 33–94, 1998. With permission.

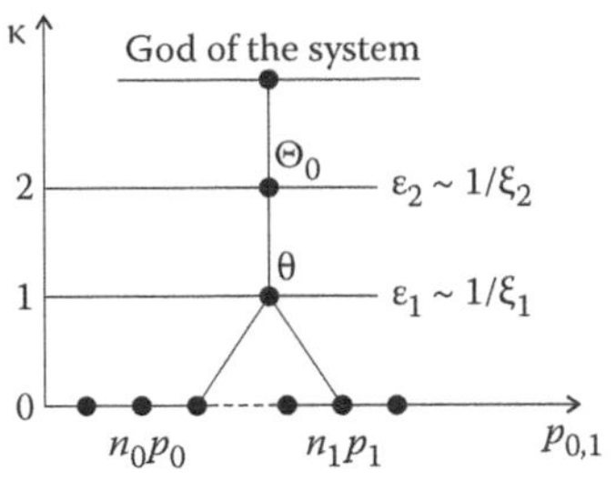

FIGURE 7.2
Hierarchic tree for the multiharmonic cyclotron resonance. Here, dark dots n_0p_0 designate the phase group of the cyclotron harmonics $n_0 = 1,2,3,\ldots$. The analogous dots n_1p_1 correspond to the phases of the electron oscillation harmonics in the field monochromatic electromagnetic wave ($m_j = m_1 = 1$). Other designations are the same as in Chapter 3.

of electron oscillations in the monochromatic electromagnetic wave n_j and, as before, with the cyclotron harmonics n_0.

According to the general theory, the next step of the calculation algorithm is the passage from the ordinary phases p_0,p_1 to the combination phases (Equations 7.26 and 7.27). In our case, they can be written as

$$\theta = n_0p_0 + \sigma_{10}n_1p_1 \tag{7.42}$$

and

$$\psi = n_0p_0 - \sigma_{10}n_1p_1, \tag{7.43}$$

where $\sigma_{01} = \pm 1$ are relevant sign functions.

In accordance with the given definitions, the case $n_0 = n_1 = 1$ is determined as the main resonance. The case of the fractional resonance takes place if the ration n_0/n_1 is a fraction. In this section, we will discuss both cases of cyclotron resonance. The case of the doubled parametric cyclotron main resonance ($j = 1,2$; $n_0 = n_1 = n_2 = \pm 1$) will be illustrated further in Section 7.6.3.

The cyclotron resonance is one of the most widespread resonant phenomena in modern applied electrodynamics [7]. Examples include accelerators of charged particles (cyclotrons, for example), electronic devices of different types (magnetrons, gyrotrons, masers at cyclotron resonance, some electronic devices of the M-type, etc.), various plasma systems (plasma-beam and plasma fusion systems, for instance), and so on [7]. However, it is known that the general physics of such systems is not as simple as it first seems. This is especially true in the case of the relativistic electron motion. Therefore, we will illustrate some mathematical aspects of the practical application of the discussed hierarchic methods, especially in the examples of the relativistic versions of the cyclotron resonance.

7.2.2 Case of the Main Cyclotron Resonance

Let us start our analysis with the simplest case of main cyclotron resonance ($n_0 = n_1 = 1$; see the hierarchic tree in Figure 7.3).

Taking into account all the accepted suppositions, it is not difficult to reduce the initial system (Equation 7.25) to the standard form similar to Equation 6.16. Then, we use the

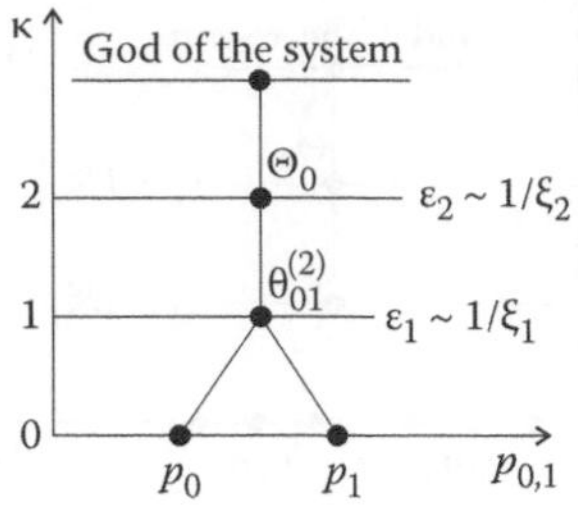

FIGURE 7.3
Hierarchic tree in the case of the main cyclotron resonance. It is the partial case of the hierarchic tree shown in Figure 7.2 for $n_0 = n_1 = 1$. Other designations are the same in Chapter 3.

compression operator on this system (see Equation 6.18 and corresponding discussion). As a result, we obtain the required truncated equations of the first hierarchic level. In the lowest (zero) approximation with respect to the large parameter (Equation 7.28), the truncated system can be obtained in the form

$$\frac{d\bar{\mathcal{H}}}{dt} = \frac{ec\omega_1}{2\bar{\mathcal{H}}} \bar{\mathcal{P}}_\perp A_1 \sin\bar{\theta},$$

$$\frac{d\bar{\mathcal{P}}_z}{dt} = \frac{eck_1}{2\bar{\mathcal{H}}} \bar{\mathcal{P}}_\perp A_1 \sin\bar{\theta},$$

$$\frac{d\bar{\mathcal{P}}_x}{dt} = -\frac{e^2 H_0}{4\bar{\mathcal{H}}} A_1 \sin\bar{\theta},$$

$$\frac{d\bar{\mathcal{P}}_y}{dt} = \frac{e^2 H_0}{4\bar{\mathcal{H}}} A_1 \cos\bar{\theta},$$

$$\frac{d\bar{p}_x}{dt} = \frac{d\bar{\mathcal{P}}_x}{dt} + \frac{eH_0\bar{p}_y}{2mc\bar{\gamma}},$$

$$\frac{d\bar{p}_y}{dt} = \frac{d\bar{\mathcal{P}}_y}{dt} - \frac{eH_0\bar{p}_x}{2mc\bar{\gamma}},$$

$$\frac{d\bar{\theta}}{dt} = \omega_1 - k_1\bar{v}_z - n_0\frac{eH_0 c\bar{\mathcal{P}}_\perp}{2\bar{\mathcal{H}}\bar{p}_\perp} - n_0\frac{e^2 H_0}{4\bar{\mathcal{H}}\bar{p}_\perp} A_1\cos\bar{\theta} + n_0\frac{e^2(\omega_1 - k_1\bar{v}_z)}{2\bar{p}_\perp c} A_1\cos\bar{\theta}, \tag{7.44}$$

where $\bar{p}_\perp = \sqrt{\bar{p}_x^2 + \bar{p}_y^2} = m\bar{v}_\perp\bar{\gamma}$, $\bar{p}_z = m\bar{v}_z\bar{\gamma}$ are the averaged transverse and longitudinal mechanical electron momentums, $\bar{\mathcal{P}}_\perp = \sqrt{\bar{\mathcal{P}}_x^2 + \bar{\mathcal{P}}_y^2}$ and $\bar{\mathcal{P}}_z = \bar{p}_z$ are the averaged transverse and longitudinal electron canonical momentums, $\bar{\gamma} = 1\Big/\sqrt{1 - \left(\bar{v}_x^2 + \bar{v}_y^2 + \bar{v}_z^2\right)/c^2} = 1\Big/\sqrt{1 - \beta^2}$

is the relativistic factor, $\bar{v}_z, \bar{v}_\perp$ are the longitudinal and transverse averaged electron velocities, and m is the electron resting mass. All other designations are already determined in this chapter.

The analytical solutions of System 7.44 can be found through Jacoby's or Weierstrass's elliptical functions [11,12]. An example of the numerical representation of the obtained solutions is illustrated in Figure 7.4.

The specific peculiarity of the zero approximation used in System 7.44 is that the decompression operator $\hat{U}^{(0)}$ in this case is equal to unity. This means that the solutions of the initial system (Equation 7.25) for the zero hierarchic level can be represented as $z \approx \bar{z}$, $\psi \approx \bar{\psi}$, i.e., in this case, we have the situation, which is similar to van der Pol's method (see Chapter 5 for details).

It should be noted that, in spite of the fact that Equation 7.44 is obtained by the lowest approximation, it satisfactorily describes the physical properties that are most important. This circumstance determines the observed relative popularity of the zero approximation in the analysis of many practical resonant-like systems. But a question arises: What is the place of the higher (first, second, etc.) approximations in practice? There are a few important aspects to the answer. Let us discuss one of the most significant.

According to the general theory, the higher the approximation, the larger the system length where the solution accuracy is guaranteed (see Equation 6.45 and relevant comments). This means that the choice of the approximation is practically determined by the physical parameters of the considered system. As a result, we sometimes obtain mathematical situations that are not so simple. First, it is because the complexity of the relevant truncated equations increases strongly with each following approximation. Let us illustrate this unpleasant peculiarity by calculating the nonlinear addends of the first approximations to Equations 7.44 (see Equation 6.18 and the corresponding discussion)

$$A_{2\mathcal{H}} = -\frac{e^2c^2k_1^2\omega_1\bar{\mathcal{P}}_\perp^2A_1^2}{8m\gamma\bar{\mathcal{H}}^2\Omega^2} + \frac{37n_0e^4\omega_1H_0A_1^3}{96\bar{p}_\perp\bar{\mathcal{H}}^2\Omega}\sin(\bar{\theta})$$

$$+\frac{3e^4n_0\omega_1H_0A_1^3}{32\bar{p}_\perp\bar{\mathcal{H}}^2\Omega}\cos(\bar{\theta}) + \frac{e^4\omega_1n_0H_0A_1^3}{32\bar{p}_\perp\bar{\mathcal{H}}^2\Omega}\cos(3\bar{\theta}),$$

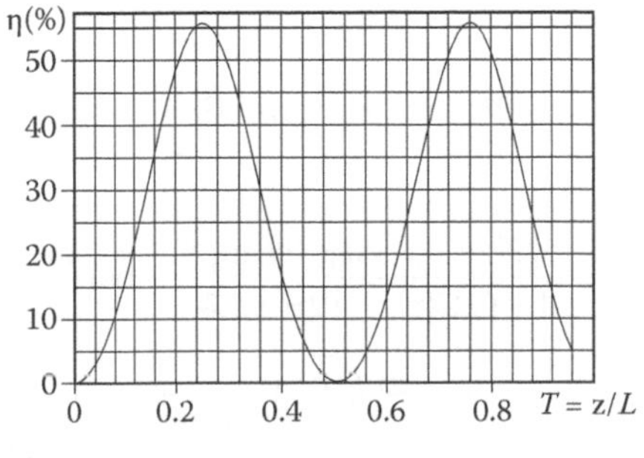

FIGURE 7.4
Dependency of the electron efficiency η on the dimensionless system length $T = z/L$ in a case of main cyclotron resonance. Here, the following system parameters are accepted for calculations: relativistic factor $\gamma = 9.45$, signal wavelength $\lambda_1 = 2$ mm, strength of the longitudinal magnetic field $H_0 = 8 \cdot 10^3$ Gs, strength of the electric field of the signal wave $\mathcal{E}_1 = 3.5 \cdot 10^7$ V/m, initial longitudinal component of electron velocity $v_{z0} = 2.0 \cdot 10^{10}$ cm/s, and initial transverse component $v_\perp = 2.2 \cdot 10^{10}$ cm/s, $L = 100$ cm.

$$A_{2\mathcal{P}z} = -\frac{e^2c^2k_1^3\bar{\mathcal{P}}_\perp^2 A_1^2}{8m\gamma\bar{\mathcal{H}}^2\Omega^2} + \frac{37n_0e^4k_1H_0A_1^3}{96\bar{p}_\perp\bar{\mathcal{H}}^2\Omega}\sin(\bar{\theta}) + \frac{3e^4n_0k_1H_0A_1^3}{32\bar{p}_\perp\bar{\mathcal{H}}^2\Omega}\cos(\bar{\theta})$$
$$+\frac{e^4k_1n_0H_0A_1^3}{32\bar{p}_\perp\bar{\mathcal{H}}^2\Omega}\cos(3\bar{\theta}),$$

$$A_{2\mathcal{P}x} = -\frac{e^3k_1\bar{v}_zH_0A_1^2}{4c\bar{\mathcal{H}}\Omega} + \frac{e^3H_0\omega_1A_1^2}{4c\Omega\bar{\mathcal{H}}} + \frac{e^3cn_0\bar{\mathcal{P}}_\perp H_0A_1}{8\bar{p}_\perp\Omega\bar{\mathcal{H}}^2}\sin(\bar{\theta})$$
$$+\frac{e^3ck_1\bar{\mathcal{P}}_\perp A_1^2}{8m\gamma\Omega^2\bar{\mathcal{H}}}\sin(2\bar{\theta}),$$

$$A_{2\mathcal{P}y} = \frac{e^3k_1\bar{v}_zH_0A_1^2}{4c\bar{\mathcal{H}}\Omega} - \frac{e^3H_0\omega_1A_1^2}{4c\Omega\bar{\mathcal{H}}} - \frac{e^3cn_0\bar{\mathcal{P}}_\perp H_0A_1}{8\bar{p}_\perp\Omega\bar{\mathcal{H}}^2}\cos(\bar{\theta})$$
$$+\frac{e^3ck_1\bar{\mathcal{P}}_\perp A_1^2}{8m\gamma\Omega^2\bar{\mathcal{H}}}\cos(2\bar{\theta}),$$

$$A_{2px} = -\frac{e^3k_1\bar{v}_zH_0A_1^2}{4c\bar{\mathcal{H}}\Omega} + \frac{e^3H_0\omega_1A_1^2}{4c\Omega\bar{\mathcal{H}}} + \frac{e^3cn_0\bar{\mathcal{P}}_\perp H_0A_1}{8\bar{p}_\perp\Omega\bar{\mathcal{H}}^2}\sin(\bar{\theta})$$
$$+\frac{e^3ck_1\bar{\mathcal{P}}_\perp A_1^2}{8m\gamma\Omega^2\bar{\mathcal{H}}}\sin(2\bar{\theta}) - \frac{\omega_1k_1e^3A_1^3}{4\sqrt{2}mc\Omega^2\gamma\bar{\mathcal{H}}}\sin(\bar{\theta}) - \frac{k_1^2e^3\bar{v}_zA_1^3}{4\sqrt{2}mc\Omega^2\gamma\bar{\mathcal{H}}}\sin(\bar{\theta}),$$

$$A_{2py} = \frac{e^3k_1\bar{v}_zH_0A_1^2}{4c\bar{\mathcal{H}}\Omega} - \frac{e^3H_0\omega_1A_1^2}{4c\Omega\bar{\mathcal{H}}} - \frac{e^3cn_0\bar{\mathcal{P}}_\perp H_0A_1}{8\bar{p}_\perp\Omega\bar{\mathcal{H}}^2}\cos(\bar{\theta})$$
$$-\frac{e^3ck_1\bar{\mathcal{P}}_\perp A_1^2}{8m\gamma\Omega^2\bar{\mathcal{H}}}\cos(2\bar{\theta}) + \frac{\omega_1k_1e^3A_1^3}{4\sqrt{2}mc\Omega^2\gamma\bar{\mathcal{H}}}\cos(\bar{\theta}) + \frac{k_1^2e^3\bar{v}_zA_1^3}{4\sqrt{2}mc\Omega^2\gamma\bar{\mathcal{H}}}\cos(\bar{\theta}),$$

$$A_{2\theta} = \frac{n_0^2e^4H_0^2A_1^2}{32\bar{p}_\perp^2\Omega\bar{\mathcal{H}}^2} - \frac{n_0e^2k_1^2\bar{\mathcal{P}}_\perp A_1^2}{4\bar{p}_\perp m\gamma\Omega\bar{\mathcal{H}}} - \frac{n_0e^2\omega_1k_1^2\bar{\mathcal{P}}_\perp A_1^2}{4\bar{p}_\perp m\gamma\Omega^2\bar{\mathcal{H}}}$$
$$+\frac{n_0e^2k_1^3\bar{v}_z\bar{\mathcal{P}}_\perp A_1^2}{4\bar{p}_\perp m\gamma\Omega^2\bar{\mathcal{H}}} - \frac{n_0^2\bar{\mathcal{P}}_\perp^2e^2c^2}{4\bar{p}_\perp^2\Omega\bar{\mathcal{H}}^2} - \frac{n_0e^4H_0\omega_1A_1^3}{12\bar{p}_\perp\Omega\bar{\mathcal{H}}^3}\cos(\bar{\theta})$$
$$+\frac{n_0e^2\bar{\mathcal{P}}_\perp\omega_1A_1}{2\bar{p}_\perp\Omega\bar{\mathcal{H}}}\cos(\bar{\theta}) - \frac{n_0e^2\bar{\mathcal{P}}_\perp k_1\bar{v}_zA_1}{2\bar{p}_\perp\Omega\bar{\mathcal{H}}}\cos(\bar{\theta}) - \frac{n_0e^4H_0k_1^2A_1^3}{6\bar{p}_\perp\Omega^2\bar{\mathcal{H}}^2m\gamma}\cos(\bar{\theta}) \tag{7.45}$$
$$-\frac{n_0e^2c^2k_1^2\bar{\mathcal{P}}_\perp^2A_1}{8\bar{p}_\perp\Omega^2\bar{\mathcal{H}}^2m\gamma}\cos(\bar{\theta}) + \frac{7n_0^2e^4H_0^2A_1^2}{32\bar{p}_\perp^2\Omega\bar{\mathcal{H}}^2}\cos(2\bar{\theta}) - \frac{n_0e^4H_0\omega_1A_1^3}{32\bar{p}_\perp\Omega\bar{\mathcal{H}}^3}\cos(3\bar{\theta}),$$

where

$$\Omega = \omega_1 - k_1 \bar{v}_z - \frac{n_0 e H_0}{2mc\gamma}.$$

It can be easily seen that System 7.44 with addends is essentially more complicated than System 7.45 in itself. Luckily, our experience shows that the real situation is not as depressing as it looks at first. The point is that the described mathematical procedures for the derivation of truncated equations of any approximation, as well as those for obtaining their asymptotic solutions, can be computerized without any essential mathematical difficulties. Experience shows also a possibility of generalization of the discussed computerization procedure for the case of the multihierarchic oscillation resonant models.

7.2.3 Case of the Fractional Cyclotron Resonance

Here we will discuss the case of fractional cyclotron resonance. Let us recall its main definition. For this, we rewrite Condition 7.1 in a more expressive form, which accounts for specific properties of the fractional resonance [44]

$$\frac{dp_1}{dt} \approx \frac{n_0}{n_1} \frac{dp_0}{dt}, \tag{7.46}$$

where the ratio n_0/n_1 plays a role in the fractional harmonic number of the cyclotron resonance. Taking into account the illustrative character of this calculation example, let us limit ourselves only to the study of the second cyclotron subharmonic $n_0/n_1 = 1/2$ ($n_0 = 1$, $n_1 = 2$), which is illustrated in Figure 7.5.

Performing the required set of calculations, we obtain the system of truncated equations (first hierarchic level) in the zero approximation

$$\frac{d\bar{\mathcal{H}}}{dt} = -\frac{e^2\omega_1}{4\bar{\mathcal{H}}} A_1^2 \sin\bar{\theta},$$

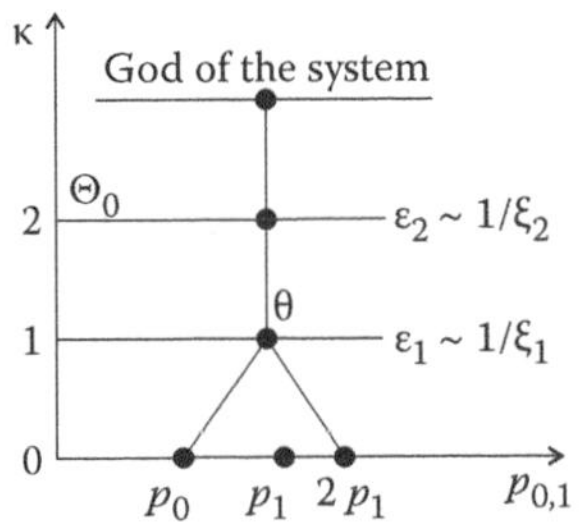

FIGURE 7.5
Hierarchic tree for the fractional cyclotron resonance at the second subharmonic ($n_0/n_1 = 1/2$). All designations are the same as given for Figures 7.2 through 7.4.

$$\frac{d\bar{\mathcal{P}}_z}{dt} = -\frac{e^2 k_1}{4\mathcal{H}} A_1^2 \sin\bar{\theta},$$

$$\frac{d\bar{\mathcal{P}}_x}{dt} = 0,$$

$$\frac{d\bar{\mathcal{P}}_y}{dt} = 0,$$

$$\frac{d\bar{p}_x}{dt} = \frac{eH_0\bar{p}_y}{2mc\gamma},$$

$$\frac{d\bar{p}_y}{dt} = -\frac{eH_0\bar{p}_x}{2mc\gamma},$$

$$\frac{d\bar{\theta}}{dt} = 2(\omega_1 - k_1\bar{v}_z) - n_0\frac{eH_0}{2mc\gamma}, \tag{7.47}$$

where all designations are given earlier in this section.

Similar to the case of the main resonance system, System 7.44, System 7.47 can also be integrated analytically through the elliptical functions. The numerical illustration of the obtained solutions is given in Figure 7.6 [44].

Comparing the materials of Figures 7.4 and 7.6, we can conclude that the real influence of the fractional resonances on the basic process (the main resonance, for example) is too weak. As a clear evident illustrative example that proves this conclusion, we give the gyro-resonant electronic devices of traditional microwave electronics [7]. However, this does not mean that we can always neglect this influence. Another situation is characteristic of the superpowerful (10^9–10^{11} W) electronic devices. Such systems began to develop during the

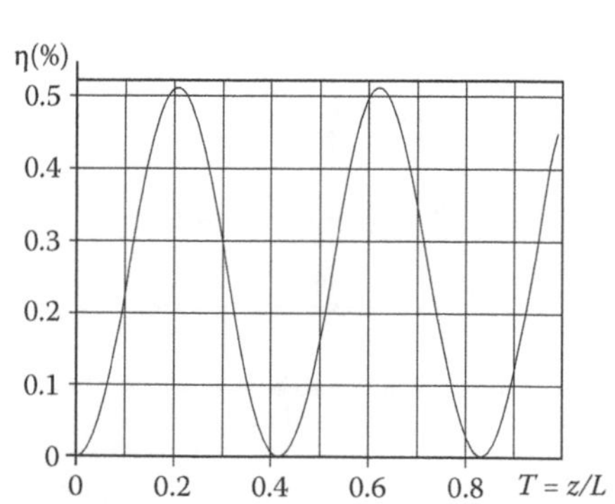

FIGURE 7.6
Dependency of electron efficiency η on the dimensionless system length $T = z/L$ in the cyclotron resonance at the second subharmonic. Here, the signal wavelength $\lambda_1 = 4$ mm; all other system parameters coincide with those that are accepted for the calculations illustrated in Figure 7.4.

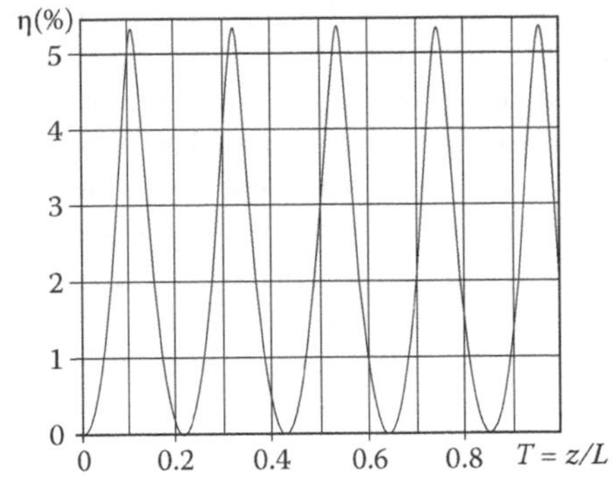

FIGURE 7.7
Dependency of electron efficiency η on the dimensionless system length $T = z/L$ in the cyclotron resonance at the second subharmonic for an amplitude higher than that shown in Figure 7.6. Here, signal wavelength λ_1 = 4 mm, and the strength of the electric field of the signal wave $\mathcal{E}_1 = 1.5 \cdot 10^8$ V/m. All other system parameters coincide with those accepted for the calculations illustrated in Figure 7.4. (From Kulish, V.V., et al., *Int. J. Infrared Millimeter Waves*, 19(1), 3–93, 1998. With permission. Kulish, V.V., *Physical Process in Parametrical Electronic Lasers (Free Electron Lasers)*, Ph.D. diss., Institute of Physics of the Academy of Sciences of Ukraine, 1985.)

time of the famous Star Wars program [13,14]. The subharmonic electromagnetic signals, as can be seen by comparison of the curves in Figures 7.6 and 7.7, become commensurable with the main working signal. This can be characterized by a rather high level of output power (10^8–10^{10}W). Obviously, this may create a lot of undesirable technological problems.

7.3 Parametric Resonances: General Case*

7.3.1 A Few Words about the Parametrical Resonances

The parametrical resonances are known in radio and electrical engineering from the twentieth century [15]. Beginning in the 1950s, this phenomenon was intensively studied in microwave electronics, nonlinear optics, plasma physics, and other fields [16–19]. However, the most promising experimental results in this area were obtained at the end of the 1970s. This was the time when the FELs first appeared as a new class of relativistic electronic parametrical devices [1,2,20–29]. Some of these first experiments indeed looked revolutionary. For instance, the power of the coherent electromagnetic signals achieved before the first FEL in the middle of the submillimeter range is characterized by magnitude ~10^{-4}W. The megawatt electromagnetic pulse at the 0.4-mm wavelength was generated in 1976 in the first experiment with the Dopplertron-type FEL [30]. Hence, the power jump in this experiment was ~10^{10}-fold.

When estimating the value of this achievement, it should be mentioned that the history of modern applied physics knows only a few examples of power jumps of such a scale. Only the nuclear explosion and the generation of optical quantum lasers can compare with this achievement. Therefore, it was not a surprise that the FELs, similar to nuclear systems

* From Kulish, V.V. et al., *Int. J. Millimeter Waves*, 19, 33–94, 1998. With permission.

and quantum lasers, have been chosen by the military as one of the key elements of the Star Wars program (see Chapter 1 for more detail) [13,14].

We can observe a no less significant impact of the phenomenon in fundamental physics. As a convincing illustrative example, we can take the mechanism of the bounded hierarchic resonant oscillations of the brane universe discussed in Chapters 2, 3, and 8. As shown, this mechanism can satisfactorily explain the process of the universe's evolution as a whole (for an example, see Figure 3.19 and corresponding discussion).

So, taking into account the intriguing aspects of the problem, let us illustrate further the fundamental ideas and concepts of the theory of parametric resonant oscillations. As before, let us do this within the framework of electrodynamics.

7.3.2 Model and Formulation of the Problem

Similar to cyclotron resonance, we will use the general model and formulation of the problem described in Section 7.1. Hence, for the initial system of equation, we chose Equation 7.25.

We are interested predominantly in the parametric pair-wise resonances where the number of oscillation phases is not larger than three (see the resonant Conditions 7.30 and 7.31). Also, we accept the supposition that all possible resonances are the main resonances.

Then, taking into consideration Equation 7.2, we introduce the following notations:

$$\mathcal{E}_{ji} = \mathcal{E}_j f_i(\hat{\psi}_i)\exp(i\chi_{ji});$$

$$f_i(\hat{\psi}_i) = \{\cos\hat{\psi}_j, (i=1)\} \quad \text{or} \quad \sin\hat{\psi}_j, (i=2), \tag{7.48}$$

where we should not confuse imaginary unit $i=\sqrt{-1}$ with the lower index i. Then, the electromagnetic wave field in the interaction region is determined by

$$E_x = -\frac{1}{c}\frac{\partial A_x}{\partial t} = \mathrm{Re}\sum_{j=1}^{n} \mathcal{E}_{j1} e^{ip_j};$$

$$E_y = -\frac{1}{c}\frac{\partial A_y}{\partial t} = \mathrm{Im}\sum_{j=1}^{n} \mathcal{E}_{j2} e^{ip_j};$$

$$A_x = -c\,\mathrm{Im}\sum_{l=1}^{n} \frac{\mathcal{E}_{j1}}{\omega_j} e^{ip_j};\; A_y = c\,\mathrm{Re}\sum_{j=1}^{n} \frac{\mathcal{E}_{j2}}{\omega_j} e^{ip_j}. \tag{7.49}$$

Then we use Equations 7.49 and 7.21 and the definition for the Hamiltonian of the electron in Equations 7.1 and 7.2

$$\mathcal{H} = \sqrt{m^2c^4 + c\left(\vec{\mathcal{P}} + \frac{e}{c}\vec{A}\right)^2} - e\varphi, \tag{7.50}$$

and pass in Equation 7.25 from differentiation with respect to time t to that with respect to coordinate z. As a result of the accomplished transformations, we can rewrite the first of the Hamiltonian equations of Equation 7.24 in the form

$$\frac{d\mathcal{H}}{dz} = -\frac{e}{\mathcal{P}_z}\left[\mathcal{P}_x \operatorname{Re}\sum_{j=1}^{n}\mathcal{E}_{j1}e^{ip_j} + \mathcal{P}_y \operatorname{Im}\sum_{j=1}^{n}\mathcal{E}_{j2}e^{ip_j}\right]$$
$$+\frac{ce^2}{\mathcal{P}_z}\left[\operatorname{Im}\sum_{j=1}^{n}\frac{\mathcal{E}_{j1}}{\omega_1}e^{ip_j}\operatorname{Re}\sum_{j=1}^{n}\mathcal{E}_{j1}e^{ip_j}\right. \tag{7.51}$$
$$\left.+\operatorname{Re}\sum_{j=1}^{n}\frac{\mathcal{E}_{j2}}{\omega_2}e^{ip_j}\operatorname{Im}\sum_{j=1}^{n}\mathcal{E}_{j2}e^{ip_j}\right],$$

where

$$\mathcal{P}_z = \mathcal{P}_z(\mathcal{H}, p_j) = \left\{\frac{(\mathcal{H}-e\varphi)^2}{c^2} - m^2c^2\right.$$
$$\left.-\left[\vec{\mathcal{P}}_\perp - \vec{e}_1 e \operatorname{Im}\sum_{j=1}^{n}\frac{\mathcal{E}_{j1}}{\omega_j}e^{ip_j} + \vec{e}_2 e \operatorname{Re}\sum_{j=1}^{n}\frac{\mathcal{E}_{j2}}{\omega_j}e^{ip_j}\right]^2\right\}^{1/2} \tag{7.52}$$

$\vec{e}_1$ and $\vec{e}_2$ are the unit vectors along the x- and y-axes, and $\vec{\mathcal{P}}_\perp = const$ is given by the motion integral of Equation 7.21. The last step is separating out the large parameter $\xi_1 \equiv \xi$ and finding which variables are components of the vector of slow variables x and which should be attributed to the vector of the fast rotating phases ψ_{vg} (Equation 7.27). In other words, we distinguish between the slowly and fast varying variables of Equation 7.52.

We have mentioned that the slowly varying variables are the electron energy $\mathcal{E} = \mathcal{H}$ and the momentum $\mathcal{P}_z$. The classification of the slow and fast variables is neither simple nor obvious because Equation 7.52 is nonlinear and some variables enter implicitly. This concerns the quantities expressed in terms of linear combinations of the oscillation phases θ_{vg} (Equation 7.26). To answer the question of which of these combination phases are slow and which are fast, one has to specify the model. In other words, imposing some conditions of slow variation for a group of combination phases physically infers specialization of the type of parametric resonance assumed to occur in the system.

Subsequently, we illustrate this assertion more clearly. In the considered problem, it is possible to introduce two combination phases (Equations 7.26 and 7.27) instead of each pair of phases p_v and p_g, (i.e., we study the model with the paired resonances only). For the simplified case $m_v = n_v = m_g = n_g = 1$—the main resonance—the expressions for the combination phases of Equations 7.26 and 7.27 can be rewritten as

$$\theta_{vg} = p_v + \sigma_{vg}p_g; \quad \psi_{vg} = p_v - \sigma_{vg}p_g, \tag{7.53}$$

where $\sigma_{\nu g} = \pm 1$ are the sign functions. Since the electron interaction mechanism is invariant under the interchange of the νth and gth waves, we have $\sigma_{\nu g} = -\sigma_{g\nu}$. It follows from Equation 7.51 that

$$\theta_{\nu g} = \sigma_{\nu g}\theta_{g\nu}; \quad \psi_{\nu g} = -\sigma_{\nu g}\psi_{g\nu}. \tag{7.54}$$

After these transformations, we rewrite Equation 7.51 with new variables $\theta_{\nu g}$ and $\psi_{\nu g}$ as parameters

$$\frac{d\mathcal{H}}{dt} = \aleph(\mathcal{H}, \psi_1, \ldots, \psi_m, \theta_1, \ldots, \theta_k, z), \tag{7.55}$$

$$\frac{d\theta_{\nu g}}{dz} = \frac{\dot{p}_\nu + \sigma_{\nu g}\dot{p}_g}{c^2\mathcal{P}_z}\sqrt{m^2c^4 + c^2\left(\vec{\mathcal{P}} + \frac{e}{c}\vec{A}\right)^2}; \tag{7.56}$$

$$\frac{d\psi_{\nu g}}{dz} = \frac{\dot{p}_\nu - \sigma_{\nu g}\dot{p}_g}{c^2\mathcal{P}_z}\sqrt{m^2c^4 + c^2\left(\vec{\mathcal{P}} + \frac{e}{c}\vec{A}\right)^2}, \tag{7.57}$$

where $m + k = n$ is the number of the initial independent phase variables (to avoid a mistake, we distinguish the index m and the similar designation for electron rest mass m),

$$\frac{dp_{\nu,g}}{dt} = \omega_{\nu,g} - s_{\nu,g}\frac{\omega_{\nu,g}N_{\nu,g}}{c}\frac{\partial}{\partial\mathcal{P}_z}\mathcal{H}(\psi_1, \ldots, \psi_m, \theta_1, \ldots, \theta_k, z, \mathcal{P}_z), \tag{7.58}$$

$\mathcal{H}(\vec{\mathcal{P}})$ is determined by Equation 7.50 and the motion integral of Equation 7.21,

$$p_\nu = (1/2)(\theta_{\nu g} + \psi_{\nu g}); \quad p_g = (\sigma_{\nu g}/2)(\theta_{\nu g} - \psi_{\nu g}), \tag{7.59}$$

and the dot denotes differentiation with respect to the laboratory time. The meaning of the function $\aleph(\ldots)$ is self-evident.

One can be sure that Systems 7.55 through 7.57, accounting for Equation 7.28, can be regarded as standard hierarchic systems like System 6.5. Hence, the general hierarchic algorithm for asymptotic integration (see Chapter 6) can be used to solve it.

7.3.3 Formation of the Combination Phases: The General Case

In this section and further in the chapter, we consider the study of the partial case of pair-wise resonances (i.e., the resonant processes with the participation of only two phases). This pair-wise illustrative case has many various methodological merits. However, experience shows that it also has shortcomings. For instance, because of its simplicity, this example cannot be useful for an illustration of the mathematical peculiarities that appear only when the number of oscillation phases p_j is larger than two. This is especially important in the formulation of a general computer algorithm for numerical solutions of multiphase problems of the type discussed. So, Sections 7.3.3 through 7.3.5 are dedicated to the discussion of these kinds of general methodological peculiarities.

In accordance with the general algorithm, an important first step is the passage from the oscillation phases p_j to the combination phases θ_{vg} and ψ_{vg}. The next, but no less important, stage is the classification of these combination phases into slow and fast. Let us discuss some general methodological features of these calculation procedures.

We assume the coordinate system K to be related to the manifold $\{p_j\}$ in the following way. We introduce an n-dimensional space, G, over the field of real numbers, i.e., $L \equiv R^n$, and describe each point of this space by an ordered set of real numbers (coordinates) $p_1, p_2, \ldots, p_n$.

Then, we put the numbers p_j (phases) in one-to-one correspondence with the functions $\exp(ip_j)$ that describe the propagation of the relevant electromagnetic waves (Equation 7.2) having complex amplitudes E_j. We regard the quantities $\exp(ip_j)$ as functions of independent variables t and z (which enter p_j). Hence, it is logical to treat the set $\{p_j\}$ as linearly independent vectors in the space L.

We are interested only in the linear nondegenerate transformations A that map the space L into itself. Instead of n linearly independent old variables, transformations (Equation 7.53) introduce $2C_n^m$ new variables (C_n^m is the number of combinations of n, m at a time), which are, generally speaking, linearly dependent. We choose n variables (θ,ψ) from this manifold in such a way that they contain each p_j of the set $\{p_j\}$ $(j = 1,2,\ldots,n)$ at least once. Then the rest $2C_n^m - n$ new variables are linear combinations of the introduced n variables and are determined by the relations

$$\psi_{ij} + \theta_{ij} = -\sigma_{lj}\psi_{lj} + \sigma_{lj}\theta_{lj}, \quad l \neq j, \quad i,j,l = 1,2,\ldots,n. \tag{7.60}$$

It is evident that not every set of the n new variables (θ,ψ) obtained in such a way is linearly independent. Therefore, the choice of variables (Equation 7.53) is not arbitrary; only the linearly independent combinations are appropriate. The condition $\det\{a_{ik}\} \neq 0$ (a_{ij} are the elements of the matrix A) makes it possible to find the group of linear nondegenerate transformations A that satisfy the requirement of the linear independence.

7.3.4 Example of Three Oscillation Phases

Let us illustrate the general scheme by three oscillation phases. According to Equation 7.53, we introduce the transformations for $\{p_j\}$ $(j = 1,2,3)$ given by

$$\begin{aligned} &\psi_1 = p_1 - \sigma_1 p_2, \quad &\psi_2 = p_2 - \sigma_2 p_3, \quad &\psi_3 = p_1 - \sigma_3 p_3, \\ &\theta_1 = p_1 + \sigma_1 p_2, \quad &\theta_2 = p_2 + \sigma_2 p_3, \quad &\theta_3 = p_1 + \sigma_3 p_3, \end{aligned} \tag{7.61}$$

where $\sigma_{1,2,3} = \pm 1$. In Equation 7.61, the symmetry of ψ_i and θ_i is used and what follow from the properties of Equation 7.54 and combinations of the type $p_3 - \sigma_3 p_1$ do not enter the formulas. It is clear that ψ_i and θ_i satisfy relations of the form of Equation 7.60, i.e.,

$$\psi_1 + \theta_1 = \psi_3 + \theta_3; \quad \psi_2 + \theta_2 = -\sigma_1(\psi_1 - \theta_1); \quad \sigma_2(\psi_2 - \theta_2) = \sigma_3(\psi_3 - \theta_3), \tag{7.62}$$

that imply that the number of the initial linearly independent variables p_j is equal to three. It may be shown that, of the total set of three-dimensional vectors whose components are phases ψ_i and θ_i, only 16 vectors formally satisfy the condition of linear independence. Indeed, let us write all the determinants corresponding to the matrices associated with the transformations A from the three variables (p_j) to the three new variables (ψ_i,θ_i) and analyze them. According to Equation 7.61, we have

$$\begin{aligned}
&\Delta(\psi_1,\psi_2,\psi_3) = -\sigma_1\sigma_2 - \sigma_3, \quad \sigma_i = \pm 1, \quad i = 1,2,3;\\
&\Delta(\theta_1,\theta_2,\theta_3) = -\sigma_1\sigma_2 + \sigma_3, \quad \sigma_i = \pm 1, \quad i = 1,2,3;\\
&\Delta(\psi_i,\psi_j,\theta_k) = s_1(\sigma_3 - \sigma_1\sigma_2), \quad s_1 = \pm 1, \quad i \neq j \neq k;\\
&\Delta(\psi_i,\theta_j,\theta_k) = -s_2(\sigma_1\sigma_2 + \sigma_3), \quad s_2 = \pm 1, \quad i \neq j \neq k.
\end{aligned} \tag{7.63}$$

All other determinants are proportional to σ_j or to the products of the type $\sigma_i\sigma_j$ and therefore do not vanish. Inasmuch as substituting σ_j instead of $-\sigma_j$ is equivalent to choosing among the values of σ_j, we can consider that all σ_j are equal to +1. Substituting $\sigma_j = +1$ in Equation 7.63, we find that there exist only four different linear degenerate transformations, i.e.,

$$p_i \to \{\theta_1,\theta_2,\theta_3\} \text{ and } p_i \to \{\psi_i,\psi_j,\theta_k\}, \quad i \neq j \neq k, \quad (i,j,k = 1,2,3). \tag{7.64}$$

The rest of the linearly independent triads of the variables (ψ,θ) can be divided into three groups with different numbers of variables ψ and θ, namely,

$$\begin{aligned}
&1.\ p_i \to \{\psi_1,\psi_2,\psi_3\};\\
&2.\ p_i \to \{\psi_i,\psi_j,\theta_k\}; \quad \text{where either } i = k \text{ or } j = k,\\
&3.\ p_i \to \{\psi_i,\theta_j,\theta_k\},
\end{aligned} \tag{7.65}$$

$i,j,k = 1,2,3$, in which the sequence order of variables in Equation 7.65 is unimportant. Then, making use of the symmetry properties of Equation 7.54, we can show that the transformations of group 2 in Formulas 7.65 are interchanged under the inversion transformations $(i \to j \to k)$ of the subscripts of the three variables (p_i). We shall refer to this property of transformations 7.53 as the invariance property with respect to the group of inverse transformations of the subscripts of the initial variables (p_i). Among the third group of the transformations, two subgroups can be separated into

1. A subgroup with $(i \neq j \neq k)$
2. A subgroup with $i = j$ or $i = k$

that are invariant with respect to each other. The first subgroup contains three transformations and the second subgroup contains six. Thus, for the case of $n = 3$, we distinguish transformations of four essentially different types described by Equation 7.61. Inasmuch as the three initial variables (p_j) have been enumerated in an arbitrary manner, we can choose any triad of variables (ψ,θ), one triad from each of the first and the second groups (Equation 7.65) and two triads from the third group of Equation 7.64, corresponding to first and second subgroups. Thus, we have

$$1.)\ \psi_1 = p_1 - p_2; \quad \psi_2 = p_2 - p_3; \quad \psi_3 = p_1 - p_3;$$

$$2.)\ \psi_1 = p_1 - p_2;\quad \psi_2 = p_2 - p_3;\quad \theta_1 = p_1 + p_3;$$

$$3.)\ \psi_1 = p_1 - p_2;\quad \theta_1 = p_1 - p_2;\quad \theta_2 = p_2 + p_3;$$

$$4.)\ \psi_1 = p_1 - p_2;\quad \theta_1 = p_2 + p_3;\quad \theta_2 = p_1 + p_3. \tag{7.66}$$

It is clear now that the transformations of Equation 7.66 may be rewritten as

$$1)\ \psi_1 = p_1 - \sigma p_2;\quad \psi_2 = p_2 - \sigma p_3;\quad \psi_3 = p_1 - \sigma p_3; \tag{7.67}$$

$$2)\ \psi_1 = p_1 - \sigma_1 p_2;\quad \theta_1 = p_2 + \sigma_2 p_3;\quad \theta_2 = p_1 + \sigma_1 p_2, \tag{7.68}$$

where $\sigma, \sigma_1, \sigma_2 = \pm 1$. Indeed, transformations 1 and 4 from Equation 7.66 are described by three ψ-variables (Equation 7.67), while transformations 2 and 3 are described by three (ψ,θ)-variables (Equation 7.68). Any interchange of the subscripts of (p_j) cannot transform Equation 7.67 into Equation 7.68 or vice versa. In view of the extension to an arbitrary n, we introduce the concept of the topological inequivalence for this and similar cases. Then, we can say that the transformations of Equations 7.67 and 7.68 are topologically inequivalent and describe the group of the allowed nondegenerate transformations from the p-variables to the (ψ,θ)-variables.

7.3.5 Separation of the Resonant Combination Phases

We have already mentioned that requiring that some combination phases be slowly varying functions is equivalent to assuming that some parametric-type resonance occurs in the system. In the mathematical formalism, this is equivalent to imposing parametric relationships for the frequencies and wave vectors of electromagnetic waves of the type of Equation 7.28 and the electron velocity in the interaction region (the conditions of parametric resonances). These relationships (resonant conditions) are usually described by equations of the form of Condition 7.30, which, in the considered case of the parametrical resonance, yield

$$\omega_i(1 - s_i N_i \beta_z) \approx -\sigma_{ij}\omega_j(1 - s_j N_j \beta_z), \tag{7.69}$$

where $s_{i,j} = \pm 1$ are the sign functions that describe the directions of the propagation of the electromagnetic waves (Equation 7.2), $N_{i,j}$ are the retardation factors of these waves in the modeling medium, $\beta_z = v_z/c$, v_z is the longitudinal electron velocity in the interaction region, and c is the light velocity in vacuum. Here, we shall follow the works of Kulish and Kochmanski and Kulish [1,10] and refer to the resonant conditions of the same type as Condition 7.69 as the conditions of synchronism.

The combination phases, whose frequencies and wave vectors are related by the synchronism conditions of the same type as Condition 7.69, belong to the components of the vector x in a relevant truncated system, while the rest of the phases are components of the vector ψ.

In Section 3.2, a diagram technique based on the idea of hierarchic trees was discussed and is destined for the graphical illustration of the different type of resonances. Let us discuss another similar technique that was also elaborated for the resonant analysis (see Figures 7.8 through 7.11) [1,10]. It should be noted that these techniques are not competitors. As will become clear, each of them has its own specific area of application.

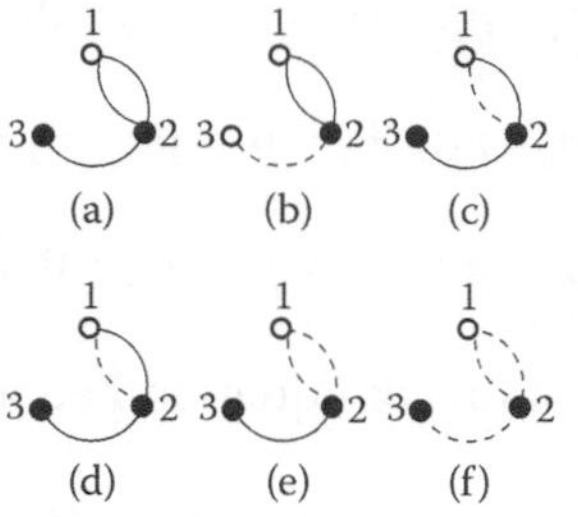

FIGURE 7.8
Graphical representation of the possible resonant modes in transformations of the types in Equation 7.68.

Thus, in order to visualize the analysis and to open the way to generalizing the results obtained in the case of arbitrary number of oscillation phases n, we introduce the graph representation for the set of probable modes of resonant interaction between electrons and the fields of the type of Equation 7.49.

Let the empty circles in Figures 7.8 through 7.11 denote the vertices associated with the ith wave (p_i), and the filled black circles correspond to $\sigma_i p_j$. The solid lines that connect the vertices p_i and p_j are associated with the fast varying angular variables, and the dashed lines are associated with the slow varying angular variables. For the sake of brevity, we shall call the former and the latter fast and slow angular variables, respectively. We shall arrange the vertices on a circle and enumerate them clockwise. Thus, it is sufficient to fix one vertex number and the numbers of all other vertices will be determined unambiguously. Two vertices, the ith and the jth connected by two lines, will be referred to as a loop (see Figure 7.10a).

Let us consider the modes described by the transformation group in Equation 7.68 in terms of the graph representation. These modes correspond to the graphs shown in Figure 7.8. It is evident from the figure that it is impossible to impose the slow variation requirements on three phases at once (this case corresponds to the graph in Figure 7.8f), i.e., the synchronism of Condition 7.69, in principle, cannot be satisfied for three phases simultaneously. Indeed, this requirement yields an equation of the type $a = -a$; its single solution is $a = 0$, which excludes the possibility of a resonance interaction. Thus, the graph in Figure 7.8f describes a forbidden mode. It is not difficult to show that this conclusion is also valid for arbitrary n. Moreover, the graph in Figure 7.8e is forbidden also. Therefore, loops like those in Figure 7.10c describe a forbidden mode. The loops in Figure 7.10b also can be excluded

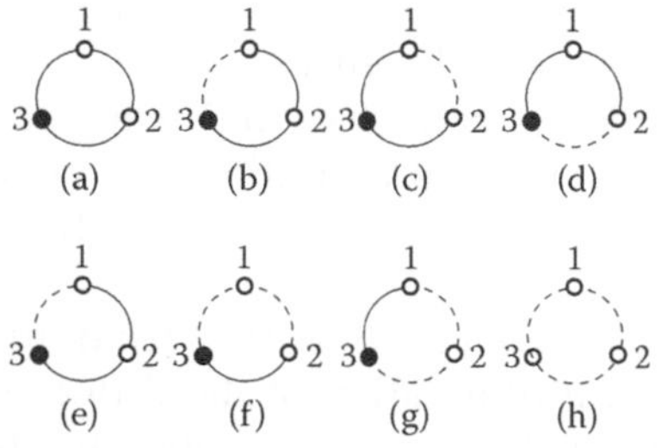

FIGURE 7.9
Graphical representation of possible resonant modes in transformations of the types in Equation 7.67.

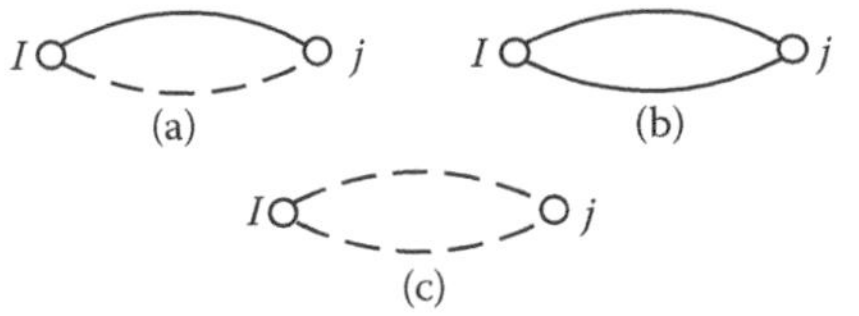

FIGURE 7.10
Graphical representation of different variants of interaction modes of oscillation phases p_i and p_j.

from consideration, though for a different reason. Our purpose is to employ the hierarchic method for analysis of the model under consideration by averaging the initial equations of Equation 7.24 over fast variables. In this method (with two or more fast variables present), the quantities inverse to the rates of change of the latter must be of the same order of smallness. It is clear, however, that the loop in Figure 7.10b changes the fast combination phases $p_i + \sigma_i p_j$ and $p_i - \sigma_i p_j$, which are of different orders of smallness. Hence, the graphs in Figure 7.8a and b should be omitted. As a result, we have to analyze the graphs in Figure 7.8c and d, which describe two essentially different cases:

1) One fast variable (ψ_1, for example) and two slow variables (θ_1 and θ_2)
2) Two fast variables (ψ_1 and ψ_2, for example) and one slow variable (θ_1).

As follows from Equation 7.68, the case when θ_1 is the fast variable does not add any new information.

Now, let us consider the transformations described by Equations 7.67. The relevant graphs are shown in Figure 7.9. Comparing the graphs of the two figures reveals essential distinctions that may be illustrated in terms of the introduced topological equivalence concept. It is clear that the graph in Figure 7.9e is forbidden. The graphs in Figure 7.9b and d describe the same mode by virtue of the interchange symmetry with respect to the subscripts of vertices 1 and 2. A similar observation corresponds to the graphs in Figure 7.9g and h. That is why we exclude the graphs in Figure 7.9d and g from consideration. As for the graph in Figure 7.9a, it does not contradict anything and, hence, describes a mode that can occur in principle.

Thus, of all the sets of graphs presented in Figures 7.8 and 7.9, only seven graphs should be analyzed: the graphs in Figure 7.8c and d and the graphs in Figure 7.9a, b, c, e, and h. We note that for $\sigma = -1$, the graphs in Figure 7.9b and c are associated with physically equivalent situations by virtue of interchange symmetry with respect to the subscripts of the vertices p_i. The same is true for the graphs in Figure 7.9e and h. For $\sigma = -1$, however, these graphs describe situations that are not physically equivalent.

Finally, we have to consider the group of so-called identical transformations whose distinctive feature is that m variables of $\{p_j\}$ are transformed into themselves ($p_j \to \psi_j$) (see Figure 7.11). The rest of $n - m$ variables are transformed according to the formulas of Equation 7.53. In the graph representation, the set of m vertices that are not involved in the transformations of Equation 7.53 gives rise to isolated vertices of types "solid line + light circle" and "dashed line + light circle" (see Figure 7.11).

The case $n = 3$ is described also by the set of graphs shown in Figure 7.11. Among them, the graphs in Figure 7.11c and d can be shown to be forbidden. The graphs in Figure 7.11a and b are associated with the modes that are of no interest in the treatment of the interaction that produces an effective particle field energy exchange. The graphs in Figure

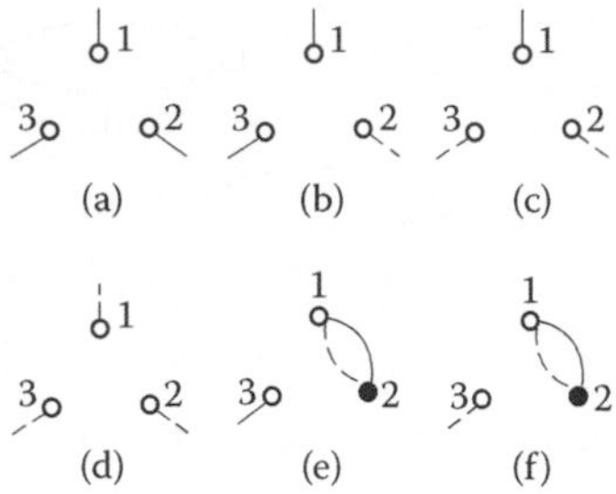

FIGURE 7.11
Graphical representation of the cases of identical transformations (a, b, and c) and the pairwise resonances (e and f).

7.11e and h will be considered further; we shall reveal their relationship with the modes described by the graphs in Figure 7.8c and d.

7.4 Case of Two Electromagnetic Waves*

7.4.1 Model and Hierarchic Tree

Let us analyze the simplest model, $n = 2$. As before, we use the formulation of the problem given in this chapter. We assume, however, that the only pair-wise resonance is realized in the model shown in Figure 7.12

$$\theta = p_1 + \sigma_{12} p_2, \ \psi = p_1 - \sigma_{12} p_2,$$

where $p_\nu \equiv p_1$, $p_g \equiv p_2$, $j = \nu, g$, $\nu = 1$, $g = 2$.

The hierarchic tree of the pair-wise main parametrical resonance is illustrated in Figure 7.13.

Now, by using the given formulation of the problem, let us analyze the dynamics of the considered resonant system.

7.4.2 Equations of the First Hierarchic Level

Let us recall that the electromagnetic waves are assumed to be arbitrary and polarized (see Definition 7.2). By using Equation 7.59, we express the fast phases $p_{1,2}$ in terms of combination phases θ and ψ. We also introduce the dimensionless longitudinal coordinate $T = z/L$, where L is the length of the system. Eventually we have

* From Kulish, V.V. et al., *Int. J. Millimeter Waves*, 19, 33–94, 1998. With permission.

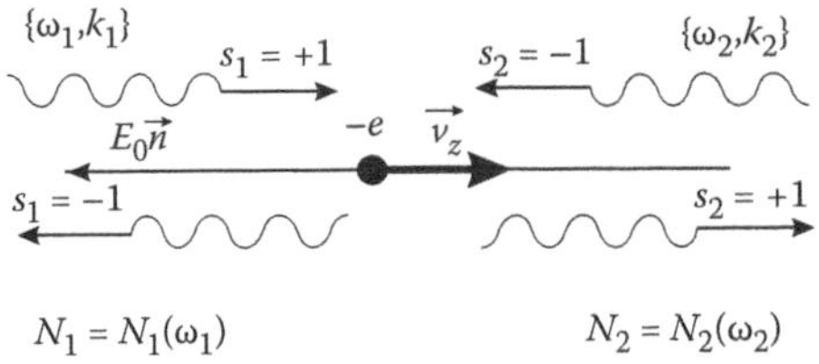

FIGURE 7.12
Arrangement of the considered model in the case of pairwise man parametrical resonance. Here, $-e$ is the electron charge, $\vec{v}_z$ is its vector of longitudinal velocity, $\{\omega_j,k_j\}$ are the frequency and wave number of the signal ($j = 1$) and the pumping ($j = 2$) arbitrary polarized electromagnetic waves, $s_j = \pm 1$ are the sign functions that describe the possible variants of the wave propagation, E_0 is the strength of the longitudinal electrostatic field (field of support), $\vec{n}$ is the unit vector along the z-axis, $N_j = ck_j/\omega_j$ are the wave retardation factors (refraction coefficients) in the modeling medium (simulated magneto-dielectric), and c is the light velocity in vacuum.

$$\frac{\mathrm{d}\mathcal{H}}{\mathrm{d}T} = -\frac{e_e L}{\mathcal{P}_z}$$
$$\times\left\{\mathcal{P}_x \operatorname{Re}\left[\mathcal{E}_{11} e^{i\frac{\psi+\theta}{2}} + \mathcal{E}_{21} e^{-i\sigma\frac{\psi-\theta}{2}}\right] + \mathcal{P}_y \operatorname{Im}\left[\mathcal{E}_{12} e^{i\frac{\psi+\theta}{2}} + \mathcal{E}_{22} e^{-i\sigma\frac{\psi-\theta}{2}}\right]\right\}$$
$$+\frac{ce_e^2 L}{\mathcal{P}_z}\left\{\operatorname{Im}\left[\frac{\mathcal{E}_{11}}{\omega_1} e^{i\frac{\psi+\theta}{2}} + \frac{\mathcal{E}_{21}}{\omega_2} e^{-i\sigma\frac{\psi-\theta}{2}}\right] \operatorname{Re}\left[\mathcal{E}_{11} e^{i\frac{\psi+\theta}{2}} + \mathcal{E}_{22} e^{-i\sigma\frac{\psi-\theta}{2}}\right]\right\}$$
$$-\frac{ce_e^2 L}{\mathcal{P}_z}\left\{\operatorname{Re}\left[\frac{\mathcal{E}_{12}}{\omega_1} e^{i\frac{\psi+\theta}{2}} + \frac{\mathcal{E}_{22}}{\omega_2} e^{-i\sigma\frac{\psi-\theta}{2}}\right] \operatorname{Im}\left[\mathcal{E}_{12} e^{i\frac{\psi+\theta}{2}} + \mathcal{E}_{22} e^{-i\frac{\psi-\theta}{2}}\right]\right\} \tag{7.70}$$

where $\sigma \equiv \sigma_{12}$, $\mathcal{P}_x$, $\mathcal{P}_y$, $\mathcal{P}_\perp = const$,

$$\mathcal{P}_z = \left\{\frac{1}{2}(\mathcal{H} + e_e\varphi)^2 - \left[\vec{\mathcal{P}}_\perp - \vec{e}_1 e_e \operatorname{Im}\left(\frac{\mathcal{E}_{11}}{\omega_1} e^{i\frac{\psi+\theta}{2}} + \frac{\mathcal{E}_{21}}{\omega_2} e^{-i\sigma\frac{\psi-\theta}{2}}\right)\right.\right.$$
$$\left.\left. + \vec{e}_2 e_e \operatorname{Re}\left(\frac{\mathcal{E}_{12}}{\omega_1} e^{i\frac{\psi+\theta}{2}} + \frac{\mathcal{E}_{22}}{\omega_2} e^{-i\sigma\frac{\psi-\theta}{2}}\right)\right]^2 - m^2c^2\right\}^{\frac{1}{2}}. \tag{7.71}$$

To avoid a mistake we temporarily introduce the special notation for the electron charge e_e. All other notations are as given before. We supplement Equation 7.70 with equations for combination phases ψ, θ and Systems 7.56 and 7.57, and separate out the large parameter of the problem

$$\xi \sim \frac{\omega_1 - \sigma\omega_2 - (s_1\omega_1 N_1 - \sigma s_2\omega_2 N_2)c\mathcal{P}_z(\mathcal{H} + e_e\varphi)^{-1}}{\omega_1 + \sigma\omega_2 - (s_1\omega_1 N_1 + \sigma s_2\omega_2 N_2)c\mathcal{P}_z(\mathcal{H} + e_e\varphi)^{-1}} >> 1. \tag{7.72}$$

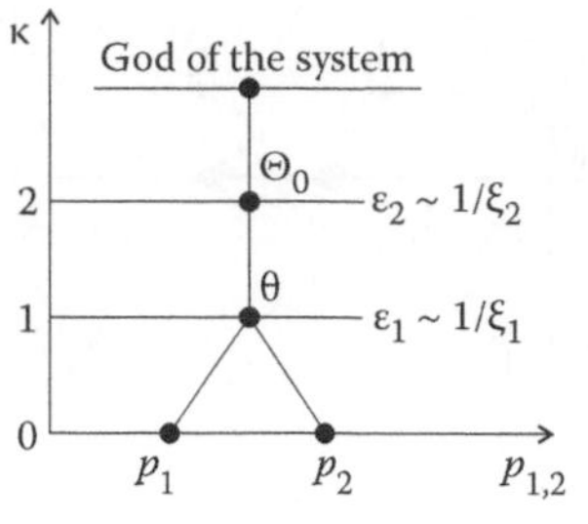

FIGURE 7.13
Hierarchic tree of the pairwise main parametrical resonance in the electron motion in the field of two collinear electromagnetic waves. Here, all designations are the same as those given in the captions for Figures 7.2 through 7.5.

The requirement $\xi >> 1$ is satisfied by virtue of parametric resonance Condition 7.69

$$\omega_1(1 - s_1\beta_z N_1) \approx -\sigma\omega_2(1 - s_2\beta_z N_2), \tag{7.73}$$

where $\beta_z = v_z/c$; $v_z = c\mathcal{P}_z/\tilde{\mathcal{H}}$ is the electron velocity along the z-axis, $\tilde{\mathcal{H}} = \mathcal{H} + e\varphi = mc^2\gamma$, and $\gamma = (1 - v^2/c^2)$. The boundary conditions can be written as

$$\mathcal{H}(T = 0), \quad \psi(T = 0) = \psi_0 = (\omega_1 - \sigma\omega_2)\tau,$$

$$\theta(T = 0) = \theta_0 = (\omega_1 + \sigma\omega_2)\tau, \tag{7.74}$$

where τ is the time of the entry of the electron into the interaction region.

We apply the required procedures of the hierarchic method (see Chapter 6). Therein, we confine ourselves to the zero approximation for the hierarchic small parameter $1/\xi$ (Equation 7.72). A system of the truncated equations of the first hierarchic level can be obtained in the form

$$\frac{d\bar{\mathcal{H}}}{dT} = \frac{ec^2\mathcal{E}_1\mathcal{E}_2\sigma(\omega_1 + \sigma\omega_2)L\hat{A}\sin(\bar{\theta} + \alpha)}{2\omega_1\omega_2\sqrt{\bar{\tilde{\mathcal{H}}}^2 - m^2c^4 - \mathcal{H}_0^2 B(\bar{\theta})}};$$

$$\frac{d\bar{\theta}}{dT} = \frac{(\omega_1 + \sigma\omega_2)L\bar{\tilde{\mathcal{H}}}}{c\sqrt{\bar{\tilde{\mathcal{H}}}^2 - m^2c^4 - \mathcal{H}_0^2 B(\bar{\theta})}} - \Phi(T), \tag{7.75}$$

where, to specialize the calculation, we assume $\omega_1 \geq \omega_2$, and the following notations for the polarization functions $\hat{A}$ are used:

$$\hat{A} = \sqrt{\cos^2(\hat{\psi}_1 + \sigma\hat{\psi}_2) + (1/2)\sin 2\hat{\psi}_1 \sin 2\sigma\hat{\psi}_2\left[\cos(\chi_{11} + \sigma\chi_{21} - \chi_{12} - \sigma\chi_{22}) - 1\right]},$$

$$\alpha = arctg\frac{\sin(\chi_{11} + \sigma\chi_{21}) + tg\hat{\psi}_1 tg\sigma\hat{\psi}_2 \sin(\chi_{12} + \sigma\chi_{22})}{\cos(\chi_{11} + \sigma\chi_{21}) + tg\hat{\psi}_1 tg\hat{\psi}_2 \cos(\chi_{12} + \sigma\chi_{22})}.$$

Other notations are

$$B(\bar{\theta}) = c^2 \mathcal{P}_\perp^2 \mathcal{H}_0^{-2} + (1/2)(\varepsilon_1^2 + \varepsilon_2^2)\varepsilon_1\varepsilon_2 \hat{A}\cos(\bar{\theta}+\alpha);$$

$$\varepsilon_{1,2} = e\mathcal{E}_{1,2}/mc\omega_{1,2}\gamma_0\,;\,\gamma_0 = \mathcal{H}_0/mc^2\,;\,\mathcal{P}_\perp = |\vec{\mathcal{P}}_\perp|;$$

$$T = z/L;\,\Phi = \frac{L}{c}(s_1\omega_1 N_1 - \sigma s_2\omega_2 N_2)\,.$$

For the accepted zero approximation $\bar{\mathcal{H}} \cong \mathcal{H}, \bar{\tilde{\mathcal{H}}} \cong \tilde{\mathcal{H}}, \bar{\theta} \cong \theta$ (hierarchic degeneration), the averaging symbol is omitted for simplicity.

7.4.3 Model with the Magnetic Undulator

In this section, the considered mechanism of the parametric resonance is used as basic in the FELs [20–33]. In FEL theory, it is known that the use of the electromagnetic wave as a pumping system (Dopplertron FEL) is only one of many possible design variants, and not the main one. It was proposed first in a USSR Patent (No. 705914) [34] and was realized experimentally in 1976 [30,31]. In this experiment, the discussed 1-MW pulse at the wavelength $\lambda_1 \cong 0.4$ mm was generated.

However, it should be mentioned that, practically simultaneously, in 1976, another first experiment with an FEL [32,33] had been accomplished. It was the infrared ($\lambda \cong 10$ μm) FEL constructed as the basis of the magnetic undulator as a pumping system. In the future, this design version became more popular. This type of system has been called the H-ubitron because its nonrelativistic versions are known in the microwave electronics from the beginning of the 1960s [35,36]. It should be mentioned that many modern researchers today, talking about FELs generally, have in mind the H-ubitron design (see, for example, the work by Schmuser et al. [26]). The main design idea of such a pumping system, which we call the H-ubitron pumping, is illustrated in Figure 7.14. In view of the discussion in this chapter, its operation principle is self-evident.

Let us stress that the general FEL theory talks about at least ten possible pumping designs, which can also be used as the FEL pumping system [1,2]. All these designs can be conditionally classified into the ubitrons and Dopplertrons [1,2]. Ubitron pumping systems are characterized by the absence of a fast time dependency of the pumping fields. In contrast, the pumping fields in the Dopplertron systems always depend on fast time. The discussed case of pumping by the electromagnetic wave can be treated as one variety of Dopplertron pumping.

The model of the electromagnetic wave propagating within a simulated (modeling) medium (magnetodielectric) used in this chapter was proposed in 1977 [37] for a uniform description of a wide spectrum of possible FEL pumping fields. Therefore, this system (System 7.75) can have much wider possibilities for practical interpretation than it might seem at the first glance. Let us illustrate one of these possibilities by passing in System 7.75 to the traditional FEL model with the magnetic ubitron pumping (H-ubitron)

$$N_2 \to \infty;\ \omega_2 \to 0;$$

$$k_2 = \lim_{\substack{\omega_2 \to 0 \\ N_2 \to \infty}} \frac{\omega_2 N_2}{c} = \frac{2\pi}{\Lambda_2} \cdot ck_2 \mathcal{E}_2 = \omega_2 \mathcal{H}_2$$

$$\mathcal{E}_2 \to 0; \lim_{\substack{\omega_2 \to 0 \\ N_2 \to \infty}} \frac{\mathcal{E}_2}{\omega_2} = \frac{\mathcal{H}_2}{ck_2} = \frac{\mathcal{H}_2 \Lambda_2}{2\pi c},$$

where $\mathcal{H}_2$ is the real amplitude of the magnetic component of the H-ubitron pumping field (see Figure 7.14), and Λ_2 is its spatial period. For more details about the theory of simulated magnetodielectric, see the earlier books by Kulish [1,2].

7.4.4 Dimensionless Variables

In computer calculations, it is suitable to rewrite System 7.75 in a dimensionless form. One of the ways to satisfy this requirement is to write equations in terms of the efficiency/phase variables. For this, we introduce the concepts of actual (true) and reduced electron efficiencies

$$\eta = \frac{\mathcal{H}_0 - \mathcal{H}}{\mathcal{H}_0 - mc^2 + e\varphi_L}; \quad w = \eta\left(1 - \gamma_0^{-1} - \Pi_L\right), \tag{7.76}$$

where $\Pi_L = \Pi(z = L)$, $\Pi = |e\varphi|/mc^2\gamma_0$, $\varphi_L = \varphi(z = L)$, and e, as before, is the magnitude of the electron charge. Also, we introduce the following notation:

$$C = -\sigma \frac{e^2 \mathcal{E}_1 \mathcal{E}_2 (\omega_1 + \sigma\omega_2)\hat{A}L}{m^2 \omega_1 \omega_2 c^3 \gamma_0^2} = -\sigma \frac{\varepsilon_1 \varepsilon_2 \mu \hat{A}}{2};$$

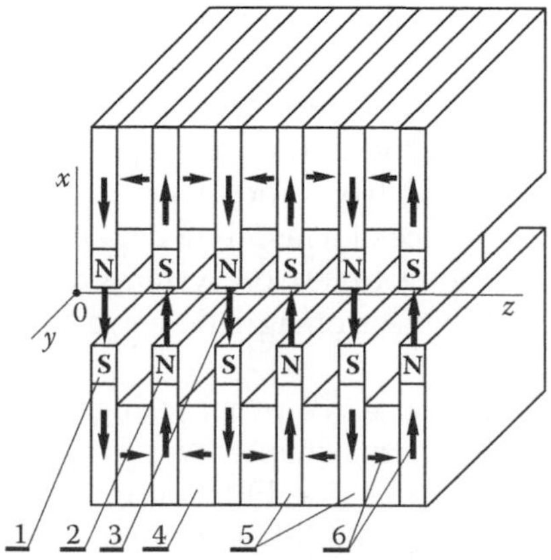

FIGURE 7.14
Design scheme of the linearly polarized magnetic undulator (H-ubitron pumping). Here, 1 are the south (S) magnetic poles, 2 are the north (N) magnetic poles, 3 are the vectors of the magnetic strength $\vec{H}_2$, 4 are the permanent magnets, and 5 are the magnetic conductors for the magnetic fluxes, which are 6.

$$\mu = (\omega_1 + \sigma\omega_2)c^{-1}L;$$

$$\theta' = \theta + \alpha;$$

$$\Phi' = \Phi - \frac{d\alpha}{dT}. \tag{7.77}$$

As a result, System 7.75 can be written in a more elegant form

$$\frac{dw}{dT} = \frac{C\sin\theta'}{\sqrt{(1-w+\Pi)^2 - \gamma_0^{-2} - B(\theta')}}; \tag{7.78}$$

$$\frac{d\theta'}{dT} = \frac{\mu(1-w+\Pi)}{\sqrt{(1-w+\Pi)^2 - \gamma_0^{-2} - B(\theta')}} - \Phi'. \tag{7.79}$$

Systems 7.78 and 7.79, truncated dimensionless nonlinear equations, describe a few possible one-dimensional models of the Dopplertron and H-ubitron types. It contains the minimum number of functional and numerical parameters, suitably normalized for practical computer calculations. It is evident that for a small magnitude of efficiency ($w, \Pi << 1$) and moderate amplitudes of electromagnetic waves $\left(\varepsilon_{1,2}^2 << 1\right)$, Systems 7.77 and 7.79 reduce to the equation of the nonlinear pendulum (for more details about the nonlinear pendulum, see Sections 3.1.4 and 3.1.5)

$$\frac{d^2\theta'}{dT^2} = -\hat{\omega}_0^2 \sin\theta', \tag{7.80}$$

where the frequency of the proper linear oscillations at the first hierarchic level is

$$\hat{\omega}_0 \cong \sqrt{\mu C/\left(1-\gamma_0^{-2}\right)}. \tag{7.81}$$

This frequency describes the FEL theory of the effect of electron oscillation in a bucket [1,2].

Let us analyze the physical meaning of the equations obtained. First, the structure of Systems 7.75, 7.78, and 7.79 suggests that parametric resonant interaction with the pumping and signal fields is of a threshold nature with respect to the electron energy $\tilde{\mathcal{H}}$, i.e.,

$$\tilde{\mathcal{H}} > \tilde{\mathcal{H}}_{th}, \tag{7.82}$$

where $\tilde{\mathcal{H}}_{th} = \sqrt{m^2c^4 + \mathcal{H}_0^2 B(\theta)}$ is the threshold energy. Therefore, for a given amplitude $\mathcal{E}_{1,2}$ and the transverse canonical momentum $\mathcal{P}_\perp$, the parametric resonance character of the energy exchange is manifested for the electron energies higher than the threshold value of Equation 7.82. The detailed analysis shows that the threshold increases by electron reverse.

The intensity of the energy exchange depends strongly on the pumping and signal wave polarizations. For $\hat{A} = 0$, the cumulative effect does not occur, i.e., $w = \eta = 0$. This situation takes place when polarization parameters ψ_{ji} and χ_{ji} are related in the transcendental equation

$$\cos^2(\hat{\psi}_1 + \sigma\hat{\psi}_2) - 1/2 \cdot \sin 2\hat{\psi}_1 \sin 2\sigma\hat{\psi}_2 \left[\cos(\chi_{11} + \sigma\chi_{21} - \chi_{12} - \sigma\chi_{22}) - 1\right] = 0 . \quad (7.83)$$

This corresponds in particular to linearly polarized waves, $\hat{\psi}_1 = 0;\ \hat{\psi}_2 = \pi/2$ or $\left(\hat{\psi}_1 = \pi/2;\ \hat{\psi}_2 = 0\right)$, or circular polarizations, $\chi_{11} = \chi_{12}$, $\chi_{21} = \chi_{22}$, $\hat{\psi}_1 = \sigma\hat{\psi}_2 = g_1(\pi/4)$, and $g = \pm 1$.

7.4.5 Isochronous Model with the Optimal Electrostatic Support

A simple analysis of Systems 7.78 and 7.79 shows that the efficiency η for the ordinary parametric resonant models is characterized by values of $\sim 10^{-1}$ or smaller. Hence, the problem of increasing the efficiency becomes practically urgent. Appropriate answers could be found by the methods of isochronization [1,2,5,11].

The electron efficiency η could be maximal in models with constant combination phases θ' (see Equation 7.79). Such a system state is referred to as isochronism. It can be practically attained in different technological ways. In this section, we discuss only two of them: the method of the optimal longitudinal electrostatic field (longitudinal support) and the method of the optimal slowly varying electromagnetic wave parameters $\mathcal{E}_{ji}, \hat{\psi}_j, \chi_j, N_j$. The method of optimal transverse undulation electric field (transversal support) will be illustrated in the next section.

The physical situation suggests the presence of an external mechanism for the provision of the optimum combination synchronism in the interaction region. In view of Systems 7.78 and 7.79, the condition $\theta' = const$ can be found. The isochronous system state can hold only if the identity

$$\mu(1 - w + \Pi) = \Phi'\sqrt{(1 - w + \Pi)^2 - \gamma_0^{-2} - B(\theta')}\ , \quad (7.84)$$

holds [1,2,5,11]. The latter makes it possible for the relevant choice of the optimum parameters of the system. In particular, using Condition 7.84 and System 7.78, we find that the electron efficiency in the isochronous model can be written in the form

$$\eta = \frac{\sin\theta'_0}{1 - \gamma_0^{-1} - \Pi_L} \int_0^T \frac{C(T')\sqrt{\Phi'^2/\mu^2 - 1}}{\sqrt{\gamma_0^{-2} + B\left(\theta'_0\right)}} dT'\ , \quad (7.85)$$

where T' is a dummy integration variable [1,2,5,11].

Let us consider the two above-mentioned special cases, which differ in the isochronization of $\theta' = \theta'_0 = const$. The first are the systems with the optimum longitudinal electrostatic support [1,2,5,11]. We assume $N_{1,2}, C, \alpha = const$, and $\gamma_0^2 B\left(\theta'_0\right) > 1$. Then, with

$$\beta_z \cong \frac{\mu}{\Phi'} \cong \beta_0\ , \quad (7.86)$$

and following synchronism condition 7.84, we find the isochronous electron efficiency [1,2,5,11]

$$\eta = \frac{C \sin\theta_0'}{\left(1 + \Pi_L - \gamma_0^{-1}\right) a\beta_z \gamma_{\|} \gamma_0^{-1}} T, \tag{7.87}$$

where $\gamma_{\|} = 1/\sqrt{1-\beta_z^2}$; $a = \sqrt{1+\gamma_0^2 B(\theta_0')}$. As follows from Equation 7.87, the efficiency η as a function of θ_0' attains maximum for $T = 1$ if

$$\theta_0' = \theta_{0\max}' = -\sigma \arccos\left[\sqrt{\left(d^2\left(\gamma_0^2 \varepsilon_1 \varepsilon_2 \hat{A}\right)^{-1} - 1\right)} - d\left(\gamma_0^2 \varepsilon_1 \varepsilon_2 \hat{A}\right)^{-1}\right], \tag{7.88}$$

where $d = 1 + \gamma_0^2 B_0$; $B_0 = B(\theta_0') - \varepsilon_1 \varepsilon_2 \hat{A} \cos\theta_0'$ for $\gamma_0^2 \varepsilon_1 \varepsilon_2 \hat{A} << 1$, $\gamma_0^2 B_2 << 1$, $\theta_{0\max}' \cong -\sigma(\pi/2)$. For $\theta_0' = 0, \pi$, we have $\eta = 0$, i.e., the cumulative effect of the interaction does not occur in these cases. For $T = 1$, in view of System 7.79, Expression 7.88 can be written as [1,2,5]

$$\eta = \frac{C \sin\theta_0'}{C \sin\theta_0' + a\beta_z \gamma_{\|} \gamma_0^{-2}(\gamma_{\|} a - 1)}. \tag{7.89}$$

By using Equation 7.84, we can derive an expression for the optimum electrostatic support

$$E_{0opt} = \frac{mc^2 C \sin\theta_0'}{eLa\beta_z \gamma_{\|} \gamma_0^{-2}} = const. \tag{7.90}$$

It can be easily seen that for the case $C \sin\theta_0' >> a\beta_z \gamma_{\|} \gamma_0^{-2}(\gamma_{\|} a - 1)$ in Expression 7.89, we obtain: $\eta \sim 1$. Therefore, in contrast to the ordinary nonisochronous models, the efficiency η in the discussed isochronous model can attain values ~1. However, it should be stressed that this result is obtained in the framework of the one-particle theory. And, therefore, it can be proved to be unrealistic in most of the multiparticle practical systems. In a real multiparticle FEL, there exist a number of causes that lead to an essential decrease in the FEL's total efficiency. Nevertheless, such a purely one-particle model result is very useful in practice.

First, it gives information about the theoretical limit of multielectron efficiency in the isochronous FEL. This limit, as a rule, is determined by the physical features of the relevant grouping mechanisms in the FEL described in Section 4.3. Indeed, one particle should always be characterized by higher efficiency than a podgy electron bunch that also contains many nonisochronous electrons (see the grouping mechanisms in Section 4.3 for details).

Second, the result suggests a very interesting technological way for constructing a highly efficient FEL. For instance, it can be klystron systems where the electron beam in the main interaction area has the form of a sequence of cold compact bunches, each of which is similar to a large particle [38]. The FEL with intermediate acceleration of the preliminarily highly modulated electron beam [1,2] can be given as a typical illustrative example of such a technological arrangement.

7.4.6 Method of Optimal Variation of the Retardation Factor

Let us discuss a second isochronization method—the method of variation of the retardation factor of the pumping electromagnetic wave [1,2,5]. This isochronization method is also known in FEL–H-ubitron technology as the method of optimal variation of the undulation period [43].

We use Systems 7.78 and 7.79: $\Pi = 0$, α, $C = const$, $N_2 = N_2(T)$. In the theory of H-Ubitron-type FEL, such a method of maintaining the synchronism $\theta' \cong \theta'_{opt}$ is also known as the method of variation of wiggler [43]. As follows from Equation 7.85, the efficiency η in this case can be described by the expression [5]

$$\eta = \frac{C \sin\theta'_0}{a\gamma_0^{-1}\left(1-\gamma_0^{-1}\right)} \int_0^T \sqrt{\Phi^2/\mu^2 - 1}\, \mathrm{d}T'. \tag{7.91}$$

Thus, the output efficiency for $T = 1$ can attain values close to one if [5]

$$\int_0^1 \sqrt{\Phi^2/\mu^2 - 1}\, \mathrm{d}T' = \frac{a\gamma_0^{-1}\left(1-\gamma_0^{-1}\right)}{C \sin\theta'_0}. \tag{7.92}$$

Theoretically, one always can ensure that Expression 7.92 holds by an appropriate choice of the retardation factors, depending on $T, N_{1,2}(T)$, or only $N_2(T)$. This confirms the conclusion that the efficiency can attain values close to 1. In practice, however, construction limitations of the sizes of the pumping system make this a difficult problem. Nevertheless, real efficiency values for the case of multiparticle interaction $\eta \sim 40\%$ (the method of variation of wiggler) were experimentally obtained long ago [43]. This means that, in this case, the individual efficiency was actually higher.

7.5 Bounded (Coupled) Parametric Resonance in the Field of Three Electromagnetic Waves*

7.5.1 Model and the Hierarchic Trees

In Chapter 3 (see Figure 3.8), we discussed the concept of the bounded resonances. In the partial case of only two bounded resonances, we spoke about coupled resonance.

The main idea of this type of resonant interaction is that at least one oscillation phase takes place in a few resonances simultaneously. The resonant Condition 7.31 describes this situation mathematically. The simplest variant of such a resonant scheme (coupled main resonances between three oscillation phases) is illustrated in Figures 7.15 and 7.16.

* From Kulish, V.V. et al., *Int. J. Millimeter Waves*, 19, 33–94, 1998. With permission.

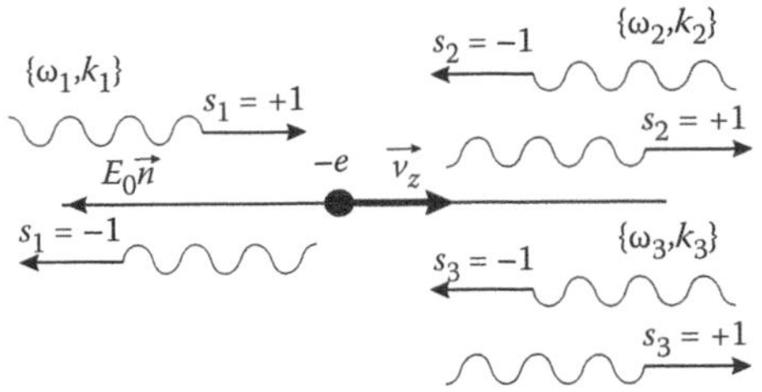

FIGURE 7.15
Arrangement of the model of a coupled parametrical resonance. Here, $-e$ is the electron charge, $\vec{v}_z$ is its vector of longitudinal velocity, $\{\omega_j,k_j\}$ are the frequency and wave number of the signal (j = 1) and the two pumping waves (j = 2,3) arbitrary polarized electromagnetic waves, $s_j = \pm 1$ are the sign functions that describe the possible variants of the wave propagation, E_0 is the strength of the longitudinal electrostatic field (field of support), and $\vec{n}$ is the unit vector along the z-axis.

Thus, we assume the Dopplertron model to be filled by the simulating medium (magnetodielectric, which is the same) with three collinear electromagnetic waves propagating through its working bulk (see Figure 7.15). For the sake of determinacy, we assume that the pumping wave has a two-frequency spectral structure ($\{\omega_2,k_2\}$ and $\{\omega_3,k_3\}$, respectively; see Figure 7.15), i.e., we have the Dopplertron FEL model with two-frequency (dichromatic) pumping. Both partial pumping waves $\left\{\omega_{2,3},\vec{k}_{2,3}\right\}$ interact simultaneously with the same signal wave $\left\{\omega_1,\vec{k}_1\right\}$ in the parametric resonant manner. Analyzing the resonant condition like Equation 7.31, one can be sure that, in this specific case, the resonant condition for both these partial waves $\left\{\omega_{2,3},\vec{k}_{2,3}\right\}$ is satisfied automatically (see Figure 7.16). Thus, all three electromagnetic waves in the discussed model interact resonantly with each other.

Then, in view of the above assumptions, we use the general formulation of the resonant problem described in Section 7.1. As a result of the required procedures, the following set of slow and fast combination phases of Equations 7.26 and 7.27 can be formulated as

$$\psi = p_1 - \sigma_{12}p_2;$$
$$\theta_{12} = p_1 + \sigma_{12}p_2;$$
$$\theta_{23} = p_2 + \sigma_{23}p_3, \tag{7.93}$$

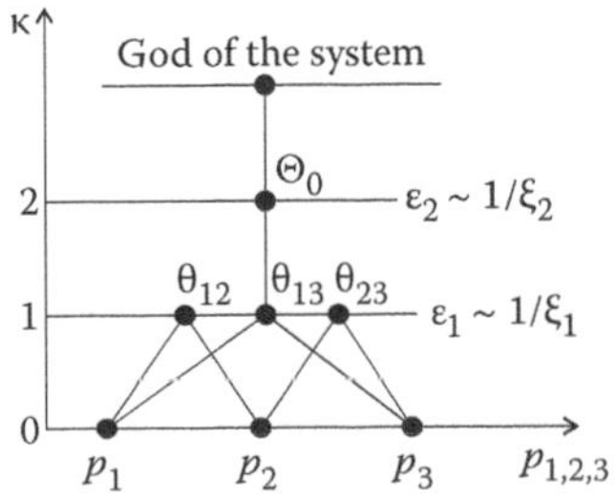

FIGURE 7.16
Hierarchic tree of bounded (coupled) parametric resonances. Here, $p_{1,2,3}$ are the phases of the zero hierarchic level, and θ_{12}, θ_{13}, and θ_{23} are the combination phases (first hierarchic level) of three coupled resonances where only two are independent. All other designations are the same as those given in the captions for Figures 7.2 through 7.4.

i.e., we take into account that the partial pumping wave $\{p_2\}$ in the considered model interacts simultaneously with the second wave component of the pumping $\{p_3\}$ and signal $\{p_1\}$ waves. We can describe the parametric resonant character of these interactions by the following resonant conditions (synchronism; see Equation 7.28):

$$\left|\frac{d\theta_{ij}}{dt}\right| << \left|\frac{d\psi_{ij}}{dt}\right|. \tag{7.94}$$

We can rewrite the conditions of the combination synchronism of Condition 7.94 as

$$\begin{aligned} &\omega_1(1 - s_1\beta_z N_1) \cong -\sigma_{12}\omega_2(1 - s_2\beta_z N_2);\ s_1\omega_1 N_1 \neq -\sigma_{12}s_2\omega_2 N_2; \\ &\omega_2(1 - s_2\beta_2 N_2) \cong -\sigma_{23}\omega_3(1 - s_3\beta_z N_3);\ s_2\omega_2 N_2 \neq -\sigma_{23}s_3\omega_3 N_3, \end{aligned} \tag{7.95}$$

where all notations are given above.

It follows from Condition 7.95 that, when $\omega_2 = \omega_3$ (but $N_2 \neq N_3$), the parametric character of reciprocal resonant interactions does not violate, even in this extraordinary case [5]. In practice, such a model can be put into practice by using two different wave modes of the same pumping monochromatic wave. In this case, the phase velocities of both modes should be different due to the different laws of dispersion. As an analysis shows, in this case, the peculiar self-consistent amplification effect can take place owing to the so-called explosive instability in FELs [39] (the explosive instability was rediscovered later by Tripathi and Lin [40]). The existence of the parametric coupling between the bichromatic pumping components and the monochromatic signal waves leads to the appearance of an additional amplification of the signal. This is the main practical advantage of the discussed type of Dopplertron systems.

7.5.2 Truncated Equations

Relevant truncated equations of the first hierarchic level (the zero approximation of the hierarchic method) can be obtained by using the general theory of Dopplertron FELs discussed in this chapter (we omit the averaging symbols because of the hierarchic degeneration: $\bar{\mathcal{H}} \cong \mathcal{H}$, $\bar{\theta}_{ij} \cong \theta_{ij}$)

$$\frac{d\mathcal{H}}{dT} = \frac{e^2cL}{2\sqrt{\tilde{\mathcal{H}}^2 - m^2c^4 - \mathcal{H}_0^2B(\theta_{ij})}} \times \sum_{i=1}^{3}\sum_{j=2,3} \mathcal{E}_i\mathcal{E}_j\hat{A}_{ij}\sigma_{ij}\frac{\omega_i + \sigma_{ij}\omega_j}{\omega_i\omega_j}\sin(\theta_{ij} + \alpha_{ij});$$

$$\frac{d\theta_{12}}{dT} = \frac{(\omega_1 + \sigma_{12}\omega_2)L\tilde{\mathcal{H}}}{c\sqrt{\tilde{\mathcal{H}}^2 - m^2c^4 - \mathcal{H}_0^2B(\theta_{ij})}} - \Phi_{12};$$

$$\frac{d\theta_{23}}{dT} = \frac{(\omega_2 + \sigma_{23}\sigma_3)L\tilde{\mathcal{H}}}{c\sqrt{\tilde{\mathcal{H}}^2 - m^2c^4 - \mathcal{H}_0^2B(\theta_{ij})}} - \Phi_{23}, \tag{7.96}$$

where

$$\Phi_{ij} = \frac{L}{c}(s_i\omega_i N_i - \sigma_{ij}s_j\omega_j N_j);\ \theta_{13} = \theta_{12} + \sigma_{12}\theta_{23};$$

$$\hat{A}_{ij} = \Big\{\cos^2\left(\hat{\psi}_i + \sigma_{ij}\hat{\psi}_j\right)$$
$$-(1/2)\sin 2\hat{\psi}_i \sin\sigma_{ij}2\hat{\psi}_j\left[\cos\left(\chi_{i1} + \sigma_{ij}\chi_{j1} - \chi_{i2} - \sigma_{ij}\chi_{j2}\right) - 1\right]\Big\}^{1/2};$$

$$\hat{A}_{32} = \hat{A}_{ii} = 0;$$

$$\alpha_{ij} = arctg\frac{\sin(\chi_{i1} + \sigma_{ij}\chi_{j1}) - tg\hat{\psi}_i\, tg\sigma_{ij}\hat{\psi}_j \sin(\chi_{i2} + \sigma_{ij}\chi_{j2})}{\cos(\chi_{i1} + \sigma_{ij}\chi_{j1}) - tg\hat{\psi}_i tg\sigma_{ij}\hat{\psi}_j \cos(\chi_{i2} + \sigma_{ij}\chi_{j2})};$$

$$\sigma_{13} = \sigma_{12}\sigma_{23},\ i \neq j;\ \sigma_{jj} = 1;$$

$$B(\theta_{ij}) = c^2\mathcal{P}_\perp^2\mathcal{H}_0^{-2} + (1/2)\sum_{i=1}^{3}\mathcal{E}_i^2 + \sum_{i=1}^{3}\sum_{j=2}^{3}\mathcal{E}_i\mathcal{E}_j\hat{A}_{ij}\cos(\theta_{ij} + \alpha_{ij}). \tag{7.97}$$

In the case $\mathcal{E}_2 = 0$, Equations 7.96 can be easily reduced to System 7.75.

It is explicit from the mathematical structure of Equations 7.97 that the above-mentioned additional parametrical coupling between both parametric resonances shown in Figure 7.16 does indeed appear. This coupling is described by the presence of the new combination phase θ_{13}. Because of Equations 7.96, the additional resonant condition

$$\left|\frac{d\theta_{13}}{dt}\right| << \left|\frac{d\psi_{13}}{dt}\right| \tag{7.98}$$

is satisfied (see Figure 7.16). Following this analysis, one can be assured that the discussed three-wave Dopplertron FEL has unusual and interesting practical properties. For instance, by simply introducing the second component of pumping (without changing the total intensity), we can obtain the growth of the efficiency of interaction for the signal wave [1,2,5].

7.6 Model with Pumping by the Crossed Magnetic and Electric Undulation Fields*

7.6.1 FEL Pumping with Crossed Magnetic and Electric Undulation Fields

In Section 7.4.3, we have discussed some aspects of the physics of the FEL with pumping by the transverse undulation magnetic field (H-ubitron pumping; see Figure 7.14). In Section

* From Kulish, V.V. et al., *Int. J. Millimeter Waves*, 19, 33–94, 1998. With permission.

7.4.5, we analyzed the isochronous FEL model with the optimal longitudinal acceleration electric field (longitudinal electric support).

In this section, we will study the physical properties of an intermediate-type model. Similar to the system illustrated in Figure 7.11, the intermediate-type model contains H-ubitron pumping and, comparable to Section 7.4.5, the model can also be isochronous. At this time, we attain the isochronous state by the use of a special transverse electric undulation field (transverse electric support). It is generated in the working bulk in such a manner that its vector of electric strength is always oriented perpendicular to the vector of the strength of the magnetic undulation field (see Figure 7.17). That type of design superimposition is called an EH-undulator [2,41–43]. We call the superposition of the crossed undulation electric and magnetic fields the EH-field.

The main idea of the EH-undulators can be illustrated by a simple modification of the design illustrated in Figure 7.14. Let us assume that instead of the permanent magnets (4), we install the magnetic core of a low-frequency electromagnet. As a result, the magnetic field between the poles [(1) and (2) in Figure 7.14] becomes adiabatically slowly varying in time. In accordance with the law of electromagnetic induction, such slow change in the time magnetic field should generate a vortex electric field in the interpole space [2,41–43]. Inasmuch as the magnetic field is the undulation (i.e., spatially periodical), the generated electric field is also spatially periodical. The described process of the formation of the EH-field is illustrated graphically in Figure 7.17.

As shown in Figure 7.17, the EH-undulator not only forms an electron undulation trajectory (6) but also provides electron acceleration. Indeed, in the space of each interpole, the electron moves in the acceleration vortex electric field (3) and (5). This means that this type of pumping design can be the basis of a new isochronous FEL.

7.6.2 Model and Formulation of the Problem

We assume that the electron moves in the superposition of the electromagnetic signal wave and the EH-undulation pumping fields, as illustrated in Figure 7.17. We also suppose that

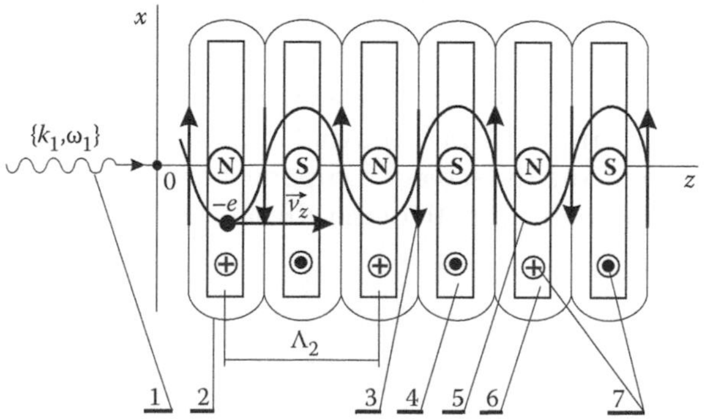

FIGURE 7.17
Arrangement of the FEL model of EH-pumping. Here, 1 is the signal wave with frequency ω_1 and wave number k_1, 2 are the strength lines of the vortex undulation electric field, 3 are the vectors of the strength of the undulation electric vortex field (2) in the interpoles spaces, 4 are the south magnetic poles, 5 is the electron trajectory in the EH-field, 6 are the north magnetic poles, 7 are the vectors of strength of the magnetic field within the gapes of the poles (4), (6) (see Figure 7.14), $-e$ is the electron charge, $\vec{v}_z$ is its vector of longitudinal velocity, and Λ_2 is the period of the magnetic and electric undulation fields. At least part of the undulation magnetic field (7), which is generated by the poles (4), (6), slowly change over time. The drawing is shown in the plane of the faces of the magnetic poles (4), (6) (see also Figure 7.11).

the longitudinal magnetic field can be introduced in this superimposition. Thus, the total electromagnetic field acting on the electron can be represented in the form

$$\vec{A} = \vec{A}_0 + \vec{A}_1' + \vec{A}_2' = \vec{A}_0 + \frac{1}{2}\sum_{j=1}^{2}\left(\vec{A}_j e^{im_j p_j} + \vec{A}_j^* e^{-im_j p_j}\right), \tag{7.99}$$

where $\vec{A}_0 = (1/2)[\vec{n}\vec{r}_\perp]$ is the vector potential of the longitudinal magnetic field, $\vec{n}$ is the unit vector along the z-axis, $\vec{r}_\perp$ is the transverse position vector, $\vec{A}_1$ is the complex amplitude of the signal wave vector potential, and $\vec{A}_2$ is the complex amplitude of the EH-pumping field.

The signal electromagnetic wave is considered plane, linearly polarized, and propagated along the z-axis

$$\vec{A}_1' = \frac{1}{2}\left(A_1(z,t)e^{ip_1} + c.c.\right)\vec{e}_x, \tag{7.100}$$

where $A_1(z,t)\vec{e}_x = \vec{A}_1$ is the complex slowly varying signal amplitude, p_1 is the phase of oscillations

$$p_1 = \omega_1 t - s_1 k_1 z, \tag{7.101}$$

ω_1 and k_1 are the circular frequency and the wave number ($k_1 = (\omega_1/c)N_1$), N_1 is the retardation coefficient, $s_1 = \pm 1$ is the sign function that determines the direction of the wave propagation along the z-axis, and $\vec{e}_x$ is the unit vector along the x-axis.

We write the EH-undulation part of the field as

$$\vec{A}_2' = \frac{1}{2}\left[A_{21}(z)e^{ip_2} + A_{22}(z,t)e^{ip_2} + c.c.\right]\vec{e}_x = \frac{1}{2}\left(A_2 e^{ip_2} + c.c.\right), \tag{7.102}$$

where $p_2 = k_2 z$ is the phase of the EH-field, $k_2 = 2\pi/\lambda_2$ is its wave number, A_{21} is the spatially complex slowly varying amplitude that represents the stationary magnetic part of the EH-field, and A_{22} represents the adiabatically slowly varying in time part of the total EH-field. According to the well-known definitions for the vector potential, the magnetic component of the acting EH-undulation pumping field can be written in the form

$$\vec{H}_2' = rot\left\{\frac{1}{2}\left[A_{21}(z)e^{ip_2} + A_{22}(z,t)e^{ip_2} + c.c.\right]\vec{e}_y\right\}, \tag{7.103}$$

and for the vortex electric field we have

$$\vec{E}_2' = -\frac{1}{c}\frac{\partial}{\partial t}\left\{\frac{1}{2}\left[A_{22}(z,t)e^{ip_2} + c.c.\right]\vec{e}_x\right\} = \frac{1}{2}(E_2 e^{ip_2} + c.c)\vec{e}_x. \tag{7.104}$$

By employing Field 7.104 in an optimal manner, we can attain the isochronous state of the considered FEL model.

Following the general formulation of the problem, we can separate out the three characteristic types of interaction depending on the value of the longitudinal magnetic field. According to the general theory set forth in Section 7.1, we obtain models with weak (Conditions 7.32 through 7.35), strong (Conditions 7.36 through 7.38), and superstrong (Conditions 7.39 and 7.40) magnetic fields. Let us limit ourselves to considering only the first two models, including the case of coupled parametric cyclotron resonance as a subtype of the strong interaction.

We formulate the resonant conditions for these models. As discussed, we chose only two types of resonances for investigation: the parametric (see Condition 7.30)

$$\left|\frac{d\theta}{dt}\right| << \left|\frac{d\psi}{dt}\right|;$$

$$\theta = p_1 + \sigma_{12}p_2; \ \psi = p_1 - \sigma_{12}p_2, \tag{7.105}$$

and the coupled parametric cyclotron (Condition 7.31) resonances

$$\left|\frac{d\theta_{0j}}{dt}\right| << \left|\frac{d\psi_{0j}}{dt}\right|;$$

$$\theta_{0j} = p_0 + \sigma_{0j}p_j; \ \psi_{0j} = p_0 - \sigma_{0j}p_2, \tag{7.106}$$

where $\sigma_{ij} = \pm 1$, ($i = 0,1$; $j = 1,2$) are the sign functions, θ and θ_{0j} are the slow combination phases, ψ and ψ_{0j} are the fast combination phases, and p_0 is the magnetic phase defined earlier (see Equation 7.22).

7.6.3 Parametric Resonance

Let us begin our study of the parametric resonant EH-model with the case of superweak magnetic fields (see Conditions 7.34 and 7.35). Similar to the discussed model of parametric resonance in two electromagnetic waves (see Section 7.4), we derive, analogous to Systems 7.78 and 7.79, a system of truncated equations in dimensionless form. For this, we introduce the concept of the true

$$\bar{\eta} = \frac{\bar{\mathcal{E}}_0 - \bar{\mathcal{E}}}{\bar{\mathcal{E}}_0 - mc^2 + W_E}, \tag{7.107}$$

and reduced

$$\bar{w} = \bar{\eta}(1 - \gamma_0^{-1} + \Pi_E) \tag{7.108}$$

average single-particle electron efficiencies. Here, $\bar{\mathcal{E}}_0 \equiv \bar{\mathcal{H}}_0$ is the initial and $\bar{\mathcal{E}} \equiv \bar{\mathcal{H}}$ is the current averaged electron energies. As is easily seen in the difference between Definitions 7.76 and 7.107, Definition 7.108 concerns only the values W_E and Π_E. In the present case, W_E is the energy acquired by the electron under the action of the vortex electric component of the EH-field. The value $\Pi_E = W_E/\mathcal{E}_0$ is the transverse support function; the value $\gamma_0 = \mathcal{E}_0/mc^2$ is the initial electron relativistic factor. Other dimensionless notations in the desired truncated system are

$$C_1 = \frac{Le^2 A_{12}}{2\mathcal{E}_0^2};\ C_2 = \frac{Le^2|A_2|}{2\mathcal{E}_0^2}E_v;\ C = 1 + E_v^2\Big/\left[E_v^2 + \left(\omega_1^2/c^2\right)|A_2|^2\right];\ \mu = \omega_1 L/c;$$

$$A_{12} = \sqrt{\frac{\omega_1^2}{c^2}|A_1|^2|A_2|^2 + |A_1|^2 E_v^2}\ ;\ E_v = |E_2|. \tag{7.109}$$

The magnetic phase of electron rotation p_0 is determined similarly to that in Equation 7.22

$$p_0 = arctg\ (\mathcal{P}_y/\mathcal{P}_x) \tag{7.110}$$

where $\mathcal{P}_{x,y}$ are the components of the transverse canonical momentum $\vec{\mathcal{P}}_\perp = \left\{\mathcal{P}_x, \mathcal{P}_y\right\}$.

We begin with the calculation in the zero approximation of the hierarchic method ($x \approx \bar{x}$). The calculations give the following system of the truncated equations (the first hierarchic level):

$$\frac{d\bar{w}}{dT} = \frac{C_1 \sin(\bar{\theta}') + C_2}{\sqrt{(1-\bar{w})^2 - \gamma_0^{-2} - B(\bar{\theta}')}};$$

$$\frac{d\bar{\theta}'}{dT} = \frac{\mu'(1-\bar{w})}{\sqrt{(1-\bar{w})^2 - \gamma_0^{-2} - B(\bar{\theta}')}} - \Phi'(T), \tag{7.111}$$

where $T = z/L$, $B(\bar{\theta}') = \left(e^2/2\mathcal{E}_0^2\right)\left[|A_1|^2 + |A_2|^2 + |A_1 A_2|\cos(\bar{\theta}')\right]$, $\mu' = C\mu$, $\bar{\theta}' = \bar{\theta} + \alpha$ is the extended combination phase, $\Phi' = L(k_1 + k_2) - d\alpha/dT$, $\mu = (\omega_1 + \sigma\omega_2)c^{-1}L$, $\alpha = \varphi_1 - \varphi_2 - arctg\left(c^2 E_v^2\big/\omega_1^2|A_2|^2\right)$, and $\varphi_{1,2} = arg\ A_{1,2}$.

As follows from the mathematical structure of System 7.111, interaction in the EH-model has the parametric resonance character only when the threshold condition for the initial electron kinetic energy $\mathcal{E}_0$ holds (see also analogous Condition 7.82)

$$\mathcal{E}_0 > \sqrt{m^2c^4 + \mathcal{E}_0^2 B(\bar{\theta}')}, \tag{7.112}$$

i.e.,

$$\mathcal{E}_0^2\left(1 - B\left(\bar{\theta}'\right)\right) - m^2c^4 > 0. \tag{7.113}$$

Only in this case, the electron, which is oscillating in the transverse plane, can move transitionally along the z-axis in the system working bulk. If

$$\mathcal{E}_0^2\left(1 - B\left(\bar{\theta}'\right)\right) - m^2c^4 < 0,$$

the reflection effect for the electrons takes place in the system input [2]. The violation of Conditions 7.112 and 7.113 within the working bulk (due to the increasing pumping amplitude, for instance) leads to realization of the capture effect [2].

7.6.4 Coupled Parametric Cyclotron Resonance: The First Approximation

From the methodological point of view, the case of the coupled (bounded) parametric cyclotron resonance is similar to the above-mentioned model with the bichromatic electromagnetic pumping (see Section 7.5). This similarity can be obviously seen by the comparison of the hierarchic trees represented in Figures 7.16 and 7.18. Figure 7.16 illustrates the bichromatic pumping, whereas Figure 7.18 corresponds to the model with the coupled parametric cyclotron resonance. However, in spite of this similarity, the physics of each type of coupled resonance is essentially different. First, it is caused by the fact that in the case represented in Figure 7.18, one of the three interaction phases—the magnetic phase p_0—is hidden. In contrast, in the model with the bichromatic pumping, all interacting phases are evident phases.

Let us illustrate this point by analyzing the energy exchanging process in the considered model. Following the discussed general calculation scheme, we can obtain a relevant set of truncated equations. However, in contrast to the preceding analogous case (see System 7.110), we accomplish the calculations in the first approximation of the hierarchic method. As a result we obtain

$$\frac{d\bar{w}}{dT}=\frac{C_1\sin(\bar{\theta}')+C_2-C_3\bar{\mathcal{P}}_\perp\sin(\bar{\theta}_{10})-C_4\bar{\mathcal{P}}_\perp\cos(\bar{\theta}_{20})}{\sqrt{(1-\bar{w})^2-\gamma_0^{-2}-B(\bar{\theta}')}}; \tag{7.114}$$

$$\frac{d\bar{\mathcal{P}}_\perp}{dT}=-\frac{C_5\sin(\bar{\theta}_{10})+C_6\sin(\bar{\theta}_{20})}{\sqrt{(1-\bar{w})^2-\gamma_0^{-2}-B(\bar{\theta}')}}; \tag{7.115}$$

$$\frac{d\bar{\theta}_{10}}{dT}=\frac{8\bar{\mathcal{P}}_\perp\mathcal{E}_0\mu'(1-\bar{w})-C_5\cos(\bar{\theta}_{10})-C_6\cos(\bar{\theta}_{20})-4LeH_0\bar{\mathcal{P}}_\perp}{8\mathcal{E}_0\bar{\mathcal{P}}_\perp\sqrt{(1-\bar{w})^2-\gamma_0^{-2}-B(\bar{\theta}')}}-\Phi_{10}; \tag{7.116}$$

$$\frac{d\bar{\theta}_{20}}{dT}=-\frac{C_5\cos(\bar{\theta}_{10})+C_6\cos(\bar{\theta}_{20})-LeH_0\bar{\mathcal{P}}_\perp}{2\bar{\mathcal{P}}_\perp\mathcal{E}_0\sqrt{(1-\bar{w})^2-\gamma_0^{-2}-B(\bar{\theta}')}}-\Phi_{20}(T); \tag{7.117}$$

$$\frac{d\bar{\theta}'}{dT}=\frac{\mu'(1-\bar{w})}{\sqrt{(1-\bar{w})^2-\gamma_0^{-2}-B(\bar{\theta}')}}-\Phi'(T), \tag{7.118}$$

where $C_3=\frac{Leck_1A_1}{2\mathcal{E}_0^2}$, $C_4=\frac{LecE_\upsilon}{2\mathcal{E}_0^2}$, $C_5=\frac{Le^2H_0A_1}{c}$, $C_6=\frac{Le^2H_0A_2}{c}$, $\Phi_{10}=Lk_1$, and $\Phi_{20}=-Lk_2$. Other values are determined already (see the preceding case of parametric resonance in System 7.111).

7.6.5 Coupled Parametric Cyclotron Resonance: The Second Hierarchic Level

In all the preceding illustrative examples, we limited ourselves to the transformation of the initial system 7.25 at the first hierarchic level only. This was motivated by the fact that the truncated equations of the first hierarchy were simple enough for physical analysis. Therefore, the first hierarchic level was considered in such situations as a level of responsibility (see this concept in Chapter 2).

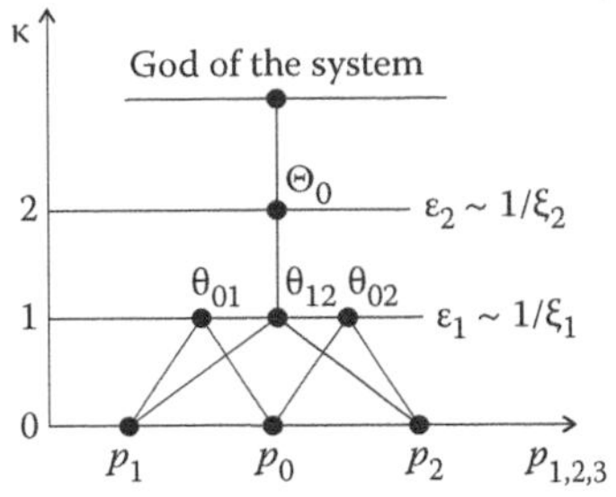

FIGURE 7.18
Hierarchic tree of bounded (coupled) parametric cyclotron resonance. Here, $p_{0,1,2}$ are the phases of the zero hierarchic level, and θ_{12}, θ_{01}, and θ_{02} are the combination phases of the first hierarchic level. There are three coupled resonances simultaneously, only two of which are independent. All designations are the same as those given in the captions of Figures 7.2 and 7.3.

Nevertheless, from a methodical viewpoint, it appears useful to illustrate the calculation peculiarities of the passage at the second hierarchic level. Just such an illustrative example, when the last (the second in our case; see Figure 7.18) hierarchic level is chosen to be a new level of responsibility, is discussed below.

The small parameter of the first hierarchy $1/\xi_1 \ll 1$ for Systems 7.116–7.118 can be separated in the form

$$1/\xi_1 \sim \left|\frac{d\theta'}{dt}\right| \Big/ \left|\frac{d\psi}{dt}\right|, \tag{7.119}$$

where $|d\theta'/dt| \sim |d\theta_{10}/dt| \sim |d\theta_{20}/dt|$. Following standard procedures, we consider the combination phases $\theta', \theta_{01}, \theta_{02}$ as the new fast oscillation phases. This means that we can determine the small parameter of the second hierarchy by the following form:

$$1/\xi_2 \sim \frac{1}{\mathcal{E}_0}\left|\frac{d\bar{\mathcal{E}}}{dt}\right| \Big/ \left|\frac{d\theta'}{dt}\right|. \tag{7.120}$$

One can be sure that both values $\xi_{1,2} \gg 1$ and, at the same time, that they satisfy the main condition of the hierarchic series (see Chapters 1, 5, and 6 for details about hierarchic series)

$$\xi_1 \gg \xi_2 \gg 1. \tag{7.121}$$

It is obvious that the accuracy of calculations at all hierarchic levels should be the same [1,2]. This automatically means that we should accomplish relevant calculations for the second level at least in the second approximation of the hierarchic method. Taking this into consideration, we find the truncated equation in the form

$$\frac{d\bar{\bar{w}}}{dT} = A_{1w}\varepsilon_1^2\varepsilon_2\varepsilon_v + A_{2w}\varepsilon_1^2\varepsilon_v + A_{3w}\varepsilon_1\varepsilon_2\varepsilon_v + A_{4w}\varepsilon_1\varepsilon_v + A_{5w}\varepsilon_1\varepsilon_2\varepsilon_v,$$

$$\frac{d\bar{\bar{\mathcal{P}}}_z}{dT} = A_{1\mathcal{P}_z}\varepsilon_1^2\varepsilon_2\varepsilon_v + A_{2\mathcal{P}_z}\varepsilon_1^2\varepsilon_v + A_{3\mathcal{P}_z}\varepsilon_1\varepsilon_2\varepsilon_v + A_{4\mathcal{P}_z}\varepsilon_1\varepsilon_v + A_{5\mathcal{P}_z}\varepsilon_2\varepsilon_v,$$

$$\frac{d\bar{\bar{\mathcal{P}}}_\perp}{dT} = A_{2\mathcal{P}\perp}\varepsilon_1^2\varepsilon_v + A_{3\mathcal{P}\perp}\varepsilon_1\varepsilon_2\varepsilon_v + A_{4\mathcal{P}\perp}\varepsilon_1\varepsilon_v + A_{5\mathcal{P}_\perp}\varepsilon_2\varepsilon_v, \tag{7.122}$$

where

$$A_{1w} = \frac{L\omega_1 c^2}{D_1\bar{\bar{\mathcal{P}}}_z v_{z0}}\left(\frac{(k_1+k_2)\Omega_{H_0}}{4v_{z0}\bar{\bar{\mathcal{P}}}_z} - \frac{k_1^2+k_1k_2}{2} + \frac{\omega_1(1-\bar{\bar{w}})(k_1+k_2)}{2v_{z0}\bar{\bar{\mathcal{P}}}_z}\right)$$

$$A_{2w} = \frac{Lw_1c}{D_1}\left(\frac{k_1\Omega_{H_0}}{8\nu_{\perp 0}\bar{\bar{\mathcal{P}}}_\perp} - \frac{k_1^2+k_1k_2}{2} + \frac{k_1^2\nu_{\perp 0}\bar{\bar{\mathcal{P}}}_\perp}{2\nu_{z0}\bar{\bar{\mathcal{P}}}_z} + \right)$$

$$+\frac{L\omega_1 c}{D_1}\left(\frac{k_1 v_{\perp 0}\bar{\bar{\mathcal{P}}}_\perp(2\omega_1(\bar{\bar{w}}-1)-\Omega_{H_0})}{4v_{z0}^2\bar{\bar{\mathcal{P}}}_z^2} + \frac{\Omega_{H_0}(2\omega_1(\bar{\bar{w}}-1)-\Omega_{H_0})}{16v_{z0}v_{\perp 0}\bar{\bar{\mathcal{P}}}_z\bar{\bar{\mathcal{P}}}_\perp}\right),$$

$$+\frac{L\omega_1 c}{D_1}\left(\frac{\Omega_{H_0}(2\omega_1(\bar{\bar{w}}-1)-\Omega_{H_0})}{4v_{z0}v_{\perp 0}\bar{\bar{\mathcal{P}}}_z^2}\right)$$

$$A_{3w} = \frac{L\omega_1 c}{D_1}\left(\frac{v_{\perp 0}\bar{\bar{\mathcal{P}}}_\perp(k_1+2k_2)(\Omega_{H_0}-2\omega_1(\bar{\bar{w}}-1))}{4v_{z0}^2\bar{\bar{\mathcal{P}}}_z^2} + \frac{k_1\Omega_{H_0}}{8v_{\perp 0}\bar{\bar{\mathcal{P}}}_\perp}\right)$$

$$+\frac{L\omega_1 c}{D_1}\left(\frac{\omega_1\Omega_{H_0}(2(\bar{\bar{w}}-1)-\Omega_{H_0})}{4v_{z0}v_{\perp 0}\bar{\bar{\mathcal{P}}}_z^2} + \frac{\Omega_{H_0}(2\omega_1(\bar{\bar{w}}-1)-\Omega_{H_0})}{16v_{z0}v_{\perp 0}\bar{\bar{\mathcal{P}}}_z\bar{\bar{\mathcal{P}}}_\perp}\right),$$

$$+\frac{L\omega_1 c}{D_1}\left(\frac{-k_1 v_{\perp 0}\bar{\bar{\mathcal{P}}}_\perp(2k_2+k_1)}{2v_{z0}\bar{\bar{\mathcal{P}}}_z}\right)$$

$$A_{4w} = \frac{L\omega_1}{D_1}\left(\frac{k_1 v_{\perp 0}^2\bar{\bar{\mathcal{P}}}_\perp^2(2\omega_1(\bar{\bar{w}}-1)-\Omega_{H_0})}{4v_{z0}^2\bar{\bar{\mathcal{P}}}_z^2} + \frac{5k_1\Omega_{H_0}}{8}\right)$$

$$+\frac{L\omega_1}{D_1}\left(\frac{\Omega_{H_0}(10\omega_1(\bar{\bar{w}}-1)-5\Omega_{H_0})+8k_1^2v_{\perp 0}^2\bar{\bar{\mathcal{P}}}_\perp^2}{16v_{z0}\bar{\bar{\mathcal{P}}}_z} + \frac{\Omega_{H_0}\bar{\bar{\mathcal{P}}}_\perp(2\omega_1(\bar{\bar{w}}-1)-\Omega_{H_0})}{4v_{z0}\bar{\bar{\mathcal{P}}}_z^2}\right),$$

$$A_{5w} = \frac{L\omega_1}{D_1}\left(\frac{k_2 v_{\perp 0}^2\bar{\bar{\mathcal{P}}}_\perp^2(\Omega_{H_0}-2\omega_1(\bar{\bar{w}}-1))}{4v_{z0}^2\bar{\bar{\mathcal{P}}}_z^2} + \frac{5k_1\Omega_{H_0}}{8}\right)$$

$$+\frac{L\omega_1}{D_1}\left(\frac{\Omega_{H0}\left(10\omega_1(\bar{\bar{w}}-1)-5\Omega_{H0}\right)-8k_1k_2v_{\perp 0}^2\bar{\bar{\mathcal{P}}}_\perp^2-8D_1}{16v_{z0}\bar{\bar{\mathcal{P}}}_z}\right),$$

$$+\frac{L\omega_1}{D_1}\left(\frac{\Omega_{H_0}\bar{\bar{\mathcal{P}}}_\perp(2\omega_1(\bar{\bar{w}}-1)-\Omega_{H_0})}{4v_{z0}\bar{\bar{\mathcal{P}}}_z^2}\right)$$

$$A_{1\mathcal{P}_z} = \frac{L\omega_1^2 c^2}{D_1 \bar{\bar{\mathcal{P}}}_z v_{z0}}\left(\frac{k_1 + k_2}{2}\right),\ A_{2\mathcal{P}_z} = -\frac{L\omega_1^2 k_1 c v_{\perp 0}\bar{\bar{\mathcal{P}}}_\perp}{2D_1 \bar{\bar{\mathcal{P}}}_z v_{z0}^2},$$

$$A_{3\mathcal{P}_z} = \frac{L\omega_1^2 c v_{\perp 0}\bar{\bar{\mathcal{P}}}_\perp}{D_1 \bar{\bar{\mathcal{P}}}_z v_{z0}^2}\left(\frac{k_1 + k_2}{2}\right),\ A_{4\mathcal{P}_z} = -\frac{L\omega_1^2 k_1 v_{\perp 0}^2\bar{\bar{\mathcal{P}}}_\perp^2}{2D_1 \bar{\bar{\mathcal{P}}}_z v_{z0}^2},$$

$$A_{5\bar{\bar{\mathcal{P}}}_z} = \frac{L\omega_1^2 k_2 v_{\perp 0}^2\bar{\bar{\mathcal{P}}}_\perp^2}{2D_1 \bar{\bar{\mathcal{P}}}_z v_{z0}^2},\ A_{2\bar{\bar{\mathcal{P}}}_\perp} = \frac{L\omega_1^2 c\Omega_{H_0}}{2D_1 \bar{\bar{\mathcal{P}}}_z v_{z0}},\ A_{3\bar{\bar{\mathcal{P}}}_\perp} = \frac{L\omega_1^2 c\Omega_{H_0}}{2D_1 \bar{\bar{\mathcal{P}}}_z v_{z0}}$$

$$A_{4\mathcal{P}_\perp} = \frac{L\omega_1^2 \bar{\bar{\mathcal{P}}}_\perp \Omega_{H_0}}{2D_1 \bar{\bar{\mathcal{P}}}_z v_{z0}},\ A_{5\mathcal{P}\perp} = \frac{L\omega_1^2 \bar{\bar{\mathcal{P}}}_\perp \Omega_{H_0}}{2D_1 \bar{\bar{\mathcal{P}}}_z v_{z0}^2},$$

$$D_1 = 4\omega_1(\omega_1(\bar{\bar{w}}^2 - 2\bar{\bar{w}} + 1) + 2k_1 v_{z0}\bar{\bar{\mathcal{P}}}_z(\bar{\bar{w}} - 1) - \Omega_{H_0}(\bar{\bar{w}} - 1))$$
$$+ 4k_1\bar{\bar{\mathcal{P}}}_z \upsilon_{z0}(k_1\bar{\bar{\mathcal{P}}}_z v_{z0}\upsilon - \Omega_{H_0}) + \Omega_{H_0}^2$$

$$\varepsilon_1 = \frac{eA_1}{mc^2\gamma_0},\ \varepsilon_2 = \frac{eA_2}{mc^2\gamma_0},\ \varepsilon_v = -\frac{ecE_v}{mc^2\gamma_0\omega_1},\ \Omega_{H_0} = \frac{eH_0}{mc\gamma},\ \mathcal{P}_{z,\perp} = \frac{\mathcal{P}_{z,\perp}}{\mathcal{P}_{z0,\perp 0}}, \qquad (7.123)$$

where $\mathcal{P}_{z0}$, $\mathcal{P}_{\perp 0}$ are the transverse and longitudinal canonical momentums.

Furthermore, let us calculate the elements of the decompression operators. After accomplishing a series of standard (for the hierarchic method) calculation procedures, we find the elements. These include the passages from the second to the first hierarchic levels, which can be obtained in the form

$$U_{1w}^{(2)} = \frac{\omega_1\varepsilon_1\varepsilon_2\varepsilon_\upsilon}{D_2}\left(\left(\frac{\cos(\bar{\theta}_{12})}{\varepsilon_\upsilon} + \frac{\sin(\bar{\theta}_{12})}{\varepsilon_2}\right) - \frac{v_{\perp 0}\bar{\bar{\mathcal{P}}}_\perp}{c}\left(\frac{\cos(\bar{\theta}_{10})}{\varepsilon_1\varepsilon_\upsilon} - \frac{\sin(\bar{\theta}_{20})}{\varepsilon_1\varepsilon_2}\right)\right)$$

$$U_{1\mathcal{P}_z}^{(2)} = \frac{c\varepsilon_1\varepsilon_2}{v_{z0}D_2}\left(c(k_1 + k_2)\cos(\bar{\theta}_{12}) + v_{\perp 0}\bar{\bar{\mathcal{P}}}_\perp\left(\frac{k_1\cos(\bar{\theta}_{10})}{\varepsilon_2} - \frac{k_2\cos(\bar{\theta}_{20})}{\varepsilon_1}\right)\right),$$

$$U_{1\mathcal{P}_\perp}^{(2)} = -\frac{c\Omega_{H_0}\varepsilon_1\varepsilon_2}{v_{\perp 0}D_2}\left(\left(\frac{\cos(\bar{\theta}_{10})}{\varepsilon_2} + \frac{\cos(\bar{\theta}_{20})}{\varepsilon_1}\right)\right),$$

$$D_2 = 2\omega_1(\bar{\bar{w}} - 1) + 2k_1 v_{z0}\bar{\bar{\mathcal{P}}}_z - \Omega_{H_0}. \qquad (7.124)$$

We find the analogous elements of the decompression operator for the passage from the first to the zero levels as

$$U^{(1)}_{1w} = -\frac{L\omega_1}{2v_{z0}\bar{\mathcal{P}}_z}\left(2\varepsilon_0\varepsilon_v\sin(p_2) + \varepsilon_1\varepsilon_v\sin(p_1+p_2) + 2\varepsilon_0\varepsilon_1\cos(p_1)\right)$$

$$+\frac{L\omega_1}{2v_{z0}\bar{\mathcal{P}}_z}\left(\varepsilon_1\varepsilon_2\cos(p_1+p_2) - \frac{v_{\perp 0}\bar{\mathcal{P}}_\perp\varepsilon_1\cos(p_1+p_0)}{c} - \varepsilon_1^2\sin(2p_1)\right),$$

$$+\frac{L\omega_1}{2v_{z0}\bar{\mathcal{P}}_z}\left(\varepsilon_2\varepsilon_v\sin(2p_2) - \frac{v_{\perp 0}\bar{\mathcal{P}}_\perp\varepsilon_v\sin(p_2+p_0)}{c}\right)$$

$$U^{(1)}_{1\mathcal{P}_z}U^{(1)}_{1\mathcal{P}_z} = -\frac{Lc^2}{2v_{z0}^2\bar{\mathcal{P}}_z}\left(2k_2\varepsilon_0\varepsilon_2\cos(p_2) - 2k_1\varepsilon_1\varepsilon_0\cos(p_1+p_2)\right)$$

$$+\frac{Lc^2}{2v_{z0}^2\bar{\mathcal{P}}_z}\left(-k_1\varepsilon_1^2\cos(2p_1) + k_2\varepsilon_1\varepsilon_2\cos(p_1+p_2) + k_2\varepsilon_2^2\cos(2p_2)\right)$$

$$+\frac{Lc^2}{2v_{z0}^2\bar{\mathcal{P}}_z}\left(-k_1\varepsilon_1\varepsilon_2\cos(p_1+p_2) - \frac{k_2 v_{\perp 0}\bar{\mathcal{P}}_\perp\varepsilon_2\cos(p_2+p_0)}{c}\right)$$

$$+\frac{Lc^2}{2v_{z0}^2\bar{\mathcal{P}}_z}\left(\frac{k_1 v_{\perp 0}\bar{\mathcal{P}}_\perp\varepsilon_1\cos(p_1+p_0)}{c}\right),$$

$$U^{(1)}_{1\mathcal{P}_\perp} = -\frac{Lc\Omega_{H_0}}{4v_{z0}v_{\perp 0}\upsilon\bar{\mathcal{P}}_z}\left(\varepsilon_1\sin(p_1+p_0) + \varepsilon_2\sin(p_2+p_0)\right)$$

$$+\frac{Lc\Omega_{H_0}}{4v_{z0}v_{\perp 0}\bar{\mathcal{P}}_z}\left(2\varepsilon_0\sin(p_0) - 2\frac{v_{\perp 0}\bar{\mathcal{P}}_\perp\sin(p_0)}{c}\right),$$

$$U^{(1)}_{1\theta_1} = -\frac{Lc\Omega_{H_0}}{4v_{z0}v_{\perp 0}\bar{\mathrm{P}}_z\bar{\mathrm{P}}_\perp}\left(\varepsilon_1\sin(p_1+p_0) + \varepsilon_2\sin(p_2+p_0) + 2\varepsilon_0\sin(p_0)\right),$$

$$U^{(1)}_{1\theta_2} = -\frac{Lc\Omega_{H_0}}{4v_{z0}v_{\perp 0}\bar{\mathrm{P}}_z\bar{\mathrm{P}}_\perp}\left(\varepsilon_1\sin(p_1+p_0) + \varepsilon_2\sin(p_2+p_0) + 2\varepsilon_0\sin(p_0)\right).$$

$$U^{(1)}_{1\theta_1} = 0. \tag{7.125}$$

The rest of the elements, $U_2^{(i)}$ and $U_3^{(i)}$, can be found analogously.

By comparing Equations 7.114 through 7.118 and Equation 7.122, it is not difficult to see that the truncated equations of the second hierarchic level (Equation 7.122) have a much simpler mathematical structure. Analogously, Equations 7.114 through 7.118 are much simpler than the initial system 7.25. In other words, it can be readily seen that the principle of hierarchic compression (see Chapter 2) holds in this illustrative example. However, the above-mentioned simplification of the mathematical structures does not automatically indicate their compactness. This follows directly from the mathematical structures of the corresponding set of coefficients (Equation 7.123) for Equations 7.122. It is not difficult to make certain that the truncated equations, being simple, are found to be rather cumbersome owing to these coefficients. However, this circumstance is important only in the case of purely analytical calculations. The noted merit of the hierarchic methods (simplification of mathematical structures in the process of hierarchic transformations) becomes their main advantage in the case of computer calculations. The point is that the calculation algorithms described in Chapters 5 and 6 can be computerized without excessive mathematical difficulties.

Let us illustrate this by the analytical asymptotic solutions of System 7.24. They are simple with respect to mathematical structure and, at the same time, are rather cumbersome with respect to the form of representation

$$\bar{\bar{\mathcal{P}}}_z = 1 + B_{1\mathcal{P}_z}\varepsilon_1^2\varepsilon_2\varepsilon_v + B_{2\mathcal{P}_z}\varepsilon_1^2\varepsilon_v + B_{3\mathcal{P}_z}\varepsilon_1\varepsilon_2\varepsilon_v + B_{4\mathcal{P}_z}\varepsilon_1\varepsilon_v + B_{5\mathcal{P}_z}\varepsilon_2\varepsilon_v,$$

$$\bar{\bar{\mathcal{P}}}_\perp = 1 + B_{2\mathcal{P}_\perp}\varepsilon_1^2\varepsilon_v + B_{3\mathcal{P}_\perp}\varepsilon_1\varepsilon_2\varepsilon_v + B_{4\mathcal{P}_\perp}\varepsilon_1\varepsilon_v + B_{5\mathcal{P}_\perp}\varepsilon_2\varepsilon_v,$$

$$\bar{\bar{w}} = B_{1w}\varepsilon_1^2\varepsilon_2\varepsilon_v + B_{2w}\varepsilon_1^2\varepsilon_v + B_{3w}\varepsilon_1\varepsilon_2\varepsilon_v + B_{4w}\varepsilon_1\varepsilon_v + B_{5w}\varepsilon_2\varepsilon_v \tag{7.126}$$

where

$$B_{1w} = \frac{L\omega_1 c^2}{Mv_{z0}}\left(\frac{(k_1+k_2)\Omega_{H_0}}{4v_{z0}} - \frac{k_1^2 + k_1k_2}{2} + \frac{\omega_1(k_1+k_2)}{2v_{z0}}\right),$$

$$B_{2w} = \frac{L\omega_1 c}{M}\left(-\frac{k_1 v_{\perp 0}(2\omega_1 + \Omega_{H_0} + 2k_1 v_{z0})}{4v_{z0}^2} - \frac{\Omega_{H_0}(10\omega_1 + 5\Omega_{H_0})}{16 v_{z0} v_{\perp 0}}\right),$$

$$B_{3w} = \frac{L\omega_1 c}{M}\left(\frac{v_{\perp 0}(k_1 + 2k_2)(\Omega_{H_0} + 2\omega_1)}{4v_{z0}^2} + \frac{k_1\Omega_{H_0}}{8v_{\perp 0}}\right) + \frac{L\omega_1 c}{M}\left(\frac{k_1^2 v_{\perp 0}}{2v_{z0}} - \frac{\Omega_{H_0}(10\omega_1 + 5\Omega_{H_0})}{16 v_{z0} v_{\perp 0}}\right),$$

$$B_{4w} = \frac{L\omega_1}{M}\left(-\frac{k_1 v_{\perp 0}^2(\Omega_{H_0} + 2\omega_1)}{4v_{z0}^2} + \frac{5k_1\Omega_{H_0}}{8}\right) + \frac{L\omega_1}{M}\left(-\frac{\Omega_{H_0}(10\omega_1 + 5\Omega_{H_0}) - 8k_1^2 v_{\perp 0}^2}{16 v_{z0}}\right),$$

$$B_{5w} = \frac{L\omega_1}{M}\left(\frac{k_2 v_{\perp 0}^2(\Omega_{H_0} + 2\omega_1)}{4v_{z0}^2} + \frac{k_1(2\omega_1 + \Omega_{H_0}) - 16k_1^2 v_{z0})}{8}\right) + \frac{L\omega_1}{M}\left(-\frac{\Omega_{H_0}(10\omega_1 + 5\Omega_{H_0}) + 32\omega_1^2 + 8k_1k_2 v_{\perp 0}^2}{16 v_{z0}}\right),$$

$$B_{1\mathcal{P}_z} = \frac{L\omega_1^2 c^2}{Mv_{z0}^2}\left(\frac{k_1+k_2}{2}\right),\ B_{2\mathcal{P}_z} = -\frac{L\omega_1^2 k_1 c v_{\perp 0}}{2Mv_{z0}^2},\ B_{3\mathcal{P}_z} = \frac{L\omega_1^2 c v v_{\perp 0}}{Mv_{z0}^2}\left(\frac{k_1+k_2}{2}\right),$$

$$B_{4\mathcal{P}_z} = -\frac{L\omega_1^2 k_1 v_{\perp 0}^2}{2Mv_{z0}^2},\ B_{5\mathcal{P}_z} = \frac{L\omega_1^2 k_2 v_{\perp 0}^2}{2Mv_{z0}^2},\ B_{2\mathcal{P}_\perp} = \frac{L\omega_1^2 c\Omega_{H_0}}{2Mv_{z0}v_{\perp 0}},$$

$$B_{3\mathcal{P}_\perp} = \frac{L\omega_1^2 c\Omega_{H_0}}{2Mv_{z0}v_{\perp 0}},\ B_{4\mathcal{P}_\perp} = \frac{L\omega_1^2 \Omega_{H_0}}{2Mv_{z0}},\ B_{5\mathcal{P}_\perp} = \frac{L\omega_1^2 \Omega_{H_0}}{2Mv_{z0}^2},$$

$$D_1 = (4\omega_1(\omega_1 - 2k_1 v_{z0} + \Omega_{H_0}) + 4k_1 v_{z0}(k_1 v_{z0} - \Omega_{H_0}) + \Omega_{H_0}^2)/T\,. \tag{7.127}$$

For the sake of simplicity we have omitted in Equation 7.127 the terms $\sim T^2$ and higher. Then by substituting Equations 7.126 and 7.127 in Equation 7.124 and so on, we eventually find complete analytical solutions of the zero hierarchic level.

7.6.6 Isochronous FEL Model with the EH-Pumping

As before (see Section 7.4.5), we determine the isochronous state of the system as the constancy of the extended combination phase θ'

$$\theta' \approx \theta_0' = const. \tag{7.128}$$

In contrast to the methods of isochronization studied in Section 7.4.5, in this case, we consider the same result we obtained due to the transverse undulation electric field. We will use the general scheme of analysis described in Section 7.4.5.

By using the second of Equations 7.111, let us rewrite Condition 7.128 in the form

$$\mu'(1-\bar{w}) = \Phi'(T)\sqrt{(1-\bar{w})^2 - \gamma_0^{-2} - B(\theta_0')}\,. \tag{7.129}$$

It is not difficult to see that Expression 7.129 coincides with the analogous condition 7.84 if we accept $\Pi = 0$ (in Condition 7.84) and $\mu' = \mu$ (in Expression 7.129). After noncomplex transformations, it is not difficult to obtain from Expression 7.111 and Expression 7.129 the relationships for the reduced

$$\bar{w} = 1 - \frac{\gamma_0^{-1}\sqrt{1+\gamma^2 B(\theta_0')}}{\sqrt{1-(\mu'/\Phi')^2}} \approx w \tag{7.130}$$

and the actual (true)

$$\bar{\eta} = \frac{1}{(1-\gamma_0^{-1})}\int_0^T \frac{(C_1\sin\theta_0' + C_2)\sqrt{(\Phi'/\mu')^2 - 1}}{\sqrt{\gamma_0^{-2} + B(\theta_0')}}\,dT \approx \eta \tag{7.131}$$

efficiencies. Then, similar to the case of the longitudinal electric support (Section 7.4.5), we accept the supposition that the considered model is homogeneous: $C, C_1, C_2, \Phi' = const.$

$$\beta_z/c = \beta_z = \beta_{0\|};\ \frac{\mu'}{\Phi'} = C\beta_{0\|} = const, \tag{7.132}$$

By using Equations 7.130 and 7.131, we obtain

$$\eta = \frac{\left(C_1 \sin\theta_0' + C_2\right)\sqrt{C^{-2}\beta^{-2} - 1}}{\left(1 - \gamma_0^{-1}\right)\sqrt{\gamma_0^{-2} + B(\theta_0')}} T. \tag{7.133}$$

Analyzing Equation 7.133, we find that efficiency η attains its maximum in the case when

$$\theta_0' = \theta_{0\max}' = -\delta \arccos\left[-d\left(\gamma_0^2 \varepsilon_1 \varepsilon_2\right) + \sqrt{d^2\left(\gamma_0^2 \varepsilon_1 \varepsilon_2\right)^{-2} - 1}\right], \tag{7.134}$$

where $d = 1 + \gamma_0^2 + B_0$, $B_0 = B(\theta_0') - \varepsilon_1 \varepsilon_2 \cos\theta_0'$, and $\varepsilon_{1,2} = eA_{1,2}/mc^2\gamma_0$. In the simplest case, $\gamma_0^2 \varepsilon_1 \varepsilon_2 << 1$, we obtain: $\theta_{0\max}' \cong -\sigma\left(\pi/2\right)$.

7.6.7 Single-Particle Numerical Analysis

Comparing Equations 7.134 and 7.88, we can conclude that the isochronization actions of the transverse electrical vortex support and the longitudinal electrostatic support are actually equivalent. Indeed, in both cases, we can completely suppress the nonlinear violation of the resonant state of the system and, as a result, to overcome the process of the saturation of the signal amplification. At first sight, the difference is only technological. Indeed, if we introduce a special system of electrodes (for creation of a longitudinal field) into the working bulk, ,we can use part of the magnetic undulation pumping—constructed on the basis of low-frequency electromagnets—for the same purpose.

The relevant numerical calculations, illustrated in Figure 7.19, completely confirm this observation. As seen in Figure 7.19, introducing the optimal transverse electric support allows increasing the single-particle electron efficiency from ~5% (curve 2) to ~73% (curve 1). And analysis shows that by increasing the system length, we can attain the efficiency approaching unit. Of course, this result carries only a theoretical model character. However, it gives a hope that, in practice, this type of isochronous system will be characterized by a relatively high level of efficiency.

The other type of result is represented in Figure 7.20. There the dynamic of electron efficiency is illustrated for a case where the electron relativistic factor is reduced from 36.1 (Figure 7.19) to 8 (Figure 7.20). Herewith, the signal wavelength is reduced from λ_1 = 10.5 μm to λ_1 = 0.214 mm, and the pumping amplitudes also are lower. It is not difficult to see that the maximal electron efficiency in this model equals to 38% (curve 2 in Figure 7.19). As an introduction into the interaction bulk, the strong magnetic field allows to increase the electron efficiency from 38% to 76% (curve 1 in Figure 7.19).

7.6.8 Some Qualitative Peculiarities of the Multiparticle Dynamics

It should be noted that the real situation in the FEL systems is not as simple as indicated by the analysis. These results concern only the single-particle electron wave interaction,

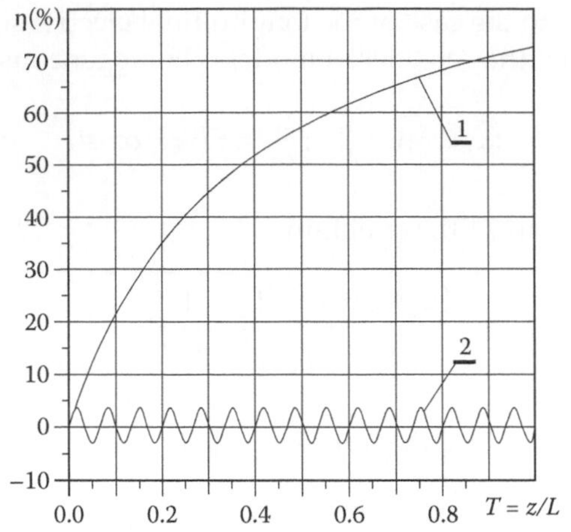

FIGURE 7.19
Dependencies of the electron efficiency η on the dimensionless longitudinal coordinate $T = z/L$ in the EH-ubitron and an equivalent H-ubitron FEL. Here, curve 1 corresponds to the isochronous model, whereas curve 2 illustrates the ordinary H-ubitron model. The parameters of the Los Alamos experimental unit [45] are accepted as the basis of the calculations: the model length L = 10 m, the signal wavelength λ_1 = 10.5 μm, the period of undulation of the pumping system Λ_2 = 2.73 cm, the relativistic factor γ_0 = 36.1, and the amplitude of undulation magnetic field B_2 = 3.1 kGs. The amplitude of the vortex electric field of support is accepted as $\varepsilon_2 \sim \varepsilon_v = 10^{-6}$.

whereas we must always deal with a multiparticle interaction. The semiqualitative physical picture of such multiparticle mechanisms is discussed in detail in Section 4.3. As shown, the coherent character of such multiparticle interactions with the signal wave is provided due to the realization of a peculiar self-organization process. In vacuum physical electronics [7], this process is treated as a realization of the grouping mechanisms, which consist

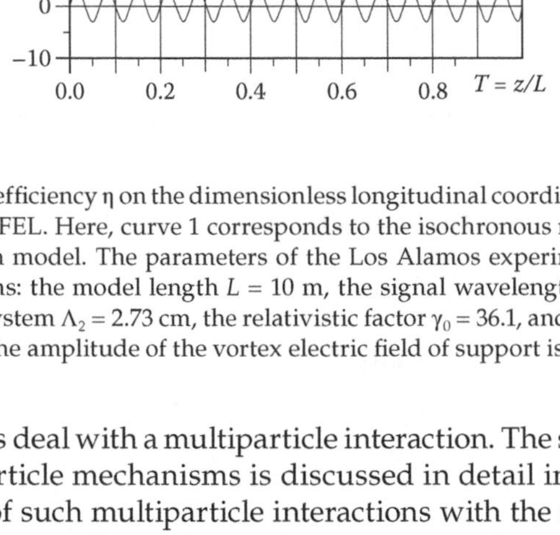

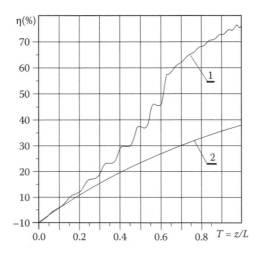

FIGURE 7.20
Dependencies of the electron efficiency η on the dimensionless longitudinal coordinate $T = z/L$ in the parametric EH-ubitron with weak and strong longitudinal magnetic fields. Here, curve 1 corresponds to the model with the strong magnetic field (B_2 = 3.1 kGs) and curve 2 illustrates the parametric EH-ubitron model with a weak magnetic field. It is assumed that $\gamma_0 = 8$, λ_1 = 214 μm.

of the formation of specific electron clusters (bunches). The initially homogeneous electron beam transforms into a periodical sequence of electron bunches. We can also treat this process as the excitation of electron waves. We interpret the resonant interaction of electron waves with the superimposition of pumping and signal waves as the wave parametric resonance. The reciprocal Coulomb electron interactions can therefore play an important role (see Section 4.3 for details).

The formation of bunches always starts with the modulation of the electron beam with respect to longitudinal electron velocities. This modulation at velocities transforms into modulation at the beam density. It is important to notice that, owing to this transformation (grouping) process, the electrons with different velocities meet within the same bunch. The velocity differences are, as a rule, rather small in comparison with the averaged longitudinal velocity of the beam as a whole. Nevertheless, the fact of their physical presence plays a very important role in the general bunch dynamic (Section 4.3).

Thus, the main idea of the above-mentioned self-organization process is that each electron bunch interacts coherently with the signal electromagnetic wave like a large charged particle. All these large particles are distributed periodically along the beam length. The single-particle conditions for each bunch of this periodical system are equivalent to the multiparticle resonant conditions in the system electron beam + electromagnetic signal wave. Owing to this, the multiparticle energy exchange in the discussed system carries a coherent character. In this case, the kinetic energy of the electron beam transforms into the energy of the signal electromagnetic wave.

In view of this, the practical applicability of the discussed results of the single-particle theory can be regarded in two ways. The first way involves the sufficiently compact (on the scale of the signal wavelength) electron bunches. In this situation, we can consider each electron bunch as a point-like large charged particle with dynamics that can be described within the framework of the developed single-particle theories. This approach can be useful for the study of the energy-exchange processes in the terminal sections of FEL klystrons.

The second case occurs when the bunches are not sufficiently compact. In such a situation, each bunch can be treated as an ensemble of separate electrons. Each electron of this ensemble interacts with the signal wave individually, in accordance with the developed single-particle theories. The total result of the energy exchange between the beam and the signal electromagnetic wave carries an integral character. This means that one group of the bunch electrons can absorb the signal-wave energy while the electrons of the other group simultaneously transfer their kinetic energy into the signal wave. The eventual balance of a superimposition of these processes determines the sign of the terminal result of the exchanging process: plus in the case of the resonant amplification of the signal wave and minus for its resonant absorption. In the first case, we have a typical FEL amplifier (or generator; see Section 4.3 for details), and the second model corresponds to the FEL accelerator (the so-called serfotron).

Let us mathematically illustrate the discussed multiparticle qualitative physical picture using the simplest Compton [1,2] model of the EH-FEL (see the corresponding classification in Section 4.3). We will use the method of large particles as the basic approach to the problem.

The essence of the method involves the representation of the electron bunches as an aggregate of relatively compact large particles—each can be considered as a point-like charge. The total charge of the bunch in such equal subdivisions equals the sum of the individual charges of all large particles.

7.6.9 Example of Quantitative Analysis of the Multiparticle EH-Model

The possibility of radiating the coherent electromagnetic radiation in any FEL is determined by the quality (including the compactness and the dynamic imbalance) of the electron bunches. Therefore, let us further introduce the function of bunch quality, R, which describes the discussed bunch compactness and the dynamic imbalance

$$R = \max_{\tau}[\rho(\tau) - a\sigma(\tau)], \tag{7.135}$$

where $\tau \in [0,1]$ is the dimensionless time of particle motion in the system working bulk, and a is a scale parameter. The functions $\rho(\tau)$ (phase density) and $\sigma(\tau) = c\delta(\tau)$ (dynamic spread within the bunch) are

$$\rho(\tau) = \frac{k(\tau)}{2^n \hat{r}(\tau)}; \tag{7.136}$$

$$\sigma(\tau) = c\delta(\tau) = \left[\frac{1}{k(\tau)-1}\sum_{i=1}^{k(\tau)}\left[v_i(\tau) - \hat{v}(\tau)\right]^2\right]^{1/2}, \tag{7.137}$$

where $v_i(\tau)$ is the single-particle velocity, and $\hat{v}(\tau)$ is the bunch velocity as a whole

$$\hat{v}(\tau) = \frac{1}{k(\tau)}\sum_{i=1}^{k(\tau)} v_i(\tau), \tag{7.138}$$

the quantity

$$\hat{r}(\tau) = r(\tau)/l(\tau = 0) \tag{7.139}$$

is the normalized spatial size of the bunch, and the value

$$l(\tau) = z_{\max}(\tau) - z_{\min}(\tau) \tag{7.140}$$

is the interval between the first [coordinate $z_{\min}(\tau)$] and the last [coordinate $z_{\max}(\tau)$] particles of the bunch. The interval $l(\tau)$ is divided into a fixed number of subintervals with the length $r(\tau)$. Therein, quantity $k(\tau)$ is a number of particles with coordinates in the interval $r(\tau)$. The quantity 2^n corresponds to the number of large particles of the electron beam flying in the interaction volume at the interval from θ_0' to $\theta_0' + 2\pi$ (θ_0', is the extended slow combination particle of the ith large particle in the system input).

We assume that the fields of the signal and the pumping are given, and the Coulomb interactions between the bunch electrons are neglected. In this case, the collective efficiency for the electron beam can be determined as

$$\hat{\eta} = \frac{1}{2\pi}\int_0^{2\pi} d\theta_0' \int_0^{\infty} \eta(\theta_0', \gamma_0)\rho_\theta(\theta_0')\rho_\gamma(\gamma_0)d\gamma_0, \tag{7.141}$$

where ρ_θ and ρ_γ are the distribution functions with respect to the initial combination phases θ_0' and the relativistic factor γ_0, and $\eta(\theta_0', \gamma_0) \cong \bar{\eta}(\bar{\theta}_0', \bar{\gamma}_0)$ is the single-particle efficiency calculated in the zero order of the hierarchic method. For the functions ρ_θ and ρ_γ we choose the normalization

$$\frac{1}{2\pi}\int_0^{2\pi} \rho_\theta(\theta_0')d\theta_0' = 1; \int_0^{\infty} \rho_\gamma(\gamma_0)d\gamma_0 = 1\,. \tag{7.142}$$

Let us assume that $\rho_\theta(\theta_0') = 1$, $\rho_\gamma(\gamma_0) = \delta(\gamma_0 - \gamma_{00})$, where $\delta(\gamma_0 - \gamma_{00})$ is the Dirac delta function. This supposition corresponds to the model of the initially nonmodulated (with respect to density and velocity at the same time) and the monoenergetic electron beam. So, let us accept that the beam is initially nonmodulated, monoenergetic, and transversely isotropic.

We use the truncated single-particle equations 7.111 as the basis for our numerical calculations of the collective efficiency $\hat{\eta}(T)$. The results of these calculations are represented in Figure 7.21. It is evident that the isochronous EH-ubitron FEL can be characterized by a relatively high collective operating efficiency. However, comparing the curves of Figures 7.19 and 7.21, we see that the collective efficiency in the considered model is essentially lower than the analogous value in the single-particle case. This is a very intriguing observation because essential physical causes should exist for the decreasing efficiency from ~73% (curve 1 in Figure 7.19) to ~23% (Figure 7.21). As analysis shows, such a cause really exists and is related to the so-called cooling effect that can be realized in the EH-pumping (for the elementary theory of this phenomenon see *Undulative Electrodynamic Systems* by Kulish [2]). Let us discuss this topic in more details.

The required modulation of electron beams in FEL occurs due to the bunching (grouping) mechanism. This mechanism is closely related to the discussed mechanism of electron modulation with respect to velocities. Hence, as mentioned, the electron bunches in any FEL consist of particles of different velocities. We can interpret the latter as an affirmation

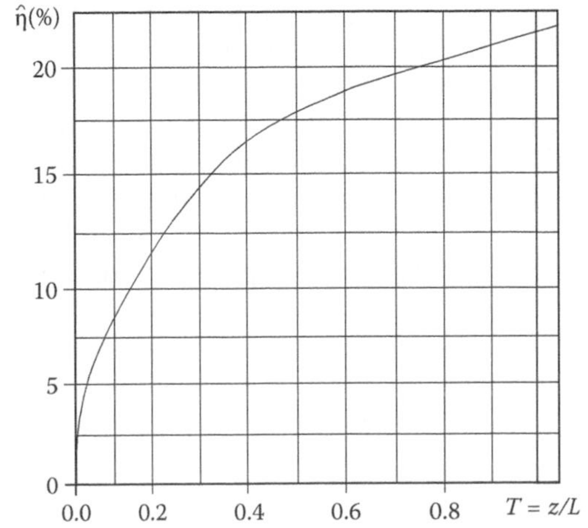

FIGURE 7.21
Dependencies of the collective efficiency $\hat{\eta}(T)$ on the dimensionless coordinate $T = z/L$. The characteristic model parameters are given in Figure 7.18.

of the fact that the electron bunches in FEL are also hot ensembles of electrons. However, as shown in the work by Kulish [2], such hot electron ensembles can be cooled in the EH system due to the cooling effect. However, we know that the cooling effect is always accompanied by an increasing bunch length [2]. We interpret this effect as the demodulation (debunching) of the modulated electron beam because its realization decreases the quality of bunches (see Definition 7.135). This physical phenomenon explains the above-mentioned effect of reducing collective efficiency, which is disclosed by the comparison of the materials in Figures 7.19 and 7.21.

Thus, in the discussed linearly polarized EH-FEL, we have the simultaneous action of the bunching and debunching (cooling) competitive mechanisms. The decrease of the collective electron efficiency as it is clarified with respect to the single-particle efficiency is the result of the action of the cooling mechanism. Hence, we can conclude that the linearly polarized EH-FEL section could be promising only for the klystron-type devices, because here the above-mentioned harmful influence of the cooling effect can be reduced to a minimum. Moreover, it could be useful because it can suppress the overgrouping effect, which is one of the main causes limiting the collective efficiency of the isochronous FEL.

In the terminal section of the klystron-type EH-FEL, the electron beam should enter in the form of a succession of almost cooled bunches [2]. However, this is not so because bunches have been formed from the electrons of different velocities. With time, the small velocity spread leads to the violation of the initial bunch compactness. In this case, the cooling effect works as a peculiar mechanism of conservation of the bunch quality. In other words, the discussed cooling effect indeed can be useful in FEL in some specific situations.

However, analysis shows that we can obtain much more promising results by studying the EH-FEL with circularly polarized real EH-pumping [2]. The point is that the real EH models are characterized by the presence of an explicitly expressed transverse inhomogeneity of the EH-field amplitudes. In a case of linearly polarized EH-pumping, this circumstance does not play any essential role. Electrons in the transverse plane move along the x-axis predominantly, whereas the mentioned inhomogeneity appears only along the y-axis [2]. In contrast, the electrons in the transverse plane of circularly polarized EH-pumping move in both coordinates x and y simultaneously. As a result, the transverse inhomogeneity exacts an essential influence on the general dynamics of the electron bunch as a whole. In particular, we can observe a peculiar phenomenon of suppression of the cooling effect in such systems. The paradox of the discussed situation is that the phenomenon of the suppression is, on one hand, bad for the EH-coolers [2] and, on the other hand, good for the isochronous EH-FEL. Indeed, in the latter case, we have an additional possibility for isochronization of the multiparticle interaction without the appearance of the discussed debunching effect. As a result, the multiparticle efficiency in such FEL models can attain essentially higher levels. Concerning the EH-coolers, we can say that peculiar coaxial designs are used for overcoming the effect of suppression of the beam cooling.

References

1. Kulish, V.V. 1998. *Methods of averaging in non-linear problems of relativistic electrodynamics*. Atlanta, GA: World Scientific Publishers.
2. Kulish, V.V. 2002. *Undulative electrodynamic systems*, Vol. 2 of *Hierarchic methods*. Dordrecht: Kluwer Academic Publishers.

3. Sokolov, A.A., and Ternov, I.M. 1974. *Relativistic electron*. Moscow: Nauka.
4. Redmond, P.J. 1965. Solution of the Klein-Gordon and Dirac equations for a particle with a plane electromagnetic wave and a parallel magnetic field. *J. Math. Phys.* 6(7):1163.
5. Kulish, V.V., Dzedolik, I.V., and Kudinov, M.A. 1985. Relativistic electron motion in periodic reversible fields. Deposited manuscript. Deposited in Ukrainian Scientific Research Institute of Technical Information, Kiev. 1490, Uk-85.
6. Kochmanski, S.S., and Kulish, V.V. 1985. Electron parametric resonance interaction with the intense electromagnetic wave field in the presence of longitudinal magnetic field. *Acta Phys. Polonica* A68(5):725.
7. Trubetskov, D.I., and Khramov, A.E. 2003. *Lectures on microwave electronics for physicists*, Vol. 2. Moscow: Fizmatlit.
8. Kochmanski, S.S., and Kulish, V.V. 1985. On the nonlinear theory of free electron lasers. *Acta Phys. Polonica* A68(5):749.
9. Kochmanski S.S., and Kulish, V.V. 1984. On the classical one-particle theory of free electron lasers. *Acta Phys. Polonica* A66(6):713.
10. Kochmanski, S.S., and Kulish, V.V. 1982. Parametric resonance under the relativistic electron motion in the field of three transverse waves. Deposited manuscript. Deposited in Ukrainian Scientific Research Institute of Technical Information, Kiev. 4792, UK-82.
11. Zhurakhovsky, V.A., Kulish, V.V., and Chemerys, V.T. 1980. Energy generation by an electron beam in the field of two transverse electromagnetic waves. *Preprint of Electrodynamics Inst.*, No. 218, Kiev.
12. Zhurakhovsky, V.A. 1972. *Nonlinear electron oscillations in magneto-directed beams*. Kiev: Naukova Dumka.
13. Velikhov, E.P., Sagdeyev, R.Z., and Kokoshyna, A.A. 1986. *Space weapons: The security dilemma*. Moscow: Mir.
14. U.S. Department of Defense (DOD). 1984. *The Strategic Defense Initiative. Defense Technology Study*. Washington, DC: DOD, April 1984.
15. van der Pol, B. 1935. *Nonlinear theory of electric oscillations*. Russian translation. Moscow: Swiazizdat.
16. Bloembergen, N. 1965. *Nonlinear optics*. Singapore: World Scientific Publishing.
17. Sukhorukov, A.P. 1988. *Nonlinear wave interactions in optics and radiophysics*. Nauka, Moscow.
18. Weiland, J., and Wilhelmsson, H. 1977. *Coherent nonlinear interactions of waves in plasmas*. Oxford: Pergamon Press.
19. Zvelto, O. 1976. *Principles of lasers*. New York: Plenum.
20. Marshall, T.C. 1985. *Free electron laser*. New York: MacMillan.
21. Brau, C. 1990. *Free electron laser.* Boston: Academic Press.
22. Kulish, V.V. 2002. *Hierarchic methods. Hierarchy and hierarchic asymptotic methods in electrodynamics*, Vol. 1. Dordrecht, the Netherlands: Kluwer Academic Publishers.
23. Freund, H.P., and Antonsen, T.M. 1996. *Principles of free electron lasers,* 2nd ed. Heidelberg, Germany: Springer-Verlag.
24. Saldin, E.L., Schneidmiller, E.V., and Yurkov, M.V. 2000. *The physics of free electron lasers. Advanced texts in physics*. Heidelberg, Germany: Springer-Verlag.
25. Shozawa, T. 2004. *Classical relativistic electrodynamics: Theory of light emission and application to free electron lasers*. Heidelberg, Germany: Springer-Verlag.
26. Schmuser, P., Ohlus, M., and Rossbach, J. 2008. *Ultraviolet and soft X-ray free electron lasers: Introduction to physical principles, experimental results, technological challenges*. Heidelberg, Germany: Springer-Verlag.
27. Williams, R.E. 2004. *Naval electric weapons: The electromagnetic railgun and free electron laser.* Master's thesis, Naval Postgraduate School, Monterey, CA. http://www.dtic.mil/cgi-bin/GetTRDoc?AD=ADA424845&Location=U2&doc=GetTRDoc.pdf.
28. Allgaier, G.G. 2004. *Shipboard employment of a free electron laser weapon system*. Master's thesis, Naval Postgraduate School, Monterey, CA. http://www.dtic.mil/cgi-bin/GetTRDoc?AD=ADA420299&Location=U2&doc=GetTRDoc.pdf.

29. Mitchell, E.D. 2004. *Multiple beam directors for naval free electron laser weapons.* Master's thesis, Naval Postgraduate School, Monterey, CA. http://www.dtic.mil/cgi-bin/GetTRDoc?AD=AD A422357&Location=U2&doc=GetTRDoc.pdf.
30. Davis, G.R. 1976. Navy researchers develop new submillimeter wave power source. *Microwave* 12:12,17.
31. Sprangle, P., Granatstein, V.L., and Barker, L. 1975. Stimulated collective scattering from a magnetized relativistic electron beam. *Phys. Rev. A.* 12:1697.
32. Elias, L.R., Madey, J.M.J., Smith, T.I., Schwettman, H.A., and Fairbank, W.M. 1976. The free-electron transverse laser: 10.6 gain measurements. *Opt. Commun.* 18:129.
33. Elias, L.R., Fairbank, W.M., Madey, J.M.J., Schwettman, H.A., and Smith, T.I. 1970. Observation of stimulated emission of radiation by relativistic electrons in a spatially periodic transverse magnetic field. *Phys. Rev. Lett.* 36(13):717–720.
34. Silin, R.A., Kulish, V.V., Klymenko, Ju.I. 1991. Soviet Inventors Certificate, SU No 705914, Priority 18.05.72, 15.05.91. *Inventions Bulletin* 26.
35. Phillips, R.M. 1960. The ubitron, a high-power travelling-wave tube based on a periodic beam interaction in unload waveguide. *IRE Trans. Electron. Devices* 7(4):231.
36. Phillips, R.M. 1988. History of the ubitron. *Nucl. Instrum. Methods Phys. Res.* A272:1.
37. Kalmykov, A.N., Kotsarenko, N.Y., and Kulish, V.V. 1977. Parametric generation and amplification of electromagnetic waves with frequencies higher than the pump wave frequency in electron beams. *Izv. Vyssh. Uchebn. Zaved. Radioelectron.* [*Soviet Radioelectronics*] 10:76.
38. Vinokurov, N.A., and Skrinsky, A.N. 1977. On the limiting power of an optical klystron based on an electron storage device. *Preprint of the Institute for Nuclear Research*, Siberian Division of the USSR Ac.Sci., No. 77-59, Novosibirsk.
39. Berezhnoi, I.A., Kulish, V.V., and Zakharov, V.P. 1981. On the explosive instability of relativistic electron beams in the fields of transverse electromagnetic waves. *Zh. Tekh. Fiz.* [*Sov. Phys.-Tech. Phys.*] 51:660.
40. Tripathi, V.K., and Lin, C.S. 1988. Explosive free-electron-laser instability with an electromagnetic wiggler. *Phys. Lett.* A132(1):47.
41. Kulish, V.V., Kosel P.B., and Kailyuk, A.G. 1998. New acceleration principle of charged particles for electronic applications: Hierarchic description. *Int. J. Infrared Millimeter Waves* 19(1):3–93.
42. Kulish, V.V., Kosel P.B., and Kailyuk, A.G., and Gubanov, I.V. 1998. New acceleration principle of charged particles for electronic applications: Quantitative analysis. *Int. J. Infrared Millimeter Waves* 19(2):106–170.
43. Kulish, V.V., Kosel, P.B., Krutko, O.B., and Gubanov, I.V. 1998. Hierarchic method and its application peculiarities in nonlinear problems of relativistic electrodynamics. Theory of EH-Ubitron accelerator of charged particles. *Ukrainian Phys. J.* 43(2):133–138.
44. Kulish, V.V., and Kailyuk, A.G. 1997. Generation at sub-harmonics in the maser at cyclotron resonance. *Radio Phys. Radio Astron.* [*Sov. Radiofizika i Radioastronomiya*] 2(3):347–352.
45. Newnam, B.E., Warren, R.W., Sheffield, R.L., Goldstein, J.C., and Brau, C.A. et al. 1984. The Los Alamos free electron laser oscillator: Optical performance. *Nucl. Instrum. Methods Phys. Res.* A237(1–2):187–198.

8

Hierarchic Oscillations and Waves: The Foundation of the World?

The number of the characteristic schemes and regularities in the universe is limited and the set of these schemes and regularities is universal for all manifestations of the world.

V.V. Kulish

Readers who find this chapter while looking through the book will, more than likely, be surprised. What relationship can this exotic material have to such basic and everyday subjects like FELs? The author agrees that, strictly speaking, there is not a direct relationship. However, an indirect connection certainly exists and will be explored in this chapter. Nevertheless, it seems that in accepted serious science, using standard logic, this chapter would not be included in this type of book, but here it is. The explanation is fairly simple: the author has taken the advantage of poetic license. The author could not resist the temptation to share with readers the extraordinary discoveries in other areas besides that of ordinary oscillations and waves and he sincerely believes that readers will also find it as interesting as he did when he wrote these lines.

So, welcome to the country of extraordinary ideas, professional dreams, and crazy new physical hypotheses.

8.1 Tree of Life: The Ancient Cosmogonic Concept and Method of Investigation

8.1.1 A Few Words of Introduction

As mentioned in Chapter 2, the concept of hierarchic resemblance as a peculiar conservation of system topology is one of the most intriguing aspects of the more general concept of the tree of life. Today, the true origin of the basic idea of the tree of life is not clear. First, this is because the time of its birth is much more ancient than written human history [5,18]. Its first material tracks have been found in the ancient archeology of Egypt [18]. The concept of the tree of life has been represented there as a partial case of the even more general concept of the flower of life [18]. Actually, we obtained most of our knowledge about this subject from written documents of ancient Middle Eastern cultures (Sumer, Judea, and others). It is well known that this region is the birthplace of the three main world religions: Judaism, Christianity, and Islam. Hence, it should not be surprising that the main fundamental concepts of ancient cosmogony take a worthy place in the basic doctrines of all these religions. The Old Testament gives obvious evidence for the thesis.

The concepts of hierarchic resemblance, in general, and the tree of life, in particular, received different amounts of attention during the evolution of the different religions [1–8].

However, it received the most essential and fundamental development in Judaism. The treatment of the mystical and cosmogonic aspects of the Old Testament forms the basis of the Kabbalah (Quabalah, Kabala, Cabal, Cabbala) [2,4–8]. It is believed that the Archangel Mettatron is the true author of the Kabbalah, which is used as

1. The philosophical and theological cosmogonic doctrine
2. The key element of the religious system
3. The object of spiritual practices

It is obvious that we are interested in the first item only. We should note that the tree of life doctrine is one of the most complex and deep cosmogonic doctrines that have ever been elaborated by a human being.

The historical European destiny of the tree of life doctrine is not so simple and rectilinear. It is, in many aspects, similar to the relationship between ancient Judaism and Christianity and their having a common holy text: the Old Testament. Born as a mystical and cosmogonic aspect of the Old Testament, the Kabbalah, as well as the concept of the tree of life, was subsequently developed in a few directions. The European branch of the Kabbalah [5] appeared historically as one of the results of the stimulating action of the Crusades. Through them, the texts of the Kabbalah found their way into the hands of medieval Christian scientists and they had an essential influence on the form of its presentation, as well as on the treatment of many aspects of its basic concepts, including the tree of life itself. The Kabbalah, by absorbing the most important achievements of medieval European science (including alchemy and astrology), transformed into the "western yoga" [5] in the nineteenth and twentieth centuries.

The temporal politics of the church are responsible for the fact that knowledge of the Kabbalah is not more widespread in Europe. One result of those politics is that the Kabbalah sometimes existed only in the form of verbal tradition because owners of hard copies were cruelly persecuted. In this connection, we can draw a parallel with the well-known stories about the scientific heritage of Copernicus, Bruno, or Galileo. The only difference is that this heritage has been appreciated for its true value by physicists and astronomers. In contrast, the Kabbalah, including the concept of the tree of life, still awaits translation into the traditional scientific languages. Such work has been accomplished for the ancient Indian, Greek, Chinese, and Japanese philosophies [1,3].

All the above make the general situation around the concept of the tree of life very complicated, delicate, and, at the same time, sensitive with respect to various inexactitudes. Realizing this, the author hopes for the forbearance of readers belonging to different creeds.

8.1.2 Tree of Life

The principle of hierarchic resemblance was briefly discussed in Chapter 2. Let us refresh the information given there and expand on it.

In accordance with this principle, our surrounding world is organized in a peculiar self-resemblance manner. This means that every subject in nature (including society), any hierarchic natural system as a whole, and each level of such a system have the same general arrangement in principle. The universal geometrical pattern, which adequately reflects this resemblance, is referred to as the tree of life (see Figure 8.1). Thus, we can treat the tree of life as a peculiar universal (but not elementary) module that plays the role of a fundamental topological basis of everything existing in the world.

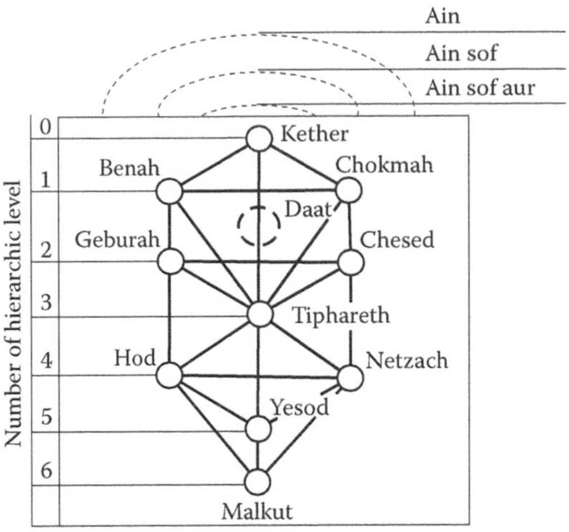

FIGURE 8.1
Tree of life. Here the circles correspond to the Sefiroth and the straight lines designate the Ways.

It is widely known that all modern natural sciences are founded on the basis of the following hypothesis: the more elementary the natural structural unit is, the simpler it should be. This hypothesis ultimately determines our modern dominant general civilization paradigm: Any complex object always consists of many simpler, more elementary structural units. Unfortunately, real scientific practice does not always conform to this obvious statement. For instance, in many aspects, the physics of the so-called elementary particles turned out not to be simpler than the physics of the galaxy.

The Kabbalah proposes a completely different approach. The above-mentioned property of structural elementarity is not as important as it is now considered. Much more important is the topological elementarity of different systems composed of elementary structural units. In accordance with the Kabbalah, all natural objects—very small (elementary) as well as very large—are hierarchically arranged in accordance with the same topological pattern. The tree of life doctrine gives the topological description of this universal elementary pattern.

As in Figure 8.1, the tree of life is a glif (composite symbol) that can be represented in the form of ten circles distributed in a relevant order and connected by twenty-two lines. These circles are called the Holy Sefiroth (plural; the singular is Holy Sefirah) and the lines are called the Ways. The Sefiroth are objective representatives of the natural forces. The Ways are subjective. They are connected with states of consciousness. In other words, the Sefiroth could be regarded as the macrocosmic subjects and the Ways as the microcosmic ones. Each Sefirah can be also treated as a relevant phase of the evolution, while the Ways represent the phases of subjective appearance.

Each Sefirah is responsible for some specialized function, manifesting in various specific aspects (potentials) that are common for all Sefiroth. Only one of these aspects (which are proper for this Sefirah) is open completely in its proper place. The potentials of other Sefiroth also are presented there. However, contrary to the proper potentials, they play a secondary role in this place.

In contrast to other versions of the Kabbalah, in the European Kabbalah, God is not regarded as a worker that fulfils, step-by-step, the process of the Creation. Instead, it is considered that God does not exert a direct influence upon subjects. Creating the world,

He has foreseen this in its inherent arrangement, operation principles, and program of action. Therewith, only deviations from relevant standards are corrected by higher hierarchic levels (i.e., different hierarchic aspects of God). In terms of the modern theory of systems, we can formulate this as an indicative method of operation.

Thus, in accordance with the European Kabbalah, The God–Creator can be regarded as the Super-Programmer who has created all-embracing algorithms and programs for the functioning of our surrounding world in all its manifestations. At the same time, He can be considered as Designer and Operator of all world mechanisms that practically realize these genial algorithm and programs. This appears at any time, there and everywhere, in all natural structures, processes, details, events of human and individual histories, etc. The most unexpected thing about this situation is the fact that the idea and the principle of evolution are the key points of the algorithms and programs. This means, in particular, that the official modern science may forget about its imaginary priority for the invention of evolution as the absolutely new modern idea and fundamental physical principle. Cabbalists knew about this for at least the last six thousand (or more) years. Darwin understood rightly the key role of the evolution in nature. However, he and his followers treated this fact incorrectly. First of all, they, in contrast to the Cabbalists, did not understand that evolution is only one of many of God's tools that provide the workability of our great universe–supercomputer–machine as a whole. Some of the properties of this incredible subject will be discussed here.

8.1.3 Evolution Process in Terms of the Tree of Life

The following picturesque model has been described as an illustration of the discussed process of evolution in the framework of the European Kabbalah [5]. Let us assume that all Sefiroth are represented as volumes, which are filling successively with a liquid (power). Let us assume also that the filling-up process begins from the upper Sefirah Kether (Figure 8.2). The liquid power fills the Kether volume, which could be treated as a process of appearance (development).

Then, after the Kether volume is overfilled, the liquid power flows down into the Chokmah volume, and so on. As mentioned, the Kether comprises all potentialities of all

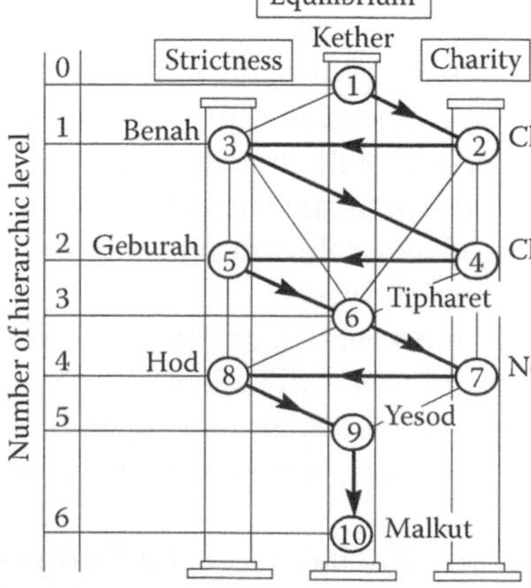

FIGURE 8.2
Descending power (evolution) in the tree of life.

other nine Sefiroth, revealing only one (proper) of them. The Chokmah does not have the Kether's potentiality, but, revealing its potentiality, Chokmah contains the potentialities of the rest of Sefiroth.

The described flowing-down process (which is treated as the descending power) is shown in Figure 8.2. The concept of the tree columns is also illustrated there. Different versions of names of these columns (strictness, equilibrium, and charity) are known in other ancient philosophical systems. They correspond to the three flows (channels) of prana (Ida, Pingala, and Sushumna) in Yoga and both side columns correspond to the concepts In' and Yan' of Eastern philosophy.

In the Kabbalah, it is considered that the central column represents the equilibrium state, whereas the right and left columns describe the positive (conditionally) and negative factors. Influences of these factors systematically disturb the equilibrium state during the evolution process. It is easily seen that the discussed concept of the three columns reflects the key idea of Hegel's dialectics, which are a basis for modern natural sciences. The side columns indicate the presence of contradictions, which are the main driving forces in the surrounding world. These forces have a different nature (physical, social, economical, psychological, etc.) and they are the main causes of motion, including the evolution process in nature.

8.1.4 Metric of the Tree of Life

The Sefiroth, as noted before, have a four-dimensional metric. In other words, they are in a peculiar four-dimensional continuum. The dimensions of this peculiar space are the four worlds: Atsilut, Beri'ah, Yetsirah, and Asiah (Figure 8.3). Their other names are the World of Archetypes, the World of Creation, the World of Forms and Angels, and the World of Making, respectively (see again Figure 8.3).

The physical meaning of these worlds is too complex for a traditional perception. It becomes especially complicated for readers of physical/mathematical books. Therefore, we note only that the worlds have a maximally generalized character and that this character has a different appearance in different systems of the surrounding world.

Hidden world	1	Ain
	2	Ain sof
	3	Ain sof aur
Manifested world	4	Atsilut (world of archetypes)
	5	Beri'ah (world of creation)
	6	Yetsirah (world of forms and angels)
	7	Asiah (world of making)

FIGURE 8.3
Four dimensions (worlds) of the manifested world and three dimensions of the hidden being (hidden world).

8.1.5 Manifested and Hidden Worlds

Apart from the ordinary surrounding world, the so-called manifested world, a concept of a hidden world also exists in the Kabbalah (Figure 8.3) and it also has its dimensionality. In this case, we have to deal with the so-called three dimensions of the Hidden Being: the Ain, Ain Sof, and Ain Sof Aur. God is beyond the manifested as well as the hidden dimensions. Therefore, the hidden dimensions are also called the Three Veils of God. The three hidden and the four manifested worlds, as dimensions, form the seven-dimensional world of the tree of life. Thus, the tree of life contains seven hierarchic levels, which exist in a seven-dimensional continuum.

Returning to Figure 8.2, we can formulate the following question: What source provides the liquid power for the first Sefirah (Kether)? The source is localized in the hidden world. The Cabbalists say, "The manifested Kether is the hidden Malkut" (the tree of life of an upper-lying world precursor). This also means that the Malkut of our world is the Kether for a lower-lying world, i.e., in terms of the tree of life doctrine, our world can be treated as a world predecessor for all other lower-lying worlds. Or, from the point of view of our personal place, our world can be regarded as an intermediate link of a chain of worlds of an incomprehensible external Something (superuniverse). In other versions of the Kabbalah, other approaches also exist to the problem of the worlds and their treatment.

8.1.6 Tree of Life as a Subject and as a Method of Investigation

We should note that the tree of life is not only a subject of study, but is also regarded as a highly efficient method of investigation. Let us discuss this interesting topic in detail.

One of the ancient researchers of Kabbalah said that an Angel descending to the Earth should acquire a human form to enable him to talk with people. The tree of life is just such humanized form of the representation of the Angel's extremely complex knowledge (Cabbalists consider that the concept of the tree of life has been written under an Angel's dictation). It is the knowledge of the surrounding Something, the description for which we lack ordinary words, definitions, and associations. It is a diagram where all forces and factors that appear in the universe and the human soul are taken into account. It is a map where we can find reciprocal positions and mutual connections between these forces and factors. This is a book where all possible scientific, psychological, philosophical, and theological knowledge is set forth. The only problem is with the book reader, who understands poorly, and sometimes not at all, the writings on the book's pages. This is the tree of life as a subject.

The method of the tree of life looks rather paradoxical to a modern civilized scientist. Indeed, how does this civilized scientist act in his professional work? He begins with the accumulation of synthesizing concepts. The method of experiment is the main methodological basis of all natural sciences. It illustrates the typical approach of a true (civilized) researcher to the investigation of a scientific problem.

It is widely known that the experiment is the "king" of modern physics. But what is the experiment itself? It is a set of special observations that are performed at different experimental units, in different places, and by different experimentalists. Only coinciding scientific results that have been obtained in such a manner are considered to be true. The key point of the experimental method is that the experimentalist is always separated from the subject of the study. Because of this, we can consider the obtained result as objective, i.e., independent of the experimentalist's personality. Synthesizing a huge number of experimental results, the experimentalist formulates the theoretical regularities. Those among them that have a general character for all of the observed world are called the fundamental laws (principles). The theorist describes the experimental results (and the fundamental laws) mathematically introducing (inventing) a lot of various abstract concepts by constructing mathematical models of the considered phenomena. Analyzing such abstract theoretical models gives new ideas and directions for further experiments, and so on. The machine of all natural sciences works in just such a manner.

Then, how does a researcher of the tree of life behave? The answer appears paradoxical: in a completely opposite manner. The researcher has all the required abstract information initially in the form of the tree of life. The problem is only to understand its essence. Meditating, the researcher analyzes the given abstract concepts (Sefiroth and Ways). In other words, the researcher reads an opened book prepared by somebody else without being interested in its origin and the author. This also means that, similar to the true experimentalist, the researcher accomplishes relevant observations (due to the meditations). But, in contrast to the true experimentalist, he uses himself (or, more precisely, his spirit) as a peculiar experimental unit for such observation. In principle, the experimentalist and his experimental unit are not separated.

Thus, at the first glance, the methods of modern physics and of the tree of life doctrine exclude each other. However, the real situation is not as depressing as it seems. The point is that

1. The time of the tree of life observation is at least six thousand (or more) years long. Thousands of generations and hundreds of thousands of researchers have studied it, developing, updating, and deepening its theory. During all this time, all of them observed basically the same general physical picture.
2. Other known spiritual traditions [1–3,5,19–22] also observe a similar picture [5]. Of course, they have essentially different terminologies and ideological approaches, and were elaborated by different human races. However, the main key of the observations and the derived conclusions (incredibly) coincide, basically, with those that were obtained by the method of the tree of life.

All this, in the author's opinion, could be considered as sufficient compensation for the above-mentioned methodological drawback of the absence of a separation of the experimentalist from the experimental units. Hence, today we all have reasons for taking seriously the scientific' results of the tree of life doctrine in solving the most urgent fundamental problems of modern physics. The first step in this direction is to construct a little bridge between them. Let us do this first step in the next sections.

8.2 Hierarchy, Oscillations, Modern Physics, and the Tree of Life

8.2.1 Seven Levels of Hierarchy of the Material Universe

We begin the not-so-simple work of constructing the above-mentioned "little bridge." Some results of such a construction are discussed further in the example of our material universe as a complex hierarchic dynamic system later in the chapter.

The hierarchic tree of the material universe is shown in Figure 2.6 and has been discussed in Section 2.4.3. A similar drawing in terms of the tree of life doctrine is represented in

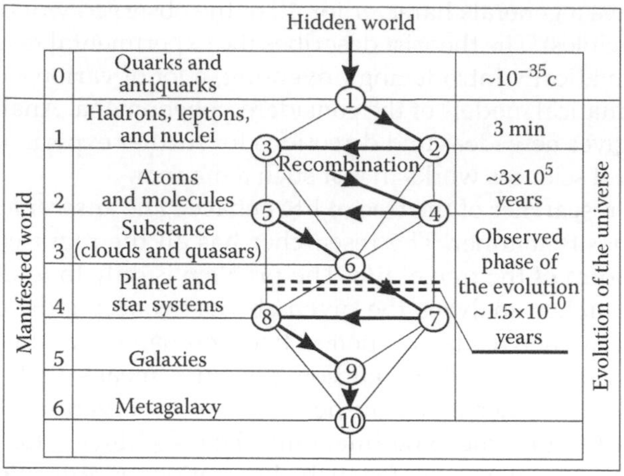

FIGURE 8.4
Tree of life as an illustration of the evolution process of the material universe. Here the circles with numbers (Sefiroth; see Figure 8.1) designate the physical types of interactions that are represented in Table 2.1. 1 is the Kether, 2 is the Chokmah, 3 is the Benah, 4 is the Chesed, 5 is the Geburah, 6 is the Tiphareth, 7 is the Netzach, 8 is the Hod, 9 is the Yesod, and 10 is the Malkut.

Figure 8.4. By comparing the drawings in Figures 2.6, 8.4, 8.8, and 8.12, we can be confident that the number of hierarchic levels (seven) coincides in all cases. At the same time, it can be easily seen that the directions of the level numbering are mutually opposite. However, this difference really is not important for the crux of the matter. The point is that the choice of directions of the numbering could be arbitrary. Therefore, this distinction does not lead to any mistakes.

Furthermore, we turn to the discussion of the physical meaning of the materials illustrated in Figure 8.4. It is an attempt to illustrate the modern scientific concept of the universe's evolution (big bang theory) in terms of the tree of life doctrine. However, let us refresh the traditional point of view that is accepted by modern physics.

8.2.2 Oscillation-Wave Nature of the Material Universe

Let us start the discussion with the title of this chapter: *Hierarchic Oscillations and Waves: The Foundation of the World?* In this connection, let us ask: What are our reasons for choosing this title? In other words, are the oscillations and waves actually the foundation of the surrounding world and how did this occur physically?

It is fascinating to realize the little known historical fact that the so-called atom epoch began more than 2500 years ago. The point is that the concept of the atom was used in the fourth century BC by ancient Greek philosophers Democritus and Anaxagoras. It should be noted that, in those times, this knowledge was already ancient. They considered the atom to be the most elementary object in the universe. Later, it was also considered that the number of the nonidentical types of atoms is rather limited. Consequently, the extremely rich variety of the surrounding world is due to the ability of this limited variety of atoms to form a huge diversity of different combinations. We meet these combinations everyday in the forms of molecules, crystals, cells, etc. However, in the twentieth century, it became obvious that the atom is not the most elementary object in nature and, in fact, consists of other smaller parts: electrons, protons, and neutrons. Following the established tradition, these smallest parts have been called the elementary particles.

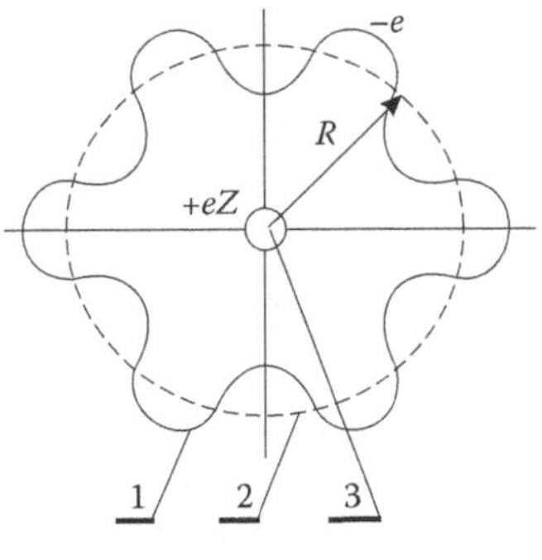

FIGURE 8.5
Electron at an orbital in the hydrogen-like atom as a de Broglie wave. Here 1 is the de Broglie's electron wave, 2 is the electron orbital with the radius R, 3 is the nucleus of a hydrogen-like atom with the charge $+eZ$, e is the absolute electron charge, and Z is the number of protons in the nucleus.

It was also found that all so-called elementary particles have a dual nature. They are simultaneously corpuscular and wave objects. Such waves have been called de Broglie's waves (see the illustration in Figure 8.5). Figure 8.5 illustrates an electron rotating in an atom in a wave-like orbital form around a nucleus. As is readily seen, the illustrated electron wave motion process can also be treated as a propagation of de Broglie's wave along a closed line (the electron orbital).

Thus, we found the oscillation-wave nature of the substance at the level of atoms, molecules, and elementary particles. When we take into consideration the oscillation-wave character of the arrangement at all other higher hierarchic levels (see Figure 2.6 and Table 2.1), we can conclude that the surrounding material world indeed has an oscillation-wave nature. Hence, the choice of chapter title was indeed motivated.

However, the history of the problem considered does not end with the discovery of de Broglie's waves. The oscillations and waves also exist at an even more elementary hierarchic level of nature. The elementary particles are found to be not as elementary as considered earlier. For instance, experiments reveal that hadrons consist of quarks. But what is the physical structure that represents quarks on this more elementary hierarchic level? Modern physics talks about different particles (including quarks) as being different modes of elementary oscillations. They are the oscillations of specific subjects, which have been referred to as the p-branes [34] (see also the relevant discussion of Figure 2.9). Let us discuss this intriguing topic in detail.

The oscillations are specific types of motion in time, when an observed object returns systematically (periodically, in particular) to its initial state (see Chapters 3 and 4). Waves are a more general type of oscillation that occur in time and space simultaneously. Considering such processes in four-dimensional (or more) spaces where time is regarded as one of the coordinates, we can ignore the difference between these concepts. For the sake of simplicity, we will talk about oscillations only, taking in view both of these processes.

We accept the definitions for oscillations and waves using the concepts of time and space. But what is time and what is space? These and other similar questions have disturbed the human mind for as long as mankind has existed. Newton has defined three-dimensional space as a "receptacle of all things" and time as the "reading of a clock." He considered that time and space can exist independently. Placing a body (material point, for example) in space, we obtain a possibility of introducing a reference frame for the mathematical description of this space.

As can be easily seen, the concept of time as distinct from the concept of three-dimensional space seems dull and incomprehensible to ordinary perception. Indeed, what is the clock in the above-mentioned definition of time? Unfortunately, we do not have any better definitions for time today. It is generally impossible to tell what time really is. We can only describe mathematically some observed properties of time. It is interesting that in the framework of the tree of life doctrine we can treat time as the world Atsilut, which is not much clearer. Such treatment is essentially richer and gives hope for clarification of this problem in the future. Now, let us turn to the main line of our discussion again.

Albert Einstein has shown that time, space, and matter energy cannot exist independently. Our three-dimensional space and our scalar time form, the so-called four-dimensional space–time (the special theory of relativity). In turn, this four-dimensional space–time has a physical meaning only in the case where it contains matter–energy. This means, in particular, that our space–time has been born with the birth of our universe. Neither space nor time existed before this (the big bang theory [23–25,34]).

Furthermore, it was found that the properties of space–time are closely connected with the properties of matter–energy. In particular, the presence of matter–energy bends the surrounding space–time that is manifested in reality as the appearance of the gravitation forces (the general theory of relativity and the theory of gravitation). If the bending process occurs systematically (periodically), we can talk about gravitation waves. On the other hand, the matter–energy in itself is a result of the manifestation of the other above-mentioned elementary oscillation properties of the p-brane in space–time [34].

We can find the origin of the theory of these elementary oscillation properties in the so-called string theory [34]. Let us assume that we have an ordinary musical string with fastened ends (see the model in Figure 8.6). The string in the initial state has a linear-like one-dimensional form (see Figure 8.6a). Perturbing it, we can obtain its different oscillation modes (spatial harmonics; see Figure 8.6b through d) The main idea of string theory is that different elementary particles correspond to different oscillation modes of strings, which move in four-dimensional space–time.

8.2.3 Complete Dimensionality and Information Properties of the Universe

The next step in the development of such representations was a generalization of the concept of string with a more complex class of objects. The simplest linear-like (i.e., one-

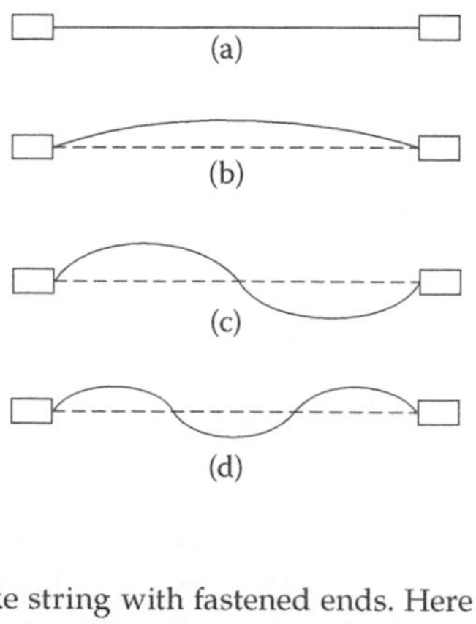

FIGURE 8.6
Oscillation-wave modes of a linear-like string with fastened ends. Here variant (a) corresponds to the resting state of the string, variant (b) to its first (main) mode of oscillation, variant (c) to the second mode, and variant (d) to the third mode.

dimensional and not closed) string model (shown in Figure 8.6) represents only one partial case of this class. However, it was discovered that the string can also have a closed form (circular, elliptical, etc.), which is similar to an atom orbital with an electron de Broglie wave (shown in Figure 8.5). The string also can be two-dimensional (closed and half-opened surfaces). It can be a p-dimensional closed surface that envelops a $(p + 1)$-dimensional volume, for example. Such a surface is called a p-brane. It is obvious in this connection that the case $p = 1$ corresponds to a one-dimensional string; in the case $p = 2$, we have an ordinary two-dimensional surface, and so on.

The models with supersymmetry [34] allow us to generalize the concept of the p-brane even more. Different researchers have used different specific details in their theories. As a result, different groups of authors have obtained five different variants of the string (superstring) theory [34].

Another direction of the development of the theory of elementary oscillations is based on the idea of a combination of the general theory of gravitation with quantum theory. The resulting theory is called the theory of supergravitation. All possible p-branes in this case can be found as solutions of the relevant equations of this theory in 10- or 11-dimensional space. The main idea of such multidimensional space is that, as before, we have to deal mainly with the traditional ordinary four-dimensional space–time. The additional six or seven dimensions are hidden. They are rolled up to such a small value that one hardly notices them [34].

It was found further that all these discussed theories are really not competitive and each of them describes this or that fragment of the total object of study (our universe). Such a general unified theory can be constructed after a unification of all these theories. This unified theory has been called the M-theory [34]. Today, this theory remains very far from completion.

The brane, in accordance with the M-theory, exists in 10- or 11-dimensional space. Six or seven of its coordinates should be rolled up to a very small value. However, the last theories [34] show that some of these rolled up coordinates (three, most probably) could really be macroscopic. This means that our world is actually seven-dimensional. Three of these coordinates are hidden (rolled up) and they not directly manifested in our universe. The remaining four dimensions could be treated as the manifested dimensions and our observed world exists only in this four-dimensional continuum. Let us remember this scientific fact because it will play a very important role in our further discussions.

In accordance with modern physics, we actually live on a four-dimensional surface brane that envelops some enigmatic five-dimensional volume. The other three macroscopic coordinates are hidden and their role in our life has an indirect character.

In Chapter 2, we stated that all natural hierarchic dynamic systems (including our universe) can be treated as peculiar information operation machines. It was found that, in the framework of modern physics, the treatment of such information is connected in a rather unexpected manner with the principle of hierarchic resemblance. Hawking [34] gave the information treatment of the universe brane. He proposed to consider it as a hologram of something that is placed within the five-dimensional volume enveloped by the four-dimensional universe brane. But by what means the hologram in the given specific case? This means that the same complete information about this inner unknown object is written within each structural element of the brane and, at the same time, within the universe brane, as a whole. This interpretation originated from the well-known properties of ordinary laser holograms. In other words, each universe hierarchic level, sublevel, and their more elementary structural elements, like the author, reader, and universe as a whole, should have the same general information arrangement and the same topology. However, this is nothing but the formulation of the principle of hierarchic resemblance given in Chapter 2. Moreover, this coincides with the main properties of the tree of life discussed in Section 8.2.1.

8.2.4 Our Universe as a Multiresonant Oscillation-Wave Hierarchic System

Unfortunately, today we do not know the physical essence of this enigmatic five-dimensional volume. We only know that this volume, together with our surface brane, is extending in nature [34]. We also suspect that different macroscopic modes of the brane surface oscillations must also be manifested in some form in our material world. But one more question arises in this connection: What are the physical forms of such possible macroscopic manifestations in the universe? The tree of life doctrine allows us to formulate the following hypothesis as an answer to that question.

As previously mentioned, the microscopic brane oscillations in the initial singularities have brought forth the first elementary particles: quarks and antiquarks. But why should this important physical peculiarity (birth of matter) be limited by the lowest hierarchic level (elementary particles)? It is because, due to the effect of brane universe expansion, we observe various similar material realizations of these oscillations at all higher macroscopic levels: the atoms and molecules, molecular clouds and quasars, planets and star systems, and so on (Figure 8.4). It is interesting to notice that the principle of hierarchic resemblance confirms this idea. The birth of all higher hierarchic levels must be fulfilled in accordance with a hierarchically appearing, oscillation-like physical mechanism. In other words, the large-scale brane oscillations should also bring into the world large-scale material objects like atoms, molecules, quasars, planets, star systems, and so on.

However, this answer to the question raises another far-from-trivial question: What are the basic operation principles of the physical mechanisms of generation of these large-scale material objects? Of course, the problem looks extremely complicated and it is impossible to give its complete solution in only one book. Moreover, the description of this kind of solution is not the main purpose of this book. Therefore, let us confine ourselves only to short qualitative comments.

Thus, we observe today the large-scale universe brane oscillations as the appearance of the matter–energy of different hierarchic ranks (particles, atoms and molecules, planet and star systems, etc.; see Figure 8.4). Moreover, we consider that the manifestation of the brane oscillations in the macroscopic material forms is realized via the plural hierarchic resonances of different oscillations and their harmonics. In other words, we treat the process of the formation of each next hierarchic level as the development of the next complex group of hierarchic resonances. This concept was discussed in detail in Chapter 3 (see Figure 3.19 and corresponding comments). Therefore, we shall limit ourselves only to the discussion of a general philosophy of this complex phenomenon.

The proposed hypothesis has the following substantiation. In accordance with the big bang theory, our initial singularity began as a result of an external action. It should be mentioned that its physical nature is still not understood today. However, we can accept the logical supposition that this action has played the role of a source of the initial multifrequency and multiharmonic oscillations for the initial universe brane. These initial brane oscillations have given birth to the first elementary particles (quarks and antiquarks; see the next sections for details). We treat this process as the first stage of the formation of the zero hierarchic level (Figure 8.4).

We assume also that at least some of these initial multioscillations can interact with each other in a hierarchic-resonant manner. The observed complete oscillation-resonant hierarchic structure of the universe (Figure 2.6) appears to owe its development to this physical mechanism. Herewith, the extension of the universe as a whole systematically extends the oscillation periods that have been demonstrated, in particular, by the effect of relict radiation.

Hence, the totality of pairs, threes, etc., of the initial oscillation phases forms a totality of nonlinear resonances (Chapter 3). However, let us recall that any nonlinear resonance is always accompanied by the appearance of so-called resonant detuning (mismatches). As a result, a number of new oscillation phases (and their harmonics), having the frequencies of this detuning, appear in the systems. It is important to note that the periods of these new detuning oscillations and their harmonics are much larger than the periods of the initial oscillations. In accordance with string theory, the appearance of these detuning oscillations means the generation of new elementary particles (hadrons in our case, the second stage of the formation of the zero hierarchic level; see Figure 8.4).

Then the described process repeats cyclically a few times and some of the detuning phases and their harmonics interact with each other due to the realization of the same hierarchic-resonant mechanism. This, similar to the preceding cycle, gives the birth to other (new) phases of a new superdetuning and their harmonics (the first part of the first hierarchic level; see Figure 8.4). The hierarchic cascade-like chain of generations of new detuning phases is finished by the formation of the highest resonance and, respectively, the highest (sixth in our case; see Figures 2.6, 8.4, and 3.19) universe hierarchic level. The presence of many harmonics and modes of the detuning oscillations stipulates a rich diversity of material objects at each macroscopic hierarchic level resonance.

Summarizing this discussion, we formulate the following oscillation-resonant algorithm of the evolution of the universe: The initial oscillations, within the initial singularity, are manifested as the appearance of the first elementary particles. Furthermore, these oscillations interact resonantly with each other and, owing to this, a number of new (detuning) oscillation phases and their harmonics are generated. Some of these phases again form the new detuning resonances and the new superdetuning phases, and so on. We observe the process of the cyclic formation of the oscillation-resonant hierarchic levels as the successive appearance in the universe of the increasingly larger scale objects: elementary particles, nuclei, atoms, molecules, quasars and space clouds, star and galaxy systems, and so on. All this is happening in the background of the effect of the extension of the universe surface brane.

Thus, the accomplished analysis shows that hierarchic oscillations and waves should actually be regarded as the foundations of our world. Herewith, any hierarchic level of the universe has the nonlinear resonant hierarchic-oscillation nature.

8.2.5 About Modern Physics and Its Theories of Everything

We should mention that modern physics does not unfortunately distinguish itself by excessive modesty, declaring that its theories (like the *M*-theory) are the theories of everything [34]. It is not difficult to convince oneself that all this is actually not so because, first, all the theories are incomplete and not really universal. They describe only a small part of the observed physical picture of our material surroundings and even such a real picture is far from complete comprehension. Second, modern physics ignores other important (nonphysical) forms of the appearance of the world. They are, for example, the biological, ethnosocial, political, and humanitarian forms, and so on. These forms also have an explicitly expressed oscillation-wave nature and exert very significant influence on the observed physical world (see Sections 8.3 and 8.4). In other words, the problem is that physics, as a natural science, long ago lost its integrity as the universal philosophy of nature, whereas our biosphere and sociosphere are the obvious inherent parts of this nature. Unfortunately, in the present day, applied trends of physics predominantly do not give any real hope for a natural restoration of this lost integrity.

It is obvious that we strongly need a new universal scientific paradigm. We need a paradigm common to all natural sciences and humanities. The author considers that the only way to integrate modern natural sciences and humanities with the tree of life doctrine is to make this paradigm a reality. Let us discuss further some illustrational examples of this type that can prove this opinion. We begin these illustrations with a comparative analysis of the evolution of the universe in terms of modern physics and the tree of life doctrine.

8.3 Evolution of the Universe in Terms of the Tree of Life Doctrine

8.3.1 Hypothesis about the Existence of Subtypes of the Fundamental Interactions

First, let us recall the physical meaning of the concept of Sefiroth. In accordance with the given explanations, they describe generalized natural forces. The first Sefirah (Kether) comprises inherent aspects (properties) of all these forces but only one of the aspects is developed (manifested) at the Kether's hierarchic level (see Table 8.1). The properties of the other nine Sefiroth do not play a key role here. The second Sefirah (Chokmah, the next hierarchic rank) also manifests its personal (main) aspect of the natural forces (see again Table 8.1). Relevant properties of the next eight Sefiroth do not play a main role here.

We can attribute the natural forces to the fundamental types of interaction (see Table 8.1). Modern physics knows four types: strong, weak, electromagnetic, and gravitational [23–25]. Apart from that, the unified field theory proposes the idea of the great unified interaction (GUI). In spite of the fact that it represents all four of the simple interactions as their sum, in our case the GUI should be regarded as an independent combined interaction. It is, therefore, that the physical processes in the initial singularity (which corresponds to the Sefirah Kether; see Figure 8.4 and Table 8.1) have essentially nonlinear characters. This means that the principle of addictiveness does not work in such situations.

Hence, we can separate five different natural forces that correspond to four simple interactions and one combined fundamental interaction. On the other hand, the tree of life doctrine tells us that we should have ten types of fundamental interactions. Because of this, according to the Kabbalah, we have ten Sefiroth as fundamental natural forces (Figures 8.4 and 8.5 and Table 8.1). The conclusion is that the remaining five types of interactions should be found by modern physics.

Basically, a few possible variants could be realized in this situation, but one of them looks like the most probable. In the author's opinion, at least two of the four simple fundamental interactions can be classified as degenerating interactions. This means that we have two groups of very close subtypes of interactions: the electromagnetic and the gravitational interactions, instead of the two traditional nondegenerating interactions.

Thus, each of the degenerating interaction could really be considered as the totality of a few partial variants of the same types of interactions (see Table 8.1). We find a relevant analogy in the quantum theory of atoms. Here, the presence of the degenerated states of system is an ordinary phenomenon. Some external electromagnetic field (magnetostatic,

TABLE 8.1

Supposed Types and Subtypes of the Fundamental Interactions in Accordance with the Tree of Life Doctrine

No.	Interaction	Quantum Field	Quantum Mass (GeV)	Interaction Radius (cm)	Source	Sefiroth
1	General unified interaction					Kether
2	Weak	Intermediate bosons $W\pm$, Z^0	82, 93	10^{-18}	Weak charge	Chokmah
3	Strong	Gluon	0	$\leq 10^{-15}$	Color charge	Benah
4	Electromagnetic I	Photon? (plasmon?)	0	∞	Electric charge	Chesed
5	Electromagnetic II	Photon	0	∞	Electric charge	Geburah
6	Gravitational I (substance level)	Graviton	0	∞	Mass	Tiphareth
7	Gravitational II (planet level?)	Graviton?	???	???	Mass?	Netzach
8	Gravitational III (star level?)	Graviton?	???	???	Mass?	Hod
9	Gravitational IV (galaxy level?)	Graviton?	???	???	Mass?	Yesod
10	Gravitational V (metagalaxy level?)	Graviton?	???	???	Mass?	Malkut

electrostatic, or electromagnetic wave) acting on an atom as a degenerating quantum system can remove this degeneration. As a result, we obtain a few close energy sublevels (the so-called thin splitting) instead of each degenerate energy level [24].

Here, let us recall the sequence of the principle of the hierarchic resemblance and the method of the hierarchic resemblance (Sections 2.2.3 and 2.2.4). We also recall that these sequences and methods, as the principle of hierarchic resemblance in itself, originated from the tree of life doctrine (Chapter 2).

Let us then assume that we know some general regularity of the discussed atom processes. In accordance with hierarchic principle and its sequence, we can expect to find basically the same general regularity in other objects of this hierarchic level, as well as in objects of other universe hierarchic levels. Of course, the forms of the manifestations of this general regularity can be essentially different in each such case. However, each time, this resemblance should be expressed clearly (Chapter 2). As an instructive example of this idea, we propose the spatial arrangement of atoms, star and galaxy systems, etc., which are discussed in Section 2.2.4.

Thus, we can use the discussed hierarchic resemblance for an analysis of the problem of subtypes of fundamental interactions. As a result, we find two (instead of one) subtypes of electromagnetic interaction and five subtypes of gravitational interaction (see Table 8.1 for details).

The evolution process in the tree of life is illustrated in Figure 8.2. The same process, but for the partial case of the material universe (which evolved from the big bang [23,24]), is shown in Figure 8.4. Let us discuss this partial case in detail, taking into consideration the following:

1. The proposed hypothesis about the existence of degeneration in electromagnetic and gravitation fundamental interactions (Table 8.1)
2. The general tree of life model for the liquid power descending (evolution) (Figure 8.2)
3. The properties and physical interpretation of the Sefiroth
4. The modern standard scenario of the evolution of the universe (the big bang theory) [23,24,34]

8.3.2 Metric of the Universe Space: The Universe Zero Hierarchic Level

We will begin the promised discussion of the evolution of the universe with the problem of the metric of the Sefiroth (Figure 8.3). According to the tree of life doctrine (Figure 8.3), the manifested metric of our universe should be four-dimensional. This coincides with the results of Einstein's general theory of relativity (three spatial coordinates plus one time coordinate; see Figure 8.7). But, apart from the dimensions, which represent the so-called manifested world, we also have the three dimensions of the hidden world (Figures 8.3 and 8.7). How can we treat the hidden world in terms of traditional physics? Let us clarify this and other similar intriguing questions.

Let us once again refresh the physics of the Sefirah Kether and the formulation: The manifested Kether is the hidden Malkut. This assertion could be reformulated in other words: The Kether is a frontier and, at the same time, a common territory between the neighboring (closest to the Kether) world and ours. And the Kether is a boundary territory between the manifested and hidden worlds (see Figures 8.4 and 8.7). Figuratively speaking, we can imagine the Kether as a peculiar object, whose head (three dimensions) is situated in the hidden world and feet (the remaining four dimensions) we see in the manifested world. All this means, automatically, that the Kether should demonstrate its general properties in

Hidden world	1	A hidden coordinate 1 (Ain)	Three-dimensional space
	2	A hidden coordinate 2 (Ain sof)	
	3	A hidden coordinate 3 (Ain sof aur)	
Manifested world	4	Time coordinate (ct) (Atsilut)	Four-dimensional space
	5	Spatial coordinate (x) (Beri'ah)	
	6	Spatial coordinate (y) (Yetsirah)	
	7	Spatial coordinate (z) (Asiah)	

FIGURE 8.7
Four dimensions (worlds) of the manifested world and the three dimensions of the hidden being (the hidden world) in the material universe.

the seven-dimensional space (4 + 3 = 7; see Figure 8.7). But just such a representation of the initial brane surface (singularity) was discussed in the preceding section. And we did it strictly in the framework of modern theoretical physics [34]. Hence, the Sefirah Kether can be interpreted as the initial singularity of the big bang theory.

Let us recall once again the properties of Sefiroth as the natural forces. We also take into consideration the corresponding discussion in Section 8.1. Then, we can treat the process of the beginning of the formation of the Kether as the start of its filling by the liquid power. The source of this liquid power, as mentioned in Section 8.1, is placed in a preceding hidden Yesod. At the same time, in terms of the big bang theory, this start of the filling also can be interpreted as the beginning of the formation of the initial singularity. It should be stressed that the modern physical theory cannot give any satisfactory scenario for this process before the time point zero to the time instance $\sim 10^{-43}$s [34]. It is considered only that unknown and incomprehensible physical processes and laws took place [34] at this stage of the universe's evolution.

We obtain a somewhat clearer physical picture by starting from the time moment $\geq 10^{-43}$s. The general unified interaction (Table 8.1) plays the main role here. The most characteristic event of this stage of evolution is a gradual disturbance of the matter–antimatter balance in favor of matter. The process of the Kether filling-up is finished within the time moment $\sim 10^{-35}$s. The singularity evolves into the initial quark–antiquark fireball, where the electroweak interaction plays the main role [34]. The quarks in this fireball begin to bind with each other, forming the hadrons (Table 8.1). In other words, the process of the transformation of the quark–antiquark fireball into the hadronic fireball is the characteristic feature of the end of the overfilling of the Sefirah Kether.

The fact that the described processes occur just in the Sefirah Kether explains, in particular, the well-known confinement problem for the quarks. The essence of this problem is the following. Numerous experiments reveal that quarks can exist only within the volume of elementary particles and never in a free state. The explanation is that that the Kether, where quarks are born, belongs also to the other, the hidden world. It is the three-dimensional world primogenitor of our four-dimensional manifested world. Our Sefirah Kether is the Sefirah Malkut of that hidden world primogenitor. Therefore, quarks, belonging to this hidden world, cannot exist in our manifested world in any manifested form. They are hidden inside of hadrons, where the preceding hidden three-dimensional space is conserved in our world.

Thus, the evolution process in the Kether is finished by the formation of the hadronic plasma fireball. In the preceding sections, we have treated the similar process as the Kether overfilling. At the same time, we can interpret it as the beginning of the process of filling of the next Sefirah (Chokmah). In the physical reference, this stage is called the epoch of hadrons and leptons (Table 8.1) [34].

8.3.3 First and Second Universe Hierarchic Levels

The significant event of the Chokmah filling process is the appearance of the leptons (Table 8.2). They are born as a result of the so-called lepton-hadronic decays. The main type of interaction in this process is the weak interaction. Other types of the fundamental interactions (see Table 8.1) are present at this stage in the second best (i.e., not main) forms. In other words, they are present physically, but do not play a key role in the background of the main (weak) interaction.

The processes of the filling of the Chokmah (Figures 8.2 and 8.4) are finished by annihilation of the electron–positron pairs ($e\tilde{e}$). It happens approximately at the third second of

TABLE 8.2
Classification of Elementary Particles

Classes	Types		Particles	
1	Photons		Photon	
2	Leptons		Electrons	
			Electron-neutrino	
			Muon	
			Muon-neutrino	
			Taon	
			Taon-neutrino	
3	Hadrons	Mesons	Pion	
			Kaon	
			η-meson	
		Barions	Nucleons	Proton
				Neutron
			Hyperons	Λ
				Σ
				Ξ
				Ω
	And many others			

the universe's evolution. The surviving particles are represented mainly by electrons and hadrons; at this time, antimatter is not present in large quantities in the universe.

The above-mentioned processes are accompanied by further expansions of the universe brane surface and a decrease of its temperature. As a result, the density of the hadronic-lepton plasma fireball decreases, too. Conditions begin to be appropriate for the formation of deuterium, helium, and lithium nuclei. The process of filling the Sefirah Benah begins (the time instant from the big bang $\tau \sim 5$ min; see Figures 8.2 and 8.4). The strong interaction plays the main role here. The universe is opaque; the substance and radiation are bounded [34].

The next stage of the evolution process is characterized by the formation of strongly ionized electron–ion plasmas. We treat this stage as the filling of the Sefirah Chesed. The electromagnetic I (collective) interaction plays the main role here. The characteristic feature of this stage is the domination of a collective electromagnetic interaction over the so-called single-particle interaction. As before, the universe is opaque for the electromagnetic radiation.

The recombination process leads to the formation of atoms and molecules from the nuclei and electrons. This corresponds to the filling of the Sefirah Geburah ($\tau \sim 3 \cdot 10^5$ years). The electromagnetic II (single-particle) interactions (see Table 8.1) play the main role at this stage of the universe's evolution.

8.3.4 Third Hierarchic Level and the Modern State of the Universe's Evolution

The formation of atoms and molecules opens a door to the formation of a quasineutral space substance. This occurs in the Sefirah Tiphareth. Let us turn our attention to the following interesting circumstance. The gravitational interaction is present at all of the stages of the universe's evolution as a secondary interaction. Therefore, it really does not exert considerable influence on the main processes. It can be explained by the fact that its action is too small in the powerful background of the main interactions: general unified, weak, strong, and electromagnetic interactions. The further evolution of the universe is accompanied by an increasing role of the recombination processes. As a result, the universe becomes transparent for electromagnetic radiation and gravitational interaction begins to play a more important role. Little by little, the universe transforms from the plasma-like state into the state of a neutral gas. The role of gravitational interaction I (Table 8.1) begins to strengthen.

The common action of the gravitational field, the large-scale brane oscillations, and various statistical fluctuations lead to the appearance of the first clusters of a substance (space clouds and quasars) and, further, of the proto-galaxies, the proto-stars, and the proto-planets. All these processes are accompanied by the appearance of the subjects of repeated heating. This, in turn, results in the appearance of the ordinary forms of the present-day electrically neutral substance. We treat this stage of evolution as the completion of the formation of the third hierarchic level of the universe. In terms of the tree of life doctrine, we regard it as the overfilling of the Sefiroth Tiphareth by liquid power (Figures 8.2 and 8.4).

It should be mentioned that we cannot precisely determine the present-day phase of the universe's evolution. It looks as if the Tiphareth filling process (Figure 8.4) is finished. We can observe the filling of the next Sefirah (forming the planetary systems; the Sefirah Netzach). Hence, we are situated now at the formation of the fourth hierarchic level (the fifth stage of evolution).

Everyday observations might appear as a contradiction of the results of the tree of life doctrine because we can now observe the already formed planets. Moreover, we can see in the night sky the star (Sefirah Hod) and galaxy systems (Sefirah Yesod). However, all these subjects are presented in Figure 8.4 as further independent stages of the universe's evolution.

For the explanation of the noted imaginary contradiction, let us recall the physical peculiarities of the preceding stages of evolution. At these stages, we had the complete set of the fundamental (GUI, weak, strong, and electromagnetic) interactions, but each time only one of them played a main role. The same rule should also hold in the subsequent stages of evolution.

Gravitational interaction exists at all these preceding stages as the secondary, not the primary, interaction. However, presented physically, it always exerts a relevant influence at all stages of the evolution. We can observe the results of this influence in the form of the existing planet, star, and galaxy systems.

This also means that the process of formation of the planetary systems is today far from complete. It can be interpreted as an affirmation of this that the formation of the planetary systems (including the planet biosphere) is the main process in the present universe. Of course, we also observe the star and galaxy systems in the universe; however, they will play the analogous main roles in the future.

8.3.5 Five Subtypes of Gravitational Interaction: Can This Be Real?

Let us recall that, according to the tree of life doctrine, we should distinguish in nature five subtypes of gravitational interaction, whereas in physics we know only one subtype (gravitation I; see Table 8.1). How can we interpret this circumstance physically? Let us discuss this interesting topic in detail.

The point is that the rest of the four subtypes will play a main role at future stages of the universe's evolution (the formation of the star and galaxy systems, and the metagalaxy as a whole; Figure 8.4). We can treat these stages as the filling processes of the Sefiroth Netzach, Hod, Yesod, and Malkut, respectively. Similar to the general situation at other stages of evolution, we can say that the remaining subtypes are also present in the present universe, although they do not play a remarkable role there. Their effect is masked by the powerful background of the first subtype (gravitational interaction I), which is now the main interaction in the universe's evolution.

Today, we do not have enough scientific information about these future processes to be rigorous and we do not have reliable knowledge about the physical scenario of further evolution of the universe. However, based on the tree of life doctrine, we can draw relevant conclusions about possible scenarios and character of the universe's evolution.

It should be mentioned that the present theory of gravitation is not a completed branch of physics. Therefore, we do not have reliable reasons to assert that the gravitational field, at different hierarchic levels of the universe, should be governed strictly by the same physical regularities. Moreover, we have more of a motive to say that these regularities most likely could be different. First, let us recall that the main sources of the gravitational field are matter and energy [34]. They bend the space of the universe and we take physically this process as the gravitational field.

In our case, matter is represented by the universe's space substance and fields. The key words in this situation are "closed universe." In the framework of the dominant physical theories, we must regard the universe as a closed dynamic system. Therefore, if it is so, we indeed should accept the supposition that the same gravitational law should work at each hierarchic level of the universe.

However, the universe really is not a closed system. We live on a four-dimensional brane surface. The total mass of the matter in this brane surface and the features of its space geometry determine the form of known gravitational law. In accordance with the bulb theory [34], our brane surface exists in a surrounding of other neighboring brane surfaces. The electromagnetic field, as it is known [34], is bounded with each of the brane surfaces. That is why we cannot visually discern its presence in our night sky in the form of luminous objects. In contrast, the gravitational field is not bounded with each brane surface [34]. Therefore, the field of one brane surface can penetrate another neighboring brane surface. Hence, the resulting gravitational field within our universe is really a superimposition of the proper and external components. In other words, the influence of external gravitational fields should not be neglected.

Thus, the external component of the acting total gravitational field is caused by the influence of the external brane universes. This influence depends on their total masses, geometries, and reciprocal distances. In accordance with the tree of life doctrine (see Section 8.3.8), all other brane universes were born simultaneously with our universe. Therefore, these external universes also should be evolved synchronously with our universe. Consequently, the reciprocal distances during the common process of evolution (inflation) can be different at different evolution stages. This, however, means that the main regularities that stipulate the configuration of the gravitational law within our universe can change during the process of evolution. Hence, we can conclude that four other subtypes of the fundamental gravitational interaction can indeed exist and the hypothesis about their existence does not look as crazy as it seems at first sight.

8.3.6 "The King Is Dead, Long Live the King!"

At the present time, we are within the intermediate stage of the universe's evolution: The filling process of the Tiphareth has been finished, and the filling of the Netzach has already started. The tree of life doctrine predicts three additional stages. They are the filling of the Sefiroth Hod, Yesod, and Malkut. But what will happen when the process of filling the Sefirah Malkut is accomplished?

This question is urgent because modern physics cannot give us an appropriate scenario without obvious internal contradictions. A few such possible scenarios are discussed in this chapter's references—see, for example, Hawking's *The Universe in a Nutshell* [34] and others. Let us shortly comment on two of them that are regarded today as the most probable.

The first is the model involving a cyclic extension–compression of the universe during its evolution. It is considered that the universe will be extended to some limit. Then, the extension process will transform into a compression. The compression, in turn, will be finished by restoring the initial singularity, and so on. This model is very attractive for modern researchers because it gives a very logical, comprehensible, and complete scenario of the universe's evolution.

But only one small difficulty mars this pleasant scenario: It, unfortunately, is not confirmed by modern astrophysical observations. The point is that, according to this model, the extension phase of the evolution should be characterized by the deceleration law of the extension. It is determined by action of the gravitational forces that are responsible for further possible compression. Real observations, however, reveal the reverse. The expansion continues to accelerate and this acceleration has increased during the last stages of the universe's evolution. It means that gravitational forces do not prevail in this situation. Because the gravitational forces are proportional to the product of masses, this phenomenon has been labeled as the problem of the universe's mass insufficiency. During the past fifty years, astrophysicists were preoccupied with the search for this lost mass. Many modern researchers consider that only ~4% of the universe's total mass is represented in the form of ordinary substance, ~74% is so-called dark energy, and ~22% is dark matter. Unfortunately, even such essential additions do not solve the problems of the decompression–compression model. Thus, today we do not have essential fundamental reasons (experimental observations, mainly) for accepting this model as depicting reality. Instead, we have only a vague hope.

Let us turn our attention to the second of the above-mentioned models. It is the so-called model of the unlimited inflated universe [34]. It is obvious that it is neither as attractive nor as logical and comprehensible as the expansion–compression model. It looks paradoxical, but the worst aspect in this situation is that it correlates well with astrophysical observations. Why, in this situation, do we use the word "worst?" The point of this paradox is that its key idea is an unlimited universe extension. It contradicts the observed symmetry of all real processes in nature. Until now, we have never had to deal with unlimited real physical objects or processes. We always knew that if something has a beginning (the initial singularity, for example), it also should have an end.

Thus, the key point of the physical problem is its internal contradictions. By accepting the big bang theory, the physicists should accept simultaneously the obvious fact of the openness of our universe as a dynamic system. Instead, they constructed their physical scenarios on the basis of the hypothesis of the closed universe.

Let us appeal to the scenario that gives us the tree of life doctrine. As can be readily seen in Figure 8.4, the process of overfilling the Sefirah Malkut finishes the universe's evolution. This point is characterized by two events. The first event is the death of our universe (i.e., the process of the universe's evolution is limited in principle). The second event is the birth of a new neighboring universe after the Malkut. As we mentioned before, "A manifested Malkut is the hidden Kether." But it will be a new Kether in a new real five-dimensional universe that will be born by our Malkut: "The king is dead, long live the king!" Hence, in reality, the universe extension process should be limited.

8.3.7 Once Again about the Problem of the End of the Universe

But what is the physical realization of this limit? In other words, what is the physical meaning of the overfilling of the Sefirah Malkut? Is this the beginning of a compression stage of the universe's evolution, or does it correspond to a different scenario? Unfortunately, modern physics does not have a synonymous answer. On the other hand, the tree of life doctrine allows us to see the general scheme of the required and not discordant scenario. We talk about the scenario of the universe's evolution, which can be referred to as the limited-inflated universe. This model, in contrast to the decompression–compression model, does not have a compression phase of the evolution. On the other hand, contrary to the similar model of an unlimited inflated universe, its extension stage is limited.

For the sake of a solution to the considered problem, we will use the previously discussed sequence of the principle of hierarchic resemblance: "The number of the characteristic schemes and regularities in the universe is limited and the set of these schemes and regularities is universal for all manifestations of the world" (Section 2.2.3). We will also apply the method of hierarchic resemblance, which originates from this sequence (Section 2.2.4).

In accordance with the bulb theory [34] (see also Figure 2.9 and relevant discussion), our universe as a four-dimensional brane bulb is not singular in the superuniverse. It is surrounded by other universe brane bulbs. Furthermore, we note that the bulb theory also tells us that the universe brane bulbs can interact with one another. As a result of these interactions, some spatial structures—clusters of the bulb universes—can form. This unknown type of interbrane gravitation interaction can play the main role in the formation of such brane clusters.

Then, we use the method of the hierarchic resemblance regarding the arrangement of the atomic nucleus as the required analogous characteristic scheme. The mentioned brane clusters can be treated as a peculiar supernucleus. The only difference is that here we have to deal with the brane universes instead of the true nucleons. Continuing this analogy, we can suppose that the nuclear forces play the role of the interuniverse gravitation-like forces. In other words, we assume that the characteristic regularities (see again the mentioned sequence) in both cases are hierarchically similar.

Following the method of hierarchic resemblance, we can suppose that this gravitation-like interaction has a quasiexchange character. Remember that one of the physical mechanisms of the realization of nuclear forces in real nuclei is the exchange of nucleons by π-mesons. As a result, a proton catastrophically transforms into a neutron, the neutron in turn transforms again into a proton, and so on. Taking into account the possibility of the existence of a similar quasiexchange interaction in the case of the superuniverse, we can conclude that every such interaction of the brane universes should be accompanied by catastrophic (transformative) events. In our case, it can be the birth of a new universe with a higher dimensionality (see also the relevant explanations in the next section).

But, this is not the only possible scenario for the interaction. In nuclear physics, the decay, collision, and other types of particle interactions are also known. Correspondingly, because of the principle of hierarchic resemblance, their relevant analogs can exist in the superuniverse, too. However, independent of the type of interaction, each time we should have an essential change (catastrophical transformation) in the forms of the subjects of the interaction. In this connection we can generalize the discussion in the form of the following conclusion: During a reciprocal interaction, the brane universes undergo an eventually catastrophic change of form each time. We interpret the beginning of the change process as overfilling the Sefirah Malkut, or, equivalently, the completion of our universe's evolution is caused by a catastrophic change of its form as the result of interactions with other brane universes. Such catastrophic transformation can look like a birth of a new singularity and, consequently, as a realization of a new big bang.

It should be mentioned that we can observe similar catastrophic processes in the modern universe: collisions of galaxies. It is believed that the largest galaxies undergo approximately four to five such catastrophic collisions during their evolution. A new galaxy appears after each of these collisions.

The hierarchic analogy in nuclear physics has another side. Indeed, following the same scheme of analysis, we can conclude that the elementary particles are the peculiar microscopic worlds. These worlds, in contrast to our four-dimensional brane universe, should have a lower dimensionality. As shown in the next section, each microuniverse can be represented as a three-dimensional surface brane that envelops a four-dimensional volume. Let us recall that our four-dimensional brane universe envelops a five-dimensional volume. Analogous to our universe, oscillations of this particle brane surface generate matter in such a universe. Let us discuss this and similarly intriguing problems in more detail.

8.3.8 The External World: What Does It Look Like?

All above-mentioned analogies suggest that they do not have a solely figurative meaning. This follows directly from the principle of hierarchic resemblance. It appears that its framework and the tree of life doctrine as a whole are much wider than previously assumed. If this is indeed so, we obtain a very effective methodological tool (the method of hierarchic resemblance) for constructing relevant extraordinary physical theories, including the theory of the superuniverse itself. Let us illustrate the possible contours of such a theory, some results of which are represented graphically in Figure 8.8.

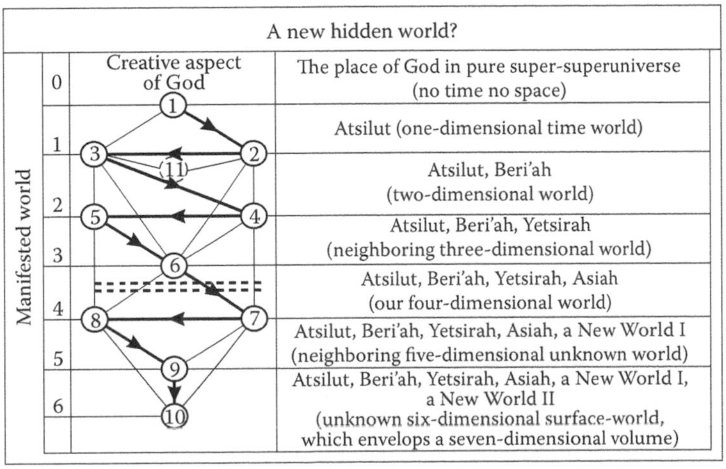

FIGURE 8.8
Illustration of the evolution scenario of our superuniverse in accordance with the tree of life doctrine. Here the circles with numbers denote the Sefiroth: 1 is the Kether, 2 is the Chokmah, 3 is the Benah, 4 is the Chesed, 5 is the Geburah, 6 is the Tiphareth, 7 is the Netzach, 8 is the Hod, 9 is the Yesod, 10 is the Malkut, and 11 is the Daat (see Figure 8.4). The two parallel dash lines correspond to the present state of the evolution process. See Figures 8.1, 8.4, and 8.12 for comparison.

Studying the relevant references dedicated to the theological and scientific cosmogony of the universe and the superuniverse, we find significant confusion of two types. The first appears in attempts at a theological treatment of some results of the physical big bang theory. The scenario of the universe's evolution, which has been constructed in the framework of this theory, cannot be reconciled directly with the Bible's scenario of the creation of our world. It is because, as a detailed comparative analysis shows, the Bible's scenario completely corresponds to the evolution of the superuniverse (Figure 8.8), whereas our universe is only a structural element of the fourth hierarchic level of the superuniverse. These scenarios are different in many specific details. However, they are governed by the same general regularities that originate from the principle of hierarchic resemblance.

The second type of confusion involves the number of worlds (dimensions; see Figures 8.7 and 8.8) in nature. The Old Testament and the Kabbalah talk about four such worlds: the Atsilut, the Beri'ah, the Yetsirah, and the Asiah, respectively. This affirmation is applicable to the description of our universe (Figure 8.7) as well as to the superuniverse (Figure 8.8). It should be stressed that this circumstance serves today as a main source of the mentioned confusion. The point is that it is very often considered that only these four worlds can exist in nature, in general, and that any other additional worlds cannot, in principle, exist. However, this is true only in the case of our universe (Figure 8.7). Being four-dimensional, it has only the four numbered world dimensions.

We have another situation in the case of the superuniverse (Figure 8.8). Until now, the worlds appeared, one after the other, during the superuniverse evolution beginning with the Atsilut and finishing with the Asiah. It is important to emphasize that, similar to the universe, the superuniverse is situated today at the stage of formation of the fourth hierarchic level (the fifth stage of evolution). Consequently, we now live in the times of the formation of the fourth world, Asiah. The question arises: Is this the last of all possible worlds in nature? The tree of life doctrine says, "No!" First, the evolution process of the superuniverse will not be finished at the present fourth hierarchic level (Figure 8.8). In accordance with the tree of life doctrine, we should expect the formation of two more unknown worlds (Figure 8.8). Taking this into consideration, let us discuss the other interesting aspects of the hypothetical scenario of the superuniverse's evolution.

In accordance with the material in Section 8.1, the Sefiroth reflect, in particular, the properties of the natural forces. In the case of our universe, it allows giving physical meaning to the Sefiroth as the types of fundamental interactions. However, in the case of the superuniverse, we have another situation principally. We cannot give an analogous attribution to the Sefiroth because we do not have enough of the required scientific information. This concerns the fact of the existence of such objects, as well as their possible physical properties. Therefore, in this section, we will confine ourselves to discussion of the general arrangement of the superuniverse as a whole, which follows from the basic principles of the tree of life doctrine.

It is not difficult to find that, similar to our universe, the superuniverse exists in seven-dimensional space. However, in contrast to the universe, we observe different correlations between the hidden and manifested dimensions at different hierarchic levels of both systems. The known combination of "3 hidden + 4 manifested" is specific only for the fourth hierarchic level of the superuniverse (the level of our universe; see Figure 8.8). During evolution, the number of the hidden dimensions changes from seven (the zero hierarchic level—Kether) to one only (the sixth hierarchic level—Malkut).

Different hierarchic levels of the system, illustrated in Figure 8.8, represent brane surfaces with different dimensionalities in the manifested world. The zero level is characterized by a zero brane, which has zero manifested dimensionality. In this case, we do not have time and space in the manifested world. The superuniverse is represented there by a point. We can treat this point as the place of the creative aspect of God in our superuniverse. We can find that our superuniverse in the Kether, being zero-dimensional in the manifested world, is completely situated in the seven-dimensional hidden world. It is because the hidden world is the world of the Malkut of a preceding superuniverse. It becomes more comprehensible if we recall that "The manifested Kether is the hidden Malkut."

The formation of the first hierarchic level (the filling of the Sefiroth Chokmah and Benah) corresponds to the birth of the world Atsilut. As is noted above, the physical aspect of the Atsilut can be treated as time. At this stage of evolution, the superuniverse in the manifested world can be represented by a one-dimensional time brane. As follows from Figure 8.8, its formation is a two-stage process (Sefiroth Chokmah and Benah). Unfortunately, we do not know the physical essence of these stages.

Relevant analysis of the process at other stages of the superuniverse's evolution can be continued in an analogous way. However, the main regularity is the same: The higher the system's hierarchic number, the higher the manifested dimensionality of the corresponding brane universes at the levels. Thus, the formation of each following superuniverse level is accompanied by the birth of a new manifested spatial dimension. It means that all spatial dimensions do not possess equal rights. This circumstance can be treated in the following manner: The complete superuniverse (in the complete hidden + manifested seven-dimensional continuum) has a specific seven-dimensional geometrical form (possibly egg-like). Different spatial dimensions of such a complete continuum play a different role in the superuniverse's everyday life. These roles, unfortunately, are not comprehensible for us now.

Similar to the case of our universe, this evolution process will stop with the filling of the Sefirah Malkut. This stage is characterized by the formation of a six-dimensional brane surface that envelops a seven-dimensional volume. Hence, at this stage the manifested world is six-dimensional and the hidden world is represented by one dimension only. But, what is the physical meaning of this hidden dimension? The Kabbalah says that it is the dimension connected with the enigmatic Sefirah Daat (Figures 8.1, 8.8, and 8.13). Let us discuss this topic in detail.

8.3.9 The Living Superuniverse

The three upper Sefiroth of the tree of life (see Figures 8.1, 8.4, and 8.8) in the Kabbalah are called the Highest Trinity. All these Sefiroth have a few other names or synonyms. Most often, they are the Crown (Kether), Wisdom (Chokmah), and Mind (Benah). All these Sefiroth are situated in the manifested world. The Sefirah Daat (knowledge) is also connected with this Trinity. This connection is fulfilled through the Chokmah and the Benah (Figures 8.8 and 8.13). It is considered that the Daat is placed outside of the manifested world. We can conditionally consider that it is placed in the hidden world.

Such a situation also takes place for all partial universes of all hierarchic levels shown in Figure 8.8. One of the hidden dimensions is always occupied by the Daat. It is possible that it is not the true hidden dimension, but at the present stage of our discussion, this is not important. We note that the connections (Ways) between the Sefiroth Chokmah, Benah, and the Daat projection on the tree of life surface form a right pyramid (see Figure 8.13).

Some Cabbalists consider that the Daat represents the idea of the perceiving and the consciousness. This and the names of the Sefiroth Chokmah (Wisdom) and Benah (Mind) allow giving a physical interpretation for the Sefirah Daat in the case of the superuniverse, as well as in the case of our universe.

In Chapter 2, we said that all natural hierarchic dynamic systems have the evidently expressed information exchange (Figure 2.10). Let us define life as a state of matter (substance and field) that is characterized by the possibility of accumulating, generating, accepting, transmitting, and processing information. In this context, we can ask: Which natural object in the world is responsible for the existence of similar processes with information? The tree of life doctrine answers, "It is the Sefirah Daat."

It should be mentioned that such ideas are not new in modern physics. For example, Hawking asserts that the four-dimensional brane surface of our universe can be treated as an information hologram of objects placed within the five-dimensional volume (let us remember that our brane surface is closed and envelops this volume). It can be readily seen that Hawking's information treatment is indeed similar to the information interpretation of the Daat: The Sefirah Daat is the main information center of the superuniverse, which is reflected holographically on our brane surface. Taking into account the given definition of life, we can say also that the superuniverse, similar to our universe, should also be a living object and the Sefirah Daat, together with the Sefiroth Chokmah and Benah, are most responsible for life in the superuniverse, as well as in our universe.

Thus, the accomplished comparative analysis confirms the correctness of the tree of life doctrine. It is done in the illustrative example of the evolution of the material universe, which shows that the proposed version of a new general theory can serve as a source of a number of very interesting and sensible physical hypotheses.

It looks as if we can finally use the tree of life doctrine effectively. It is the beginning of the twenty-first century, when human civilization has accumulated enough scientific knowledge. Never before in the nineteenth century or earlier did scientists have such great possibilities. This means that today we have a chance to lay down the foundation for a new, in principle, general scientific paradigm, the true theory of everything.

Let us illustrate some contours of the other possibilities that this new paradigm can provide in the future. Emphasizing its general character, we will demonstrate it by examples that are far removed from today's physics: human ethnosocial genesis, including the hierarchic structure of the human ethnopolitical organization, and the biosphere as a hierarchic dynamic system. Of course, the author is aware of the fact that, contrary to physics, at the present time the modern state of the sciences responsible for these topics cannot provide the required level of quality of relevant experimental and theoretical materials. However, he considers that this is not the crucial point of the formulated illustrational purpose.

8.4 Hierarchic Cycles (Oscillations) in Earth Systems

8.4.1 A Few Words of Introduction

As shown in the preceding section, our four-dimensional brane universe could be imagined as a gigantic four-dimensional hierarchic-oscillation system. The three-dimensional components are represented in the surrounding world in the form of a rich variety of various spatially periodical and quasiperiodical natural objects: crystals, planets, stars, galaxies, etc. The temporal component determines the cyclic life of these spatial hierarchic systems in time: various rotation motions, oscillations, resonances, precessions, and so on.

After being acquainted with the materials of the preceding sections, a reader can conclude that this type of ideas is principally a new branch of advanced human scientific thought. Indeed, the superstring and the *M*-theory, the representation of the evolution of the universe as a successive cascade of bounded hierarchic resonances, etc., look like absolutely modern subjects of present-day fundamental science. However, this is not so. The idea of the hierarchic-oscillation arrangement of the surrounding world has been known as far back as written human history exists. In various ancient sources [18–22,27–29,44–47], we find information about the natural temporal cycles of various periods: 11, 15–16, 19, 30, 33, 52, 100, 125, 300, 400, 500, 600, 700, 1000, 1200, 1500, 2160, and tens, hundreds, and even billions of years.

Analyzing this series, one can be sure that all known periods (cycles) can be conditionally divided into two different groups. The first group is the hierarchy of the different periods (cycles): 11, 15–16, 19, 30, 52, 100, 125, etc. The second group is characterized by the presence of periods that are harmonics (or subharmonics) of the first cycles, for example, 11 and 33, or, 100, 300, 400, 500, and so on. In other words, the periods (cycles) of the first group are connected with the resonances, i.e., with the birth of relevant detuning phases of oscillations (see Section 8.2.4 and relevant discussion of Figure 3.19). The periods of the second group are correlated with the oscillation harmonics of these detuning phases. Basically, it correlates well with the discussed hypothesis about the resonant-wave arrangement of the universe discussed in Section 8.2.

We find that a part of the discussed periods (cycles) was discovered during investigations of not purely physical (human-related, in particular) phenomena. Apart from the astrophysical cycles of different scales, we also find cycles in the ethnopolitical, historical, and sociocultural processes, and in the biological evolution of the earth and the universe as a whole [18–22,27–29,32,33,44–47]. Researchers have also found these human-related cycles while studying purely physical systems—of astronomical objects, for example.

All these intriguing facts, together with the material discussed in the preceding sections, recall that the ancients perceived the integrity of our universe as a whole. They deeply understood its hierarchic oscillation-wave nature and intuitively felt the world as an object where all processes and structural elements are bounded by myriads of connections, and where everything is interlaced with each other. It is interesting that the ancients realized this filling practically in the form of such extraordinary products of intellectual activity as astrology and calendars. And, similar to the discussed situations in the Old Testament, the most enigmatic aspect of this ancient knowledge is the sources of the incredible scientific information that they used for the construction of their cosmogonical concepts and theories. Let us briefly illustrate this nontrivial side of the problem in this and the following sections.

8.4.2 Ancient Indian Hierarchic Cycles

The first enigma that we meet by starting to study the ancient Indian cosmogony is the scale of units for measuring time intervals. The smallest unit was the truty, which equals one-fifth of a second. In Europe, researchers began to measure such time intervals only in the nineteenth century. The Indians, as it follows from the references, used it 3000–4000 years ago. How could they use it practically? Unfortunately, we do not know. No less incredible is the largest time interval: the ancients have handled numerical data of $3.1104 \cdot 10^{15}$ earth years (Table 8.3).

The ancient Indian cosmogonic doctrine includes a rather developed theory of cycles of different kinds. From the point of view of modern physics, the presence of a system of

TABLE 8.3
System of Megacycles in Ancient Indian Doctrine

No. of Hierarchic Cycle		Name and Duration	
0	Mahayuga	Kaliyuga: $0.432 \cdot 10^7$ E.y.	
1	Mahayuga	Dvaparayuga: $0.864 \cdot 10^7$ E.y.	
2	Mahayuga	Tretayuga: $1.296 \cdot 10^7$ E.y.	
3	Mahayuga	Kritayuga: $1.728 \cdot 10^7$ E.y.	
4		Brahma's day (kalpa):	1 000 Mahayugas ($4.32 \cdot 10^{10}$ E.y.)
5		Brahma's year:	720 Kalpas ($3.1104 \cdot 10^{13}$ E.y.)
6		Brahma's century:	100 Brahma's years ($3.1104 \cdot 10^{15}$ E.y.)

Note: Here the notation E.y. denotes earth years.

megacycles exerts the strongest influence on our imagination. The variant of the system of such cycles is illustrated in Table 8.3.

Relevant analysis shows that this system practically corresponds to the discussed scheme of the evolution of our superuniverse (see Figure 8.8 and corresponding discussion). This scheme was obtained in the process of physical treatment of the Old Testament (Kabbalah). Comparing both schemes, we find that the materials of Table 8.3 and Figure 8.8 describe different sides of the same object. The material in Figure 8.8 displays qualitative aspects of the general scheme of the superuniverse's evolution. At the same time, we can regard the content of Table 8.4 as a very interesting numerical addendum to this qualitative scheme.

TABLE 8.4
Chmykhov's Historical Cycles

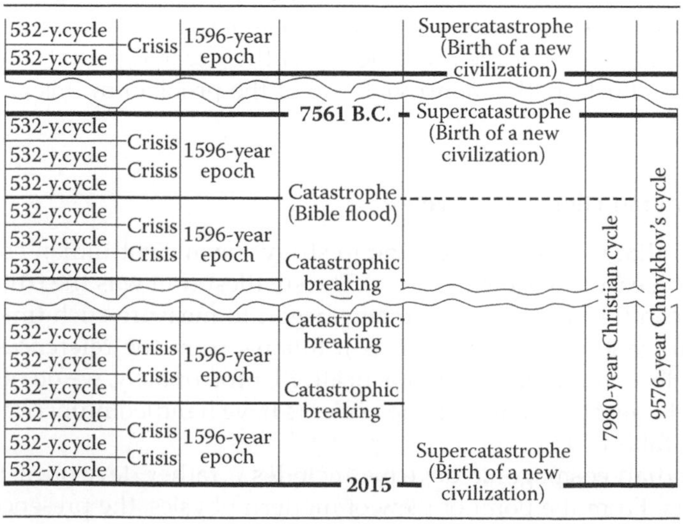

By using Table 8.3, we can determine, in particular, the duration of the formation of each hierarchic level of the superuniverse and the superuniverse's time-life as a whole.

Unfortunately, we also have some contradictions with the ancient Indian evolution scenarios with those that originate from the tree of life doctrine. Let us discuss the most important of them.

According to ancient Indian doctrine, our universe exists only during one Brahma's day (Kalpa) (see level 4 in Figure 8.8 and Table 8.3). The world disappears catastrophically at the end of Brahma's day and the beginning of Brahma's night. It is considered that the duration of Brahma's night and that of Brahma's day are equal. Hence, the world of the following (fourth) hierarchic level can appear only after such a "night break." Whereas in the general theory of tree of life we have continuous transformation of one superuniverse hierarchic level into the following level (see the model with the liquid power descending illustrated in Figures 8.2, 8.4, and 8.8). At the same time, in the framework of the ancient Indian doctrine, the analogous processes have a principally discontinuous character. The first four hierarchic levels (see Table 8.3) evolve without any catastrophic Brahma's breaks. The first Brahma's break happens at the end of the fourth (i.e., our) level. The evolution of the last two hierarchic levels (see Figure 8.8) should be accompanied by numerous catastrophic Brahma's breaks.

However, the presence of catastrophes in ancient Indian doctrine does not mean the existence of the above-mentioned contradictions. In the tree of life doctrine, the evolution process, being continuous in principle, also has a catastrophic character. The point is that the birth of any hierarchic level of the superuniverse is accompanied by the birth of a new dimension (see Figure 8.8 and corresponding discussion). For example, the birth of our universe (in the form of the initial singularity, the fourth hierarchic level) was accompanied by the birth of a new (fourth) dimension (see Figure 8.8). Inasmuch as the preceding hierarchic world-level was three-dimensional, we treat this phenomenon as the birth of a new four-dimensional space. We perceive this process as the catastrophic initial big bang.

Thus, the key contradiction concerns the ancient Indian idea about the complete disappearance of the superuniverse during the Brahma's night and its subsequent restoration during the following Brahma's day. It should be noted that, from a physical point of view, such a scenario looks rather improbable. First of all, in accordance with the principle of hierarchic resemblance, such a general scenario should be reproduced in the evolution process at all of the hierarchic levels and sublevels. This, in particular, should concern the processes of evolution on our earth. However, we do not find anything like this in the observed world. On the other hand, we see that the general regularities of the evolution process that are given by the tree of life doctrine are systematically reproduced at any hierarchic level and sublevel of any natural hierarchic system of any hierarchic rank.

It should be noted that the idea of the Brahma's night also contradicts the fundamental principles of modern physics. According to these principles, time–space can exist only under the condition of the existence of matter (Section 8.2). In other words, modern physics forbids the existence of an empty space. Hence, the fact of the disappearance of the matter of the world during the Brahma's night automatically means the disappearance of the time dimension. However, in this case, the following question arises: What does the affirmation "The duration of the Brahma's night equals the duration of the Brahma's day" mean? Unfortunately, it is not clear within the framework of the traditional interpretations of ancient Indian texts. In the framework of modern physics, the duration of the Brahma's night in the manifested world should be equal to zero. In reality, this means that no time breaks should exist between any two Brahma days. As a result, we return to the continuously catastrophic tree of life model, which is represented in Figure 8.8.

Thus, by comparing the two ancient scenarios of the evolution of the superuniverse, we can see that both have a catastrophic character. However, if in the scenario that adheres to the tree of life doctrine, the catastrophes do not violate the continuous character of the superuniverse evolution. In the ancient Indian scenario, we have a systematic interruption of the evolution process. And, as it follows from the principles of modern physics, this interruption cannot really exist in nature. But is this contradiction insurmountable? The author thinks that it is not. Most likely, it is a result of inaccuracies in the modern interpretation of the relevant ancient texts. Therefore, taking into account the existing level of contemporary knowledge, we propose simply to ignore this contradiction.

This is motivated also by the fact that we observe a few very important points. The seven levels of hierarchy and the sensible proportions of the times of level formation create a possibility of making a more complete general evolution scenario, which is represented in Figure 8.8. Additionally, we can determine the theoretical duration of the present stage of the superuniverse's evolution (between the Sefiroth Tiphareth and Netzach). In accordance with Table 8.3, it could be estimated as $\sim 4.32 \cdot 10^{10}$ years. In view of the above, it is also the complete lifetime of our universe (Figures 8.4 and 8.8), because the total lifetime of all previous universes is much smaller (Table 8.3).

Let us determine the present time point of the universe's evolution line (Figure 8.4). Modern physics gives estimations for the present age of our universe from $\sim 1.3 \cdot 10^{10}$ to $\sim 3 \cdot 10^{10}$ years. For simplicity, we can accept the mean value of $\sim 2 \cdot 10^{10}$ years. The ancient Indian doctrine gives similar data: $2.20 \cdot 10^{10}$ years. It is approximately half of the complete universe lifetime. Thus, the reader need not worry. Our universe will exist for $2.12 \cdot 10^{10}$ more years.

8.4.3 Ancient Mesoamerican Hierarchic Cycles

The ancient Mesoamericans (Maya and Aztecs) also had a deep knowledge of the universe's hierarchic cycles. But, in contrast to the ancient Indians, they evaluated this knowledge mainly from a practical point of view. The Mesoamericans are known as creators of the most precise calendar among all others known, including the modern.

The Maya's calendar contains a 13-day week, a 20-day month, and a 365- or 366-day year. Each 52 years, they jointed into a cycle. This cycle contains two half-cycles, which in turn form the basis of the main Mesoamerican 5125-year cycle. The last such cycle started on 13 August, 3113 B.C. and its end is expected in December of 2012. The ancient Maya considered that each end of a cycle is accompanied by a series of catastrophes, which they considered the end of the world. The ancient Mesoamericans, similar to modern researchers, considered each such end of the world as a radical renewal of the world order. This renewal involves all components of the hierarchic earth besom, beginning with the climate and finishing with the social and political arrangement of human civilization. We will have a good chance to examine the correctness of this theory in practice very soon. However, another interpretation of this ancient scheme gives us a more optimistic date: December 2115.

The Mesoamericans also knew large-scale cycles. Similar to the ancient Indians and the creators of the Old Testament and Kabbalah, they believed that we live in the fifth stage (the number of our hierarchic level + the zero level) of the universe's evolution (compare with the material in Figure 8.4). They are called the preceding four stages

1. First stage: the Four Jaguars
2. Second stage: the Four Winds
3. Third stage: the Four Water
4. Fourth stage: the Four Earthquakes

They also knew about the four manifested worlds and the principle of hierarchic resemblance. The Mesoamericans also applied these four universe evolution stages to the earth's evolution.

It is interesting that, besides the 5125-year cycle, the Mesoamericans also used the 8000-year main cycle. With respect to the duration, this cycle is very close to the well-known Christian 7980-year cycle (Section 8.4.5). It should be taken into account that, in accordance with Christian tradition, this cycle should end in 2015. The above-mentioned Mesoamerican cycle has another starting point and should end in 4846. This means that, in spite of the closeness, both of these cycles have essentially different physical natures.

Concerning the dates 2012 and 2015, it is possible that they are not a true end of the world from a physical point of view. In accordance with the above-mentioned Mesoamerican doctrine, it could be a break of the existing world order. We should expect a series of crisis phenomena that will be manifested as a profound change in all main aspects of the general arrangement of the surrounding world. In other words, a serious shock will almost simultaneously touch all spheres of our everyday lives: the global and local climate, political, social, economic, and financial systems, the ethnocultural and biogeological global processes, etc. Global warming, increasing tectonic activity, the series of bloody wars in Asia, Africa, and Europe, as well as the world financial and economic crisis that started in June of 2008 speak to this very expressively. The other side of this incredible process of crises is the increasing rate of change in human nature, which has been especially distinctive during the past two decades [5]. Taking into account the large-scale character of the discussed cycles (5125 and 7980 years), we can expect that the transitional period of the present global catastrophic crisis can be estimated to be at least 100 (or more) years. Therefore, we should add to this list the critical events of the twentieth century: both World Wars, the Russian revolutions, the series of cruel civil wars, the terrible large-scale famines and holocausts, etc. What shall we expect in the near future? Welcome to the world of global change.

8.4.4 Other Ancient Hierarchic Cycles

All large ancient civilizations had relevant knowledge of the natural and human-related cycles. Similar to the Mesoamericans, they used this knowledge as the basis of their calendars. The ancient Egyptians had the moon calendar, which had 12 months of 29 or 30 days; every 2–3 years they added a 13th month. To determine when this extra month should be added, they used astronomical observations of the star Sirius. They knew about the existence of the so-called Sirius cycle with a duration of 1460 years and believed that this cycle governed all natural and human-related processes in the surrounding world.

The ancient Indo-Aryans had knowledge about the ecliptic. This is the great circle of the sky sphere, which the sun completes during one year. The ecliptic plane transverses the sky equator at the points of the vernal and autumnal equinoxes.

The Sumerians divided the ecliptic into 12 parts and gave animal names to each part. The Sumerians also discovered the 'saros'—the 19-year cycle of the repetition of eclipses of the sun and moon. But their most important discovery involves their disclosure of the so-called zodiacal circle. The duration of the zodiacal cycle is 259,200 years. It was also divided into 12 zodiacal epochs of 21,600 years each.

Similar to other ancient civilizations, the Sumerians knew the principle of hierarchic resemblance. They translated the large-scale zodiacal principle into other small-scale earth hierarchic cycles (individual and collective cycles). This system of small-scale cycles was established as the basis of astrology, which is widely used in our everyday life. (Do you have your personal astrological horoscope?)

The ancient Chinese also knew about cycles and they used them as a basis for their calendars. They used a 60-year cycle that contains five branches, with each branch then divided into 12 sections/animals. The complete period of Jupiter's rotation and doubled periods of Saturn's rotation were used as the basis of this 60-year cycle. This calendar has been used for more than 2600 years.

8.4.5 Chmykhov's Theory of Historical Cycles

The investigation of natural and human-related cycles remains an important direction of modern science. As an example of such research, we propose the theory of the space-social cycles, developed by Ukrainian scientist M.O. Chmykhov (1953–1994) [29,45]. In contrast to ancient doctrines, the novelty of Chmykhov's theory is that it was constructed on the modern basis of the numerous reliable historical, archeological, astronomical, and geological materials. Let us discuss this interesting topic in detail.

Similar to the ancients, Chmykhov considered that all our planetary cycles involve the totality of hierarchic branches of the earth as a gigantic complex hierarchic dynamic system. In terms of the method of the hierarchic tree described in Chapter 2 (see Section 2.4), this can be represented graphically by using the concept of the hierarchic besom (see Figure 2.7 and corresponding explanations). This means that the astrophysical cycles disclosed as a result of relevant astronomical observations should exist parallel to the biological, human-related, and other cycles in the all-earth hierarchic system. We can also treat this as one more manifestation of the principle of hierarchic resemblance.

Thus, strictly speaking, we cannot regard any partial earth hierarchic system (human society, for example) separately. First, it is because all systems of the united hierarchic earth besom are always bounded by numerous so-called horizontal reciprocal connections (see Figure 2.7). Hence, analyzing the behavior of any singular hierarchic branch, we cannot simply neglect the influence of other branches without estimating the importance of their influence. Chmykhov proved this hypothesis experimentally by studying the known astronomical, climatic, historical, geophysical, and other earth cycles in one chosen Earth territory. He obtained the best evidence in favor of the correctness of his theory by the discovery of the so-called watch-cycle effect [29,45], which works only in this territory.

Traditionally, the territory around the Black Sea—the Balkans with Greece, Asia Minor (including the Middle East), Caucasus, Ukraine, Romania, and Bulgaria—has been referred to as the circumpontic zone. It is characterized by an incredible concentration of historical enigmas. Let us mention old European civilization, which included the ancient archeological cultures Cucuteni-Trypolye, Lengyel, and Vincha. These cultures existed in Southeast Europe from 5400 to 2750 B.C. [48,49]. We also remember the ancient Aryans, Scythians, Sumerians, Assyrians, Babylonians, and Judeans. And it should be noted especially that all main world religions (Judaism, Christianity, and Islam) were born in this area. Some historical properties of the Circumpontic zone actually look mystical. Very often, it is called the "center and omphalos of the earth." The above-mentioned watch-cycle effect has been named specifically for this territory.

The essence of this effect is as follows. During the past 10,000 years (Table 8.4), the circumpontic zone has served as a peculiarly powerful generator of cultural pulses. These pulses have propagated out in all possible directions, with the effect becoming weaker and more delayed with increasing distance. In other words, the process explicitly expressed a wave-like nature. Generally, we can talk about an unusual oscillation-wave hierarchic structure of historical processes in this region. The historical cycles (oscillations) occur there synchronously with some natural cycles: astronomical, climatic, biological, and others. Thus, we do indeed have a peculiar hierarchic besom of many closely connected natural–historical cyclic branches.

The Bible tells us that our world began in 7561 B.C. (Table 8.4). Some people consider this date funny. However, well-educated people know that it is only the birth date of the present-cycle variety of the world with specific mankind, climate, biological, and geophysical peculiarities that are intended for this world. In accordance with the relevant calculations, this world should end in 2015 (see also the discussion in Section 8.4.3).

This so-called Christian 7980-year cycle is an important part of Chmykhov's general theory. However, the theory also contains another 9576-year Chmykhov's cycle, which its author considers the Main Cycle of the circumpontic zone. Both cycles are constructed on the basis of the same 1596-year cycle that is referred to as the epochs. In accordance with Chmykhov, the Christian cycle contains five epochs, whereas the Main Cycle consists of six epochs. Each epoch is a composite of three 532-year cycles (see again Table 8.4). Each such 532-year cycle consists of two 266-year half-cycles, and each half-cycle contains two 133-year sub-half-cycles. This complex analysis allows us to ascertain that all these cycles, epochs, half-cycles, and sub-half-cycles really exist in the global natural and human-related earth hierarchic systems [29,45].

The last three 532-year cycles of the last epoch, in particular, are the following:

- Cycle 1 (419–951)
- Cycle 2 (951–1483)
- Cycle 3 (1483–2015)

Each 532-year cycle is characterized by four phases of development, the durations of which are equal to the 133-year sub-half-cycle. The end of each 133-year cycle is always characterized by some violation of the existing world order. The end of each 532-year cycle is always accompanied by a systemic crisis (political, economic, demographic). The crisis is deeper if it coincides with the end of an epoch, and much deeper (catastrophic) at the end of the Christian 7980-year cycle. The crisis has the character of a global supercatastrophe if it is at the end of the 133-year sub–half-cycle, the 532-year, the 1596-year, the Christian's 7980-year cycles, and the 9576-year Chmykhov's cycle. Just such a situation, as it is readily seen in Table 8.4, is expected for realization in 2015.

Of course, human psychology basically is pessimistic; people more willingly believe bad news than good news. Without any connections with the theories of cycles discussed here, many modern scientists during the past few decades have been busy elaborating various dark prognostications. Most of these prognoses concern global warming and the crisis states of the modern world's political and economic systems (Chapter 1). Today the Internet is saturated with such material. For example, a typical prognosis for the sequence of the global warming is that the velocity of the change in mean global temperatures will increase rapidly after 2013. This will cause an acceleration of glacial thawing. This will lead to the flooding of many territories in Europe, Asia, and America. In 2040, many people will be forced to leave their normal dwellings; Holland will be completely flooded. The same fate is expected for northern Germany, northern Poland, and the Baltic countries. Three new islands will appear instead in what is today Great Britain: Scotland, Wales, and the Canterbury Mounting. Many famous cities such as New York and Saint Petersburg will disappear. All these natural cataclysms will be accompanied by no less cataclysmic political, social, economic, demographic, and humanitarian upheavals. Many more optimistic prognoses are dedicated to these crises. But, let us go beyond this not so funny topic.

8.5 Integrity of the Surrounding World as a Totality of Seven-Level Hierarchic Besoms

8.5.1 Once More about the Earth and the Universe as Hierarchic Besoms

Let us continue to illustrate the process of the specific features of a few human-related (i.e., nonphysical) hierarchic structures in terms of the discussed new hierarchic theory.

We begin the discussion with the hierarchic trees represented in Figures 8.9 through 8.11. These are the human ethnogenetic system (Figure 8.9), the system of human ethnopolitical organization (Figure 8.10), and the biosphere of the universe (Figure 8.11). The first two correspond to the earth as a hierarchic system. The third one, together with Figure 2.6, is related to the universe as a whole. The proposed examples describe different aspects of the universe as a whole and its separate hierarchic levels (the earth, in this case). Comparing all these illustrative examples, we find two interesting facts.

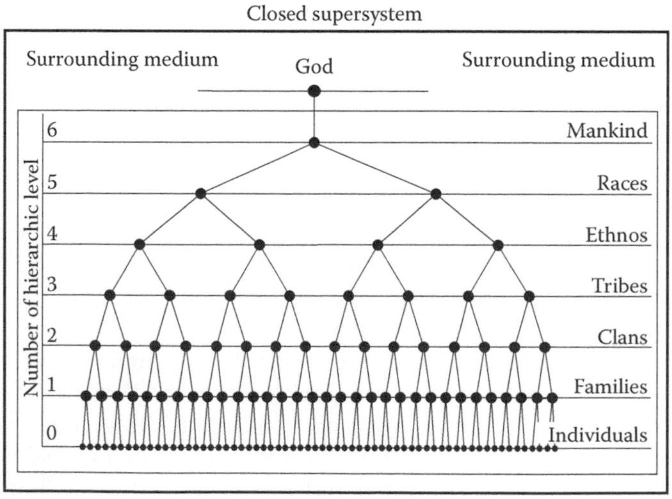

FIGURE 8.9
Hierarchic tree for the human ethnogenetic system (see Figure 2.6 for comparison).

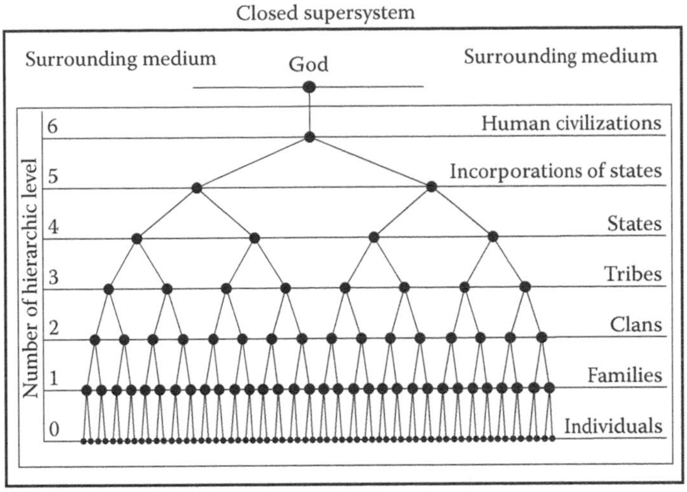

FIGURE 8.10
Hierarchic tree for the system of human ethnopolitical organization (see also Figures 2.6, 8.9, and 8.11 for comparison).

1. In spite of their obviously expressed differences in physical nature and the rank of the hierarchy, all analyzed systems have the same number of hierarchic levels (seven). This could be regarded as one more evident manifestation of the principle of hierarchic resemblance in the surrounding world.
2. Each of the pairs of hierarchic trees (Figures 2.6 and 8.11 and Figures 8.9 and 8.10, respectively) can be encompassed in two hierarchic supersystems and each of them can be referred to as the *hierarchic beson* (see Figure 2.7 and corresponding explanations).

Thus, each of the two separate hierarchic trees represented in Figures 2.6 and 8.11 could be treated as a branch of the hierarchic universe besom. Analogously, the separate

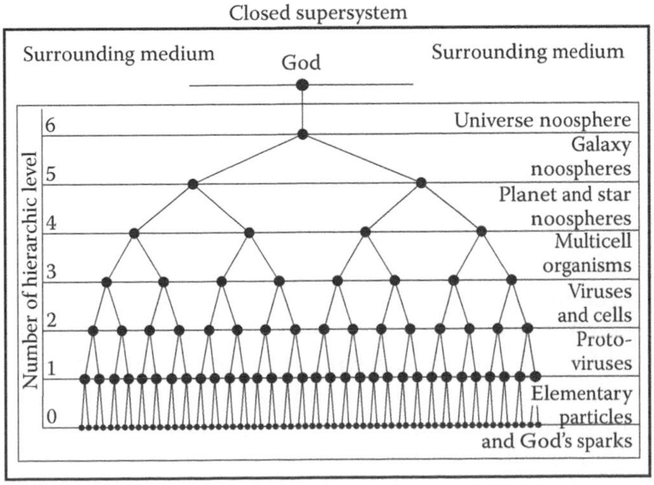

FIGURE 8.11
Hierarchic tree for the biosphere of the universe (see also Figures 2.6, 8.4, and 8.9 for comparison).

hierarchic trees in Figures 8.9 and 8.10 can be regarded as the branches of the hierarchic earth besom. It is obvious that these hierarchic besoms are systems of unimaginable complexity, containing many other similar branches. However, let us limit the discussion to the study of the proposed simplified partial models, each of them containing only two branches.

Comparing the material in Figures 2.6, and 8.9 through 8.11, we can also make the following observations. First, all the common levels of different branches in each hierarchic besom are connected by vertical connections. The operation principles of these connections as information channels are governed by the hierarchic principles discussed in Sections 2.1 through 2.5. Second, the horizontal connections should also exist between the different branch levels of the complete universe or earth as hierarchic besoms. Let us emphasize that the horizontal connections act between objects of the same level of different branches (subsystems) only.

8.5.2 Vernardsky's Doctrine of the Earth Biosphere

It should be mentioned that the idea of hierarchic besoms and the horizontal connections between their branches, as follows from the material of the preceding section, is not new. In 1927, V.I. Vernardsky provided its modern scientific formulation [27]. He proposed the doctrine of the biosphere as a specific biogeological earth shell. This shell includes the other earth subshells (living world, lithosphere, hydrosphere, atmosphere, and stratosphere) as the hierarchic branches. In other words, in the term of hierarchic theory, we can say that the totality of these subshells forms the hierarchic earth besom.

It was found, in particular, that the earth's evolution 3.3 billion years ago changed from a purely chemical form into a biological form. The influence of the living world on the geological evolution became prevalent during this period. The existing horizontal connections in the real hierarchic biosphere earth besom now became obvious to modern science. It is clear today that mankind is considered only as one of many branches of the general hierarchic earth besom. Today, human activity is one of the most important geological and biological factors that act in the modern earth biosphere. The well-known global ecological problems, in particular, speak to this rather expressively.

The noosphere (the sphere of the intellect) is such a stage of the earth besom evolution, when the human collective intellect will become the main factor of the biosphere evolution. This idea is very important for a better understanding of the essence of the present and future stages of the biosphere evolution. Some scientists treat the appearance of the noosphere as the formation of a superorganism—living earth. They consider that it possesses its own superconsciousness. It is also believed that the end of the formation of this earth superorganism means, simultaneously, the end of the formation process of the universe hierarchic level of planetary systems (see Figures 8.4 and 8.12).

Vernardsky's ideas about the importance of horizontal connections have evolved a great deal during the past eight decades and are remarkable because they play an important role in many essentially different modern scientific directions simultaneously. We can observe this in the ethnogenesis of human civilization and its ethnopolitical evolution, in astrobiophysics and biogeology, and in ecology and philosophy, as well.

Today, as before, Vernardsky's doctrine is true for modern natural sciences as well as for the humanities. The further expansion of the concepts of the biosphere and noosphere from the scale of the earth to the universe (see Figures 8.11 and 8.12, for example) may

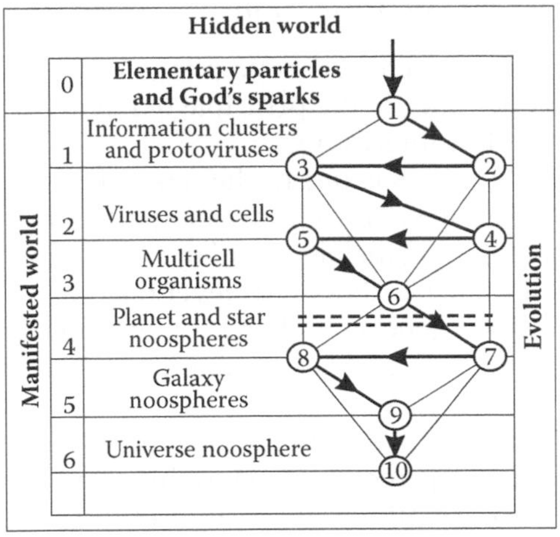

FIGURE 8.12
Tree of life glif and an illustration of the evolution process of the universe biosphere (see also Figures 8.2 and 8.4 for comparison). Here the circles with numbers (Sefiroth; see Figure 8.1) designate the acting natural force (1 is the Kether, 2 is the Chokmah, 3 is the Benah, 4 is the Chesed, 5 is the Geburah, 6 is the Tiphareth, 7 is the Netzach, 8 is the Hod, 9 is the Yesod, and 10 is the Malkut). The two parallel dash lines correspond to the present state of the evolution process (see Figures 8.1 and 8.8 for comparison).

serve as an evident illustration of this affirmation. We will discuss some modern aspects of such a generalization of Vernardsky's ideas and doctrine.

8.5.3 Gumiliov's Theory of Ethnogenesis

First, we should note the revolutionary theory of passionarity, which has been elaborated by Russian historian L.N. Gumiliov in the middle of the twentieth century [28,32]. He has found, in particular, that the ethnosocial and ethnopolitical processes in earth history have always been governed by very specific regularities that are connected strongly with the properties of the surrounding natural landscape and (unknown for the time being) space factors. It is interesting that some of these regularities (the passionary explosion, for example) are very similar to the bifurcations in the theory of nonlinear oscillations. Gumiliov has disclosed the very intriguing correlations between the passionarity of historical processes and astrophysical factors. Herewith, the passionarity process seizes confined earth territories only. It looks as if a fantastic space searchlight illuminates this territory with an unknown radiation. The territory boundaries are always outlined precisely. The Hun and Mongolian invasions, the Arabian, Spanish, British, and Russian expansions, and many other historical events are given as the best illustrative examples of the historical phenomenon of passionary explosions. Gumiliov has found a close connection between the phenomenon of passionarity and the discussed phenomenon of historical cycles. Based on this similarity, he has given a few historical prognoses, some of which (the swift rise of China, for example) have already had vivid confirmations.

It should be noted that modern science does not have any rational scientific explanations of this observed historical phenomenon. On the other hand, it is obvious that such explanations might be found in the framework of the discussed concept of horizontal connections in the universe and earth hierarchic besoms and the theory of hierarchic oscillations and waves.

8.5.4 Hierarchic Resemblance in Ethnogenetics

Let us continue the discussion of ethnogenetic phenomena. Other results of this kind, which illustrate the real workability of the principle of hierarchic resemblance in modern ethnogenetic systems, are no less interesting.

Let's remember that ethnogenetics is one of the new sciences that study regularities of origination and evolution of ethnos. The ethnos is regarded as a biosocial organism with a general structure and operation principles similar to an individual. In other words, the ethnos is a personality of a higher taxonomical level that has its own consciousness and subconscious, intellect, memory, genotype, character, age, and destiny. The end of the formation of the earth's noosphere is regarded as the final stage of the global earth ethnic evolution.

Similar to the preceding illustrative examples (Sections 8.2 and 8.3), the ethnogenetic variant of the manifestation of the principle of hierarchic resemblance is also rather ancient. Plato spoke about the "Person small" (individual) and the "Person large" (ethnos). The novelty of the modern situation is that the methods of the study of ethnos genealogy are similar to the modern (including genetic) methods of the study of individuals. Figuratively speaking, each ethnos has its specific genetic map, its father and mother, grandfathers and grandmothers, and so on. The wedding of "men-ethnos" and "women-ethnos" can, similar to people, produce an "ethnos-baby." That is interesting, especially as all the conditions for a successful wedding are also very similar to the human condition [which are well-known (the author hopes) for everybody, at least from TV and some Internet sites]. In other words, we have an obvious triumph of the principle of hierarchic resemblance in human surroundings. Such triumphs look somehow incredible.

Other manifestations of the principles of ethnogenetics look no less incredible. Let us discuss some political sequences of Gumiliov's theory of ethnogenesis, which are especially urgent today. We talk about the widely known (in the relevant expert medium) Euro-Asian concept [32] that directly originates from the theory of passionarity. The government of Russian President V.V. Putin has made this theory the ideological basis of the new strategic state concept of Russia. As follows from Gumiliov's theory, today's Russia should be considered a natural historical successor to Genghis Khan's ancient Mongolian empire [28,32,33]. (Sorry, this is not fantasy, but a serious branch of today's historical political science.) First, this is because the Moscoviya (Russia's name until the seventeenth century) was legally an "ulus" (province) of the Golden Horde (the western part of Genghis Khan's empire) from the thirteenth through the sixteenth century. Second, the Russian Tsars—from Alexander Nevskiy to Ivan the Terrible—were the ritual relatives of the Genghisides [28,32,33]. Third, the Tatar–Finn component of the population is prevailing (~70%) in the genotype of the modern Russian nation. According to Gumiliov, these facts explain the specific Euro-Asian ethnic psychology, as well as the natural political tradition of modern Russia. That is why, in accordance with the opinion of leading Moscow politologists [32], only the Euro-Asian (traditionally Mongol) vector of Moscow politics was historically natural. Therefore, it should play the main role in Russia's political future. With time, the purely western political vector will become secondary.

What connection does this interesting topic have to do with the main topic of the present chapter? The point is that ethnogeneticists use this concept as a very important argument for the substantiation of the so-called theory of reincarnation of political nations. The concept of individual reincarnation is well known in ancient Indian philosophy [81]. It illustrates, in particular, the oscillation-like (cyclic) nature of the evolution of the human spirit during the process of mankind's development. The key point of the proposed new theory is that the same idea could be applied to the ethnogenesis of political nations. Such a reincarnation process can be treated as a cyclical birth of modern political nations in the form (and approximately in the same boundaries) of their old analogs. It happens with the same spirit of the nation but in another ethnic and political body.

The theory of historical cycles and the principle of hierarchic resemblance serve as the basic argument for the development of this extraordinary hypothesis. The ethnogeneticists of this school of thought regard modern Russia (as well as the tsar's empire and the former Soviet Union) as a modern reincarnation of Genghis Khan's empire. The other historical illustration of this theory is the European Union. The adepts of this theory regard it as a modern reincarnation of the Roman Empire. We can also mention China, which is considered to be the modern reincarnation of the ancient Chinese empire. Of course, similar to other ethnogenetic ideas, the discussed theory appears too extraordinary; however, modern science cannot give convincing arguments for its complete rejection, whereas the arguments of the proponents of this theory sometimes look very reasonable.

8.5.5 Concept of the Living Universe

The most interesting and intriguing ideas of modern bioastrophysics concern the theory of a living universe (see Medvedyev's *The Space Creation of Human Beings* [31], for example). It actually can be regarded as a relevant extension of Vernardsky's doctrine of the universe as a whole (see Figures 8.11 and 8.12).

The problem is that, today, we do not have any cause for considering life on earth generally—and mankind in particular—as unique in the universe. Moreover, most of the inhabitants of earth (probably under Hollywood's influence) are convinced that our civilization is one among many in the universe.

There exists a group of scientists whose opinions have not been formed solely under Hollywood's influence. Their most shocking hypothesis is that the big bang (in our traditional understanding; see Figure 8.4) has been accompanied by a "big biological bang" in open space [31] (see Figure 8.12). The key foundations for their hypothesis are the following.

- Life is an all-universe phenomenon and it started with the birth of our manifested world (i.e., with the appearance of the initial singularity).
- All sectors of the observed universe developed practically synchronously, since all processes within these sectors have a common beginning (i.e., the same initial conditions in the form of the common initial singularity); as a result, the modern universe as a whole is practically homogeneous and different sectors should have a similar evolutional destiny.
- Life, analogous with the ordinary substance (inert matter), also originated from the same initial singularity; the role of the initial living particles is referred to as God's sparks (see Figures 8.11 and 8.12).
- The first living matter (information clusters) appeared as a result of the combination of subjects of the inert matter (substance and fields) with God's sparks, in accordance with unknown to us quantum laws.

In accordance with these somewhat revolutionary ideas, the general scenario of the evolution of the living universe in terms of the tree of life doctrine could be imagined as is shown in Figure 8.12. Let us briefly discuss some key points of this peculiar hybrid of the tree of life and Vernardsky's doctrine, on one hand, and the big bang theory, on the other.

First, we should note that the main difference between inert and living matter involves information. We have defined living matter as matter that can accumulate, generate, accept, transmit, and process information. The concept of living matter is a key point for understanding the true essence of today's "information universe."

Second, apart from the known properties of the general unified physical interactions, it appears that it should also contain unknown aspects that involve their information properties. These aspects are responsible for the appearance in the initial singularity of the unknown "holders of living properties." We call the hypothetical component that provides these properties the God's spark interaction. In other words, the initial quark–antiquark fireball in the Sefirah Kether also included specific God's spark quarks. Unfortunately, we do not know today which of the twelve quarks and antiquarks are responsible for this role. We also cannot say what type of initial particles (or other substance) played the role of the God's spark particles.

8.5.6 Daat as an Information Aspect of God

Thus, two kinds of initial particles should exist in the framework of the modernized Vernardsky's doctrine. The first represents the well-known type of matter—inert matter. The second could be treated as the God's spark particles. Only the latter do we consider as the probable candidate for the first living matter. In this connection, it should be mentioned that, being objects of an unknown physical nature, these new particles might not really be true particles. Therefore, we should regard the introduced terminology as conditional only.

The birth of both kinds of particles could be treated as the overfilling of Kether by the liquid power. In turn, the birth of the first true living matter occurs in the next two stages of the living universe evolution. In accordance with the tree of life doctrine, this happens in the Sefiroth Chokmah and Benah. These Sefiroth are also called the wisdom and the mind (or the understanding), respectively [4–8]. The Sefirah Daat (knowledge; see Figures 8.1 and 8.13) plays the key role. The specific feature of this Sefirah is that it is situated out of

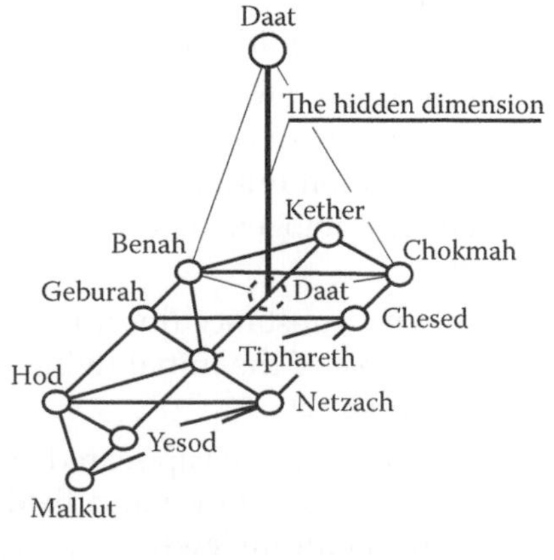

FIGURE 8.13
Sefirah Daat (knowledge) in the tree of life (see also Figure 8.1 for comparison). According, to the tree of life doctrine, Daat does not belong to our manifested universe (the plane of the tree of life) and is connected only with the Sefiroth Chokmah (wisdom) and Benah (understanding).

the plane (or more precisely out of the surface) of the tree of life (see Figure 8.13). The Daat plays some type of role as a universe information center. Therefore, we can also treat Daat as the information aspect of God.

It is interesting that such an interpretation has something in common with Hawking's theory [34]. By taking its main ideas into consideration, we can conclude that Daat is placed within the five-dimensional volume, which is enveloped by our universe brane surface. It is because of the fact that the plane of the tree of life is connected with a four-dimensional brane surface, which in turn is closed and envelops a five-dimensional volume including Daat.

According to Hawking's theory, the brane surface (and our world, consequently) can be regarded as an information hologram of an object contained within the five-dimensional volume. This brane hologram has an information exchange with an information store processor within this volume.

The appearance of the information clusters should be closely connected with the action of the weak interaction within the Sefirah Chokmah (see Figures 8.4, 8.5, and 8.9 through 8.13, and Table 8.2 and corresponding discussion). The fundamental regularities of this process are not clear for us for the time being. It is clear only that each such combination of matter + information could be stable only in the case where the relevant quantum conservation laws (like the conservation laws for the baryon charge, color, smell, charm, strangeness, etc., in the modern theory of elementary particles) are satisfied.

The Sefirah Daat (see Figures 8.1 and 8.13) fills the information clusters (Sefirah Chokmah) and the protoviruses (Sefirah Benah) by the relevant programming information. This means that the first stage of the discussed process of reviving the inert matter happens in the Sefirah Chokmah during the formation of the primary life information matrix. The second stage (Sefirah Benah; protoviruses) is connected with the activation of the first primitive consciousness.

Thus, the Sefirah Daat plays the role of a peculiar external life programmer for the evolution of the living universe (see Figures 8.1, 8.4, and 8.11 through 8.13). This programmer determines an algorithm of action (evolution) for the living matter at each hierarchic level of the universe. It is believed that the living part of the universe is the main part, whereas the inert material plays a secondary role. The inert material universe is created only for the sake of the fulfillment of the given program by the living universe. It should also be mentioned that, in the case of the superuniverse, each hierarchic level has an individual program for its living part. In other words, similar to a human child, universes of any superuniverse hierarchic level are always born with their individual program, which should be realized during their evolution (Section 8.3.9). Therefore, in summary, we can determine that the main purpose of the existence of the inert universe is to provide material resources for the successive accomplishment of the main mission by the living part.

8.5.7 Angels, Archangels, Spirits, and Others as Physical Subjects: Why Shouldn't They Exist?

It is well known that matter in nature exists in two main forms: substance and field. We do not have any guarantee that the form of living matter—based on substance (living substance) has been created only during the evolution process. Theoretically, an analogous type of living matter could also be formed parallel to the form of a living field. Such living forms of fields could undergo further evolution until the present time. Today our living universe could be treated as a complex hierarchic dynamic structure. It should include the objects of living substance as well as objects of living fields as inherent structural elements.

On the other hand, we do not have any cause for the assertion that both of these forms should always be spatially separated and that they mutually do not interact in the common life of the universe. The possibility of the existence of symbiotic forms (i.e., living substance + living field) of life in the universe is discussed by many experts. In contrast to animals and other simpler forms of life, man is considered as a bright representative of such symbiotic form.

Some researchers believe that the fact of the presence of objects in the living field (including symbiotic forms) in our world can explain many so-called anomalous phenomena in present-day human life, as well as some enigmas of human nature. On the other hand, their accounting in the framework of future extraordinary physical theories can allow agreement with modern physics and theology. We talk about the well-known difference of the physical pictures of the observed surrounding world, which give these two directions of human intellectual activity. Indeed, angels, archangels, soul, spirit, and so on as physical subjects—why shouldn't they exist?

8.5.8 Vikramsingh's Living Space Clouds

Unfortunately, this version of the first steps of the evolution of life has only a range of promising hypotheses. The first material tracks of real living space matter (living substance) have been fixed experimentally by astrophysicist Vikramsingh [31]. These tracks, in terms of the tree of life doctrine, correspond to the level of Sefirah Geburah. Vikramsingh has observed a significant amount of an organic dust-like substance in space in the form of huge carbon clouds. Physical and chemical spectral analysis shows that these clouds are nothing but clouds of charred (dead) biological cells.

These living clouds appear in the universe on the level Sefirah Geburah. The processes of repeated heating of matter (Section 8.2) have been accompanied by numerous explosions of very new stars. We can observe these processes in the present universe, too. Such explosions are always characterized by an emission of fantastically powerful ionized radiation. This radiation could kill all living beings within a huge sphere with the point of explosion at its center. This radiation is regarded today as one of the possible causes of the death of Vikramsingh's living space clouds [31].

Thus, the following revolutionary conclusion could be formed on the basis of Vikramsingh's discovery: Life was born in open space; the Earth is only a bosom for its cultivation [31].

8.5.9 Planetary, Star, Galaxy, and All-Universe Noospheres

We mentioned that we live today in an intermediate stage of the universe's evolution. This is the stage when the Sefirah Tiphareth (substance in Figure 8.4 and multicell organisms in Figure 8.12, respectively) is overfilled by the liquid power. The process of the filling of the Sefirah Netzach (planetary systems in Figure 8.4 and planet noospheres in Figure 8.12) continues. But how can we treat the formation of the planet noospheres as the present stage of the universe's evolution? As an answer, we propose the following hypothesis that originates directly from Vernardsky's doctrine and the hierarchic resemblance principle.

The end of the formation of the earth noosphere is an evolution stage in which the human collective intellect and the human collective will become the main factor of the biosphere evolution [27]. It will be manifested as the formation of a collective consciousness of the living earth. In other words, the present stage of the universe's biosphere evolution is the formation of a united all-planet noosphere with their collective consciousnesses. Taking into account the hierarchic resemblance principle, we can draw similar conclusions for other evolution stages of the universe. We can extend this hypothesis to the Sefiroth Hod (the formation of the noospheres of star systems), Yesod (the formation of the galaxy noospheres), and Malkut (the formation of the all-universe noosphere; see Figure 8.12).

But, "The manifested Malkut is the hidden Kether." This means automatically that the main source of God's sparks for the next universe (see Figure 8.8) will be the collective being of our all-universe noosphere. Using this as an analogy, we can formulate the following hypothesis: The collective being of the preceding all-universe noosphere served as the source of the initial God's sparks for our Kether (our living initial singularity).

8.5.10 Other Civilizations: Can They Really Exist?

At last, let us add a few words about the possibility of the existence of other civilizations in the universe. Researchers have accomplished the following calculations for the substantiation of the thesis of the multiplicity of developed civilizations in the universe [31]. Approximately 10^{11} stars exist in our galaxy alone. They should comprise ~10^9 planets with life. Relevant estimations show that only ~$5.5 \cdot 10^7$ planets among them can have developed civilizations. As mentioned previously, we have basically the same initial starting conditions as other potential civilizations. At first sight, this enunciates our equality with respect to a chance to reach the same level of development as potential competitors. Unfortunately, this does not automatically mean an absolutely synchronous character of this process. The main cause of this is the existence of various fluctuations in the universe's structure as a stochastic system. Relevant calculations of the characteristic values of these fluctuations show the existence of a scattering with respect to the actual starting times of evolution for different civilizations. This value can be estimated as ~10^7 years [31]. We draw the following conclusion: The modern united civilizations of the universe (UCU) can simultaneously comprise the planet civilizations, the ages of which can, in principle, differ by ~10^7 years. For our civilization, for which civilization experience is not more than 100,000 years, it does not look very optimistic.

8.6 Instead of a Conclusion

The author is a professional physicist who has served as the head of departments of physics for 28 years, the last twenty of which were as the head of a department of theoretical physics. When writing this chapter, he mentally heard the indignant cries of his colleagues: "How could you?" Of course, the presence of God, the tree of life, angels, archangels, souls, spirits, and God's sparks in the temple of modern high physics is unaccustomed, to say the least. But to this, the answer is, "Unfortunately, we have, in reality, no alternative."

Modern physics is undergoing a deep systemic crisis, which can be classified as a crisis of integrity. Formerly united as a philosophy of nature, physics has disintegrated into hundreds and even thousands of separate branches and subbranches. Long ago, most of scientists lost the feeling of integrity of physics. We dig deeper and deeper and, more and more, we only increase the numbers of petty new facts and details, estranging us from one another. All of this gives only an extensive (quantitative) growth of the sum of our knowledge, where we wish urgently for a revolutionary qualitative growth.

As a result, the realities of the current situation are very unpleasant. Practically, we have not moved very far forward from the nineteenth century with respect to true fundamental achievements. Indeed, the main set of the fundamental ideas (the concept and theory of fields, discovery of the electron as a first elementary particle, etc.) that became the basis of the modern high technology revolution were created in the nineteenth century. The last outstanding fundamental discovery in the required critical range has been the disclosure of the possibility of discrete (quantum) processes in nature. This is the discrete emission of light portions (photons) that were discovered by M. Plank in 1900. Modern quantum physics and technologies (from nuclear to laser) should be regarded only as successful further development and practical implementation of this key idea. This is the same situation we have with other fundamental achievements: We use only the fundamental treasures accumulated predominantly in the nineteenth century. Unfortunately, the modern intellectual achievements like the *M*-theory and others can be only regarded as bright mind-games. First, they do not contain any revolutionary ideas and concepts. Second, in spite of the complexity and deepness of modern physical theories, they are extremely pretentious and do not promise the required level of novelty, universality, and integrity for the near future.

It looks as if modern fundamental science has reached a deadlock. Therefore, any new unifying ideas (even crazy ones like those proposed in this chapter) that can overcome the present problem of excessive fragmentation are worthy of attention and relevant discussion. We cannot allow ourselves to become too intellectually haughty and too quickly and easily label ideas as nonsense or delirium. We have an extreme need for a flow of principally new ideas and concepts. It is natural that most of these ideas and concepts should sometimes look extraordinary. However, they look so only because of the present-day established point of view. Only such extraordinary ideas can provide a truly radical break in modern scientific thought. The discussed examples of the physical interpretation of the tree of life doctrine show that such ideas can give, at least, the level of "revolutionarity" and "extraordinarity" that is currently required.

It is obvious that the traditional scientific paradigm has exhausted itself already. Historically physicists look for the most elementary element in nature, where they should be searching for the most elementary scheme of nature's organization. In other words, they should look for a most elementary topology. This is the key point proposed in this book: a new physical approach for creating a new physical paradigm. In the author's opinion, the most important part of this new paradigm should be the introduction of the principle of hierarchic resemblance in physics as a new fundamental law. Similar to the known conservation laws, it can be treated as a peculiar conservation of hierarchic topology.

Unfortunately, today we can see only weak signs of this new paradigm, but we desire it with great impatience. We dream about superpowerful sources of a gratis energy (instead of the now existing oil and gas), superengines for interstellar voyages, a means for global climate control, and other marvelous solutions. However, only a new-in-principle paradigm can realize these dreams. It should be a paradigm that will encompass the dispersed body of the present-day natural sciences and humanities. We desire a paradigm that will join in the framework of a unified doctrine of all intellectual achievements of modern as well as ancient researchers. It should be a paradigm that will allow formulating a new and clear program for further development of our fundamental science. In the author's opinion, the tree of life doctrine is a very promising candidate for the core role of this new extraordinary and sufficiently unifying paradigm. The author's modest experience with the practical application of the hierarchic resemblance principle [9–16,40,41,43] allows confirmation that the fundamental ideas of the tree of life doctrine could be very effective.

And a few last words. The reader might believe or not the general philosophy set forth in Chapters 2 through 4 and 8. However, this need not have any effect on the practical workability and effectiveness of the algorithms and mathematical methods (see Chapters 5 through 7) developed on the basis of this philosophy. The material of Part II is proposed as evident proof of this statement.

References

1. Flood, G.D. 1996. *Introduction to Hinduism*. Cambridge: Cambridge University Press.
2. Nataf, A. 1988. *Dictionary of the occult*. Herttordshire: Wordworth Editions Ltd.
3. Kapra, F. 1994. *Dao of physics*. St. Petersburg: ORIS.
4. Laitman, M. *Kabbalah*. 1984. Printed in Israel.
5. Fortune, D. 1995. *The mystical Qabalah*. New York: Alta Gaia Books.
6. Garbo, N. 1978. *Cabal*. New York: W. W. Norton & Company.
7. Ponce, C. *Kabbalah*. 1995. Wheaton, IL: Quest Books and Adyar, Madras, India: The Thesophical Publishing House.
8. Parfitt, W. 1995. *The new living Qabalah*. Queensland: Element.
9. Kulish, V.V., and Lysenko, A.V. 1993. Method of averaged kinetic equation and its use in the nonlinear problems of plasma electrodynamics. *Fizika Plazmy (Sov. Plasma Physics)* 19(2):216–227.
10. Kulish, V.V. 1991. Nonlinear self-consistent theory of free electron lasers. Method of investigation. *Ukrainian Physical Journal* 36(9):1318–1325.
11. Kulish, V.V. 1997. Hierarchic approach to nonlinear problems of electrodynamics. *Visnyk Sumskogo Derzshavnogo Universytetu* 1(7):3–11.
12. Kulish, V.V., Kosel P.B., and Kailyuk, A.G. 1998. New acceleration principle of charged particles for electronic applications: Hierarchic description. *The International Journal of Infrared and Millimeter Waves* 19(1):3–93.
13. Kulish, V.V. 1998. Hierarchic method and its application peculiarities in nonlinear problems of relativistic electrodynamics. General theory. *Ukrainian Physical Journal* 43:483–499.
14. Kulish, V.V., Kosel, P.B., Krutko, O.B., and Gubanov, I.V. 1998. Hierarchic method and its application peculiarities in nonlinear problems of relativistic electrodynamics. Theory of EH-Ubitron accelerator of charged particles. *Ukrainian Physical Journal* 43(2):133–138.
15. Kulish, V.V., and Kayliuk, A.G. 1998. Hierarchic method and its application peculiarities in nonlinear problems of relativistic electrodynamics. Single-particle model of cyclotron-resonant maser. *Ukrainian Physical Journal* 43(4):398–402.
16. Kulish, V.V., Kuleshov, S.A., and Lysenko, A.V. 1993. Nonlinear self-consistent theory of superheterodyne and free electron lasers. *The International Journal of Infrared and Millimeter Waves* 14(3):451–568.
17. Kahanetz, I. 1997. *Psychological aspects in management: The Young's typology, the socionics, and the psycho-informatics*. Kiev: A.S.K.
18. Melchizedek, D. 2002. *The Ancient secret of the flower of life*, Vol. I and Vol. II, Kiev: Sofiya Publishing House and Helios.
19. Blavatskaya, E.P. 2003. *The secret doctrine*, in three volumes. Moscow: Folio Kharkov.
20. Arguelles, J. 1987. *The Mayan factor. Path beyond technology*. Santa Fe, NM: Bear & Company.
21. Andreyev, D.L. 1998. *The rose of world*. Herndon, VA: Lindisfarne Books.
22. Kastaneda, C. 1992. *Collected works in ten volumes*. Kiev: Sofiya.
23. Green, M.B., Schwartz, J.H., and Witten, E. 1995. *Superstring theory*. Cambridge, UK: Cambridge University Press.
24. Kuzmichov, V.E. 1989. *Laws and formulas of the physics*. Kiev: Naukova Dumka.

25. Sivukhin, D.V. 1989. *General course of the physics. Atomic and nuclear physics*, Vol. V-2. Moscow: Nauka.
26. Arshynov, V., Lightman, M., and Svirskiy, Ya. 2007. *Sephiroth of cognition*. Moscow: URSS.
27. Vernardsky, V.I. 1988. *The naturalist's philosophical thoughts*. Moscow: Nauka.
28. Gumiliov, L.N. 2004. *Ethno-genesis and the earth biosphere*. Moscow: Irys Press.
29. Chmykhov, M.O., Kravchenko, N.M., and Chernjakov, I.T. 1992. *Archeology and the ancient history of Ukaraine*. Kiev: Publishing House of Kievan-Mohyliansky Academy.
30. Conte, O. 1908. *Course de Philosophie positive*. Paris.
31. Medvedyev, V. 2003. *The space creation of human beings*. Moscow: Piligrim Press.
32. Gumiliov, L.N. 2005. *The rhythms of Euro-Asia. Epochs and civilizations*. Moscow: AST Publishing House.
33. Gumiliov, L.N. 2003. *The ancient Russ' and the great step*. Moscow: Irys Press.
34. Hawking, S. 2001. *The universe in a nutshell*. New York: Bantam Books.
35. Graeber, D. 2007. *Possibilities: Essay on hierarchy, rebellion, and desire*. Oakland, CA: AK Press.
36. Harding, D.E. 1979. *The hierarchy of heaven and earth: A new diagram of man in the universe*. Gainesville, FL: University Press of Florida.
37. Bailey, A.A. 1983. *The externalization of the hierarchy*. New York: Lucis Pub.
38. Kaivarainen, A. 2008. *The hierarchic theory of liquids and solids: Computerized application for ice, water and biosystems*. New York: Nova Science Publishers, Inc.
39. Rammal, R., Toulouse, G., and Virasoro, M.A. 1986. Ultrametricity for physicists. *Reviews of Modern Physics* 58(3):765–788.
40. Kulish, V.V. 1998. Hierarchic theory of oscillations and waves and its application to nonlinear problems of relativistic electrodynamics. In *Causality and locality in modern physics*. Dordrecht, the Netherlands: Kluwer Academic Publishers.
41. Kulish, V.V. 1997. Hierarchic oscillations and averaging methods in nonlinear problems of relativistic electronics. *The International Journal of Infrared and Millimeter Waves* 18(5):1053–1117.
42. Von Betralanffy, L. 1976. *General system theory: Foundations, development, applications*. New York: George Braziller.
43. Kulish, V.V. 2002. *Hierarchy and hierarchic asymptotic methods in electrodynamics*, Vol. 1 of *Hierarchic methods*. Dordrecht: Kluwer Academic Pusblishers.
44. Cyclic character of the social systems (round table). 1992. *Sociological Researches* (Ukraine) 6.
45. Chmykhov, M.O. 1994. *Ancient culture*. Kiev: Publishing House of Kievan-Mohyliansky Academy.
46. Sorokin, P. 1992. *Human being*. Moscow: Civilization Society.
47. Tokariov, S.A. 1991. *Myths of the world's people*. Moscow: Iris Press.
48. Gimbutas, M. 1991. *The civilization of the goddess—the world of Old Europe*. San Francisco: Thames&Hudson.
49. Videiko, M.Yu. 1996. Die Crobsiedlungen der Tripoll'e-Kultur in der Ukraine. *Eurasia Antiqua* 1:46–80.
50. Carroll, L., and Tober, J. 1999. *The indigo children. The new kind have arrived*. Carlsbad, CA: Hay House, Inc.

Part II

High-Current Free Electron Lasers

As is mentioned in the Preface, Part II consists of six chapters. Chapters 9 and 10 are written as an introduction to the theory of high-current FELs (HFELs). The subsequent two chapters (Chapters 11 and 12) are dedicated to the weak-signal and nonlinear theory of ordinary parametrical HFEL. Chapter 13 occupies a special position in Part II because they contain the theory of two-stream superheterodyne FELs (TSFELs). Let us recall that just these systems are most promising as a main design basis of the cluster systems (see Chapter 1 in detail). These include the monochromatic (one-frequency) and multiharmonic TSFEL discussed in Chapter 13. Finally, the superheterodyne FELs (SFELs) of other types (plasma-beam and parametrical electron-wave SFELs) are discussed in Chapter 14.

It should be mentioned that FELs [1–9], as a new class of electronic devices, have appeared in the first two experiments [1,2]. Both experiments, as is well known, were accomplished in 1976. It is important to note that two types of the FELs have been realized in those experiments: the weak-current high-relativistic (Stanford University [2]) and the high-current moderate-relativistic (Navy Research Laboratory [1]) FELs. It is interesting to note that further historical destiny of both these types turned out to be unexpectedly different.

With time, the first device [2] (with the pumping in the form of a transverse magnetic undulator—H-Ubitron) became an academic version of the FEL in spite of the fact that its working up was financed by the military. Today, this type of FEL is most widely known among educated people. For instance, when talking about FELs, most physicists keep actually in mind just the H-Ubitron. Its academic status is explained, first of all, by the fact that it has very interesting fundamental physics. Indeed, being a purely classical device, the FEL, at the same time, works in the optical range, which is traditional for the ordinary quantum lasers.

The second (high-current) type of the first experimental FEL [1] historically has been constructed and studied in the framework of programs like the Star Wars (Chapter 1). Besides the mentioned high-current electron beam, the FEL realized in the mentioned experiment [1] has once more specific peculiarity: it had the pumping system in the form of a microwave waveguide with powerful traveling electromagnetic wave (Dopplertron pumping). It happened that such system never was considered as an academic device because its military nature did not provoke any doubt. The satellite method of basing (Chapter 1), for example, requires the FEL with moderate length and, at the same time, a maximum level of power. The use of so-called two-stage design schemes (where the Dopplertron has been used at the second stage—see below) allows to effectively solve this problem. After the end of the Star Wars program, a real interest to the HFELs, generally, and the Dopplertron, particularly, strongly reduced. As a result, the physical community with time forgot practically this interesting variety of FEL. As a result, during the past 20 years, the main attention (and the main financing, correspondingly) in this field was paid mainly to investigation of its competitor: the weak-current high-relativistic academic FEL [10–12].

The author's opinion is that the discussed situation is not fair. First, it is because real fundamental physics of the HFEL is no less interesting than the weak-current ones. Moreover, being plasma-like devices, the high-current systems demonstrate a much richer (and, consequently, more interesting) general physical picture. It is caused by the fact that the particle–wave interactions, as well as the wave–wave interactions, are possible in the high-current systems, depending on the FEL work mode. In contrast, the only particle–wave interaction mainly determines the basic physics of the weak-current systems.

The other side of the problem is that, as was already said in Chapter 1, the application possibilities of the HFEL were not appreciated at their true value until today. First and foremost, it concerns the cluster femtosecond systems, whose main ideas have been discussed in Chapter 1. The ordinary parametric, two-stream, plasma-beam FEL and all totality of the cluster multiharmonic FEL, whose operation principles have purely plasma (i.e., high-current) nature, can serve as convenient examples for confirmation of this thought (Chapter 1). Therefore, one of the main purposes of Part II is to illustrate this opinion by various authors' materials in the field of the theory of HFEL.

References

1. Davis, G.R. 1976. Navy researchers develop new submillimeter wave power source. *Microwave* 17(12):12.
2. Deacon, D.A.G., Elias, L.R., and Madey, J.M.J. 1976. First operation of a free electron laser. *Phys. Rev. Lett.* 38:892.
3. Marshall, T.C. 1985. *Free electron laser*. New York: MacMillan.
4. Brau, C. 1990. *Free electron laser*. Boston: Academic Press.
5. Luchini, P., and Motz, U. 1990. *Undulators and free electron lasers*. Oxford: Clarendon Press.
6. Kulish, V.V. 1998. *Methods of averaging in nonlinear problems of relativistic electrodynamics*. Atlanta: World Scientific Publishers.
7. Saldin, E.L., Schneidmiller, E.V., and Yurkov, M.V. 2000. *The physics of free electron lasers (Advanced texts in physics)*. Berlin: Springer.
8. Kulish, V.V. 2002. *Hierarchic methods. Undulative electrodynamic systems*. Vol. 2. Dordrecht: Kluwer Academic Publishers.
9. Kulish, V.V. 1990. The physics of free electron lasers. General principles. *Deposited in Ukrainian Scientific Research Institute of Technical Information* on 05.09.90, Kiev, No. 1526, Uk-90, 192 pp.
10. Freund, H.P., and Antonsen, T.M. 1996. *Principles of free electron lasers*. Berlin: Springer.
11. Toshiyuki, S. 2004. *Classical relativistic electrodynamics: Theory of light emission and application to free electron lasers*. Berlin: Springer.
12. Schmuser, P., Ohlus, M., and Rossbach, J. 2008. *Ultraviolet and soft X-ray free electron lasers: Introduction to physical principles, experimental results, technological challenges*. Berlin: Springer.

9

Free Electron Lasers for the Cluster Systems

The general concepts of the cluster systems, constructed on the basis of FELs of different types, are discussed in the ideological Chapter 1. Let us continue discussion of some of their aspects in this chapter. However, at this time, we will be interested, mainly, in the FEL classification, history, and their key technological basis.

The FELs are actually unique objects, from both the purely scientific as well as technological points of view. As shown in Chapter 1, their most interesting practical application is connected with a possibility of constructing the so-called cluster FEL systems (CFEL). As is shown in Figure 9.1, all CFELs can be classified into three main classes:

1. The CFEL on the basis of harmonic FEL
2. The CFEL on the basis of multiharmonic Cluster FEL
3. The combined CFEL

Relevant illustration examples of the CFEL of different classes, which are shown in the classification in Figure 9.1, were given and discussed already in Chapter 1 (see Figures 1.15 through 1.17). Therefore, let us turn our further attention to the classification of the types of FEL known today, which can be used as the CFEL technological basis.

As shown in Chapter 1, many of the key design peculiarities of the FEL, as main design elements, depend strongly on the chosen type of arrangement and destination of the CFEL. In turn, this circumstance gives birth to a rich variety of FEL designs, which can be used in the CFEL of different types. This observation is evidently illustrated in Figures 9.2 and 9.3.

The FEL classification with respect to the basic working physical mechanisms is given in Figure 9.2. As is readily seen, the number of different variants of possible types of FEL is actually very rich. Herewith, we should take into consideration the circumstance that we did not include in the classification in Figure 9.2 of many other possible FEL types, which are investigated up to now very feebly. For instance, they are the FELs on the basis of parametric electron-waves resonances, whose dispersion is determined by the beam transverse size effects, the FELs on the Cherenkov's instability, and so on.

The classification, which is shown in Figure 9.3, represents all possible arrangements of FEL designs known today. As we can see, they might be classified into only three types of the design:

- The monotones (devices that have only one design section),
- The klystrons (systems that contain more than one section, which are placed successively along the same electron beam),
- The multistage devices (systems that, similarly with the klystrons, also have more than one section, but they are arranged with respect to the signal wave).

Just the above given classifications are put in the basis of the general structure of Part II, where each of these FEL types will be discussed.

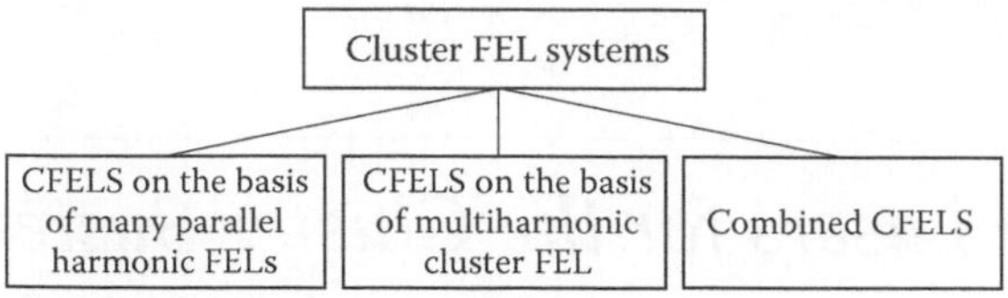

FIGURE 9.1
General classification of the cluster FEL systems (CFEL).

9.1 Parametrical Free Electron Lasers: Most General Information*

9.1.1 Priority Problem

A specific feature of the science of FELs is that here, the problem of priorities is extremely complicated. In the author's opinion, the main cause of such situations is that the here-occurring essentially different partial priorities were not distinguished. Making clearer the system of different priorities, we obtain a possibility to harmonize the general historical situation in this field. Therefore, let us discuss shortly an example of systems of such partial priorities.

As is well known [1–8], the work of FELs, as the electronic devices, is based on a specific combination of two different physical mechanisms:

1. The mechanism of stimulated electron radiation in undulation-like electromagnetic fields
2. The Doppler-up transformation of frequency of electromagnetic radiation emitted by this electron

As a result, the priority for this or that variant of such combination appears as one of the FEL priorities. Besides that, researchers have proposed each or other FEL designs and their practical applications. At the same time, other researchers have accomplished first experiments, and so on. Hence, we can conclude that only one absolute general priority cannot be found in the field of FEL. Therefore, it looks more right to search for a system of partial priorities instead of one absolute. Basically, they could be

1. The priorities for each of the mentioned basic mechanisms (radiation and Doppler transformation)
2. The priority of the idea of combining a chosen radiation mechanism with the Doppler's effect
3. The priorities for first designs of FELs
4. The priorities for first experiments with FELs
5. The priorities for their first practical applications

Let us start our discussion with the partial priorities for basic mechanisms and first designs.

* From Kulish, V.V., *Hierarchic Methods. Undulative Electrodynamic Systems,* Kluwer Academic Publishers, Dordrecht/Boston/London, 2002. With permission.

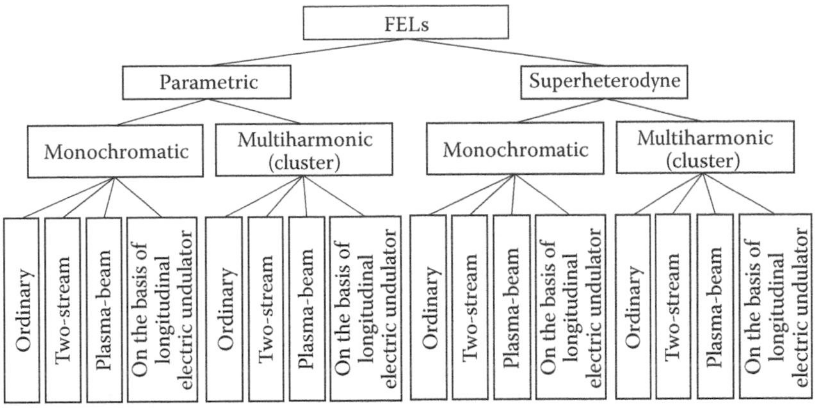

FIGURE 9.2
General classification of the FELs with respect to the basic physical mechanisms.

9.1.2 Basic FEL Physical Mechanisms: Kapitza–Dirac and Doppler Effects

Thus, as is mentioned above, the operation principles of FEL are based on the use of a specific combination of a radiation mechanism and the Doppler's effect. The first general idea for obtaining optical radiation in such a way belongs to Soviet Academician Vitalij Ginzburg (1947) (see reference [9]). However, he did not offer concrete working radiation mechanism for its experimental realization. Pantell et al. did it in 1968 [10]. Their design scheme is represented in Figure 9.4. Here, the bulk of relativistic electron beam 6 has been used as a quantum active laser medium. Electron beam 6 passes through superconducting resonator 3 along the optical axis of optical resonator 5. The powerful microwave electromagnetic wave of pumping ω_1 is excited in superconducting resonator 3. The generated (amplified) electromagnetic wave of signal ω_2 is excited in the work bulk of optical resonator 6 by an emission mechanism.

The essential relativism of the electron beam provides realization of the large-scale Doppler's effect for both the signal as well as pumping waves simultaneously. The Kapitza–Dirac effect (1933) [11] had been proposed firstly to be used as a main working radiation mechanism. The general idea of the Kapitza–Dirac effect [11] is as follows (see Figure 9.5). An initially immobile electron 3 oscillates under action of the straight 2 and reflected 1 light waves with the same frequency ω. The electron $-e$ absorbs photon of the straight wave and then emits it in the reflected wave. In the framework of the FEL terminology, straight wave 2 plays a role of the pumping wave and the reflected wave can be treated as the initial signal wave. Superposition of the reflected and emitted waves forms the generated (amplified) signal wave because both these waves are coherent and have the same polarization, that is, they are identical.

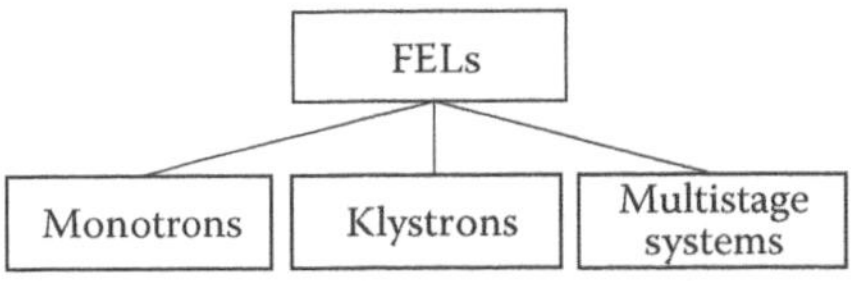

FIGURE 9.3
General classification of the FELs with respect to the basic design arrangements.

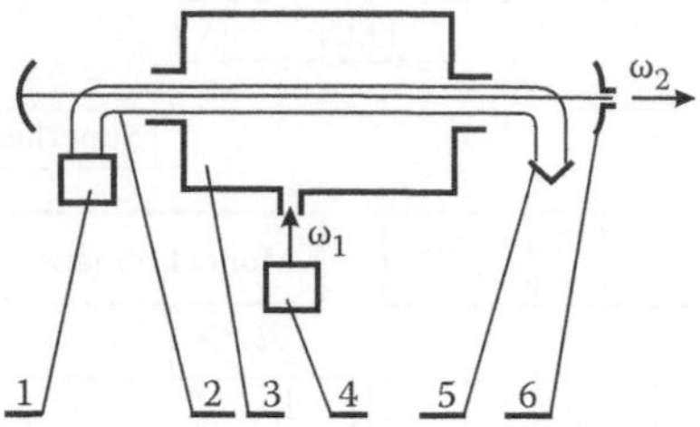

FIGURE 9.4
First scheme of free electron laser (Pantell's FEL). Here, 1 is the accelerator, 2 is the relativistic electron beam, 3 is the superconducting microwave resonator, 4 is the generator of microwave pumping, 5 is the optical resonator, 6 is the optical resonator for the generated signal wave, ω_1 is the pumping microwave signal, and ω_2 is the generated optical signal. (From Kulish, V.V., *Hierarchic Methods. Undulative Electrodynamic Systems*. Kluwer Academic Publishers, Dordrecht/Boston/London, 2002. With permission.)

Thus, in our case, the Kapitza–Dirac effect [11] can be regarded as one of the so-called mechanisms of light amplification by stimulation of emission of radiation. Peculiarity of this mechanism is that it is analogous to the well-known similar phenomenon, which is characteristic for atoms. In other words, an electron, moving in the pumping wave, forms a peculiar quantum energy system. Absorption or emission of photons accompany the electron transactions up or down between the system energy levels. Below we will discuss this physical picture in detail (Figure 9.10). But here let us confine ourselves to the statement that the electron behavior in the Kapitza–Dirac effect is similar to an electron in an atom. As the analysis shows, this mechanism has some very interesting fine physical peculiarities, which are connected with the Doppler's effect [24–28].

Essence of the Doppler's effect (Figure 9.6) is well known and has been described in numerous references [12,13]. Therefore, let us discuss shortly only some of its most general properties.

We consider that an electron (–*e*) (see item 2 in Figure 9.6) oscillates with the frequency ω. Simultaneously, it moves along some spatial axis with velocity $\vec{v}$ (see item 4 in the same place). Therein, the electron radiates other electromagnetic waves because it moves with acceleration. Let us choose a pair of radiated waves that are propagating in reciprocally opposite directions. For instance, they could be the waves collinearly propagated along the same axis that, in turn, is directed under the angle θ with respect to the *z*-axis (see Figure 9.6). According to the theory of the Doppler's effect, in general, the frequencies ω′ and ω″ of both radiated waves should be different [12,13]

$$\omega' = \omega \frac{\sqrt{1 - v^2/c^2}}{1 - (v/c)\cos\theta}; \quad \omega'' = \omega \frac{\sqrt{1 - v^2/c^2}}{1 + (v/c)\cos\theta}. \tag{9.1}$$

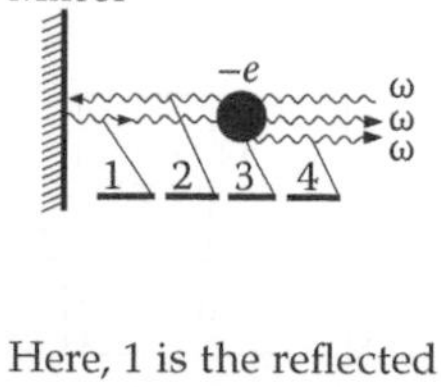

FIGURE 9.5
General scheme of the Kapitza–Dirac effect. Here, 1 is the reflected light wave, 2 is the straight light wave, 3 is the electron (–*e*), 4 is the radiated light wave, and ω is the frequency of the straight, reflected, and emitted light waves, respectively. (From Kulish, V.V., *Hierarchic Methods. Undulative Electrodynamic Systems*. Kluwer Academic Publishers, Dordrecht/Boston/London, 2002. With permission.)

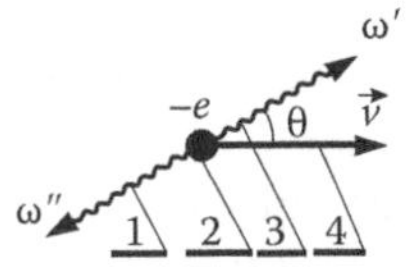

FIGURE 9.6
General scheme of the Doppler's effect. Here, 1 is the electromagnetic wave with frequency ω″, 2 is the electron (−*e*), 3 is the electromagnetic wave with frequency ω′, $\vec{v}$ is the vector of the electron velocity, and θ is the angle between directions of propagation of the waves 1,3 and the velocity vector $\vec{v}$. (From Kulish, V.V., *Hierarchic Methods. Undulative Electrodynamic Systems*. Kluwer Academic Publishers, Dordrecht/Boston/London, 2002. With permission.)

The following relation for the frequencies ω′ and ω″ can be derived from Equation 9.1:

$$\omega' = \omega'' \frac{1+(v/c)\cos\theta}{1-(v/c)\cos\theta}. \tag{9.2}$$

Analyzing Equation 9.2, we can obtain the following simplified formula in the particular case θ = 0 (i.e., in the case when both waves are propagated collinearly to the *z*-axis):

$$\omega' = \omega'' \frac{1+v/c}{1-v/c}. \tag{9.3}$$

This means that in the case of a moving electron ($v \neq 0$), we always have ω′ > ω″. This phenomenon usually is treated as the Doppler's effect.

Furthermore, let us accept that in the Pantell's FEL, the wave with frequency $\omega' = \omega_2$ is the signal wave (emitted by the electron) and the wave with frequency $\omega'' = \omega_1$ corresponds to the pumping wave (absorbed by the electron) (Figure 9.4). Besides that, we consider also that both these waves are propagated along the *z*-axis in reciprocally opposite directions (i.e., the angle θ in Figure 9.6 equals zero).

Then, it should be noted that the peculiar feature of all FELs is that the Doppler's effect is used here two times. The first is when we try to satisfy a relevant resonance condition, which is specific for the chosen basic mechanism of stimulated emission (see Chapters 3, 4, and 7). The second takes place when we transform the signal frequency $\omega' = \omega_2$ up and, simultaneously, the pumping frequency $\omega'' = \omega_1$ down.

Let us discuss this double role of the Doppler's effect beginning with the first of the mentioned mechanisms. The point is that a possibility of realization of any known variety of stimulated radiations [13] essentially depends on the presence of a frequency mismatch [13] between the electron oscillation frequency, on one hand, and the frequency of signal (amplified) wave, on the other hand. In other words, the Kapitza–Dirac effect can be realized in a case only when the frequencies of straight and reflected waves are each shifted with respect to one another (see Figure 9.5) in the proper electron reference frame. But, a nontrivial question arises: What is the physical mechanism that provides required frequency shift in the case of the Kapitza–Dirac effect under condition, when the electron is initially immobile and both frequencies are equal?

The matter is that these frequencies are equal in the chosen laboratory reference frame only that is obviously illustrated in Figure 9.5. Therefore, the right answer to the formulated question is the following: The required frequency shift really appears owing to common action of two physical effects in the proper electron reference frame. The first from these physical effects is the quantum recoil (Compton recoil effect in this case). The second is the above-mentioned Doppler's effect. The preliminary immobile electron gets some small longitudinal velocity $\vec{v}'$ just due to the Compton recoil effect after the absorption of the photon of the straight wave (see Figure 9.5). This leads to the fact that in the proper reference frame, the electron begins to see the reflected (see item 1 in Figure 9.5) as well as the straight (see item 2) waves as the waves with different (shifted) frequencies (see formula 9.3). Then the electron emits a photon in the reflected wave. Due to this, it obtains oppositely the directed Compton recoil. The immobile status quo is restored. Hence, the difference of frequencies in the proper reference system, caused by the Compton recoil effect, provides presence of the required frequency mismatch.

Thus, it is the first use of the Doppler's effect in the discussed working mechanism of the Pantell's FEL. The second use is directly connected with the utilization of relativistic electron beam in the Pantell's scheme (see item 2 in Figure 9.4). It should be noted that, generally, in any FEL, relativistic electron beams always play a role of a peculiar large-scale Doppler-transformer of the pumping and signal waves. Therein, the frequency of the pumping wave transforms to down, and the signal wave frequency transforms to up. We have also the same situation in the case of Pantell's FEL: its main FEL idea consists of the use of large-scale transforming of a low-frequency resonator pumping wave into the high-frequency signal wave. This obviously demonstrates formula 9.3. Indeed, in the relativistic case, $v \approx c$ (i.e., for $(1 - v/c) << 1$) $\omega_2 >> \omega_1$ (let us recall that $\omega' \equiv \omega_2$ and $\omega'' \equiv \omega_1$ in the considered case—see Figure 9.4).

Then, we note that, generally, apart from the FEL with microwave electromagnetic pumping (FEL-Dopplertron—see Pantell's FEL in Figure 9.4), a wide class of FELs with pumping by periodically reversed (including undulation) electric and magnetic fields could be constructed also. For instance, they are the FELs with pumping by transverse undulation magnetostatic field [14,15] (H-Ubitron FEL—see below in detail), FEL with pumping by oscillating on time electric and magnetic fields, or their combinations [16–20] (E-, H-, and EH-Dopplertrons), the FEL with pumping by transverse crossed magnetic and electric undulation fields (EH-Ubitron and E-Ubitron FEL) [21], and many others. However, in spite of this imaginary variety, the main operation principles of all these FELs are, basically, the same with the FEL-Dopplertrons, that is, the FEL with the pumping by an electromagnetic wave traveled toward the relativistic electron beam [22,23]. It is because of the fact that any electromagnetic field, which transforms in the proper electron reference frame into an oppositely directed (with respect to the signal wave) electromagnetic wave, could be used as a FEL pumping field. This affirmation has been referred to as the theorem about the FEL pumping [8]. Correspondingly, any electromagnetic field, which in a proper electron coordinate frame transforms into the directly propagated electromagnetic wave, can play the role of a FEL signal wave [8].

9.1.3 Is the Pantell's Device the First True FEL?

It should be mentioned that, in spite of the very clear physics and bright applied prospects, the Pantell's FEL (Figure 9.4) was not constructed until today. Correspondingly, the following question appears: Why did it happen?

The problem is hidden in some specific physical features of the above-mentioned Kapitza–Dirac effect. However, before discussing these features, let us discuss preliminarily the arrangement of the energy-level system in the ordinary Bohr's atom. It is well

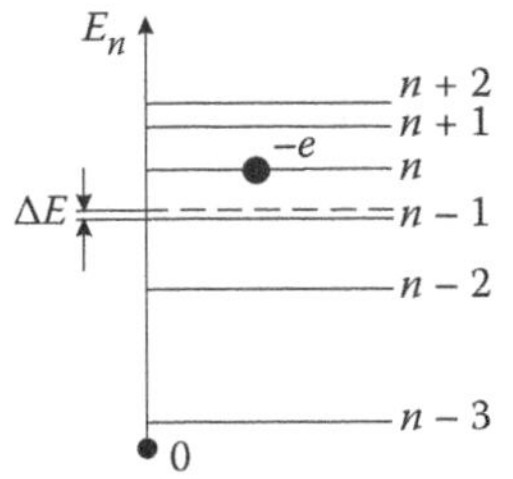

FIGURE 9.7
Scheme of the energy levels in the Bohr's atom. Here, E_n is the electron energy on the nth energy level, ΔE is the magnitude of the nonequidistance of energy levels, and n is the level number. (From Kulish, V.V., *Hierarchic Methods. Undulative Electrodynamic Systems.* Kluwer Academic Publishers, Dordrecht/Boston/London, 2002. With permission.)

known that the concept of equidistance of the energy level plays a key role in the work of usual quantum lasers.

The main idea of the concept of equidistance is illustrated in Figure 9.7. We can see that in the considered case, the distance $E_{n+1} - E_n$ between any pair of neighboring energy levels is different. This phenomenon was called the nonequidistance of energy levels. A possibility to work most of the quantum lasers is determined specifically by this phenomenon. Let us assume that an electron in an atom emits a photon or absorbs it in a stimulated way. This can happen under the action of another external photon (Figure 9.8). Because of the nonequidistance, the frequencies of radiation and absorption (that correspond to the stimulated transition down and up) are different. It is important to note that owing to this, the probabilities of these stimulated transitions turn out to be essentially different, too (Figure 9.8).

For illustration of this phenomenon, let us assume that an electron is located on an n + 1 energy level. Besides that, we accept that the energy of the external perturbing photon is equal to the energy difference $E_{n+1} - E_n$ (Figure 9.8). It is easy to see that, owing to the explicitly expressed nonequidistance of energy levels, this down energy difference $E_{n+1} - E_n$ is not equal to the upper energy difference $E_{n+2} - E_{n+1}$. As a result, the probability of the electron-stimulated radiation in this case remarkably prevails under the probability of stimulated absorption.

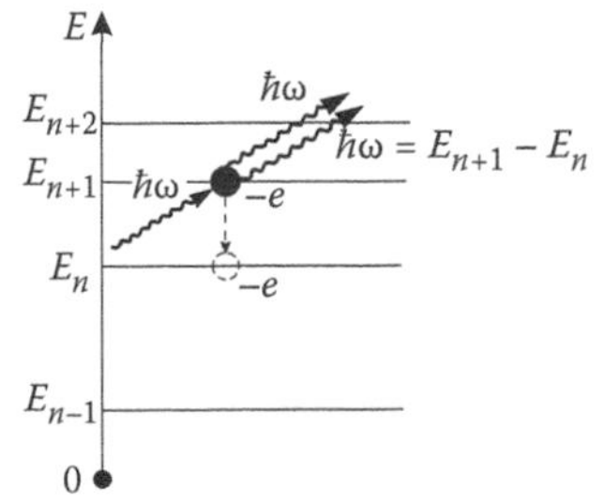

FIGURE 9.8
Stimulated radiation of an electron in a nonequidistant *stationary* quantum system. Here, $-e$ is the electron, E_j is the electron energy on the jth quantum level (where $j = n, n \pm 1, n \pm 2, \ldots$ are the numbers of energy levels), and $\hbar\omega = E_{n+1} - E_n$ is the energy of radiating photon, which is equal to the energy of the perturbing external photon $\hbar\omega$. (From Kulish, V.V., *Hierarchic Methods. Undulative Electrodynamic Systems.* Kluwer Academic Publishers, Dordrecht/Boston/London, 2002. With permission.)

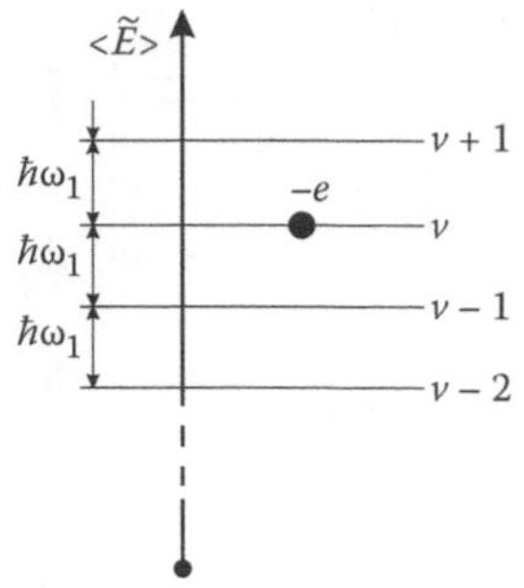

FIGURE 9.9
Quantization of the statistically averaged energy of electron in the field of pumping electromagnetic wave. Here, $\langle\tilde{E}\rangle$ is the statistically averaged electron energy (i.e., the averaged energy of some quantum ensemble of electrons in the field of an electromagnetic wave), ν is the number of the νth energy level, and $\hbar\omega_1'$ is the distance between two neighboring energy levels [at the same time, it is the energy of quantum of the electromagnetic wave (pumping)]. (From Kulish, V.V., *Hierarchic Methods. Undulative Electrodynamic Systems.* Kluwer Academic Publishers, Dordrecht/Boston/London, 2002. With permission.)

Let us turn again to the topic about the Pantell's FEL physics. It is known that analogous to the ordinary atoms, the averaged (statistically) energy of a quantum electron ensemble in the field of the electromagnetic (pumping, for instance) wave is also quantized [12,13] (Figure 9.9). However, as is readily seen in Figure 9.9, the system of averaged energy levels in the case of the electromagnetic wave, in contrast to the situation with the atom (Figure 9.8), is quite equidistant (Figure 9.9). Hence, it seems that in such a system, the stimulated radiation is impossible because of the equality of the probabilities of radiation and absorption. However, it is correct only in the case when relevant calculations are accomplished in the lowest (first) order of quantum theory of radiation. The point is that in accordance with basic physics of the Kapitza–Dirac effect, the signal photon, owing to the Compton recoil, slightly perturbs the structure of the quantum system electron in the field of the pumping wave (Figure 9.10). Therefore, a small nonequidistance appears in the higher orders of the quantum theory of radiation [13]. As a result, the probabilities of radiation and absorption, as well as their frequencies, again become different. But, at this time, contrary to the case of Bohr's atom (Figure 9.7), the equidistance is expressed rather slightly.

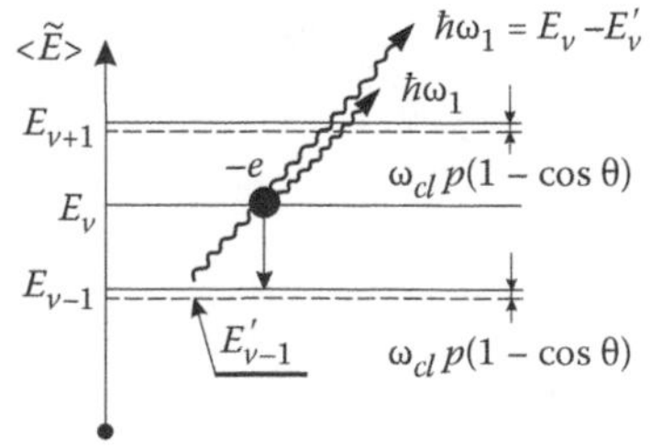

FIGURE 9.10
Stimulated radiation of an electron in the field electromagnetic pumping wave. Here, $\omega_{cl}p(1-\cos\theta)$ is the amplitude of nonequidistance, ω_{cl} is the radiation frequency obtained in the framework of classical theory of radiation, and p is some quantum parameter related with the Compton wavelength. Other definitions are given above. (From Kulish, V.V., *Hierarchic Methods. Undulative Electrodynamic Systems.* Kluwer Academic Publishers, Dordrecht/Boston/London, 2002. With permission.)

Thus, the appearance of the equidistance means that we again come to the situation signifying resemblance with the Bohr's atom (see Figure 9.8 and corresponding discussion). The main difference of both these models is that in the Pantell's FELs, magnitude of nonequidistance $\omega_{cl}p(1 - \cos\theta)$ (see Figure 9.10) turns out to be too small.

It should be noted that the last conclusion has very important practical sequences. It is clear that in any such experiment we should have a source of the microwave pumping wave with extremely narrow half-width of the spectral line. In this case, only the signal photon has a possibility to distinguish the frequencies of radiation or absorption. Practically, only the utilization of superconductive microwave resonators can provide the required extremely narrow half-width of the pumping spectral line.

As was mentioned earlier, just this version of the Kapitza–Dirac effect had been proposed by Pantell et al. [10] as a suitable physical basis for their FEL (Figure 9.4). Unfortunately, numerical estimation shows that, in view of the above-accomplished analysis, the direct practical application of the discussed idea is related with a number of essential technological difficulties. The main difficulty from them is connected with the above-mentioned requirement for half-width of the pumping spectral line. Let us repeat once more this key point because of its importance.

Thus, in the discussed Pantell's case, the difference of frequencies of radiation (signal) and absorption (pumping) is extremely small. Or, more precisely, in the electron reference frame, we obtained the practical difficulty to distinguish both of these frequencies. In accordance with the main idea of the design given in Figure 9.4, this distinguishing should be provided by the pumping microwave resonator 3. However, the problem is that the required value of such frequency difference cannot be achieved practically in the case of ordinary warms resonators. Only in the case, when special superconductive pumping microwave resonators (see item 3 in Figure 9.4) are used for this purpose, the basic Kapitza–Dirac effect of the light amplification by stimulated emission of radiation can be technologically realized. Therefore, just this design solution has been put by Pantell et al. [10] in the basis of their FEL version. And just owing to this, the proposed design is found so hard for practical realization. It is the main cause why the Pantell's FEL, which, actually, is historically the first system of the FEL type, is not constructed experimentally until today.

9.1.4 FEL Is a Classical Device

Let us mention that at times essentially other radiation mechanisms have been known also in the FEL theory [24,25]. The characteristic feature of these mechanisms is that they allow avoidance of the above-mentioned main technological difficulties of the Pantell's FEL. The key point of such FEL is that they, in spite of the use of quantum method of its description, in reality carry explicitly expressed nonquantum (i.e., classical) nature. Two such classical mechanisms of this type were discussed in the theory of FEL.

It is interesting to point out that, according to the improved theory of the Kapitza–Dirac effect, the classical version of the radiation mechanisms is also possible [24–30]. For the sake of illustration of this fact, we will study the FEL model with relative wide spectral lines of signal and pumping waves: the half-width of these lines are larger than the value of nonequidistance (see item 2 in Figure 9.11). For the sake of illustration of the key physical peculiarity of such model, we write the general expression for electron powers of stimulated radiation W^{+}_{stim} and absorption W^{-}_{stim}. Then we expend it in a series in powers of the normalized Plank constant $\hbar$:

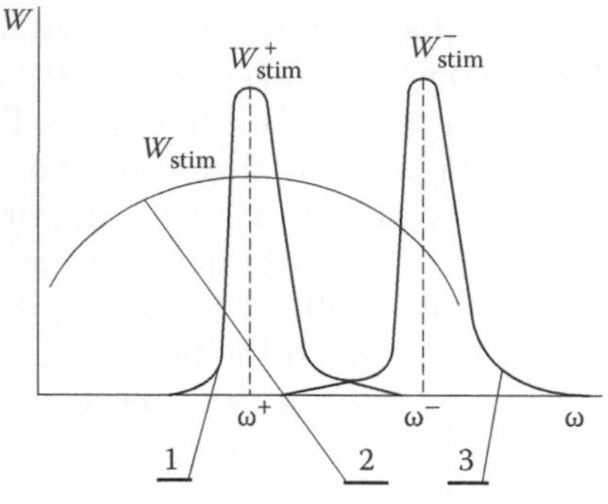

FIGURE 9.11
Quadratic quasiclassical mechanism of the collective stimulated radiation in FEL. Here, 1 is the radiation spectral line, 2 is the spectral line of the signal wave, 3 is the absorption spectral line, $W^{\pm}_{stim}$ is the power of the stimulated processes (the signs "+" and "–" correspond to radiation and absorption, respectively), W_{stim} is the power of collective radiation; in view of the above said, the meaning of ω^+ and ω^- are obvious. (From Kulish, V.V., *Hierarchic Methods. Undulative Electrodynamic Systems.* Kluwer Academic Publishers, Dordrecht/Boston/London, 2002. With permission.)

$$W^{\pm}_{stim} \cong \left(\frac{1}{\hbar} a_0 \pm a_1 + \hbar a_2 \pm \hbar^2 a_3 + \ldots\right), \tag{9.4}$$

where a_k (k = 0,1,2,...) are known coefficients. As the analysis shows, the first term in Equation 9.4 plays the main role in the traditional version of Kapitza–Dirac effect as a quantum phenomenon. It is obvious that here we have to do with a purely quantum effect because in this case the first term in Equation 9.4 is proportional to $1/\hbar$.

Then we remember the above-accepted supposition: the considered resonant system does not distinguish radiation and absorption frequencies. This means that both processes of stimulated radiation and absorption occur there simultaneously. The resulting (collective) radiation power W_{stim} in this case can be represented as

$$W_{stim} \cong W^+_{stim} - W^-_{stim} \cong 2a_1 + 2\hbar^2 a_3 + 0(\hbar^4), \tag{9.5}$$

that is, we see that the first terms in Equation 9.4 are reciprocally eliminated in sum (Equation 9.5). In this situation, the term $2a_1$ which is not proportional to the Plank constant $\hbar$ plays the main role. This means that the discussed improved version of the Kapitza–Dirac effect is not quantum. In this case, the resulting power W_{stim} attains maximum if the central frequency of such spectrally wide signal coincides with the frequency ω^+ (Figure 9.11).

Thus, we have a very unusual situation, that is, superimposition of two essentially quantum effects (stimulated radiation and absorption, respectively—see Equation 9.4) gives a classical radiation effect (Equation 9.5). It is explained mathematically by the fact that the term $2a_1$ becomes the main in the series in Equation 9.5 due to the summing procedure. It is important that this term does not depend on Plank constant $\hbar$. Just this circumstance we treat as an evidence of the described effect indeed carries the classical (quasiclassical) nature. Then, as analysis shows [24,25], the coefficient a_1 in Equation 9.5 is proportional to square of the signal wave amplitude. Therefore, this mechanism is referred to as the so-called quadratic quasiclassical mechanism of stimulated radiation.

However, it should be noted that the discussed mechanism of the stimulated radiation is not the basic working mechanism of real FEL because it is characterized by too low effectiveness. It was found that, apart from the quadratic radiation mechanism, some other more effective quasiclassical mechanisms are known in the FEL theory [25,32]. Here, we have in view the so-called linear (quasilinear) mechanism of stimulated radiation [25,32]. In this case, the power of electron radiation is linearly proportional to the signal wave amplitude.

It should be recognized that its main physical idea until today looks too unwanted for the traditional quantum laser experts. Unfortunately, the number of similar out-of-date professional stereotypes exists in the modern quantum electronics until today. Therefore, it does not seem strange that the real authors of the first FEL design [10] have missed this so promising historical opportunity.

9.1.5 Linear Quasiclassical Mechanism of Amplification in FEL

To understand their psychological difficulties, let us consider a solid-state quantum laser (e.g., the first ruby laser). Here, only temporal energy distribution in a spatially quasihomogeneous nonmodulated medium is the main object of study traditionally. Indeed, it is difficult to imagine that atoms in such a solid medium can freely move under the action of a stimulated radiation and form by this way some periodically modulated spatial structures. This means automatically that any small volume of the medium is not different with respect to other analogous small volumes: all these volumes are practically equivalent.

Analyzing the design scheme of the Pantell's laser (Figure 9.4), one can be easily convinced that its authors were captives of this obsolete paradigm. The ideology, which is illustrated in Figures 9.10 and 9.11, corresponds to such type of professional psychology. The revolution in the creation of a new FEL ideology has begun with radically destroying the described traditional quantum laser paradigm. The main novelty of the general situation in modern FEL is that the active laser medium, which is represented in this case by a relativistic electron beam, is spatially modulated in the periodical way [25,32] (see illustration in Figure 9.13). This creates a new principle of physical situation, which is characteristic of the modern FEL theory.

At first, we discuss the model where a nonmodulated electron beam moves in a field of oppositely directed electromagnetic pumping wave $\vec{A}_1(\xi)$ (see Figure 9.12a). Each electron in this model oscillates in the transverse plane under the action of the pumping field. Because of this, all beam electrons move with acceleration and, consequently, they become sources of the signal electromagnetic radiation [32]. It is important to note that initial phases of the electron radiations are determined by the initial phases of the pumping wave in electron spatial points (see Figure 9.12a).

In what follows, let us recall that the radiation of any electron occurs with relevant Doppler-shifted frequency (see expression 9.1 and Figure 9.6). Then, we choose randomly some control points *A*, *B*, *C*, . . . in the electron beam bulk. These points have a random character and they are also characterized by the random nature of the observed initial phases of oscillation.

We register radiation from the beam points *A*, *B*, *C*, . . . in the observation point *M*. The total electromagnetic field of radiation is formed there as a result of superimposition of radiation fields of all electrons. It is important to emphasize that, in general, this total field is noncoherent because the initial phases of oscillations of radiation fields of different electrons have random character.

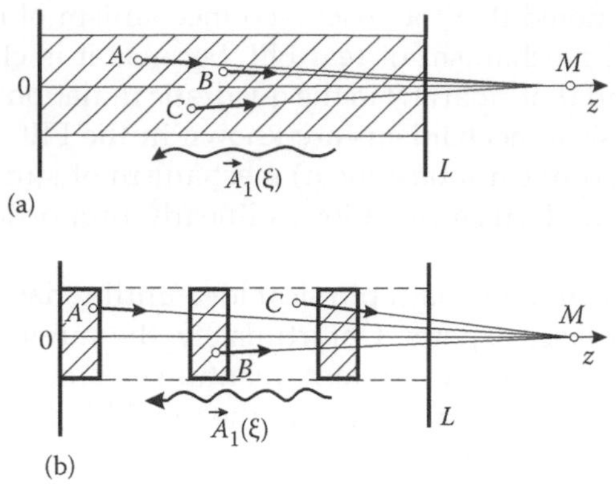

FIGURE 9.12
Linear quasiclassical mechanism of collective stimulated radiation in FEL. Here, *A*,*B*,*C* are the chosen radiating points within electron beam, *M* is an observation point, $\vec{A}_1(\xi)$ is the vector-potential of the opposite (pumping) electromagnetic wave field, $\xi = \omega_1(t + z/v_{1ph})$, v_{1ph} is its phase velocity, ω_1 is the cyclic frequency of the pumping wave, *z* is the longitudinal coordinate, and *L* is the length of the system. The upper drawing (a) illustrates noncoherent radiation of the nonmodulated electron beam; the lower drawing (b) corresponds to a case of coherent radiation of the modulated electron beam. (From Kulish, V.V., *Hierarchic Methods. Undulative Electrodynamic Systems*. Kluwer Academic Publishers, Dordrecht/Boston/London, 2002. With permission.)

Thus, the superimposition of different electron radiation fields in the considered homogeneous model (Figure 9.12a) gives the noncoherent total signal wave field in the observation point *M*.

Then, we consider the spatially modulated electron beam model shown in Figure 9.12b (for an example of the FEL modulation mechanism, see Section 4.3). Similarly to the abovementioned nonmodulated case (Figure 9.12a), we again have the superposition of radiation fields of different electrons in the observation point *M*. However, this time, the result of this superimposition is essentially different [25,31]. Let us discuss it in detail.

As is readily seen in Figure 9.12b, the electrons of type *A* or *B* only (i.e., the electrons that are members of relatively narrow neighboring electron bunches) take part in the formation of a total radiation field in point *M*. Let us recall once more that the electron initial phases in the same spatial point are determined by the initial phase of the pumping wave [32]. Therefore, the initial phases of all electron oscillations in any separate narrow bunch are approximately the same (see Figure 9.12b). As a result of this, each such electron bunch radiates coherently as a single large particle. At the same time, we do not have any radiation from the points that are analogous to the points like *C* in Figure 9.12b [25,31].

Or, generalizing, we can say that we do not have radiation from the points within the spatial intervals between any two electron bunches. Also, we can state that the total observed radiation in the point *M* is emitted by bunches and each bunch emits coherently.

In the case when the beam modulation is periodical, the difference between the initial phases of radiation of any two bunches is the same. In such a situation, we can choose the modulation period in such a way that in the observation point *M*, these phase differences are equal to $2\pi n$ (where $n = 0,\pm1,\pm2,\ldots$). Relevant mathematical expression for the chosen modulation period, written through other system parameters, has the physical meaning of FEL resonant condition.

We also can regard the discussed resonant process as a peculiar physical mechanism of coherentization of the initially noncoherent beam radiation. It should be emphasized especially that, as shown above, it can occur only due to the periodic modulation of the radiated electron beam. Because of the additive physical nature of this collective radiation mechanism, it is linear with respect to the signal wave amplitude.

However, it is only in the framework of the above-chosen terminology where we are interested mainly in the investigation of a reaction of the considered system to the signal wave action. However, let us repeat that this considered system, apart from the discussed radiation mechanism, in reality contains also the self-consistent beam modulation processes. As shown in Chapter 4, the radiated signal wave plays a very important role. As a result, in the framework of another possible terminology, which accounts for this circumstance, the complete interaction FEL mechanism can be classified as nonlinear. It looks as an inner terminological contradiction. But, actually, we have not any contradiction in this case. Let us illustrate this statement in detail.

The point is that we have also in the considered model the pumping field, which, together with the signal wave, is responsible for the mentioned modulation process. It is important to recall that this pumping field is characterized by the pumping wave amplitude. We may understand all this in the following way: in general, together with the signal wave, the system reaction to the pumping wave action is also investigated. As a result, we can say that the complete (doubled) system reaction depends on the product of both these amplitudes (Chapter 4). Hence, the complete interaction mechanism indeed can be treated as quadratic-nonlinear with respect to common action of both amplitudes: the signal and pumping wave amplitudes.

Below we will use both terminologies. However, as will be shown below, in reality, it does not lead to any confusion.

In the traditional vacuum electronics, the process of longitudinal electron-beam modulation is regarded as an excitation of the longitudinal electronic wave (the SCW). In the discussed model, this means that we have three interacting waves: the two transverse electromagnetic (signal and pumping) waves and one longitudinal SCW. Therefore, the above-mentioned physical mechanism of collective coherent radiation in FEL can be also treated as the effect of three-wave parametrical resonance (Chapter 4).

Thus, the main working mechanism of the modern FEL has obviously classical nature, in spite of the fact that it, basically, can be described within the framework of the quantum theory, too. A specific feature of this mechanism is that it can be realized in the case only when the relativistic electron beam is modulated. Therein, a characteristic feature of the above-mentioned mechanism is that, in reality, the beam modulation process has a peculiar self-consistent character. The required modulation appears in the beam in the quadratic approximation of the theory as result of its quadratic-nonlinear interaction with the superimposition of the pumping and signal waves. The physical details of such modulation and radiation mechanisms were discussed earlier in Chapter 4 on a semiqualitative level. Relevant qualitative theories are discussed below in the following chapters.

9.1.6 First Designs of the Free Electron Lasers

Madey in 1973 (Figure 9.13) [15] and Silin, Kulish et al. in 1971/1972 (Figure 9.14) [22] have proposed the first design schemes of FEL based on the above-mentioned linear (quasilinear) classical radiation mechanism. Just these designs were realized in the first experiments with FELs [23,33].

The pumping system in the Madey's scheme (Figure 9.13) [15] is performed in the form of a periodically reversed transverse magnetic system (H-Ubitron pumping). It is interesting to note that the weak-relativistic (hundreds of kilovolts) version of the device shown in Figure 9.13 was well known in the traditional microwave electronics long before the

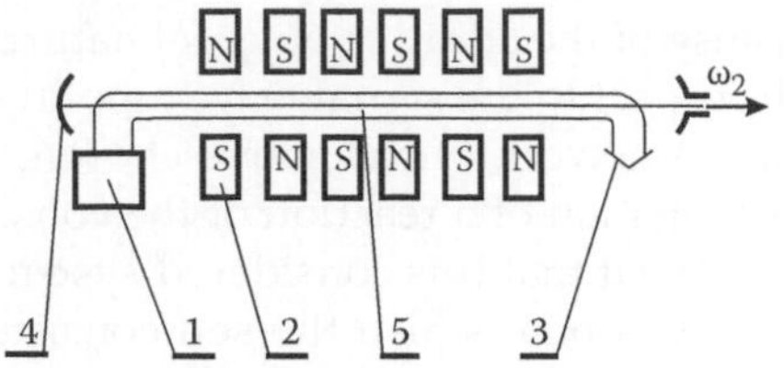

FIGURE 9.13
The first design scheme of FEL of the H-Ubitron type [15]; the scheme is realized in the first experiments with H-Ubitron FEL performed in 1976 [33,34] by Elias's group. Here, 1 is the electron accelerator, 2 is the magneto-undulation pumping system (H-Ubitron pumping), 3 is the electron collector, 4 is the optical resonator, 5 is the relativistic electron beam, and ω_2 is the generated optical signal. (From Madey, J.M.J., et al., *IEEE Trans. Nucl. Sc.*, 20, 980–983, 1973. With permission. From Deacon D.A.G., et al., *Phys. Rev. Lett.*, 38, 892, 1976. With permission. From Elias, L.R., et al., *Phys. Rev. Lett.*, 36, 717–720, 1976. With permission.)

appearance of the works [15,33,34]. An analogous device has been created by Phillips in 1960 and it was called the Ubitron [35,36]. Its basic difference from the Madey's scheme is the design of the source of the electron beam. In the case of Phillips's Ubitron, it was a weak-relativistic electron gun, while an electron accelerator (see item 1 in Figure 9.13) plays the same role in the Madey's FEL. Owing to this, the old microwave Ubitron [35,36] obtained a possibility to work in the optical range.

The first FEL design scheme proposed by Silin, Kulish et al. [22] contains the Dopplertron type pumping (Figure 9.14). Or, more concretely, the pumping system here was performed in the form of a smooth waveguide with attached powerful microwave oscillator. In contrast to the H-Ubitron FEL, the proposed design had not any precursors in the microwave electronics. It is because in the case of nonrelativistic version, the proposed design [22] cannot have any applied significance: the signal and pumping frequencies become too close to one another. It should be noted, however, that in further similar designs [39–43], the possibility of the use of even weak-relativistic electron beams has been legalized. In this regard, we can mention the

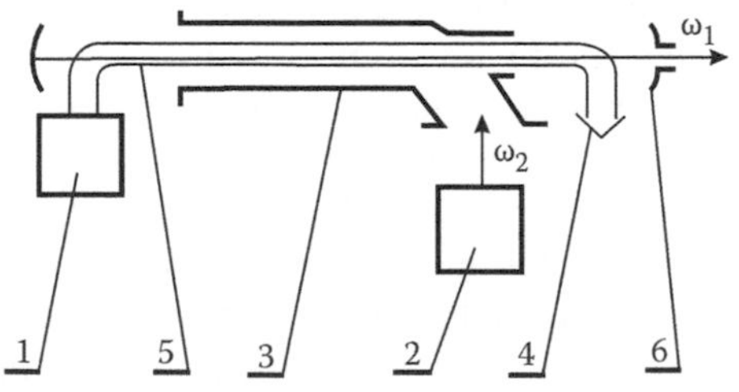

FIGURE 9.14
The first design scheme of FEL of the Dopplertron type [22]; the scheme has been realized in the first experiments with FEL by Granatstein's group in 1976 [23,37,38]. Here, 1 is the accelerator, 2 is the power microwave pumping generator, 3 is the smooth wave-glide, 4 is the electron collector, 5 is the relativistic electron beam, 6 is the optical system (resonator or lens-type), ω_1 is the microwave pumping electromagnetic signal, and ω_2 is the amplified electromagnetic signal. (From Silin, R.A., Kulish, V.V., et al., Patent of USSR SU N705914. Electronic device. Priority of 18.05.72. Published in non-secret press of USSR after taking off of the relevant stamp of secrecy: 15.05.91, Inventions Bulletin 26, 1991. From Davis, G.R., *Microwave*, 12, 12–17, 1976. With permission. From Granatstein, V.L., Mechanisms for Coherent Scattering of Electromagnetic Wave from Relativistic Electron Beams. *2nd Int. Conf. and Winter Sch. 'Sub-millimeter waves and their application'.* San Juan, Puerto Rico; New York, 87–89, 1976. With permission. From Granatstein, V.L., and Sprangle, P., *IEEE Trans. Microwave Theory Techn.*, 25, 6, 545–550, 1977. With permission.)

so-called Dopplertron FEL with the retarded electromagnetic pumping, whose physics we will widely discuss below.

As is mentioned above, one of the general key FEL features is that in the proper reference frame, the beam electrons see the H-Ubitron magnetic pumping as an oppositely directed electromagnetic pumping wave [6,8]. This means that, from the physical point of view, the Dopplertron [22,23,37,38] and H-Ubitron FELs [15,33,34] are basically the same devices. All observed differences actually have only the technological character.

The above-formulated affirmation is known as the theorem about the pumping fields in FELs [6,8]. Let us illustrate it performing the Lorentz' transformations for the pumping and signal fields in the design shown in Figure 9.13

$$E_x = \frac{E'_x + \beta_0 H'_y}{\sqrt{1-\beta_0^2}};$$

$$E_y = \frac{E'_y - \beta_0 H'_x}{\sqrt{1-\beta_0^2}};$$

$$E_z = E'_z;$$

$$H_x = \frac{H'_x - \beta_0 E'_y}{\sqrt{1-\beta_0^2}};$$

$$H_y = \frac{H'_y + \beta_0 E'_x}{\sqrt{1-\beta_0^2}};\ H_z = H'_z, \tag{9.6}$$

where $E_{x,y,z}$ and $H_{x,y,z}$ are the components of strength vectors of electric and magnetic fields, $E'_{x,y,z}$ and $H'_{x,y,z}$ are analogous quantities in the moving (proper) reference frame, and $\beta_0 = v_0/c$, v_0 is the velocity of the motion of the proper reference frame and, at the same time, the electron beam. Indeed, accepting the harmonic law for the H-Ubitron pumping

$$\vec{H}_{c2} = \frac{1}{2}\left(\vec{H}_2 e^{-ik_2 z} + c.c.\right);\ \vec{E}_{c2} = 0, \tag{9.7}$$

after relevant elementary transformations for the electric component of the pumping field in the proper reference frame, we obtain

$$\vec{E}'_{c2} = \vec{e}_1 \frac{1}{2}\left(-\beta_0 H'_{2y}\right) e^{i(\omega'_2 t + k'_2 z)} + \vec{e}_2 \frac{1}{2}\left(\beta_0 H'_{2x}\right) e^{i(\omega'_2 t + k'_2 z)} + c.c., \tag{9.8}$$

where $\vec{e}_{1,2}$ are the unit vectors along the x- and y-axes. Hence, H-Ubitron field (Equation 9.7) electrons that are evidently expressed in the laboratory reference frame indeed see (in the moving reference frame) as an oppositely directed nonproper electromagnetic wave with the new cyclic frequency and wave vector

$$\omega_2' = \frac{k_2 v_0}{\sqrt{1-\beta_0^2}};$$

$$k_2' = \frac{2\pi}{\Lambda'} = \frac{k_2}{\sqrt{1-\beta_0^2}}. \tag{9.9}$$

Thus, we became convinced that the difference between the schemes shown in Figures 9.13 and 9.14 is not related with the basic working physical mechanisms. As is noted already, this difference concerns mainly the FEL design arrangement. Hence, the theoretical model with pumping in the form of an electromagnetic wave is more general because it includes the H-Ubitron pumping fields as a particular case (see Equations 9.7 and 9.9). This circumstance will be widely used below.

As was mentioned already, the H-Ubitron FEL [15] has been realized in the first experiments by Elias's group [33,34]. Until today, this experiment is considered as an academic realization of the main FEL idea because it engaged in optical (infrared) range, which is traditional for the quantum lasers.

The Dopplertron design scheme of FEL (Figure 9.14) [22] was realized practically simultaneously by other research groups (Granatstein, Sprange, and others) in other series of the first experiments with FEL [23,37,38]. There, a superpowerful microwave pumping wave ω_1 propagates oppositely within the working bulk of smooth wave-guide 3. The main peculiarity of these experiments is that they were accomplished within the submillimeter range [23,37,38]. In these experiments, the record-breaking power results were achieved for the submillimeter range: 1 MWt for the 0.4-mm signal wavelength.

Furthermore, a number of new designs of FEL were proposed, for example, the FEL with retarded electromagnetic pumping wave [39–43], superheterodyne [43–47] and multistage [48–50] FEL, FEL with the laser [51] and EH-Ubitron pumping [8,52], and some other designs [53–59]. All these types will be illustrated below by a series of relevant FEL design variants.

It should be mentioned that, besides the enumerated types of pumping, a number of other varieties have also been proposed [8]. They are, for instance, the pumping by the oscillating on time magnetic, electric, and crossed electromagnetic fields (H-, E-, and EH-Dopplertrons, respectively), and so on. However, their practical value is not clear until today. Therefore, we will omit their discussion below in this book.

9.2 Two-Stream Superheterodyne Free Electron Lasers: History and Typical Design Schemes*

9.2.1 History of the Two-Stream Problem

As shown in Chapter 4, the effect of superheterodyne amplification takes place in the case only when some additional longitudinal mechanism for the electron-beam wave

* From Kulish, V.V., *Hierarchic Methods. Undulative Electrodynamic Systems,* Kluwer Academic Publishers, Dordrecht/Boston/London, 2002. With permission.

amplification is introduced in the FEL work bulk [3]. On the other hand, as is well known, traditional difficulty of the classical microwave vacuum electronics [36] is the search of methods to suppress various types of inherent (proper) beam instabilities [61–65]. All these instabilities, in principle, could be used as such addition mechanisms. Thus, the paradox point here is that in the traditional vacuum electronics, such inherent (proper) beam instabilities are suppressed for the sake of excitation of other instabilities in the same beam, which are considered as useful (the Cherenkov, transient, gyro-resonance, parametric, etc.). A natural question arises: Can inherent instabilities of the beam be also directly used to amplify and generate electromagnetic waves?

It should be noted that researchers have faced this question at the early stage of microwave electronics and have even found reasonable answers, in particular, the electronic devices on the basis of plasma-beam [66] and two-stream instabilities [67] (it should be mentioned that the two-velocity electron beams really could be treated as the simplest model of the electron beam with a transversally inhomogeneous distribution of the beam density [61–64]). For example, the first two-stream electronic device (the two-stream electron-wave tube) was designed in 1948 (in the former USSR, it was called the Haeff tube [36,67–70]). The scheme of this device is shown in Figure 9.15.

Nonrelativistic two-velocity electron beams 1, 2 (with partial velocities v_1 and v_2, respectively) move through two segments of retarding systems (3 and 7) that are spatially separated by transit space 5. The first retarding system 3 serves as an input of the device. Its function is to transform the input electromagnetic signal ω_3 into the longitudinal electrostatic electron-beam wave (SCW). Then the transformed electron-wave form signal is amplified in transit 5 space due to the developing two-stream instability. Such devices (the electron-wave tubes) were a subject of intensive theoretical and experimental studies in the 1950s [67]. However, during the next six decades, they attracted far less interest and were eventually forgotten for two main reasons.

The first reason is associated with the working physical mechanism of two-stream instability. Maximal amplification of the signal could be attained in the case when the signal frequency satisfied the following condition:

$$\omega_3 = \omega_{3\text{opt}} \cong \frac{\sqrt{3}\omega_p}{\delta} = \frac{\sqrt{3}\omega_p}{\Delta v} v_0, \tag{9.10}$$

where $\delta = \Delta v / v_0$, $\Delta v = v_1 - v_2$, $v_0 = (v_1 + v_2)/2$, $v_{1,2}$ are the partial electron beam velocities, and $\omega_{3\text{opt}} = \omega_{\text{opt}}$ is the so-called optimal frequency for the two-stream instability. Putting, say, $\delta \sim 10^{-3}$–10^{-2}, $\omega_p \sim (10^8$–$10^9)\text{s}^{-1}$, we see that such simple devices can (at the first sight) work

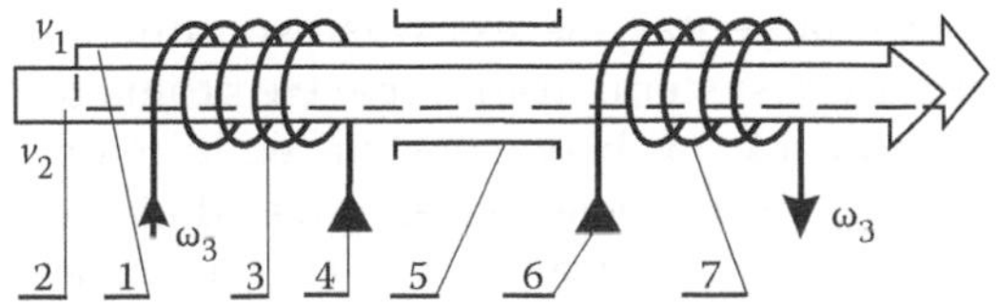

FIGURE 9.15
Design scheme of the electron-wave tube (Haeff's tube). Here, 1 and 2 are partial nonrelativistic electron beams; 3 is the first retarding system; 4 is the first microwave power absorber; 5 is the drift space; 6 are second microwave power absorbers; 7 is the second retarding system; and ω_3 is the amplified signal frequency. (From Kulish, V.V., *Hierarchic Methods. Undulative Electrodynamic Systems*. Kluwer Academic Publishers, Dordrecht/Boston/London, 2002. With permission.)

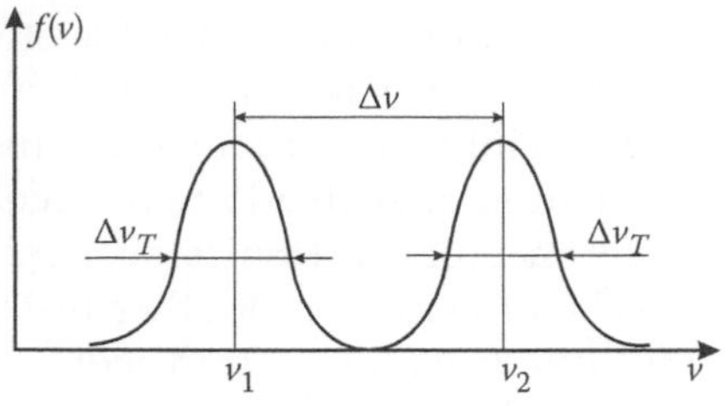

FIGURE 9.16
The velocity distribution function $f(v)$ of a two-velocity beam. Here, Δv is the velocity difference and Δv_T is the thermal velocity spread of the partial beams. (From Kulish, V.V., *Hierarchic Methods. Undulative Electrodynamic Systems*. Kluwer Academic Publishers, Dordrecht/Boston/London, 2002. With permission.)

in the millimeter and even in the shorter wavelength ranges. Unfortunately, researchers in practice did not manage to go beyond the centimeter range. The main difficulty was that desired numerical values of the parameter δ could not be obtained for standard kilovolt beams used at that time.

The essence of this technological difficulty is illustrated in Figure 9.16. It shows the electron velocity distribution function of the two-velocity beam. We see that velocity difference Δv cannot be smaller than beam thermal spread Δv_T (otherwise, the beam really becomes one-velocity). This requirement, along with Equation 9.10, results in the above restriction for frequency ω_3. Moreover, even when this difficulty is avoided (e.g., employing special cooled cathodes), prospects for creating the electron-wave devices of this type for the millimeter and infrared ranges would be doubtful for other technological reasons. Because fragments of the retarding systems (as in Figure 9.19) or the klystron resonators were used there as input elements, the electron-wave devices discussed automatically inherited all main shortcomings of competing microwave devices (traveling wave tube, klystron, etc.). Among these shortcomings are the restrictions for the frequency range associated with the design of the input elements. As a result, the considered electron-wave devices turned out to be noncompetitive. With time, they became subjects of historical reviews only in manuals on the physical electronics [36,68–70], like the Barkhausen–Kurtz generator, the Arsent'eva–Hail tube, Adler's lamp, and other historical devices [68].

It should be stressed, however, that the status quo in the vacuum electronics considerably changed during the past four decades. First, the technology of relativistic electron beams (REBs) has been developed. Second, new experimental methods for transforming the millimeter- and visible-range electromagnetic waves into the electron-beam wave form are proposed. These methods are based on involving the three-wave parametric resonance (Chapters 4, 11, and 12). Both innovations historically made the basis for parametrical (ordinary) FEL. Fundamental success was experimental realization of the first FEL. By this, the transformation systems transverse electromagnetic waves → longitudinal electron-beam waves and back has been constructed in reality. The strongest impression from the practical realization of these systems was that they operated in the whole frequency range from millimeter to optical waves. This opens a lot of new prospects for revision of many old ideas, which, at this time, could be fulfilled on the basis of the newly appeared technological possibilities. Unfortunately, in spite of such promising prospects, most of the FEL researchers do not see these prospects clearly. Until today, as was noted above, most experimental works of such kind are devoted to just one type of the FEL—the ordinary systems with H-Ubitron pumping. Herewith, it is well known that most of such FELs until now are characterized by relatively low amplification per unit of length; they

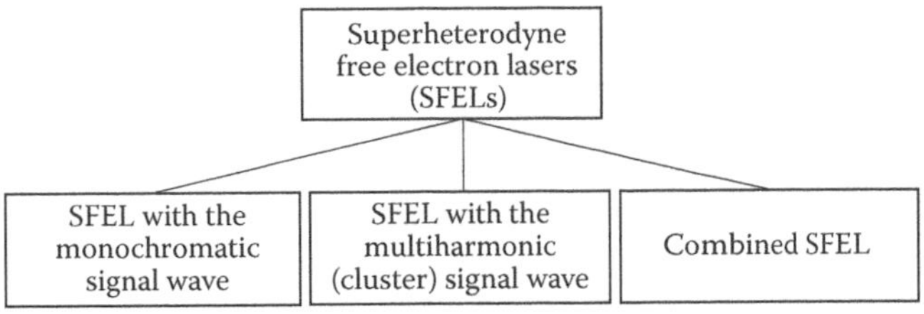

FIGURE 9.17
General classification of SFEL with respect to the signal spectrum.

are too expensive, cumbersome, and heavy. Assistance to overcoming this professional thinking and professional conservatism is also one of the main purposes of Part II.

9.2.2 Types of the SFEL

As is noted above, a lot of various types of the inherent beam instabilities are known in electrodynamics [61–70]. Theoretically, most of them can be used as the above-mentioned additional mechanisms of amplification in SFELs.

However, in reality, only a few such mechanisms can be successful in this role. In spite of this, the real diversity of possible SFEL designs, which can be constructed on their basis, is rich enough. Apart from that, a number of design combinations of such design solutions with the other traditional FEL elements can also be rather large. This even more increases this diversity. So, in the described methodological situation, the necessity of introduction of some SFEL classifications becomes obvious. However, elaborating such classifications, we should keep in mind that it depends essentially on the choice of the key classification criteria. Let us choose two such criteria: the character of the signal wave spectrum (see Figure 9.17) and the physical mechanism of the addition amplification (Figure 9.18).

As shown in Figure 9.17, all SFELs, with respect to the signal spectrum, could be classified into the monochromatic, multiharmonic (cluster), and combined devices.

As is noted in Chapter 1, the main subject of attention in this book is the FEL for the cluster systems (CS). From such a point of view, the optimal utilization of the monochromatic SFEL is their use in the CS with external blocks of harmonic generation (see Figures 1.15

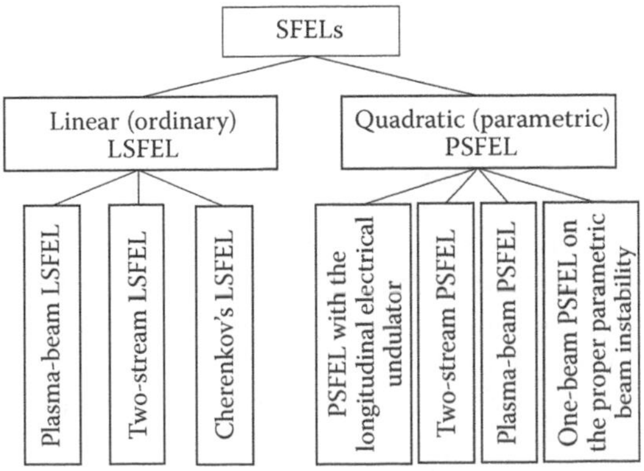

FIGURE 9.18
General classification of SFEL with respect to the basis physical mechanisms.

and 1.16) as harmonic partial amplifiers. In accordance with the key idea of this CS type (Chapter 1), each such external block is accomplished as a composed cluster oscillator with an external multiharmonic excitation. The oscillator, in turn, consists of the many harmonic partial amplifiers and the so-called master oscillator, which is common for all these amplifiers (see relevant explanations for Figures 1.15 and 1.16). A specific design feature of these partial amplifiers is that each of them is tuned in to its personal signal harmonic. Below in the book, the different types of such type of monochromatic SFEL amplifiers are described and analyzed in detail.

In the case of the CS on the basis of SFEL with the internal generation of the signal wave harmonics (Figure 1.17 in Chapter 1), the cluster amplifier is accomplished in the form of a multiharmonic SFEL. In the case of the construction of the combined systems, the monochromatic as well as multiharmonic SFEL amplifiers are used for forming different parts of the signal spectrum.

Example of the SFEL classification on the used physical mechanisms is illustrated in Figure 9.18. As is readily seen, all possible designs of the proposed SFEL can be classified into two classes only: the linear (quasilinear) and quadratic (parametric) SFEL, respectively. In turn, each of these classes contains a number of more specific SFEL types. Let us shortly discuss the most general characteristics and the history of these SFEL types.

It should be noted that two types of the SFEL addition physical mechanisms are known today: the linear (quasilinear) and quadratic (parametric) correspondingly. The mechanisms of the first type (linear) are characterized by the appearance of the effect of the wave amplification in the first (linear) order in relevant SFEL theory. Besides that, other types of mechanisms (quadratic, parametric) are also known in theory. In contrast to the linear mechanisms, there the amplification appears only in the second (quadratic) order. In this regard, as is noted already, all SFELs, with respect to the discussed properties of the addition physical mechanisms, are classified into the linear (LSFEL) and parametric (PSFEL) SFEL (Figure 9.18).

9.2.3 Short History of the SFEL

Let us begin the discussion of the classification in Figure 9.18 with a short review of the history of the SFEL.

The first SFEL, generally, and the LSFEL [44], particularly, have been proposed to be constructed on the basis of the plasma-beam instability [43] (plasma-beam superheterodyne FELs—PBSFEL; Chapter 14) [53–55,57]. The second type of SFEL is the two-stream SFEL (TSFEL; Chapter 13) [72].

First, the idea to use the two-stream instability in FEL (for increasing their gain factor) has been proposed by Bekefi and Jacobs [71]. However, it was not yet the true TSFEL because the two-stream instability there was used as an electron-beam prebuncher only for a next traditional parametrical FEL section. In other words, the main amplification mechanism in the FEL section there has actually the ordinary parametric-wave nature. This means that it prevailed there under the two-stream amplification mechanism. Whereas, as shown in Chapter 4, in the case of the superheterodyne amplification, we have a completely opposite situation: the two-stream mechanism should prevail under the parametric mechanism. Hence, strictly speaking, the work [71] cannot be considered as a first work on the SFEL because it is sconcerned to essentially another (nonsuperheterodyne) physical idea.

The first physical idea of the true TSFEL, as the superheterodyne device, was proposed and analyzed in reference [72]. In what follows, it has been developed further in many works (see references [58,59,72,74–91] and others), including the first TSFEL design-schemes that were described in patents [47]. The first concept and design scheme of the TSFEL-klystrons have been proposed in the patent [47] and reference [75]. The first nonlinear TSFEL theory has been constructed in references [78,90]. The first cluster TSFELs were proposed in references [85–89] and so on.

It should be mentioned that, from the physical point of view, the two-stream systems could be treated as a more general case of the plasma-beam systems. Indeed, transferring into the moving reference frame, we obtain the transformation of the initially immobile (as a whole) electron plasmas into a moving electron beam. Hence, in this case, the initial plasma-beam system transforms into the two-beam system. However, in spite of this physical similarity, technologically both systems (two-stream and plasma-beam) are essentially different and should be regarded as separate types of the systems for the additional amplification.

The Cherenkov's effect is widely known in the modern relativistic electronics (see, for example, reference [63]). It is obvious that relevant SFELs (Cherenkov's SFELs—see classification in Figure 9.18) can be constructed, too. However, corresponding works on the theory of such devices are not known for the author.

Furthermore, let us discuss shortly the history of the quadratic (parametrical) variety of the considered kind of devices: the parametrical superheterodyne FELs (PSFELs, see Figure 9.18).

The PSFELs with longitudinal electric undulator [56] are based on the very old microwave electronic idea (so-called amplifier on jumps of potential) [69]. It is well known that electron-beam waves can be amplified in the case, when the beam moves along the longitudinal undulation electric field (i.e., through the jumps of potential). In reality, it is a variety of the effect of three-wave parametric resonance for the electron-beam waves. The undulation electric field is created within the work bulk of PSFEL by means of special periodical systems on the basis of magnetic inductors with electrodes or without [56]. Another evident variant—the high-voltage source connected with similar system of electrodes—is also possible, in principle. But, the first type of designs is much preferred because of many technological causes.

The next types of PSFEL are also based on different variants of the effect of three-wave parametric resonance [parametric electron-wave SFEL (PTSFEL); Chapter 14]. Such PTSFELs have been proposed firstly in references [46,76,77]. From the physical point of view, the difference with the preceding PSFEL type [56] concerns only the method of generation of the above-mentioned pumping longitudinal undulation electric field (jumps of potential). Let us recall that such field in the preceding case is generated by means of some external electric undulator, which generates the nonproper pumping electric wave-like field. The functionally similar field in the discussed case is the proper electron-beam wave generated within one of two partial electron beams taking into account its modulation (Chapter 14).

One of the plasma waves also can be used as the similar longitudinal electric pumping in the case of parametric PBSFEL (see the classification in Figure 9.18). Unfortunately, such systems are not studied until today, although its basic physical properties are clear because of above-mentioned similarity to the parametric electron-wave two-stream SFEL (Chapter 14) [46,76,77].

Apart from them a few other types of the parametric SFEL can be, basically, realized, too. For instance, the system based on the mechanism of the three-wave instability in the transversally limited electron beam (Figure 9.18). A characteristic feature of these systems is that the resonance conditions in this case are satisfied to account for the specific properties of electron-wave dispersion in the transversely limited beams. They are well known in traditional nonrelativistic microwave electronics [65]. However, relativistic versions of the systems of such kind are not studied until today as well as the systems on the basis of four-wave, five-wave, etc. parametric resonances [4–6].

Finally, let us make a few comments to the above. The main physical mechanism of the parametric (Quadratic) SFEL can be treated also as the phenomenon of coupled two three-wave resonances [91]. Indeed, we have in this case two connected three-wave systems. The first system contains only the electron-beam longitudinal waves (it provides their additional amplification), whereas the second system has two transversal electromagnetic waves and one longitudinal electron-beam wave (the typical FEL configuration). The connection between both systems is fulfilled through the common electron-beam wave. Owing to such doubled belonging, the common wave is amplified due to the two parametric amplification mechanisms simultaneously. Herewith, the first mechanism (three longitudinal waves) already predominates under the second mechanism (two transversal and one longitudinal). The transformation of the common amplified electron-wave into the electromagnetic signal waves in all cases occurs owing to traditional (the second in our case) parametric FEL mechanism.

Thus, all SFELs could be classified (see Figures 9.17 and 9.18) with respect to the basic additional amplification mechanism for the common electron-beam wave. In turn, we classified all additional mechanisms into the linear (quasilinear) and quadratic (parametric) ones. Correspondingly, all SFELs also can be classified into the quasilinear and quadratic ones. In this regard, let us recall once more some key points of the accepted classification system.

What are the quasilinear and quadratic mechanisms in themselves? The quasilinear additional mechanisms are characterized by the fact that an amplification effect appears in the first approximation, that is, in the framework of the linear (quasilinear) theory. We treat the two-stream, plasma-beam, Cherenkov's, gyro-resonant, and other similar instabilities as the quasilinear (linear) mechanisms. It is readily seen that all of them are *ordinary* for the traditional vacuum electronics amplification mechanisms. It is natural to call such type of SFELs as the ordinary SFEL (Figure 9.18). The quadratic mechanisms are described by truncated equations obtained at in the quadratic approximation (Chapter 4). They, as a rule, have the explicitly expressed parametric wave-resonant physical nature. Therefore, furthermore, we will use more often the definition parametric SFEL than the quadratic SFEL.

9.2.4 Examples of the Design Schemes of the Monochromatic TSFEL

In what follows, let us discuss shortly the variants of general design arrangement of the monochromatic TSFEL. We classify additionally all TSFEL designs into the TSFEL generators (oscillators) and the TSFEL amplifier, correspondingly. The sketch of a simplest TSFEL generator is represented in Figure 9.19. Here, relativistic electron beams 1 and 4, which have different initial energy, are formed by means of accelerators 2, 3 [block of these accelerators could be accomplished, for instance, in the form of a two-beam undulation induction accelerator (UNIAC)—see Section 9.5]. Then both beams are merged into only two-velocity (two-stream) electron beam 5. The beam 5, in turn, is directed in the working bulk of pumping system 6. Electron collector 6 absorbs the worked out beam 5. Optical resonator 8 is placed by such a manner that two-velocity electron beam 5 moves within the pumping system 6 along its optical axis. Generation of the initial signal wave $\omega_1\vec{k}_1$ within resonator 7 begins from some random initial fluctuation. Then this signal wave $\omega_1\vec{k}_1$ is amplified owing to the superheterodyne mechanism passing many times the same

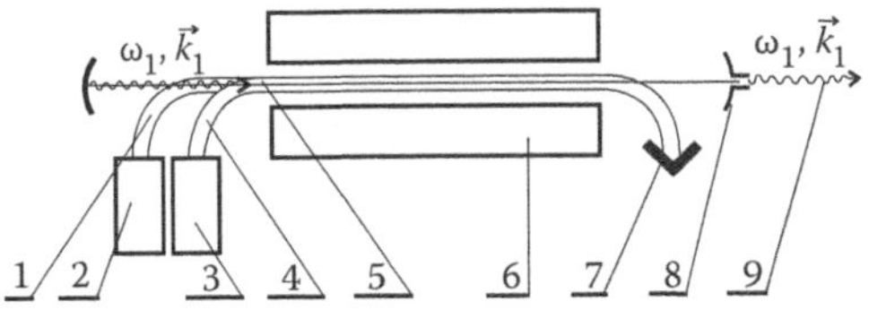

FIGURE 9.19
Design scheme of the TSFEL generator (oscillator). Here, 1 is the first one-velocity relativistic electron beam, 2 is the first electron accelerator, 3 is the second electron accelerator, 4 is the second electron beam, 5 is the two-velocity electron beam, 6 is the pumping system, 7 is the electron collector, 8 is the optical resonator, and 9 is the generated signal with frequency ω_1 and wave vector $\vec{k}_1$. (From Kulish, V.V., *Hierarchic Methods. Undulative Electrodynamic Systems*. Kluwer Academic Publishers, Dordrecht/Boston/London, 2002. With permission.)

working bulk between the resonator 7 mirrors. The generated (output) signal 9 leads out from the output window in one of the resonator 7 mirrors.

In view of the discussion in Chapter 1, the TSFEL generators look essentially lesser interesting than the TSFEL amplifiers. That is why in this book, we will pay further attention to the TSFEL amplifiers. Let us discuss the design peculiarities of the two types of the TSFEL amplifiers, which, as is mentioned above, also could be treated as generators with external excitation.

The first is the one-section (monotron) TSFEL amplifier. Its design scheme is shown in Figure 9.20. The main difference from the system shown in Figure 9.19 is that the initial amplifying signal 1 enters through system input 2. Correspondingly, the signal wave $\omega_1, \vec{k}_1$ is amplified only by first passing the working bulk. Nonlinear interaction of the signal wave $\omega_1, \vec{k}_1$ with the pumping field in plasmas of electron beam 7 leads to excitation of the electron stimulated wave with combination (in general) frequency ω_3 (see Chapter 4 in more detail). The effect of superheterodyne amplification of the signal $\omega_1', \vec{k}_1'$ appears maximally if the frequency of electron-beam stimulated wave ω_3 equals to the optimal frequency of two-stream instability ω_{opt}. Nonlinear superposition of the parametric and two-stream instabilities, as was mentioned already, is treated just the same as the effect of superheterodyne amplification. Output amplified signal 11 leads out through output system 10.

The second is the klystron TSFEL amplifier. Its main design idea is similar to the discussed earlier idea of the Haeff lump (see Figure 9.15 and corresponding comments). Two main distinctions are the following. The first is that short sections of the FEL-pumping systems (see below items 8 and 10 in Figure 9.21) are used instead of the retarding systems (see items 3 and 7 in Figure 9.15). The second distinction is the use in the TSFEL relativistic

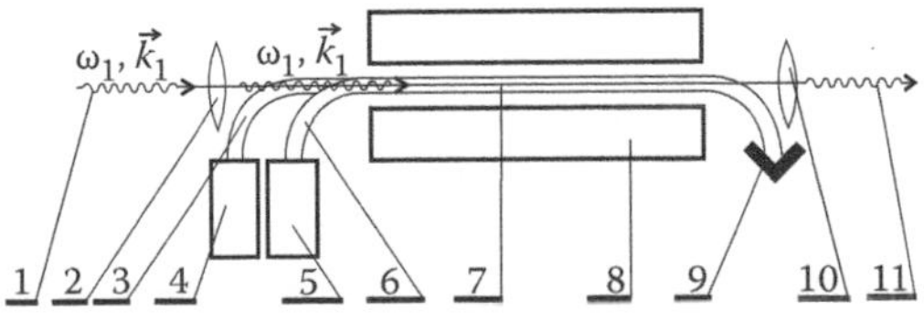

FIGURE 9.20
Design scheme of the monotron TSFEL amplifier. Here, 1 is the input wave signal, 2 is the signal input system, 3 is the first one-velocity relativistic electron beam, 4 is the first electron accelerator, 5 is the second electron accelerator, 6 is the second electron beam, 7 is the two-velocity electron beam, 8 is the pumping system, 9 is the electron collector, 10 is the optical output system, and 11 is the generated signal with frequency ω_1 and wave vector $\vec{k}_1$. (From Kulish, V.V., *Hierarchic Methods. Undulative Electrodynamic Systems*. Kluwer Academic Publishers, Dordrecht/Boston/London, 2002. With permission.)

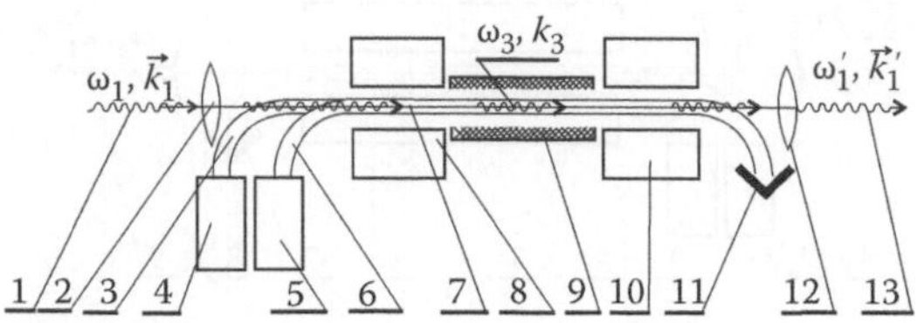

FIGURE 9.21
Simplest variant of design scheme of the klystron TSFEL amplifier. Here, 1 is the input wave signal $\omega_1, \vec{k}_1$; 2 is the signal input system; 3 is the first one-velocity relativistic electron beam; 4 is the first electron accelerator; 5 is the second electron accelerator; 6 is the second electron beam; 7 is the two-velocity electron beam; 8 is the first pumping system; 9 is the transit system; 10 is the second pumping system; 11 is the electron collector; 12 is the output signal system; 13 is the output amplified signal $\omega_1', \vec{k}_1'$; and ω_3, k_3 is the amplifying SCW within the transit section 9. (From Kulish, V.V., *Hierarchic Methods. Undulative Electrodynamic Systems*. Kluwer Academic Publishers, Dordrecht/Boston/London, 2002. With permission.)

electron beams. The klystron-TSFEL amplifiers occupy a special place among the devices of the discussed class owing to uniqueness of number of working characteristics. The simplest variant of the scheme of the klystron TSFEL amplifier is shown in Figure 9.21. Taking into account the above, let us discuss its operating principles in more detail.

The most characteristic feature of the klystron TSFEL amplifier, in comparison with the monotron (one-sectional) TSFEL, is the presence of the transit section 9 (see Figure 9.21). The interaction scenario in the first part of the system (items 1–8) is analogous to the above-mentioned interaction mechanism in the one-sectional TSFEL. The difference is only that first pumping system 8 in Figure 9.21 has relatively smaller length than in the above-mentioned one-sectional case. As a result, the input signal $\omega_1, \vec{k}_1$ within the working bulk of pumping system 8 amplifies rather feebly. The main result of this interaction is the modulation of doubled electron beam 7 on density. In other words, the excitation of electron waves with frequency ω_3 in beam 7 occurs in the first section.

Then electron beam 7 enters in transit section 9. The two-stream instability continues to evolve within doubled electron beam 7. The worked-up initial signal wave $\omega_1, \vec{k}_1$ is absorbed in the transit section 9. This means that the input signal $\omega_1, \vec{k}_1$ further exists in section 9 in the specific form of SCW (electron-beam waves) with frequency ω_3. As a result of evolving the two-stream instability, the depth of the beam 7 modulation strongly increases.

Then the strongly modulated electron beam enters in the input of second pumping system 10. The generation (restoration) of output signal 13 and its further amplification occurs within just this pumping system. This goes on due to the nonlinear superheterodyne interaction of strongly modulated electron beam 7 with the undulation pumping field in the working bulk of system 10. We can say that the back transformation of the amplifying signal (from the electron wave-form into the electromagnetic one) happens in system 10. The worked off electron beam collects by electron collector 11. The amplified electromagnetic signal $\omega_1, \vec{k}_1$ 13 goes out through system output 12. The main advantages of the klystron-SFEL are the high level input–output decoupling and a possibility to work in modes with transforming up the initial signal frequency ($\omega_1' > \omega_1$—see Figure 9.21).

9.2.5 Theoretical Models One-Sectional and Klystron TSFEL

Previously in Section 9.2, it was made clear that a lot of various pumping systems could be used in FELs. This means that the above-mentioned TSFEL represented in Figures 9.19

through 9.21 can have various design realizations differing by performance of the pumping sections. Let us confine ourselves in this chapter by studying the designs only, which are based on use of the H-Ubitron pumping and the Dopplertrons with retarded pumping, correspondingly. Herewith we assume that the explosive instability is the main working mode in the Dopplertron. All varieties of further studied models are presented in Figures 9.22 through 9.27.

Forestalling further theoretical analysis, let us turn the reader's attention to the fact that the Dopplertron pumping plays a much more important role in the case of SFEL technology than it takes place in the tradition ordinary parametrical FEL. As was written before in Section 9.2, two different experimental models of the parametric FEL have been realized in the historically first experiments. They are the H-Ubitron and Dopplertron FEL, respectively. However, then the H-Ubitron FEL became most popular in all subsequent experiments. The main cause of this is the technological problem of creation of required amplitude of the electromagnetic pumping wave. It should be mentioned that suitable technological solutions in the case of H-Ubitrons (amplitude of the magnetic undulation field ~hundreds–thousands Gauss) were found rather quickly. In contrary, the situation in the case of Dopplertron pumping (tens–hundred megawatts of electromagnetic wave power) turned out to be essentially worse—any obvious competitive technological solutions were not found.

However, the general situation began to change strongly with the appearance of the first SFEL designs. As will be shown below, the same (comparing with the parametrical FEL) levels of amplification can be attained in the TSFEL for at least a hundred times lower of the Dopplertron pumping power. This makes much more soft technological requirements for the SFEL-Dopplertron pumping systems. Indeed, constructing the pulsed system with powerful electromagnetic waves on the level hundreds of kilowatts to units of megawatts is not a too difficult problem for the modern microwave technology. The situation can be simplified additionally in the case of the use of special resonator-like design solutions for the pumping. On the other hand, the Dopplertron pumping, as relevant project analysis shows, have a lot of additional design and technological advantages in comparison with the H-Ubitron Parametrical FEL. The main of them is a possibility to construct the designs with the multistage transformation of the signal frequency up [58]. In many cases, it makes today the Dopplertron-SFEL even more competitive than the FEL with the traditional H-Ubitron pumping.

The model of the one-sectional H-Ubitron SFEL amplifier is shown in Figure 9.22. Here, the two-velocity relativistic electron beam 5 consists of two partial one-velocity electron beams 2 and 3, moving within the H-Ubitron pumping system 3. Amplifying electromagnetic signal 1 enters in the system input and then propagates along the two-velocity beam

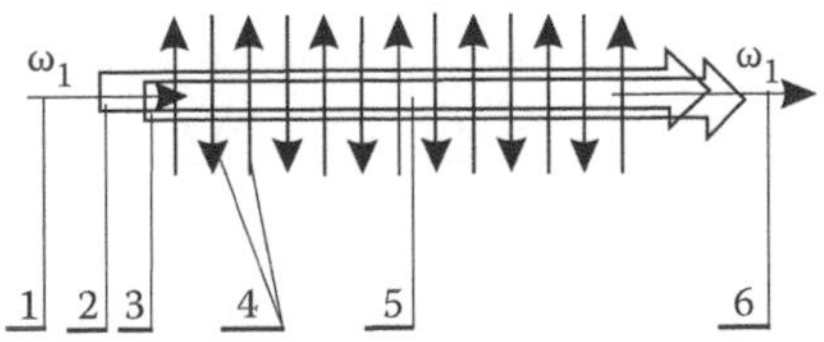

FIGURE 9.22
Model of the one-sectional H-Ubitron SFEL amplifier. Here, 1 is the amplifying signal wave with frequency ω_1, 2 and 3 are one-velocity relativistic electron beams, 4 is the H-Ubitron pumping, 5 is the doubled (two-velocity) electron beam, and 6 is the amplified signal with the same frequency ω_1. (From Kulish, V.V., *Hierarchic Methods. Undulative Electrodynamic Systems.* Kluwer Academic Publishers, Dordrecht/Boston/London, 2002. With permission.)

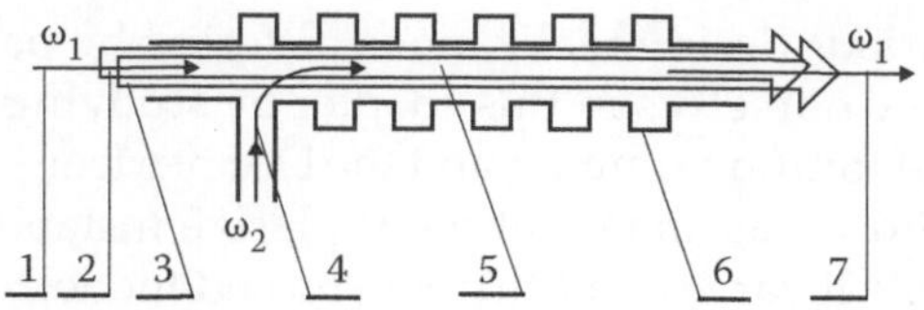

FIGURE 9.23
Model of the one-sectional TSFEL Dopplertron-amplifier with retarded wave of pumping. Here, 1 is the electromagnetic signal with the frequency ω_1; 2 and 3 are one-velocity partial relativistic electron beams, 4 is the retarded pumping wave with the frequency ω_2; 5 is the two-stream relativistic electron beam; 6 is the retardation system for the pumping wave 4; and 7 is the output amplified signal wave with frequency ω_1. (From Kulish, V.V., *Hierarchic Methods. Undulative Electrodynamic Systems.* Kluwer Academic Publishers, Dordrecht/Boston/London, 2002. With permission. From Kulish, V.V., Physical Process in Parametrical Electronic Lasers (Free Electron Lasers), Thesis for the Scientific Degree of Doctor of Physical-Mathematical Sciences, Institute of Physics of Academy of Sciences of Ukraine, 1985.)

5. The signal frequency ω_1 is chosen equal to the optimal frequency of the two-stream instability ω_{opt}. Owing to this, the electron wave with the frequency $\omega_3 = \omega_1$ (which is excited within electron beam 5) and the signal wave with the same frequency ω_1 are amplified due to the superheterodyne mechanism. Amplified signal wave 6 goes out from the system output.

Another version of the one-sectional TSFEL amplifier model (see Figure 9.20) is given in Figure 9.23. This is the Dopplertron amplifier with pumping by retarded electromagnetic waves [39–42]. There, similarly to the preceding case, two-stream relativistic electron beam 5, which is formed by the merging of two one-velocity beams 2, 3, is injected into the working bulk of retardation system 6. The amplifying electromagnetic signal 1 with frequency ω_1 and the retarded electromagnetic pumping wave ω_2 propagate in the same longitudinal direction within the system working bulk. Realization of the parametric interaction mechanism for the signal and pumping fields in plasma of beam 5 leads to excitation of the electron wave with combinative frequency $\omega_3 = \omega_1 + \omega_2$. Similarly to the above case, the frequency ω_3 is chosen equal to the optimal frequency ω_{opt}. As a sequence, the amplification of all three waves occurs due to the explosive version of the superheterodyne amplification mechanism. The meaning of all the rest of the items in Figure 9.23 is the same with Figure 9.22.

The devices, which models are shown in Figures 9.24 through 9.27, represent different types of TSFEL-klystrons [47,74,75,78,83,84].

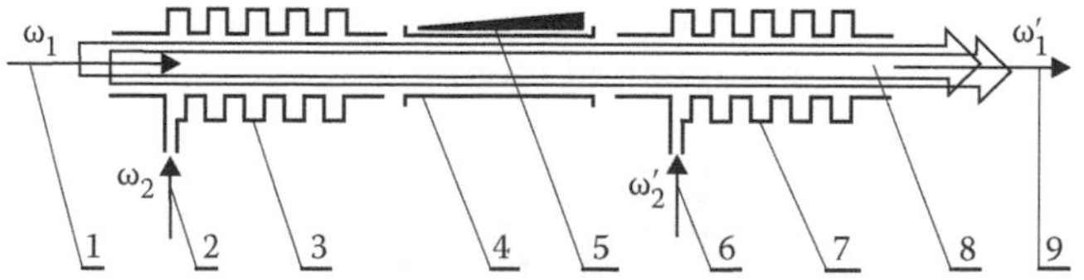

FIGURE 9.24
Model of the TSFEL-amplifier-klystron of the Dopplertron type. Here, 1 is the input amplifying signal wave, 2 is the first retarded pumping wave with frequency ω_2, 3 is the first retardation system, 4 is the transit section, 5 is the absorber for signal and pumping waves, 6 is the second retardation system, 7 is the second retarded pumping wave with frequency ω'_2, 8 is the doubled (two-velocity) relativistic electron beam, and 9 is the output amplified wave signal frequency ω'_1. (From Kulish, V.V., *Hierarchic Methods. Undulative Electrodynamic Systems.* Kluwer Academic Publishers, Dordrecht/Boston/London, 2002. With permission. From Kulish, V.V., Physical Process in Parametrical Electronic Lasers (Free Electron Lasers), Thesis for the Scientific Degree of Doctor of Physical-Mathematical Sciences, Institute of Physics of Academy of Sciences of Ukraine, 1985.)

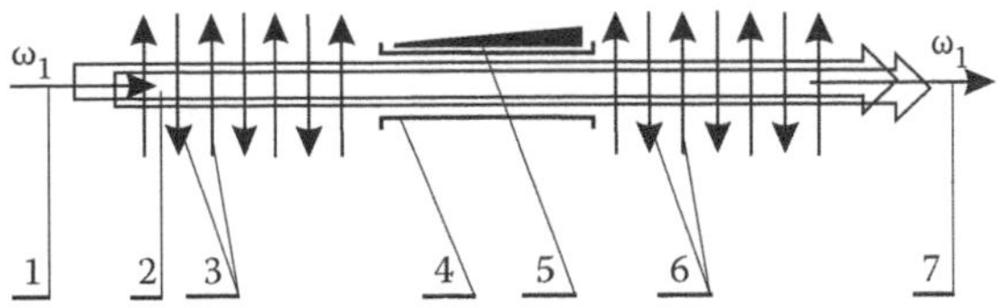

FIGURE 9.25
Model of the TSFEL-amplifier-klystron of the H-Ubitron type. Here, 1 is the amplifying wave signal with the frequency ω_1, 2 is the is the doubled (two-stream) relativistic electron beam, 3 is the first H-Ubitron magnetic pumping system, 4 is the transit section, 5 is the absorber for the electromagnetic signal wave 1, 6 is the second H-Ubitron pumping system, and 7 is the amplified signal wave with the frequency ω_1. (From Kulish, V.V., *Hierarchic Methods. Undulative Electrodynamic Systems.* Kluwer Academic Publishers, Dordrecht/Boston/London, 2002. With permission. From Kulish, V.V., Physical Process in Parametrical Electronic Lasers (Free Electron Lasers), Thesis for the Scientific Degree of Doctor of Physical-Mathematical Sciences, Institute of Physics of Academy of Sciences of Ukraine, 1985.)

The model of TSFEL-klystron-Dopplertron is given in Figure 9.24. A characteristic feature of this design version is that both pumping sections (see Figure 9.21) are made on the basis of Dopplertron pumping. A specific feature of this design is that the frequencies of both pumping sections in general could be different ($\omega_2 \neq \omega_2'$). The frequencies of input ω_1 and output ω_1' signals also turn out to be not coinciding. This allows to increase additionally the level of input–output electromagnetic decoupling. The operation principles as well as the meaning of corresponding items are self-evident in view of the above discussions relating to Figures 9.22 and 9.24.

Comparing the one-sectional TSFELs and traditional parametric FELs, we can find that the main advantage of the first of them is the extremely high levels of signal amplification. However, just this advantage leads to the main drawback of the one-section TSFEL. We take in view their inclination to self-excitation because the system input is really always weakly coupled with the system output. As the analysis shows (see below in this chapter), the klystron TSFEL solves this problem radically. Owing to the presence of the transit section (see Figures 9.21 and 9.24 through 9.27), the klystron TSFELs are characterized by very reliable input–output decoupling. This allows us to suppress strongly the self-excitation effect. The input–output decoupling could be done even stronger due to the shifting of the output signal frequency ω_1' with respect to the input signal frequency ω_1. Herewith, the combination

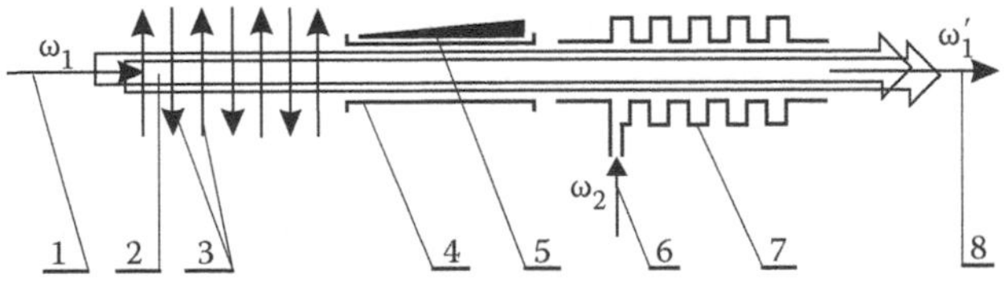

FIGURE 9.26
Model of the TSFEL-amplifier-klystron of the mixed H-Ubitron–Dopplertron type. Here, 1 is the amplifying wave signal with frequency ω_1, 2 is the doubled (two-velocity) relativistic electron beam, 3 is the H-Ubitron pumping system, 4 is the transit section, 5 is the absorber for the electromagnetic signal wave 1, 6 is the Dopplertron pumping wave with frequency ω_2, 7 is the retardation system, and 8 is the amplified signal wave with the frequency ω_1'. (From Kulish, V.V., *Hierarchic Methods. Undulative Electrodynamic Systems.* Kluwer Academic Publishers, Dordrecht/Boston/London, 2002. With permission. From Kulish, V.V., Physical Process in Parametrical Electronic Lasers (Free Electron Lasers), Thesis for the Scientific Degree of Doctor of Physical-Mathematical Sciences, Institute of Physics of Academy of Sciences of Ukraine, 1985.)

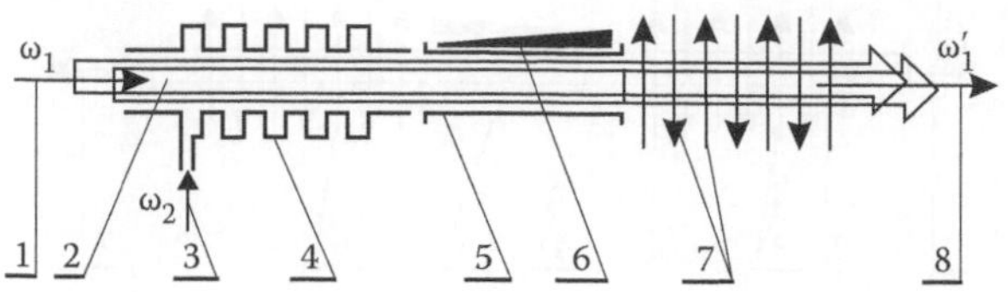

FIGURE 9.27
Model of the TSFEL-amplifier-klystron of the mixed Dopplertron–H-Ubitron type. Here, 1 is the amplifying wave signal with frequency ω_1, 2 is the doubled (two-velocity) relativistic electron beam, 3 is the Dopplertron pumping wave with frequency ω_2, 4 is the retardation system, 5 is the transit section, 6 is the absorber for the signal and pumping waves, 7 is the H-Ubitron pumping system, and 8 is the amplified signal wave with the frequency ω'_1. (From Kulish, V.V., *Hierarchic Methods. Undulative Electrodynamic Systems.* Kluwer Academic Publishers, Dordrecht/Boston/London, 2002. With permission.)

electron-wave frequency ω_3 is the same for the input and output sections. Technologically, such possibility can provide all design schemes of the TSFEL-klystrons containing the Dopplertron sections (i.e., the systems shown in Figures 9.24, 9.26, and 9.27).

Besides that, each type of the TSFEL-klystron has more specific advantages. For example, the systems similar to that are shown in Figures 9.25 and 9.27 (i.e., containing the H-Ubitron pumping in the terminal sections) have, as a rule, higher gain factor than the systems in Figures 9.24 and 9.26. At the same time, the system containing the Dopplertron pumping in the first section can work in a specific superheterodyne mode. In this case, the system has a possibility to change the input frequency ω_1 without changing frequency of the output signal ω'_1. Herewith, the signal frequency ω_1 as well as the pumping frequency ω_2 changes synchronously in the way when the combination frequency $\omega_3 = \omega_1 + \omega_2$ is constant.

Finally, let us turn the reader's attention at one very interesting applied prospect of systems in Figures 9.25 and 9.27. As is well known, attaining the possibility to work on higher signal frequencies always is connected in the case of traditional (parametrical) FEL with increasing the beam energy (see Section 9.2). The following question arises: do we have other easier possibilities for attaining the same result without the beam energy increasing? The answer is yes.

Basically, we have three such possibilities

1. The use of higher electron harmonics of the two-velocity beam [4,6,25, 59,85–89]
2. The use of the multistage signal transformation [4,6,25,49–51,58]
3. The use of retarded electromagnetic waves as the Dopplertron pumping [4,6, 25,39–43,53–56]

Let us begin with the first of these possibilities. It could be realized practically because of unique multiharmonic properties of the two-stream instability [59,85–89]. The main idea of these properties consists of that the amplitudes of higher electron-beam (SCW) harmonics can be essentially higher than the amplitudes of lower harmonics [59,85–89]. This means that the klystron-TSFEL could be used as a peculiar multiplier of the signal frequency. The first (pumping) and second (transit) sections in this case play the role of a generator and an amplifier of the high electron-beam harmonics. The last pumping system is used as a system for selection of the working harmonic (of a few harmonics simultaneously) and, at the same time, for amplification of the chosen harmonic (harmonics) of the output signal.

9.2.6 Example of TSFEL with Multistage Transformation Up of Signal Frequency

Let us turn to the second of the mentioned possibilities, which is illustrated in Figure 9.28. The key point of this multistage design is that there, the output signals of one TSFEL section are used as a Dopplertron pumping for other TSFEL sections.

At first, the general concept and example of designs of such kind (multistage PFEL) have been proposed and analyzed in references [50,51]. The main advantage of this design solution is that we have a possibility to attain high levels of output signal frequencies using moderately relativistic electron beams.

The following is the explanation of this effect. Let us assume for simplicity that the pumping system 3 (in Figure 9.28) is performed in accordance with the Dopplertron scheme with pumping by a nonretarded electromagnetic wave. We regard such TSFEL (items 1–5 in Figure 9.28) as the first stage of frequency transformation. We have in view the transformation of the pumping electromagnetic wave with frequency $\omega_2^{(1)}$ into the signal wave with frequency $\omega_1^{(1)}$. These frequencies are bound (see, for example, Section 9.2) by the following relationship:

$$\omega_1^{(1)} \approx \frac{1+\beta_{01}}{1-\beta_{01}}\omega_2^{(1)} \approx 4\gamma_{0i}^2\omega_2^{(1)}, \tag{9.11}$$

where β_{01} is the averaged velocity of the two-velocity beam for the first stage of transformation, $\gamma_{0i} = \left(1-\beta_{0i}^2\right)^{-1/2}$ is the averaged relativistic factor of the ith two-velocity electron beam. Then, let us assume that this signal wave is used further (see Figure 9.28) as the Dopplertron pumping for the next TSFEL section (second transformation stage), and so on. It is not difficult to obtain the signal frequency of the last (nth) transformation stage $\omega_1^{(n)}$, which could be written in the form

$$\omega_1^{(n)} \cong \omega_2^{(1)}\prod_{i=1}^{n}\left(\frac{1+\beta_{0i}}{1-\beta_{0i}}\right) \approx \omega_2^{(1)}\prod_{i=1}^{n}\left(4\gamma_{0i}^2\right). \tag{9.12}$$

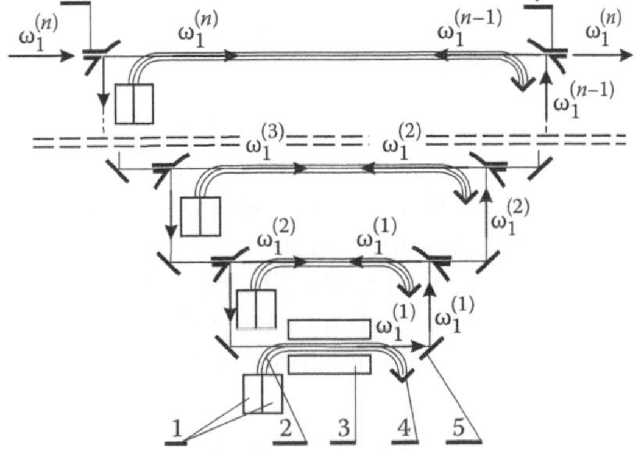

FIGURE 9.28
General idea of the two-stream SFEL amplifier with multistage transformation of the signal frequency. Here, 1 are two-beam electron injectors, 2 are two-velocity electron beams, 3 is the pumping system, 4 are electron collectors, 5 is the system of bounded ring optical resonators, 6 is the signal input, and 7 is the signal output.

It is obvious from Equation 9.12 that using the multistage design scheme, we really obtained a possibility to amplify rather high-frequency electromagnetic signals with the use of moderately relativistic electron beams only.

As is mentioned already, the practical realization of the ordinary (i.e., purely parametric) multistage Dopplertron FEL is always complicated by technological problems: they are the problems related with the too small signal gain factors, which are characteristic of the PFEL-Dopplertrons. The introduction of the two-stream superheterodyne amplification into the mechanism of multistage transformation allows avoiding these difficulties. It becomes possible because the superheterodyne gain factor is essentially higher than the parametric gain factor for the same pumping amplitude and beam parameters. As a result, the multistage TSFELs on the basis of Dopplertron pumping, in contrast to their parametrical analogy, become competitive to the traditional H-Ubitron TSFELs in many important applications. It concerns especially the cluster systems that use the satellite method of basing.

Let us accomplish some numerical estimation for illustration of practical prospects of the multistage SFEL. We accept, for instance, the three-stage TSFEL scheme shown in Figure 9.28, the averaged energies of all two-velocities beams at all stages are equal ($E_0^{(1)} = E_0^{(2)} = E_0^{(3)} = 1$ MeV), the frequency of Dopplertron pumping for the first stage $\omega_2^{(1)} = 10^{10}\,s^{-1}$ (that corresponds to wavelength $\lambda_2^{(1)} \approx 18.84$ cm). Using formula 9.12, after noncomplex calculations, we obtain: $\omega_1^{(1)} \approx 3.5 \cdot 10^{11}$ s^{-1} ($\lambda_1^{(1)} \approx 5.3$ mm—the first stage), $\omega_1^{(2)} \approx 1.22$ s^{-1} ($\lambda_1^{(2)} \approx 150$ μm—the second stage), $\omega_1^{(3)} \approx 4.29 \cdot 10^{14}$ s^{-1} ($\lambda_1^{(3)} \approx 4.28$ μm—the third stage), and the total transformation factor $\xi = \omega_1^{(3)}/\omega_2^{(1)} \approx 4.29 \cdot 10^4$. The same transformation factor, but in the one-stage case only, could be obtained for the beam energy $E_{0i} \approx 69.6$ MeV ($\gamma \approx 137.5$). The simplest project analysis shows that the used three parallel 1 MeV three-channel high-current linear induction accelerators (HLIACs) have much more smaller longitudinal sizes (~1.5 m) than one 70 MeV HLIAC (80–100 m). Moreover, as will be shown below (see Sections 9.5 and 9.6), the use of such too relativistic HLIAC provides essential addition and reduction of the TSFEL factor. It is because the increment growth for the two-stream instability strongly falls down with the beam energy increasing (see further Chapter 13 for details).

9.3 Cluster Klystron SFEL: The Main Design Schemes and Operation Principles

As was already indicated in Chapter 1 and above in this chapter, today the cluster klystron superheterodyne FELs (CKSFELs) represent the most interesting class of FEL both in physical and purely practical aspects. By this reason, special attention is paid to these systems further in the book. In the light of the above-mentioned, we shall consider this section as a sort of an introduction to the theory of CKSFEL. However, we shall consider only some of the most general theoretical issues, such as basic schemes, operating principle, etc. We shall discuss the physical problems of the systems of this type in detail in Chapter 14.

9.3.1 General Scheme of the Cluster Klystron SFEL

We begin the discussion with the analysis of the general block diagram of the version of the cluster klystron shown in Figure 9.29 [58]. The operating principle of the installation is as follows.

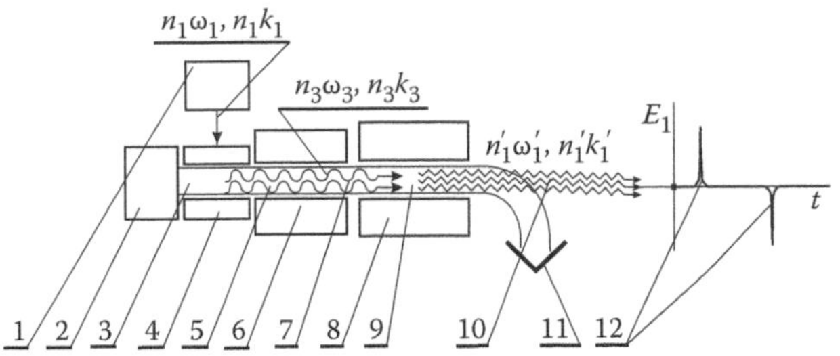

FIGURE 9.29
Example of the general block diagram of the cluster multiharmonic Klystron FEL. Here, 1 is the source of multiharmonic input signal $n_1\omega_1,n_1k_1$ (where ω_1 is its cyclic frequency, k_1 is the wave vector, n_1 are harmonic numbers); 2 is the source of relativistic electron beam; 3 is the electron beam emitted by the source 2; 4 is the multiharmonic modulator; 5 is the multiharmonically modulated relativistic electron beam; 6 is the transit section consisting, in general, of active and passive parts; 7 are harmonics of the SCWs of the beam 5; 8 is the pumping system of the terminal FEL section; 9 is the electron beam 5 in the zone of interaction of the FEL section 8; 10 is the multiharmonic output signal generated by the FEL section in the form of a sequence of femtosecond electromagnetic clusters (signal cluster wave) 12; 11 is the recuperation system and collector of the used electron beam 5, 8; 12 is the cluster wave of the signal; $n_3\omega_3,n_3k_3$ is the multiharmonic wave packet of the SCW, where ω_3 is the frequency, k_3 is the wave vector of the SCW, n_3 are numbers of harmonics; $n_1'\omega_1',n_1'k_1'$ is the multiharmonic wave packet of the output signal, where ω_1' is the cyclic frequency of the first harmonic, k_1' is the wave number, n_1' is the harmonic number; in general, we consider that the frequency of the output signal, ω_1', is shifted relative to the frequency of the input signal, ω_1.

The relativistic electron beam 3 (in general, is a multispeed beam) is generated by the source 2. Some examples of design schemes of the source 2 are discussed in the next section below. Then the beam 3 is fed into the entrance of the multiharmonic modulator 4 (see examples of the design schemes shown in Figures 9.30 and 9.31). The input signal, or in general, multiharmonic signal $(n_1\omega_1,n_1k_1)$ generated by the source 1, also enters the modulator 4. An arrangement of a combined type (consisting, for example, of many parallel monochromatic FEL or other devices) or another multiharmonic FEL may be used as such a source.

The signal $(n_1\omega_1,n_1k_1)$ interacts with the electron beam 3 in the working volume of the modulator 4. According to a simple project analysis, the practical realization of an enough big number of diagram variants for modulator technological versions is possible in this case. For example, in models of cluster klystrons studied earlier [58], it was supposed to fabricate

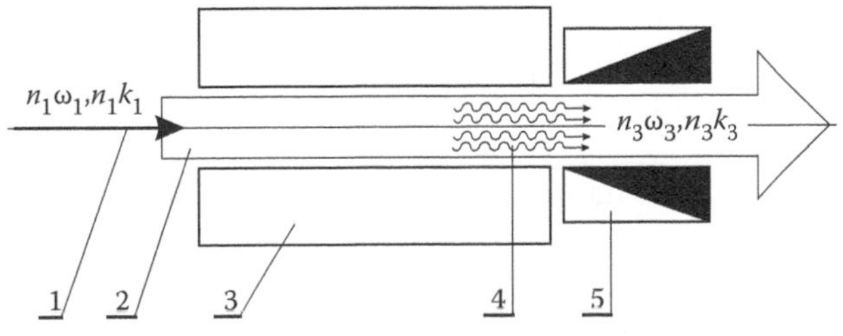

FIGURE 9.30
Design scheme of the electron beam resonance (parametric) modulator. Here, 1 is the multiharmonic input signal (signal cluster wave with $n_1\omega_1,n_1k_1$ frequency spectrum, where ω_1 is the first harmonic cyclic frequency, k_1 is the wave number, n_1 is the harmonic number); 2 is the nonmodulated electron beam; 3 is the transversal undulator; 4 is the modulated electron beam with the packet of electron waves (SCW) (cluster SCW with $n_3\omega_3,n_3k_3$ frequency spectrum, where ω_3 is the frequency, k_3 is the wave vector of the SCW, n_3 is the harmonic numbers); and 5 is the absorber of the signal 1 cluster wave energy.

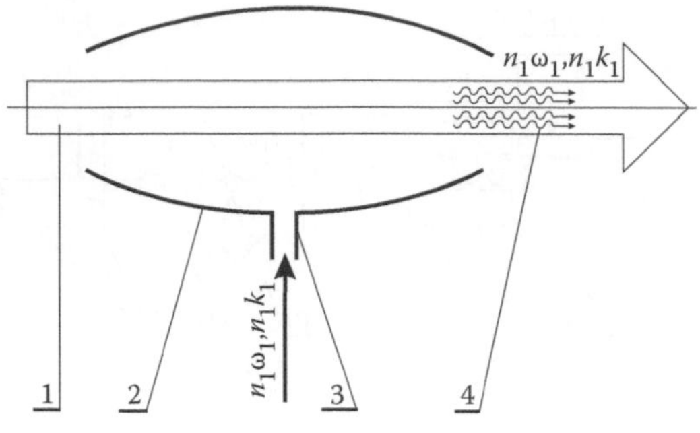

FIGURE 9.31
Design scheme of the optical (quasioptical) modulator of the electron beam. Here, 1 is the nonmodulated input electron beam; 2 is the dome-shaped multiharmonic optical (or quasioptical) resonator; 3 is the entrance of the resonator 2 for cluster electromagnetic signal with $n_1\omega_1, n_1k_1$ frequency spectrum (where ω_1 is the first harmonic cyclic frequency, k_1 is the wave number, n_1 is the harmonic number); and 4 is the modulated electron beam at the exit from the system (i.e., the cluster SCW with the same $n_1\omega_1, n_1k_1$ frequency spectrum (as the signal) is generated).

a modulator as an input harmonic (or multiharmonic, in general) section of the SFEL (see further discussion related to Figure 9.31), that is, principally resonance systems. However, it was found later that at high amplification levels, which are typical, for instance, for two-stream and plasma-beam superheterodyne modulator sections 4, such a solution seems to be technologically redundant. This conclusion is based, first, on the fact that to get sufficient modulation, a much more simple nonresonance modulator can be used with the same success. One of the examples of this sort is shown below in Figure 9.31. It is worth to note that, in general, independent of the technological solution for the modulator 4, at the system exit we always have a (multiharmonically, as well) modulated beam 5, which passes into the active part of the transit section 6. The formation of the intensive cluster SCW of the electron beam 9 takes place in it. It was already mentioned in Chapter 1 that this cluster wave is the result of a nonlinear synthesis of various harmonics of the SCW 7 in the course of development of an additional mechanism of amplification of longitudinal electron waves. These mechanisms may constitute both quasilinear (plasma-beam or two-stream) and parametric (electron wave and others) beam instabilities (see the previous section for more details).

After passing the transit section 6, the modulated beam is directed into the zone of interaction of the multiharmonic FEL section 8. Similar to the modulator case, here we also have a rather wide variety of particular scheme variants. However, the general principle of their operation appears to be the same, that is, the condition of the parametric resonance with like harmonics of cluster pumping waves $(n_2\omega_2, n_2k_2)$ and electromagnetic signal $\left(n_1'\omega_1', n_1'k_1'\right)$ is true for each of the harmonics of the multiharmonic cluster SCW $(n_3\omega_3, n_3k_3)$ (see also Chapter 1).

Therefore, the physical effect named as multiple three-wave parametric resonance for harmonics [58,85–89] is laid as the basis of the discussed type of electronic systems, and this effect is far from the traditional one. Regulating the shape of the cluster SCW spectrum at the modulator 4 exit by using any method, we get the real opportunity to control the parameters of electromagnetic clusters at the exit of the FEL section 8. As we can easily see, the main principal difference of such a multiresonant FEL section from the "ordinary" one (see Sections 9.2 and 9.3) consists only of the use of the multiharmonic, that is, cluster pumping system. It may be realized both in the form of a magnetic undulator (wiggler—see example shown in Figure 9.33) and intensive electromagnetic wave pumping (Dopplertron pumping—see example shown in Figure 9.34 below).

9.3.2 Modulators

It was already mentioned that it is possible to use two principally different variants of electron beam modulators—resonance and nonresonance, in the cluster FEL klystrons considered herein. The resonance modulators include all the diversity of any possible design solutions based on the principles of resonance interaction of the input electromagnetic signal with relativistic electron beam. From the theoretical standpoint, it must be a very big number of examples of the design solutions of this type. First and foremost, because it seems that it is possible to take them directly from the classical microwave electronics, plasma electronics, etc., in principle, they could be some sections of traveling wave tubes (TWTs), klystrons and cyclotron type systems, or other similar arrangements designed to excite intrinsic instabilities in the beam.

However, the more detailed project analysis indicates that in reality, this visible variety does not mean too much for the relativistic electronics practice discussed herein. The main reason is that any relativistic system is characterized by a number of rather specific technological and physical peculiarities. Hence, more complete analysis results in a conclusion that only one type of technologically accepted resonance systems that satisfy all above-mentioned specific peculiarities exists, and it is FEL sections of various types. These sections may be both ordinary parametric FEL (Section 9.2) and super heterodyne two-stream and plasma-beam FEL sections (Chapter 14).

An example of the general design scheme of this type of modulators is shown in Figure 9.30. The multiharmonic electromagnetic signal 1 (signal cluster wave with $n_1\omega_1,n_1k_1$ frequency spectrum) enters into the multiharmonic undulator 3. Originally nonmodulated electron beam 2 also enters the undulator. Then the cluster wave of SCW, which is characterized by $n_3\omega_3,n_3k_3$ frequency spectrum, is generated as a result of the effect of multiple three-wave parametric resonance [85–89]. Then this cluster wave enters the working volume of the next (transit) section (see Figure 9.29).

Note that the variant of the FEL section with monochromatic input signal and pumping [85–89] may be also used for the above-formulated purpose of multiharmonic modulation. In this case, the appearance of multiharmony in SCW spectrum results from the nonlinear nature of the FEL basic working mechanism. It is expressively revealed in cases of quasi-monochromatic resonance modulators built on the basis of superheterodyne two-stream and plasma-beam FEL [53–55,57,85–89].

The scheme shown in Figure 9.30 may have two principle design versions. The first suggests the amplified component arrangement of the FEL as modulator, when signal 1 (Figure 9.30) passes through the zone of interaction of the section only once. In this case, the worked-out input signal is absorbed by the absorber 5 after passing the section. In the second case, an additional open resonator for signal 1 is inserted into the construction.

In general, different design modifications of resonance modulators may also differ by electron beam structure, pumping type used in undulators, etc. As to the beam structure, we can say that the most often found proposal in well-known works consists of the use of both single-speed and two-speed rectilinear beams. In principle, as the preliminary analysis shows, the use of three- and more speed beams is possible. However, as of today, these versions of femtosecond FEL are not investigated yet, though the respective semiqualitative analysis shows their potentially great practical perspective. As to design variants of the undulator 3 in Figure 9.30, the diversity of possible design realizations seems to be even richer. Among others, it may be magnetic (H-Ubitron) undulators, undulators with crossed magnetic and electric fields (EH-undulators), undulators on the basis of intensive electromagnetic waves (Dopplertron systems) [4,6,58], etc. At this point, all the variants may be of both quasiharmonic and essentially multiharmonic versions. Two examples of such FEL multiharmonic pumping are shown further in Figures 9.33 and 9.34.

In case of nonresonance undulators, another excitation principle for the SCW is used. It is known that at any periodic excitations of the electron speeds in the beam, the intrinsic waves appear in it, including longitudinal waves, which are of special interest to us [4,6]. The real efficiency of this "electromagnetic wave—SCW transformation" is relatively low compared with the resonance case. However, in practice, this circumstance does not always play the primary part. For example, in design versions of cluster FEL klystrons with high amplification of the SCW in the transit section (in particular, due to two-stream and plasma-beam instabilities), the use of nonresonance modulators has technological preferences, first and foremost because constructions of these systems are always simpler than for their resonance analogs.

An example of the design of the optical (quasioptical) nonresonance modulator of the electron beam of this type is shown in Figure 9.31. In this figure, the electron beam 1 enters into the interaction zone of the resonator 2, where the field of the multiharmonic cluster signal with $n_1\omega_1, n_1k_1$ frequency spectrum is generated. The longitudinal component of the field modulates the beam. The modulated beam 4 then passes into the transit section (see Figure 9.29).

9.3.3 Transit Section

In general, as was already said, the transit section of the cluster FEL klystron (see Figure 9.29) may contain both active 3 and passive 5 parts (see Figure 9.32). In its turn, the analysis shows that from the technological standpoint, the most promising are those variants where the active part is based on two-stream, plasma-beam, as well as parametric electron-wave systems (including the systems with the longitudinal electric pumping undulator) [56].

Two key physical processes simultaneously take place in the active part of the transit section. One of them is the amplification of electron harmonics of the beam 5 (SCWs), and thus formation of the multiharmonic spectrum of the SCW. The other one is the process of optimal shaping of the cluster SCW spectrum. This process is realized due to the additional generation (amplification) of the SCW higher harmonics. It is worse to note that in case of quasiharmonic (i.e., single-frequency) modulators, the two processes are physically easily distinguishable. It is clearly revealed when performing the respective quantitative analysis. However, in case of "originally multiharmonic" modulators (similar to those that are shown in Figures 9.30 and 9.31), the second of these two effects seems to be hidden by the effect of the multiharmonic amplification, though in reality, the observed harmonic amplification mainly is a mere process of their generation.

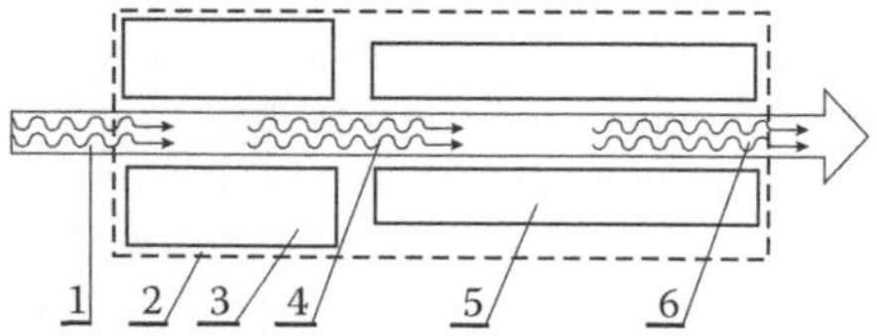

FIGURE 9.32
Block diagram of the transit section, general case. Here, 1 is the modulated electron beam from the modulator; 2 is the transit section as a whole; 3 is its active part; 4 is the multiharmonically modulated electron beam; 5 is the passive part of the transit section 2; and 6 is the outgoing multiharmonically modulated electron beam.

The passive part of the transit section 5 may be made in the form of a magnetic compression (decompression) system or as an intermediary acceleration unit for electron beam 4. The technological reason for inserting the passive part of the transit section into the construction consists, first of all, in getting the optimal configuration of the beam 6 prior to its passing into the next (terminal) FEL section. In the latter, as was already said, the transformation of electron wave (beam) clusters into signal electromagnetic clusters takes place. The thing is that the optimum conditions for two-stream or plasma-beam instabilities differ, as a rule, from the optimal operating conditions for the next terminal FEL section. Therefore, in general, it is not rare to get design contradiction in optimal adjustment of transit and terminal sections for systems without passive parts. Insertion of the passive part allows us to remove this type of contradiction.

Of all the variety of analyzed variants of possible construction solutions for section 3 passive parts, we can choose three most interesting variants as to their practical implementation. For the possibility to vary beam parameters, such as, for example, plasma frequency, parameters of waves present in the working volume (for instance, dispersion law for the extraordinary pumping wave in plasma-beam systems [43,53–55,57], etc.) are provided in systems of the first type. Technologically, one can realize it by beam compression (or decompression) during its motion in accompanying, slowly changing and focusing, magnetic field. In the physical aspect, such a situation is realized when effects of plasma-beam, two-stream, longitudinal parametric (among others, that with the use of the longitudinal electric pumping undulator) and other similar instabilities are used to amplify longitudinal SCW. As was already mentioned, the beam parameters within the active part of the transit section are taken optimal for mechanisms of SCW-cluster formation, while within the passive part, these parameters are readjusted to be optimal for the terminal FEL section.

Systems of the second type, in addition, provide for change in beam energy by means of beam acceleration or deceleration. The thing is that, as a rule, any increments of growth of the above-mentioned longitudinal instabilities are much more sensitive to the grade of electron beam relativity compared with the traditional transverse–longitudinal parametric FEL mechanism. As a result, enough severe constructive limitation as to the beam energy in the zone of the transit section appears in some practically important cases. At the same time, functionally similar criteria for the terminal FEL section require the use of much more relativistic beams. Introduction of the intermediary acceleration allows to noticeably moderate the described conflict of requirements [25,56]. Then the technological possibility to execute principal procedures on SCW clusters formation with relatively low energy levels (units of megaelectron volts) of the beam, and realize the procedure of energy takeoff and electromagnetic clusters formation at considerably higher energies, appears.

And finally, design variants of the third type are characterized with the simultaneous application of the both above-mentioned technological methods, that is, simultaneous change in such parameters as, for instance, plasma frequency of the beam and its energy. Such situations appear, for example, in two-stream cluster FEL klystrons, when it is reasonable to continue the process of two-stream amplification of longitudinal SCW of the beam in the process of acceleration.

9.3.4 Pumping Systems for the Terminal Sections

As the analysis shows, one of the key technological problems of the technology of the femtosecond FEL is the practical realization of multiharmonic (cluster) pumping. By analogy with ordinary FEL here, in principle, it is possible to create a rather long series of possible design variants. They are H-Ubitron, crossed EH-Ubitron, Dopplertron, etc. [4,7,8,25] multiharmonic versions of well-known techniques of FEL pumping systems. Two examples of this type are shown in Figures 9.33 and 9.16.

The example shown in Figure 9.33 represents the simplest variant of cluster (multiharmonic) H-Ubitron pumping [58]. The idea of the periodically reversible consequence of so-called magnetic clusters is realized in it. These clusters are formed in gaps between very narrow (compared with the undulation period λ_2) magnetic pole pieces 1, which are constructively connected with the magnetic poles 2 (see Figure 9.33). It is obvious that it is not difficult to get enough expressive multiharmonic spectrum broken up by wave numbers $k_2 = 2\pi/\lambda_2$ by expanding the magnitude of magnetic field induction vector $\vec{B}_2$ of this cluster pumping wave into the Fourier series. Moreover, the most characteristic physical peculiarity of the wave formed in this way is its static (magnetostatic, in this case) character. In addition, in this case, the pumping field itself is principally nonintrinsic for the working region of cluster superheterodyne FELs (CSFELs).

Versions of Dopplertron pumping are based on the phenomena of an essentially other nature. By this reason, the pumping fields in this case are, as a rule, intrinsic but not static (Section 9.2). The example of the multiharmonic Dopplertron pumping, which is shown in Figure 9.34, belongs to the class of so-called multistage systems (see also Figure 9.28 and respective comments). In this case, the stage structure consists of the following: the same electromagnetic cluster wave is the signal for the lower (first) stage 6 and, at the same time, it acts as multiharmonic (cluster) pumping in the second (upper) stage of the multistage two-stream cluster superheterodyne FEL-klystron (MTCSFEL-Klystron).

The key design element of the scheme shown in Figure 9.34 is the ring resonator 1. The resonator 1 has two optical axes, the generator 6 of multiharmonic (cluster) wave 4 (with $n_1\omega_1, n_1k_1$ frequency spectrum) is located on one of them, and the former of the output femtosecond cluster wave of signal 10 (with $n'\omega_1', n'k_1'$ frequency spectrum), which consists of structural elements 2, 3, 5, 7, and 8 is located on the other axis. The pumping system operates as follows. The generator 6 generates the multiharmonic (cluster) electromagnetic wave 4, which circulates in the ring resonator 1. At this stage, as was already mentioned, this wave plays a part of the multiharmonic (cluster) signal of the first stage. This wave propagates along the optical axis of the same resonator 1 in the direction that is opposite to the two-speed electron beam 7. At the same time, it pierces through working volumes of the system of beam transportation 8 and modulator 5.

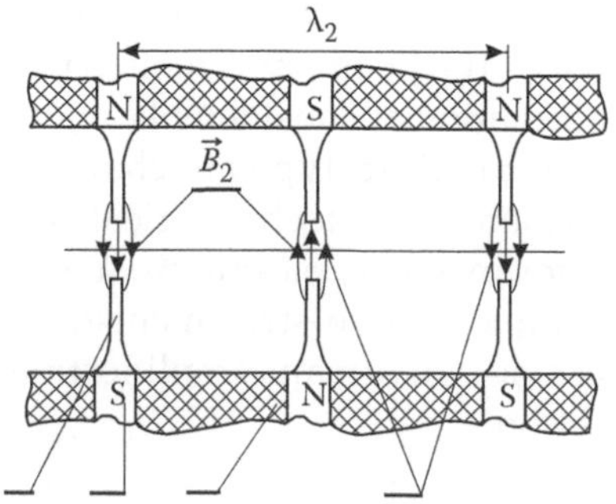

FIGURE 9.33
Design of the multiharmonic (cluster) magnetic undulator. Here, 1 are narrow magnetic pole pieces; 2 are magnetic poles; 3 is the magnetic insulator; 4 are force lines of the magnetic field of multiharmonic pumping (magnetostatic clusters); λ_2 is the undulator period (the period of the first harmonic); N is the North magnetic pole; S is the South magnetic pole; and $\vec{B}_2$ is the induction vector of the magnetic field.

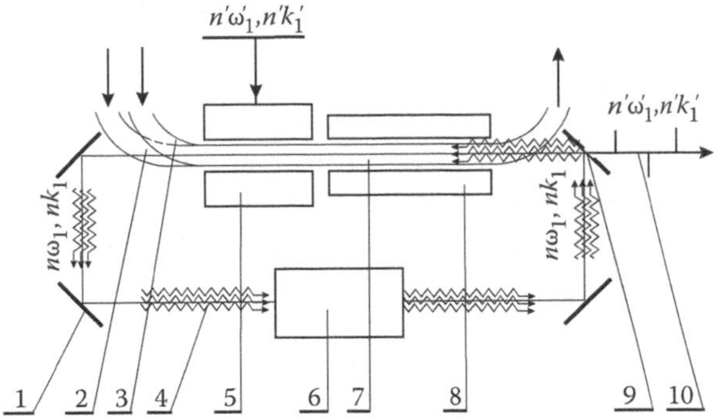

FIGURE 9.34
Example of the scheme of two-stage Dopplertron multiharmonic pumping in the femtosecond cluster MDFEL-Klystron based on the ring resonator. Here, 1 are mirrors of the ring resonator of the multiharmonic pumping; 2 and 3 are two single-speed relativistic electron beams; 4 is the electromagnetic cluster wave which is the signal (with $n_1\omega_1, n_1k_1$ frequency spectrum) for the lower stage 6 and at the same time—the pumping wave in the second stage of the MDFEL-Klystron; 5 is the modulator of the two-speed electron beam 7; 6 is the signal generator of the cluster pumping wave (the first stage of the MDFEL-Klystron); 7 is the two-speed electron beam; 8 is the transit section of the second stage of the MDFEL-Klystron (system of the beam transportation); 9 is the exit window for the generated (formed) cluster wave of the signal; 10 is the generated (formed) cluster electromagnetic signal; and $n'\omega_1', n'k_1'$ is the frequency spectrum of the generated output femtosecond signal.

Now it plays a part in the opposite multiharmonic Dopplertron pumping for the second stage of the FEL. Finally, the needed output cluster signal 10 is generated (formed) as a result of realization of the multiharmonic version of the superheterodyne amplification mechanism.

It is known that the principle advantage of the stage FEL consists of the possibility to get signals of rather high frequency while using relatively low energy electron beams. It is easy to illustrate the said statement if we use, for example, a correlation which is well known from the theory of Dopplertron FEL (Equation 9.11):

$$\omega_1' \approx 4\gamma_0^2\omega_1 >> \omega_1. \tag{9.13}$$

Here, γ_0 is the average relativistic factor of two-stream systems 2 and 3, and other designations are explained in Figure 9.34. Let us perform some illustration estimations. It is considered that the first stage in the design in Figure 9.34 is also accomplished in accordance with the Dopplertron scheme. If we assume that, for example, $\gamma_0 \sim 10$ (average energy of beams 2 and 3 in Figure 9.34 ~5 MeV), cyclic frequency of the first harmonic of the pumping wave $\omega_2 \approx 10^{10}$ s^{-1} (wavelength $\lambda_2 \approx 20$ cm); we obtain for the cyclic frequency of the first harmonic of the signal generated by the first stage: $\omega_1 \sim 4 \cdot 10^{12}$ s^{-1} (wavelength $\lambda \approx 0.5$ mm). After noncomplex calculations, we get for the output signal wave frequency: $\omega_1' \approx 1 \cdot 6 \cdot 10^{15}$ s^{-1} (i.e., the generated signal is in the optical range of frequencies). In case of the single-stage transformation equivalent as to its scale (assuming, for example, that the generator 6 in Figure 9.34 is also built on the basis of Dopplertron FEL scheme), we find that to get the same transformation coefficient, it needs to use beams of energy ~100 MeV. It is much more difficult from the technological standpoint especially, if we talk about the high-current systems. Under this assumption, we can suppose that in the future, it will be possible to create sources of sub-femtosecond clusters in visible, ultraviolet, and other short-wave ranges of wavelengths by using cluster versions of this type of two- and more stage structures, and it is possible to do it while using relatively moderate relativistic accelerating systems.

Hence, the qualitative physical-technological substantiation for the possibility of practical realization of a new wide class of cluster FEL, which are called here as cluster FLE klystrons, is given in this section. It is worth to mention that earlier, certain studies of single-section (i.e., monotron) versions of femtosecond two-stream FEL only (see the preceding section) were performed; they are known from the bibliography. In other words, the majority of the ideas and klystron principles of FEL construction exposed here are new enough. By this reason, on one hand, the demonstrated possible diversity (which is exclusively rich) of the proposed new design solutions really surprises, and on the other hand, it sets the problem of their quantitative analysis at the equivalent level. The point is that it is the quantitative analysis that can give the final answer to the questions like this: which constructive solutions of the above-mentioned variants are really prospective? Partially, the analysis of this sort is given in Chapters 13 and 14. However, if we compare the scope of results that are already received with the scope of knowledge, we can sadly state that this extremely interesting research sphere is still in its incipient state.

9.4 Linear High-Current Induction Accelerators*

The project analysis shows that the most interesting practical versions of CSFEL may be built on the basis of the sources of high-current and superhigh-current relativistic beams. In connection with this, we shall consider the examples of the constructive sources of this type.

In the conceptual aspect, the principle idea of the multichannel high-current induction accelerators (MHIACs) [92–98] consists of realization of the deep modernization of concepts and basic technological solutions elaborated during the almost 30-year period in the sphere of linear induction accelerators (LIACs) [99–102]. The rapid development of this class of technologies which was observed during the past 30 years was mainly caused by the influence of the global geopolitics factor, and, mainly, by the fact that LIAC became a key element of programs of the Star Wars type (Chapter 1). Therefore, more than lavish financing was historically invested into their realization. As a result, at the same time, great breakthroughs became possible in a number of technological spheres. The most crucial influence was caused by the drastic breakthrough in the sphere of new amorphous materials and ferrite technologies which, as is known, play a key part in the technology of modern LIAC. The second (in sequence, but not as to its significance) achievement is the phenomenal progress in the technology of getting high-current beams of charged particles (electrons and ions).

It is known that one of the principally recognized LIAC advantages is the continuous simplicity of the operating principle. As to its physical essence, this principle is similar to the principle of a common electric transformer operation. In LIAC, the voltage from any powerful impulse source of high-frequency energy is also applied to the primary windings of magnetic inductors. Electron beam, in its turn, plays a part of the secondary winding of this singular "transformer–accelerator." In other words, in this case, the electron beam accelerates due to electromagnetic induction, which is a well-known phenomenon

* From Kulish, V.V., *Hierarchic Methods. Undulative Electrodynamic Systems*, Kluwer Academic Publishers, Dordrecht/Boston/London, 2002. With permission.

in general physics. We shall analyze this analogy of "LIAC–transformer" in detail below. And now let us speak about more general issues.

Historically, the systems of multichannel high-current induction accelerator (MHIAC) [93–98] class appeared as the result of further development of LIAC technologies. First and foremost, the problem of searching for construction systems, which were more adequate for technological conditions of cluster systems, was set. On one hand, these construction systems must maintain all unique peculiarities of the LIAC, and on the other hand, they must considerably moderate their most significant shortcomings, if not remove them all. First, it is excessive longitudinal dimension (the first accelerator ATA was ~70 m long [100]). Besides, another task, which consisted of forming two-speed (and in general, multispeed) high-current electron beams, increasing the beam general current at the system exit up to 100 kA, at least, and in simultaneous acceleration of both electron and ion beams, was formulated.

9.4.1 Classification of Multichannel High-Current Induction Accelerators

Prior to considering the basic concepts and key ideas of multichannel high-current induction accelerators (MHIACs), let us briefly speak about their general classification (see Figure 1). In the previous section, we have already said that the choice of classification criteria is a problem of primary importance for any classification system, and it is important that the classifications built in this way must be independent and the simultaneous application of several of them must allow us to get the most complete idea of the respective research sphere. Following this methodological scheme, let us choose the number of acceleration channels (see Figure 9.35), shape of these channels (Figure 9.36), and their constructive location (Figure 9.37) as classification criteria.

However, let us also mention that just now the total number of the proposed scheme variants that are compatible with the above-said classification is of hundreds of variants, and it is not the end, because historically, the multichannel induction accelerators as a class of modern acceleration systems were proposed just recently [92–98]. Hence, we can expect that the list of design variants will continue to increase in the future. However, it is clear that the adequate complete description of all this list of schemes in the framework of this study which is limited by its volume is not real. Taking into account all the above-mentioned, we shall limit by the brief description of the most visual examples only.

9.4.2 Linear Induction Accelerators as a Technological Basis of the MHIAC

As to their construction, the MHIACs are almost completely based on the use of already known and practically proved LIAC technologies. As the majority of key differences

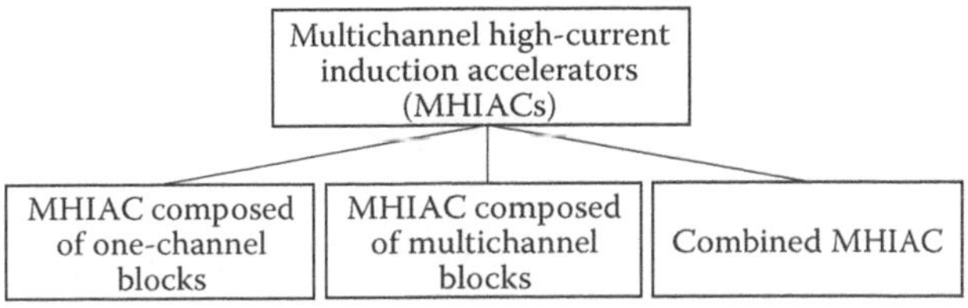

FIGURE 9.35
General classification of multichannel induction accelerators (MHIACs) by "number of acceleration channels" criterion.

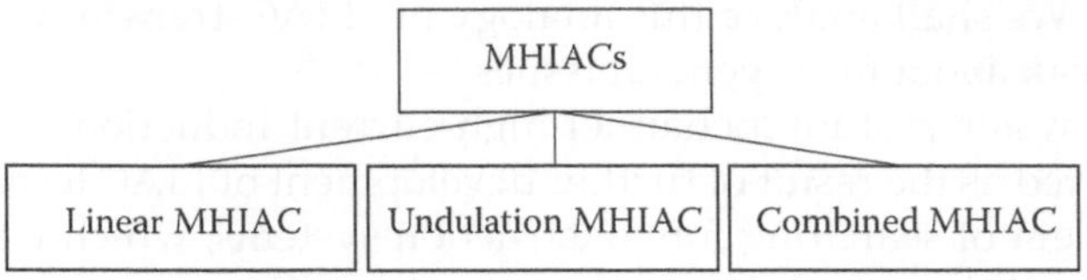

FIGURE 9.36
General classification of multichannel induction accelerators (MHIACs) by "shape of acceleration channels" criterion.

have scheme and arrangement character, then, while describing the technological principles of the LIAC, we give the description of basic technologies that serve as a ground for the whole class of systems of the MHIAC type. Hence, we shall begin the brief analysis of physical and technological peculiarities of the MHIAC with the discussion of their operating principle and some constructive peculiarities of LIAC key constructive elements and units.

The LIAC general block diagram is shown in Figure 9.38. The system operates as follows. The injector 1 generates the beam of charged particles 2, which, in its turn, passes into the acceleration unit 3. This unit consists of acceleration sections 4. The beam 2 passes through section 4 and the eddy electric field accelerates it (see more details about the physics of these processes below in this section). The acceleration unit 3 as a whole, and its sections 4 as the components, are connected to units 5 and 6. They are the unit of impulse power and auxiliary systems 5 and control unit 6. The purposes and functions of these units are clear from their denominations.

And now we continue with more detailed analysis of some of mentioned assemblies and units. Let us begin with the discussion about the operating principle and basic constructions of injectors of charged particle beams (see unit 1 in Figure 9.38).

9.4.3 Injectors

The design solution of LIAC injectors is one of their characteristic peculiarities that distinguish them from accelerators of other traditional types. In the latter, so-called electron guns or standard sources of ion beams (their operating principle is a standard one) are used as injectors. The specific feature of these injectors consists of the fact that they usually operate with relatively low-current beams of charged particles (typical beam impulse currents rarely exceed hundreds of amperes). However, it is known that practically all modern LIACs are designed to form high-current (units to tens of kiloamperes) electron or ion beams [99–102]. It is also known that the above-mentioned sources of standard design cannot form beams of such a level due to purely physical reasons.

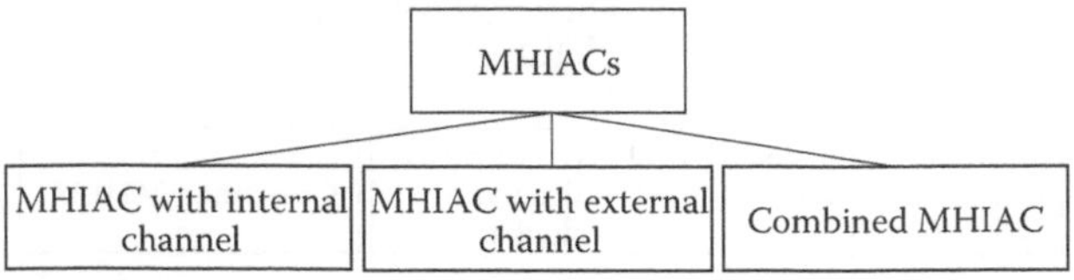

FIGURE 9.37
General classification of multichannel induction accelerators (MHIACs) by "acceleration channels locations" criterion.

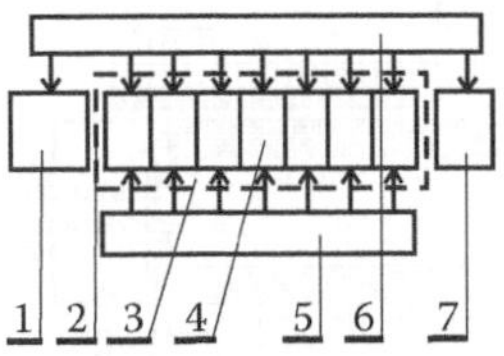

FIGURE 9.38
General block diagram of the linear induction accelerator (LIAC). Here, 1 is the injector of charged particles (electrons or ions); 2 is the beam of charged particles; 3 is the acceleration unit; 4 are acceleration sections; 5 is the impulse power supply and auxiliary systems unit; 6 is the control systems unit; and 7 is the exit system for the beam of charged particles.

The latter is explained principally by the fact that if the purveyance of the kiloampere beam is of the order of a unit (or several units), then to "take it out" of the cathode region, static (quasistatic) voltage of ~0.5 MV as minimum, or even higher voltage, needs to be applied. From the technological standpoint, it means that it is necessary to use high-voltage power systems capable to assure this level of anode voltage. The latter automatically means the appearance of rather great technological problems. They are related, mainly, to problems of transit of voltage of units of megavolts from the power source to electrodes of the electron gun, including the problem of electrical breakdown. That is, the list of emerging problems and standard ways of their solution are well known. By this reason, we shall limit ourselves by statement of the fact that traditional electron (or ion) guns seem to be of little interest for the practical systems of LIAC type (and of MHIAC type as well).

That is why the scientists took another technological way in LIAC technology. They proposed and practically realized the idea of so-called inductive injectors, where some specific properties of high-frequency fields are originally used. Two examples of the injectors of this type are shown in Figures 9.39 and 9.40, and they expressively illustrate the main specifics of this technology.

The scheme shown in Figure 9.39 operates as follows. The magnetic inductor 2 induces eddy electric field in the internal part of the shield 1. Due to skin effect, this field is completely concentrated in a relatively thin near-surface layer (skin-layer) of the shield 1 material. In other words, the potential of the external part of the shield does not change in this case and equals the earth potential (which is zero), whereas all the internal surface may have a high HF-potential. The breakage in the loop of intensity vector circulation for eddy field is realized as a gap between the central electrode 3 and the part of the electric shield

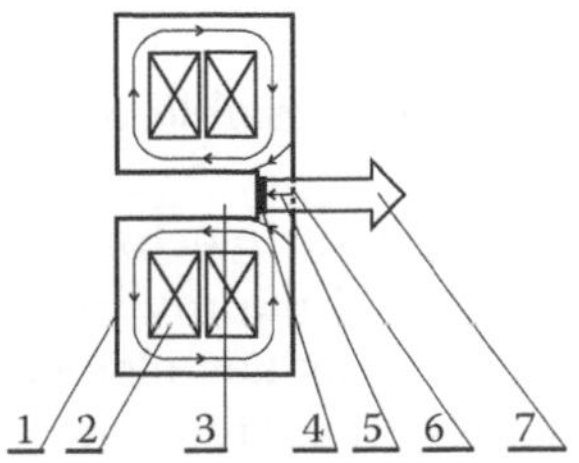

FIGURE 9.39
Example of the simplest design scheme of the induced high-current injector of electron beam. Here, 1 is the electron shield; 2 is the magnetic inductor; 3 is the central electrode; 4 is the cathode; 5 is the force line of the internal part of the field used to accelerate the beam; and 6 is the part of the electric shield 1 (e.g., anode, grid, or foil), which is transparent for the beam of electric particles 7.

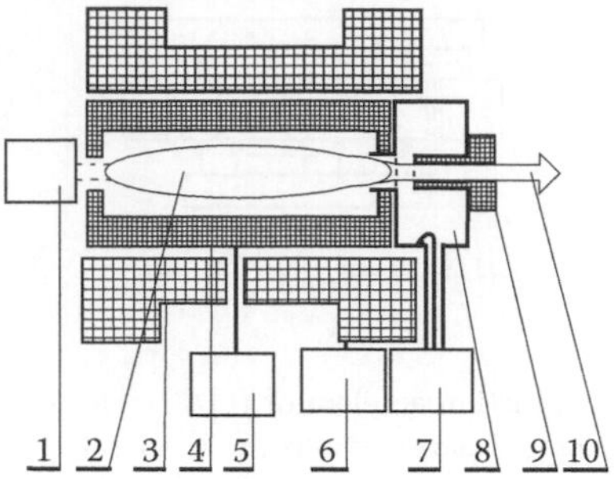

FIGURE 9.40
Example of design scheme of the injector of modulated ion beam on the basis of magnet plasma trap. Here, 1 is the gas source; 2 is the plasma filament; 3 is the magnet system to suspend the plasma filament 2; 4 is the system of HF-heating of the filament plasma 2 (HF-inductor); 5 is the power source for the system of plasma 4 HF-heating; 6 is the power supply system for the magnetic system 3; 7 is the source of microwave frequency radiation for the ion beam modulator; 8 is the microwave frequency resonator for ion beam modulator; and 9 is the magnetic system for ion beam 10 transportation.

1 which is transparent for electrons (it may be, for example, a grid or foil) 6. The force lines of the HF-field that exist on the internal surface of the shield 1 and electrode 3 are closed through vacuum in this gap, where the cathode 4 is located. Electrons ejected by this cathode experience the acceleration action of the HF-electric field 5. As a result, during the "negative" phase of the field, these electrons are "taken out" of the cathode region forming the high-current beam 7. Then this beam is directed to the input of the LIAC acceleration unit (see block diagram in Figure 9.38). During the "positive" phase, the system does not operate.

It was already indicated that the main advantage of induction injectors compared with traditional electron guns consists of the absence of static superhigh potential difference between any two pairs of electrodes in the external part of the injector. In this case, the maximum voltage is that across the primary windings of inductors 2, in practice, it usually equals ~25–50 kV if acceleration voltage inside the injector (i.e., between the cathode and anode) is of ~0.6–5.0 MV or more. However, it is worth mentioning that the physics of magnetic injectors is not always as simple as it may seem at first sight. For example, in case of superhigh voltages between cathode and anode, electrons may be ejected from the lateral surface of the central electrode 3 (autoemission phenomenon), and this effect is not desirable for constructions of this type. To neutralize this effect, the special magnetic field is formed in the region of the possible autoemission (it is so called magnetic isolation), etc. Hence, we want to emphasize that today the magnetic injector technique is a rather complicated, developed, and independent technological sphere, whose more detailed description, unfortunately, is impossible within the limited framework of this review.

One of the main shortcomings of magnetic injectors, whose example is shown in Figure 9.39, is their limited useful life. It is caused mainly by technological restrictions as to the real service life of cathodes that are used as key elements of the construction. The thing is that any solid cathode, especially a high-current one, destroys early or late and stops to normal function, independent of the type of its construction. During the past ten years, the professional dream of creating a so-called eternal cathode that never wears out existed in this and in associated spheres. In the époque of the Star Wars program, this dream was practically realized in the form of so-called plasma cathodes. The idea consists of drawing the beam of charged particles (electrons or ions) out of specially treated gas plasma that plays the part of such an eternal cathode. It is obvious that such a cathode never wears out because it is not rigid or fixed in time but constantly resumes during its operation. However, it does not mean that the system of this type may be considered as an eternal one in the direct sense of this word. Naturally, its service life is determined by factors of the next hierarchy level, such as the service life of high voltage cables isolation, operating life of thyratrons of power systems, etc. Nevertheless, the practice demonstrates that the plasma cathode technologies are extraordinary prospective for creation of injector systems with elevated term of service life.

An example of the system that illustrates the idea of the plasma cathode is shown in Figure 9.40. Here, we see the design scheme designed to form high-current impulse ion beams of industrial destination, modulated with microwave frequency (as to the density). They are used, for example, in technological systems for upgrading material surfaces, etc. The construction is based on the use of magnetic traps of "magnetic bottle" type, which, in due time, were created within the framework of the well-known program on controlled thermonuclear synthesis [104]. It was proposed [92] to use duly modified "magnetic bottle" as electron or ion plasma cathode.

The system shown in Figure 9.40 operates as follows. Gas from the gas source 1 enters the volume limited by the internal chamber of the system of plasma HF-heating 4, and the source of HF-power 5 is connected to it. HF-field of the system 4 ionizes gas in the working chamber of the system 4, which transforms in a rather dense plasma. The heating system 4, in its turn, is enclosed by the magnet system (solenoid of special construction) to suspend plasma 3; this magnet system is powered by the power source 6. Magnetic field of system 3 forms the plasma filament 2 of special shape in the working region of the system 4. The ends of this filament seem to be tied by special magnet strings (which is characteristic for the magnet traps of this type). As a result, all the plasma of the filament 2 is localized in a closed volume, or in a so-called plasma bottle [104]. The construction of the considered injector provides for one (e.g., right) end of the bottle to be in the working volume of the entrance part of the microwave resonator 8, which is connected to the microwave resonator 7. This construction may have two versions. One of them is ion, and the other one is electron version.

Let us begin the analysis with the ion version. The peculiarity of the system consists of the fact that the plasma in the region of the right "magnet string" is polarized. The special system of electrodes is also localized here; its task is to regulate the "correct" polarity and the magnitude of this polarization. To do it, the electric field created by the electrodes "pulls" the plasma electron component to the resonator 8 entrance. During positive half-periods, microwave electric field of the resonator can pull ions out of filament plasma—bottle 2 and form ion beam 10 at the system output. In this case, the output magnetic system 9 serves to adjust the shape of the output signal 10 similarly to a transit section where the velocity modulation is transformed into density modulation. As the plasma of the filament 2 is dense enough, the output beam current may be of units to tens of kiloamperes, and the beam 10 is formed as an ion bunch of several hundreds of nanoseconds duration, which is modulated with microwave frequency (as a rule, in decimeter band). In this case, the bunch has a complicated internal structure because it consists of singular microbunches. As a result of modulation, the final transformation of velocity modulation into density modulation takes place in the resonator volume 8. Then the microbunch peak current formed in the output beam 10 may reach hundreds of kiloamperes.

In the case of the injector electron version shown in Figure 9.40, we have the situation with symmetric signs. The negative voltage is applied across the electrodes and it is sufficient to "pull" the electron component to the resonator 8 entrance. Its electric field when being in the "negative" phase pulls the plasma electron component, then high-current electron beam formation and modulation takes place. The most prospective design variant for electrode power system is the magnetic inductor (for the simplicity it is not shown in Figure 9.40). It is similar to those inductors that are used in high-current linear induction accelerators (LIACs) and multichannel induction accelerators (MIACs) (see this and next sections given below).

9.4.4 Acceleration Sections

The illustration of operating principles of LIAC acceleration sections is given in Figures 9.41 and 9.42.

The scheme shown in Figure 9.41 also explains the operating principles of so-called nonshielded acceleration section (in this case, the electric shield is made for the entire acceleration unit as a whole). The scheme operates as follows. Impulse electric current in the primary winding of the magnetic inductor 1 induces variable magnetic flux in its volume. This flux (due to its changes in time) generates eddy electric field 2, 3 in the region near inductors. Item 2 corresponds to the internal and item 3—to the external part of the field. Electron beam 4 accelerates by the electric field. It is worth mentioning that the characteristic peculiarity of LIAC traditional schemes, as illustrated in Figure 9.41, is the use of the internal part of the field 2 to accelerate electron beam 4, whereas the external part 3, as a rule, is unused. We shall show later that it is possible to develop rather effective technological methods on the basis of the idea of the use of the field external part, and these technologies result in certain essential advantages for some systems of the MHIAC class.

The scheme shown in Figure 9.41 is physically visual; however, it is not a widespread scheme of the LIAC-type constructions. The construction of the shielded acceleration section whose principal idea is illustrated in Figure 9.42 is widely used. In contrast to the case illustrated in Figure 9.41, now each LIAC acceleration section is enclosed by an individual electric shield 3 (see Figure 9.42). The special gap in the shield is made in the form of the acceleration gap 2 with the internal part of the eddy electric field 1 localized between the electrodes (compare with the similar gap in the construction shown in Figure 9.39). In this case, the external part of the electric field is limited in space by the walls of the shield 1

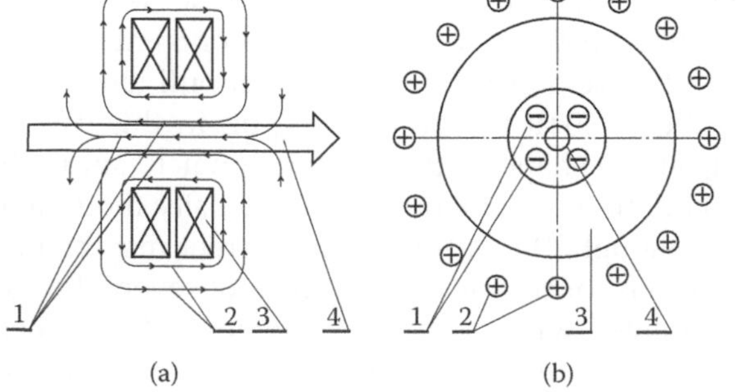

FIGURE 9.41
Illustration of basic operating principle of LIAC nonshielded acceleration section. Here, 1 are force lines of the internal part of the eddy electric field; 2 are force lines of the external part of the electric field; 3 are inductors; and 4 is the electron beam that is accelerated. We can see from the figure that the electron beam 4 is accelerated by the internal electric field 2. The left part of the figure (a) is the frontal projection, and the right part (b) is its cross projection.

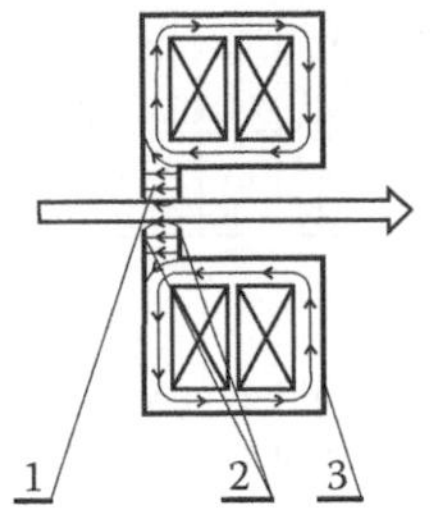

FIGURE 9.42
Illustration of the basic operating principle of the LIAC shielded acceleration section. Here, 1 are force lines of the eddy internal electric field in the acceleration gap 2; 2 is the acceleration gap; 3 is the electric shield that is partially used as a secondary winding of the "transformer-accelerator." As we can see in the figure, electron beam closes the secondary winding in the acceleration gap 2.

itself. Electron beam accelerates by the internal field 1 exclusively within the limits of the acceleration gap 2.

9.4.5 Inductors

Now it is worth to mention that two types of magnetic inductor constructions are used in the technology of traditional LIAC (see item 2 in Figure 9.39 and item 1 in Figure 9.41). The first of them is characterized with the presence of special magnetic cores made of amorphous magnetic material (of MetGlass type) or HF-ferrite. In the second case, the magnetic cores are absent. By this reason, this type of construction is often called "air-cored."

In practice, the constructions of the first type are preferential because the availability of the magnet core allows to considerably increase (by hundreds of times) the magnetic field induction (compared with the "air-cored" case) in the inductor with the same current in primary windings. On the other hand, magnetic cores become a source of considerable HF-energy losses. As a result, realization of LIAC with acceptable practical efficiency becomes reasonable only in the case when it is needed to accelerate high-current (kiloampere) beams. In addition, the magnetic cores themselves have a rather big mass, and the general weight of the accelerator as a whole critically increases. Besides, they also create great technological problems related to heat withdrawal from the working volume of the acceleration section, etc. The problems of this type become critical in case of systems of satellite basing with the accelerator. In this case, the use of both LIAC and MIAC becomes practically impossible. However, let us mention that in case of systems of ground basing, these shortcomings are not critical. The exclusions are rare; they refer to those cases when especially severe requirements are formulated with respect to mass–dimension characteristics of the accelerator and its heat losses.

The said inferior limitation as to the beam current in air-cored construction variants principally does not exist, and it is their principal advantage. Besides, they are much lighter than systems of the first type. That is why their practical use in airborne systems and systems of satellite basing seems rather prospective. However, we must emphasize that this advantage is associated with considerable increase in current (by tens and even hundreds of times) in primary windings of the inductors. The solution of associated technological and construction problems in air-cored LIAC consists of the use of special superconductor constructions for primary windings of inductors and parts of power sources. It is this circumstance that plays the role of the principal limiting factor with respect to wide practical application of air-cored designs.

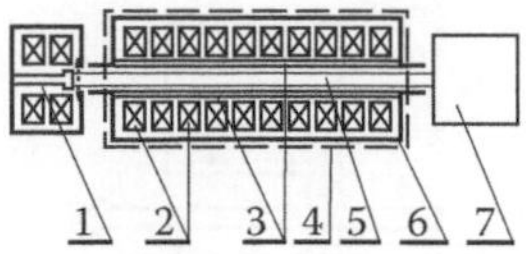

FIGURE 9.43
Example of the LIAC design scheme with nonshielded acceleration sections. Here, 1 is the injector of charged particles (electrons or ions); 2 are inductors; 3 is the acceleration channel with dielectric walls; 4 is the acceleration unit as a whole; 5 is the electron beam under acceleration; 6 is the common electric shield; and 7 is the exit system for the accelerated beam.

9.4.6 Traditional Design Schemes

The example of the LIAC-designed scheme based on nonshielded acceleration sections (similar to those shown in Figure 9.41) is shown in Figure 9.43. The scheme operates as follows. The injector of charged particles (electrons or ions) 1 forms a beam of charged particles 5, which passes into the working volume of the first acceleration section 2 that envelops the acceleration channel 3. The beam 5 accelerates in the acceleration channel 3 by the eddy electric field (see Figure 9.41 and respective comments) and passes through it into the exit system 6 or acceleration channel of the next unit. The latter may be realized in the form of a system of beam formation and control, system of beam escape from vacuum into the atmosphere, etc. The common shield 5 provides for inductors electromagnetic shielding from the medium. In addition, it resolves the problem of electromagnetic compatibility and biological protection of the personnel from the HF-field action.

The scheme shown in Figure 9.44 distinguishes from that which was already analyzed (see Figure 9.43) only by the construction of the acceleration section 2. In this case, the acceleration unit is executed as the series connection of shielded acceleration sections 2 (see Figure 9.42 and respective comments). It means that in the course of beam (formed by injector 1) motion in the acceleration channel 3, the increase in its energy (acceleration) takes place only at those moments when it passes through acceleration gaps of sections. The beam moves without any acceleration along the segment of channel 3 between the acceleration gaps. This fact offers additional possibilities to adjust its shape and other parameters. Comparing the two design schemes shown in Figures 9.43 and 9.44, we can easily see that the existing construction differences have rather conditional character, first and foremost, because any acceleration section 2 in Figure 9.44 may be interpreted as the shortened variant of the acceleration unit 4 in Figure 9.43.

Therefore, the specific peculiarity of the acceleration process in the LIAC which principally distinguishes them from those processes that take place, for example, in electrostatic accelerators, is the following. In the case of LIAC, the beam energy systematically accumulates in the course of its consecutive transit through acceleration sections. It means

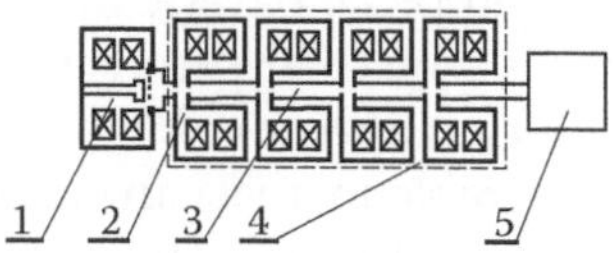

FIGURE 9.44
Example of the LIAC design scheme with individual shields for every acceleration section. Here, 1 is the injector of charged particles (electrons or ions); 2 is the shielded acceleration section; 3 is the acceleration channel; 4 is the linear single-channel acceleration unit; and 5 is the exit system for the accelerated beam.

that if the beam passes the potential difference U in every acceleration section (see, for example, Figure 9.42), then, when passing the whole linear one-channel acceleration unit 4 (see Figure 9.42), its energy becomes qNU (where q is the absolute charge of the accelerated particle and N is the number of sections 2). The key circumstance in this case is the fact that the potential difference between the internal and external surfaces of the shield for any of sections 2 (hence, between the same surfaces of electrodes in the acceleration gap) equals U. The above is visually illustrated in Figure 9.45, where we can see the equivalent scheme of the LIAC as a "transformer." All the primary windings (of inductors) are connected in parallel and each of them is under the same voltage U. It is obvious that in this situation, the voltage across the secondary winding (potential difference passed by the beam during its acceleration) equals NU, as was already mentioned.

The latter is caused by the physical peculiarities of the skin-effect which were discussed earlier; this effect is realized here for the eddy HF electric field. Let us recall that in this case, almost all the electric field in the material of the wall is concentrated in its thin (internal) near-surface layer. It means that if the wall is thick enough, then this field "does not reach" the opposite (external) surface of the material. Then there is every reason to interpret the sequence of inductors (see Figures 9.43 and 9.44) as a parallel connection of N primary windings of the LIAC transformer (see Figure 9.45).

At the same time, in the formally similar process of charges acceleration in the electrostatic accelerator [105] (that is, by the static electric field), such an effect of the mutual isolation of internal and external surfaces of the material does not occur. In this case, the skin-effect "does not work" and, as a result, potential difference between the external and internal walls of electrodes in acceleration gap equals zero. In other words, all the surface of each of the electrodes of the acceleration gap is principally equipotential. As a result, the beam gets the energy qNU equal to that being get in the LIAC, only in the case when the potential difference between the first and last electrodes of the acceleration gaps of the entire unit is N times bigger, and equals NU. When the beam energy is the same as in the case of LIAC, the maximum potential difference in case of electrostatic accelerator is N times bigger. This fact causes the LIAC principal technological advantage compared with electrostatic accelerators. The LIAC advantage compared with another traditional type of accelerators, namely, radio-frequency (linear) accelerators (LINAC) [105] is caused, first and foremost, by the unavailability of superpowerful sources of microwave energy sources in LIAC, as their creation and operation are associated with certain specific problems. On the other hand, LIAC can generate electron beams with record-breaking current (up to 30 kA and more). However, let us mention that in some cases [e.g., while creating parametric superheterodyne FEL (PSFEL), see the previous section], the use of LINAC may be reasonable, too. First, due to the fact that electron beam at the LINAC output is always

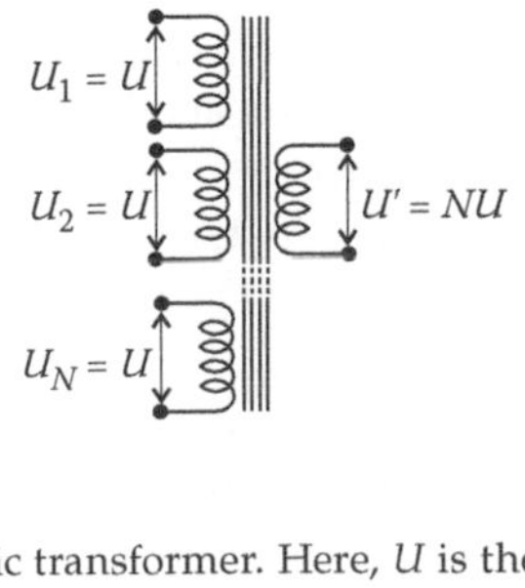

FIGURE 9.45
Equivalent scheme of the LIAC as electric transformer. Here, U is the voltage across the primary windings of inductors; N is the number of inductors; and NU is the resulting potential difference the beam passes in the course of its acceleration.

modulated. In the PSFEL case, modulation electron waves are used as additional (electron wave) pumping (see more details in Chapter 14).

It is generally known that the principal shortcoming of traditional LIAC as technological basis for cluster systems is their excessive length. Better to say, it is not the problem of the length itself, but of the ratio of characteristic longitudinal and transversal dimensions. This is the main obstacle for their wide practical use, especially for the systems of satellite basing. For example, the length of the famous ATA accelerator (built in the 1980s in Livermore laboratory in the USA within the framework of the Star Wars program) [100] was of ~70 m (with beam current of 10 kA and electron energy of 50 MeV) and characteristic transversal dimension of ~2 m. In this case, the said ratio equals ~35. However, it is well known that the value of this ratio which is ideal for practice is ~1. Besides, traditional LIACs are characterized with rather limited functional possibilities. They can accelerate only one beam (electron or ion) at the same time. All the LIACs have principal limitation as to the maximum value of current (several tens of kA). The analysis demonstrates that they cannot be used to form beams with given energy structure (e.g., two- or multispeed structure), etc. All these shortcomings, as we shall see later, may be effectively removed by using the acceleration systems of MHIAC type.

9.5 Undulation High-Current Induction Accelerators*

9.5.1 Key Design Elements

Prior to continuing with the description of constructive peculiarities of one or another MHIAC modification, consider certain basic ideas that are specific for the systems of this class, as we see later. These basic ideas are

1. The use of both internal and external parts of the eddy (vortex) electric field to accelerate charged particles [98]
2. The use of magnetic turning systems to transport beams of accelerated charged particles from output of one acceleration unit to input of another parallel acceleration unit [92,97,98]
3. Realization of multichannel principle including external [98] and internal multichannels [93,94,97,98]

Furthermore, we discuss these types of ideas and construction innovations in more detail.

We shall start with the analysis of the first above-mentioned idea. To do it, let us go back once again to the basic principles of standard accelerating systems of the LIAC class (see Figure 9.41). This issue was analyzed completely enough in the previous section. Here, we shall limit by the statement that the eddy electric field generated by magnetic inductors 3 includes both internal 1 and external 2 parts (Figure 9.41a represents the frontal projection of this scheme, and Figure 9.41b shows the same scheme but in its cross projection). It is easy to notice that among the principal differences between external and internal fields, we must mention that the vectors of their electric fields are directed in opposite directions.

* From Kulish, V.V., *Hierarchic Methods. Undulative Electrodynamic Systems,* Kluwer Academic Publishers, Dordrecht/Boston/London, 2002. With permission.

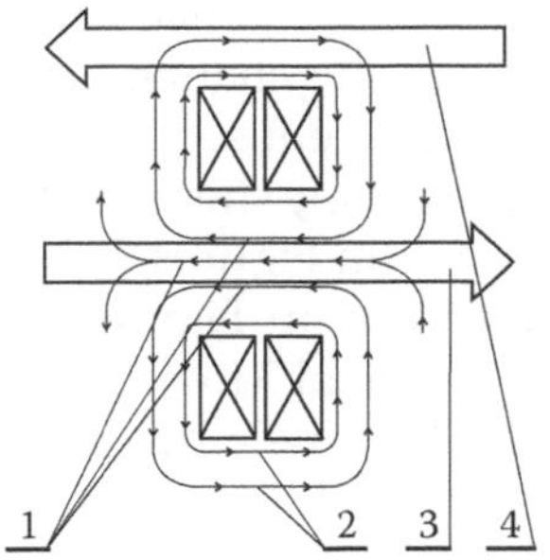

FIGURE 9.46
Illustration of the idea of simultaneous use of both internal and external parts of the electric field to accelerate beams of charged particles. Here, 1 are force lines of the internal part of the eddy electric field; 2 are force lines of the external part of the eddy electric field; 3 is the internal electron beam; and 4 is the external electron beam.

Then we note that in traditional LIAC constructions [99–103], it is only the internal part of the general field 1 (see Figure 9.41) that is used to accelerate beams. In contrast to tradition, the idea to use also the external field 2 to accelerate beams was proposed in patents [93,94,98]. The essence of the idea is illustrated in detail in Figure 9.46. It is easy to see that the idea consists of using the same inductor 3 (shown in Figure 9.41) to simultaneously accelerate no less than two beams of charged particles 3 and 4 (Figure 9.46). One of the beams (in this case, it is the beam 3, see it in the same figure) is accelerated by the internal electric field 1 (Figure 9.46), whereas the beam 4 accelerates due to the work of the external field 2 (see the same figure). In general, we speak about both electron beams (like those in Figure 9.46) and beams of positively or negatively charged ions.

It is easy to become sure that the above idea may have two partial scheme realizations that differ by the sign of accelerated beams. If the two beams have the same charge, then directions of their acceleration are opposite (the example is shown in Figure 9.46). In the other case (beams of different signs), they accelerate in the same direction.

Two technological versions to realize this idea were proposed. One of them [93,94,96,97] is illustrated in Figures 9.47 and 9.48. The characteristic peculiarity of this version is that the part of external field 3 (see Figure 9.47) generated by the inductors 6 of one acceleration section is used for additional acceleration of the beam 1 in the internal channel of the

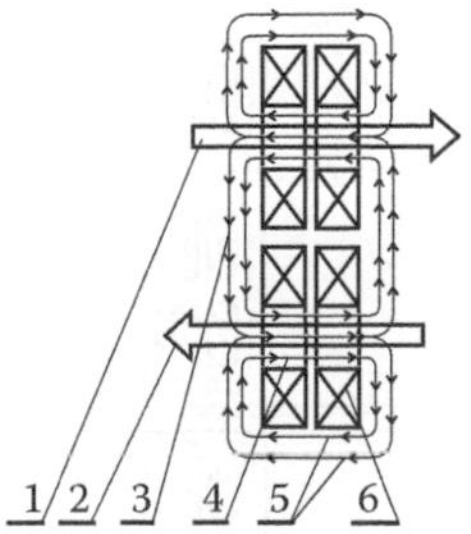

FIGURE 9.47
Illustration of the idea of the use of external parts of the electric field to accelerate particles in the adjacent acceleration channels (construction with nonshielded separate sections). Here, 1 is the first internal electron beam; 2 is the second internal electron beam; 3 is the common (for the two acceleration channels) part of the electric field outside the acceleration channels; 4 is the part of the intrinsic internal electric field in acceleration channels; 5 is the external electric field; and 6 are inductors.

section which is parallel to this one, and vice versa. The principal acceleration results from the intrinsic internal field 4 of each of the channels, and additional acceleration, which results from the external field of the adjacent section, plays an auxiliary role. As to other details, the operating principle of the scheme shown in Figure 9.47 coincides with the operating principles of constructions shown in Figures 9.41 and 9.46.

A principally similar idea is illustrated in Figure 9.48. It differs from the first design scheme (see Figure 9.47) by additional shield 3 (which is the same for two individual parallel acceleration sections). In this case, the shield has two design tasks.

The first one consists of simplification of the construction of the accelerator as a whole and increase in its fabricability. The existing experience of practical design of this type of system shows that the common shield (which is used simultaneously for several parallel acceleration sections) allows to essentially improve conditions for convection of the isolating gas (insulating gas or nitrogen) inside the shield, provides for additional possibilities in design maneuver in the volume of the acceleration section to place auxiliary units and systems there, etc.

The second task of the shield 3 (see Figure 9.48) consists of additional intensification of the total internal field in the acceleration gap of one section due to greater (compared with the previous case) component of the external field of the adjacent section and vice versa. The essence of this amplification effect is as follows. In Figure 9.47 (system without a shield), we can see force lines of the external electric field 5, which further become closed in the acceleration channel of the same acceleration section. However, it is only one variant of the space localization of the given field. It is easy to understand that the same force lines may have other variants of space closing. According to one of these variants, force lines 5 (which are localized at essentially large distances from the inductor) close in the acceleration channel of the adjacent (and not the same, as shown in Figure 9.47) acceleration section. However, in this case, the field is opposite to the principal acceleration field in the channel and somewhat weakens it. By this reason, the gain due to the effect of resultant field increase, which is discussed here, becomes partly underestimated in systems with nonshielded acceleration sections. In the variant with the shield, a part of this weakened field is particularly shielded by the shield 3 (see Figure 9.48).

The idea of another technological variant of acceleration of charged particles beams by external fields is shown in Figure 9.49 [98]. In this case, the acceleration section, in addition to the internal acceleration gap 1 includes one or several external acceleration gaps 4 located at electrodynamical gaps of the shield 3. Hence, the electron beam 6 in this construction

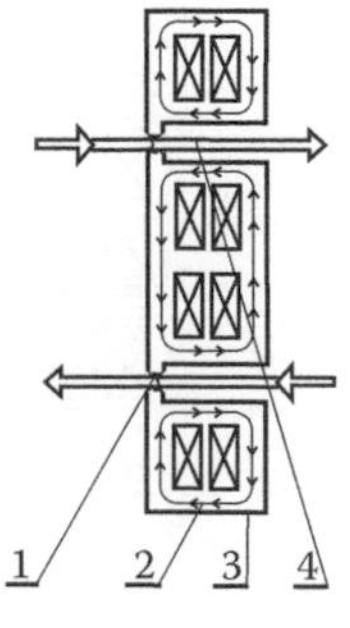

FIGURE 9.48
Design scheme of MHIAC twin acceleration section with common (for two individual acceleration sections) electric shield. Here, 1 are force lines of the field in the acceleration gap; 2 is the part of the external electric field inside the shield 3; and 4 is the accelerated electron beam in the adjacent acceleration channel.

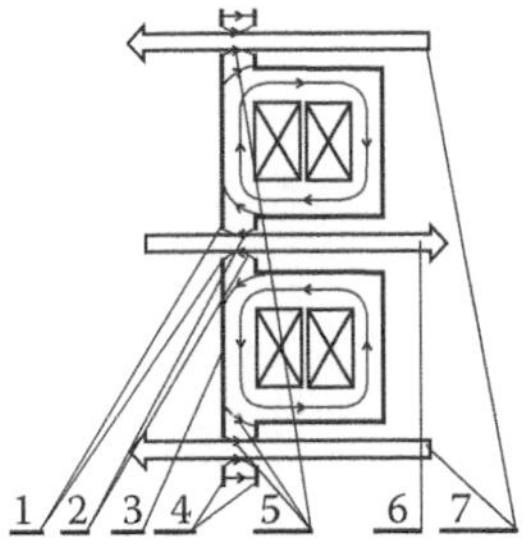

FIGURE 9.49
Design scheme of MHIAC acceleration section with external acceleration channels. Here, 1 is the internal acceleration gap; 2 are force lines of the internal accelerating electric field; 3 is the electric shield; 4 are external acceleration gaps; 5 are force lines of the external accelerating electric field; 6 is the electron beam that accelerates in the internal channel; and 7 are electron beams that accelerate in external channels.

accelerates by the internal field 2, whereas beams 7 accelerate by the external field 5, and the directions of acceleration of both beams are opposite.

The elementary numerical estimations allow us to easily verify that one of the principal physical differences between internal and external fields consists of the very high strength of the former compared with the strength of the latter. Then automatically we realize that, in general, the rate of beams acceleration in external acceleration channels must be very small compared with the rate of beams acceleration in the internal channels. This fact may have either positive or negative significance in practice. For example, if the goal consists of forming intensive relativistic beams of quasineutral plasma, a complicated problem of measuring speeds of electron and ion components of the output plasma beam arises. The existence of this problem is explained by the fact that the speed of the ion beam is always considerably less than that of the electron beam at the same energy because the order of magnitude of the ion mass is three times and over greater compared with the mass of an electron. The idea of the simultaneous use of both internal and external channels is rather prospective (as analysis shows) to form this kind of beams of quasineutral plasma. The thing is that in this case, the directions of acceleration of both components of plasma beam are equal. Besides, due to the above-mentioned difference in acceleration rates of external and internal channels, we get the opportunity to form two beam components with essentially different output energies (but close speeds). When the electron component of the plasma beam accelerates in external channels and the ion component accelerates in internal channels, we get the desirable speed difference for the two components at the exit of the accelerator.

However, the case of acceleration of beams that have the same sign is much more actual. Then the above-mentioned significant difference in acceleration rates in external and internal channels is extremely undesirable. In these cases, the use of special technological methods is proposed; their goal consists of considerable increase in typical acceleration rate of beams in external channels. An example of the design scheme of this type is shown in Figure 9.50 [98], where two methods are illustrated at the same time.

The first of the said methods consists of selection of the special shape for the shield because the shield operates as a concentrator of electric field force lines [98]. The idea is to maximally approximate the density of electric field force lines in the external channel to the density of lines in the internal channels. To do it, it is necessary to concentrate the majority of force lines in one (or two, as maximum) external acceleration gap (see the shape of shields in a design shown in Figure 9.50, as an example). It is worth mentioning

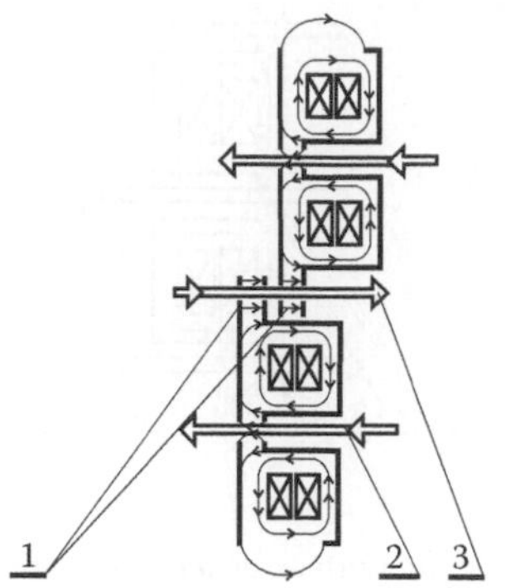

FIGURE 9.50
Design scheme of parallel connection of two acceleration sections with external acceleration channels. Here, 1 are two external acceleration gaps with parallel connection; 2 are internal beams, which accelerate; and 3 is the external beam, which accelerates.

that more or less precise calculation of shields-concentrators of this type, in practice, is a complicated problem. However, rough, semiqualitative physical estimations allowed getting certain results that look rather encouraging. They prove that, in fact, it is principally possible to get field intensity in (one) external channel at the level of ~70%–80% of the field intensity in the internal channel when selecting the special shape of shield concentrators.

The other of two above-mentioned technological methods used in the scheme shown in Figure 9.50 consists of the use of series connection of external gaps 1, which form part of two (or more) different parallel acceleration sections [98]. As a result, one external acceleration channel with doubled (or even greater) acceleration rate is formed. In principle, according to estimations, acceleration rate for the beam 3 in such a common channel may even exceed the acceleration rate of beam 2 in internal channels of the system shown in Figure 9.50.

9.5.2 Concept of Undulation Accelerators

It was already mentioned in the previous section that one of the principal shortcomings of traditional LIACs, which considerably complicates the possibility of their wide practical use, is their excessive length. As an example, we took the ATA system of linear induction accelerators, which is the most known system [100–103]. Well, speaking of this length problem, we take into account not the length itself (the length is not informative enough) but the ratio of maximum and minimum characteristic dimensions of the acceleration system. In the chosen ATA example, the maximum dimension (length) equals ~70 m, whereas the minimum dimension (diameter) equals only ~2 m. Therefore, the ratio of two dimensions is ~35, and it is extremely unfavorable. We know quite well that, from the design standpoint, the ideal value is ~1.

A series of designs of undulation MHIAC [97,98] was proposed to solve this problem of dimension optimization of the construction (their classification is given in Figures 9.35 through 9.37). The basic idea of this type of spatial arrangement is illustrated by the simplest design scheme shown in Figure 9.51. Acceleration channels 5, 6 of two independent parallel acceleration sections (see Figure 9.48 and respective comments) are connected by the turning system 7. As a result, the accelerated beam 1 passes through the first acceleration gap, enters the turning system 7 where it changes its direction by 180°, passes through the second acceleration gap and goes out of the system as the accelerated beam 2, but now it has opposite direction. The beam path has a characteristic U-like shape. Then, in general,

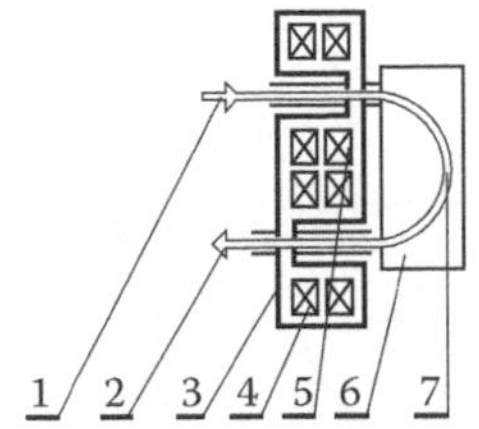

FIGURE 9.51
Design scheme of connection of two acceleration sections with the use of the turning system. Here, 1 is the accelerated input beam; 2 is the accelerated output beam; 3 is the common electric shield for two parallel acceleration sections; 4 are magnetic inductors of the second (along the electron beam path) acceleration unit; 5 are magnetic inductors of the first acceleration unit; 6 is the turning system; and 7 is the beam in the turning system.

the output beam 2, as a rule, is directed into the internal channel of the other parallel (with respect to the first two) acceleration section (unit). As a result, the path of the beam acquires the expressed undulator (that is, wave-like) shape. By this reason, the acceleration systems built on the basis of this principle are called UNIACs.

However, another version of the UNIAC is also possible; it characterizes with the use of both internal and external acceleration channels. For example, there are design variants where the beam, being accelerated in the internal channel, passes the turning system and then it is channeled into the external acceleration channel of the same section [98]. In this case, the same inductors are used twice for the same beam acceleration.

The other possible design variant of a one-section accelerator is shown in Figure 9.52; the same inductors are used three times. The operating principle does not differ too much from that described in the comments to Figure 9.49. The difference consists of connection of external beams 7 and internal 6 beam (see Figure 9.49); in the scheme shown in Figure 9.52, they are connected by means of the turning systems, similar to the connection of the input 1 and output 2 beams in Figure 9.51. As a result, the trajectory of the beam 1 in Figure 9.52 acquires an expressive undulator shape.

As the analysis shows, these design versions become extremely efficient in case of simultaneous use of the method of series connection of external channels, too, which was illustrated earlier in Figure 9.50. For example, it is the system of four parallel acceleration units with their axes of symmetry in transversal plane located along a circular line (Figure 9.53) [98]. Each unit possesses one internal and two external channels. The external channels of each unit are connected in series in such a way that they form four common channels of the type shown in Figure 9.50. Therefore, internal and external channels are connected

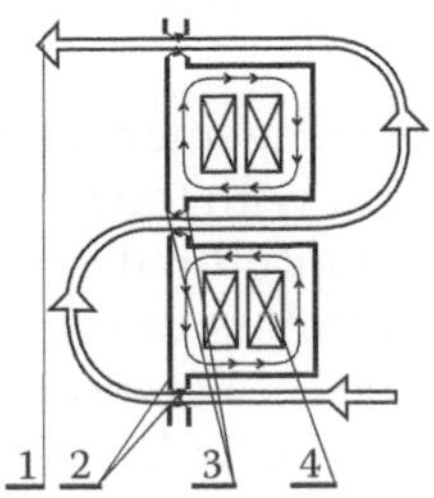

FIGURE 9.52
Design scheme of the acceleration section of undulator type with the use of both internal and external acceleration channels. Here, 1 is the output beam; 2 are external acceleration gaps; 3 is the internal acceleration gap; and 4 are magnetic inductors.

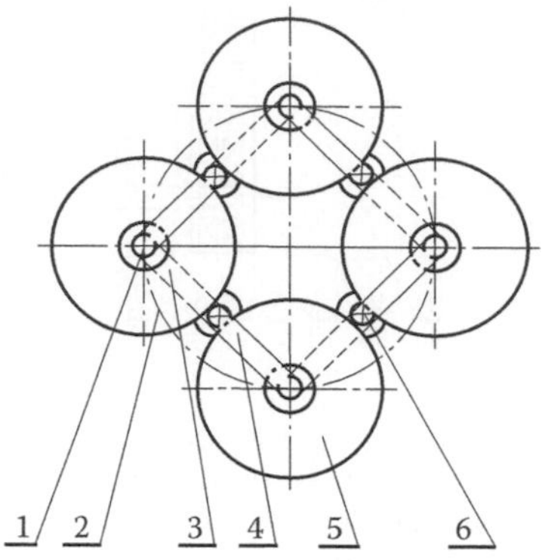

FIGURE 9.53
Design scheme of eight-channel UNIAC arrangement, transverse projection. Here, 1 are internal acceleration channels; 2 is the circular axis line crossed by the axes of symmetry of all acceleration units 5; 3 are turning systems located *under* the partial acceleration units 5; 4 are turning systems located over the partial acceleration units; 5 are partial linear acceleration units; and 6 are coupled external acceleration channels (they are connected in series—see scheme in Figure 9.50).

by seven turning systems and the same beam during its acceleration passes through all these six acceleration channels (four internal and four external channels, respectively, see Figure 9.53).

And now let us give certain elementary numerical estimations which illustrate the efficiency of design idea of turns when used in practice. We shall take the really existing ATA accelerator as a technological basis for estimations. Let us recall that its total length equals ~70 m, its maximum width equals ~2 m [15], and the energy of the output 10-kA electron beam is 50 MeV. We calculate the longitudinal dimension of the functionally equivalent eight-channel (four internal + four external channels) UNIAC which provides for the same output energy in case of the same current; its example is shown in Figure 9.53.

To get a functionally equivalent model of the considered system, let us conditionally divide the whole acceleration unit of the equivalent ATA into four parts of the same length; we get four partial linear acceleration units 5 (see Figure 9.53). Let us place these units parallel to each other with their centers of symmetry in the transversal plane lying on the same circular axis line 2. Each unit 5 is supplied with two external channels. The two adjacent external channels of different units are connected in series; hence, they form coupled external acceleration channels 6 (see the comments to the idea of such a connection in Figure 9.49). Then let us connect inputs and outputs of internal 1 and external 6 channels by turning systems 3 and 4 (see Figure 9.53); it is similar to the scheme shown in Figure 9.51.

Hence, as a result of the described rearrangement of the traditional linear induction accelerator ATA, we get the eight-channel (but with four units) UNIAC design scheme shown in Figure 9.53. Now let us estimate the longitudinal dimension of this system. Assume that due to special design of shield concentrators of the acceleration sections (see respective comments to Figure 9.51), the field in the acceleration gap of each of two external channels of each unit 5 (Figure 9.53) of the channel equals ~40% of the field in its internal channel. Recall that each external channel, as was already mentioned, has two acceleration gaps with two adjacent units (see Figures 9.49 and 9.50). Taking into account such a doubling of the acceleration voltage, the equivalent acceleration field in each common external channel 6 (see Figure 9.53) reaches ~80% of the internal field. Then we find that the construction shown in Figure 9.53 can provide for the same parameters (50 MeV, 10 kA), but the length of the system is approximately seven times less (~10–11 m) than in the ATA case (~70 m), and we get this result with the same basic technologies and design solutions that were used during creation of the ATA acceleration units.

Now, we come to another conclusion. Passing from the inductors with magnetic cores to their "air-core" variants, we somewhat lose in acceleration rate for linear acceleration channels; however, we do not obligatory lose in the longitudinal dimension of the accelerator as a whole. On the other hand, in this case, we get a possibility to essentially gain in weight characteristics which is the principal advantage of "air-core" UNIACs, and this fact makes UNIAC beyond comparison in their class of accelerators. The said conclusion becomes even more important when we speak about cluster systems of satellite basing.

Now let us note that the proposed system, as was illustrated earlier, is characterized with less dimensions and besides, it is also more technological and simple in manufacturing and operation. It is explained by the fact that its use instead of the systems of ATA-type gives the possibility to essentially simplify (and make cheaper) the systems of radiation protection, power supply, control, vacuum pumping, etc.

Based on the results, the key elements of the UNIACs are special turning magnetic systems. Three types of designs are proposed for them:

1. Systems based on two-pole magnets [6,92]
2. Systems based on segments of toroid and cylindrical solenoid [97,98]
3. Combined systems [97,98]

The analysis proves that the sphere of application of turning systems of the first type is limited by the magnitude of the impulse current of the beam (no more than tens of amperes). Systems of the second and third type are designed for constructions with elevated magnitudes of current in the beam (tens of kiloamperes and more).

9.5.3 Multichannel Concept

As was already mentioned, two types of multichannel acceleration systems, namely, internal and external systems, can be seen in the systems of MHIAC class. In its turn, the external multichannels are divided into two subtypes. In the case of the first subtype, multichannels are formed by several external channels in the same acceleration unit [98]. In the second variant, multichannels are formed by several parallel one-channel units connected in accordance with a special scheme [93,94,96,97]. The idea of both of the mentioned types of the external multichannels is expressly based on the information presented in this and the previous sections. Hence, we shall further focus on the discussion about certain peculiarities of the design schemes with internal multichannels, which are not self-evident.

It is known that classical traditional accelerators of LIAC-type [99–103] are characterized by the existence of only one acceleration channel in each acceleration unit. However, we must note that initially the idea of internal multichannels was formed in the endeavor to improve these acceleration systems. The idea to divide one high-current beam into many partial beams moving in the same channel was created among others. The further progress in the given sphere consisted of the search for constructive and technological methods to adhere specific practical forms to these ideas [93,94,96–98].

Historically, the formation of the internal multichannel idea was stimulated by the attempts to solve the well-known problem of increase of the total critical current of the beam accelerated in the channel. Let us recall that the essence of the problem consists of the following [61–63]. It is known that severe limitations as to the numerical value of current in the beam are caused by the specific mechanism of electrostatic interaction of the beam with the conducting walls of the channel. Several ways to increase the value of the critical current per series were proposed. One of them consists of application of the idea of internal multichannels: one above-critical beam is divided into a series of parallel subcritical partial beams. Then it was proposed to accelerate these beams parallel to each other in one wide acceleration channel with common focusing magnetic field. Later it was proposed to insert partial solenoids for each separate partial beam aimed at adjustment of the process of partial beams transportation to this channel.

The main shortcomings of the above-mentioned variant of formation of a system of many partial beams are as follows. First and foremost, in this case, a single-cathode high-current injector is used; it forms one "big" beam that is later divided into a series of small (partial) beams with the help of grids. As a result, the investigators received all problems typical for superpowerful injectors with elevated current of the beam [the necessity to use superhigh voltage (megavolts), the tendency of the injected beam to instability at the stage of beam formation, etc.]. All these factors essentially limit the maximum value of the output beam current in the system. Besides, the procedure of beam separation into a series of partial beams was realized with the help of special grids that were placed at the exit of the injector. In its turn, it caused significant heating of grids, and this heating also limited the maximum current of the total beam. Second, the optimal structure of partial beams was disturbed, and it was very undesirable because of complicated technological tasks, which should be solved for the long transport path of the accelerated beam in the acceleration channel and, later, in cluster systems.

Taking into account the above considerations, other, more perfect, design approaches were elaborated to practically use the idea of internal multichannels. All of them are based on construction combination of the following technological methods:

1. Formation of individual independent (isolated) acceleration channel for each partial beam [93,94]
2. Combination of designs with internal and external multichannels and undulation ideas [97,98]
3. Use of the designs of special multicathode injectors [93,94,98]
4. Creation of the external multichannels by using several parallel single-channel acceleration units [98]
5. Use of both partial beams formed by the same multibeam injector, and beams which passed the acceleration procedure in other [98] acceleration units, in the same multibeam channel

Furthermore, we give the brief description of some partial technological interpretations of the presented technological methods.

9.5.4 Multichannel Acceleration Sections

Let us begin with the discussion about the first of the above-mentioned technological methods—the idea of formation of individual isolated independent acceleration channel for each partial beam [93,94,98]. First and foremost, it means that now each acceleration section must have several acceleration gaps 2 (see Figure 9.54) at the same time, located in

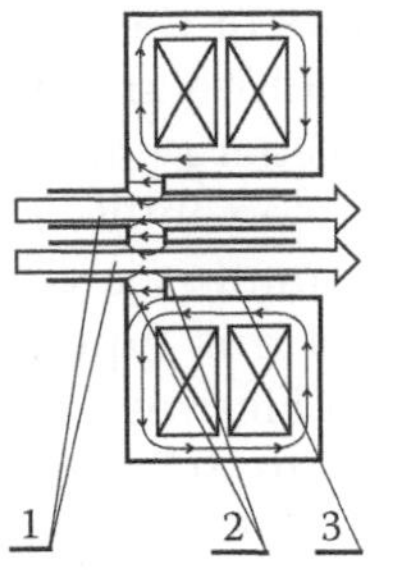

FIGURE 9.54
Design scheme of acceleration section with several separate internal acceleration channels at the same time. Here, 1 are partial beams of charged particles; 2 are acceleration gaps for partial beams 1; and 3 are transportation channels for partial beams.

the same plane [10], instead of one acceleration gap (existing in single-channel systems). Accelerated partial beams 1 move in their "personal" acceleration channels which include individual acceleration gaps 2 and transportation channels 3 (see the same figure). The specific feature of the multichannel schemes of this type is the energy of partial beams 1: it may be the same for all beams or may differ. As to partial beams, in reality, it may be the same beam but at different stages of its acceleration [98].

Figure 9.55 illustrates the idea of combination variant for internal and external multichannels with the undulation idea [98]. It is easy to notice that the use of the design schemes of this type gives an interesting possibility to form beams of charged particles with the given energy spectrum of partial beams. For example, it may be two-speed electron beams in the two-stream superheterodyne FEL (TSFEL). As we know, the energy shift (energy mismatch) for the components of such a beam in the TSFEL may be characterized by values of ~10% and, in some cases, even less (see Chapters 13 and 14 given below). At the same time, the TSFEL basic operating mechanism is extraordinary sensitive to any time change of this mismatch. In the case when independent acceleration systems are used to form each of partial beams, the problem of beams' mutual stabilization with respect to this mismatch becomes extremely critical. First and foremost, this is because the characteristic mutual instability of the energy of both beams during their acceleration may have the same (or even greater) order of magnitude due to external factors. It is known that physical processes in inductors and their power sources are the principal sources of instabilities in the systems of this type. Hence, to solve this problem, the idea of simultaneous acceleration of two

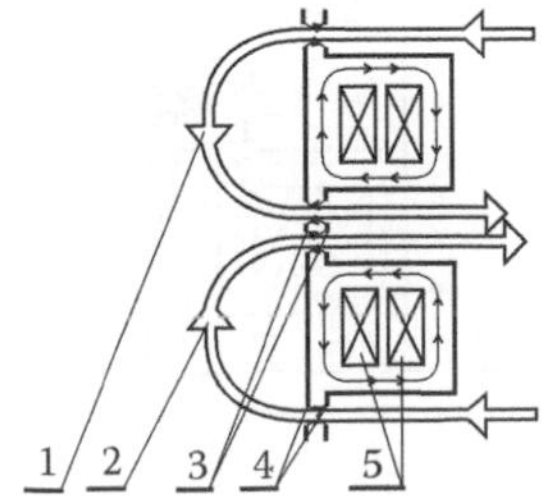

FIGURE 9.55
Example of internal and external multichannel combination with the undulation idea. Here, 1,2 are partial beams of charged particles; 3 are internal acceleration gaps; 4 are external acceleration gaps; and 5 are magnetic inductors.

beams in different channels of inductors of the same acceleration section was proposed. In this case, all possible dynamic instabilities are realized simultaneously for both beams; in other words, the given initial mutual energy mismatch practically does not change during the acceleration process. The TSFEL sensibility relative to the instability of the average energy of two-stream system is very small (see Chapters 13 and 14 for more details). It is proposed to construct the MHIAC of this type on the basis of different technological methods, including those illustrated in Figures 9.54 and 9.55, respectively. In view of the above-mentioned (see, for example, Figure 9.51 and respective comments), the operating principle of the scheme shown in Figure 9.55 may be considered as self-evident.

9.5.5 Multichannel Injectors

In technological aspect, to practically realize the idea of multichannels as it is, it is needed to previously solve the problem of construction of special multicathode injectors. Examples of such injectors are shown in Figures 9.56 through 9.59. In Figures 9.56 and 9.57, one can see the design scheme of a two-cathode (or four-cathode and more) injector for MHIAC with internal multichannels [98].

The injector shown in Figure 9.56 operates as follows. Magnetic inductors 5 generate eddy electric field inside the shield 4. Some of the force lines of this field become closed on the way through the internal surface of the field 4, external surface of the bar 1, and acceleration gaps between cathodes 3 and anode grids 6, 10. Electrons emitted by cathodes 3 experience the action of the electric field existing in these gaps, and accelerate forming partial electron beams 7, 9 which pass into acceleration channels of the respective multichannel accelerator.

The peculiarity of the design scheme shown in Figure 9.56 is that the output partial electron beams 7, 9 have a small relative energy shift, whose value is specified in respective performance specifications. To get this result, an additional anode grid 12 is inserted into the scheme. "Shift voltage" from the special external high-voltage power source is applied between the grids 10 and 12. As a result, another beam, which is formed in the acceleration gap 3–10, gains (or loses) additional shift energy. This type of injector is designed mainly for use in MHIAC acceleration units with internal multichannels; they are also used in two- or multistream SFELs.

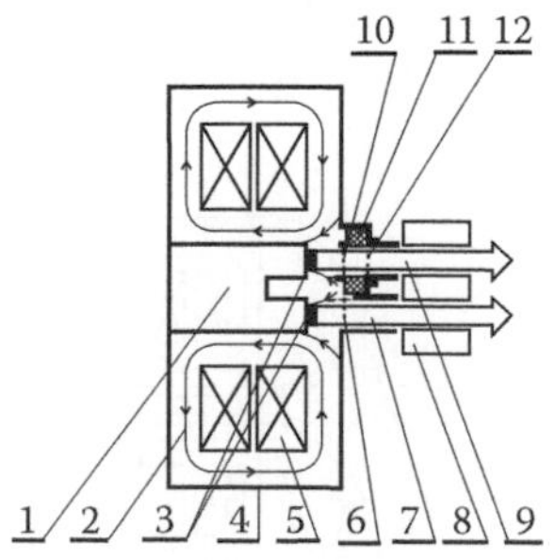

FIGURE 9.56
Design scheme of the multichannel injector designed to form two high-current electron beams with different speeds. Here, 1 is the central bar; 2 are force lines of the eddy electric field; 3 are partial cathodes; 4 is the electric shield; 5 are magnetic inductors; 6 is the second anode grid (or foil) of the first beam 7; 7 is the first electron beam; 8 is the system of beam 7 formation; 9 is the second electron beam; 10 is the first anode grid (or foil) of the second electron beam 9; 11 is the isolator; and 12 is the second anode grid (foil) of the second electron beam 9.

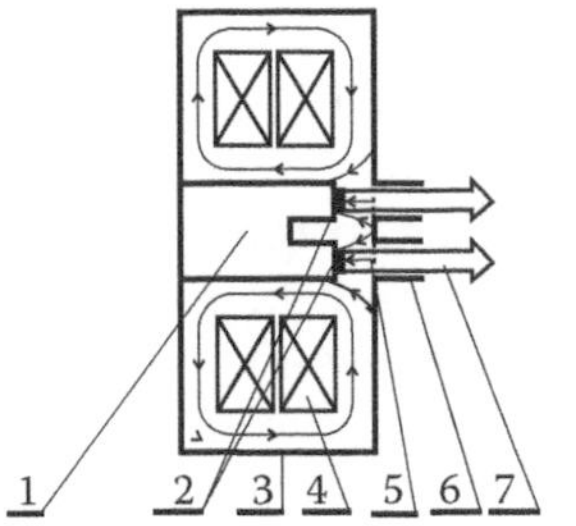

FIGURE 9.57
Design scheme of the multichannel injector designed to form two high-current electron beams with same speeds. Here, 1 is the central bar; 2 are cathodes; 3 is the electric shield; 4 are magnetic inductors; 5 are anode grids (or foils); 6 are input sections of the acceleration channels; and 7 are injected partial electron beams.

The injector whose design scheme is shown in Figure 9.57 differs from the injector shown above in Figure 9.56 only by absence of the subsystem that assures the reciprocal energy shift between two injected beams (see items 11 and 12 in Figure 9.56). The principal destination of acceleration systems with internal multichannels containing such injectors is the formation of superhigh-stream electron beams (after the respective convergence procedure). The explanation consists of a rather dense (in transversal plane) group of partial beams emitted in this type of system; such a high density is needed because further the beams are directed into the same wide, but transversely limited common acceleration channel. The ideas of some design schemes of this type are described below in this section. But now let us discuss other illustrated examples of multichannel injectors.

In contrast to the schemes shown in Figures 9.56 and 9.57, injectors, whose design schemes are shown in Figures 9.58 and 9.59, are intended mainly for their use in MHIACs with external channels. It is explained, firstly by the fact that external channels in accordance with the basic design idea (see, for example, illustration given in Figure 9.55) are always spaced in the transversal plane because later they are directed into technologically different partial channels. The most interesting design solutions are those that are intended for their use in combined electron–ion systems (Figure 9.58).

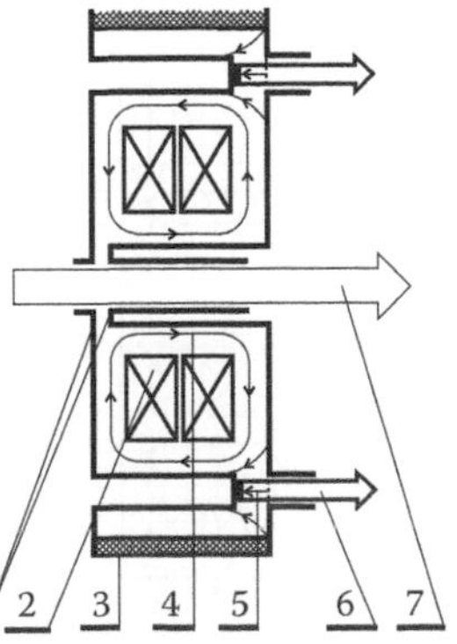

FIGURE 9.58
Combined design scheme of multicathode electron injector with external placement of cathodes and acceleration section for ion beam. Here, 1 is the acceleration gap for ion beam 8; 2 are magnetic inductors; 3 is the insulator; 4 are force lines in the central part of the inductor; 5 are acceleration gaps for electron beams; 6 are injected partial electron beams; and 7 is the high-current ion beam.

The operating principle of constructions shown in Figures 9.58 and 9.59, in principle, is similar to that described earlier (Figures 9.56 and 9.57). However, there are certain differences. Let us speak briefly about them.

Let us first settle the generality of some design ideas of constructions shown in Figures 9.58 and 9.59. The external parts of the electric field generated by internal inductors are used in both cases. Due to this fact, we get the above-mentioned opportunity to form partial beams, which are spaced in transversal plane, at the entry of external channels of multichannel acceleration units, because the diameter of external force lines is big enough. Second, the number of partial beams in this case is much bigger than the number of beams in the system with internal cathodes (similar to those shown in Figure 9.13) may generate, because the total area of cathodes in this case is much bigger. It gives good prospects for independent practical use of these injector systems. The principal destination of acceleration systems on the basis of this type of injector is cluster systems on the basis of FEL with the external block of harmonic generation (see Figures 1.15 and 1.16). However, they may have a wide commercial use too, for example, in systems of industrial smoke purification systems, electron pumping of excimer lasers, stimulation of chemical reactions by electron beams, etc.

The key difference between the designs shown in Figures 9.58 and 9.59 is the energy of partial beams. The injectors of the first type (see Figure 9.58) are designed for their use in situations when, on one hand, it is needed to guarantee a big total current (of tens of kiloamperes, for instance) and on the other hand, to assure rather low levels of beam energy (up to one megavolt). It is obvious that in virtue of well-known physics of electron injector processes, these beams cannot be formed using the standard methods. In this case, the problem of superhigh purveyance is solved by simultaneous partial generation of a number of relatively low-voltage beams with their further summation according to well-known schemes. We can mention systems intended to form combined compensated-by-charge electron-ion beams with like speeds of the components as their example. It is known that if the energy level equals several tens of megavolts, ion high-current beams (see item 7 in Figure 9.58) are practically nonrelativistic (or, to be more precise, weakly relativistic). It means that the electron component of the composite beam (see item 6 in the same figure), on one hand, must be high-current too, but on the other hand, it must be weakly relativistic. Such a peculiarity of the systems with external multichannels that the acceleration field in the internal channel is much stronger than the field in the external channels is used here.

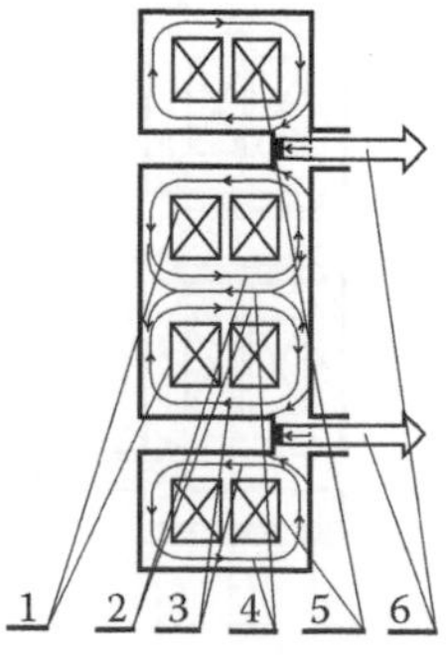

FIGURE 9.59
Design scheme of multicathode coaxial injector with combined external–internal placement of cathodes. Here, 1 are internal coaxial inductors; 2 are external coaxial inductors; 3 are force lines of the external electric field of the internal coaxial inductor 1; and 4 are force lines of the internal electric field of the external coaxial inductor 5. Diameter of inductors 5 is greater than that of inductors 1, and inductors 5 enclose them coaxially.

Multicathode injectors of the other type (see Figure 9.59) are intended for somewhat other cases of their application. They are used in those cases when it is needed to form supercritical (hundreds of kiloamperes) and, at the same time, enough energetic (tens of megaelectron volts) electron beams. It is proposed to solve this problem by simultaneous generation of many (for example, 12) subcritical beams (at the level of ~10 kA) with their further acceleration in the multichannel MHIAC (for example, up to energy values of ~50 MeV) and their further convergence into one superhigh-current (~120 kA) output beam.

Here, we take into account that the bigger the critical current of the beam, the bigger the energy of the beam [61,63]. Hence, this total superhigh-current beam, being the overcritical input beam, after accelerated partial beams convergence, may become a subcritical output beam. Therefore, in this case we can get around the mentioned problem of overcritical beam by its separation into a number of input subcritical partial beams and the convergence of output accelerated beams (or performing their convergence step by step during acceleration).

The other application is the multibeam injector of similar multibeam accelerator for a multichannel cluster system (see Chapter 1).

Now let us pay attention to the specific *coaxiality* of the general arrangement of the system that is illustrated by the scheme given in Figure 9.59. The "cathode–anode" pairs are located in circular "slit" between external 5 and internal 1 magnetic inductors. Both subsystems of inductors of the coaxial pair are characterized by opposite directions of magnetic flux circulation in their cores. Then the directions of force lines of electric fields generated by them are also opposite. As a result, both internal components of the electric field of external inductors 5 and the external component of internal inductors 1 form the accelerating field in the "cathode–anode" gaps. Hence, the accelerating action of fields on electrons is added in the gaps. It is the principal reason for getting the opportunity of simultaneous generation of a number of relatively high-current high-energy electron beams. On the other hand, both electric fields compensate each other in the gap between the inductors 5 and external electric shield. In a similar way, the fields are compensated in the internal part of the inductors 1 (see Figure 9.59). Then losses in shields and losses related to external radiation diminish. Summarizing, we can say that the coaxial design scheme shown in Figure 9.59 is one of the most interesting schemes from the standpoint of its possible use as a technological basis for cluster systems (see Chapter 1 for more details). In this case, each injected beam (after its acceleration) is used as the basis for respective harmonic or multiharmonic FEL amplifiers (see more details in Chapter 1).

Summarizing the analysis of the examples of injector schemes given in this section, it is worth mentioning that each of them only illustrates one or another design idea, and it means that in reality, when we deal with their combinations in different variants, the number of partial variations may be rather large. Besides, only two-electrode injector's construction was used in all given examples. If we pass to the well-known three-electrode construction, no new principles will be inserted into the described ideas, but the number of possible variants will increase by several times, etc.

9.5.6 One-Channel Undulation Accelerators

The simplest design version of the multichannel UNIACs (MUNIAC) may be the variant of one-turning single-beam system built on the basis of two partial single-channel linear acceleration units [97]. The example of the scheme of this type is shown in Figure 9.60.

The system consists of the injector 1 that is connected to the first acceleration units 4. We can see in the figure that these units are formed by acceleration sections 5. The characteristic

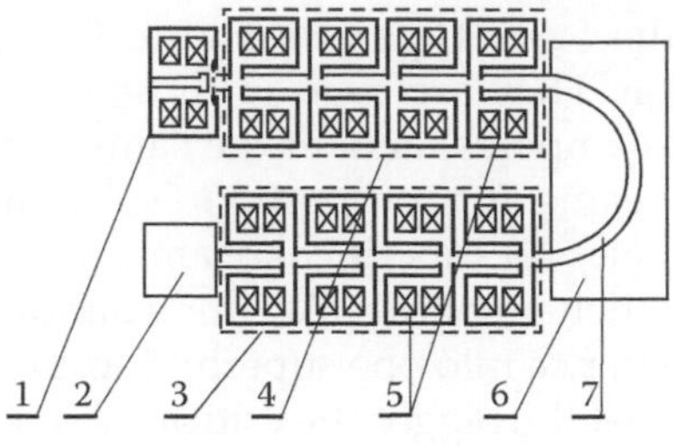

FIGURE 9.60
Design scheme of one-turning single-beam undulation induction electron accelerator built on the basis of two parallel single-channel linear acceleration units. Here, 1 is the injector; 2 is the output system; 3 is the first linear acceleration unit; 4 is the second linear acceleration unit; 5 are acceleration sections with individual electric shields; 6 is the turning mechanism; and 7 is the transportation channel for the electron beam in the turning system 6.

feature of the design of acceleration sections 5 in this case is that each of them is enveloped by individual electric shields.

The acceleration channel output of the first acceleration unit 3 is connected to the channel input of the other unit 3 by means of the channel for beam transportation, which is located in the turning system 6. The output system 2 is connected to the system 4 output; its specific construction depends on the functional destination of the accelerator. It may be, for example, the "vacuum–atmosphere" outlet window, X-ray target, FEL input, or another transfer system input as well. In the context of the said above in this chapter, the operating principles of the construction shown in Figure 9.60 is rather self-evident. However, we shall recall that directions of beam acceleration in the both acceleration units 3 and 4 are opposite. This fact, in addition to all above-mentioned technological consequences: a possibility of realization of the effect of cancellation of external electric fields of the accelerator as a whole. The point is that in spite of shields, a part of eddy electromagnetic field penetrates into the external space. By different reasons, this effect is very undesirable during practical operation of all standard induction acceleration systems, without any exclusion. However, in the case of undulator arrangement (see Figure 9.60), each of these units creates electric fields in the surrounding environment, and these fields are directed toward each other. By this reason, these electric fields cancel each other at a distance of two transversal dimensions of the inductors of the acceleration units 3, 4, as was already mentioned.

Other possible design versions of the accelerator similar to that shown in Figure 9.60 are absolutely evident. They are the variants of design with the common electric shields, which envelop several parallel acceleration sections at the same time. We can find among them systems with both planar and circular spatial arrangement of units, or systems with both one beam or many beams, which, in their turn, are formed by independent single-beam injectors or individual multibeam injectors, systems with combined turning and acceleration systems, etc. However, all the described variety of possible versions is based on the same principal design idea. It is the use of individual single-channel acceleration units as technological basis, which exist today and are successfully implemented in systems of linear induction accelerator (LIAC) type. Today it gives certain advantages to this type of MUNIAC. Unfortunately, neither adequate systems with internal multichannels nor similar systems with external multichannels were experimentally realized in the time of the Star Wars program, and today as well. On the other hand, during the past 20 years, there were no doubts that multichannel systems will make the most impressive breakthroughs in the course of creating the newest cluster systems on the basis of FEL.

However, let us continue with the principal line of our discussion.

9.5.7 Multichannel Undulation Accelerators

It was already mentioned that the principal technological obstacles in the technique of traditional superhigh-current LIAC are related, first of all, with the peculiarities of physics of "pulling" beams out of the cathode region and with the specifics of their further transportation in working channels. The main obstacle in case of the problem of "pulling" the beams out consists of the necessity of using ideally high acceleration voltages in cathode–anode gaps. In the second case, the key circumstance is that maximum possible current of the beam transported in the channel with conducting walls is limited by so-called critical current [61–63]. The latter, as is known, is almost proportional to the beam energy at a given density. In physical interpretation, it is the maximum (that is, the greatest possible) current which may be transported in the given acceleration channel.

In the view of the above-mentioned, in the framework of the Star Wars program and later, there was a lot of persistent attempts aimed at a search of new nonstandard approaches to design superhigh-current acceleration systems. The most interesting from both a technological and physical standpoint are those design solutions that are based on the idea of the synthesis of one superhigh-current electron beam basing on a series of weaker partial high-current beams, which was already discussed. However, due to special actuality of this idea for the further description, let us insert several additional comments into the above.

Hence, as was already mentioned, each of partial beams in any multichannel high-current induction accelerator (MHIAC) is independently injected and then accelerates. At every stage of the acceleration process, the convergence occurs: first, a part of beams joins "a little bit more high-current" beams, and then the convergence of these "a little bit more high-current" beams with a less number of "even more high-current" beams takes place, and so on. Therefore, at the last stage, the formation of one or several relativistic superhigh-current beams of different speeds (if it is provided for by the requirements of performance specification) takes place. Basing on the estimations of the modern state of high-current induction accelerators technique, we can find that, taking into account its actual level, it is principally possible to create acceleration systems intended to form fantastic (from the modern standpoint) beams with current of units of mega-amperes. However, both the level of technological complexity of the problem and the financing needed for its solution are very serious.

In case of multiharmonic cluster FEL (CFEL) (both two-stream, and multistream), the use of two-channel (or multichannel, respectively) design schemes of MUNIAC seems to be optimal. Efficiency of this kind of solutions becomes even more evident if a problem of creation of a superhigh-current (superpowerful) CFEL is set. The example of the MUNIAC design scheme, intended for use in this type of systems, is shown in Figure 9.61 [97].

Here, eight partial injectors 7 (see Figure 9.61) form eight electron partial beams of single-channel linear acceleration units. The single-channel units themselves are located along the generating line of an imaginary cylinder whose walls bound the central eight-channel linear acceleration unit 8. Single-channel units are connected with each of their channels of the central unit 8 by means of turning systems 1.

At the first stage of acceleration, the sum of currents of injected partial beams is much greater than the critical current of an imaginary total superhigh-current beam. In other words, at this stage of such a superhigh-current beam formation (if we really have a wish to form it), it would be essentially overcritical. Hence, its transportation in a (imaginary) single channel would be physically impossible. However, if it is divided into many partial beams, then its transportation becomes realizable.

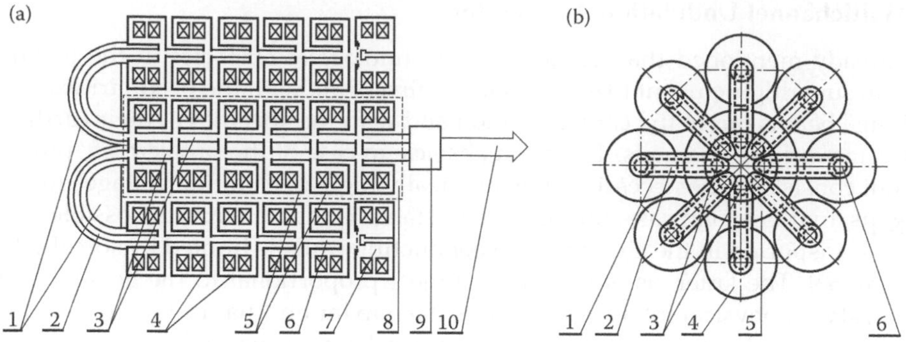

FIGURE 9.61
Design scheme of multichannel induction eight-beam accelerator—former of superhigh-current electron beam [frontal (a) and cross (b) projections]. Here, 1 are turning systems; 2 are curvilinear channels for electron beam transportation in turning systems 1; 3 are internal rectilinear channels of central eight-channel acceleration units; 4 are peripheral single-channel acceleration sections; 5 are acceleration sections of the central eight-channel linear acceleration units; 6 are linear channels of peripheral acceleration units; 7 are single-beam partial electron injectors; 8 is the central eight-channel linear acceleration unit; and 9 is the system of integration of eight accelerated high-current electron beams into one superhigh-current beam 10.

Injected partial beams accelerate by peripheral acceleration sections 4 of single-channel units. After passing the turning systems 1, each of the beams continues to accelerate in linear channels of eight-channel linear acceleration unit 8. Then all these eight high-current partial beams are joined in one superhigh-current (single-speed or multispeed) beam 10 in integration system 9. In other words, resultant electron beam with total current which is eight times bigger than any partial current, is formed. As the energy and, respectively, critical current of the total beam essentially increase, the further passage of this beam through the transportation channel becomes physically possible. In view of this purpose, the parameters of the system are chosen in such a way that after the complete acceleration cycle for all partial beams, the values of their critical currents increase eight times at least. Therefore, when electrons of partial beams acquire their total energy, the formed resultant beam 10, as was already mentioned, is not overcritical.

Hence, summarizing, we can say that the principal physical idea of the described method of beam synthesis is based on the said proportionality of the critical current to the energy of the beam. It means that when each partial beam is well accelerated, the critical current of each of them noticeably increases. The parameters of the system in the considered case are chosen in such a way that the critical current at every stage of partial beams convergence exceeds the critical threshold needed for the further acceleration of a single resultant beam.

In the author's opinion, the level of significance of special problems which may be solved by systems constructed on the basis of superpowerful cluster FEL (CFEL), in principle, may give sufficient motivation to overcome possible technological difficulties and remove any doubt with respect to the reasonability of bearing impressive financial expenses. From this standpoint, the most prospective FEL of this type are

1. Two-stream quasilinear and parametric FEL
2. Multibeam cluster two-stream superheterodyne FEL (whose theory is not elaborated yet)
3. Plasma-beam quasilinear and parametric single-beam cluster FEL

4. Parametric cluster FEL, both classical electron wave and those where the middle section is made as a longitudinal electric undulator, and others

To illustrate the potential possibility to use the design under consideration as a system that forms superhigh-current electron beams, let us make once again some numerical estimates. It is known that the LIAC technology today allows forming electron beams with currents up to ~30 kA (and somewhat more). Consider that partial (single-channel) linear acceleration systems of this type are taken as a basis in our tough experiment. Assume that the acceleration rate in their channels and partial channels of the central acceleration unit equals ~1 MeV/m, and the length of all the units is of ~10 m. Then the elementary calculations show that the output parameters of the beam are estimated as follows: current equals ~340 kA, energy equals ~25 MeV (we take into account the energy of injectors), that is, the peak power in this case equals ~8.5 TW, and the total longitudinal dimension of the system does not exceed 13 m. From the standpoint of the actual state of this type of technology, this result looks like record-breaking for both the current of the output beam and dimensions. If we use such an accelerator as the source of electron beam for cluster FEL, then at electron efficiency of ~50% (achieved experimentally 20 years ago [1]), we principally get the technological possibility to generate impulses of electromagnetic radiation of several terawatt of power (in this case, the duration of macro-impulse of radiation equals ~100–300 ns) in infrared–visible range of frequencies. As we can see from the modern bibliography review, this fact may be considered as the record-breaking result as to "power-dimensions" parameters. Let us recall that in practice, the overall dimension of powerful FEL is mainly determined by dimensions of accelerators with their power supply systems.

A radically different design idea is implemented in the construction whose typical example is shown in Figure 9.62 [98]. The main goal of this construction is to achieve high values of the electron efficiency and high value of the average acceleration rate at moderate values of longitudinal dimension. It is possible to achieve this goal by triple passage of each of the two accelerated beams through the same working volume of the four-channel acceleration unit with internal multichannels.

The design scheme of channel placement in case of another design variant is shown in Figure 9.63 [97]. Here, the inductors 2 are enveloped by the common electric shield 1. Turning systems 3 connect partial acceleration channels 4, which are located in two parallel planes (at right angles to the plane of the figure). We can easily see that the principal difference between the design variants shown in Figures 9.62 and 9.63 consists

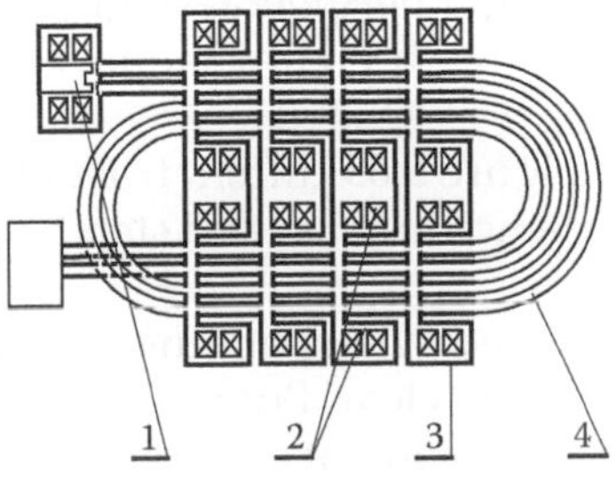

FIGURE 9.62
Design scheme of the plane two-beam undulation MUNIAC with three turns for each of the beams (frontal projection). Here, 1 is the two-beam injector; 2 are inductors of accelerations sections 3; and 4 are curvilinear channels in turning systems (they are not shown in the figure for the simplicity of the scheme).

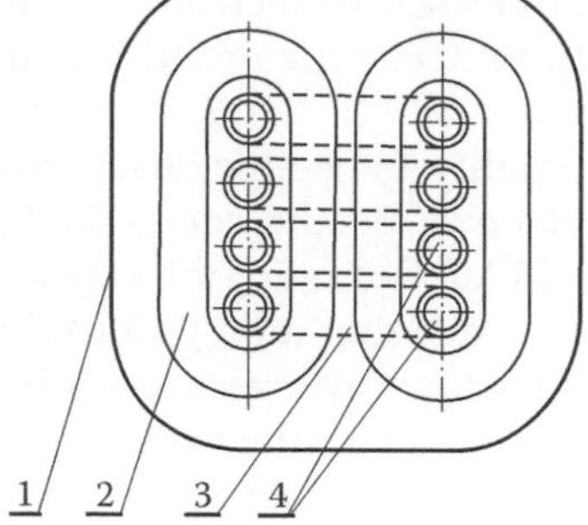

FIGURE 9.63
Three-dimensional design scheme of one-beam undulation MHIAC with seven turns of the beam (cross section). Here, 1 is the electric shield; 2 are inductors; 3 are front curvilinear channels for beam transportation in turning systems (which are not shown in the figure); and 4 are linear acceleration channels.

of reciprocal orientation of partial acceleration units. In the first case (Figure 9.62), they are located in the same plane, but in the other case (Figure 9.63), they are located in two parallel planes.

Similar to the previous case, we present numerical estimates for another possible variant of the record-breaking MHIAC built in accordance with the scheme shown in Figures 9.62 and 9.63. The number of partial channels in this case equals 8; we have 2 beams of 30 kA each, the acceleration rate in each partial channel equals ~1 MeV/m, and the length of linear acceleration units (channels) equals ~10 m. After performing elementary calculations, we get the following results: the output energy of electrons is ≥45 MeV (we take into account the energy of the injector), the total current of the output beam equals 60 kA, and the longitudinal dimension of the system as a whole is of ~13 m if the output impulse beam power is of ≥2.7 TW. Therefore, in both cases, the output impulse beam power is in the terawatt range. However, in the first case (see Figure 9.61), its record-breaking characteristic consists of the possibility to get supercritical currents in the output beam (340 kA), while in the second case (Figures 9.62 and 9.63), we have the illustration of the technological achievement of high resultant acceleration rates (4 MeV/m) at relatively low acceleration rates in each partial channel (1 MeV/m).

Construction versions, which are shown in Figures 9.64 and 9.65 [97], illustrate another idea of arrangement of such systems. It is the idea of combination of single-channel (round) and multichannel linear acceleration units. In the first case (Figure 9.64), it is the design version of the system with eight peripheral single-channel linear units 4 connected (through turning systems 5 and 6) with eight-channel acceleration unit 1 with internal multichannels.

The design version shown in Figure 9.65 differs from other versions by the additional idea of accelerated beams convergence in pairs and step-by-step (in contrast to the variant shown in Figure 9.61).

In the view of the above-mentioned, the operating principle of design versions shown in Figure 9.64 and Figure 9.65 is quite clear. But we shall mention that the advantages of each of them essentially depend on statement of the specific project problem. For example, if the principal task is to achieve supercritical values of the output beam (hundreds of kiloamperes), then the schemes of the type shown in Figures 9.61 and 9.65 are beyond comparison. But if the task is to get elevated values of the resultant acceleration rate, then the schemes whose examples are shown in Figures 9.62 through 9.64 are preferential.

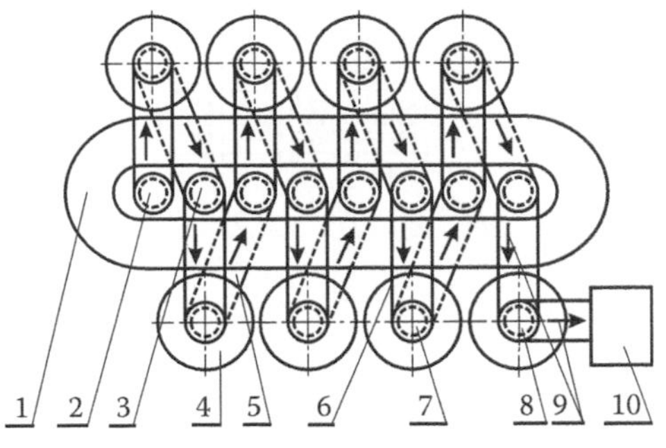

FIGURE 9.64
Design scheme of the single-beam 16-channel UNIAC with eight peripheral single-channel acceleration units and one central multichannel acceleration unit with eight internal channels (cross projection). Here, 1 is the plane (planar) multichannel acceleration unit with eight internal channels; 2 is the input acceleration channel with the injector (now shown in the figure) connected from below; 3 are intermediary internal channels; 4 are peripheral acceleration units; 5 are lower turning systems that connect internal and peripheral channels; 6 are upper turning systems; 7 are intermediary peripheral acceleration channels; 8 is the output peripheral acceleration channel; 9 are arrows that indicate directions of electron beam motion in turning systems; and 10 is the output system intended to turn the electron beam through 90°.

9.5.8 Systems on the Basis of Acceleration Blocks with External Channels

The simplest scheme of UNIAC design with external channels is shown in Figure 9.66 [98]. The electron beam formed by the injector 1 accelerates in the internal channel 4, passes through the turning system 6, and then accelerates in the external channel 6. The formed (and accelerated) beam passes to the output system 2.

The main shortcoming of the simplest design solutions of this type (as was already mentioned and explained earlier—see Figures 9.50 and 9.53) consists of a very small acceleration rate in the external channel 3 (Figure 9.66) compared with the acceleration rate in the internal channel 4, and this phenomenon is always observed (which is of principal significance). To remove this shortcoming, the designs of multiple (two-fold, three-fold, etc.) external channels (Figure 9.50) were proposed in reference [98]. An example of the design scheme of UNIAC with a two-fold acceleration channel is given in Figure 9.67. The key idea is that external channel 5 is common for two acceleration units 4, 7 at the same time, which

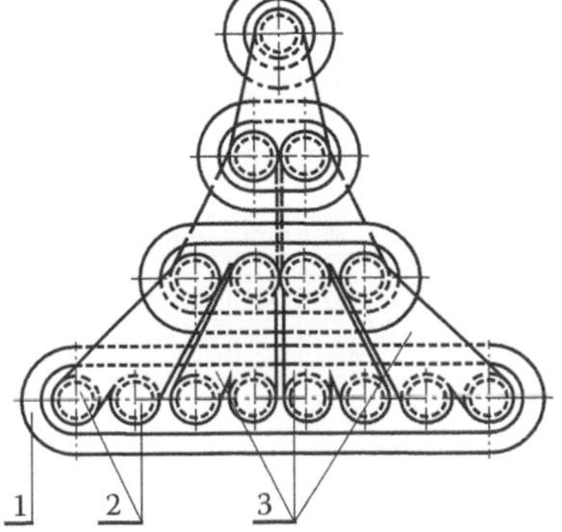

FIGURE 9.65
Example of the pyramidal design scheme of the UNIAC with systems to turn and converge beams (cross projection). Here, 1 are planar inductors with internal multichannels; 2 are acceleration channels; and 3 are systems to turn and converge beams.

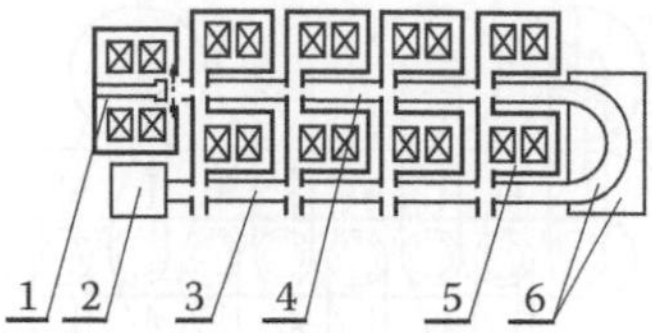

FIGURE 9.66
Design scheme of UNIAC with one internal and one external channels. Here, 1 is the injector; 2 is the input unit; 3 is the external acceleration channel; 4 is the internal acceleration channel; 5 are inductors; and 6 is the turning system.

is similar to the scheme shown in Figure 9.53. Due to this peculiarity, the average acceleration rate in channel 5 (Figure 9.67) at all identical conditions is twice as high as it is in the similar channel 3 in Figure 9.66. As to other characteristics, the operating principle of the scheme shown in Figure 9.67 is principally indistinguishable from the scheme shown in Figure 9.66.

9.5.9 Multichannel Systems on the Basis of Acceleration Blocks with External and Internal Channels

The analysis proves that those combined constructions are the most perfect (from the technological standpoint), where the best features of both systems with internal and external multichannels are organically united. The example of such a scheme (transverse projection) is shown in Figure 9.68 (see also commentaries to Figure 9.53). The earlier discussed design whose scheme is shown in Figure 9.61 serves as the basis for it. The scheme is upgraded through addition of eight external coupled channels 4 (see Figure 9.68).

Therefore, each linear peripheral acceleration unit 1 simultaneously participates in formation of the electric field in one internal 3 and two adjacent external 4 acceleration channels. It gives the unique possibility to essentially increase the energy of the total output electron beam at the cost of rather "modest" design complication (namely, through the addition of external channels 4). To illustrate this idea, let us take numerical values of the system's parameters used earlier when analyzing the scheme shown in Figure 9.61 as the basis. Hence, the current in each of eight partial beams equals 30 kA, the acceleration rate in internal channels of acceleration units of channels equals 1 MeV/m, and the length of acceleration units equals 10 m. Besides, assume that the acceleration rate in coupled external channels equals 0.8 MeV/m (see respective comments to Figure 9.53). Making

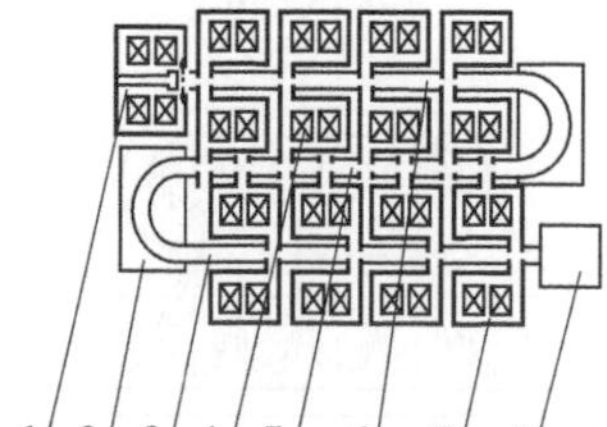

FIGURE 9.67
Design scheme of UNIAC with one two-fold external acceleration channel. Here, 1 is the injector; 2 are turning systems; 3 is the first internal acceleration channel; 4 is the first linear acceleration unit; 5 is the two-fold acceleration channel; 6 is the second internal acceleration channel; 7 is the second linear acceleration unit; and 8 is the output unit.

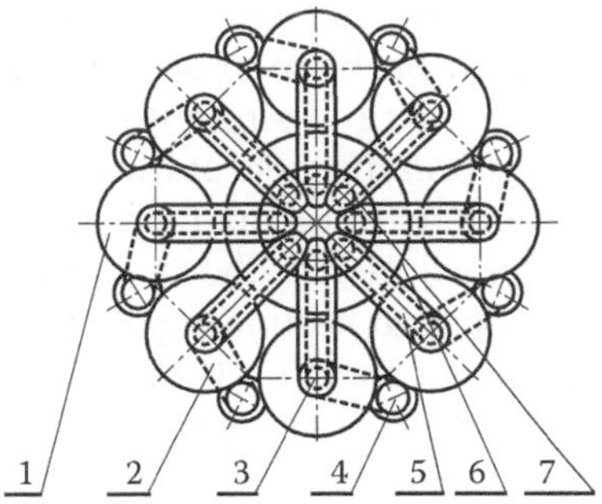

FIGURE 9.68
Example of the modernized scheme (see the simple scheme in Figure 9.61) of the 24-channel 8-beam UNIAC (transversal projection) with 8 peripheral acceleration units, each of them having one internal and two external (coupled) channels and one central 8-channel unit. Here, 1 are peripheral acceleration units, each of them includes one external and two external channels; 2 are turning systems between external and internal channels of units 1; 3 are internal channels of peripheral units 1; 4 are coupled external channels that are connected two by two in series; 5 are channels for electron beam transportation in turning systems; 6 is the central eight-channel unit with internal channels; and 7 are acceleration channels of the unit 6.

elementary calculations, we can easily get the following estimates for the output parameters of the total beam: the total current is the same, ~340 kA; energy equals ~33 MeV (together with the energy of injectors), that is, the peak power in this case equals ~11.3 TW, if the overall longitudinal dimension of the system is ~13 m. We can easily deduce that the price needed to pay for the increase in the output energy from 25 MeV (Figure 9.61) to 33 MeV (Figure 9.68), and, respectively, the increase in peak energy from 8.5 TW (Figure 9.61) to 11.3 TW (Figure 9.68) is not excessive.

9.5.10 Delaying Effect

The brilliant prospects described above are not overshadowed by potential problems that result from the basic physical principle of the MUNIAC operation itself. Among these problems, we shall especially mention the so-called problem of "accelerating bunch delay." The essence of the problem consists of the following. As we know, the high-voltage impulse sources assure power supply for systems of both the LIAC and NHIC type, as well as the UNIAC type. The typical duration of impulses (for example, in cases when standard METGLASS magnetic cores are used) equals 100–300 ns. On the other hand, the physics of the acceleration process requires for electrons to experience the same eddy electric field from the beginning to the end of the acceleration process. It implies that this time interval must not be less than half of the duration of the electric field impulse. Then it is easy to get estimates for the maximum possible length of the acceleration channel, $l_{\max}$, which results from the formulated requirement

$$l_{\max} \leq \frac{1}{2} v_b \tau_b \sim (15 - 45)m. \tag{9.14}$$

Here, $v_b \sim c$ is the beam speed; c is the light speed; $\tau_b \sim \tau_i$ is the bunch duration; and τ_i is the duration of the acceleration phase of the electric field impulse in the channel. It is obvious that the received estimation principally limits the possibility to "infinitely" increase the number of electron beam transits in the same acceleration unit with either internal or external multichannels (see, for example, design schemes in Figures 9.62 through 9.68). As a result, a complicated technological problem arises in practice. The thing is that when we try to increase the beam output energy using multichannels, we face the necessity to increase the duration of the acceleration impulse of the electric field. However, at the same time, we decrease the operating intensity of the field and, hence, the beam acceleration rate in partial channels. The radical solution of the said problem is related to the use of the method of multichannel acceleration unit sectioning. It means that one long multichannel undulation unit is divided into several short units and the criterion of Equation 9.14 type is absolutely true for each of them. The power supply of each of these "shortened" units is realized with the use of the system of impulse start delay whose schemes and designs are well known in the standard technology of the "ordinary" LIAC. The total compensation of the said "delaying" effect is possible through correct adjustment of the delay value from one multichannel section to another.

Hence, the presented brief analysis allows us to formulate the statement that today MUNIACs are a very prospective class of systems from the standpoint of their ideology and the significance of their possible practical application in FEL technologies. It is obvious that the further development of this ideology type opens real prospects to create absolutely unique (as to their characteristics) series of acceleration systems with impulse currents from tens or hundreds to several hundreds of amperes (or even bigger) in the energy range from hundreds of kiloelectron volts to hundreds of megaelectron volts and electron efficiency of ~90% and more. These systems are considered today as the most technologically adequate basis for the future version of the Star Wars program.

References

1. Marshall, T.C. 1985. *Free electron laser*. New York: MacMillan.
2. Brau, C. 1990. *Free electron laser*. Boston: Academic Press.
3. Luchini, P., and Motz, U. 1990. *Undulators and free electron lasers*. Oxford: Clarendon Press.
4. Kulish, V.V. 1998. *Methods of averaging in nonlinear problems of relativistic electrodynamics*. Atlanta: World Scientific Publishers.
5. Saldin, E.L., Schneidmiller, E.V., and Yurkov, M.V. 2000. *The physics of free electron lasers* (Advanced texts in physics). Berlin: Springer.
6. Kulish, V.V. 2002. *Hierarchic methods. Undulative electrodynamic systems*. Vol. 2. Dordrecht: Kluwer Academic Publishers.
7. Kulish, V.V. 2002. *Hierarchic methods. Hierarchy and hierarchical asymptotic methods in electrodynamics*. Vol. 1. Dordrecht: Kluwer Academic Publishers.
8. Kulish, V.V. 1990. The physics of free electron lasers. General principles. Deposited in Ukrainian Scientific Research Institute of Technical Information on 05.09.90, Kiev, No. 1526, Uk-90, 192 pp.
9. Bratman, V.L., Ginzburg, N.S., and Petelin, M.I. 1979. Nonlinear theory of stimulated radiation waves on relativistic electron beams. *Sov. J. Theory Exp. Phys.* 76:930–43.
10. Pantell, R.N., Soncini, G., and Puthoff, H.E. 1968. Stimulated photon electron scattering. *IEEE J. Quantum Electron.* 4:905.
11. Kapitza, P.L., and Dirac, P.A.M. 1933. The reflection electrons from standing light waves. *Proc. Cambridge Philos. Soc.* 29:297–300.
12. Landau, L.P., and Liftshitz, E.M. 1974. *Theory of field*. Moscow: Nauka.
13. Sokolov, A.A., and Ternov, I.M. 1974. *Relativistic electron*. Moscow: Nauka.

14. Phullips, R.M. 1960. The ubitron, a high-power traveling-wave tube based on a periodic beam interaction in unloaded waveguide. *IRE Trans. Electron. Devices* 7(4):231–241.
15. Madey, J.M.J., Schwetman, H.A., and Fairbanc, W.H. 1973. A free electron laser. *IEEE Trans. Nucl. Sci.* 20:980–983.
16. Kulish, V.V., Dzedolik, I.V., and Kudinov, M.A. 1985. Relativistic electron motion in periodically reversible electromagnetic field. Deposited in Ukrainian Scientific Research Institute of Technical Information on 23.07.85, Kiev, No. 1490, Uk-85, 110 pp.
17. Kohmanski, S.S., and Kulish, V.V. 1984. To the classic single-particle theory of free electron laser. *Acta Phys. Pol.* A66(6):713–740.
18. Kohmanski, S.S., and Kulish, V.V. 1985. To the nonlinear theory of free electron lasers. *Acta Phys. Pol.* A68(5):749–756.
19. Kohmanski, S.S., and Kulish, V.V. 1985. To the nonlinear theory of free electron lasers with multi-frequency pumping. *Acta Phys. Pol.* A68(5):741–748.
20. Kohmanski, S.S., and Kulish, V.V. 1985. Parametric resonance interaction of an electron and the field of electromagnetic waves and the longitudinal magnetic field. *Acta Phys. Pol.* A68(5):725–736.
21. Kulish, V.V., and Krutko, O.B. 1995. Amplifying properties of free electron lasers with combined crossed EH-Ubitron pumping. *Pis'ma v Zh. Tekh. Fiz. [Sov. Tech. Phys. Lett.]* 21(11):47–51.
22. Silin, R.A., Kulish, V.V., and Klimenko, Ju.I. Patent of USSR SU N705914. Electronic device. Priority of 18.05.72. Published in non-secret press of USSR after taking off of the relevant stamp of secrecy: 15.05.91, *Inventions Bulletin* N26, 1991.
23. Davis, G.R. 1976. Navy researchers develop new sub-millimeter wave power source. *Microwave* 17(12):12.
24. Kulish, V.V. 1978. Some problems of the theory of charged particle interaction with the field of electromagnetic waves. Candidate Dissertation (PhD Thesis), Kiev State University, Kiev.
25. Kulish, V.V. 1987. Physical processes in the parametrical electron lasers (free electron lasers). Doctoral Dissertation (Doctor of Sciences Thesis), Institute of Physics of Ukrainian Academy of Sciences, Kiev.
26. Kompaneyets, A.S. 1956. About the Compton stimulated scattering. Quantum approach. *Sov. J. Theory Exp. Phys.* 31:836–841.
27. Bagrov, V.G., and Khalilov, V.V. 1969. Stimulated radiations of charged particles moving in the field of plane electromagnetic waves. *Izvestiya Vuzov, Fizika* (USSR) 9:50–55.
28. Bagrov, V.G., and Khalilov, V.V. 1972. About the stimulated radiations of electrons in the field of plane electromagnetic waves. *Sov. Nucl. Phys.* 16(1):174–184.
29. Shukhatme, V.P., and Wolf, P.A. 1973. Stimulated Compton as a radiation source—Theoretical limitation. *J. Appl. Phys.* 44:2331.
30. Madey, J.M.J. 1971. Stimulated emission of bremsstrahlung in a periodic magnetic field. *J. Appl. Phys.* 42:1906–1913.
31. Fedorov, M.V. 1991. *Electron in the strong light field*. Moscow: Nauka.
32. Bunkin, F.V., Kazakov, A.B., and Fedorov, M.V. 1972. Interaction of intensive optical radiation with free electrons. *Uspekhi Fizicheskih Nauk* (USSR) 107(4)559–593.
33. Deacon, D.A.G., Elias, L.R., and Madey, J.M.J. 1976. First operation of a free electron laser. *Phys. Rev. Lett.* 38:892.
34. Elias, L.R., et al. 1976. Observation of stimulated emission of radiation by relativistic electron beam in a spatially periodic transverse magnetic field. *Phys. Rev. Lett.* 36:717–720.
35. Phullips, R.M. 1960. The ubitron, a high-power traveling-wave tube based on a periodic beam interaction in unloaded waveguide. *IRE Trans. Electron. Devices* 7(4):231–241.
36. Trubetskov, D.I., and Khramov, A.E. 2003. *Lectures on the microwave electronics for physicists.* In two volumes. Moscow: Fizmatlit.
37. Granatstein, V.L. 1976. Mechanisms for coherent scattering of electromagnetic wave from relativistic electron beams. 2nd International Conference and Winter Sch. Sub-millimeter waves and their application, San Juan, Puerto Rico, pp. 87–89.

38. Granatstein, V.L., and Sprangle, P. 1977. Mechanisms for coherent scattering of electromagnetic wave from relativistic electron beams. *IEEE Trans. Microwave Theory Technol.* 25(6):545–550.
39. Kalmykov, A.N., Kotsarenko, N.Y., and Kulish, V.V. 1979. On the theory of parametric frequency-increasing transformation in electron beams. *Radiotech. Electron. [Radio Eng. Electron. (USSR)]* 24(10):2084–2088.
40. Kalmykov, A.N., Kotsarenko, N.Y., and Kulish, V.V. 1977. Parametric generation and amplification of electromagnetic waves with frequencies higher than the pumping wave frequency in electron beams. *Izv. VUZov. Radioelectron. [Sov. Radioelectron.]* 10:76–82.
41. Berezhnoi, I.A., Kulish, V.V., and Zakharov, V.P. 1981. On the explosive instability of relativistic electron beams in the fields of transverse electromagnetic waves. *Zh. Tekh. Fiz. [Sov. Phys. Technol. J.]* 51:660–662.
42. Kulish, V.V., and Kotsarenko, N.J. Patent of USSR No. 668491 (cl. H01 J 25/00) Electronic microwave device. Priority 17.05.77.
43. Kotsarenko, N.Y., and Kulish, V.V. 1980. Superheterodyne amplification of electromagnetic waves in a beam-plasma system. *Radiotech. Electron [Radio Eng. Electron. (USSR)]* 25(11): 2470–2475.
44. Kotsarenko, N.Y., and Kulish, V.V. 1980 On the possibility of superheterodyne amplification of electron beam electromagnetic waves. *Zh. Tekh. Fiz. [Sov. Phys. Tech. Phys.]* 50:220–226.
45. Perekupko, V.A., Silivra, A.A., Kotsarenko, N.Y., and Kulish V.V. Patent of USSR No. 835259. Electronic device. Priority 28.01.80.
46. Kulish, V.V., and Storizhko, V.E. Patent of USSR No. 1837722. Free electron laser. Priority 15.02.91.
47. Kulish, V.V., and Storizhko, V.E. Patent of USSR No. 1809934. Free Electron Laser. Priority 18.07.90.
48. Kulish, V.V., and Kotsarenko, N.Y. Patent of USSR No. 711927. Electronic device. Priority 10.03.78.
49. Elias, L.R. 1979. High-power, CW efficient, tunable (UV through IR) free-electron laser using low-energy electron beams. *Phys. Rev. Lett.* 42(15):977–981.
50. Zakharov, V.P., and Kulish, V.V. 1983. Multi-stage increase of transverse electromagnetic wave frequency in electron beams. *Radiotech. Electron. [Radio Eng. Electron. (USSR)]* 27(9):1799–1807.
51. Kulish, V.V., and Kotsarenko, N.Y. Patent of USSR No. 711927. Electronic device. Priority 10.03.78.
52. Kulish, V.V., and Krutko, O.B. 1995. Amplifying properties of free electron lasers with combined crossed EH-Ubitron pumping. *Pis'ma Zh. Tekh. Fiz. [Sov. Tech. Phys. Lett.]* 21(11):47–51.
53. Kulish, V.V., Lysenko, A.V., and Koval,V.V. 2007. On the theory of the plasma-beam superheterodyne free electron laser. *Visnyk Sumskoho Universytetu* (Ukraine). *Seria Phys. Math. Mech.* 2:112–119.
54. Kulish, V.V., Lysenko, A.V., and Koval, V.V. 2009. Multi-harmonic cubic-nonlinear theory of the superheterodyne plasma-beam free electron lasers with Dopplertron pumping. *Prikladnaya Fizika* (Russia) 5:76–82.
55. Kulish, V.V., Lysenko, A.V., and Koval, V.V. 2009. On the theory of the plasma-beam superheterodyne free electron laser with H-Ubitron pumping. *Tech. Phys. Lett.* 35(8):696–699.
56. Kulish, V.V., Lysenko, A.V., Gubanov, I.V., Brusnik, A.Ju. Patent US No. 87750. Superheterodyne free electron laser with longitudinal electric undulator. Published in *Bulletin of Invention* (Ukraine) 08.10.2009, No. 15.
57. Kulish, V.V., Lysenko, A.V., and Koval,V.V. 2009. Multi-harmonic cubic-nonlinear theory of the superheterodyne plasma-beam free electron lasers with H-Ubitron pumping. *Radiotechnika I Elektronika* (Russia) 14(3):383–389.
58. Kulish, V.V., Lysenko, A.V., and Brusnik, A.Ju. 2010. Active FEL-klystrons as formers of femtosecond clusters of electromagnetic field. *J. Nanoelectron. Phys.* (Ukraine) 2(2):50–78.
59. Kulish, V.V., Lysenko, A.V., and Rombovsky, M.Yu. 2010. Effect of parametric resonance on the formation of waves with a broad multi-harmonic spectrum during the development of two-stream instability. *Plasma Phys. Rep.* (Russia) 36(7):594–600.

60. Kulish, V.V., and Pugachev, V.P. 1991. On the theory of superheterodyne wave amplification in a two-stream plasma. *Fiz. Plazmy [Sov. J. Plasma Phys.]* 17(6):696–705.
61. Rukhadze, A.A., et al. 1980. *Physics of high-current relativistic electron beams*. Moscow: Atomizdat.
62. Kuzelev, M.V., and Rukhadze, A.A. 1990. *Electrodynamics of dense electron beams in plasmas*. Moscow: Nauka.
63. Kuzelev, M.V., Rukhadze, A.A., and Strelkov, P.S. 2002. *Plasma relativistic microwave electronics*. Moscow: Publishing House of the Bauman Moscow State Technological University.
64. Kondratenko, A.N., and Kuklin, V.M. 1988. *Principles of plasma electronics*. Moscow: Energoatomizdat.
65. Zshelezovskii, B.E. 1971. *Electron-beam parametrical microwave amplifiers*. Moscow: Nauka.
66. Akhiezer, A.I., and Fainberg, Y.B. 1949. Interaction of a charged particles beam with a plasma. *Dok. Akad. Nauk SSSR [Sov. Phys. Doklady]* 69(4):555–568.
67. Lopoukhin, V.M. 1953. *Excitation of electromagnetic oscillations and waves by electron beams*. Moscow: Gostekhizdat.
68. Lebedev, I.V. 1964. *Microwave technology and devices*. Vol. II. Moscow: Energiya.
69. Haiduk, V.I., Palatov, K.I., and Petrov, D.N. 1971. *Physical principles of microwave electronics*. Moscow: Sovetskoye Radio.
70. Vainshtein, L.A., and Solntsev, V.A. 1973. *Lectures in microwave electronics*. Moscow: Sovetskoye Radio.
71. Bekefi, G., and Jacobs, K.D. 1982. Two-stream FEL. *J. Appl. Phys.* 53:4113–4121.
72. Bolonin, O.N., Kulish, V.V., and Pugachev, V.P. 1988. Superheterodyne amplification of electromagnetic waves in a relativistic electron two-stream system. *Ukr. Fiz. Zh. [Ukr. J. Phys.]* 33(10):1465–1468.
73. Botton, M., and Ron, A. 1990. Two-stream instability in FEL. *IEEE Trans. Plasma Sci.* 18(3):416–423.
74. Kulish, V.V. 1992. Physics of two-stream free electron lasers. *Moscow Univ. Bull., Phys. Astron.* 33(3):64–78.
75. Kulish, V.V. 1991. On the theory of klystron-type superheterodyne free electron lasers. *Ukr. Fiz. Zh. [Ukr. J. Phys.]* 36(1):28–33.
76. Kulish, V.V. 1991. On the theory of relativistic electron-wave free electron lasers. *Ukr. Fiz. Zh. [Ukr. J. Phys.]* 36(5):682–694.
77. Kulish, V.V. 1993. Superheterodyne electron-wave free-electron laser. *Int. J. Infrared Millimeter Waves* 14(3).
78. Kulish, V.V., Kuleshov, S.A., and Lysenko, A.V. 1994. Nonlinear self-consistent theory of two stream superheterodyne free electron laser. *Int. J. Infrared Millimeter Waves* 15(1).
79. Wilhelmsson, H. 1991. Double beam free electron laser. *Phys. Scr.* 44:603–605.
80. Davydova, T.A., and Wilhelmsson, H. 1992. Resonant and nonresonant wave excitation in a double beam free electron laser. *Phys. Scr.* 45:607–612.
81. Bekefi, G. 1992. Double-stream cyclotron maser. *J. Appl. Phys.* 71(9):4128–4131.
82. Kulish, V.V., Lysenko, A.V., and Savchenko, V.I. 2003. Two-stream free electron lasers. General properties. *Int. J. Infrared Millimeter Waves* 24(2):129–172.
83. Kulish, V.V., Lysenko, A.V., and Savchenko, V.I. 2003. Two-stream free electron lasers. Analysis of the system with monochromatic pumping. *Int. J. Infrared Millimeter Waves* 24(3):285–309.
84. Kulish, V.V., Lysenko, A.V., and Savchenko, V.I. 2003. Two-stream free electron lasers. Physical and project analysis of the multiharmonic models. *Int. J. Infrared Millimeter Waves* 24(4):501–524.
85. Kulish, V.V., Lysenko, A.V., Savchenko, V.I., and Majornikov, I.G. 2005. The two-stream free electron laser as a source of electromagnetic femtosecond wave packages. *Mater. Int. Workshop Microwave*, Radar Remote Sens. MRRS 2005, September 19–21, Kyiv, Ukraine, pp. 304–309.
86. Kulish, V.V., Lysenko, A.V., and Majornikov, I.G. 2004. A two-stream free electron laser for generation of the electromagnetic femtosecond wave packages. Abstracts of the International Conference on Infrared and Millimeter Waves & International Conference on Terahertz Electronics (THz) (IRMMW2004/THz2004), Karlsruhe, Germany, September 27–October 01, 2004.

87. Kulish, V.V., Lysenko, O.V., and Majornikov, I.G. 2004. The two-stream multi-harmonic FEL as a powerful source of femtosecond wave packages. Proceedings for the European Radar Conference, Amsterdam, Netherlands, 14–15 October 2004.
88. Kulish, V.V., Lysenko, A.V., Savchenko, V.I., and Majornikov, I.G. 2005. The two-stream free electron laser as a source of electromagnetic femtosecond wave packages. *Laser Phys.* 12:1629–1633.
89. Kulish, V.V., Lysenko, A.V., and Mayornikov, I.G. 2005. The two-stream multi-harmonic FEL as a powerful source of femtosecond wave package. Proceedings of the National Aviation University, Kyiv, No. 2, pp. 126–130.
90. Kulish, V.V, Kuleshov, S.A., and Lysenko, A.V. 1993. Nonlinear self-consistent theory of superheterodyne and parametric electron laser. *Int. J. Infrared Millimeter Waves* 14(3):451–568.
91. Bolonin, O.I., Kochmanski, S.S., and Kulish, V.V. 1989. Coupled paramagnetic resonances in FEL. *Acta Phys. Pol.* A76(3):455–473.
92. Kulish, V.V., Kosel, P.B., Melnyk, O.K., and Kolcio, N. Patent No. US 6,433,494, B1, Inductional Undulative EH-accelerator, Date of the Patent August 13, 2002, Filed April 18, 2000.
93. Kulish, V.V., and Melnyk, A.C. Patent No. US 6,653,640 B2, Multi-channel linear induction accelerator, Date of Patent Nov. 25, 2003, Filed Aug. 15, 2002.
94. Kulish, V.V., and Melnyk, A.C. Patent No. US 2001020957, Multi-channel linear induction accelerator of charged particles, Date of the Patent April 15, 2004, Filed February 13, 2001.
95. Kulish, V.V., and Melnyk, A.C. Patent No. US 2001020953, Electronic sterilizator, Date of Patent April 15, 2004, Filed February 13, 2001.
96. Kulish, V.V., Melnyk, A.C., and Langraf, A.K. Patent No. US 7,045,978 B1, Multi-channel induction accelerator, Date of Patent May 18, 2006, Filed March 10, 2004.
97. Kulish, V.V., Melnyk, A.C., and Langraf, A.K. Patent No. US 7,030,577 B2, Multi-channel induction undulation accelerator. Date of Patent April 18, 2006, Filed March 10, 2004.
98. Kulish, V.V., and Melnyk, A.C. Patent No. US 7,012,385 B1, Multi-channel accelerator with external channels. Date of Patent March 14, 2006, Filed September 24, 2004.
99. Vakhrushyn, Ju.P., and Anatskiy, A.I. 1978. *Linear induction accelerators*. Moscow: Atomizdat.
100. Redinato, L. 1983. The advanced test accelerator (ATA), a 50-Mev, 10-kA inductional LINAC. *IEEE Trans.* vNS-30(4):2970–2983.
101. Chen, Y.-J. 1992. Beam control in the ETA-II linear inductional accelerator. Digest of technical paper of 16th LINAC Conference, Ottawa, Ontario, Canada, August 1992.
102. Scarpetti, R.D., Boyd, J.K., and Earley, G.G., et al. 1998. Upgrades to the LLNL flash X-ray induction linear accelerator (FXR). Digest of technical papers of 11th IEEE International Pulsed Power Conference, Baltimore, MD, USA, June 29–July 2, 1997, Vol. 2, pp. 597–602.
103. Motojima, O., Muratov, V.I., and Shishkin, A.A. 1993. *Plasma physics in pictures*. Kharkov, Osnova.
104. Abramyan, E.A. 1988. *Industrial electron accelerators and applications*. CRC.

10

General Description of the FEL Models

The content of this chapter is mainly methodological. The material should be regarded as a peculiar theoretical introduction to the rest of the book. Here, we will formulate the key methodological concepts and ideas that form the basis of the general FEL theory. Something similar has already been done in Part 1. However, the material of Part 1 covers only the most abstract aspects of the theory. In contrast, this chapter will cover the problems that are characteristic for hierarchic electrodynamic oscillation-wave-resonant systems. They will be used for physical analysis of different types of FEL models.

10.1 General FEL Model

10.1.1 About "Precise" and "Rough" Theoretical Models

Prior to describing the theoretical model of devices, we shall note some aspects of the methodological character. First, let us note that the theoretical models in FEL theory are of two principally different types. We shall label the models of the first type as precise models, for convenience. The most characteristic peculiarity of these models involves the achievement of the maximum possible adequacy level (for the actual level of theory development) for the elaborated theoretical models and experimental systems being studied. Naturally, being the most precise models, they automatically become the most complicated and their practical application is very narrow. That is, this type of theoretical model is good for an acceptable quantitative description of the specific FEL designs with specific design solutions for pumping, signal systems, systems of electron tracts for electron beam transportation, etc. A good example of this sort of model is the theoretical model of the ordinary FEL with magneto-undulator pumping (H-Ubitrons, according to our classification; see Figure 9.2). The advantage of these precise models involves the possibility to take into account a big number of small physical and design details that specify, for example, the multimode nature of beam and electromagnetic wave processes in the zone of interaction of FELs of specific geometry. However, their shortcomings are continuations of their merits. These most complicated and informative models are very effective in those situations where the preproject analysis is already done and optimal design solutions are chosen. In this case, the task of theoretical analysis, first and foremost, consists of a maximum and precise fine-tuning of all the key physical and technological elements of the future experiment through numerical and analytical simulation of all the processes taking place in the system. It is necessary to remember that experiments in the sphere of FEL technologies are very expensive, and mistakes in the design will cost too much in the future; that is why the practical importance of precise models is high.

But what can one do in cases where the problem of preproject analysis is set? In this case, prior to starting the precise analysis, it is necessary to at least approximately estimate the perspectives of one or another design solution from the rather long list of possible pretenders and choose one that has a chance to be optimal. Obviously, it is practically unrealistic to realize this with the help of precise models because each of them is very complicated and extremely specialized. At the same time, the number of possible variants of design solutions—as mentioned in Chapter 9—is too big. As a result, the scope of work needed in this case becomes unrealistically great. Also, the large amount of detailed information received becomes excessive for practical purposes. Shall we need, for example, information about the details of mode structure of the above-mentioned electromagnetic field if we are not sure of the capacity of one or another construction to operate? In such situations, preference shall be given to more rough but much more universal models. This means that, in principle, the necessary general characteristics may be obtained in a simpler way at that stage of preproject analysis. Then, we can specify the physical and technological peculiarities of the chosen project solution by using more precise models. For example, the majority of pumping systems used in FEL (including those mentioned in Chapter 9, namely H-Ubitron, electromagnetic waveguide, laser, retarded, plasma, etc.) may be described within the framework of the unified artificial magneto-dielectric model. A general approach of this type was elaborated in Kulish's scientific school in the 1970s and 1980s [1,2], and later was widely used in many problems of preproject analysis [3–16].

However, let us note that multivariant problems of this type will be the main object of interest in the remaining chapters. The essence of each of these problems consists of an idea to make approximate quantitative and qualitative estimates of the whole range of variant design solutions for cluster FELs within the framework of a universal multifunctional model. First, within the framework of this analysis, it is necessary to clarify when and under what conditions the use of one or another design scheme is reasonable and even optimal (from the standpoint of the criteria formulated in preliminary specifications). In our case, these criteria may be, for example, the instant power of the output femtosecond cluster, the average power of the generated cluster wave, the electron efficiency, or the dimensions of the FEL part of the analyzed system. It is obviously impossible to make such a complete analysis for all possible design schemes described in Chapters 1 and 9—where the total number may exceed hundreds—within the framework of this book. For this reason, in this and the following chapters, we shall be limited to studying several of them only. In particular, we shall study those design schemes that seem to be the most promising for cluster systems practice using the ideology of the approximate, but rather universal approach.

10.1.2 Rough Model of a Wide Electron Beam

Today the majority of experts believe that the transversely unlimited model of the relativistic electron beam is obsolete, that it does not adequately describe the physics of real processes in FELs. However, long time ago T.C. Marshall stated in his monograph [17], with a certain astonishment, that in spite of the obvious roughness of these sorts of simplified models, they "by certain reason" describe rather well many key peculiarities that are received in the course of experiments with the basic operating mechanism in FELs. More careful analysis shows that this stereotype is true only in part. In case of relatively narrow weak-current but high-voltage beams in FELs, which are the most popular objects of study in this sphere today [1–21], the model of transversely unlimited beams is applied little. However, in cases of high-current and moderate-relativistic systems with wide beams, the situation is not so single-valued and obvious. There is a wide range of beam parameter combinations (its plasma density, transversal dimension and geometrical configuration, range of frequencies of the SCW, etc.) when the use of more complete and perfect transversely limited high-current models becomes practically unreasonable. The main reason is that in spite of a dramatic increase in the general volume of calculations, we do not get equivalent compensation in the form of principally new knowledge. It is possible to get more practically important results by a much easier method using a much simpler transversely unlimited model [3–5] that is sufficiently reasonable from the standpoint of its criteria.

Now, we will outline key benchmarks of the criterion analysis that allows us to legalize the application of transversely unlimited models to study the physics of the class of processes in the high-current FELs that are of interest to us. It is known that the impact of the phenomenon of the beam's real transversal limitation consists, mainly, in the appearance of three basic effects [3,22–24]. The first is the appearance of the beam's Coulomb field sagging beyond the limits of its transversal border and the appearance of an associated transversal shear of electron energy along its radius. In turn, this results in the appearance of radially nonuniform attenuation of the SCW field inside the beam. In traditional microwave electronics [25–28], this phenomenon is known as beam wave depression. It is convenient to describe it using the so-called reduced plasma frequency ω_{pr}.

The second effect involves the appearance of a multimode spectrum of beam waves; in general, this is reduced to the appearance of additional (with respect to longitudinal Langmuir waves) transversal and mixed longitudinal–transversal electron waves. Finally, the third effect is related to a certain deformation of dispersion laws for longitudinal Langmuir waves, which are of interest to us due to border influence.

It is known that the reduced frequency dependence on the equilibrium radius R_b of the beam with compensated charges may be roughly estimated in the following way [3]:

$$\omega_{pr} \approx R_r \frac{\omega_p}{\gamma^{1/2}} \tag{10.1}$$

here, for the plasma reduction coefficient $R_r(R_b)$, we can write

$$R_r \sim 1 - exp\{-0.7\omega_3 R_b/v_0\}, \tag{10.2}$$

$\omega_p/\gamma^{1/2} = \sqrt{4\pi e^2 n_e/m\gamma}$ is the relativistic plasma frequency of the beam, e and m are the electron charge and its rest mass, v_0 is its undisturbed axial velocity, n_e is the electron plasma density of the equilibrium beam, $\gamma = E/mc^2$ is the common relativistic factor, and E is the beam energy. We can easily see that if

$$R_b >> 1/k_3 \approx v_0/\omega_3, \tag{10.3}$$

where $k_3 \approx \omega_3/v_0$ is the wave number of the SCW in the beam, we can neglect the impact of the plasma reduction effect. Numerical estimations show that in situations typical for high-current FELs, Condition 10.3 may be true, for example, in cases of uniform cylindrical beams [22–24].

The specific character of the model chosen for this work is that in cases where Condition 10.3 is true, one can also neglect the impact of the above-mentioned beam multimode effect. This results from the fact that, for example, the cluster FEL, which is a system with multiple three-wave parametric resonances in the range of energy take-off, simultaneously operates as a singular active filter for the set of operating SCWs. And in this case, the operating waves are only longitudinal low and fast SCW of ω_3 frequency and their harmonics. Therewith, the effect of wave excitation of transversal and transversal–longitudinal types has a nonresonance character. This does not guarantee that such a noticeable relation will not appear in the framework of the theory of the higher approximations due to wave nonlinear relations and excessive width of resonance bands. However, the analysis shows that these effects are not significant in cases with a reasonable choice of parameters for the model under study in the framework of the cubic approximation used further in Chapters 12 through 15.

Finally, let us say a few words about the impact of the effect of dispersion properties deformation for longitudinal SCW, which is caused by the existence of transversal boundaries. A simple analysis shows that respective deformation adjustments to the dispersion laws formulated in the framework of transversally unlimited beams are proportional to ω_p [22–24,29,30]. As a rule, in the case of wide high-current beams, these adjustments are less than (or, at least, do not exceed) the separation Δ' for wave numbers of low and fast SCWs. On the other hand, from FEL theory we know (see, for example, Chapter 4) that the influence of this separation on the processes in FELs is physically observed only in cases where the half-width of the SCW resonance line Δk becomes less than the separation itself ($\Delta k < \Delta'$) in the process of resonance interaction. This case corresponds to the so-called Raman mode of interaction in FELs (see Chapter 4) [3–5]. Otherwise, when the half-width is greater than the separation Δ', the system seems not to notice (i.e., it does not see the fast and low SCWs separately). It is a Compton mode of FEL operation (see again Chapter 4) [3–5].

Therefore, we come to the conclusion that if the other discussed criteria are true, then at least in the case of Compton mode realizations, we can neglect the impact of the effect of dispersion law deformation for longitudinal SCWs. The Raman mode of FEL operation is more sensible to the impact of this deformation effect; however, in this case, we can choose rather wide ranges for parameters, which allows us to use the dispersion laws obtained in the framework of the theory of transversally unlimited beams. It is reasonable to add that the width of the Compton resonance line must not be too big: it must be less than the distance to the nearest transversal and longitudinal–transversal beam modes.

As for other criteria, we choose the assumptions that are well known in FEL theory [2]: The transversal dimension of the beam is much greater than the Debye screening radius, and the period for plasma oscillations is much less than the interval of time for electron transit through the interaction zone (for more details, see Section 10.1.4). We assume that the beam propagates in the axial uniform magnetic field, the strength of which is small (it is the case of the so-called weak magnetic field) in accordance with the classification given in Chapter 7. The latter physically means that the cyclotron frequency of rotation is small compared with SCW frequencies.

10.1.3 Isochronous Models

Dispersion properties of SCWs strongly depend on constant (averaged) drift velocity v_0 of the REB. Hence, with time, the FEL-resonant conditions can violate due to radiation losses of beam energy (trapping mechanism of saturation of amplification works; see Chapters 11 through 15).

Another possible scenario is that, in the course of parametric instability development, the velocities of interacting waves appreciably differ from initial ones. As a result, the wave characteristics $\omega_{j,q}, k_{j,q}$ become slowly varying functions of coordinate z and time t and, again, the resonant conditions are violated (saturation mechanism of nonlinear frequency shift) [10,11]. Several ways are known to suppress these nonlinear effects [7,10]. Some of them were used in the framework of the single-particle theory described in Chapter 7. We employ the method first proposed in the work by Zhurakhovski et al. [16]—a specially fitted supporting electrostatic field

$$\vec{E}_0 = -E_0\vec{n}\,, \tag{10.4}$$

is applied to the interaction region to compensate for the violation of the resonance state of the system. Moreover, the pumping wave parameters can slowly vary along the system length (wiggler variation) [6–8].

10.1.4 Some Criteria

Any charged particle beam may be treated, in principle, as a flow of drifting plasma. However, in many cases, collective plasma properties are weak compared to the background of numerous processes occurring in the system. That is why it is important to introduce criteria that will enable us to estimate when and to what extent plasma properties of an electron beam should be taken into account. Such criteria are well known in electrodynamics [31].

The *Debye* radius is associated with the geometry of the system. The scale is determined by the quantity

$$r_D = \sqrt{k_B T_e / 4\pi e^2 n_e}\,, \tag{10.5}$$

which is referred to as the Debye screening radius. Here, k_B is the Boltzmann constant, and T_e and n_e are the beam electron temperature and density, respectively. For the collective properties of the beam plasma to be manifested appreciably, the inequality

$$r_D << d \tag{10.6}$$

must be satisfied, where d is the characteristic beam dimension. Most often it is the transverse dimension—the radius (if the beam is cylindrical), thickness (for a strip beam), etc. Common sense suggests that collective effects must be fairly pronounced in high-current beams. This is not so clear as far as moderate-current beams (of the order of amperes) are concerned. Therefore, it seems to be useful to estimate Criterion 10.6 in view of characteristic conditions of FEL experiments. To specialize the problem, we assume the parameters to be those given by Elias in his historical experiment with the FEL on the basis of an electrostatic accelerator [32]. In this case, we take the beam current to be equal to 2A, and the dimensionless velocity thermal spread $\beta_T \approx 5.6 \cdot 10^{-3}$ ($\beta_T = v_T/c$, v_T is the electron thermal velocity). We assume that, in the FEL interaction range, the beam is compressed to the diameter $d \cong 2$ mm. Then, it is not difficult to obtain $r_D \cong 0.3$ mm $<< d \cong 2$ mm. Thus, Criterion 10.6 is satisfied, though rather weakly. This infers that the possibility that collective properties of electron plasma can be manifested should be borne in mind, even for ampere-level beams.

Criterion 10.6 suggests rather than provides manifestations of the collective plasma properties in a beam. The necessary condition is that electrons must travel within the interaction region for a time longer than the plasma oscillation period, i.e.,

$$t_{tr} > T_p = 2\pi/\omega'_p, \tag{10.7}$$

where

$$\omega_p' = \sqrt{4\pi e^2/m\gamma} = \omega_p/\sqrt{\gamma} \tag{10.8}$$

is, as before, the plasma (Langmuir) frequency given by

$$\omega_p = \sqrt{4\pi e^2/m}, \tag{10.9}$$

$\gamma = E/mc^2$ is the relativistic factor, and E is the total energy of an electron. For this numerical example, it is not difficult to find that Criterion 10.7 is satisfied for $L > 2.5$ m as the interaction region length. We remind the reader for comparison that, in Elias's experiments, the interaction region length was ~6 m, i.e., Criterion 10.7 was basically satisfied. As follows from these estimates, situations in which electron beams manifest collective plasma properties are fairly typical for relativistic electrodynamics. Thus, we can accept that the electron beam is cold, transversally unlimited, homogeneous, relativistic, and high current. Apart from that, we consider that all the above-mentioned criteria are satisfied.

10.2 Formulation of the General FEL Problem

10.2.1 Equations for the Electromagnetic Field

Equations that describe electromagnetic fields in the long time interaction devices (LTIDs) in terms of the general self-consistent theory must take into account both external and internal sources. Thus we have

$$rot\vec{H} = \frac{1}{c}\frac{\partial\vec{D}}{\partial t} + \frac{4\pi}{c}\left(\vec{j} + \vec{j}_0\right); div\vec{B} = 0;$$

$$rot\vec{E} = -\frac{1}{c}\frac{\partial\vec{B}}{\partial t}; div\vec{D} = 4\pi\rho + 4\pi\rho_0, \tag{10.10}$$

where $\vec{E}$ and $\vec{D}$, and $\vec{H}$ and $\vec{B}$ are electric and magnetic field strengths and inductions, respectively, $\vec{j}$ and ρ are current and charge densities produced by the fields in the beam plasma, $\vec{j}_0$ and ρ_0 are current and charge densities of external sources, t is the laboratory time, and c is the light velocity in vacuum. As a rule, the method of REB production, i.e., the nature of external sources, is of no interest for the LTID theory. Devices of the magnetron type with appreciable secondary emission are an exception. Nevertheless, it is not difficult to extend the conceptual approaches described here, even to such cases. The input parameters $\vec{j}_0$ and ρ_0 are usually regarded as given. On the contrary, the densities $\vec{j}$ and ρ are self-consistent quantities and calculating these is an important part of the general problem.

10.2.2 Beam Current and Space Charge Densities

Under physical conditions occurring in the LTID, it is convenient to write the densities $\vec{j}$ and ρ in terms of the one-particle distribution functions for the beam plasma particles [31,33], i.e.,

$$\rho = \sum_{\alpha} e_{\alpha} \int_{-\infty}^{\infty} f_{\alpha}\left(t,\vec{r},\vec{\mathcal{P}}\right) d^3\mathcal{P}; \tag{10.11}$$

$$\vec{j} = \sum_{\alpha} e_{\alpha} \int_{-\infty}^{\infty} \vec{v} f_{\alpha}\left(t,\vec{r},\vec{\mathcal{P}}\right) d^3\mathcal{P}, \tag{10.12}$$

where the subscript α corresponds to the charged particle species (electrons and ions, respectively), $\vec{v}$ and $\vec{\mathcal{P}}$ are the charged particle coordinate, Lagrange velocity, and canonical momentum. It is obvious that the set of Equations 10.10 through 10.12 is closed and only provided that we know the material relations and the relation between the distribution function f_{α} and the field strength $\vec{E}$ and induction $\vec{B}$. The latter are determined by kinetic equations of various types.

The densities ρ and $\vec{j}$ can be determined in other ways. For example,

$$\rho = \sum_{\alpha} e_{\alpha} n_{\alpha}; \tag{10.13}$$

$$\vec{j} = \sum_{\alpha} e_{\alpha} n_{\alpha} u_{\alpha}, \tag{10.14}$$

where n_{α} is the density, and u_{α} is the Eulerian velocity of a particle of species α averaged over the ensemble. In this case, in order to close the set of Equations 10.10, 10.13, and 10.14, we should know the dependencies of the density n_{α} and velocity u_{α} on the field strength $\vec{E}$ and induction $\vec{B}$. Such dependencies can be derived in terms of the hydrodynamic description of beam plasmas [29–31,33].

10.2.3 Kinetic Equations

In the first part of this chapter, we discussed the semiqualitative one-particle theory of operation mechanisms for some LTID types (TWT and FEL). In principle, knowledge of the laws of individual particle motion is sufficient for describing the dynamics of the whole system. Klimontovich formalism [33] provides the clearest evidence confirming this assertion. We describe the state of the system in terms of the microscopic distribution function [29–31,33]

$$N_{\alpha}\left(t,\vec{p},\vec{r}\right) = \sum_{i=1}^{N_0^{\alpha}} \delta\left(\vec{r} - \vec{r}_{\alpha i}\right) \delta\left(\vec{p} - \vec{p}_{\alpha i}\right), \tag{10.15}$$

where N_0^α is the number of particles of species α in the system, $\vec{r}_{\alpha i}$ and $\vec{p}_{\alpha i}$ are particle coordinates and mechanical momenta, and δ is the Dirac delta function. The equation for $N_\alpha(t,\vec{p},\vec{r})$ follows from the beam plasma continuity in the phase space. We have

$$\frac{\partial N_\alpha}{\partial t}+\frac{d\vec{r}}{dt}\frac{\partial N_\alpha}{\partial \vec{r}}+\frac{d\vec{p}}{dt}\frac{\partial N_\alpha}{\partial \vec{p}}=0, \tag{10.16}$$

where

$$\frac{d\vec{r}}{dt}=\vec{v};\ \frac{d\vec{p}}{dt}=e_\alpha\left\{\vec{E}_M(\vec{r},t)+\frac{1}{c}\left[\vec{v}\vec{H}_M(\vec{r},t)\right]\right\}=\vec{F}, \tag{10.17}$$

and $\vec{E}_M$ and $\vec{H}_M$ are the microscopic strengths of the electric and magnetic fields, $\vec{v}$ is the particle velocity, and $\vec{F}$ is the force that acts on the particle.

The velocity $\vec{v}$ and the momentum $\vec{p}$ satisfy the relation

$$\vec{v}=\frac{\vec{p}/m_\alpha}{\left(1+p^2/m_\alpha^2c^2\right)^{1/2}}=\frac{\vec{p}}{m_\alpha\gamma}, \tag{10.18}$$

where γ is the relativistic factor. Hence, the expressions for the microscopic charge and current densities may be written as

$$\rho_\alpha^M=e_\alpha\sum_{i=1}^{N_0^\alpha}\delta(\vec{r}-\vec{r}_{\alpha i})=e_\alpha\int_{-\infty}^{\infty}N_\alpha(t,\vec{r},\vec{p})d^3p; \tag{10.19}$$

$$\vec{j}_\alpha^M=e_\alpha\sum_{i=1}^{N_0^\alpha}\vec{v}\delta(\vec{r}-\vec{r}_{\alpha i})=e_\alpha\int_{-\infty}^{\infty}\vec{v}N_\alpha(t,\vec{r},\vec{p})d^3p, \tag{10.20}$$

whereas the microscopic fields are determined by the Maxwell–Lorentz system of equations

$$rot\vec{H}_M=\frac{1}{c}\frac{\partial \vec{E}_M}{\partial t}+\frac{4\pi}{c}\sum_\alpha e_\alpha\int_{-\infty}^{\infty}\vec{v}N_\alpha(t,\vec{r},\vec{p})d^3p;$$

$$div\vec{H}_M=0;$$

$$rot\vec{E}_M=-\frac{1}{c}\frac{\partial \vec{H}_M}{\partial t};$$

$$div\vec{E}_M=4\pi\sum_\alpha e_\alpha\int_{-\infty}^{\infty}N_\alpha(t,\vec{r},\vec{p})d^3p; \tag{10.21}$$

Equations 10.15 through 10.21 form a closed system that describes an electron beam that moves in a superposition of proper and external fields. This system is too complicated to be solved directly. That is why simpler approaches are employed, based on the Boltzmann and Vlasov kinetic equations (or quasihydrodynamic equations) and the self-consistent field concept.

Let us derive Boltzmann and Vlasov kinetic equations from Equations 10.15 through 10.21. We carry out the statistical averaging of Equations 10.16 and 10.17 and thus obtain the kinetic Boltzmann equation [29–31,33] given by

$$\hat{D}\langle N_\alpha\rangle = \left\{\frac{\partial}{\partial t} + \vec{v}\frac{\partial}{\partial \vec{r}} + e_\alpha\left(\vec{E} + \frac{1}{c}\left[\vec{v}\vec{B}\right]\right)\frac{\partial}{\partial \vec{r}}\right\} f_\alpha\left(t,\vec{r},\vec{p}\right) = \varsigma I^{\alpha}_{col}, \qquad (10.22)$$

where $\hat{D}$ is the differential operator, $f_\alpha = (\langle n_\alpha\rangle)\langle N_\alpha\rangle$, the symbol $\langle\ldots\rangle$ implies statistical averaging, $\vec{E} = \langle \vec{E}_M\rangle$, $\vec{B} = \langle \vec{H}_M\rangle$, I^{α}_{col} is the collision term, and ς is the scale factor. In our case, the collision term is described by the expression

$$\varsigma I^{\alpha}_{col} = -\frac{e_\alpha}{\langle n_\alpha\rangle}\left\langle \delta\vec{E} + \frac{1}{c}\left[\vec{v}\delta\vec{B}\right]\right\rangle \frac{\partial \delta N_\alpha}{\partial \vec{p}}, \qquad (10.23)$$

with the fluctuations being defined as

$$\delta N_\alpha = N_\alpha\left(t,\vec{r},\vec{p}\right) - \left\langle N_\alpha\left(t,\vec{r},\vec{p}\right)\right\rangle;$$

$$\delta\vec{E} = \vec{E}_M\left(\vec{r},t\right) - \vec{E}\left(\vec{r},t\right);$$

$$\delta\vec{B} = \vec{H}_M\left(\vec{r},t\right) - \vec{B}\left(\vec{r},t\right). \qquad (10.24)$$

Equation 10.23 shows that, distinct from the neutral gas, the nature of the collision mechanism in plasmas is electromagnetic, i.e., it is determined by the long-range forces. The term collisions, in this case, is due to the tradition of neutral gases in kinetic theory. We note that the description of the system in terms of Equation 10.22 requires knowledge of the high-order moments. The latter are derived in plasma theory by means of the well-known calculation procedure [34]. So we shall not consider this aspect in detail. Moreover, collision mechanisms have a minor effect on the regimes occurring in the LTID (except some plasma beam and other similar systems). In particular, it is not difficult to show that if the gas parameter is small [31],

$$\varepsilon_k = (r_m/r_D)^3 \ll 1, \qquad (10.25)$$

then the collision term on the right-hand side of Equation 10.22 may be disregarded. In Equation 10.25, r_m is the mean distance between particles, and r_D is the Debye radius (see Criterion 10.6). Thus, assuming that the collision term vanishes, we obtain the Vlasov equation [29–31,33],

$$\left\{\frac{\partial}{\partial t}+\vec{v}\frac{\partial}{\partial \vec{r}}+e_{\alpha}\left(\vec{E}+\frac{1}{c}\left[\vec{v}\vec{B}\right]\frac{\partial}{\partial \vec{p}}\right)\right\}f\left(t,\vec{r},\vec{p}\right)=0. \tag{10.26}$$

Comparing Equations 10.16 and 10.17 to Equations 10.22 and 10.26, we see that the Boltzmann and Vlasov equations disregard the fine structure of electric and magnetic fields that Equation 10.21 associated with the individual particle motion. Therefore, vectors $\vec{E}$ and $\vec{B}$ describe macroscopic electromagnetic fields acting on large ensembles of charged particles. So, on one hand, solutions for the distribution function f_α correspond to the given values of $\vec{E}$ and $\vec{B}$, and on the other hand, these vectors $\vec{E}$ and $\vec{B}$ are determined by the distribution function (see Equations 10.10 through 10.12). Thus, the electromagnetic field entering Equations 10.22 and 10.26 is self-consistent. That is why the Boltzmann equation (Equation 10.22) and the Vlasov equation (Equation 10.26) are referred to as kinetic equations with self-consistent fields.

In the high-current FEL theory discussed here, we show that in many cases it is convenient to employ the canonical (Hamiltonian) form of the kinetic equations. In particular, this procedure reduces the Boltzmann equation to the canonical form given by

$$\left\{\frac{\partial}{\partial t}+\frac{\partial \mathcal{H}}{\partial \vec{\mathcal{P}}}\frac{\partial}{\partial \vec{r}}+\frac{\partial \mathcal{H}}{\partial \vec{r}}\frac{\partial}{\partial \vec{\mathcal{P}}}\right\}f_{\alpha}\left(t,\vec{r},\vec{\mathcal{P}}\right)=\varsigma I_{col}^{\alpha}, \tag{10.27}$$

where $\mathcal{H}=\sqrt{m_{\alpha}^{2}c^{4}+c^{2}\left(\vec{\mathcal{P}}-\frac{e_{\alpha}}{c}\vec{A}\right)^{2}}+e_{\alpha}\varphi$ is the Hamiltonian of a charged particle of species α in the electromagnetic field described by the vector-potential $\vec{A}$ and the scalar potential φ, and $\vec{\mathcal{P}}$ is the canonical momentum.

10.2.4 Quasihydrodynamic Equation

The quasihydrodynamic equation is employed for the description of beams with relatively high densities [22–24,29–31,33], when inverse inequality (Equation 10.25) is satisfied. To derive this method, we make use of the Chapman–Enskog equation [29,30]. We multiply both the right- and left-hand sides of Equation 10.22 by an arbitrary function $\Phi\left(\vec{p}\right)$ and integrate over $\vec{p}$ to find

$$\int \Phi\left(\vec{p}\right)\hat{D}\left\langle N_{\alpha}\right\rangle d^{3}p=n_{\alpha}\Delta\left\langle \Phi\right\rangle, \tag{10.28}$$

where

$$\Delta\left\langle \Phi\right\rangle=\varsigma\frac{1}{n_{\alpha}}\int \Phi\left(\vec{p}\right)I_{col}d^{3}p.$$

Then, we introduce the mean velocity and momentum, $\vec{u}_\alpha$ and $\vec{P}_\alpha$,

$$\vec{u}_{\alpha}=\frac{1}{n_{\alpha}}\int \vec{v}\left\langle N_{\alpha}\left(t,\vec{r},\vec{p}\right)\right\rangle d^{3}p; \tag{10.29}$$

$$\vec{P}_\alpha = \frac{1}{n_\alpha}\int \vec{P}\langle N_\alpha(t,\vec{r},\vec{p})\rangle \mathrm{d}^3p. \tag{10.30}$$

We put successively $\Phi(\vec{p}) = 1, \vec{p}$ and obtain from Equation 10.28 the equations for particle balance and momentum density, i.e.,

$$\frac{\partial u_\alpha}{\partial t} + \frac{\partial}{\partial t}(n_\alpha \vec{u}_\alpha) = \varsigma\int I_{col}^\alpha \mathrm{d}^3p;$$

$$\frac{\partial}{\partial t}(n_\alpha P_{i\alpha}) + \frac{\partial}{\partial r_i(n_\alpha P_{i\alpha j\alpha}u)} + \frac{\partial R_{ij}}{\partial r_j} - e_\alpha n_\alpha\left(\vec{E} + \frac{1}{c}\left[\vec{u}_\alpha \vec{B}\right]\right)_i = \varsigma\int p_i I_{col}^\alpha \mathrm{d}^3p. \tag{10.31}$$

Here, $\hat{R}$ is the symmetric pressure tensor whose components are given by

$$R_{ij} = \int \delta p_i \delta v_j \langle N_\alpha(t,\vec{r},\vec{p})\rangle \mathrm{d}^3p, \tag{10.32}$$

$\delta\vec{p} = \vec{P}_\alpha - \vec{p}$; $\delta\vec{v} = \vec{u}_\alpha - \vec{v}$. Then, we put $\Phi = p^2/2m$ and, in the same manner, derive equations for the tensor $\hat{R}$. This time, however, the equation contains the components of the thermal flux vector, $\hat{Q}$, to be found in turn, and so on. To break this chain, we have to introduce some additional assumptions. In physical situations under consideration, it is reasonable to neglect variations of particle kinetic energy within an elementary volume due to the difference between kinetic energies of particles that enter and leave this volume. The latter assumption is equivalent to disregarding the terms proportional to $\hat{Q}$. Thus, the chain is broken. Moreover, for the sake of simplicity, we assume that the components of the pressure tensor $\hat{R}$ are described by the Boltzmann distribution law.

Let us analyze Equations 10.31. The right-hand side of the first equation determines the change of particle density under collisions. It must evidently vanish, since we have assumed that interaction is not accompanied by charge creation or annihilation. The right-hand side of the second equation of Equation 10.31 describes the momentum transfer under collisions. Calculating it requires knowledge of the form of the collision term ςI_{col}^α. We have already mentioned that collisions do not influence the processes occurring in the FEL, so we take, for example, the model Bhatnaghar–Gross–Krook collision term [31]. With regard to the above-mentioned speculations, we find that

$$\varsigma\int \vec{p} I_{col}^\alpha \mathrm{d}^3p = -\nu n_\alpha \vec{p}, \tag{10.33}$$

where ν is the effective collision frequency. We notice by the way that in the case of a sufficiently cold beam, ν satisfies the criterion

$$\nu \ll \omega_j, \tag{10.34}$$

where ω_j are the characteristic oscillation frequencies of the system. In view of the above assumptions, Equation 10.31, after some simple transformations, may be written in the form

$$\left(\frac{\partial}{\partial t}+\vec{u}_\alpha\frac{\partial}{\partial\vec{r}}+\frac{\nu}{\gamma_\alpha^2}\right)\vec{u}_\alpha=\frac{q_\alpha}{m_\alpha\gamma_\alpha}\left\{\vec{E}+\frac{1}{c}\left[\vec{u}_\alpha\vec{B}\right]-\frac{\vec{u}_\alpha}{c^2}\left(\vec{u}_\alpha\vec{E}\right)-\frac{v_T^2}{n_\alpha\gamma_\alpha}\left[\frac{\partial n_\alpha}{\partial\vec{r}}-\frac{\vec{u}_\alpha}{c}\left(\vec{u}_\alpha\frac{\partial}{\partial\vec{r}}\right)n_\alpha\right]\right\}, \quad (10.35)$$

where v_T is the root-mean-square particle thermal velocity. Equation 10.35 is obviously analogous to the Euler equation (to the accuracy of the terms $\sim\nu$ and v_T), which is well known in hydrodynamics. This is the reason why this equation is referred to as the quasi-hydrodynamic equation, and for $\nu = 0$, $v_T = 0$ as the hydrodynamic equation. The terminological distinctions are insignificant for our analysis and we shall ignore them in what follows.

10.2.5 Fields and Resonances

The external focusing magnetic field is assumed to be negligibly weak. Suppose an electromagnetic pumping wave $\{\omega_2,\vec{k}_2\}$ and a signal wave $\{\omega_1,\vec{k}_1\}$ propagate in REB drift space, collinearly to the z-axis. Wave polarization is arbitrary, and wave direction can be both similar and opposite. The dispersion and impedance properties of the waves (as well as the properties of stationary transverse reversible pumping fields) are described in terms of the method of simulating magnetodielectric (SM). The electric field strength vectors of signal ($j = 1$) and pumping wave ($j = 2$) are of the form

$$\hat{\vec{E}}_j=\frac{1}{2}\sum_{n_j,m_j}\left(\vec{E}_{n_jm_j}e^{ip_{n_jm_j}}+c.c.\right), \quad (10.36)$$

where $\vec{E}_{n_jm_j}$ are the slowly varying vector complex amplitudes of partial waves, and p_{m_j} are the partial wave phases given by

$$p_{m_j}=n_jm_j\omega_jt-s_jn_jm_jk_jz, \quad (10.37)$$

where $n_j = \pm1,\pm2,\ldots$ are the harmonic numbers of electron oscillations in fields, $m_j = \pm1,\pm2,\ldots$ are the wave harmonic numbers, ω_j are the frequencies, $k_j=\left|\vec{k}_j\right|$ are the wave numbers, $\vec{k}_j$ are the wave vectors of transverse partial waves, $s_j = sign\left\{\vec{k}_j\vec{n}\right\}$ are the sign functions, $\vec{n}$ is the unit vector along the z-axis, t is the laboratory time, and $j = 1,2$. In view of Equations 10.36 and 10.37, definitions for the magnetic component of acting electromagnetic wave fields are obvious.

Fields (Equation 10.36) excite electron waves in the plasma of REBs. Among these, both the stimulated and proper waves are present (see Chapters 4 and 5, Part I). Some of the stimulated waves have a combination nature [9], i.e.,

$$\vec{E}_{st}=E_{st}\vec{n}=\frac{1}{2}\sum_{n,m}\left(E_{\theta nm}e^{i\theta_{nm}}+E_{\psi nm}e^{i\psi_{nm}}+c.c.\right)\vec{n}, \quad (10.38)$$

where $E_{\theta_{n,m},\psi_{n,m}}$ are the complex amplitudes of the stimulated wave, and θ_m, ψ_m are relevant combination phases to be specified later (see above Chapters 4 and 5). The proper waves can be driven both by fields (Equation 10.36) and by external sources at the interaction range input [9]. They are given by

$$\vec{E}_i = E_i\vec{n} = \frac{1}{2}\sum_{l,q}\left(E_q e^{ip_{lq}} + c.c.\right)\vec{n}, \tag{10.39}$$

where E_q is the complex amplitude of electron waves with phase $p_{lq} = l\omega_q t - lk_q z$, $l,q = 1,2,\ldots$. We showed in Chapters 4 and 5 that any wave resonance, including the parametric wave resonance, occurs in the system if the stimulated wave phase is close to some proper wave phase. This can be formulated as

$$p_{n_1m_1} + \sigma_{12}n_2p_{n_2m_2} \approx p_{lq}, \tag{10.40}$$

or equivalently,

$$\begin{aligned} &n_1m_1\omega_1 + \sigma_{12}n_2m_2\omega_2 \approx l\omega_q;\\ &n_1m_1s_1k_1 + \sigma_{12}n_2m_2s_2k_2 \approx lk_q, \end{aligned} \tag{10.41}$$

where $n_j = \pm1,\pm2,\ldots$ are the harmonic numbers of electron oscillations in fields, and $m_j = \pm1,\pm2,\ldots$ are the wave harmonics. Apart from that, $\sigma_{12} = \pm1$ is the sign function. If Expressions 10.40 and 10.41 are satisfied, the induced wave (Equation 10.38) becomes indistinguishable from a proper wave (Equation 10.39). These waves merge and form a unified wave, a SCW.

Expressions 10.41 are referred to as the parametric wave coupling conditions. They give only necessary parametric resonance conditions. A conclusion about whether the parametric resonance occurs in the system (sufficient conditions) can be drawn only by means of the amplitude analysis.

10.2.6 Reduction of the Maxwell Equation to the Standard Form

In accordance with the general theory given in Chapter 5, the first step in constructing the general FEL theory is reducing the initial equation for fields (Equation 10.10) to the so-called standard form like Equation 5.26

$$A\frac{\partial U}{\partial t} + (BP)U + CU = R(U, \partial U/\partial t, (PU), z, t), \tag{10.42}$$

where A,B,C are the square matrices of size $n \times n$, $U = U(z,t)$ is some vector function in Euclidean n-dimensional space R^n with coordinates $\{z_1,z_2,\ldots,z_n\}$, i.e., $\forall z \in R^n$ $z = (z_1,z_2,\ldots,z_n)^T$, $\forall z_I \in (-\infty, +\infty)$, $I \in (1,2,\ldots,n)$, P is some linear differential operator in the space R^n, $R(\ldots)$ is a given weak nonlinear vector function, and t is a scalar variable (laboratory time, for example). Let us illustrate the reduction procedures of this type.

We start with the Maxwell equations (Equation 10.10), supplemented by a kinetic equation (Equation 10.26 or 10.27) and relevant definitions (Equations 10.11 and 10.12).

Comparing Equations 10.10 through 10.12 and 10.42, we see that the Maxwell equations (Equations 10.10) in their original form do not satisfy the standard equation requirement (Equation 10.42). But this impression is wrong. Let us illustrate it by standardizing the set of Equations 10.10 through 10.12.

Let us represent the current density vector $\vec{J}$ and the charge density ρ in Equations 10.11 and 10.12 as sums of linear and nonlinear parts, respectively

$$\vec{J} = \vec{J}^{(1)} + \vec{J}^{(2)}, \quad \rho = \rho^{(1)} + \rho^{(2)}. \tag{10.43}$$

In the most general case, the linear parts $\vec{J}^{(1)}$ and $\rho^{(1)}$ can be written as

$$\begin{pmatrix} J_x^{(1)} \\ J_y^{(1)} \\ J_z^{(1)} \end{pmatrix} = \begin{pmatrix} \sigma_{11} & \sigma_{12} & \sigma_{13} \\ \sigma_{21} & \sigma_{22} & \sigma_{23} \\ \sigma_{31} & \sigma_{32} & \sigma_{33} \end{pmatrix} \begin{pmatrix} E_x \\ E_y \\ E_z \end{pmatrix} + \begin{pmatrix} d_{11} & d_{12} & d_{13} \\ d_{21} & d_{22} & d_{23} \\ d_{31} & d_{32} & d_{33} \end{pmatrix} \begin{pmatrix} B_x \\ B_y \\ B_z \end{pmatrix}; \tag{10.44}$$

$$\rho^{(1)} = (g_1\, g_2\, g_3) \begin{pmatrix} E_x \\ E_y \\ E_z \end{pmatrix} + (h_1\, h_2\, h_3) \begin{pmatrix} B_x \\ B_y \\ B_z \end{pmatrix}. \tag{10.45}$$

Explicit expressions for the matrices $[\sigma_{ij}]$, $[d_{ij}]$, $[g_{ij}]$, $[h_{ij}]$ can be easily derived from the Maxwell equations (Equation 10.10). Then, we arrange the components of the vectors $\vec{E}$, $\vec{H}$, $\vec{D}$, and $\vec{B}$ as a column vector

$$U = \begin{pmatrix} \vec{E} \\ \vec{H} \\ \vec{B} \\ \vec{D} \end{pmatrix}. \tag{10.46}$$

We have for the operator $\hat{P}$

$$\hat{P} = \hat{\vec{\nabla}}. \tag{10.47}$$

Maxwell Equations 10.10 in terms of the above notations could be reduced to a form like Equation 10.42

$$A\frac{\partial U}{\partial t} + (B\hat{P})U + CU = R, \tag{10.48}$$

where

$$A = \begin{pmatrix} [0] & [0] & \left[\frac{F}{c}\right] & [0] \\ [0] & [0] & [0] & \left[-\frac{F}{c}\right] \\ (0) & (0) & (0) & (0) \\ (0) & (0) & (0) & (0) \end{pmatrix}; B = \begin{pmatrix} \vec{k} & [0] & [0] & [0] \\ [0] & \vec{k} & [0] & [0] \\ (0) & (0) & \vec{b} & (0) \\ (0) & (0) & (0) & \vec{b} \end{pmatrix};$$

$$C = \begin{pmatrix} [0] & [0] & [0] & [0] \\ \left[-\frac{4\pi}{c}\sigma_{ij}\right] & [0] & \left[-\frac{4\pi}{c}d_{ij}\right] & [0] \\ (0) & (0) & (0) & (0) \\ (-4\pi g_i) & (0) & (-4\pi h_i) & (0) \end{pmatrix};$$

$$R = \begin{pmatrix} \{0\} \\ \left\{\frac{4\pi}{c} j_i^{(2)}\right\} \\ 0 \\ 4\pi\rho^{(2)} \end{pmatrix}; b = (\vec{e}_x \vec{e}_y \vec{e}_z); \vec{k} = \begin{pmatrix} 0 & -\vec{e}_z & \vec{e}_y \\ \vec{e}_z & 0 & -\vec{e}_x \\ -\vec{e}_y & \vec{e}_x & 0 \end{pmatrix}, F = \begin{pmatrix} 1 & 0 & 0 \\ 0 & 1 & 0 \\ 0 & 0 & 1 \end{pmatrix}, \quad (10.49)$$

[0] are 3 × 3 zero matrices; (0) and {0} are three-dimensional zero-row vectors and zero-column vector, respectively. The function R can be represented as $R = \sum_{n=1}^{\infty} \frac{1}{\xi^n} R^{(n)} = \sum_{n=1}^{\infty} \varepsilon^n R^{(n)}$.

In case the quasihydrodynamic equation is used, the difference from previous kinetic cases is only that the vector U includes the space charge density ρ, as a component

$$U = \begin{pmatrix} \vec{E} \\ \vec{H} \\ \vec{B} \\ \vec{D} \\ \rho \end{pmatrix}. \quad (10.50)$$

Accomplishing the sequence of these calculative procedures, we get the following expressions for the matrices A, B, and C, and the vector function R:

$$A=\begin{pmatrix} [0] & [0] & (F/c) & [0] & [0] \\ [0] & [0] & [0] & (-F/c) & [0] \\ (0) & (0) & (0) & (0) & (0) \\ (0) & (0) & (0) & (0) & (0) \\ (0) & (0) & (0) & (0) & F \end{pmatrix};B=\begin{pmatrix} \vec{k} & [0] & [0] & [0] & [0] \\ [0] & \vec{k} & [0] & [0] & [0] \\ (0) & (0) & \vec{b} & (0) & (0) \\ (0) & (0) & (0) & \vec{b} & (0) \\ \vec{b}(\sigma_{ij}) & (0) & \vec{b}(d_{ij}) & (0) & (0) \end{pmatrix};$$

$$C=\begin{pmatrix} [0] & [0] & [0] & [0] & [0] \\ (-4\pi/c)[\sigma_{ij}] & [0] & (-4\pi/c)[d_{ij}] & [0] & [0] \\ (0) & (0) & -4\pi(h_i) & (0) & (0) \\ (0) & (0) & (0) & (0) & -4\pi \\ (0) & (0) & (0) & (0) & (0) \end{pmatrix};R=\begin{pmatrix} \{0\} \\ (4\pi/c)\vec{j}_i^{(2)} \\ 0 \\ \{0\} \\ \hat{N}_i \end{pmatrix};b=(\vec{e}_x\vec{e}_y\vec{e}_z);$$

$$\vec{k}=\begin{pmatrix} 0 & -\vec{e}_z & \vec{e}_y \\ \vec{e}_z & 0 & -\vec{e}_x \\ -\vec{e}_y & \vec{e}_x & 0 \end{pmatrix};F=\begin{pmatrix} 1 & 0 & 0 \\ 0 & 1 & 0 \\ 0 & 0 & 1 \end{pmatrix}, \tag{10.51}$$

where $\hat{N}_i = \hat{\vec{\nabla}}\vec{j}_i^{(2)}$ are known predetermined functions. As will be shown in the following chapters, the beam motion equations (kinetic and quasihydrodynamic) can also be represented in another standard form. Common properties of these forms are that both can be asymptotically integrated by the use of hierarchic algorithms described in Chapters 5 and 6.

10.2.7 Free Electron Laser as a Hierarchic Oscillation System

Taking into consideration the hierarchic book ideology (see Part I), let us look at the FEL as a hierarchic oscillation system. One can be sure that the considered FEL model could be regarded as a four-level hierarchic oscillation-wave system. A scheme of this system is shown in Figure 10.1. A similar scheme has been discussed in Chapter 4. Let us continue this discussion.

The zero level of the considered hierarchic system is formed by the phases $p_{p'}$, where p_1 is the phase of the signal wave, and p_2 and p_{lq} describe the phases of pumping and SCWs (see Definitions 10.37 and 10.39 and corresponding comments).

The first hierarchic level is determined by the stimulated combination phase $\theta = n_1 p_{m_1} + \sigma_{12} n_2 p_{m_2}$ and the phase of the SCW $p_{lq} = l\omega_q t - lk_q z$ (see Definitions 10.38 and 10.39 and corresponding comments). The combination phase of the next hierarchic level appears as a result of realization of the wave resonance $\bar{\theta} \approx \bar{p}_{lq}$ (see Expressions 10.40 and 10.41) on the first hierarchic level. Namely, the mismatch $\bar{\theta} - \bar{p}_{lq} = \bar{\Theta}$ forms this phase $\bar{\Theta}$ on the second hierarchic level. The third hierarchic level is the maximal one for the considered hierarchic system. According to the material set forth in Chapters 2 through 4, the corresponding phase for the maximal level should be constant: $\bar{\bar{\Theta}} = const$ (the hierarchic analogy with the third low of the thermodynamics; Chapter 2).

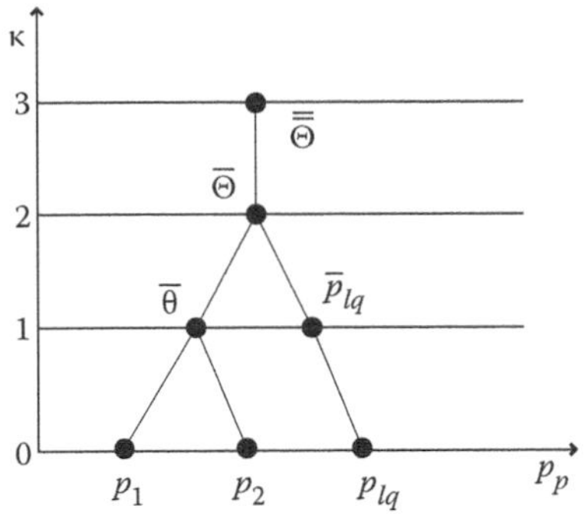

FIGURE 10.1
Parametrical FEL as a hierarchic oscillation system. Here, $p_{1,2}$ are the phases of the signal and pumping waves, p_{lq} is the phase of the l_qth electron wave ($l_q = 1$ and $l_q = 2$ correspond to slow and fast SCWs, respectively), κ is the number of the hierarchic level, $\theta = n_1 p_{m_1} + \sigma_{12} n_2 p_{m_2}$ is the combinative single-particle phase of the first hierarchic level (coinciding with Euler's stimulated phase and corresponding comments), $\bar{\Theta}$ is the combinative phase of the second hierarchic level (it can be determined as a result of averaging the combinative phase $\Theta = \bar{\theta} - \bar{p}_{lq}$, where $\bar{p}_{lq}$ is the averaged wave phase p_{lq}), and $\bar{\bar{\Theta}} = const$ is the averaged combinative phase of the third hierarchic level (see Chapters 3 and 4 for details).

Furthermore, in this chapter, we will illustrate the practical meaning of the system represented in Figure 10.1 at various two- and three-level hierarchic calculative schemes.

In view of the discussed cluster ideology (Chapter 1), we can ask: how does the graph in Figure 10.1 change with the effect of multiharmonic (including the cluster) FELs? In this case, as follows from Figure 10.1, many parallel three-wave resonances work simultaneously. The result of this multiplication is illustrated in Figure 10.2. The hierarchic three for the three-wave multiharmonic cluster resonances can be represented as a system of numbers of hierarchic trees with each of them being the hierarchic three for three resonant harmonics with the same number (Figure 10.1).

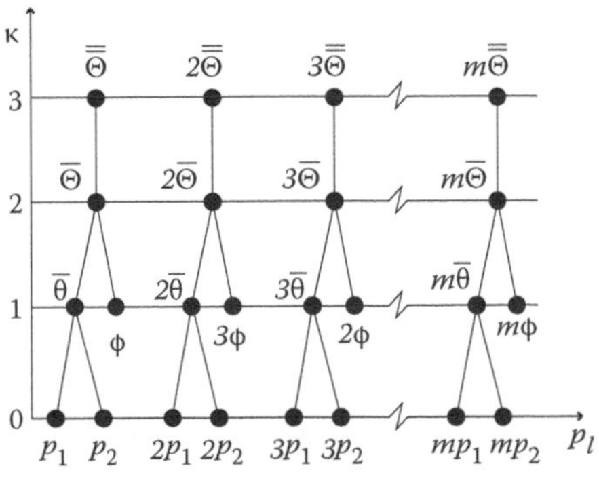

FIGURE 10.2
Multiharmonic two-stream superheterodyne FEL as a hierarchic wave-oscillation system with m parallel three-wave resonances. Here, $p_{1,2}$ are the phases of the signal and pumping waves (zero hierarchic level), $\phi = \mathrm{Re}\{\bar{p}_{3,1}\}$, p_{3j} is the phase of a jth electron-beam wave, κ is the number of the hierarchic level, $\bar{\theta}$ is the averaged slow combinative phase $\theta = p_1 - p_2$ (first hierarchic level), $\bar{\Theta}$ is the averaged combinative phase $\Theta = \bar{\theta} - \mathrm{Re}\{\bar{p}_{3j}\}$ (second hierarchic level where $\mathrm{Re}\{\bar{p}_{3j}\}$ is the real averaged part of the phase p_{3j}), $\bar{\bar{\Theta}} = const$ is the combinative phase of the third (highest) hierarchic level, m is the number of harmonics, and l is the current number of harmonic.

10.3 Method of Simulating the FEL Pumping Fields

10.3.1 Modeling the Pumping System Using the Method of Simulated Magnetodielectric

Many types of FEL pumping systems can be rather large. However, according to the formulated pumping theorem, all these systems can have the same (with the Dopplertron-type) field configuration in the proper reference frame. Two related questions arise. Is there a similar universal description of the pumping field in the laboratory reference frame only? If the answer to that question is, "yes," what are ways to solve this problem? In this section, we discuss a possible answer to both of these questions. We will consider the method of SM [1–16].

The main idea of SM originates from the observation that dispersion of electromagnetic waves is associated with any of the following two factors:

1. Atoms and molecules occurring in the region of wave propagation (in our case a continuous medium).
2. Extended interfaces (e.g., vacuum–metal, vacuum–dielectric, etc.). The distances between boundaries (there is more than one boundary in an arbitrary system) and other characteristic dimensions must be of the same order of magnitude as the wavelength.

The design and technological parameters of second-type systems can be purposefully fitted to provide similar dispersion and impedance properties with the system of the first type. Therefore, each fixed electromagnetic wave propagating in a complex configuration wave-guide (the system of the second type) can correspond to a similar electromagnetic wave in a continuous medium (the system of the first type). However, we should take into consideration that the characteristic feature of the second type of system are, as a rule, rather complex boundary conditions. At the same time, the systems of the first type can be characterized by simpler boundary conditions for the approximately equal dispersion and impedance properties. In other words, the boundary conditions are much simpler in the first case and, thus, the electrodynamic problem here is considerably less involved. This forms the basis of the method of SM. The essence of the latter consists of simplifying the mathematical problem by replacing actual electrodynamic system by a simulating medium.

Simulation procedures based on the comparison of dispersion and impedance characteristics were addressed in detail in microwave technology long ago [35,36]. Since simulated systems can be regarded as linear passive ones (i.e., they do not contain any electron beams) with fairly low active losses, we can replace the simulating systems with a continuous nonconducting linear passive media. In general case, the latter can be anisotropic and inhomogeneous [36]. For simplicity, in this chapter, we assume that the simulating systems are transversely homogeneous unbounded media.

Introducing a simulating media is only a convenient methodological technique. Such an artificial medium can be attributed with properties absent in nature. In particular, we can replace the electrodynamic part of an LTID (pumping or signal systems in FEL, for example) by an SM satisfying the following requirements. The SM imitates a usual continuous medium for electromagnetic fields and, at the same time, has vacuum properties with respect to the electron beam, i.e., it does not prevent electron propagation through the interaction region. It is obvious that, strictly speaking, such systems cannot be observed in nature. However, introducing the SM only as a convenient calculative approach does not cause any physical contradictions. Here we will give some definitions and useful information on electrodynamics of continuous simulating media [2,3] and discuss real media, bearing in mind that the latter is the assumed SM.

Let us consider the propagation of an electromagnetic wave in a continuous medium, discussing first the medium response to static fields. The medium in this case is described by constitutive equations

$$\vec{D} = \varepsilon \vec{E}; \quad H^{-1} = \mu^{-1}\vec{B}; \quad \vec{j} = \sigma \vec{E}, \tag{10.52}$$

where $\vec{B}$ and $\vec{D}$ are the magnetic and electric induction vectors, $\vec{E}$ and $\vec{H}$ are the field strength vectors, $\vec{j}$ is the current density vector, ε is the dielectric permittivity, μ is the magnetic permeability, and σ is the conductivity of the medium. In static cases, the vectors $\vec{D}$, $\vec{H}$, and $\vec{j}$, at any instant and at any space point, are determined by the force characteristic of electromagnetic fields $\vec{E}$ and $\vec{B}$ at the same time and point up to constants ε, μ, and σ. All changes of intrinsic motion velocities and structure of the substance forming the medium produced by the external quasistatic alternating field occur almost without retardation.

The situation changes as the field rate of change increases. We note that mechanisms of intrinsic motions of atoms and molecules include certain time lags. Here, to avoid a mistake, let us remember that, according to the definition, the SM has an uncertain intrinsic physical structure. Therein, the processes in real mediums are not connected with any processes in the modeled vacuum electrodynamic system. We present them only as convenient illustrations of physical mechanisms that cause that or other dispersion and impedance properties of real electromagnetic mediums.

Because of the above-mentioned mechanism of time lag (related to the intrinsic motions of atoms and molecules), polarization of the medium at any given point depends on the field at other points and other time instants. In other words, the nonlocality of interaction, retardation, and anisotropy are manifested. In this case, the medium is described by tensor quantities ε_{ij}, μ_{ij}, and σ_{ij}, rather than by scalars ε, μ, and σ, so constitutive Equations 10.52 take the form

$$D_i(t,\vec{r}) = \int_{-\infty}^{t} dt' \int \varepsilon_{ij}(t,t',\vec{r},\vec{r}')E_j(t',\vec{r}')dr';$$

$$H_i(t,\vec{r}) = \int_{-\infty}^{t} dt' \mu_{ij}^{-1}(t,t',\vec{r},\vec{r}')B_j(t',\vec{r}')dr';$$

$$j_i(t,\vec{r}) = \int_{-\infty}^{t} dt' \sigma_{ij}(t,t',\vec{r},\vec{r}')E_j(t',\vec{r}')dr', \tag{10.53}$$

where all notations are self-evident. Formulas 10.53 give the most general system of constitutive equations where the nonlocality, retardation, and anisotropy are taken into account.

In principle, these equations can form a basis for the theory of SM. The problem is to verify the adequacy of the considered real electrodynamic system and the chosen simulating medium described by parameters ε_{ij}, μ_{ij}, and σ_{ij}.

We will confine ourselves to a demonstration of efficiency of the SM idea for the particular case of models with transverse electromagnetic waves. The expediency of this special choice is justified for three reasons. First, and most important, formal calculation is simplified. It becomes easy to illustrate pumping and signal wave dispersion influence on the operation of FELs. Second, in many cases, real fields are close to pure transverse waves propagating in SM. Third, this approach allows, in principle, generalization of the obtained theoretical results. Indeed, according to the known Brillouin condition (see Figure 10.3), the field of an arbitrary electromagnetic wave can be the superposition of many normal plane wave fields. Hence, the model of SM with transverse waves propagating in a bulk filled by the simulating medium can be universal under moderate complications of description.

Thus, we write the wave fields in an LTID as a superposition of transverse electromagnetic waves propagating in the SM bulk. We disregard active losses $\sigma_{ij} = 0$ and assume that properties of the medium are time-independent. For the time being, we do not discuss anisotropy aspects. In this case, the parameters of the medium, i.e., $\varepsilon_{ij} \equiv \varepsilon$, $\mu_{ij} \equiv \mu$, and $\sigma_{ij} \equiv \sigma$, are scalar quantities depending on coordinate difference $\vec{r} - \vec{r}' = \vec{R}$ and time difference $t - t' = \tau$.

We describe fields in the Fourier representation, i.e.,

$$\vec{E}(t,\vec{r}) = \int_{-\infty}^{+\infty} d\vec{k} \int_{-\infty}^{+\infty} d\omega \vec{E}(\omega,\vec{k}) e^{i(\omega t - \vec{k}\vec{r})};$$

$$\vec{D}(t,\vec{r}) = \int_{-\infty}^{+\infty} d\vec{k} \int_{-\infty}^{+\infty} d\omega \vec{D}(\omega,\vec{k}) e^{i(\omega t - \vec{k}\vec{r})}. \tag{10.54}$$

With these assumptions, we find that field amplitude Fourier components satisfy relations

$$\vec{D}(\omega,\vec{k}) = \varepsilon^{-1}(\omega,\vec{k})\vec{E}(\omega,\vec{k})$$

$$\vec{H}(\omega,\vec{k}) = \mu^{-1}(\omega,\vec{k})\vec{B}(\omega,\vec{k}). \tag{10.55}$$

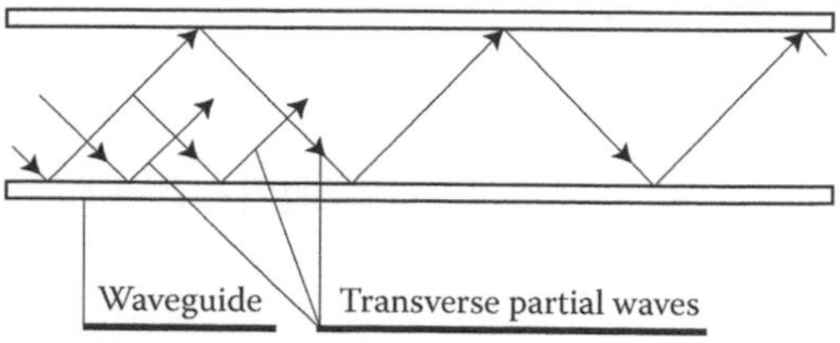

FIGURE 10.3
Illustration of the Brillouin condition.

Comparing Equations 10.52 and 10.54, we see that their mathematical structure is similar. The latter observation is crucial for the theory of SM. The spectral dielectric permittivity $\varepsilon(\omega,\vec{k})$ and the inverse spectral magnetic permeability $\mu^{-1}(\omega,\vec{k})$ are constants and (in our case $\sigma = 0$) completely describe the dispersion and impedance properties of a simulating medium. They are fitted to provide the equivalence of the given medium and real electrodynamic system; that is, they are controlled parameters within the theory of SM. On the other hand, the analogy to the static case (Equation 10.52) where ε, μ^{-1}, and σ are also constants makes it possible to simplify the calculation since all dependencies on t and $\vec{r}$ enter only exponential functions of $(\omega t - \vec{k}\vec{r})$. It is clear that all of these simplifications of the equation structure are obtained owing to the Fourier expansions (Equation 10.53), which is the main reason for the spectral approach to be employed. We notice that if active damping is allowed for $\sigma \neq 0$, then analogous results can be obtained by Laplace or Carson transformations.

Thus, the general model parameters $\varepsilon(\omega,\vec{k})$ and $\mu^{-1}(\omega,\vec{k})$ are functions of frequency ω and wave vector $\vec{k}$ only. These dependencies are called frequency dispersion and spatial dispersion. The frequency dispersion occurs in many electrodynamic systems. The spatial dispersion is a rather rare phenomenon. It occurs in magnetron-like resonator systems with links, in retarding systems operating at higher oscillation modes, in plasma wave-guides, etc. That is why, in the analysis that follows, we consider mainly frequency dispersion.

We turn to the concepts of dispersion function and dispersion equation. Let us write the Maxwell equations for a plane monochromatic wave with frequency ω_j and wave vector $\vec{k}_j$ propagating in an SM. After some transformations we have

$$D(\omega,\vec{k}) = \left(\omega_j^2\varepsilon_j - c^2k_j^2\mu_j^{-1}\right)\omega_j^{-1}, \tag{10.56}$$

where

$$\varepsilon_j = \varepsilon(\omega_j,\vec{k}_j); \quad \mu_j^{-1} = \mu^{-1}(\omega_j,\vec{k}_j). \tag{10.57}$$

The equation

$$D(\omega_j,\vec{k}) = 0 \tag{10.58}$$

is referred to as the dispersion equation. Its solutions describe proper waves (for given electromagnetic systems) whose dispersion law is

$$\vec{k}_j = \vec{n}_j\frac{\omega_j}{c}\sqrt{\varepsilon_j\mu_j} = \vec{n}_jk_j = \vec{n}_j\frac{\omega_j}{c}N_j, \tag{10.59}$$

where $\vec{n}_j$ is the unit vector along the wave propagation direction, and $N_j = \sqrt{\varepsilon_j\mu_j}$ is the retardation factor. If the wave propagates along the z-axis, the unit vector $\vec{n}_j = s_j\vec{n}$. Here we introduce the sign function s_j characterizing one of the two possible directions of wave propagation along the z-axis, i.e.,

$$s_j = \vec{n}_j\vec{n} = \pm 1. \tag{10.60}$$

Using the concept of magnetodielectric, there is a rather simple method of modeling different versions of the grouping and phase velocities. In the given case, the phase velocity can be defined as

$$\vec{v}_{ph} = \frac{\omega_j}{\vec{k}_j}, \tag{10.61}$$

and the group velocity

$$\vec{v}_{gr} = \frac{\partial \omega_j}{\partial \vec{k}_j}. \tag{10.62}$$

It is not difficult to verify that, in general, the directions of the vectors of phase and group velocities can be arbitrary (the angle between the vectors can be from 0 to 2π, including π/2 and π). In the case of collinear propagation, both situations can be realized: $(\partial\omega_j/\partial k_j > 0)$, when the vectors are oriented in the same direction, as well as $(\partial\omega_j/\partial k_j < 0)$, when they are oriented reciprocally opposite. For a description of the latter version, the concept of the dispersion sign is introduced

$$\zeta_j = sign\{\vec{v}_{gr}\vec{v}_{ph}\} = sign\{\partial\omega_j/\partial k_j\} = \pm 1. \tag{10.63}$$

Using Equations 10.54a through 10.58 and 10.62, we can determine the group velocity for the considered model

$$\vec{v}_{gr} = |M_j/L_j|, \tag{10.64}$$

where

$$L_j = \frac{\partial(\omega_j^2 \varepsilon_j)}{\partial \omega_j} - c^2 k_j^2 \frac{\partial \mu_j^{-1}}{\partial \omega_j}; \; M_j = c^2 \frac{\partial(k_j^2 \mu_j^{-1})}{\partial k_j} - \omega_j^2 \frac{\partial \varepsilon_j}{\partial k_j}. \tag{10.65}$$

If only frequency dispersion occurs, we have

$$M_j = 2c^2 k_j \mu_j^{-1}. \tag{10.66}$$

To this point, we have dealt with constant Fourier amplitudes of the fields. In practice, such a case takes place within the theory of a given field. Different versions of the theory are given in Chapter 7. In Chapters 11 through 14, we will study the model with slowly varying amplitudes (Chapters 3 and 4). The fact that the amplitudes of electromagnetic waves are slowly varying functions changes the situation as a whole. However, after relevant modification, the method of SM can also be applied. Only relations such as Equation 10.54a should be specified. For instance, in the simplest one-dimensional model, expressions for material relations are

$$\vec{D}_j(z,t) \cong \left(\varepsilon_j - i\frac{\partial \varepsilon_j}{\partial \omega_j}\frac{\partial}{\partial t} + is_j\frac{\partial \varepsilon_j}{\partial k_j}\frac{\partial}{\partial z} + \dots \right)\vec{E}_j(z,t);$$
$$\vec{H}_j(z,t) \cong \left(\mu_j^{-1} - i\frac{\partial \mu_j^{-1}}{\partial \omega_j}\frac{\partial}{\partial t} + is_j\frac{\partial \mu_j^{-1}}{\partial k_j}\frac{\partial}{\partial z} + \dots \right)\vec{B}_j(z,t). \tag{10.67}$$

When comparing to Equation 10.54, we see that the corresponding operators in a case of slowly varying amplitudes should replace the relevant spectral electric permittivity and magnetic permeability. The concrete examples of such calculation schemes are given in Chapters 12 and 13.

10.3.2 Types of Modeled Pumping Fields

At last, we discuss some results of the application of SM in modeling electromagnetic fields in FEL. It is considered that the plane transversal electromagnetic wave $\{\omega_j, k_j\}$ propagates through some isotropic magnetodielectric. The following nine versions of transversal pumping systems are obtained:

1. $\varepsilon_j \to \infty$, $\omega_j \to 0$, $\mu_j \neq 0$ (and is finite). These parameters correspond to H-Ubitron pumping [17–21,37]. For the electric field strength, in this case, we obtain

$$E_j = \lim_{\substack{\varepsilon_j \to \infty \\ \omega_j \to 0}} \frac{\omega_j}{ck_j} B_j = 0\,, \tag{10.68}$$

since the wave number is not equal to zero

$$k_j = \lim_{\substack{\varepsilon_j \to \infty \\ \omega_j \to 0}} \frac{\omega_j}{c}\sqrt{\varepsilon_j \mu_j} = \frac{2\pi}{\Lambda} \neq 0\,, \tag{10.69}$$

and is finite for the finite magnetic field

$$\vec{H}_j = \mu_j^{-1} B_j\,. \tag{10.70}$$

It is assumed that

$$D_j = \lim_{\substack{\varepsilon_j \to \infty \\ \omega_j \to 0}} (\varepsilon_j E_j) = 0\,. \tag{10.71}$$

Thus, in the given case, the field of a plane electromagnetic wave transforms into a periodic transverse magnetostatic (H-Ubitron) field.

2. $\mu_j \to \infty, \omega_j \to 0, \varepsilon_j \neq 0$ (and is finite). Such a set of parameters corresponds to E-Ubitron pumping [38]. Analogous to the above discussion, the wave number is nonzero, too

$$k_j = \frac{1}{c} \lim_{\substack{\omega_j \to 0 \\ \mu_j \to \infty}} \omega_j \sqrt{\varepsilon_j \mu_j} = \frac{2\pi}{\Lambda} \neq 0\,, \tag{10.72}$$

and finite. It is not difficult to verify that as $H_j \to 0$ and $B_j \to 0$, the plane wave in SM is transformed into a periodic electric (E-Ubitron vortex) field [2]. Just such a field is used in the models of EH systems as a component of the total EH-field (see Chapter 7).

3. $\varepsilon_j \to 0$, $\mu_j \neq 0$ (and is finite). In this case, $k_j \to 0$, $H_j \to 0$, we have one of the Dopplertron-type pumping (E-Dopplertron vortex pumping) [2,3,12]. The electric field is oscillating only in time. This field could be realized near a wide core of a high-frequency electromagnet. It should be noted that the E-Dopplertron systems have no applications in FEL technologies and are interesting only from a theoretical point of view.
4. $\varepsilon_j = \mu_j = 1$, the field of an electromagnetic wave propagating in vacuum (a variety of Dopplertron-type pumping) [39]. Such a pumping field is of interest because a possibility exists for generating short-wave radiation (ultra-violet and higher).
5. $\varepsilon_j\mu_j > 1$, decelerating (retarded) electromagnetic wave pumping (a variety of the Dopplertron-type pumping) [1–17]. This is a very promising type of pumping that allows a number of unique designs of FEL (FEL with the energy of electron beams but that can work in the optical range, the explosive-type FEL, for examples; see Chapters 11 through 14 for more detail).
6. $\varepsilon_j\mu_j < 0$, accelerating electromagnetic pumping wave (a variety of the Dopplertron-type pumping) [39]. This version of the pumping system was used in the first experiments with FELs in the submillimeter range [40,41].
7. $\varepsilon_j \to \infty$, $\mu_j \to \infty$, $(\varepsilon_j/\mu_j) \neq 0$ (and is finite), a variety of the EH-ubitron pumping [2,3,42] (see Chapter 7).
8. $\mu_j \to 0$, $\varepsilon_j \neq 0$ (and is finite). This is H-Dopplertron pumping [2,3,42]. A field of this type can be realized in the gap of a high-frequency electromagnet.
9. $\mu_j \to 0$, $\varepsilon_j \to 0$, $(\varepsilon_j/\mu_j) \neq 0$ (and is finite). It is an EH-Dopplertron pumping field [2,3,42]. Such a field can also be realized in the specific inhomogeneous gap of a high-frequency electromagnet.

Thus, the method of SM provides a unified description of a wide class of periodically reversible electromagnetic fields. The discussed simulation allows us to model a framework of the same theoretical approach to various existing electromagnetic waves (retarded and accelerated, with positive and negative energy and dispersion, etc.), and electric and magnetic fields with different undulation. A generalization of the described version of SM lies in modeling the transversely nonuniform fields containing longitudinal components. This can be made passing from the scalar to the simulated tensors of dielectric permittivity and magnetic permeability using transversely bounded smooth wave-guides. However, use of the discussed modeling of electric and magnetic pumping FEL fields has specific peculiarities. Some of them relate to uncertainties of type $0/0$ or $0 \cdot \infty$. Considering the main purpose of this book, in the following chapters, we use only models without any such uncertainties. As an example, we take the magnetodielectric model universally describing all Dopplertron-type systems with electromagnetic waves and the H-Ubitron model.

References

1. Kalmykov, A.N., Kotsarenko, N.Y., and Kulish, V.V. 1977. Parametric generation and amplification of electromagnetic waves with frequencies higher than the pump wave frequency in electron beams. *Izv.Vyssh.Uchebn.Zaved. Radioelectron.* [*Soviet Radioelectronics*] 10:76.
2. Kulish V.V. 1990. *The physics of free electron lasers. General principles.* Deposited manuscript. Deposited in Ukrainian Scientific Research Institute of Technical Information, Kiev. 1526 Uk-90:192.
3. Kulish, V.V. 1998. *Methods of averaging in non-linear problems of relativistic electrodynamics.* Atlanta, GA: World Scientific Publishers.
4. Kulish, V.V. 2002. *Hierarchic methods. Undulative electrodynamic systems,* Vol. 2. Dordrecht, the Netherlands: Kluwer Academic Publishers.
5. Kulish, V.V. 2002. *Hierarchy and hierarchic asymptotic methods in electrodynamics,* Vol. 1 of *Hierarchic methods.* Dordrecht: Kluwer Academic Publishers.
6. Kalmykov, A.N., Kotsarenko, N.Y., and Kulish, V.V. 1979. Concerning the theory of frequency parametric transformation 'up' in electron beams *Radiotekhika I electronika* [*Sov. Radio Eng. and Electron.*] 24(10):2084–2088.
7. Berezhnoi, I.A., Kulish, V.V., and Zakharov, V.P. 1981. On the explosive instability of relativistic electron beams in the fields of transverse electromagnetic waves. *Zh.Tekh.Fiz.* [Sov.Phys.-Tech. Phys.] 51:660.
8. Kulish, V.V. 1997. Hierarchic oscillations and averaging methods in nonlinear problems of relativistic electronics. *The International Journal of Infrared and Millimeter Waves* 18(5):1053–1117.
9. Kulish, V.V., and Lysenko, A.V. 1993. Method of averaged kinetic equation and its use in the nonlinear problems of plasma electrodynamics. *Fizika Plazmy* (*Sov. Plasma Physics*) 19(2):216–227.
10. Kulish, V.V., Kuleshov, S.A., and Lysenko, A.V. 1993. Nonlinear self-consistent theory of superheterodyne and free electron lasers. *The International Journal of Infrared and Millimeter Waves* 14(3):451–568.
11. Kulish, V.V., Dzedolik, I.V., and Kudinov, M.A. 1985. Relativistic electron motion in periodic reversible fields. Deposited manuscript. Deposited in Ukrainian Scientific Research Institute of Technical Information, Kiev. 1490, Uk-85.
12. Kochmanski S.S., and Kulish, V.V. 1984. On the classical one-particle theory of free electron lasers. *Acta Phys.Polonica* A66(6):713.
13. Kochmanski, S.S., and Kulish, V.V. 1985. On the nonlinear theory of free electron lasers. *Acta Phys.Polonica* A68(5):749–756.
14. Kohmanski, S.S., and Kulish, V.V. 1985. To the nonlinear theory of free electron lasers with multi-frequency pumping. *Acta Phys. Polonica* A68(5):741–748.
15. Kohmanski, S.S., and Kulish, V.V. 1985. Parametric resonance interaction of electron in the field of electromagnetic waves and longitudinal magnetic field. *Acta Phys. Polonica* A68:6525–6736.
16. Zhurakhovski, V.A., Kulish, V.V., and Cheremis, V.T. 1980. Generation of energy by a flow electron in the field of two wave of transverse electromagnetic type. Institute of Electrodynamic Academy of Science of Ukraine, Kiev. *PrePrint No. 218.*
17. Marshall, T.C. 1985. *Free electron laser*. New York: MacMillan.
18. Brau, C. 1990. *Free electron laser.* Boston: Academic Press.
19. Luchini, P., and Motz, U. 1990. *Undulators and free electron lasers*. Oxford: Clarendon Press.

20. Saldin, E.L., Schneidmiller, E.V., and Yurkov, M.V. 2000. *The physics of free electron lasers. Advanced texts in physics*. Heidelberg, Germany: Springer-Verlag.
21. Wu, J., Bolton, P.R., Murphy, J.B., and Zhong, X. 2007. Free electron laser seeded by IR laser driven high-order harmonic generation. *Appl. Phys. Lett.* 90(2):1109.
22. Ruhadze, A.A., Bogdankevich, L.S., Rosinkii, S.E., and Ruhlin, V.G. 1980. *Physics of high-current relativistic beams*. Moscow: Atomizdat.
23. Kuzelev, M.V., and Rukhadze, A.A. 1990. *Electrodynamics of dense electron beams in plasmas*. Moscow: Nauka.
24. Kuzelev, M.V., Rukhadze, A.A., and Strelkov, P.S. 2002. *Plasma relativistic microwave electronics*. Moscow: Bauman Moscow State Technological University.
25. Lebedev, I.V. 1964. *Microwave technology and devices*, Vol. II. Moscow: Energiya.
26. Gaiduk, V.I., Palatov, K.I., and Petrov, D.M. 1971. *Principles of microwave physical electronics*. Moscow: Soviet Radio.
27. Vainstein, L.A., and Solnzev, V.A. 1973. *Lectures on microwave electronics*. Moscow: Soviet Radio.
28. Trubetskov, D.I., and Khramov, A.E. 2003. *Lectures on microwave electronics for physicists*, Vols. 1 and 2. Moscow: Fizmatlit.
29. Davidson, R.C. 1974. *Theory of nonneutral plasmas*. Reading, MA: W.A. Benjamin.
30. Davidson R.C. 1990. Physics of nonneutral plasmas. Reading, MA: Addison-Wesley.
31. Alexandrov, A.F., Bogdankevich, L.S., and Ruhadze, A.A. 1978. *Principles of plasma electrodynamics*. Moscow: Vyschja Shkola.
32. Elias, L.R. 1987. Electrostatic accelerator FEL. *Society of Photo-Optical. Instrumentation Engineers* 738:28–35.
33. Klimontovich, J.L. 1980. *Kinetic theory of electromagnetic processes*. Moscow: Nauka.
34. Sytenko, A.G. 1980. *Fluctuations and nonlinear wave interactions in plasmas*. Moscow: Nauka.
35. Silin, R.A., and Sazonov, V.P. 1966. *Retarding systems*. Moscow: Soviet Radio.
36. Silin, R.A. 1960. On the dispersion of two- and three-dimensional periodic systems (simulating dielectric). *Sov. Radio Eng. and Electron.* 5(4):688–674.
37. Phillips, R.M. 1960. The ubitron, a high-power travelling-wave tube based on a periodic beam interaction in unload waveguide. *IRE Trans. Electron. Devices* 7(4):231.
38. Bekefi, G. 1980. Electrically pumped in relativistic free-electron wave generator. *Appl. Phys.* 51(6):2447–2452.
39. Silin, R.A., Kulish, V.V., Klymenko, Ju.I. 1991. Soviet Inventors Certificate, SU No. 705914, Priority 18.05.72, 15.05.91. *Inventions Bulletin* 26.
40. Davis, G.R. 1976. Navy researchers develop new submillimeter wave power source. *Microwave* 12:12–17.
41. Granatstein, V.L., and Sprangle, P. 1977. Mechanisms for coherent scattering of electromagnetic waves from relativistic electron beams. *IEEE Trans., Microwave Theory and Techn.* 25(6):545–550.
42. Kulish, V.V., Dzedolik, I.V., and Kudinov, M.A. 1985. Relativistic electron motion in periodic reversible fields. Deposited manuscript. Deposited in Ukrainian Scientific Research Institute of Technical Information, Kiev. 1490, Uk-85.

11

Parametrical (Ordinary) Free Electron Lasers: Weak-Signal Theory

Every professional in FEL theory knows that in 40 years since its development, this branch of science has become a boundless ocean of knowledge and information. But, not having a system of dependable beacons, one can easily get lost or even drown in this ocean. Therefore, a demanding question arises for any author who aims to write this type of book and that is the question of what beacons he intends to use. This author was not exempt from this question. As mentioned in Chapters 1 and 9, one of the basic beacons is the phrase "cluster high-current FEL." These words fully represent the basic trends of this book as a whole, as well as this and the next chapter, which are dedicated to the theory of the parametrical (ordinary) FELs.

According to the classification presented in Figure 9.1, the position of the given class of the devices in the cluster system branch can be determined to be the CFEL on the basis of parallel harmonic FEL (see examples in Figures 1.15 and 1.16). This is the case of the final evaluation of and answer to the question of the beacons.

The choice of the mode of interaction is the subsequent choice of the beacon question. Among the two basic interaction modes that are possible in the parametric FELs, contrary to tradition, we dedicated significantly more attention to the Raman mode. This is based on two circumstances. The first depends on the fact that, traditionally, most of the existing literature is dedicated to the study of the Compton mode FELs because this mode is characteristic for the moderate current FELs that are the most popular today. Consequently, the basic physics of the Raman FEL is comparatively much less studied. On the other hand, the shortage of basic physical information might, in principle, result in the occurrence of significant mistakes in the process of preproject analysis. It is important to stress that this especially concerns the cluster systems, for which the choice of basic demands for FEL is far above the estimated limit.

Furthermore, one should mention that even in high-current FELs, the experimentally practical realization of the Raman mode is significantly more complex technologically than the Compton mode. This is related to the more strict demands for the quality of the REBs, which appear in such situations (Chapter 9). Simultaneously, the accumulated experience with the formation of the high-quality high-current beams gives reason to believe that the Raman FEL might be experimentally quite efficient. Let us recollect that the Raman FEL exhibits high levels of amplification and lower levels of electromagnetic noise. This in its totality means, even in the preproject stage, that the parametrical Raman FEL shell should not be excluded from the list of the most promising candidates for the technological basis of the cluster systems.

There exists one more question on the subject of beacons and the next two chapters devote significant attention to it. It is the physics of the polarization phase effects. Traditionally, the investigators disregarded these effects. However, one cannot disregard them in the case of the cluster systems in general and in the CFEL specifically on the basis of many parallel parametrical harmonic FELs. It is their key peculiarity—the interferential nature (effect of multiharmonic interference)—that is the very process of the synthesis of cluster waves (Chapter 1). This automatically means an elevated sensitivity of the cluster system to the phase and polarization characteristics of the cluster signal harmonics, whose totality forms the cluster wave. By taking this information into account in the subsequent chapters, we will give a detailed analysis of the evolution of the initial phases and polarization properties in the process of the interactions in the parametrical FEL.

Finally, the last beacon, to which we dedicate special attention, is the choice of the type of pumping. As mentioned many times, in the beginning of FEL technology development as the basis of the Star Wars program (Chapter 1), the FEL with the magneto-undulation (H-Ubitron) pumping obtained absolute domination. Simultaneously, as shown in Chapters 9 and 10, many other types of pumping might be used in FELs. However, analysis reveals that most of them cannot compete with the traditional H-Ubitron systems. The systems with powerful electromagnetic wave pumping, the Dopplertron-type pumping, are the only exclusion. Even in this case, investigators of the practical realization of FELs traditionally gave preference to the H-Ubitron-type systems. This is primarily because, considering the equivalent action on electrons, here the field amplitude of pumping can be achieved by a technologically more simple method. However, in a case of the cluster systems, where the question of assurance of the mutual coherence of fields (including the pumping field harmonics) is technologically acute, the situation changes essentially. A project analysis reveals that, in many cases, only the parametric FEL with the retarded Dopplertron pumping opens a series of perspective design solutions for the cluster systems.

For example, the frequency transformation coefficient for some types of Dopplertron amplifiers can be increased significantly without an increase of the beam energy and the pumping field frequency. It can be done only because of the increase in the retardation of the pumping electromagnetic wave (see Equations 11.28, 11.36, and 11.37 and related discussions). What is especially important is that the increase can reach a few tens times (see numerical estimations for Expressions 11.36 and 11.37). This, in turn, opens very interesting technological ways for creating the relatively compact superpowerful satellite-based cluster systems. This is illustrated in Figure 11.1, where one of the many possible examples of this type of cluster system is illustrated.

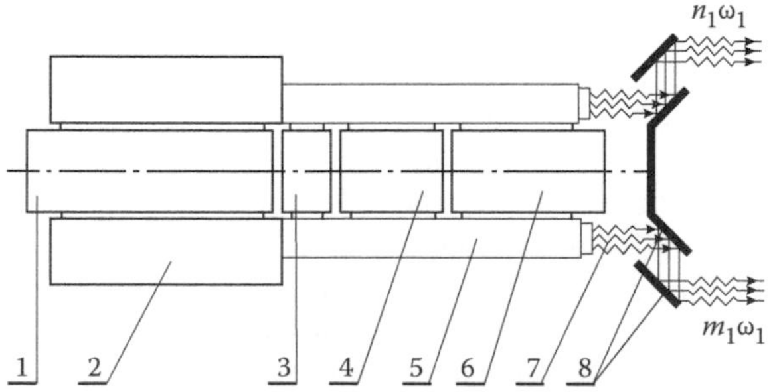

FIGURE 11.1
Illustrative example of the arrangement of the cluster FEL on the basis of many parallel harmonic Dopplertron FELs. Here, 1 is the driving and operating systems of the multichannel and multibeam induction accelerator, which is 2, 3 is the block of multiharmonic input signals, 4 is the microwave pumping oscillator for the Dopplertrons, 5 are the monochromatic partial Dopplertron-amplifiers with retarded microwave pumping, 6 are the driving and operating systems of the Dopplertron (5), 7 are the beams of output Dopplertron radiation with the frequencies $n_1\omega_1$, $m_1\omega_1$ of the output cluster signal (which is not shown here), and $n_1 = 1,2,\ldots$, $m_1 = 1,2,\ldots$ are the numbers of harmonics of the master oscillator frequency ω_1, $n_1 \neq m_1$.

Its characteristic peculiarity is that all the partial Dopplertron amplifiers (5) have the same energy as the electron beams and are connected to a common monochromatic pumping generator (4). Simultaneously, all the amplifiers also have a similarly designed pumping system: each such design is built on the basis of the magnetized plasma-filled wave-guide. As will be shown, the Dopplertron signal wave frequency in this case can strongly depend on the retardation coefficient of the magnetized plasma. The difference between the amplifiers consists of the peculiarity of tuning of these retardation coefficients for each partial pumping system. Namely, they are chosen in such a way that the difference in the signal frequencies for two neighboring amplifiers coincides with the first (or second) harmonic of the master oscillator ω_1, which is within block (3). As a result, we obtain the required multiharmonic spectrum for the totality of signal harmonics that is required for further synthesizing the cluster wave (Chapter 1).

In correlation with all the above-mentioned facts, the theoretical models of the monochromatic partial Dopplertron amplifiers with the retarded microwave pumping are objects of special value in this and in the next chapters. However, the above-mentioned practical side of the matter is not the only reason for its choice. As mentioned in the previous chapters, such theoretical models are also the most generalized. By changing the parameters of the Dopplertron pumping within the framework of the artificial magneto-dielectric model, we obtain different cases of its technological realization in different types of FEL pumping. In particular, the traditional case of the H-Ubitron (magneto-undulation) pumping appears there as one of partial realization. Therefore, many results given in this chapter appear to be useful in the description of the classical H-Ubitron FEL, as well as in the nonclassical Dopplertron-type FELs.

11.1 Self-Consistent Truncated Equations: Simplest Example

As noted in Chapter 6, the method of averaging characteristics can be very effective in cubic nonlinear FEL theory. However, as experience shows, application of this method to the weak-signal theory is not rational. The use of the simplified version of the slowly varying amplitude method [1–7] (see the illustrative example in Chapter 4) allows us to obtain the same results in a simpler way. Accounting for this, in this chapter, we will use this theoretical approach to study the weak-signal parametric FEL model, using as a basis the general model described in Chapter 10 and we conserve all designations that are accepted there.

11.1.1 Formulation of the Quadratic Nonlinear Problem

Signal and pumping waves (Equation 10.36) are linearly polarized in the same plane ($E_y = H_x = 0$) and can be retarded: $N_j = ck_j/\omega_j \geq 1$, and dispersions of signal and pumping electromagnetic waves $dN_j/d\omega_j \cong 0$, $dN_j/dk_j \cong 0$ are neglected. The motion of the beam is described by quasihydrodynamic Equation 10.35 in the simplest case of a charge-compensated (as a whole) cold electron beam $\left(\nu = 0, v_T = 0, q_\alpha = -e, m_\alpha = m, \vec{u}_\alpha = \vec{u} = \{u_x, u_y, u_z\}\right)$. Assuming a small parameter amplitude

$$\varepsilon_j = e\left|E_j\right|/mc\omega_j\gamma_0 << 1, \tag{11.1}$$

we accept that the electron beam velocity $\vec{u}$ can have the form of series in powers of ε_j

$$\vec{u}=\sum_{n=0}^{\infty}\varepsilon^n\vec{u}^{(n)},\quad \varepsilon_1\sim\varepsilon_2\sim\varepsilon. \tag{11.2}$$

It is obvious that $\vec{u}^{(0)}$ ($n = 0$) describes free motion (i.e., without any influences of pumping or signal waves) of the electron beam in the working bulk because Equation 11.1 gives $\varepsilon\sim|E_j|$. Correspondingly, for the oscillatory part of the motion ($n > 0$), we can write

$$\tilde{\vec{u}}=\sum_{n=1}^{\infty}\varepsilon^n\vec{u}^{(n)}=\sum_{n=1}^{\infty}\tilde{\vec{u}}^{(n)}. \tag{11.3}$$

11.1.2 Initial Equations

Then we write Maxwell equations (Equation 10.10) and quasihydrodynamic equations (Equation 10.35 for $v = 0$, $v_T = 0$) with Equations 10.13 and 10.14 as

$$-\frac{\partial H_y}{\partial z}=\frac{\hat{\varepsilon}}{c}\frac{\partial E_x}{\partial t}-\frac{4\pi e}{c}nu_x;\qquad 0=\frac{\hat{\varepsilon}}{c}\frac{\partial E_z}{\partial t}-\frac{4\pi e}{c}nu_z;$$

$$\frac{\partial E_z}{\partial z}=-\frac{4\pi e}{c}(n-n_0);\qquad \frac{\partial E_x}{\partial z}=-\frac{1}{c}\frac{\partial H_y}{\partial t};$$

$$\frac{\partial u_x}{\partial t}+u_z\frac{\partial u_x}{\partial z}=-\frac{e}{m}\sqrt{1-\frac{\vec{u}^2}{c^2}}\left\{E_x-\frac{1}{c}u_zH_y-\frac{1}{c^2}u_x(u_xE_x+u_zE_z)\right\};$$

$$\frac{\partial u_z}{\partial t}+u_z\frac{\partial u_z}{\partial z}=-\frac{e}{m}\sqrt{1-\frac{\vec{u}^2}{c^2}}\left\{E_z+\frac{1}{c}u_xH_y-\frac{1}{c^2}u_z(u_xE_x+u_zE_z)\right\}, \tag{11.4}$$

where $n = n_\alpha$, $n_0 = n_{0\alpha}$, and $\hat{\varepsilon}$ is the dielectric permittivity of the simulated magnetodielectric. For simplicity, let the magnetic permeability of the latter be $\hat{\mu}=1$.

11.1.3 Truncated Equations in the Complex Form

Furthermore, with Equations 11.1 through 11.3, we expand the roots in the right-hand side of the last two equations of Equation 11.4 and take the first-order terms (n = 0,1) with respect to $\tilde{u}^{(n)}_{x,y,z}$

$$\sqrt{1-\frac{\vec{u}^2}{c^2}}\cong(1/\gamma_0)-\beta_0\gamma_0(\tilde{u}/c), \tag{11.5}$$

where $\gamma_0=(1-\beta_0)^{-\frac{1}{2}}$, $\beta_0=u_0/c$, $u_0=(\vec{u}^{(0)}\vec{n})$, $\vec{n}$ is the unit vector along the z-axis. After substitution of Equation 11.5 into Equation 11.4 and accounting for quadratic nonlinearity only, we reformulate Equation 11.4 as

$$\frac{\partial H_y}{\partial z}+\frac{\hat{\varepsilon}}{c}\frac{\partial E_x}{\partial t}-\frac{4\pi e n_0}{c}\tilde{u}_x=\hat{\varepsilon}\frac{\tilde{u}_x}{c}\frac{\partial E_z}{\partial z};$$

$$\frac{\partial E_z}{\partial t}+u_0\frac{\partial E_z}{\partial z}-\frac{4\pi e n_0}{\hat{\varepsilon}}\tilde{u}_z=0;$$

$$\frac{\partial E_x}{\partial z}+\frac{1}{c}\frac{\partial H_y}{\partial t}=0;$$

$$\left(\frac{\partial}{\partial t}+u_0\frac{\partial}{\partial z}\right)\tilde{u}_x+\frac{e}{m\gamma_0}(E_x-\beta_0 H_y)=-\tilde{u}_z\frac{\partial \tilde{u}_x}{\partial z}+$$

$$+\frac{e\beta_0\gamma_0}{mc}\tilde{u}_z(E_x-\beta_0 H_y)+\frac{e}{mc\gamma_0}(\tilde{u}_z H_y+\beta_0\tilde{u}_x E_z);$$

$$\left(\frac{\partial}{\partial t}+u_0\frac{\partial}{\partial z}\right)\tilde{u}_z+\frac{e}{m\gamma_0^3}E_z=-\frac{e}{mc\gamma_0}\tilde{u}_x(H_y-\beta_0 E_x). \tag{11.6}$$

Then, we look at the problem from a hierarchic point of view. According to the general calculation scheme (see Chapter 5), System 11.6 should be presented in the so-called Rabinovich standard form. It is seen that, in principle, it is not difficult to construct relevant vector U, vector function $R^{(n)}$, and matrices A, B, and C (Chapter 10). The next stage of this calculation scheme is accomplishing the hierarchic transformations to the following (first) hierarchic level. For this, we find solutions for linearized System 11.6—where all terms higher than the first order are neglected—that can be represented in the form

$$E_{xj}^{(1)}=\frac{1}{2}E_{js_j}e^{i(\omega_j t-s_j k_j z)}+c.c.;$$

$$E_3^{(1)}=\frac{1}{2}E_{3r}e^{i(\omega_3 t-k_{3r}z)}+c.c.;$$

$$H_y^{(1)}=s_j\frac{ck_j}{\omega_j}E_{xj};$$

$$u_x^{(1)}=\frac{eE_{js_j}}{2m\omega_j\gamma_0}e^{i(\omega_j t-s_j k_j z)}+c.c.;$$

$$u_z^{(1)}=-r\frac{eE_{3r}}{2m\omega_p\gamma_0^{3/2}}e^{i(\omega_3 t-k_{3r}z)}+c.c., \tag{11.7}$$

where dispersion relations $kq(\omega_q)$ (given by solutions of relevant dispersion equations; see Chapter 4) have the form

$$k_j = \frac{N_j}{c}\sqrt{\omega_j^2 - \frac{\omega_p^2}{N_j^2\gamma_0}};$$

$$k_{3r} = k_3 = \frac{\omega_3}{v_0} + r\frac{\omega_p}{v_0\gamma_0^{3/2}}, \tag{11.8}$$

where $r = \pm 1$ is the sign function ($r = \pm 1$ corresponds to the slow SCW of plasmas of electron beam, and $r = -1$ corresponds to the fast SCW), $s_j = sign\{\vec{k}_j\vec{n}\}$ are the sign functions, $\vec{n}$ is the unit vector along the z-axis, $\omega_p = \sqrt{4\pi e^2 n_0/m}$ is the plasma frequency, amplitudes $E_{js}^{(1)}$, $E_{3r}^{(1)}$ correspond to electromagnetic waves ($j = 1,2$) and SCWs ($r = \pm 1$), respectively, N_j are the retardation coefficients for the electromagnetic waves, and $v_0 \equiv u_0$ is the unperturbed electron beam velocity. The other notations are obvious or given already in Chapter 10.

It can be seen from Equation 11.7 that the magnitudes of magnetic field $H_y^{(1)}$ and velocities $u_{x,z}^{(1)}$ are linearly related to the intensities of electric field $E_{xj,z}^{(1)}$. Hence, for further description of the system, the use of three quantities $E_{xj,z}^{(1)}$ is enough. Thus, following the general calculation scheme of a slowly varying amplitude method [1–7], we can write the straight and backward hierarchic transformations as

$$E_x = \frac{1}{2}\left\{\sum_{j,s_j} E_{js_j}(z,t)e^{i(\omega_j t - s_j k_j z)} + c.c.\right\};$$

$$E_z = \frac{1}{2}\left\{\sum_{r} E_{3r}(z,t)e^{i(\omega_3 t - k_{3r} z)} + c.c.\right\};$$

$$H_y = \frac{1}{2}\left\{\sum_{j,s_j} s_j\frac{ck_j}{\omega_j}E_{js_j}(z,t)e^{i(\omega_j t - s_j k_j z)} + c.c.\right\};$$

$$u_x = \frac{1}{2}\left\{\sum_{j,s_j}\frac{eE_{js_j}(z,t)}{m\omega_j\gamma_0}e^{i(\omega_j t - s_j k_j z)} + c.c.\right\};$$

$$u_z = -\frac{1}{2}\left\{\sum_{r} r\frac{eE_{3r}(z,t)}{m\omega_p\gamma_0^{3/2}}e^{i(\omega_3 t - k_{3r} z)} + c.c.\right\}, \tag{11.9}$$

where $E_{js_j}(z,t)$ and $E_{3r}(z,t)$ are the corresponding slowly varying complex amplitudes.

Let us use the following notation for the electromagnetic waves and SCW parameters: $E_{js} \equiv E_j$, $E_{3r} \equiv E_3$, and $k_{3r} = k_3$. We assume that the parametric wave-resonant condition (Equation 10.41) is satisfied. It is accepted that $n_j = m_j = l = 1$ (the case of main resonance; see Chapters 6 and 7) and $\sigma_{12} = \sigma = -1$; $s_1 = -1$, $s_2 = +1$, i.e., we have the model with interactions of the first harmonics of waves only; SCW frequency ω_3 approximately equals the difference between electromagnetic wave frequencies ω_j (see Equation 10.41)

$$\omega_3 \approx \omega_1 - \omega_2; \quad k_3 \approx k_1 + k_2, \tag{11.10}$$

i.e., we accept here for simplicity $\sigma_{12} = -1$, n_j, $m_j = +1$. The direction of propagation of the pumping wave is opposite to the z-axis direction ($s_2 = -1$) and, at the same time, opposite to the signal wave ($s_1 = +1$), too (see again Equation 10.41). Also, we assume that the chosen model is quasistationary (i.e., $\partial E_{j,3})/\partial t = 0$). This means that all transition processes in the system are finished already, i.e., $E_j = E_j(z)$, $E_3 = E_3(z)$ only.

Now, let us again employ the initial set of System 11.6. We remind that nonlinearity and dispersions are small in this model. Therefore, derivatives of slowly varying amplitudes with respect to z can be estimated as

$$\frac{\partial}{\partial z} \to \varepsilon_j; \quad \frac{\partial E_q}{\partial z} \to \frac{dE_q}{dz} \sim \varepsilon_j^2; \quad \frac{\partial^2 E_q}{\partial z^2} \sim \varepsilon_j^3, \tag{11.11}$$

where index $q = j,3$. We substitute System 11.9 into System 11.6, bearing in mind all the given assumptions. Therein, we equate the coefficients of similar exponential functions in the left- and right-hand parts and use parametric resonance Condition 11.10. After some calculations, we obtain the required set of truncated equations for slowly varying complex wave amplitudes, i.e., equations of the first hierarchy

$$\frac{dE_1}{dz} = -B_1 E_2 E_3; \quad \frac{dE_2}{dz} = -B_2 E_1 E_3^*; \quad \frac{dE_3}{dz} = -rB_3 E_1 E_2^*, \tag{11.12}$$

where linear coefficients $B_{1,2,3}$ are called the matrix elements

$$B_1 = \frac{e\omega_1}{4mc^3\omega_2\gamma_0 k_1}\left(ck_3 + r\beta_0\omega_p\gamma_0^{1/2}\right);$$

$$B_2 = \frac{e\omega_2}{4mc^3\omega_1\gamma k_2}\left(ck_3 - r\beta_0\omega_p\gamma_0^{1/2}\right); \quad B_3 = \frac{e\omega_p(ck_3 - \beta_0\omega_3)}{4mc\gamma_0^{1/2}v_0\omega_1\omega_2}. \tag{11.13}$$

System 11.12 can be solved analytically. Hence, substituting these solutions into Equation 11.9, we can obtain the complete analytical solutions of the studied problem. It should be mentioned, however, that in practice, the solutions for amplitudes $E_{1,2,3}$ have an independent interest. This is because the important characteristic of resonant-wave interaction in this class of electrodynamic problems is the gain of signal wave

$$K_1 = |E_1(z = L)| / |E_1(z = 0)| \tag{11.14}$$

where L is the system length. Therefore, in physical analysis, we could be more interested in solutions of truncated systems like System 11.12, and not in relevant total solutions like System 11.9.

11.1.4 Truncated Equations in Real Form

One can be sure that a few motion integrals can be obtained for System 11.12. For this, let us separate out the real and imaginary parts of complex amplitudes

$$E_q = \mathcal{E}_q e^{i\varphi_q}, \tag{11.15}$$

where $q = j,3$, $\mathcal{E}_q = |E_q|$ are the real amplitudes and $\varphi_q = \arg E_q$ are the phases of complex amplitudes E_q. (Do not confuse oscillation phases p_q with phases of complex amplitudes φ_q.) Because of separation, the scalar equations could be obtained

$$\frac{d\mathcal{E}_1}{dz} = -B_1\mathcal{E}_2\mathcal{E}_3\cos\phi;$$

$$\frac{d\mathcal{E}_2}{dz} = -B_2\mathcal{E}_1\mathcal{E}_3\cos\phi;$$

$$\frac{d\mathcal{E}_3}{dz} = -rB_3\mathcal{E}_1\mathcal{E}_2\cos\phi;$$

$$\mathcal{E}_1\frac{d\varphi_1}{dz} = B_1\mathcal{E}_2\mathcal{E}_3\sin\phi;$$

$$\mathcal{E}_2\frac{d\varphi_2}{dz} = -B_2\mathcal{E}_1\mathcal{E}_3\sin\phi;$$

$$\mathcal{E}_3\frac{d\varphi_3}{dz} = -rB_3\mathcal{E}_1\mathcal{E}_2\sin\phi, \tag{11.16}$$

where $\phi = \varphi_1 - \varphi_2 - \varphi_3$ is the phase mismatch of the interacted waves.

11.1.5 Motion Integrals

Then, combining three phase equations and three amplitude equations from the set of equations in Equation 11.16, we can obtain

$$\left(B_1\frac{\mathcal{E}_2\mathcal{E}_3}{\mathcal{E}_1} + B_2\frac{\mathcal{E}_1\mathcal{E}_3}{\mathcal{E}_2} + rB_3\frac{\mathcal{E}_1\mathcal{E}_2}{\mathcal{E}_3}\right)\sin\phi = \frac{d\phi}{dz}; \tag{11.17}$$

$$-\frac{1}{\mathcal{E}_1\cos\varphi}\frac{d\mathcal{E}_1}{dz} - \frac{1}{\mathcal{E}_2\cos\varphi}\frac{d\mathcal{E}_2}{dz} - \frac{r}{\mathcal{E}_3\cos\varphi}\frac{d\mathcal{E}_3}{dz} =$$

$$= B_1\frac{\mathcal{E}_2\mathcal{E}_3}{\mathcal{E}_1} + B_2\frac{\mathcal{E}_1\mathcal{E}_3}{\mathcal{E}_2} + rB_3\frac{\mathcal{E}_1\mathcal{E}_2}{\mathcal{E}_3}. \tag{11.18}$$

It is easy to obtain from Equations 11.17 and 11.18 the expression,

$$\frac{d\phi}{dz}=\left(-\frac{1}{\mathcal{E}_1}\frac{d\mathcal{E}_1}{dz}-\frac{1}{\mathcal{E}_2}\frac{d\mathcal{E}_2}{dz}-r\frac{1}{\mathcal{E}_3}\frac{d\mathcal{E}_3}{dz}\right)\frac{\sin\phi}{\cos\phi}, \tag{11.19}$$

that can be transformed into the form

$$\frac{d}{dz}\left(\ln\frac{1}{\mathcal{E}_1\mathcal{E}_2\mathcal{E}_3}\right)=\frac{d}{dz}\left(\ln\left(\sin\phi\right)\right). \tag{11.20}$$

The integral of motion obviously follows from Expression 11.19:

$$\mathcal{E}_1\mathcal{E}_2\mathcal{E}_3\sin\phi = C = const. \tag{11.21}$$

It can be seen that for $\phi = n\pi$ (n = 0,1,2,...) or if one of the initial amplitudes $\mathcal{E}_{1,3}(z = 0)$, $\mathcal{E}_2(z = L)$ is equal to zero, then $C = 0$. In this case, Equations 11.16 can be written as

$$\frac{d\mathcal{E}_1}{dz}=-B_1\mathcal{E}_2\mathcal{E}_3;\frac{d\mathcal{E}_2}{dz}=-B_2\mathcal{E}_1\mathcal{E}_3;\frac{d\mathcal{E}_3}{dz}=-rB_3\mathcal{E}_1\mathcal{E}_2; \tag{11.22}$$

$$\varphi_{1,2,3} = const, \tag{11.23}$$

where we accept $n = 0$. Equations 11.22 could be easily integrated. But we will discuss this problem later. Here we again return to the discussion of the problem of motion integrals.

Combining the first three equations of Equations 11.16, we obtain the following three additional integrals of motion:

$$\frac{\mathcal{E}_2^2}{B_2}-\frac{\mathcal{E}_1^2}{B_1}=C_1=const;$$

$$\frac{\mathcal{E}_2^2}{B_2}-r\frac{\mathcal{E}_3^2}{B_3}=C_2=const;$$

$$\frac{\mathcal{E}_1^2}{B_1}-r\frac{\mathcal{E}_3^2}{B_3}=C_3=const. \tag{11.24}$$

One can be sure that only two from motion integrals Equation 11.24 can be considered as independent ones.

11.1.6 Raman and Compton Modes

Performing a separation of the quadratic terms in equations to obtain Equation 11.12, we accepted the following nonevident supposition:

$$\left|\frac{d^2E_3}{dz^2}\right| << \Delta k_3\left|\frac{dE_3}{dz}\right|, \tag{11.25}$$

where $\Delta k_3 = k_3\big|_{r=-1} - k_3\big|_{r=+1} = 2\dfrac{\omega_p}{v_0\gamma_0^{3/2}}$ is the width of the split between the slow and fast SCWs. The validity of Condition 11.25 determines the Raman interaction mode (see Chapter 4). This condition describes the interaction of wave–wave types and is a characteristic feature of the collective mechanism of wave interactions in REB plasmas. In the second (opposite) case, we deal with interactions of the wave–particle type (Compton mode) (see Chapter 4). The system as a whole demonstrates the explicitly expressed single-particle nature of the interaction mechanism. Relevant examples of the Raman and Compton FEL models are discussed in this chapter.

11.2 Kinematical Analysis

11.2.1 Model of Cold Electron Beam

We start with the case of a cold electron beam. Contrary to the preceding case, let us assume that we have a more general model: $s_j = \pm 1$. We assume $n_j, m_j, l = 1$, $\sigma_{12} \equiv \sigma = \pm 1$ (see the definitions for Formulas 10.37 and 10.41 and the scheme of the parametrical model described in Chapter 7). The dispersion law for interacting waves can be written analogously to Equation 11.8

$$k_j = \frac{\omega_j N_j}{c}\sqrt{1 - \frac{\omega_p^2}{\omega_j^2 N_j^2 \gamma}}; \qquad k_3 = \frac{\omega_3}{v_0} + r\frac{\omega_p}{v_0\gamma^{3/2}}, \tag{11.26}$$

where all designations are as given before. Substituting Equation 11.26 into the condition of parametric wave resonance Equation 10.41

$$\omega_1 - \sigma\omega_2 = \omega_3; \qquad s_1k_1 - \sigma s_2 k_2 = k_3 \tag{11.27}$$

we can rewrite the latter as

$$\omega_1 = \sigma\omega_2\frac{s_2N_2\beta - 1}{1 - s_1N_1\beta} - \frac{r\omega_p}{(1 - s_1N_1\beta)\gamma^{3/2}}, \tag{11.28}$$

where we use the notations: $\beta = v_0/c$, $\gamma = (1-\beta^2)^{-1/2}$.

Parametric resonance (combination synchronism) Condition 11.28 illustrates the dependence of interaction development on wave parameters. Therein, we separate two characteristic dependencies on sign function s_j, σ and r

1. The dependency $\omega_1(r)$
2. The dependency $\omega_1(s_j, \sigma)$.

The first shows that the formal resonant conditions for slow ($r = +1$) and fast ($r = -1$) beam SCWs are different. Both resonance frequencies shift as

$$\Delta\omega_{r1} = \pm\omega_p/(1 - s_1N_1\beta)\gamma^{3/2} \tag{11.29}$$

with respect to the central resonant frequency

$$\omega_{c1} = \sigma\omega_2(s_2N_2 - 1)/(1 - s_1N_1\beta). \tag{11.30}$$

It is widely known that the resonance conditions of any type indicate only about a possibility of realization of a resonance. The final answer to the question of a true realization of the resonance can be obtained only from the relevant amplitude analysis. Such analysis will be performed in this chapter. Let us continue the analysis of resonant conditions.

Let us discuss the characteristic dependency of the second mentioned type, i.e., the dependency $\omega_1(s_j, \sigma)$. Taking into consideration that in practice $\Delta\omega_{r1} \ll \omega_{c1}$, we neglect the difference $\Delta\omega_{r1}$ in Condition 11.28 and we consider

$$\omega_1 \cong \omega_{c1} = \sigma\omega_2 \frac{s_2N_2 - 1}{1 - s_1N_1}. \tag{11.31}$$

Bearing in mind all the accepted suppositions and using Condition 11.31, we can derive the sufficient condition for parametric resonance in the form: $\omega_1 > 0$, i.e.,

$$\sigma(s_2N_2\beta - 1)(1 - s_1N_1\beta) > 0. \tag{11.32}$$

11.2.2 Anomalous Doppler Effect in the Dopplertron FEL

Criterion 11.32 imposes restrictions on all theoretically possible combinations of the sign functions σ and s_j, depending on combinations of parameters $N_j\beta$. It is not difficult to verify that the number of probable interaction configurations can be rather large. Analysis of Condition 11.28 accomplished with respect to the variations of $N_j\beta$ in the case $\omega_1 \approx \omega_{c1}$ shows the following. If even one of the phase velocities of electromagnetic waves $v_{phj} = c/N_j$ is smaller than the electron beam velocity $v_0 = c\beta$, very specific effects can occur. Here, we take into account a similar phenomenon known in the theory of cyclotron resonance: the so-called anomalous Doppler effect (ADE). Accounting for this in the above-mentioned situation with the Dopplertron FEL, when $v_{phj} < v_0$, we also regard this as a variety of ADE. Therefore, we refer to the case $N_j\beta = v_0/v_{phj} > 1$ as the ADE interaction mode in the Dopplertron FEL.

It should be emphasized that there are essential distinctions between both versions of this phenomenon. The pure anomalous Doppler effect occurs in the case of cyclotron resonance when the signal wave propagates in the direction of the electron beam progressive motion or, at least, has a velocity component along this direction. In our Dopplertron model, the directions of electromagnetic wave propagation do not determine immediately the possibility of the ADE interaction mode occurring. The condition, analogous to the condition of cyclotron resonance, can be obtained here for the combinative wave, but not with respect to the signal and pumping waves. The phenomenon of the combination wave was discussed in Chapter 4. In accordance with the above-mentioned facts, this combination wave in our case should be directed along the positive direction of the z-axis

$$sign\{\vec{v}_{cw}\vec{n}\} = sign\left\{\frac{\sigma\omega_1 + \omega_2}{s_1k_1 + \sigma s_2k_2}\right\} = +1, \tag{11.33}$$

where $\vec{v}_{cw}$ is the velocity vector of the combination wave, and $\vec{n}$ is the unit vector along the z-axis. Let us remember that we have previously accepted $\vec{v}_0 = v_0\vec{n}$ for the direction of the velocity of the electron beam. Within Condition 11.31, we find that three interaction modes with anomalous Doppler effects could be realized in the considered Dopplertron model:

1. $N_2\beta > 1, N_1\beta < 1\,(s_2\sigma = +1)$, the ADE for the wave with frequency ω_2 (pumping wave)
2. $N_1\beta > 1,\ N_2\beta < 1\ \left(s_1\sigma = +1\right)$, the ADE for the wave with frequency ω_1 (signal wave)
3. $N_2\beta > 1,\ N_1\beta > 1\ \left(s_1 s_2\sigma = -1\right)$, the ADE for both electromagnetic waves

In what follows, we show that some interesting physical effects accompany ADE modes. The explosive instability of the electron beam is most interesting [1–3,7–11].

11.2.3 Passing to the Case of H-Ubitron Pumping

We remember the discussed universal character of the FEL model with the retarded pumping electromagnetic wave. Let us illustrate methodological possibilities of such model-taking as a convenient illustrative example of the H-Ubitron models.

Here, according to Formulas 10.68 through 10.71 in Chapter 10, $N_2 \to \infty$, $\omega_2 \to 0$. Passing to the limit in Condition 11.28, we obtain [8,9,12]

$$\omega_1 = \mu \frac{ck_2\beta}{1 - s_1 N_1\beta}, \tag{11.34}$$

where $\mu = \sigma s_2 = \pm 1$ is the number of signal wave harmonic involved in the interaction (minus 1 or plus 1 in our simplest model). For the case $N_1\beta > 1$ (AED mode), it is found that $-\mu s_1 = +1$, i.e., the interaction occurs both for $\mu = +1$ (if $s_1 = -1$) and $\mu = -1$ (if $s_1 = +1$). In the case $N_1\beta < 1$ (normal case), $\mu = +1$ always.

11.2.4 Dopplertron FEL Models with Retarded Pumping

We turn to other versions of Dopplertron models. The traditional Dopplertron model with nonretarded pumping follows from Condition 11.28, if $N_1 \backsim 1$. In this case, we have the Dopplertron FEL with laser pumping wave [13] or a microwave in a smooth wave-guide [14]. However, from the application point of view, the case $N_2 > 1$ (Dopplertron with retarded pumping wave) is the most interesting [7–12,15–28]. It is readily true that, for $N_2\beta \gg 1$, sufficiently large values of the frequency transformation coefficients $K_{12} = \omega_1/\omega_2$ can be obtained using moderately relativistic electron beams, even $\gamma \backsim 1$ (we will ignore the problem of generating such high-current beams in this methodological example). In the discussed Dopplertron model, the expression for the transformation coefficient follows from Condition 11.31, i.e.,

$$K_{12} \cong \sigma \frac{s_2 N_2\beta - 1}{1 - s_1 N_1\beta}. \tag{11.35}$$

Among interaction modes described in terms of Condition 11.35, only those with large coefficient K_{12} are of practical interest. Inasmuch as the working frequencies of FEL output signals usually belong to the millimeter ultraviolet range, the signal wave retardation factor N_1, in practice, is near 1. Therefore, the requirement that K_{12} is maximum immediately infers $s_1 = +1$. At the same time, the most suitable pumping for the Dopplertron model is provided by electromagnetic waves with $N_2 \gg 1$. Experimentally, these are produced either by microwave retarding systems [29] or by extraordinary plasma waves [30]. In the latter case, the retardation factor is known to attain especially large values. We carry out some numerical estimates to illustrate this assertion. The dispersion law for the extraordinary waves in magnetized plasma is [31]

$$k_2 = \frac{\omega_2}{c}\sqrt{1 - \frac{\omega_{pp}^2}{\omega_2(\omega_2 - \Omega_0 - i\nu)}}, \tag{11.36}$$

where ω_{pp} is the present gas–plasma frequency, Ω_0 is the cyclotron frequency, and ν is the electron collision frequency. We assume $\nu << \omega_{pp}, \omega_p/\gamma^{3/2} << \omega_{pp}, \omega_{pp} >> \Omega_0 >> \omega_2$. Taking these suppositions into account, we can obtain from Equation 11.36 the simplified expression for N_2

$$N_2 \approx \frac{\omega_{pp}}{\sqrt{\omega_2 \Omega_0}}. \tag{11.37}$$

Then, we put $\omega_2 \sim 10^{10}s^{-1}$, $\Omega_0 \sim 10^{11}s^{-1}$, $\omega_{pp} \sim 3 \cdot 10^{12}s^{-1}$. For these parameters, Expression 11.37 yields $N_2 \backsim 100$. If we put $\beta \backsim 0.82$ (which corresponds to the accelerating $U \backsim 375kV$), then we find from Condition 11.28 that $\omega_1 \backsim 5 \cdot 10^{12}s^{-1}$, and for $\beta \backsim 0.95$, $\omega_1 \backsim 2 \cdot 10^{12}s^{-1}$, and so on. For a Dopplertron with a nonretarded pump wave ($N_2 \backsim 1$), such values of transformation coefficient can be attained for electron beam energies $\mathcal{E} \backsim 5$ Mev and $\mathcal{E} \backsim 22$°MeV.

Let us discuss the above-performed kinematical analysis. At this time, we turn to the design version of cluster systems, which can be constructed on the basis of the Dopplertron FEL with retarded pumping (see Figure 11.1 and relevant comments). According to the material presented in the introduction to this chapter, here we should accept the following set of assumptions:

1. We have a totality of many parallel Dopplertron amplifiers arranged into the cluster system in the manner given in Figure 11.1.
2. All Dopplertrons have the high-current REB with the same energy and also have a common source of the pumping wave.
3. All Dopplertrons amplify the monochromatic signal waves, the frequencies of which coincide with the frequencies of the odd harmonics of the formed cluster wave.
4. All Dopplertrons have pumping systems that are made as retarding systems on the basis of extraordinary plasma waves.
5. The retardation coefficients of all pumping systems are chosen in such a way that Condition 11.31 is satisfied for each odd harmonic of the cluster wave.

In the case of the obtained estimation for the maximal retardation coefficient $N_2 = N_{2max} \backsim 100$, we obtain an estimation for the maximal number of signal harmonics of $\backsim 50$. This means that, in the design shown in Figure 11.1, we can create the following technological situation: Varying only the retardation coefficients for the extraordinary plasma waves in different pumping systems, we can provide the number of parallel Dopplertron amplifiers on the level of ~50 items. As other estimations show, it is enough for the synthesis of rather narrow superpowerful femtosecond electromagnetic clusters. This practical prospect historically determined that the essential attention that our research group paid to the study of the Dopplertrons with retarded pumping for the past thirty years was justified [1–3,7–12,15–28].

The Dopplertron model with retarded pump wave in the case $\omega_1 = \omega_2 = \omega$ is of individual interest. It can be treated as a model with the parametric interaction of oscillations of two electromagnetic waves with frequency ω of different types with SCWs with the frequency 2ω. In this case, we obtain $\sigma = +1$ and synchronism Condition 11.28 reduces to

$$\beta \cong \frac{2}{s_1 N_1 + s_2 N_2} \leq 1. \tag{11.38}$$

Such systems can be used as an active electron-beam modulator for the klystron FELs of different types (see the relevant examples in Chapter 9).

In what follows, we show that amplitude analysis reveals the most interesting case, $s_1 = s_2 = +1$, associated with the explosive instability of electron beams. The case of combination synchronism reveals promising prospects for new designs of powerful microwave devices. The E-, H-, EH-Dopplertron models discussed in Chapter 10 are associated with $N_2 = 0$. Then, the combination synchronism Condition 11.28 takes the form

$$\omega_1 \cong \frac{\sigma\omega_2}{s_1 N_1 \beta - 1}. \tag{11.39}$$

We see that frequency transformation coefficient K_{12} is somewhat lower than in the case of Dopplertron with retarded pumping wave. However, in some cases, this is not a shortcoming for practical applications. In some situations, the requirement of maximum attainable beam power can contradict the given moderate value of coefficient K_{12} (e.g., for millimeter-range superhigh-powered devices). In such situations, systems of these types can be of a certain practical significance.

11.2.5 Model of Thermalized Electron Beam

At last, we briefly discuss a role of the thermal electron beam spread in the FEL resonance conditions. In this chapter, we mainly consider the physical situations for which electron thermal spread is irrelevant. That is why we employ cold electron beam approximation. In the simplest case, the validity criterion for this assumption can be derived for initial Maxwell electron distribution. After a series of calculations, the synchronism condition analogous to Condition 11.28 can be given by [11,22]

$$\omega_1 = \sigma\omega_2 \frac{s_2 N_2 \beta - 1}{1 - s_1 N_1 \beta} - \frac{r\omega_p}{(1 - s_1 N_1 \beta)\gamma^{3/2}}$$

$$\times \left[1 + \frac{3 v_T \omega_1^2}{2\omega_p c^2 \gamma} \left(1 - \sigma - \beta\left(s_1 N_1 + \sigma s_2 N_2\right)\right)^2 \right], \tag{11.40}$$

where v_T is the electron thermal velocity. Using Conditions 11.14 and 11.40, we find the required criterion [11]

$$\frac{v_T}{c} << \frac{\omega_p \gamma^{1/2}}{\omega_2\left[1-\sigma-(s_1N_1+\sigma s_2N_2)\beta\right]}. \tag{11.41}$$

Analyzing Formula 11.41 to reveal the influence of thermal effects, we find that the latter to a considerable extent are determined by the electromagnetic wave retardation factors $N_{1,2}$. In particular, for $\sigma = s_2 = -1$, $s_1 = +1$, Formula 11.41 yields [11]

$$N_1 + N_2 << \frac{\omega_p \gamma^{1/2}}{\omega_2 \beta v_T}. \tag{11.42}$$

For $\omega_p \backsim 5 \cdot 10^{10} s^{-1}$, $\beta \backsim 0.8$, $\omega_2 \backsim 10^{11} s^{-1}$, $N_1 = 1$, and $v_T/c \backsim 10^{-3}$, we find $N_2 \ll 10^3$, while for $v_T/c \backsim 10^{-2}$ and similar values of other parameters, we have $N_2 \ll 10^2$. Therefore, restrictions on the thermal spread of beam electrons in a Dopplertron model with a retarded pumping wave—see comments to Equations 11.35 through 11.37—can be more stringent than in the case of a nonretarded pumping wave.

11.3 Amplitude Analysis

We begin the amplitude analysis with the simplest version of the FEL Dopplertron model discussed in Section 11.1. The system of truncated equations for the slowly varying amplitudes (Equation 11.12) (the first hierarchic level)

$$\frac{dE_1}{dz} = -B_1E_2E_3; \quad \frac{dE_2}{dz} = -B_2E_1E_3^*; \quad \frac{dE_3}{dz} = -rB_3E_1E_2^* \tag{11.43}$$

serve here as initial equations. All notions in Equation 11.43 are given in the explanation for Equation 11.12. Therein, two characteristic physical situations are studied:

1. The case of the given pumping field $E_2 \approx E_2(z = 0) = E_{20} = const$ (i.e., the rate of varying the pumping wave amplitude is negligibly small compared to varying the other two waves; this interaction mode is called the parametrical stage of interaction.
2. A case of the self-consistent change of all slowly varying amplitudes simultaneously.

Similar to Section 11.1, we again confine ourselves to studying only the simplest model: $s_1 = +1$, $s_2 = -1$, $\sigma = -1$.

11.3.1 Approximation of the Given Pumping Field

We begin the amplitude analysis with the approximation of the given pumping field. Let us differentiate the last of Equations 11.43, taking into account the first of Equations 11.43

for E_1 and the accepted assumption $E_2 \approx E_2(z = 0) = E_{20} = const$. The result of such a mathematical transformation is the following:

$$\frac{d^2E_3}{dz^2} = rB_1B_3\mathcal{E}_{20}^2E_3, \tag{11.44}$$

where $\mathcal{E}_{20} = |E_{20}|$. We can write the solution of Equation 11.44 as

$$E_3 = \hat{C}_1e^{\alpha_1 z} + \hat{C}_2e^{\alpha_2 z}, \tag{11.45}$$

where $\hat{C}_{1,2}$ are integration constants, and $\alpha_{1,2}$ are determined as roots of characteristic equation

$$\alpha^2 - rB_1B_2\mathcal{E}_{20}^2 = 0, \tag{11.46}$$

i.e.,

$$\alpha_{1,2} = \pm\mathcal{E}_{20}\sqrt{rB_1B_3} = \pm\alpha. \tag{11.47}$$

The quantity α is called the increment or spatial growth rate at the parametric stage of parametric interaction. The boundary conditions are chosen in the form

$$E_1(z=0) = E_{10}; \quad E_3(z=0) = 0, \tag{11.48}$$

so, the considered beam model could be classified as an initially nonmodulated electron beam. Then, the solutions for amplitudes E_1 and E_3 can be found

$$E_1 = -B_1E_{20}\int_0^z E_3(z)\,dz = E_{10}ch\alpha z; \tag{11.49}$$

$$E_3 = -\frac{\alpha E_{10}}{B_1E_{20}}sh\alpha z = -E_{10}\sqrt{\frac{rB_3}{B_1}}\exp(i\varphi_{02})sh\alpha z, \tag{11.50}$$

where $\varphi_{02} = \arg E_{20}$.

Analyzing Solutions 11.49 and 11.50, we see that the studied variety of parametric resonant interaction is characterized by the following two nontrivial features:

1. Amplification of signal waves $\{\omega_1, k_1\}$
2. Excitation of SCWs $\{\omega_3, k_3\}$ with the difference frequency $\omega_3 = \omega_1 - \omega_2$

Amplification takes place only in the case where the SCW is the slow SCW (SSCW) (i.e., $r = +1$; see comments to Equation 11.8). It is interesting to analyze the physical nature of the SSCW. Such analysis shows that it can be classified as a longitudinal wave with negative energy (see Chapter 4). Physical features of waves of this type are unusual. For example, the amplitude (i.e., energy) of an SSCW becomes higher as it loses more energy. In reality, the main source of the spent energy in this situation is the kinetic beam energy.

In other words, taking into consideration the energy conservation law, we can conclude that the main energy source for amplifying the signal wave in the Dopplertron FEL model is the kinetic energy of the longitudinal motion of the electron beam. In other words, some physical mechanism of transforming the kinetic energy into the SSCW energy should exist in the considered system. It is characteristic that this mechanism cannot be developed in a framework of the quadratic theory discussed here. It will be shown further that the mechanisms of such a type should have at least a cubic nonlinear nature. It will be shown also that the pumping wave in the Dopplertron model in some cases can serve as a second (auxiliary) energy source.

Thus, we can write the total solutions of the problem using inverse transformations of Equation 11.7. However, the obtained solutions for slow varying amplitudes (Equations 11.49 and 11.50), in practice, play much more important roles than the total solutions. It is explained that slow varying wave amplitudes are very often more convenient for the description of the interactions than analogous total characteristics. For instance, amplification properties of the physical mechanism can be described based on the definition of amplitude (see Equation 11.24). In our concrete case for the gain factor (coefficient of amplification) of the signal wave, we have the definition

$$K_1 = \frac{\mathcal{E}_1(z=L)}{\mathcal{E}_{10}} = ch\alpha z, \tag{11.51}$$

where L is the total length of the interaction region, $\mathcal{E}_1 = |E_1|$. Later in this chapter, Definition 11.51 will be used for the study of the effects of polarization and phase discrimination, explosive instability, superheterodyne amplification, and others.

11.3.2 Self-Consistent Model: Integration Algorithm

We turn to the technological peculiarities of analysis for the models with self-consistent change of all three amplitudes. As a first step in this direction, let us find exact solutions of System 11.43. We accept the boundary conditions of Equation 11.48 as the same (see Section 11.1).

Let us pass to the real amplitudes $\mathcal{E}_j = |E_j|$, $j = 1,2,3$. In this case, System 11.43 can be rewritten in a form similar to Equation 11.22

$$\frac{d\mathcal{E}_1}{dz} = -B_1\mathcal{E}_2\mathcal{E}_3; \quad \frac{d\mathcal{E}_2}{dz} = -B_2\mathcal{E}_1\mathcal{E}_3; \quad \frac{d\mathcal{E}_3}{dz} = -rB_3\mathcal{E}_1\mathcal{E}_2. \tag{11.52}$$

Then, we express the amplitudes $\mathcal{E}_1$ and $\mathcal{E}_3$ from the first of two integrals of Equation 11.24

$$\frac{\mathcal{E}_2^2}{B_2} - \frac{\mathcal{E}_1^2}{B_1} = C_1 = const;$$

$$\frac{\mathcal{E}_2^2}{B_2} - r\frac{\mathcal{E}_3^2}{B_3} = C_2 = const;$$

$$\frac{\mathcal{E}_1^2}{B_1} - r\frac{\mathcal{E}_3^2}{B_3} = C_3 = const \tag{11.53}$$

and substitute obtained results into the second of Equations 11.52. Introducing the notation

$$\frac{1}{y} = \frac{\mathcal{E}_2}{\sqrt{B_2 C_2}} \tag{11.54}$$

we reduce the second of Equations 11.52 to the form

$$\frac{dy}{dz} = \sqrt{B_1 B_2 B_3 C_2}\sqrt{\left(1-y^2\right)\left(1-\frac{C_1}{C_2}y^2\right)}. \tag{11.55}$$

Let us estimate the ratio C_1/C_2. For this we use the boundary condition $\mathcal{E}_3(z = 0) = 0$, substituting the latter into motion integrals (Equations 11.24 and 11.53). As a result, we find $C_2 = \mathcal{E}_2^2(z=0)/B_2 > C_1$, i.e., $C_1/C_2 < 1$. In this case, Expression 11.55 can be represented in the form of normal Legendre's elliptical integral of the first kind [32] with the integrand

$$\frac{dy}{\sqrt{(1-y^2)\left(1-\frac{C_1}{C_2}y^2\right)}} = \mathcal{E}_2(0)\sqrt{B_1 B_3}\,dz.$$

Furthermore, we invert this integral and pass to the Jacobi elliptical functions in Legendre's form [32]

$$y = sn(\mathcal{E}_2(0)\sqrt{B_1 B_3}\,z + K_0) = sn(w|\mu), \tag{11.56}$$

where K_0 is the constant of integration,

$$w = \mathcal{E}_2(0)\sqrt{B_1 B_3}\ z + K_0;\ \mu = \sqrt{1 - \frac{B_2\mathcal{E}_{01}^2}{B_1\mathcal{E}_2^2(0)}}. \tag{11.57}$$

The parameter μ is called the module of elliptic function. It should be emphasized especially that the constant $\mathcal{E}_2(0) = \mathcal{E}_2(z = 0)$ is an unknown value at this stage of calculation. Its method of determination is discussed later (the point here is that the pumping wave in the chosen model oppositely propagates to the *z*-axis, i.e., $s_2 = -1$). The required explicit solutions of Equations 11.52 can be found using Solution 11.56, Notation 11.54, and the rest of motion integrals 11.53

$$\mathcal{E}_1 = \mathcal{E}_2(0)\sqrt{\frac{A_1}{A_2}}\frac{dn\left(\mathcal{E}_2(0)\sqrt{B_1 B_3}\,z + K_0\right)}{sn\left(\mathcal{E}_2(0)\sqrt{B_1 B_3}\,z + K_0\right)};$$

$$\mathcal{E}_2 = \mathcal{E}_2(0)\frac{1}{sn\left(\mathcal{E}_2(0)\sqrt{B_1B_2}\,z + K_0\right)};$$

$$\mathcal{E}_3 = \mathcal{E}_2(0)\sqrt{\frac{A_3}{A_2}}\frac{cn\left(\mathcal{E}_2(0)\sqrt{B_1B_3}\,z + K_0\right)}{sn\left(\mathcal{E}_2(0)\sqrt{B_1B_3}\,z + K_0\right)}, \tag{11.58}$$

where $cn(w|\mu)$ and $dn(w|\mu)$ are the mentioned Jacobi elliptical functions in Legendre's form. The function $cn(\ldots)$ is called the elliptical cosine and the function $dn(\ldots)$ is the determinant of amplitude [32]. The constant K_0 can be found from the boundary condition $\mathcal{E}_3(z = 0) = 0$

$$cn(K_0) = cn(2m + 1)K = 0,$$

$$K_0 = (2m + 1)K,\ m = 0,\pm1,\pm2,\ldots, \tag{11.59}$$

where K is the complete elliptical integral (K is the real period of elliptical function) [32]. Furthermore, for convenience, we consider $m = 0$, i.e., $K_0 = K$.

At the next stage of our calculative procedure is the determination of the unknown constant $\mathcal{E}_2(0)$. For this, we use the circumstance that the pumping amplitude $\mathcal{E}_2$ is determined on the boundary $z = L$. In addition, we use the second of Solutions 11.58 for the plane $z = L$. As a result, it is easy to obtain

$$\mathcal{E}_2(z = L) = \mathcal{E}_{20} = \mathcal{E}_2(0)\,sn\left(\mathcal{E}_2(0)\sqrt{B_1B_3}\,L + K\right). \tag{11.60}$$

In principle, solving Equation 11.60, we can determine the required constant $\mathcal{E}_2(0)$. Unfortunately, it can be performed only numerically and that is inconvenient for our analysis. Therefore, we use another (nondirect) way to solve the discussed problem [16]. This way is based on some specific mathematical structures of Solutions 11.58. It allows solving the peculiar inverse problem. Including for the known constants $\mathcal{E}_2(0)$ and $\mathcal{E}_{20}$, we can find the length on which such a distribution of amplitude $\mathcal{E}_2(z)$ is realized.

Thus, the main idea of the considered boundary problem is the following. We consider the values $\mathcal{E}_2(0)$, $\mathcal{E}_{20}$, $\mathcal{E}_{10}$ as given quantities. At the same time, we assume that the system length L is an unknown function, which should be determined. Solving the corresponding equation with respect to the unknown length L, we really solve the discussed problem of determining the complete set of integration constants.

Let us illustrate the formulated idea in more detail. For convenience, we introduce the following dimensionless parameters:

$$\hat{p} = \frac{\mathcal{E}_2(0)}{\mathcal{E}_{20}};\quad \hat{q} = \frac{\mathcal{E}_{10}}{\mathcal{E}_{20}}\sqrt{\frac{B_2}{B_1}}, \tag{11.61}$$

where parameter $\hat{p}$ can be treated as the coefficient of exhaustion of the pumping wave. Then, using the second of Solutions 11.58, we can write

$$\mathcal{E}_{20} = \hat{p}\mathcal{E}_{20}\frac{1}{sn(\alpha\hat{p}L + K)} = \hat{p}\mathcal{E}_{20}\frac{dn(\alpha\hat{p}L)}{cn(\alpha\hat{p}L)}, \tag{11.62}$$

where the value α is determined by Expression 11.47. Furthermore, let us use some known properties of elliptic functions [32]. Accomplishing not complex calculations, we can reduce Expression 11.62 to the form

$$sn(\alpha\hat{p}L) = \frac{1-\hat{p}^2}{1-\hat{p}^2+\hat{q}^2}. \tag{11.63}$$

Inverting the elliptical sinus in Equation 11.63, we eventually obtain the required solution for system length

$$L = \frac{1}{\alpha\hat{p}} F(\hat{\varphi}, \mu), \tag{11.64}$$

where $\hat{\varphi} = (1-\hat{p}^2)/(1-\hat{p}^2+\hat{q}^2)$.

Thus, we can write self-consistent solutions of the discussed problem as

$$\mathcal{E}_1 = \mathcal{E}_{10}\frac{1}{cn(\alpha\hat{p}z)};$$

$$\mathcal{E}_2 = \hat{p}\mathcal{E}_{20}\frac{dn(\alpha\hat{p}z)}{cn(\alpha\hat{p}z)};$$

$$\mathcal{E}_3 = \mathcal{E}_{10}\sqrt{\frac{B_3}{B_1}\frac{sn(\alpha\hat{p}z)}{cn(\alpha\hat{p}z)}};$$

$$L = \frac{1}{\alpha\hat{p}} F\left(\frac{1-\hat{p}^2}{1-\hat{p}^2+\hat{q}^2}, \mu\right); \tag{11.65}$$

where the length L is regarded as one more (fourth) determining value, the module of elliptic function $\mu = \sqrt{1-\hat{q}^2/\hat{p}^2}$. Solutions 11.65 say the following: For an initial signal amplitude $\mathcal{E}_{10}$, the given magnitudes of pumping amplitudes $\mathcal{E}_2(0)$ and $\mathcal{E}_2(L)$ in two points $z = 0$ and $z = L$, at the same time, can be realized on the obtained system length L (see the last equation of Solutions 11.65).

11.3.3 Self-Consistent Model: Passage to the Case of Given Pumping Field

In the particular case, $\mathcal{E}_2 \approx \mathcal{E}_{20}$, we obtain Solutions 11.49 and 11.50, obtained earlier in the approximation of the given pumping field. Indeed, taking into consideration the following properties of elliptical functions [32], for $\mu \to 1$ (i.e., for $\hat{q} \to 0$ and $\hat{p} \to 1$, because in this case $\mathcal{E}_{20}^2 >> \mathcal{E}_1^2, \mathcal{E}_3^2$) the following passing is satisfied

$$sn(\alpha\hat{p}z) \to th\alpha z; \quad cn(\alpha\hat{p}z) \to 1/ch\alpha z; \quad dn(\alpha\hat{p}z) \to 1/ch\alpha z.$$

So, performing relevant transformations in Solutions 11.65, we obtain

$$\mathcal{E}_1 \cong \mathcal{E}_{10} ch\alpha z; \quad \mathcal{E}_2 \cong \mathcal{E}_{20} \frac{ch\alpha z}{ch\alpha z} = \mathcal{E}_{20};$$

$$\mathcal{E}_3 \cong \mathcal{E}_{10} \sqrt{\frac{B_3}{B_1}} th\alpha\hat{p} \cdot ch\alpha\hat{p} = \mathcal{E}_{10} \sqrt{\frac{B_3}{B_1}} sh\alpha\hat{p}. \tag{11.66}$$

The signal wave gain for self-consistent interactions can be obtained from the first equation of Solutions 11.65

$$K_1 = \frac{\mathcal{E}_1(L)}{\mathcal{E}_{10}} = \frac{1}{cn(\alpha\hat{p}L)}. \tag{11.67}$$

Comparing Solutions 11.51 and 11.67, we conclude that amplification in the self-consistent model is characterized by dynamics that are much more complex than in the first case. Processes of such type are discussed in detail later in this chapter based on more general analogous self-consistent models.

11.4 More General Dopplertron Model: Explosive Instability

Thus, analyzing the Dopplertron model with the pumping wave oppositely propagated to the electron beam, we obtain a technologically unpleasant result: the effect of the pumping wave exhausting (see Expressions 11.65). Practical realization of this effect means that the pumping amplitude reduces its amplitude during the signal amplification process. In accordance with Expression 11.47, this automatically means the reduction of the Dopplertron gain factor (Equation 11.51). In the theory of Dopplertrons, this type of the interaction mode is called the asymptotic mode.

In view of the above, it is obvious that the asymptotic interaction mode is not too promising for the cluster system represented in Figure 11.1. Hence, we should look for more technologically interesting modes. For a solution to this problem, we will study a more general Dopplertron FEL model that includes an essentially larger number of interaction schemes. All self-consistent modes possible in such a model can be conditionally classified into three characteristic groups:

1. Explosive instability modes. The amplitudes of interacting waves grow infinitely at a finite length of the interaction region [1–3,7–11,18–28].
2. Asymptotic modes. Wave amplitudes asymptotically tend to some fixed values.
3. Oscillatory modes. Amplitudes of interacting waves depend on coordinate z in an oscillatory manner [1–3,7–11,15–28].

The explosive instability modes are most interesting for the realization of the Dopplertron cluster FEL system, the design scheme of which is shown in Figure 11.1. However, the question arises: How can we find these most interesting model configurations among the many other possible arrangements? Let us discuss this more general Dopplertron model with retarded pumping wave in detail.

11.4.1 Truncated Equations

We begin with a cold linearly polarized Dopplertron model. Then, by performing relevant transformations, corresponding truncated equations can be found in the form

$$\frac{d\mathcal{E}_1}{dz} = s_1\sigma B_1\mathcal{E}_1\mathcal{E}_2\cos\phi;$$

$$\frac{d\mathcal{E}_2}{dz} = s_2 B_2\mathcal{E}_1\mathcal{E}_3\cos\phi;$$

$$\frac{d\mathcal{E}_3}{dz} = r\sigma B_3\mathcal{E}_1\mathcal{E}_2\cos\phi;$$

$$\frac{d\phi}{dz} = -\left(\sigma s_1 B_1\frac{\mathcal{E}_2\mathcal{E}_3}{\mathcal{E}_1} + \sigma s_2 B_2\frac{\mathcal{E}_1\mathcal{E}_3}{\mathcal{E}_2} + \sigma r B_3\frac{\mathcal{E}_1\mathcal{E}_2}{\mathcal{E}_3}\right)\sin\phi, \tag{11.68}$$

where $E_j = E_j\, exp\{i\varphi_j\}$; $E_3 = \mathcal{E}_3\exp\{i\varphi_3\}$, $\phi = \varphi_1 + \sigma\varphi_2 - \varphi_3$ is a phase mismatch. In the particular case $s_1 = +1$, $s_2 = \sigma = -1$, Equation 11.68 is evidently reduced to Equations 11.16 and 11.17. Analogous to the calculation scheme above, we obtain the generalization of motion integrals in Equations 11.24 and 11.53

$$s_1\sigma\frac{\mathcal{E}_1^2}{B_1} - s_2\frac{\mathcal{E}_2^2}{B_2} = C_1 = const;$$

$$s_2\frac{\mathcal{E}_2^2}{B_2} - r\sigma\frac{\mathcal{E}_3^2}{B_3} = C_2 = const;$$

$$\mathcal{E}_1\mathcal{E}_2\mathcal{E}_3\sin\phi = C = const, \tag{11.69}$$

where all notations are given above. We simplify the initial conditions, assuming the initial phase mismatch in Equation 11.68 vanishes, i.e., $\phi_0 = \phi(0) = 0$. Then, taking into account the second motion integral in Equation 11.69, Equation 11.68 simplifies (compare with Equations 11.22 and 11.53)

$$\frac{d\mathcal{E}_1}{dz} = \sigma s_1\sigma B_1\mathcal{E}_2\mathcal{E}_3;$$

$$\frac{d\mathcal{E}_2}{dz} = s_2 B_2\mathcal{E}_1\mathcal{E}_3;$$

$$\frac{d\mathcal{E}_3}{dz} = r\sigma B_3 \mathcal{E}_1 \mathcal{E}_2. \tag{11.70}$$

In the following, we employ the set of System 11.70 to analyze wave explosive instability in FEL-Dopplertron.

System 11.70 describes the explosive instability, provided that right-hand (nonlinear) parts have similar signs, i.e., $\sigma s_1 = s_2 = r\sigma$. Actually, the frame of applicability of System 11.70 is essentially wider as it could appear at first sight. We have in mind the case of the Dopplertron models with pumping by waves with negative dispersion [9,21]. As can be readily seen, System 11.70 does not immediately contain the sign of the pumping wave dispersion $\xi_2 = \pm 1$. Nevertheless, this does not mean that it does not exert any influence on the processes under consideration. The key point in this case is hidden in the method of formulation of the boundary conditions. For example, if we put $s_2 = -1$ and impose the condition $E_{20} = E_2(z = 0)$, we can thrust the pumping wave negative dispersion $\xi_2 = -1$ on the system. In this case, as noted in Chapter 4, the direction of the wave group velocity (i.e., the direction of wave energy propagation) is oppositely directed to the direction of the wave phase velocity.

The energy analysis of the considered system [1–3,7,11,21] allows finding that explosive instability occurs if

a) $\xi_j = s_j = \sigma = r = +1$, i.e., the interaction of electromagnetic waves with similar propagation directions and electron beam SSCW; the directions of the group velocities of all three waves are the same [1–3,7,11,21].
b) $\xi_2 = s_2 = \sigma = -1$, $\xi_1 = s_1 = r = +1$, i.e., interaction of SSCWs with different frequencies, pumping waves with negative dispersion, and signal waves with positive dispersion; the directions of the group velocities of all three waves are the same [1–3,21].

11.4.2 Kinematical Analysis

For type-a explosive instability, the frequency transformation coefficient within the context of Equation 11.35 is given by [1–3,7,11,21]

$$K_{12} = \frac{N_2\beta - 1}{1 - N_1\beta}. \tag{11.71}$$

As follows from Equation 11.71, the explosive instability can occur only when retarding is provided for at least one of the electromagnetic waves. Since, $N_1 \backsim 1$, this implies in most cases of practical interest $N_2\beta > 1$. In the case of type-b explosive instability, the frequency transformation coefficient is

$$K_{12} = \frac{1 + N_2\beta}{1 - N_1\beta}. \tag{11.72}$$

The latter expression does not require any restrictions to retardation coefficient N_2. These arise, however, as we analyze conditions $\xi_2 = -1$, $v_0 < c$. We then have

$$\frac{dN_2}{d\omega_2} < 0; \quad \left|\frac{dN_2}{d\omega_2}\right| < \frac{1 + N_2}{\omega_2}. \tag{11.73}$$

For $N_2\beta \gg 1$, both instability types are equivalent in view of the requirement that the frequency transformation coefficient must be large, $K_{12} \gg 1$. For $N_2\beta \backsim 1$, type-b is preferable for practical applications. The type-b explosive instability effect underlies the new design of Dopplertron-type FELs [33]. The type-a explosive instability is employed in the retarded pump wave FEL, the design of which was proposed in the Refs. [15,16]. However, in the practice case, type-a is much more realistic than type-b. On the other hand, type-b is more interesting from a physical point of view.

In view of the practical interpretations of the type-b mode, one must bear in mind the following points. The negative dispersion waves in conventional retarding systems [34,35] have no negative energies, i.e., strictly speaking, they do not satisfy the above requirements. Nevertheless, this does not contradict our reasoning since these waves are not transverse but have surface nature [34,35]. However, the question arises whether theoretical and realistic models are adequate; in other words, how to accomplish the type-b mode in practice? The analysis shows that the desired result can be obtained for active gaseous media (for instance, gas laser inverted media) with negative dispersion [1–3].

11.4.3 Amplitude Analysis

Furthermore, let us carry out the amplitude analysis of the effect. For boundary conditions, we take

$$z = 0, \quad \mathcal{E}_1 = \mathcal{E}_{10}, \quad \mathcal{E}_2 = \mathcal{E}_{20}, \quad \mathcal{E}_3 = \mathcal{E}_{30}. \tag{11.74}$$

It should be noted that Condition 11.74 has an essentially wider application frame than the simplified version used in Section 11.3. Here we obtain a possibility to describe the ordinary one-sectional (monotron) Dopplertron designs (Section 11.3), as well as the klystron versions, too.

We use the solution algorithm set forth in Section 11.2.1. As a result of relevant calculations, we can be convinced that the solution form is essentially determined by signs of the constants C_q in Equation 11.69 (where $q = j,3$; $C_3 = C_1 + C_2$). For instance, if $C_q > 0$ (i.e., $\mathcal{E}_{20}^2 B_1 > \mathcal{E}_{10}^2 B_2$, $\mathcal{E}_{10}^2 B_3 > \mathcal{E}_{30}^2 B_1$, $\mathcal{E}_{10}^2 B_3 > \mathcal{E}_{30}^2 B_1$) solutions can be written as

$$\mathcal{E}_1 = \mathcal{E}_{10} X_1 dc(\alpha_2' z + H_1);$$

$$\mathcal{E}_2 = \mathcal{E}_{20} X_2 nc(\alpha_2' z + H_1);$$

$$\mathcal{E}_3 = \mathcal{E}_{30} X_3 sc(\alpha_2' z + H_1), \tag{11.75}$$

where $\alpha_2' = \sqrt{\mathcal{E}_{20}^2 B_1 B_3 - \mathcal{E}_{30}^2 B_1 B_2}$, elliptic function module $\mu_1 = \alpha_2^{-1}\sqrt{\mathcal{E}_{20}^2 B_1 B_3 - \mathcal{E}_{30}^2 B_2 B_3}$, and $H = F\left(\hat{\varphi}_1 \middle| \mu_1\right)$ is a normal elliptic integral of the first type of the argument $\hat{\varphi}_1 = Arc\sin \left[\left(\mathcal{E}_{30}^2 / \mathcal{E}_{10}^2\right)\sqrt{B_1/B_3}\right]$ and the module μ_1; $X_1 = \sqrt{1 - \mathcal{E}_{30}^2 B_1 / \mathcal{E}_{10}^2 B_3}$, $X_2 = \sqrt{1 + \mathcal{E}_{30}^2 B_2 / \mathcal{E}_{20}^2 B_3}$, $X_3 = \sqrt{\mathcal{E}_{10}^2 B_3 / \mathcal{E}_{30}^2 B_1 - 1}$, $nc(x) = 1/cn(x)$, $dc(x) = dn(x)/cn(x)$, $sc(x) = sn(x)/cn(x)$, and $cn(x)$, $sn(x)$, $dn(x)$ are the Jacobi elliptic functions of module μ_1. Analogous expressions can be found for other combinations of constants C_q signs.

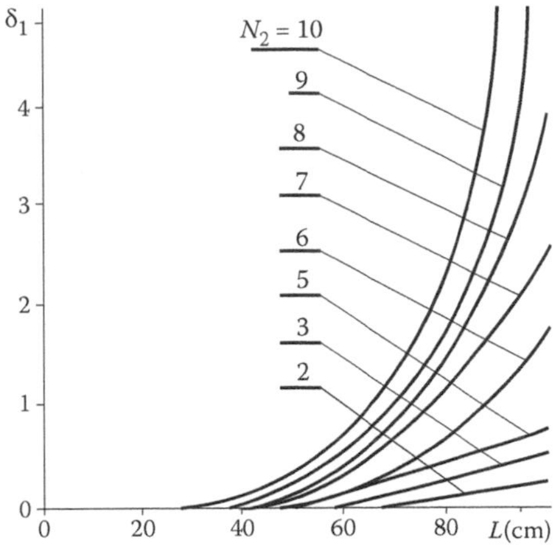

FIGURE 11.2
Dependency of the normalized signal amplitude $\delta_1 = \mathcal{E}_1 \times 10^{-6}$ V/m on the system length L for the explosive instability mode at different magnitudes of the pumping wave retarding coefficient N_2. Here, $\mathcal{E}_{20} = 1.23 \cdot 10^6$ V/cm, $\mathcal{E}_{10} = 10^{-2}\mathcal{E}_{20}$, $\mathcal{E}_{30} = 10^{-2}\mathcal{E}_{10}$, $\omega_2 = 10^{11}s^{-1}$, $N_1 = 1$, and $\omega_p = 0.56 \cdot 10^9 s^{-1}$.

As can be readily seen in Figure 11.2, the peculiar feature of the solutions obtained is their explosive nature. If the argument of the elliptic function in Solutions 11.75 tends to complete the elliptic integral K, the amplitudes grow infinitely, $\mathcal{E}_j \to \infty$, because $cn(K) = \infty$. Therein, the interaction region length L_{cr}, for which the spatial explosion mode is accomplished, is finite,

$$L_{cr} = \frac{K - H_j}{\alpha'_j}. \tag{11.76}$$

The above kinematical analysis suggests that explosive instability occurs only under the anomalous Doppler effect for one of electromagnetic waves (see Section 11.2.4), for example, the wave with frequency ω_2 that is conventionally referred to as the pump wave.

Figure 11.2 shows the variations of signal wave amplitude $\mathcal{E}_1(z)$ on the interaction region of length L = 100 cm in the case of explosive instability under the anomalous Doppler effect for the pumping wave with frequency ω_2.

Figure 11.3 shows the curves for spatial growth rate α'_2 and frequency transformation coefficient K_{12} (see Equation 11.71) corresponding to the amplitude curves in Figure 11.2. As previously mentioned, frequency transformation coefficient K_{12} increases with the retardation factor N_2. As can be readily seen in Figures 11.2 and 11.3, in such a situation the gain factor also increases.

11.4.4 Case of Degeneration on the Wave Frequencies

An interesting version of explosive instability is associated with the case $\omega_1 = \omega_2 = \omega$ (a necessary condition for this version of parametric resonance is given by Equation 11.38). It can be interpreted as the interaction of two different modes of electromagnetic waves with equal frequencies ω and the SSCW with a frequency that is twice as large. The solutions for wave amplitudes can be obtained in the manner similar to $\mathcal{E}_{10} = \mathcal{E}_{20} = \mathcal{E}_0$. Thus, we have

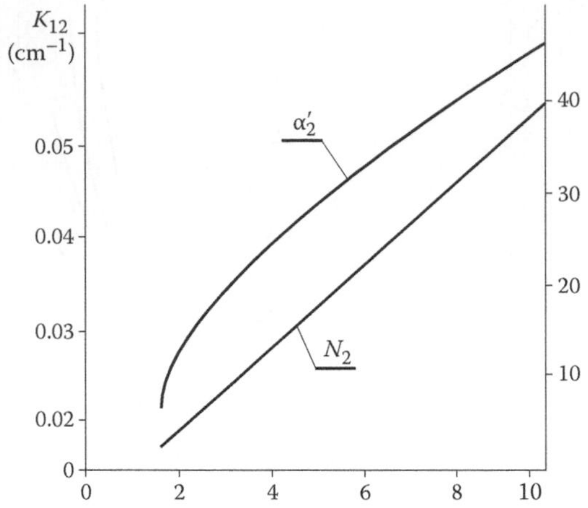

FIGURE 11.3
Dependencies of the spatial growth rate (increment) α_2' and the transformation coefficient K_{12} on the pumping retardation coefficient N_2. All parameters are the same as in Figure 11.2.

1. $z = 0$, $\mathcal{E}_{30} = 0$ (initially nonmodulated electron beam), $N_2 > N_1$ (the opposite case can be obtained by inverting the subscripts) [9]

$$\mathcal{E}_1 = \mathcal{E}_0 nc(\tilde{\alpha}_2 z);$$

$$\mathcal{E}_2 = \mathcal{E}_0 dc(\tilde{\alpha}_2 z);$$

$$\mathcal{E}_3 = \mathcal{E}_0 \left(\frac{\omega_p N_1}{\omega\beta\gamma^{3/2}} \right) sc(\tilde{\alpha}_2 z), \tag{11.77}$$

where $\tilde{\alpha}_2 = \left(e\mathcal{E}_0/mc^2\beta\gamma^{7/4}\right)\left(\sqrt{\omega}/\omega_p\beta N_<\right)^{\frac{1}{2}}$, the elliptic function module $\tilde{\mu} = (1 - N_</N_>)^{\frac{1}{2}}$, and $N_<$, $N_>$ are the greatest and the smallest coefficients $N_{1,2}$, respectively.

2. $z = 0$, $\mathcal{E}_3 = \mathcal{E}_{30}$ (initially modulated electron beam), $N_2 > N_1$ [9]

$$\mathcal{E}_1 = \mathcal{E}_0 \kappa_1 nc\left(\tilde{\tilde{\alpha}}_2 z + H\right);$$

$$\mathcal{E}_2 = \mathcal{E}_0 \kappa_2 dc\left(\tilde{\tilde{\alpha}}_2 z + H\right);$$

$$\mathcal{E}_3 = \mathcal{E}_{30} \kappa_3 sc\left(\tilde{\tilde{\alpha}}_2 z + H\right), \tag{11.78}$$

where $\kappa_{1,2} = \left[1 - \left(\mathcal{E}_{30}^2 \beta\gamma^{3/2}\omega/\mathcal{E}_0^2 N_{1,2}\omega_p\right)\right]^{\frac{1}{2}}$; $\tilde{\tilde{\alpha}}_2 = \tilde{\alpha}_2 \kappa_>$; $\kappa_3 = [(\mathcal{E}_0^2 \omega_p N_2/\mathcal{E}_{30}^2 \beta\gamma^{3/2}\omega) - 1]^{\frac{1}{2}}$; elliptic function module $\tilde{\tilde{\mu}} = \tilde{\mu}\kappa_>^{-1}$; $\kappa_>$ is the largest of $\{\kappa_1, \kappa_2\}$; $H = F\left(\tilde{\varphi}\,|\,\tilde{\tilde{\mu}}\right)$; $\beta = v_0/c$; $\gamma = (1 - \beta^2)^{-\frac{1}{2}}$; and $\tilde{\varphi} = Arc\sin\left[\left(\mathcal{E}_{30}/\mathcal{E}_0\right)\sqrt{\omega_p N_</\gamma^{3/2}\omega\beta}\right]$.

Thus, we see that the character of interaction in both cases of initially nonmodulated and modulated electron beams is mainly determined by the retardation coefficients $N_{1,2}$. For example, the growth rate $\tilde{\alpha}_2$ with the same initial values as those given in Figure 11.2 (except $\mathcal{E}_{10} = \mathcal{E}_{20} = \mathcal{E}_0 = 1.2 \cdot 10^6$ V/cm) is equal to 0.27 cm^{-1}. For $N_2 = 10$, we have $N_2 \to N_1$ and solutions of Equations 11.77 and 11.78 are expressed in terms of trigonometric functions. Therefore, in this case, the process is passive oscillatory and the wave with frequency ω is not amplified.

11.4.5 Influence of Dissipation of SCWs

At last, we discuss the influence of dissipation of SCWs on the development of explosive instability [18–20,22]. The dissipation of interacting waves in the system under consideration is associated with the kinetic properties of electron beam plasma. In the simplest model, particularly, it can be caused by the initial thermal spread of electron beams. We already mentioned this problem in the kinematics analysis of parametric resonance conditions (see Equations 11.40 and 11.41). Under the weak signal amplitude analysis, the most obvious manifestation of the thermal spread is the threshold nature of the explosion process. Namely, the explosive amplification occurs only for pumping field amplitudes greater than the threshold value

$$\mathcal{E}_{20} > (\mathcal{E})_{th}, \tag{11.79}$$

where $(\mathcal{E})_{th}$ is the threshold pumping amplitude. In terms of Equation 11.108, we obtain for linear polarizations the estimate for $(\mathcal{E})_{th}$ as [19,22]

$$(\mathcal{E})_{th} = \frac{K}{\left|B_1 B_3^*\right|^{1/2}} \frac{\sqrt{\pi}\omega_p^2 \left|\omega_3 - v_0 k_3\right|}{k_3 v_T v_0} \tilde{\xi} \exp\left\{-\tilde{\xi}^2\right\}, \tag{11.80}$$

where $\tilde{\xi} = \frac{\omega_p c}{\sqrt{2} v_T \gamma^{3/2} \omega_2} \left|\left[\beta(s_1 N_1 + s_2 \sigma N_2) - 1 - \sigma\right](1 - s_2 N_2 \beta)\right|^{-1}$, and K is the complete Legendre elliptic integral of module $\left(1 - \left(\mathcal{E}_{10}^2 / \mathcal{E}_{20}^2\right)\left|B_1/B_2\right|\right)^{\frac{1}{2}}$. As follows from Equation 11.80, reasonable practical values of $(\mathcal{E})_{th}$ can be obtained for $\tilde{\xi} >> 1$ that are equivalent to

$$\frac{\Delta Q_T}{Q_0} << \left|\frac{2\beta\gamma^{-1/2}}{\left[\beta(s_1 N_1 + s_2 \sigma N_2) - 1 - \sigma\right](1 - s_2 N_2 \beta)(\gamma - 1)}\right| \frac{\omega_p}{\omega_2}, \tag{11.81}$$

where Q_0 is the electron beam energy density, and ΔQ_T is the electron beam mean thermal energy density. For example: $\omega_2 \cong 10^{11} s^{-1}$, $\omega_p \cong 10^{10} s^{-1}$, $\beta \cong 0.9$, $N_1 \backsim N_2 \backsim 1$, and $\sigma = +1$, $(\Delta Q_T)/Q_0 << 2.5 \cdot 10^{-3}$.

11.5 Arbitrarily Polarized Dopplertron Model: Truncated Equations

In previous sections, we discussed particular FEL models only. We considered that electromagnetic pumping and signal fields are linearly polarized and that polarizations of both these fields are oriented in the same polarization plane. This means that a number of polarization effects that, in principle, could take place here during the interaction processes were not accounted for in this case. On the other hand, these effects, as will be shown in what follows, can be very interesting from a physical point of view as well as from a practical one. This especially concerns the cluster systems like those illustrated in Figure 11.1, which is constructed on the basis of the effect of multiharmonic interference (see Chapter 1).

11.5.1 Formulation of the Problem

Let us further generalize the considered model, which contains

1. Arbitrary polarization of pumping and signal waves
2. Kinetic properties of the electron beam
3. Quadratic terms only in the right parts of relevant truncated equations (i.e., we consider the so-called quadratic approximation)

Also, the model is filled by the nonmagnetic ($\hat{\mu}_j = 1$) SM having frequency dispersion $\hat{\varepsilon}_j(\omega_j)$ only.

Analogous to the example discussed previously in this chapter, we use the simplified slowly varying amplitudes method. Here we choose the kinetic Boltzmann equation (Equation 10.27) as an initial motion equation for the electron beam.

We found the fields of pump and signal waves

$$E_x = \frac{1}{2}\sum_{j=1}^{2}\left(E_{j1}(t,z)e^{ip_j} + c.c.\right);\; E_y = \frac{-1}{2}\sum_{j=1}^{2}\left(iE_{j2}(t,z)e^{ip_j} + c.c.\right), \tag{11.82}$$

where $E_{j1,2}$ are the complex amplitudes and $p_j = \omega_j t - s_j k_j z$ are the wave phases. For the magnetic component of the field we can write

$$H_x = \frac{c}{2}\sum_{j=1}^{2}\left(\frac{is_j k_j}{\omega_j}E_{j2}(t,z)e^{ip_j} + c.c.\right);\; H_y = \frac{c}{2}\sum_{j=1}^{2}\left(\frac{s_j k_j}{\omega_j}E_{j1}(t,z)e^{ip_j} + c.c.\right). \tag{11.83}$$

The field of the SCW is

$$\hat{E}_3 = \frac{1}{2}\left(E_3(t,z)e^{ip_3} + c.c.\right), \tag{11.84}$$

where the phase p_3 is determined as before

$$p_3 = \omega_3 t - k_3 z. \tag{11.85}$$

11.5.2 Truncated Equations for Wave Amplitudes

Substituting Equation 11.82 through 11.84 into Maxwell equations (Equation 10.10) and averaging over all fast oscillation phases $p_{j,3}$ (according to previously described procedures of the simplified slowly varying amplitudes method), we obtain the shortened truncated equations

$$\left(\xi_j\frac{\partial}{\partial t}+s_jv_{grj}\frac{\partial}{\partial z}+i\frac{\hat{\varepsilon}_j\omega_j^2-c^2k_j^2}{\hat{\varepsilon}_j\omega_j}\right)E_{jk}=\frac{i}{\hat{\varepsilon}_j\omega_j}q_{jk};\tag{11.86}$$

$$\left(\frac{\partial}{\partial t}+v_0\frac{\partial}{\partial z}+i\Omega_{30}\right)E_3=\frac{i}{\Omega_{30}}q_{33},\tag{11.87}$$

where $\Omega_{30}=\Omega_3(v_z=v_0)$; $\Omega_3=\omega_3-k_3v_z$, $v_0=v_z\big|_{z=0}$, v_z is the longitudinal electron velocity, $\hat{\varepsilon}_j=\hat{\varepsilon}(\omega_j)$ is the permittivity of the simulated magneto-dielectric at frequency ω_j,

$$q_{jk}=2\pi ei^kn_0\omega_jQ_{jk},\quad \text{if } k=1,2;$$

$$q_{jk}=2\pi ein_0\Omega_{30}Q_{jk},\quad \text{if } k=3;\tag{11.88}$$

$\xi_j=sign\left\{\vec{v}_{gj}\vec{n}\right\}=\pm1$ is the dispersion sign for the vector of group velocities $\vec{v}_{gj}=d\omega_j/dk_j$ (see Chapter 4), $\vec{k}_j=s_jk_j\vec{n}$ is the wave vector of the jth electromagnetic wave, k_j are the wave numbers, $v_{gj}=\left|\vec{v}_{gj}\right|$, $\vec{n}$ is the unit vector along the z-axis, n_0 is the nonperturbed density the of electron beam, e is the electron charge,

$$Q_{jk}=\left\langle v_kf\left(t,\vec{r},\vec{p}\right)e^{-ip_j}d\vec{p}\right\rangle\Big|_{p_j},\tag{11.89}$$

$\left\langle\ldots\right\rangle\big|_{p_j}$ is the averaging symbol over phases p_j, $f\left(t,\vec{r},\vec{p}\right)$ is the distribution function solving the Boltzmann kinetic equation, $\vec{p}=\{p_x,p_y,p_z\}$ is the mechanical electron momentum, $v_k=v_x$, if $k=1$, $v_k=v_y$, if $k=2$, and $v_k=v_z$ if $k=3$.

11.5.3 Solving the Kinetic Equation by Successive Approximations

We begin by solving the electron beam motion problem. Taking into consideration the explicit expressions for Fields 11.82 through 11.84, we write the Boltzmann kinetic equation (Equation 10.27) as (collisions are neglected)

$$\left(\frac{\partial}{\partial t}+v_z\frac{\partial}{\partial z}-\hat{M}\right)f\left(t,\vec{r},\vec{p}\right)=0,\tag{11.90}$$

where

$$\hat{M} = \sum_{m=1}^{3} \hat{a}_m e^{ip_m} + k.c., \; \hat{a}_m = E_{m1}\hat{a}_{mx} + E_{m2}\hat{a}_{my}, \text{ if } m = 1,2; \; \hat{a}_m = E_3\hat{a}_z, \text{ if } m = 3; \tag{11.91}$$

$$\hat{a}_{mx} = \frac{e}{\omega_m}\left(\Omega_m \frac{\partial}{\partial p_x} + s_m k_m v_x \frac{\partial}{\partial p_z}\right); \; \hat{a}_{my} = \frac{ie}{\omega_m}\left(\Omega_m \frac{\partial}{\partial p_y} + s_m k_m v_y \frac{\partial}{\partial p_z}\right); \; \hat{a}_z = -e\frac{\partial}{\partial p_z}. \tag{11.92}$$

Therein, we consider $v_{x,y} \ll v_z$ (i.e., any transverse electron motions are considered nonrelativistic).

We find solutions of Equation 11.90 using successive approximations. For this, we represent the distribution function in the form of series

$$f(t,\vec{r},\vec{p}) \cong f^{(0)}(t,\vec{r},\vec{p}) + f^{(1)}(t,\vec{r},\vec{p}) + \ldots, \; \hat{a}_z = -e\frac{\partial}{\partial p_z}, \tag{11.93}$$

where each term satisfies the convergence condition

$$\left|f^{(n+1)}(t,\vec{r},\vec{p})\right| << \left|f^{(n)}(t,\vec{r},\vec{p})\right|. \tag{11.94}$$

Then, we substitute Equation 11.93 into Equation 11.90, equalizing terms of the same order in the left- and right-hand sides of the equation. Therefore, Equation 11.90 can be written as

$$\left(\frac{\partial}{\partial t} + v_z \frac{\partial}{\partial z}\right) f^{(0)}(t,\vec{r},\vec{p}) = 0$$

$$\left(\frac{\partial}{\partial t} + v_z \frac{\partial}{\partial z}\right) f^{(n)}(t,\vec{r},\vec{p}) = \hat{M} f^{(n-1)}(t,\vec{r},\vec{p}); \; n = 1,2,3,\ldots. \tag{11.95}$$

We choose the nonperturbed distribution function $f^{(0)}$, satisfying the first equation from Equation 11.95 in the form $f^{(0)} = f^{(0)}(\vec{p})$ with normalization

$$\int_{-\infty}^{+\infty} f^{(0)}(\vec{p}) d^3 p = 1. \tag{11.96}$$

Taking into account the explicit expressions for operator $\hat{M}$ (Equation 11.8), we represent $f^{(n)}$ as

$$f^{(n)} = \sum_{k=j,3} f_{kn} e^{ip_k} + c.c. \tag{11.97}$$

Averaging with respect to phases p_k in Equation 11.95, we obtain solutions for complex amplitudes f_{kn}

$$f_{kn} = -\frac{i}{\Omega_{k0}}\left\langle \sum_{k'} \hat{M} f_{k',n-1} e^{i(p_{k'}-p_k)} \right\rangle\Big|_{p_k}. \tag{11.98}$$

The explicit form of functions f_{kn} (Equations 11.98) in the first approximation can be obtained by

$$f_{11} = f_{11}^{(1)} E_{11} + f_{11}^{(2)} E_{12};\ f_{21} = f_{21}^{(1)} E_{21} + f_{21}^{(2)} E_{12};$$

$$f_{31} = f_{31}^{(1)} E_3, \tag{11.99}$$

where

$$f_{j1}^{(1)} = \frac{\hat{a}_{jx}}{i\Omega_j} f^{(0)};\quad f_{j1}^{(2)} = \frac{\hat{a}_{jy}}{i\Omega_j} f^{(0)};\quad f_{31}^{(1)} = \frac{\hat{a}_z}{i\Omega_3}. \tag{11.100}$$

In the second approximation they are

$$f_{12} = f_{12}^{(1)}\left(E_3^* E_{21}\delta_{\sigma,-1} + E_{21}^* E_3 \delta_{\sigma,+1}\right) + f_{12}^{(2)}\left(E_{22}E_3^*\delta_{\sigma,-1} + E_3 E_{22}^*\delta_{\sigma,+1}\right);$$

$$f_{22} = f_{22}^{(1)}\left(E_3 E_{11}\delta_{\sigma,-1} + E_3 E_{11}^* \delta_{\sigma,+1}\right) + f_{22}^{2}\left(E_3 E_{12}\delta_{\sigma,-1} + E_3 E_{12}^*\delta_{\sigma,+1}\right);$$

$$f_{32} = f_{32}^{(1)}\left(E_{21} E_{11}^*\delta_{\sigma,-1} + E_{21} E_{11} \delta_{\sigma,+1}\right) + f_{32}^{(2)}\left(E_{22}E_{11}^*\delta_{\sigma,-1} + E_{22} E_{11}\delta_{\sigma,+1}\right)$$

$$+ f_{32}^{(3)}\left(E_{21} E_{12}^*\delta_{\sigma,-1} + E_{21} E_{12} \delta_{\sigma,+1}\right) + f_{32}^{(4)}\left(E_{22}E_{3}^*\delta_{\sigma,-1} + E_{22} E_{12}\delta_{\sigma,+1}\right); \tag{11.101}$$

where $\delta_{\sigma,\pm1}$ are the Kronecker symbols,

$$f_{12}^{(1)} = \frac{\sigma}{\Omega_1}\left(\hat{a}_z \frac{1}{\Omega_2}\hat{a}_{2x} - \hat{a}_{2x}\frac{1}{\Omega_3}\hat{a}_z\right) f^{(0)};\ f_{12}^{(2)} = \frac{1}{\Omega_1}\left(\hat{a}_{2y}\frac{1}{\Omega_3}\hat{a}_z - \hat{a}_z \frac{1}{\Omega_2}\hat{a}_{2y}\right) f^{(0)};$$

$$f_{22}^{(1)} = \frac{\sigma}{\Omega_2}\left(\hat{a}_z \frac{1}{\Omega_1} a_{1x} - \sigma\hat{a}_{1x}\frac{1}{\Omega_3}\hat{a}_z\right) f^{(0)};\ f_{22}^{(2)} = \frac{\sigma}{\Omega_2}\left(\hat{a}_{1y}\frac{1}{\Omega_3}\hat{a}_z - \sigma\hat{a}_z \frac{1}{\Omega_1}\hat{a}_{1y}\right) f^{(0)};$$

$$f_{32}^{(1)} = -\frac{1}{\Omega_3}\left(\hat{a}_{1x}\frac{1}{\Omega_2}\hat{a}_{2x} + \sigma\hat{a}_{2x}\frac{1}{\Omega_1}\hat{a}_{1x}\right) f^{(0)};\ f_{32}^{(2)} = -\frac{1}{\Omega_3}\left(\hat{a}_{1x}\frac{1}{\Omega_2}\hat{a}_{2y} + \sigma\hat{a}_{2y}\frac{1}{\Omega_1}\hat{a}_{1x}\right) f^{(0)};$$

$$f_{32}^{(3)} = -\frac{1}{\Omega_3}\left(\hat{a}_{2x}\frac{1}{\Omega_1}\hat{a}_{1y} + \sigma\hat{a}_{1y}\frac{1}{\Omega_2}\hat{a}_{2y}\right) f^{(0)};\ f_{32}^{(4)} = -\frac{1}{\Omega_3}\left(\hat{a}_{2y}\frac{1}{\Omega_1}\hat{a}_{1y} + \sigma\hat{a}_{1y}\frac{1}{\Omega_2}\hat{a}_{2y}\right) f^{(0)}. \tag{11.102}$$

Therefore, an analogously relevant solution of the third approximation can be written.

11.5.4 Again, Truncated Equations for Wave Amplitudes

Furthermore, we specify definitions for the fields. It is known that any arbitrary plane electromagnetic wave can be represented by two possible methods [36]:

1. As a sum of two reciprocally perpendicular linearly polarized waves
2. As a sum of two circularly polarized waves rotating in reciprocally opposite directions

Here, we choose the second description method as the basic one

$$E'_{j1} = E_{j1} + E_{j2}; \quad E'^{*}_{j2} = E_{j1} - E_{j2}, \tag{11.103}$$

where E'_{j1} describes the slowly varying complex amplitudes of circularly counterclockwise-polarized waves, and E'_{j2} corresponds to the clockwise polarized waves. In the case $E'_{j1} = 0$ (or $E'_{j2} = 0$), we have the circularly polarized waves. For $E'_{j1} = E'^{*}_{j2} = E'_{j}/2$ we have the linearly polarized waves, and so on [36]. Furthermore, for convenience we neglect the sign of torch in complex amplitudes E'_{j1}, E'_{j2}.

At the next stage of calculation, we substitute the obtained solutions for the distribution function into truncated Equations 11.86 and 11.87 and take into account Notations 11.103. Because of transformations, we reduce Equation 11.86 to the form (Raman interaction mode)

$$\left(\xi_1 \frac{\partial}{\partial t} + s_1 v_{g1} \frac{\partial}{\partial z}\right) E_{11} = \sigma A_{11}\left(E_{21}E_3\delta_{\sigma,-1} + E_{22}E_3\delta_{\sigma,+1}\right);$$

$$\left(\xi_1 \frac{\partial}{\partial t} + s_1 v_{g1} \frac{\partial}{\partial z}\right) E_{12} = \sigma A_{12}\left(E_{22}E_3^{*}\delta_{\sigma,-1} + E_{21}E_3^{*}\delta_{\sigma,+1}\right);$$

$$\left(\xi_2 \frac{\partial}{\partial t} + s_2 v_{g2} \frac{\partial}{\partial z}\right) E_{21} = A_{21}\left(E_{11}E_3^{*}\delta_{\sigma,-1} + E_{12}E_3\delta_{\sigma,+1}\right);$$

$$\left(\xi_2 \frac{\partial}{\partial t} + s_2 v_{g2} \frac{\partial}{\partial z}\right) E_{22} = A_{22}\left(E_{12}E_3\delta_{\sigma,-1} + E_{11}E_3^{*}\delta_{\sigma,+1}\right);$$

$$\left(\frac{\partial}{\partial t} + v_0 \frac{\partial}{\partial z} + \Delta'\right) E_3 = r\sigma A_3\left[\left(E_{12}^{*}E_{11} + E_{22}E_{12}^{*}\right)\delta_{\sigma,-1} + \left(E_{21}E_{12}^{*} + E_{22}^{*}E_{11}\right)\right], \tag{11.104}$$

where $A_{11} = \dfrac{ek_3\omega_1\kappa}{m|L_1|\omega_2} = A_{12}^{*}$; $A_{21} = \dfrac{ek_3\omega_2\kappa}{m|L_2|\omega_1} = A_{22}^{*}$; $A_3 = \dfrac{ek_3\Omega_{30}\kappa}{2m\omega_1\omega_2}$; $\kappa = \dfrac{m\omega_p^2}{k_3}\displaystyle\int_{-\infty}^{+\infty}\frac{1}{\gamma_z\Omega_3}\frac{\partial g(p_z)}{\partial z}dp$;
$\Delta' = (\omega_3 - v_0k'_3)\,\mathrm{Im}\{\kappa\}$; $L_j = \partial D_j/\partial\omega_j$; $k'_3 = \mathrm{Re}\{k_3\}$; $\gamma_z = (1 - v_z^2/c^2)^{-\frac{1}{2}}$; $\Omega_3 = \omega_3 - v_zk_3$, $\Omega_{30} = \omega_3 - v_0k_3$; and $p_z = m\gamma_zv_z$; D_j are the dispersion functions for the electromagnetic waves. Dispersion laws $k_j(\omega_j)$ are determined by the solutions of the dispersion equations

$$D_j = (-i)\left(k_j^2 - \varepsilon_j \frac{\omega_j^2}{c^2} + \frac{\omega_p^2}{c^2} \int_{-\infty}^{+\infty} \frac{g(p_z)}{\gamma_z} dp_z \right) = 0. \tag{11.105}$$

Therein, the linear dispersion law $k_3(\omega_3)$ is found from the dispersion equation for the SCW

$$D_3 = (-1)\left(k_3 + \omega_p^2 m \int_{-\infty}^{+\infty} \frac{1}{\omega_3 - k_3 v_z} \frac{\partial g(p_z)}{\partial p_z} dp_z \right) = 0, \tag{11.106}$$

where $r = \pm 1$ is the sign function ($r = -1$ corresponds to the SSCW of the electron beam plasmas, and $r = +1$ corresponds to the fast SCW); $\omega_p = \sqrt{4\pi e^2 n_0 / m}$ is the plasma frequency (see comments to Equation 11.8). Similar to the previous case, we assume the initial electron spread is oriented only along the z-axis

$$f^{(0)}(\vec{p}) = \delta(p_x)\delta(p_y)g(p_z). \tag{11.107}$$

11.5.5 Stationary Version of the Truncated Equations for Wave Amplitudes

System 11.104 can be completely solved analytically only in some special cases [10]. That is why, similar to the previous cases, we restrict consideration to the particular case of a stationary model bounded in one dimension. Assuming the system to be in the stationary state, we omit the derivative $\partial/\partial t$ in System 11.104. Then we have [8–9,18–20]

$$\begin{aligned}
\frac{dE_{11}}{dz} &= \sigma s_1 B_{11}\left(E_{21}E_3\delta_{\sigma,-1} + E_{22}E_3\delta_{\sigma,+1} \right); \\
\frac{dE_{12}}{dz} &= \sigma s_1 B_{12}\left(E_{22}E_3^*\delta_{\sigma,-1} + E_{21}E_3^*\delta_{\sigma,+1} \right); \\
\frac{dE_{21}}{dz} &= s_2 B_{21}\left(E_{11}E_3^*\delta_{\sigma,-1} + E_{12}E_3\delta_{\sigma,+1} \right); \\
\frac{dE_{22}}{dz} &= s_2 B_{22}\left(E_{12}E_3\delta_{\sigma,-1} + E_{11}E_3^*\delta_{\sigma,+1} \right); \\
\frac{dE_3}{dz} + \Delta E_3 &= \sigma r B_3\left[\left(E_{21}^*E_{11} + E_{22}E_{12}^* \right)\delta_{\sigma,-1} + \left(E_{21}E_{12}^* + E_{22}^*E_{11} \right)\delta_{\sigma,+1} \right],
\end{aligned} \tag{11.108}$$

where $B_{jk} = (1/v_{gj})A_{jk}$; $k = 1,2$; $B_3 = (1/v_0)A_3$; $\Delta = \Delta'/v_0$.

In the case of linearly polarized waves with polarizations in the same plane, $\Delta = 0$ (cold electron beam), and $\sigma = -1$ (SCW with difference frequency) Equations 11.108 reduce to Equations 11.12.

The structure of Equations 11.104 and 11.108 demonstrates the peculiarities of the discussed model concerning polarization properties. For instance, considering Equations 11.104 and 11.108 in the case of linearly polarized electromagnetic waves with arbitrary orientation of polarization planes ($E_{j11} = E^*_{j2} = E_j/2$), we can be sure that interaction does not occur if the polarization planes are orthogonal and if $E_{30} = E_3(z = 0)$. An analogous result for circularly polarized waves can be obtained for $\sigma = +1$, if the electric field strength vectors rotate in the same direction and for $\sigma = -1$ if these vectors rotate in opposite directions. These effects are discussed later in detail.

11.5.6 Motion Integrals

In the case of a relatively cold electron beam when SCW dissipation can be neglected ($\Delta = 0$, $B_{j1} = B_{j2} = B_j$), Equations 11.108 have integrals of motion generalizing motion integrals 11.21 and 11.24 for the case of the arbitrarily polarized Dopplertron model

$$s_2\frac{\mathcal{E}_{21}^2}{B_2} - s_1\sigma\frac{\mathcal{E}_{11}^2\delta_{\sigma,-1} + \mathcal{E}_{12}^2\delta_{\sigma,+1}}{B_1} = C_1; \tag{11.109}$$

$$s_2\frac{\mathcal{E}_{21}^2}{B_2} - s_1\sigma\frac{\mathcal{E}_{12}^2\delta_{\sigma,-1} + \mathcal{E}_{11}^2\delta_{\sigma,+1}}{B_1} = C_2; \tag{11.110}$$

$$s_2\frac{\mathcal{E}_{21}^2\delta_{\sigma,-1} + \mathcal{E}_{22}^2\delta_{\sigma,+1}}{B_2} - r\sigma\frac{\mathcal{E}_3^2}{B_3} = C_3; \tag{11.111}$$

$$\mathcal{E}_3\mathcal{E}_{21}(\mathcal{E}_{21}\delta_{\sigma,-1} + \mathcal{E}_{22}\delta_{\sigma,+1})\sin\hat{\psi}_1$$

$$-\mathcal{E}_3\mathcal{E}_{12}\left(\mathcal{E}_{22}\delta_{\sigma,-1} + \mathcal{E}_{21}\delta_{\sigma,+1}\right)\sin\hat{\psi}_2 = C_4, \tag{11.112}$$

where $E_{jk} = \mathcal{E}_{jk}\exp(i\varphi_{jk})$, $(k = 1,2)$; $E_3 = \mathcal{E}_3\exp(i\varphi_3)$; C_l $(l = 1,2,3,4)$ are constants, and $B_{jk} = B_j$; $\hat{\psi}_1 = \varphi_{12} - \varphi_3 - \varphi_{21}\delta_{\sigma,-1} - \varphi_{22}\delta_{\sigma,+1}$, $\hat{\psi}_2 = \varphi_{12} + \varphi_3 - \varphi_{22}\delta_{\sigma,-1} - \varphi_{21}\delta_{\sigma,+1}$ are the combination polarization phases.

In the case of waves linearly polarized in different planes, there is one more integral of motion, i.e.,

$$\mathcal{E}_1\mathcal{E}_2\sin\hat{\psi} = C_5, \tag{11.113}$$

where $\mathcal{E}_{11} = \mathcal{E}_{12} = \mathcal{E}_1/2$ $\mathcal{E}_{21} = \mathcal{E}_{22} = \mathcal{E}_2/2$; $\hat{\psi} = (1/2)(\hat{\psi}_1 + \hat{\psi}_2)$ is the angle between the polarization planes (i.e., between the major axes of polarization ellipses). For $\hat{\psi} = 0$ (both waves are polarized in the same plane), the integral of motion (Equation 11.112) takes the form

$$\mathcal{E}_1\mathcal{E}_2\mathcal{E}_3\sin\phi = C_6, \tag{11.114}$$

where $\phi = (1/2)(\hat{\psi}_1 - \hat{\psi}_2)$ is the phase mismatch of interacting waves. It is a generalization of Motion Integral 11.21 (with exchange $\varphi \to \phi$) for the case $s_j = \pm1$, $\sigma = \pm1$.

Motion Integrals 11.109 through 11.113 illustrate the dynamics of polarizations and oscillation phases of interacting waves. In particular, we see that $(n = 0,1,2,\ldots)$ (see Motion Integral 11.113) only as $\mathcal{E}_j \to \infty$, i.e., electromagnetic waves initially linearly polarized in different planes cannot become polarized in the same plane in the course of interaction.

11.6 Arbitrarily Polarized Kinetic Model: Approximation of Given Pumping Field in the Case of Raman Mode

The simplest amplitude analysis to study amplification in FELs was performed previously in this chapter. In that case, the simplification consisted of accepting a number of additional assumptions. They are the linear polarizations of both wave fields in the same plane, neglecting the kinetic effects, and the model of an initially nonmodulated electron beam. Also, the dispersion of the electromagnetic wave was positive ($\xi_j = +1$). In this section, we set forth the result of the amplitude analysis (in the approximation of a given pumping field) for a more general FEL model. Namely, contrary to the mentioned cases, we choose the arbitrarily polarized kinetic Dopplertron model with an initially modulated electron beam. We will suppose that SCWs possess the sum or difference frequencies ($\sigma = \pm 1$), and that the electromagnetic wave can have a positive as well as negative dispersion ($\xi_j = \pm 1$).

Thus, we assume the pumping wave amplitude remains nearly unchanged during the interaction (it is the above-mentioned approximation of a given pumping field or the so-called parametric approximation). As it is mentioned above, this condition is valid when this amplitude is much greater than the two other wave amplitudes at any point of the interaction region. Then we say that, for energetic reasons, its interaction with the two weaker waves cannot have an appreciable effect on the pumping field intensity, provided the length of the system is moderate.

11.6.1 Types of Instabilities That Are Possible in the Dopplertron FEL

First, we find out the complete set of instabilities that can be realized in the model. We employ the instability criteria derived in the works by Kotsarenko and Fedorchenko [37] and Akhiezer and Polovin [38]. As a result of the accomplished work, we find the state of the system is

1. Absolutely unstable if
   $r\xi_1 > 0, \quad s_1\xi_1 < 0$ (pumping by the wave with frequency ω_2)
   $\sigma r\xi_2 > 0, \quad s_2\xi_2 < 0$ (pumping by the wave with frequency ω_1)
   $\xi_1\xi_2\sigma > 0, \quad s_1 s_2\xi_1\xi_2 < 0$ (pumping by the SCW)
2. Convectively unstable if
   $r\xi_1 > 0, \quad s_1\xi_1 > 0$ (pumping by the wave with frequency ω_2)
   $\sigma r\xi_2 > 0, \quad s_2\xi_2 > 0$ (pumping by the wave with frequency ω_1)
   $\xi_1\xi_2\sigma > 0, \quad s_1 s_2\xi_1\xi_2 > 0$ (pumping by the SCW)
3. Stable if
   $r\xi_1 < 0$ (pumping by the wave with frequency ω_2)

$\sigma r \xi_2 < 0$ (pumping by the wave with frequency ω_1)
$\xi_1 \xi_2 \sigma < 0$ (pumping by the SCW)

Let's remember the accepted assumption concerning the frequencies ω_j: $\omega_1 > \omega_2$. It is obvious that this physical model allows a great variety of probable interaction modes. Let us consider the models only with pumping by the wave with frequency ω_2.

11.6.2 Boundary Conditions

In what follows, we turn to solving the truncated Equations 11.108 [22–25]. We assume the field of the pumping wave with frequency ω_2 is strong and its amplitude, initial oscillation phase, and polarization remain unchanged in the course of interaction (the approximation of the given pumping field). It is convenient to introduce the concept of the intermittent length of the system

$$L_{sj} = (1/2)(1 - s_j)L, \tag{11.115}$$

where L is the true system length. If $s_j = +1$, we have the intermittent length $L_{sj} = 0$, and in the case $s_j = -1$, $L_{sj} = L$. Hence, by such a method, we can describe both models with directly propagating electromagnetic waves ($s_j = +1$) and those with oppositely propagating waves $s_j = -1$.

Then we take boundary conditions in the form

$$\begin{aligned} z = L_{s2},\quad & E_{21} = E_{210},\quad E_{22} = E_{220}; \\ z = L_{s1},\quad & E_{11} = E_{110},\quad E_{12} = E_{120}; \\ & z = 0,\quad E_3 = E_{30} \end{aligned} \tag{11.116}$$

11.6.3 Solutions

Analogous to the case of Equation 11.49, we obtain solutions in a similar form

$$E_{12} = E_{120} + s_1\sigma \frac{B_1^*}{\alpha_2^*}\left(E_{220}\delta_{\sigma,-1} + E_{210}\delta_{\sigma,+1}\right)E_{30}^*\left(sh\left(\alpha_2^* z\right)e^{-\frac{\Delta}{2}z} - sh\left(\alpha_2^* L_{s1}\right)e^{-\frac{\Delta}{2}L_{s1}}\right)$$

$$+\frac{\left(\mathrm{E}_{220}^2\delta_{\sigma,-1} + \mathrm{E}_{210}^2\delta_{\sigma,+1}\right)E_{120} + E_{210}E_{220}E_{110}^*}{\mathrm{E}_{110}^2 + \mathrm{E}_{120}^2} \times \left[\left(ch\left(\alpha_2^* z\right) + \frac{\Delta}{2\alpha_2^*} sh\left(\alpha_2^* z\right)\right)e^{-\frac{\Delta}{2}z}\right.$$

$$\left. - \left(ch\left(\alpha_2^* L_{s1}\right) + \frac{\Delta}{2\alpha_2^*} sh\left(\alpha_2^* L_{s1}\right)\right)e^{-\frac{\Delta}{2}L_{s1}}\right];$$

$$E_{11} = E_{110} + s_1\sigma \frac{B_1}{\alpha_2}\left(E_{210}\delta_{\sigma,+1} + E_{220}\delta_{\sigma,-1}\right)E_{30}^*\left(sh\left(\alpha_2 z\right)e^{-\frac{\Delta}{2}z} - sh\left(\alpha_2 L_{s1}\right)e^{-\frac{\Delta}{2}L_{s1}}\right)$$

$$-sh\left(\alpha_2 L_{s1}\right)e^{-\frac{\Delta}{2}L_{s1}} + \frac{\left(\mathrm{E}_{210}^2\delta_{\sigma,-1} + \mathrm{E}_{220}^2\delta_{\sigma,+1}\right)E_{110} + E_{210}E_{220}E_{120}^*}{\mathrm{E}_{210}^2 + \mathrm{E}_{220}^2} \times \left[\left(ch\left(\alpha_2 z\right) + \frac{\Delta}{2\alpha_2} sh\left(\alpha_2 z\right)\right)e^{-\frac{\Delta}{2}z}\right.$$

$$-\left(ch(\alpha_2 L_{s1})+\frac{\Delta}{2\alpha_2}sh(\alpha_2 L_{s1})\right)e^{-\frac{\Delta}{2}L_{s1}}\Bigg];$$

$$E_3 = E_{30}\left(sh(\alpha_2 z)-\frac{\Delta}{2\alpha_2}ch(\alpha_2 z)\right)e^{-\frac{\Delta}{2}z}$$

$$+\ \sigma r_3\frac{B_3}{\alpha_2}\left[\left(E^*_{210}\delta_{\sigma,-1}+E^*_{220}\delta_{\sigma,+1}\right)E_{110}+\left(E_{120}\delta_{\sigma,-1}+E_{210}\delta_{\sigma,+1}\right)E_{120}\right]\times sh(\alpha_2 z)e^{-\frac{\Delta}{2}z}, \quad (11.117)$$

where

$$\alpha_2 = \sqrt{\frac{\Delta^2}{4}+s_1 r B_1 B_3\left(\mathcal{E}^2_{210}+\mathcal{E}^2_{220}\right)} \quad (11.118)$$

is the spatial growth rate at the parametric stage of interaction (increasing increment).

The structure of Equations 11.117 and 11.118 indicates that the signal wave that can be amplified only provided $s_1 r = +1$; the amplification is considerable only if the pumping wave amplitude exceeds the threshold value [18], namely,

$$\left(\mathcal{E}^2_{210}+\mathcal{E}^2_{220}\right) >> \frac{\Delta^2}{4B_1B_3^*\cos(2\vartheta)} = \left(\mathcal{E}^2_{210}+\mathcal{E}^2_{220}\right)_{th}, \quad (11.119)$$

where $\vartheta = arctg(\mathrm{Im}\{\kappa\}/\mathrm{Re}\{\kappa\})$, κ is determined by notations for Equation 11.104. The latter inequality suggests that the greater the electron beam thermal spread Δ, the higher the threshold amplitude value $\left(\mathcal{E}^2_{210}+\mathcal{E}^2_{220}\right)_{th}$.

11.6.4 Passage to the Arbitrarily Polarized H-Ubitron Model

By the SM method (Chapter 10), we describe both the true Dopplertron models as well as the H-Ubitron model. In particular, to derive expressions for the H-Ubitron model, we pass to limits of Equations 10.68 through 10.71

$$\mathcal{E}_{2i}\to 0;\ \ \omega_2\to 0;\ \ \lim_{\omega_2\to 0}\frac{\mathcal{E}_{2i}}{\omega_2}=\frac{\mathcal{H}_{2i}}{ck_2}, \quad (11.120)$$

where $\mathcal{H}_{2i}$ is the H-Ubitron magnetic field strength amplitude, $i = 1,2$, $k_2 = 2\pi/\Lambda$, and Λ is the field undulation period.

11.6.5 Phase Effects

One can be sure that if $E_{j1}=E_{j2}=E_j/2$, $\hat{\psi}=0$ (see (Equation 11.113)), $\sigma = -1$, $s_1 = +1$, $\Delta = 0$, then Solutions 11.117 and Increment 11.118 reduce to 11.49 and 11.50. Using Solutions

11.117, let us generalize the linearly polarized model discussed in Section 11.2.1. For this, we transform Expression 11.117 supposing $\sigma = \pm 1$, $s_1 = \pm 1$, $\Delta = 0$ for the two particular cases

1. Initially nonmodulated beam ($E_{30} = 0$)

$$E_1 = E_{10}\, ch(\alpha_2 z);$$
$$E_3 = r\sigma \frac{B_3}{\alpha_2} \mathcal{E}_{20}\mathcal{E}_{10} sh(\alpha_2 z)\exp\{i(\varphi_{10} + \sigma\varphi_{20})\}; \tag{11.121}$$

2. Initially nonmodulated beam ($E_{10} = 0$; $E(z = 0) = E_{30}$)

$$E_1 = s_1\sigma \frac{B_1}{\alpha_2} \mathcal{E}_{20}\mathcal{E}_{30} sh(\alpha_2 z)\exp\{i(\varphi_{30} - \sigma\varphi_{20})\};$$
$$E_3 = E_{30} ch(\alpha_2 z), \tag{11.122}$$

where $E_q = \mathcal{E}_q \exp\{i\varphi_q\}$, $q = j, 3 = 1,2,3$, $E_{q0} = E_q(z = 0)$.

As follows from Equations 11.121 and 11.122, the SCW and the signal wave with zero initial amplitude always arise with initial oscillation phases giving phase mismatch $\phi = \phi_0 = n\pi$ ($n = 0, \pm 1, \pm 2, \ldots$) (see Equation 11.16). This holds for the arbitrary electromagnetic wave polarizations, too. This means that the state of the system with the mismatch phase $\phi = n\pi$ is the most advantageous with respect to the energy, and therefore the system always aims for this state. This feature is an obvious manifestation of the purposefulness principle (Chapter 2) in the case of the FEL model. It is well known that this principle is very close to the concept of self-organization (Chapter 2). To illustrate this, we consider the expression for phase mismatch ϕ easily obtained from Equation 11.117 for the particular case $E_{j1} = E_{j2} = E_j/2$, $\hat{\psi} = 0$, $\sigma = \pm 1$, $s_1 \pm 1$, $\Delta = 0$

$$ctg\phi = ctg\phi_0\,(ch^2(\alpha_2 z) + sh^2(\alpha_2 z))$$
$$+ \left(s_1\sigma \frac{B_1\mathcal{E}_{30}^2 + B_3\mathcal{E}_{10}^2}{2\sqrt{B_1 B_3}\,\mathcal{E}_{10}\mathcal{E}_{30}}\, sh(2\alpha_2 z) \right) \cos ec\phi, \tag{11.123}$$

where all initial amplitudes are nonvanishing. As $z \to \infty$, ϕ always tends either to zero or to π depending on the sign of $\sigma s_1 = \pm 1$.

11.6.6 Polarization Effects

The possibility of studying various polarization effects is the most important advantage of the considered model. Therefore, let us then illustrate relevant analysis technology for the case of arbitrary polarizations. As follows from Equations 11.117, the polarizations of the electromagnetic wave are slowly varying in the course of interactions along with the initial phases. So, for initially nonmodulated electron beams, $E_{30} = 0$ (Equation 11.117), for $s_1 r = +1$ we can obtain the following expressions for polarization parameters:

$$\delta_1(z)=\delta_{20}\left\{\frac{\mathcal{E}_{110}+\mathcal{E}_{120}}{\mathcal{E}_{210}\mathcal{E}_{220}}ch(\alpha_2 z)+\frac{\mathcal{E}_{110}^2+\mathcal{E}_{120}^2}{\mathcal{E}_{210}^2+\mathcal{E}_{220}^2}\left[ch(\alpha_2 z)-1\right]+R^2(z)\right\}^{\frac{1}{2}}\left[\frac{\mathcal{E}_{110}+\mathcal{E}_{120}}{\mathcal{E}_{210}+\mathcal{E}_{220}}+R(z)\right]^{-1},\tag{11.124}$$

$$tg\left(2\hat{\psi}(z)\right)=\mathcal{E}_{110}\mathcal{E}_{120}ch(\alpha_2 z)\sin\left(2\hat{\psi}_0\right)\Bigg/\Bigg(\mathcal{E}_{210}\mathcal{E}_{220}R'(z)\left[ch(\alpha_2 z)-1\right]$$
$$+\cos\left(2\hat{\psi}_0\right)\mathcal{E}_{110}\mathcal{E}_{120}\left\{ch(\alpha_2 z)+2\delta_{10}\frac{\left(\mathcal{E}_{210}+\mathcal{E}_{220}\right)^2}{\mathcal{E}_{210}^2+\mathcal{E}_{220}^2}\left[ch(\alpha_2 z)-1\right]^2\right\}\Bigg),\tag{11.125}$$

where $\delta_j = \delta_j(z)$ is the jth wave polarization ellipse eccentricity, δ_{j0} is its limiting value, $\hat{\psi} = \hat{\psi}(z)$ is the angle between major semiaxes of polarization ellipses (see the comments on Equation 11.114), $\hat{\psi}_0 = \hat{\psi}(0)$, and $R(z)$ and $R'(z)$ are some auxiliary functions

$$R(z)=\frac{\left(\mathcal{E}_{220}\delta_{\sigma,-1}+\mathcal{E}_{210}\delta_{\sigma,+1}\right)\mathcal{E}_{120}+\left(\mathcal{E}_{210}\delta_{\sigma,-1}+\mathcal{E}_{220}\delta_{\sigma,+1}\right)\mathcal{E}_{110}}{\mathcal{E}_{210}^2+\mathcal{E}_{220}^2}\times\left(ch(\alpha_2 z)-1\right);\tag{11.126}$$

$$R'(z)=\frac{\mathcal{E}_{110}^2+\mathcal{E}_{120}^2}{\mathcal{E}_{210}^2+\mathcal{E}_{220}^2}+R(z).\tag{11.127}$$

For sufficiently long systems ($\alpha_2 L \gg 1$)

$$\delta_1(z\to L)\to\delta_{20},\quad \hat{\psi}(z\to L)\to 0,\tag{11.128}$$

independent of initial wave polarizations.

Thus, polarization of the weaker electromagnetic wave (signal) in the course of interaction always matches the polarization of the stronger (pumping) wave. As it is marked above, the final state of the system does not depend on boundary values δ_{j0} and $\hat{\psi}_0$; the latter determines the length L at which polarization matching occurs. One should bear in mind, however, that the matching never becomes exact; the process is asymptotic. For linear polarization, this is illustrated by integral of motion Equation 11.113. In the view set forth earlier in Chapter 2, we treat the discussed behavior of wave polarizations as one more illustration of the purposefulness principle.

An interesting detail is that the character of the tendency $\delta_1 \to \delta_2$ is not always monotonic. Now we discuss the change of rotation directions of vectors $\vec{E}_j$ in the course of interaction. We consider the waves with circular initial polarizations. This case is of considerable methodological interest since the assumption that the initial rotation directions of vectors $\vec{E}_j$ are conserved in the course of interaction is often postulated in literature without appropriate substantiation.

We consider the interaction of a pumping wave with clockwise circular polarization ($E_{220} = 0$) and a signal wave with counterclockwise circular polarization ($E_{110} = 0$) with SCWs of difference frequency ($\sigma = -1$). As follows from the mathematical structure of Equation 11.108 and the accepted boundary conditions, at the first stage field derivatives vanish, $dE_{21}/dz = dE_{12}/dz = 0$, i.e., $\mathcal{E}_{21}(z \backsim 0)$ and $\mathcal{E}_{12}(z \backsim 0)$ conserve their initial values. However, the amplitudes grow, since their derivatives are not zero, and this violates the initial wave circular polarization. This character of the process is confirmed by Solution 11.117 that, in the considered case, is ($\Delta = 0$)

$$\mathcal{E}_{11} = \sqrt{\frac{B_1}{B_3}}\mathcal{E}_{30} sh(\alpha_2 z); \quad \mathcal{E}_{12} = \mathcal{E}_{120}; \quad \mathcal{E}_3 = \mathcal{E}_{30} ch(\alpha_2 z);$$

$$\delta_1(z) = 2\left(\sqrt{\frac{B_1}{B_3}}\frac{\mathcal{E}_{30}}{\mathcal{E}_{120}} sh(\alpha_2 z)\right)^{\frac{1}{2}} \Bigg/ \left(\sqrt{\frac{B_1}{B_3}}\frac{\mathcal{E}_{30}}{\mathcal{E}_{120}} sh(\alpha_2 z)\right). \tag{11.129}$$

The growth of amplitudes $\mathcal{E}_{11}$ and $\mathcal{E}_{12}$ at the initial stage of interaction leads to the transformation of the signal wave with counterclockwise circular polarization into an elliptically polarized wave. At the point

$$z = z_0 = \frac{1}{\alpha_2} Arsh\left(\sqrt{\frac{B_1}{B_3}}\frac{\mathcal{E}_{30}}{\mathcal{E}_{120}}\right) \tag{11.130}$$

the amplitude $\mathcal{E}_{11}$ attains the value $\mathcal{E}_{120}$ and signal wave polarization becomes linear. Then, for $\mathcal{E}_{11} > \mathcal{E}_{120}$, the wave transforms into an elliptically polarized wave with clockwise polarization. We see that $\delta_1 \to 0$ as $z \to \infty$, i.e., the signal wave polarization tends to become clockwise (i.e., opposite to the initial direction of rotation) circular.

11.6.7 Effects of Phase and Polarization Discrimination

Now, we discuss the very interesting effects called phase and polarization discrimination. The analysis of Solutions 11.117 shows that the general picture of signal wave amplification in the FEL can be accompanied with two background phenomena:

1. Phase discrimination, i.e., the dependence of the signal wave gain factor on its initial oscillation phase
2. Polarization discrimination, i.e., analogous dependence on the initial polarizations of interacting waves

The phase discrimination can be suppressed by a special choice of the boundary conditions (assuming some initial wave amplitude to be zero). On the other hand, with the initial wave polarization fixed, polarization discrimination has no sense because, for this effect, slowly varying polarizations are necessary. Thus, both these phenomena can be treated independently.

We define the signal wave gain factor in a standard manner (see, for example, Equation 11.14), i.e.,

$$K_1 = \sqrt{\frac{\mathcal{E}_{11}^2\left(L'_{s1}\right) + \mathcal{E}_{12}^2\left(L'_{s1}\right)}{\mathcal{E}_{110}^2 + \mathcal{E}_{120}^2}}, \tag{11.131}$$

where $L' = (1/2)(1 + s_1)L$ is the modernized intermittent length (compared to the intermittent length Equation 11.115).

First, we consider the phase discrimination. We assume the waves to be linearly polarized in the same plane. Then, within the context of Equations 11.117 and 11.131, we find the gain factor K_1 to be ($s_1 = +1$)

$$K_1 = \left(ch^2(\alpha_2 L) + 2\sigma\sqrt{r\frac{B_1}{B_3}\frac{E_{30}}{E_{10}}} sh(\alpha_2 L) ch(\alpha_2 L)\cos\phi_0 + r\frac{B_2}{B_3}\frac{E_{30}^2}{E_{10}^2} sh^2(\alpha_2 L) + r\frac{B_2}{B_3}\frac{E_{30}^2}{E_{10}^2} sh^2(\alpha_2 L) \right)^{1/2} \tag{11.132}$$

With φ_{02} and φ_{03} fixed, dependence $K_1(L, \phi_0)$ is similar to $K_1(L, \varphi_{01})$, i.e., phase discrimination of signal wave occurs in the system. The possibility that the discrimination of the electron beam SCW phase can occur can be demonstrated similarly.

As follows from Equation 11.132, the gain factor as a function of ϕ_0 attains its maximum for $\phi_0 = \phi_{0\max} = (2n + \delta_{\sigma,-1})\pi$ and minimum for $\phi_0 = \phi_{0\min} = (2n + \delta_{\sigma,+1})\pi$. The phase discrimination is manifested to the greatest extent for $B_1\mathcal{E}_{30}^2 = B_3\mathcal{E}_{10}^2$. In this case

$$K_1 = \sqrt{ch^2(\alpha_2 L) + 2\sigma sh(\alpha_2 L)\cos\phi_0 ch(\alpha_2 L) + sh^2(\alpha_2 L)}, \tag{11.133}$$

for $\alpha_2 L \gg 1$, $ch(\alpha_2 L) \cong sh(\alpha_2 L)$, $K_1(\phi_{0\max}) \gg K_1(\phi_{0\min})$.

Thus, the phase discrimination depth is close to 100% (see Figure 11.4). The change of initial oscillation phases in the course of nonlinear interaction can be interpreted as well as being due to z-dependence of phase velocities of (hot) interacting waves.

Now let us consider the polarization discrimination. We assume the initial electromagnetic wave polarizations to be linear and lie in different planes $\hat{\psi}_0 \neq 0$. Then gain factor K_1 (for $s_1 r = +1$) is given by

$$K_1 = \left| \sqrt{\frac{B_1}{B_3}\frac{E_{30}}{E_{10}}} sh^2(\alpha_2 L) + s_1\sigma\cos\hat{\psi}_0 ch(\alpha_2 L) \right|. \tag{11.134}$$

$K_1\left(\hat{\psi}_0\right)$ is appreciable only if the electron beam is weakly modulated $\mathcal{E}_{30} \to 0$. In this case, the gain factor attains its maximum for

$$\hat{\psi}_{0\min} = Arc\cos\left(-s_1\sigma\sqrt{\frac{B_1}{B_3}\frac{\mathcal{E}_{30}}{\mathcal{E}_{10}}} \right) \tag{11.135}$$

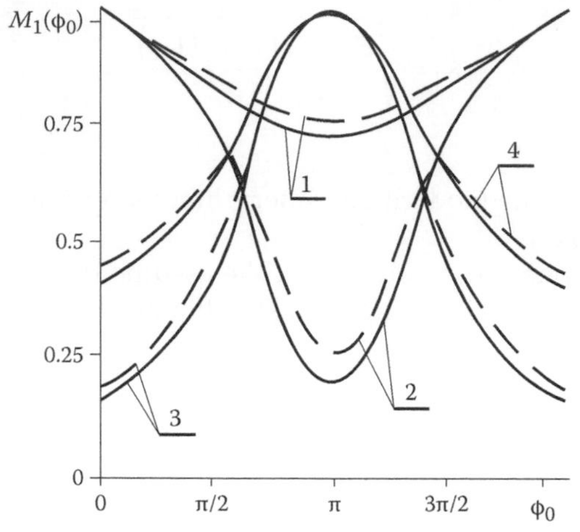

FIGURE 11.4
Illustration of the effect of phase discrimination [the dependency of the reduced gain factor $M_1 = K_1(\phi_0)/K_1(\phi_{0max})$ on the initial phase mismatch ϕ_0]. Here, the continuous lines correspond to solutions obtained in the given field approximation, the dotted lines illustrate the same exact dependencies, curves 1 and 4 correspond to the case $a = \sqrt{B_1/B_3}\left(\mathcal{E}_{30}/\mathcal{E}_{10}\right) = 1$, curves 2 and 3 correspond to the case $a = 0$, curves 1 and 2 are calculated for $\sigma = +1$, and at calculation of curves 3 and 4, it is accepted that $\sigma = -1$.

(see Figure 11.5). If the electron beam is strongly modulated initially (terminal klystron section), then transient process $\hat{\psi}_0 \to 0$ occurs very quickly and $K_1\left(L, \hat{\psi}_0\right) \cong K_1(L, 0)$.

In general, the gain factor depends on the eccentricities of electromagnetic wave polarization ellipses being stronger. For $E_{30} = 0, \;\; \hat{\psi}_0 = 0$ we find from Equation 11.117 ($\Delta = 0$) that

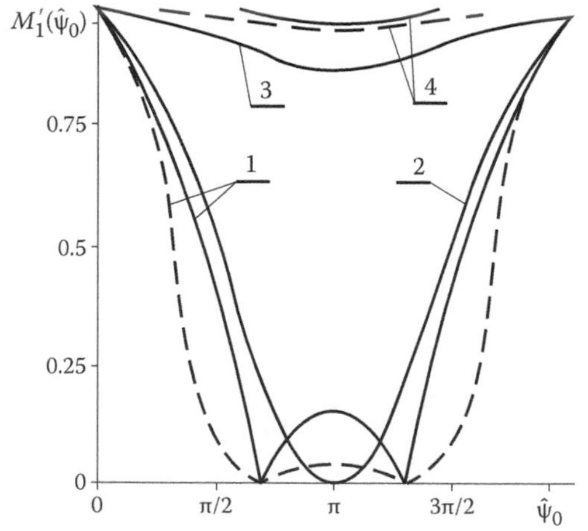

FIGURE 11.5
Illustration of the effect of polarization discrimination [dependency of the reduced gain factor $M_1' = K_1(\hat{\psi}_0)/K_1(0)$ on the initial angle between the polarization planes $\hat{\psi}_0$]. Here, the continuous lines correspond to the solutions obtained in the given field approximation, the dotted lines illustrate the same exact dependencies, curve 1 corresponds to the case $a = \sqrt{B_1/B_3}\left(\mathcal{E}_{30}/\mathcal{E}_{10}\right) = 0.7$, curve 2 corresponds to the case $a = 1$, curve 3 corresponds to the case $a = 10$, and curve 4 corresponds to the case $a = 10^2$; curves 1 and 2 are calculated for $\sigma = +1$, and at calculation of curves 3 and 4, it is accepted that $\sigma = -1$.

$$K_1 = ch(\alpha_2 L)\left[(a_+b_- + b_+d)^2 + (a_-b_+ + b_-d)^2\right], \tag{11.136}$$

where

$$a_\pm = \left[1 + \left(\frac{1 \pm \sigma p_2\sqrt{1-\delta_{20}^2}}{\delta_{20}}\right)^4\right]^{-\frac{1}{2}}; \quad b_\pm = \left[1 + \left(\frac{1 \pm \sigma p_1\sqrt{1-\delta_{10}^2}}{\delta_{10}}\right)^4\right]^{-\frac{1}{2}}; \quad d = \frac{\delta_{20}^2}{2\left(2-\delta_{20}^2\right)}, \tag{11.137}$$

and $p_j = \pm 1$ is the sign function ($p_j = +1$ and $p_j = -1$ correspond to clockwise- and counter-clockwise-polarized jth electromagnetic wave, respectively). The dependence of the reduced gain factor $M_1'(\delta_{10}, \delta_{20}) = K_1(L, \delta_{10}, \delta_{20})/K_{1\max}$ on eccentricities δ_{j0} is shown in Figure 11.6. One can verify that, for the direction of the wave going around polarization ellipses being governed by the condition $p_1 p_2 \sigma = -1$, the gain factor is maximum if the pump wave polarization coincides with the initial polarization of the signal wave.

This analysis suggests that the stage during which phase and polarization discriminations are manifested actually is the establishment of favorable initial oscillation phases and polarizations of the waves. If the signal wave amplitude vanishes, then it arises with the polarization corresponding to $\hat{\psi}_0 = 0$, $\delta_{10} = \delta_{20}$, $p_1 = -\sigma p_2$ and initial oscillation phase $\varphi_{10} = \phi_{0\max} + \varphi_{30} - \sigma\varphi_{20}$, i.e., in all cases the system tends to accomplish maximum gain factor. So, we again have the manifestation of purposefulness principle.

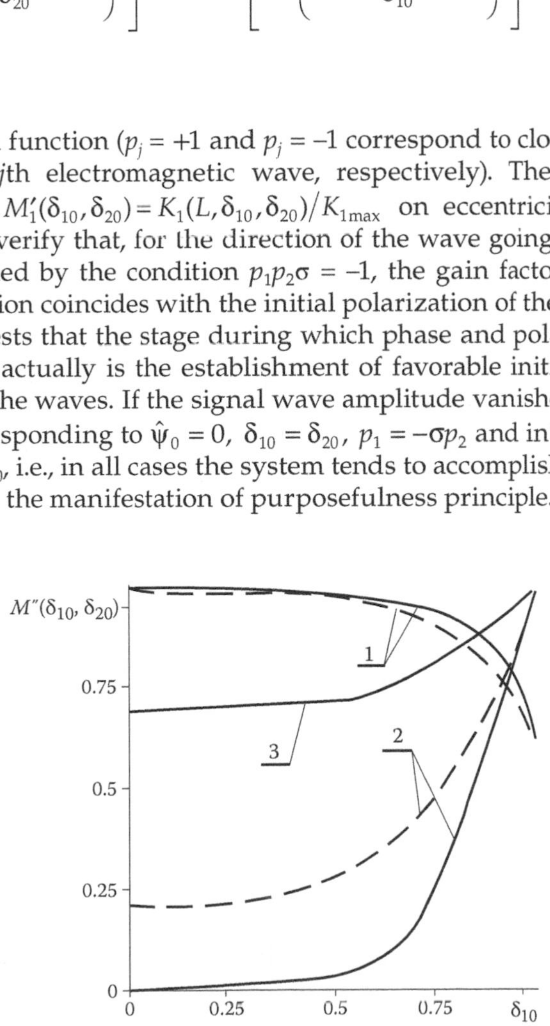

FIGURE 11.6
Illustration of the effect of polarization discrimination [the dependency of the reduced gain factor $M'' = K_1(\delta_{10}, \delta_{20})/K_{\max}$ on the initial signal wave polarization ellipse eccentricity δ_{10}]. Here, the continuous lines correspond to the solutions obtained in the given field approximation, the dotted lines illustrate the same exact dependencies, curve 1 corresponds to the case $\delta_{20} = 0$, $p_1 p_2 \sigma = -1$, curve 2 corresponds to the case $\delta_{20} = 1$, $p_1 p_2 \sigma = +1$, and curve 3 corresponds to the case $\delta_{10} = 1$.

11.6.8 Role of the Pumping Wave Retardation in the Amplification Process

At last, we discuss once more the influence of pumping wave retardation on the parametric amplification of the signal wave [1–3,15,16]. One of the conclusions is that pumping wave retardation causes the increase of frequency transformation coefficient K_{12} (see Equation 11.35) and corresponding comments. Now, let us show that growth rate (increment) α_2 increases in the arbitrarily polarized model in accordance with the same (with the linearly polarized case) regularity.

For simplicity, we put $s_2 = \sigma = -1$, $s_1 = r = +1$, $N_1 = 1$, $\Delta = 0$. As before, the wave with frequency ω_2 is taken for the pumping wave. Then, considering accepted assumptions, we find from Equation 11.117, the expression for growth rate α_2, i.e.,

$$\alpha_2 = \frac{e(N_2+1)}{2mc^2\beta}\sqrt{\frac{2\omega_p}{\omega_2}\frac{\beta(1+\beta)}{(1+N_2\beta)}}\gamma^{1/2}\left(\mathcal{E}_{210}^2+\mathcal{E}_{220}^2\right). \tag{11.138}$$

We see that for $N_2\beta \gg 1$, the growth rate α_2 depends on the pumping wave retardation degree as $\backsim N_2^{1/2}$. Let us estimate α_2 for some cases of practical interest. For the model used by V.L. Granatstein and P. Sprangle in their experiment [39,40]: $K_{12} \backsim 50$; $\omega_p/\omega_2 \backsim 1$; $\beta \backsim 0.96$; $\gamma \backsim 3.5$ $N_2 \backsim 1$; $\sqrt{\mathcal{E}_{210}^2+\mathcal{E}_{220}^2} \sim 10^5$ V/cm; $\omega_2 \backsim 10^{11}\,\text{s}^{-1}$. The relevant estimate for α_2 is $\alpha_2 \cong 0.15$ cm^{-1}. We retain the value of coefficient $K_{12} \cong 50$, assuming the pumping wave is retarded, $N_2 \cong 10$. This is equivalent to the decrease of the electron beam relativism degree to $\beta \cong 0.8$. Moreover, for the same pumping wave field intensity, we find $\alpha_2 \cong 0.23$ cm^{-1}. If the electron beam energy is fixed, $\beta \cong 0.96$, and then for retardation $N_2 \cong 10$, the frequency transformation coefficient grows to $K_{12} \cong 260$. It is peculiar that the growth rate value remains reasonable, $\alpha_2 \cong 0.12$ cm^{-1}.

11.7 Arbitrarily Polarized Kinetic Model: An Approximation of a Given Pumping Field in the Case of the Compton Mode

11.7.1 Truncated Equations and Boundary Conditions

Let us consider the main characteristics of the wave parametric resonance phenomenon in the Compton modification (see Criterion 11.25) [1–3]. We assume the electron beam is cold ($\Delta = 0$). Then the equation for the SCW can be written as

$$\frac{d^2E_3}{dz^2} = -i\sigma\tilde{B}_3\left[\left(E_{21}^*E_{11}+E_{22}E_{12}^*\right)\delta_{\sigma,-1}+\left(E_{21}E_{12}^*+E_{22}^*E_{11}\right)\delta_{\sigma,+1}\right], \tag{11.139}$$

whereas equations for electromagnetic waves reproduce (Equation 11.108). Here the Compton matrix element is

$$\tilde{B}_3 = ek_3\omega_p\kappa/2mv_0^2\omega_1\omega_2\gamma^3, \tag{11.140}$$

where all notations are as given above.

As before, we take wave $j = 2$ for the pumping wave: $E_{21} = E_{210}$; $E_{22} = E_{220}$. The boundary condition is ($s_1 = +1$)

$$E_{11}(0)=E_{110};\quad E_{12}(0)=E_{120};\quad E_{30}(0)=E_{30};\quad dE_3(0)/dz=E'_{30}=0;$$

$$\frac{d^2E_3(0)}{dz^2}=E''_{30}=-i\sigma\tilde{B}_3\left[\left(E^*_{210}E_{110}+E_{220}E^*_{120}\right)\delta_{\sigma,-1}+\left(E_{210}E^*_{120}+E^*_{220}E_{110}\right)\delta_{\sigma,+1}\right]. \quad (11.141)$$

11.7.2 Solutions

Solutions for normalized amplitudes

$$U_{jk}=\sqrt{B_mB^*_n}E_{jk}\ \ (m\neq n\neq j);\ U_3=\sqrt{B^*_1B_2}E_3, \quad (11.142)$$

are

$$U_{11}=U_{110}+iu_{20}^{-2}\left[\sigma\left(U_{210}\delta_{\sigma,-1}+U_{220}\delta_{\sigma,+1}\right)\sum_{n=1}^{3}\alpha_n^2R_n\exp\{\alpha_nz\}\right.$$
$$\left.+iU_{210}U_{220}U^*_{120}\right]-u_{20}^{-2}U_{110}\left(u^2_{210}\delta_{\sigma,-1}+u^2_{220}\delta_{\sigma,+1}\right);$$

$$U_{12}=U_{120}-iu_{20}^{-2}\left[\sigma\left(U_{220}\delta_{\sigma,-1}+U_{210}\delta_{\sigma,+1}\right)\sum_{n=1}^{3}\alpha_n^{*2}R^*_n\exp\left\{\alpha^*_nz\right\}\right.$$
$$\left.+iU_{210}U_{220}U^*_{110}\right]+u_{20}^{-2}U_{120}\left(u^2_{210}\delta_{\sigma,-1}+u^2_{220}\delta_{\sigma,+1}\right);U_3=\sum_{n=1}^{3}R_n\exp\left\{\alpha_nz\right\}, \quad (11.143)$$

where

$$\alpha_n=-iu_{20}^{3/2}\beta_n;\ \beta_n=\exp\left\{2i(n-1)\frac{\pi}{3}\right\};\ u_{ij}=\left|U_{ij}\right|;\ u_{j0}=\sqrt{u^2_{j10}+u^2_{j20}};$$
$$R_n=(-1)^n\frac{\alpha_l-\alpha_n}{\Gamma_n}\left(E''_{30}+\alpha_l\alpha_nE_{30}\right),\ \ (m\neq n\neq l),\ \ l>m;$$
$$\Gamma_n=\alpha_1^2(\alpha_3-\alpha_2)+\alpha_2^3(\alpha_1-\alpha_3)+\alpha_3^2(\alpha_2-\alpha_1).$$

The condition for the Compton mechanism to be realized in this case is

$$u_{20}^{2/3}>>\omega_p/\gamma^{3/2}v_0. \quad (11.144)$$

The solutions obtained infer that if the length of the system is sufficiently large, i.e.,

$$L\geq u_{20}^{-2/3}, \quad (11.145)$$

then it is sufficient to account for only one of the three electron waves, namely wave $n = 2$. Then, expressions for the complex amplitudes are simplified considerably. We have

$$U_{11} = \frac{1}{3} u_{20}^{-2} \left(U_{210} \delta_{\sigma,-1} + U_{220} \delta_{\sigma,+1} \right) \hat{p} \exp\left\{ \frac{\sqrt{3}}{2} u_{20}^{2/3} z \right\};$$

$$U_{12} = \frac{1}{3} u_{20}^{-2} \left(U_{220} \delta_{\sigma,-1} + U_{110} \delta_{\sigma,+1} \right) \hat{p} \exp\left\{ \frac{\sqrt{3}}{2} u_{20}^{2/3} z \right\}, \tag{11.146}$$

where

$$\hat{p} = \left[\left(u_{210} u_{120} + u_{220} u_{110} \exp\{2i\hat{\psi}_0\} \right) \delta_{\sigma,-1} + \left(u_{210} u_{110} \exp\{2i\hat{\psi}_0\} \right. \right.$$

$$\left. \left. + u_{220} u_{120} \right) \delta_{\sigma,+1} \right] \exp\left\{ i \left(\varphi_{210} \delta_{\sigma,+1} + \varphi_{220} \delta_{\sigma,-1} - \varphi_{120} \right) \right\} \times \exp\left\{ i \left(u_{20}^{2/3} \frac{z}{2} + \pi \right) \right\}$$

$$+ u_{20}^{4/3} u_{30} \exp\left\{ i \left(\varphi_{30} + \sigma \frac{\pi}{2} + \frac{\pi}{3} \right) \right\} \times \exp\left\{ i \left(u_{20}^{2/3} \frac{z}{2} + \pi \right) \right\}; \varphi_{ji} = \arg U_{ji};\ \varphi_3 = \arg U_3;\ \hat{\psi}_0 = \hat{\psi}(z = 0);$$

and $\hat{\psi} = (1/2)(\varphi_{12} + \varphi_{22} - \varphi_{11} - \varphi_{21})$ is the angle between the major semiaxes of polarization ellipse.

11.7.3 Phase and Polarization Effects

As follows from Equation 11.146, in the interaction region determined by Equation 11.144, eccentricities of wave polarization ellipses

$$\delta_j = \frac{2\sqrt{u_{j1} u_{j2}}}{u_{j1} + u_{j2}} \tag{11.147}$$

nearly coincide. Therefore, in the Compton case, the wave polarizations are equalized rather quickly. The comparison to Equation 11.124 shows that the analogous process in the Raman interaction mechanism develops much more slowly. Moreover, in the Raman case, the initial wave oscillation phases—including the mismatch phase ϕ (see, for example, Equation 11.123)—tend to certain limits, whereas in the Compton case, they grow monotonously. The reason is that in the Raman mechanism, steady hot wave phase velocities are constant (the growth rate α_2 in Equations 11.118 and 11.117 is real), whereas in the Compton mechanism, they vary along the z-direction (the growth rate α_2 in Equation 11.143 is complex). In both cases, the output amplitudes of the signal wave and SCW depend considerably on the initial oscillation phases, the initial mismatch phase in particular.

$$\phi_0 = \varphi_{30} - \frac{1}{2}\left[\varphi_{110} + \varphi_{120} + \sigma\left(\varphi_{210} + \varphi_{220}\right)\right], \tag{11.148}$$

and on the initial angle $\hat{\psi}_0 = \hat{\psi}(0)$ and the wave amplitudes. Let us illustrate this for linearly polarized waves. We introduce the gain factor in the standard form

$$K_1 = \frac{u_1(L)}{u_1(0)}. \tag{11.149}$$

Then, for electromagnetic waves polarized in the same plane we have

$$K_1 = \frac{1}{3}\left[1 + u_{20}^{2/3} u_{10}^{-1/3} u_{30}^{-1/3} - 2\sigma u_{20}^{1/3} u_{10}^{-1} u_{30}^{2/3} \sin\left(\phi_0 + \frac{\pi}{4}\right)\right]^{\frac{1}{2}} \times \exp\left(\frac{\sqrt{3}}{2} u_{20}^{2/3} L\right). \tag{11.150}$$

We see that phase discrimination in the Compton mechanism is manifested as clearly as in the Raman case. The maximum of gain factor K_1 is attained for

$$\phi_0 = \phi_{0\max} = \frac{\pi}{6}(12n - 2 - 3\sigma), \tag{11.151}$$

and the minimum corresponds to

$$\phi_0 = \phi_{0\min} = \frac{\pi}{6}(12n + 4 - 3\sigma). \tag{11.152}$$

The phase discrimination depth is the greatest (close to 100%) for initial amplitude

$$u_{20} = \left(u_{10}^{2/3} u_{30}^{-1/3}\right)^3. \tag{11.153}$$

The dependence of the gain factor K_1 on the eccentricities of electromagnetic wave polarization ellipses is more complicated. For example, if the beam is nonmodulated initially, $E_{30} = 0$, then we have

$$K_1 = \frac{1}{3}\left[\frac{S_1^2 + S_2^2 + 2S_1S_2\cos\hat{\psi}_0}{\left(1 + S_1^2\right)\left(1 + S_2^2\right)}\right]^{\frac{1}{2}} \exp\left\{\frac{\sqrt{3}}{2} u_{20}^{3/2} L\right\}, \tag{11.154}$$

where $S_j = \delta_j^2 \Big/ \left(1 + \hat{p}_j \sigma \sqrt{1 - \delta_j^2}\right)$; $\hat{p}_j = \pm 1$ ($\hat{p}_j = +1$ and $\hat{p}_j = -1$ correspond to counterclockwise and clockwise polarized waves, respectively). As follows from Equation 11.154, the gain factor is maximum for the similar directions of going around wave polarization ellipses in the case $\sigma = -1$ and for the opposite directions if $\sigma = +1$.

As in the Raman mechanism, the dependence of the gain factor on initial wave polarization parameters is due to the fact that signal wave polarization is matched to pump wave polarization in the course of interaction. It is natural that the dependence of K on S_j is manifested at greater lengths for greater differences of initial pump and signal wave polarization. For example, for $\psi_0 = \phi_o = 0$ we have

$$\frac{u_{11}}{u_{12}} = \frac{u_{210}}{u_{220}}\delta_{\sigma,-1} + \frac{u_{220}}{u_{210}}\delta_{\sigma,+1}. \tag{11.155}$$

Then, polarization sign parameters are

$$\hat{p}_j = sign\left\{\frac{u_{j1}}{u_{j2}} - 1\right\}. \tag{11.156}$$

Thus, irrespective of initial polarization parameters, the rotation directions of E-vectors in the interaction region output are similar for $\sigma = -1$ and opposite for $\sigma = +1$.

11.8 Arbitrarily Polarized Dopplertron Model: Explosive Instability in the Raman Model

The physical picture of processes associated with the self-consistent arbitrarily polarized Dopplertron models is of considerable interest from a methodological point of view. In this section, we demonstrate the latter assertion by means of the explosively unstable stationary Raman modes.

Exact integration of truncated Equations 11.43 (for the linearly polarized model) is described in Section 11.1. However, this calculative algorithm is not acceptable directly for the arbitrarily polarized case because of a difference in initial truncated equations. Therefore, below we will use a modernized algorithm that is suitable for treating more generally arbitrarily polarized stationary systems similar to System 11.104. Let us confine ourselves to the case of a cold electron beam $\Delta = 0$.

11.8.1 Truncated Equations in the Real Form

At first, following the principal scheme performed in Section 11.1, we rewrite Equations 11.108 (considering $\Delta = 0$) in terms of real variables [1–3], i.e.,

$$\frac{d\mathcal{E}_{11}}{dz} = s_1\sigma B_1(\mathcal{E}_{21}\delta_{\sigma,-1} + \mathcal{E}_{22}\delta_{\sigma,+1})\mathcal{E}_3\cos\hat{\psi}_1;$$

$$\frac{d\mathcal{E}_{12}}{dz} = s_1\sigma B_1\left(\mathcal{E}_{22}\delta_{\sigma,-1} + \mathcal{E}_{21}\delta_{\sigma,+1}\right)\mathcal{E}_3\cos\hat{\psi}_2;$$

$$\frac{d\mathcal{E}_{21}}{dz} = s_2 B_2(\mathcal{E}_{11}\cos\hat{\psi}_1\delta_{\sigma,-1} + \mathcal{E}_{12}\cos\hat{\psi}_2\delta_{\sigma,+1})\mathcal{E}_3;$$

$$\frac{d\mathcal{E}_{22}}{dz} = s_2 B_2(\mathcal{E}_{12}\cos\hat{\psi}_2\delta_{\sigma,-1} + \mathcal{E}_{11}\cos\hat{\psi}_1\delta_{\sigma,+1})\mathcal{E}_3;$$

$$\frac{d\mathcal{E}_3}{dz} = r\sigma B_3\left[\left(\mathcal{E}_{21}\delta_{\sigma,-1} + \mathcal{E}_{22}\delta_{\sigma,+1}\right)\mathcal{E}_{11}\cos\hat{\psi}_1 + \left(\mathcal{E}_{12}\delta_{\sigma,-1} + \mathcal{E}_{11}\delta_{\sigma,+1}\right)\mathcal{E}_{22}\cos\hat{\psi}_2;\right.$$

$$\frac{d\hat{\psi}_1}{dz} = -\left[s_1\sigma B_1\frac{\left(\mathcal{E}_{21}\delta_{\sigma,-1} + \mathcal{E}_{22}\delta_{\sigma,+1}\right)\mathcal{E}_3}{\mathcal{E}_{11}}\right.$$

$$\left. + s_2 B_2\frac{(\mathcal{E}_{21}\delta_{\sigma,-1} + \mathcal{E}_{22}\delta_{\sigma,+1})\mathcal{E}_3}{\mathcal{E}_{12}} + r\sigma B_3\frac{(\mathcal{E}_{21}\delta_{\sigma,-1} + \mathcal{E}_{22}\delta_{\sigma,+1})\mathcal{E}_{11}}{\mathcal{E}_3}\right]$$

$$\times\sin\psi_1 + \frac{(\mathcal{E}_{12}\delta_{\sigma,-1} + \mathcal{E}_{11}\delta_{\sigma,+1})\mathcal{E}_{22}}{\mathcal{E}_3}\sin\hat{\psi}_2;$$

$$\frac{d\hat{\psi}_2}{dz} = -\left[s_1\sigma B_1\frac{(\mathcal{E}_{11}\delta_{\sigma,-1} + \mathcal{E}_{12}\delta_{\sigma,+1})\mathcal{E}_3}{\mathcal{E}_{22}} + s_2 B_2\frac{(\mathcal{E}_{11}\delta_{\sigma,-1} + \mathcal{E}_{12}\delta_{\sigma,+1})\mathcal{E}_3}{\mathcal{E}_{21}} + \right.$$

$$\left. r\sigma B_3\frac{(\mathcal{E}_{12}\delta_{\sigma,-1} + \mathcal{E}_{11}\delta_{\sigma,+1})\mathcal{E}_{22}}{\mathcal{E}_3}\right]\times\sin\hat{\psi}_2 + r\sigma B_3\frac{(\mathcal{E}_{21}\delta_{\sigma,-1} + \mathcal{E}_{22}\delta_{\sigma,+1})\mathcal{E}_{11}}{\mathcal{E}_3}\sin\hat{\psi}_1, \tag{11.157}$$

where $E_{ji} = \mathcal{E}_{ji}\exp\{i\varphi_{ji}\}$; $\hat{\psi}_1 = \varphi_{11} - \varphi_{21}\delta_{\sigma,-1} - \varphi_{22}\delta_{\sigma,+1} - \varphi_3$; $\hat{\psi}_2 = \varphi_{12} - \varphi_{22}\delta_{\sigma,-1} - \varphi_{21}\delta_{\sigma,+1} + \varphi_3$.

Apart from the constants $C_1 - C_4$ determined by Motion Integrals 11.109 through 11.112, in the case where $C_4 = 0$, System 11.104 (and, consequently, System 11.157 also) possesses two more motion integrals

$$\mathcal{E}_{11}(\mathcal{E}_{21}\delta_{\sigma,-1} + \mathcal{E}_{22}\delta_{\sigma,+1})\sin\hat{\psi}_1 = const = \Gamma_1;$$

$$\mathcal{E}_{12}(\mathcal{E}_{22}\delta_{\sigma,-1} + \mathcal{E}_{21}\delta_{\sigma,+1})\sin\psi_2 = const = \Gamma_2. \tag{11.158}$$

11.8.2 Functions *u*(*z*) and *R*(*z*) and the Nonlinear Potential

Let us introduce the functions

$$u(z) = \mathcal{E}_{21}^2 \quad \text{and} \quad R(z) = \mathcal{E}_{22}^2. \tag{11.159}$$

Then, using motion integrals 11.109, 11.111, and System 11.157, we obtain the following system

$$\frac{du}{dz} = 2s_2\left\{s_1 r B_1 B_3(u + R - s_2 B_2 C_3)\left[u(u - s_2 B_2 C_1) - s_1 s_2 \sigma\Gamma_1\right]\right\}^{\frac{1}{2}};$$

$$\frac{dR}{dz} = 2s_2\left\{s_1 r B_1 B_3(u + R - s_2 B_2 C_3)\left[R(R - s_2 B_2 C_2) - s_1 s_2 \sigma\Gamma_1\right]\right\}^{\frac{1}{2}}, \tag{11.160}$$

where

$$\Gamma_1 = \Gamma^2 \left| L_1 v_{gr1} / L_2 v_{gr2} \right| K_{12}^{-2}; \vec{v}_{gr} = \left| M_j / L_j \right|;$$
$$L_j = \partial\left(\omega_j^2 \hat{\varepsilon}_j\right) / \partial\omega_j - c^2 k_j^2 \left(\partial\hat{\mu}_j^{-1} / \partial\omega_j\right);$$
$$M_j = c^2 \left[\partial\left(k_j^2 \hat{\mu}_j^{-1}\right) / \partial k_j\right] - \omega_j^2 \left(\partial\hat{\varepsilon}_j / \partial k_j\right)$$

(see Equations 11.113 and 11.114);

$$K_{12} = \omega_1/\omega_2 \cong \sigma(s_2\beta N_2 - 1)/(1 - s_1\beta N_1);$$

constants C_l (k = 1,2,3) are determined by Equations 11.109 through 11.111. Making use of System 11.160, we find R as a function u, i.e.,

$$R(u) = \frac{1}{4AG(u)}\left[\left(AG(u) + s_2 B_2 C_2\right)^2 + 4 s_1 s_2 \sigma \Gamma_1\right], \tag{11.161}$$

where

$$G(u) = 2u - s_2 B_2 C_1 + 2\sqrt{u(u - s_2 B_2 C_1) - s_1 s_2 \sigma \Gamma_1};$$
$$A = G^{-1}(u_0)\left[2R_0 - s_2 B_2 C_2 + 2\sqrt{R_0(R_0 - s_2 B_2 C_2) - s_1 s_2 \sigma \Gamma_1}\right], \tag{11.162}$$

$$u_0 = u(z=0) = \mathcal{E}_{210}^2, \quad R_0 = R(z=0) = \mathcal{E}_{220}^2.$$

We employ Systems 11.161 and 11.162 to reduce the set of equations in System 11.160 to the form [41]

$$\frac{1}{2}\left(\frac{du}{dz}\right)^2 + \pi(u) = 0, \tag{11.163}$$

where nonlinear potential $\pi(u)$ can be written as

$$\pi(u) = -2 s_2 r B_1 B_2 \left[u\left(u - s_2 B_2 C_1\right) - s_1 s_2 \sigma \Gamma_1\right]\left(u + R(u) - s_2 B_2 C_3\right). \tag{11.164}$$

11.8.3 Analytical Solutions

It is not difficult to reveal the formal analogy of Function 11.163 with the equation of motion of a material point in a potential field. That is why Function 11.164 is referred to as the nonlinear potential [41]. We employ the Cardano formula [32] to find roots $(u_1, u_2, u_3,)$ of equation $\pi(u) = 0$. Then the solution of Equation 11.163 can be written as

$$z = \pm\frac{1}{\sqrt{2}}\int_0^u \frac{du}{\sqrt{s_1 s_2 r(u-u_1)(u-u_2)(u-u_3)}}. \quad (11.165)$$

Thus, the form of polynomial $\pi(u)$ determines the form of the desired solution.

In our special case, equation $\pi(u) = 0$ is satisfied for three real roots, as shown in Figure 11.7a and b. For example, in the case $s_1\sigma = s_2 = \sigma r$, two roots can be positive in $u = \mathcal{E}_{21}^2 > u_1$. The oscillations range from the positive root u_1 to the negative one u_2. This suggests the possibility that energy exchange in a three-wave system can have an oscillatory character for an accepted combination of sign functions. In the second case (Figure 11.7b), amplitudes can grow infinitely that correspond to the explosive instability mode discussed for the linearly polarized model. Here, we continue using the arbitrarily polarized model as the basic object.

We return to Solutions 11.163. Inversion of the elliptic integral in Equation 11.165 yields the explicit form of the solution, i.e.,

$$u(z) = (u_2 - u_1)sn^2\left[\left((u_1-u_2)^{\frac{1}{2}}z + \hat{\psi}\right)\middle|\tilde{\mu}\right], \quad (11.166)$$

where $\tilde{\mu} = \sqrt{(u_1-u_2)/(u_1-u_3)}$ is the elliptic function module, $u_1 \geq u_2 \geq u_3$,

$$\hat{\psi} = Arc\cos\left\{\left[u_1/(u_1-u_2)^{1/2}\right]\right\}. \quad (11.167)$$

Furthermore, substituting Solution 11.166 into Equation 11.161 and using motion integrals 11.109 through 11.111, we find a total set of solutions for Problem 11.157

$$\mathcal{E}_{11}(z) = \left[s_1 s_2 \sigma K_{12}^2\left|\frac{L_2 v_{gr2}}{L_1 v_{gr1}}\right|\left(\mathcal{U}(z)\delta_{\sigma,-1} + \mathcal{R}(z)\delta_{\sigma,+1}\right)\right]^{\frac{1}{2}};$$

$$\mathcal{E}_{12}(z) = \left[s_1 s_2 \sigma K_{12}^2\left|\frac{L_2 v_{gr2}}{L_1 v_{gr1}}\right|\left(\mathcal{R}(z)\delta_{\sigma,-1} + \mathcal{U}(z)\delta_{\sigma,+1}\right)\right]^{\frac{1}{2}} \mathcal{E}_{21}(z) = u^{1/2}(z); \mathcal{E}_{22}(z) = R^{1/2}(z);$$

$$\mathcal{E}_3(z) = \left[s_2\frac{B_2}{B_2}(u+R) - \sigma B_3 C_3\right]^{\frac{1}{2}}; \hat{\psi}(z) = \frac{1}{2}(\hat{\psi}_1 + \hat{\psi}_2) = \frac{1}{2}Arc\sin\left\{\sqrt{s_1 s_2 \sigma \Gamma_1\left[uR\mathcal{U}(z)\mathcal{R}(z)\right]^{-1}}\right.$$

$$\left.\times\left[\sqrt{u\mathcal{U}(z) - s_1 s_2 \sigma \Gamma_1} + \sqrt{R\mathcal{R}(z) - s_1 s_2 \sigma\Gamma_1}\right]\right\};$$

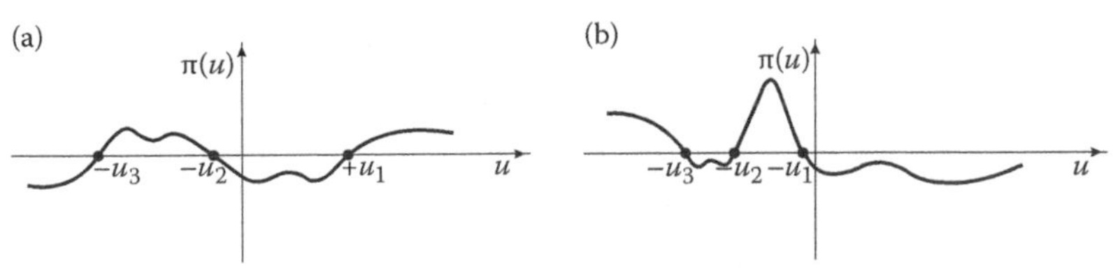

FIGURE 11.7
Characteristic examples of the function $\pi(u)$ for two typical situations: a) corresponds to the oscillation interaction mode and b) describes the explosive instability.

$$\phi(z) = \frac{1}{2}(\hat{\psi}_1 - \hat{\psi}_2) = \frac{1}{2} Arc\sin\left\{\sqrt{s_1 s_2 \sigma \Gamma_1 \left[uR\mathcal{U}(z)\mathcal{R}(z)\right]^{-1}}\right.$$

$$\left.\times\left[\sqrt{R\mathcal{R}(z) - s_1 s_2 \sigma\Gamma_1} - \sqrt{u\mathcal{U}(z) - s_1 s_2 \sigma\Gamma_1}\right]\right\}, \quad (11.168)$$

where $\mathcal{U}(z) = u - s_2B_2C_1$; $\mathcal{R}(z) = R - s_2B_2C_2$, $\hat{\psi}$ is the angle between polarization planes (i.e., between major axes of polarization ellipses), and ϕ is a phase mismatch of interacting waves. The other notations here are the same as those introduced above. These solutions completely describe the dynamics of polarization effects under explosive instability as well as under asymptotic modes.

Solutions 11.168 yield an expression for the critical explosion length (see also Expression 11.76)

$$L_{cr} = \frac{K - H}{\sqrt{B_1 B_3 (u_2 - u_3)}}. \quad (11.169)$$

It is not difficult to verify that if $z \to L_{cr}$ then $\mathcal{E}_{ji}$, $\mathcal{E}_3 \to \infty$. Therein, we have $\hat{\psi} \to 0$, $\phi \to n\pi$ ($n = 0,1,2,\ldots$).

11.8.4 Polarization Effects

Let us consider the evolution of electromagnetic wave polarizations in the course of interaction.

The eccentricity of the signal wave polarization ellipse is

$$\delta_j = \frac{2\sqrt{\mathcal{U}(z)\mathcal{R}(z)}}{\sqrt{\mathcal{U}(z)} + \sqrt{\mathcal{R}(z)}}. \quad (11.170)$$

Bearing in mind that $u,R \to \infty$ as $z \to L_{cr}$, we find that

$$\delta_1(z) \to 2\sqrt{uR}\left(\sqrt{u} + \sqrt{R}\right)^{-1} = \delta_2(z), \quad (11.171)$$

i.e., the eccentricities of ellipses tend to become closer in given field approximation (see relevant discussion for the second equation of Expression 11.129).

Unfortunately, the analytic solutions of Equations 11.157 for the case $C_4 \neq 0$ have not been found. That is why we involve results obtained by numerical methods. As before, we assume the electron thermal spread to be small ($\Delta = 0$, cold beam). For all the cases considered, we take the frequency to be $\omega_2 \cong 10^{11}\ s^{-1}$ and the accelerating potential of the electron beam to be ~375 kV, which corresponds to $\beta \cong 0.82$. The electromagnetic wave dispersion is assumed to be normal, i.e., we assume the type-a explosive instability mode. The length of the interaction region was found from the condition that signal wave intensity must attain the value $I_{1out} = K_{12}I_{20}$, where, as before, $K_{12} = \omega_1/\omega_2$ is the frequency transformation coefficient, and I_{20} is the input magnitude of pumping wave intensity. This condition follows from Motion Integrals 11.109 and 11.111. It is known as one of the possible forms of the Manly–Row relations equivalent to the condition of conservation of the total number of falling (pumping) and radiated (signal) photons.

The transformation picture typical to electromagnetic wave polarization is shown in Figure 11.8. It is clear that eccentricities of electromagnetic wave polarization ellipses become closer (compared with the results shown in Figure 11.6), and, as expected, the polarization of the pumping wave that is stronger remains nearly unchanged. The picture shown in Figure 11.8 is basically similar for any boundary conditions. However, the interaction region length at which polarizations are equalized depends considerably on initial amplitudes ratio, polarizations, and phases of interacting waves. This assertion is illustrated by plots of polarization ellipse eccentricities δ_j for various amplitude boundary values and phase mismatch ϕ.

It should be mentioned that eccentricities δ_j sometimes are not very close at a given interaction region length; however, the trend is clear. For commensurable initial wave amplitudes, the establishment of polarizations is hindered by their mutual influence. For opposite directions of going around electromagnetic wave polarization ellipses ($p_1p_2\sigma = -1$), eccentricities δ_j monotonically tend to asymptotic value δ_0 (for relevant definitions for p_j, see the notation in Equation 11.137). For similar directions ($p_1p_2\sigma = +1$), the eccentricity of the weaker wave first tends to one (linear polarization). Then, after changing the direction of going around the ellipse, it tends to the asymptotic δ_0 given by

$$\delta_0 = \delta_2(L_{cr}) = \frac{2A^{1/4}}{1 + A^{1/2}}, \tag{11.172}$$

where A is determined by the second of Expressions 11.162. If the initial amplitude of some wave (say, pumping wave) is larger, then $A \sim R_0/u_0$. Figure 11.9 shows a normalized interaction region length, for which the signal wave attains the value $I_{1out} = K_{12}I_{20}$ as a function of the initial pumping wave polarization ellipse eccentricity for the same values of wave amplitudes as in Figure 11.2.

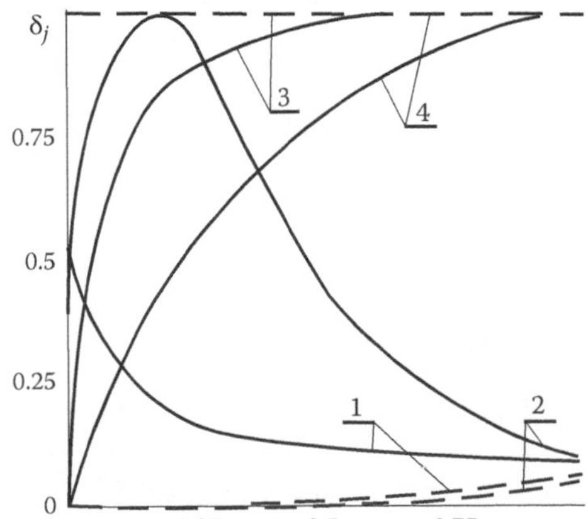

FIGURE 11.8
Dependencies of the wave eccentricities δ_j on the undimensional coordinate z/L for the explosive instability mode. Here, the solid curve corresponds to the eccentricity δ_1, the dotted line corresponds to δ_2, curves 1, 2, and 3 are calculated for $\mathcal{E}_{20}/\mathcal{E}_{10}/\mathcal{E}_{30} = 1/10^{-2}/10^{-4}$ and curve 4 for $\mathcal{E}_{20}/\mathcal{E}_{10}/\mathcal{E}_{30} = 1/10^{-1}/10^{-4}$; it is accepted for curves 1, 3, and 4 that $p_1p_2\sigma = -1$ and for curve 2 that $p_1p_2\sigma = +1$.

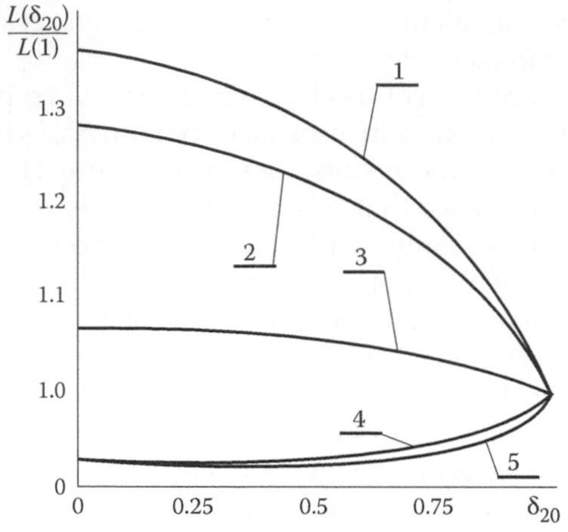

FIGURE 11.9
Dependencies of the normalized optimum length of the interaction region $L(\delta_{20})/L(1)$ on the initial eccentricity of the pumping field δ_{20}. Here, $L(1)$ is the optimum length for the case of linear polarization, curve 1 corresponds to the case $\delta_{10} = 0$, $p_1p_2\sigma = +1$, curve 2 corresponds to $\delta_{10} = 0.5$, $p_1p_2\sigma = +1$, curve 3 corresponds to $\delta_{10} = 1$; $p_1p_2\sigma = \pm1$, curve 4 corresponds to $\delta_{10} = 0$; $p_1p_2\sigma = -1$, and curve 5 corresponds to $\delta_{10} = 0.5$, $p_1p_2\sigma = -1$.

For opposite directions of going around the polarization ellipses, $p_1p_2\sigma = -1$, the optimum length is almost independent of the eccentricity δ_2, whereas for $p_1p_2\sigma = +1$, it decreases by 1.3 times (for $\delta_{10} = 0$).

In the course of interaction, not only eccentricities of wave polarization ellipses are changed but also each polarization ellipse turns as a whole (see Figures 11.9 and 11.10). The results of the relevant numerical computations are given in Figure 11.9.

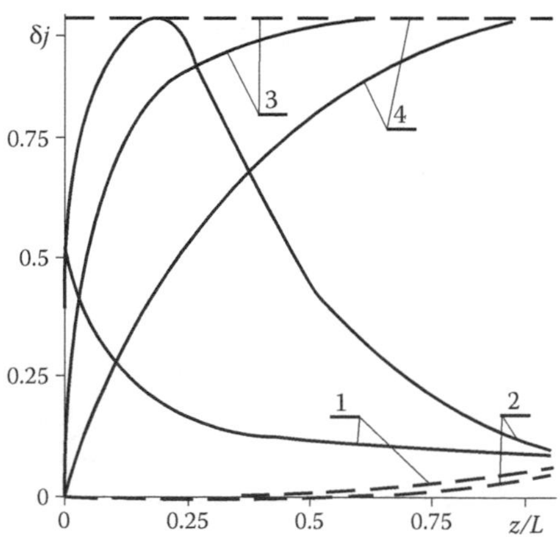

FIGURE 11.10
Dependencies for the angle between the major axes of the polarization ellipses $\hat{\psi}$ on the undimensional longitudinal coordinate z/L at the explosive instability mode. Here, $\omega_2 = 10^{11}\ s^{-1}$, $\beta = 0.82$, $\omega_p = 10^{10}\ s^{-1}$, $N_1 = 1$; $N_2 = 5$, curve 1 corresponds to the case $\mathcal{E}_{20}/\mathcal{E}_{10}/\mathcal{E}_{30} = 1/10^{-2}/10^{-5}$, curve 2 corresponds to $\mathcal{E}_{20}/\mathcal{E}_{10}/\mathcal{E}_{30} = 1/10^{-2}/10^{-4}$, curve 3 corresponds to $\mathcal{E}_{20}/\mathcal{E}_{10}/\mathcal{E}_{30} = 1/10^{-2}/10^{-3}$, and curve 4 corresponds to $\mathcal{E}_{20}/\mathcal{E}_{10}/\mathcal{E}_{30} = 1/10^{-2}/10^{-2}$.

The qualitative regularities are in accordance with the conclusions of Section 11.4: The closer the initial amplitudes of interacting waves, the greater the quantitative discrepancies between the results.

11.9 Explosive Instability in the Linearly Polarized Compton Model

The wave theory developed here makes it possible to describe parametric resonance in terms of the Compton version of explosive instability [41]. Let us consider some physical aspects of the mechanism. We take the first four equations of System 11.108 for the initial system and supplement them with the equation for SCWs (Equation 11.139)

$$\frac{dE_{11}}{dz} = \sigma s_1 B_{11}(E_{21}E_3\delta_{\sigma,-1} + E_{22}E_3\delta_{\sigma,+1});$$

$$\frac{dE_{12}}{dz} = \sigma s_1 B_{12}(E_{22}E_3^*\delta_{\sigma,-1} + E_{21}E_3^*\delta_{\sigma,+1});$$

$$\frac{dE_{21}}{dz} = s_2 B_{21}(E_{11}E_3^*\delta_{\sigma,-1} + E_{12}E_3\delta_{\sigma,+1});$$

$$\frac{dE_{22}}{dz} = s_2 B_{22}(E_{12}E_3\delta_{\sigma,-1} + E_{11}E_3^*\delta_{\sigma,+1});$$

$$\frac{d^2E_3}{dz^2} = -i\sigma\tilde{B}_3\left[(E_{21}^*E_{11} + E_{22}E_{12}^*)\delta_{\sigma,-1} + (E_{21}E_{12}^* + E_{22}^*E_{11})\delta_{\sigma,+1}\right], \qquad (11.173)$$

where all definitions are given in relevant comments for Equations 11.108 and 11.139. The electron beam is considered cold ($\Delta = 0$).

We assume electromagnetic wave polarizations to be linear and lie in the same plane. Then the system of equations possesses integrals of motion

$$s_2u_2^2 - s_1\sigma u_1^2 = C_1' = const;\ k_1u_1^2 + k_2u_2^2 + 2k_3u_3\frac{d\varphi_3}{dz} = C_2' = const, \qquad (11.174)$$

where $u_q = |U_q|, q = j,3, U_1 = \sqrt{B_2\tilde{B}_3}E_1, U_2 = \sqrt{B_1\tilde{B}_3}E_2, U_3 = \sqrt{B_1B_2}E_3$, and $E_q = \mathcal{E}_q\exp\{i\varphi_q\}$.

Expressions 11.174 reflect the energy conservation of interacting waves. The other notation is the same as in Equations 11.108 and 11.140 for linearly polarized electromagnetic waves (see Section 11.2.4). As follows from these equations, all wave amplitudes can grow simultaneously if

$$\frac{d\varphi_3}{dz} < 0. \qquad (11.175)$$

We already mentioned that the explosive instability occurs if energies of the interacting waves are transferred in the same direction, and the sign of the energy of the wave with the highest frequency must be opposite to the energy signs of the other two waves. For SCW energy density

$$W_3 = \frac{v_0 \gamma^{3/2}}{2\omega_p^2} \frac{d\varphi_3}{dz} E_3 E_3^*, \tag{11.176}$$

to be negative, the second of Conditions 11.174 must hold. Therefore, similar to the Raman mode, the Compton explosive instability mode occurs particularly for $s_1 = s_2 = \sigma = +1$ (type-a mode; see the corresponding classification in Section 11.5.4).

Solutions of initial truncated equations can be found within the context of their symmetry with respect to continuous transformation groups [43]. According to Davydova et al. [43], we find that

$$u_{1.2} = \frac{\sqrt{5}}{(L_{cr}^C - z)^{3/2}}; \quad u_3 = \sqrt{\frac{5}{2}} \frac{1}{(L_{cr}^C - z)};$$

$$\varphi_{1,2} = \tilde{\psi}_{1,2} \frac{1}{2} \ln(L_{cr}^C - z); \quad \varphi_3 = \tilde{\psi}_3 + \ln(L_{cr}^C - z), \tag{11.177}$$

where L_{cr}^C is the Compton explosive length (critical Compton length).

In Equations 11.177, constants $\tilde{\psi}_q$ must satisfy the condition

$$tg(\tilde{\psi}_3 - \tilde{\psi}_1 - \tilde{\psi}_2) = \frac{1}{3}, \tag{11.178}$$

and the critical Compton length, at which amplitudes grow infinitely, is equal to

$$L_{cr}^C \cong 5^{1/3} u_{20}^{-2/3}, \tag{11.179}$$

where $u_{20} = u_2$ $(z = 0)$. Comparing Equation 11.177 to the Raman version of the explosive instability, Equation 11.75, we see that SCW amplitude growth rates are approximately equal, whereas electromagnetic wave amplitudes grow much faster in the Compton case. We also note that the Compton critical length L_{cr}^C is much shorter than the Raman critical length L_{cr} (see Equation 11.76)

$$L_{cr}^C / L_{cr} \sim u_{20}^{-1/3} < 1. \tag{11.180}$$

11.10 Effect of Generation of the Transverse H-Ubitron Field

The idea of the effect of magnetic field generation in a FEL was first formulated in a paper by Kulish and Miliukov [44]. An electron beam periodically oscillating in a transverse direction (in the pumping field, for example) can be attributed with a transverse periodic reversible current. According to the Biot–Savart–Laplace law, such a current should produce a periodic reversible magnetic field similar to the H-Ubitron field. If the process occurs in the H-Ubitron interaction range, then the generated H-Ubitron field is superposed with a basic field and, thus, a common self-consistent H-Ubitron pump field is formed. Later this physical mechanism was studied in more detail [45–49] (see Chapter 12). It has been clarified that the effect can be considerable in the FEL. Apart from that, it was shown that its influence should necessarily be taken into account, especially in terms of nonlinear theory. However, it should be noted that many physical aspects of the nonlinear theory of this phenomenon are still not clear today.

11.10.1 Two Modes of the Effect of Generation of Additional Magnetic Field

The detailed nonlinear analysis of the nonlinear stage of the effect is given in the following sections. Now, let us discuss some aspects of the weak-signal theory of the effect. Two basic generation mechanisms are distinguished

1. Diamagnetic
2. Wave nonlinear

The first mechanism is realized under electron beam motion in any external H-, E-, or EH-Ubitron fields. The electron trajectories are of a slalom shape, i.e., oscillating in a transverse plane. Furthermore, we confine ourselves to only the H-Ubitron version of this phenomenon. It should draw our attention to the fact that a generated H-Ubitron magnetic field is completely determined by the pumping and beam parameters. The influence of signal wave and SCW fields on the generation mechanism in this case can be insignificant. Some quantitative analyses of the process will be given in the following sections. Furthermore, in this section, we consider the second above-mentioned mechanism only.

The fundamental distinction of the second mechanism as compared to the first is that it is a wave-resonant nonlinear one. The basic process here is parametric resonance interaction of a signal electromagnetic wave with an SCW of a relativistic electron beam. The idea of the mechanism is as follows.

First, all trajectories of electrons (with the same stationary boundary conditions) in any external H-Ubitron field coincide. Alternatively, independent of the time of electrons entering the system input, all follow along the same trajectories. We turn to the model without any pumping, i.e., we assume that the electrons move in the field of the electromagnetic signal only (Chapter 7). It is obvious that owing to the nonstationary nature of the signal field, the trajectory of all electrons must be different because of the nonstationarity of the boundary conditions. However, we can observe that in view of periodicity of the boundary conditions of time, the trajectories of electrons are repeated systematically in a periodical manner.

Let us assume then that the electron beam is initially modulated on density, i.e., it can be represented as a sequence of electron bunches. Also, we assume that the conditions for parametric resonance are satisfied. In this case, the trajectories of electrons belonging to different bunches turn out to be approximately coinciding, i.e., all bunches, as electron aggregates, follow along the same trajectories in the signal electromagnetic field. Therefore, the aggregate of synchronous oscillating bunches can be treated as a transverse undulation current. Correspondingly, this transverse current generates a transverse undulation magnetic field. The latter, in turn, generates the undulation vortex electric field, and so on. But, the situations are possible when the magnitude of this electric field is too small on the background of accepted accuracy of the considered problem (we will discuss such a situation later). As a result, we obtain the field whose configuration is very similar to the pumping field in the usual H-Ubitron FEL. The difference, however, is that the H-Ubitron field in the discussed case is generated as a result of a specific wave-resonant nonlinear interaction in the system.

But the question arises: What is the basic physical cause of this difference? As analysis shows [46], the main cause is the proper and improper nature of the field generated in both discussed models. In the first (diamagnetic) case, the generated magnetic undulation field is an improper field of the H-Ubitron system. In the second (wave-nonlinear) case, the generated magnetic field is proper. Let us consider both modes in terms of the weak signal theory.

11.10.2 Wave-Nonlinear Mechanism

Let us discuss one of the situations where we can neglect the influence of the generated vortex electric field. The idea can be interpreted by the well-known fact that the electron does not feel the difference between a low-frequency anomaly retarded electromagnetic wave $N_2 >>1$ (see Equation 11.37 and relevant comments) and a true static (or, more exactly, quasistatic) H-Ubitron field [46]. The key point is that relevant terms in the equation of electron motion in such a field can be of the order of magnitude beyond the acceptable accuracy. So, the electron sees only the magnetic component of the total electromagnetic field.

Let us illustrate the above as an example of the Raman quadratic-nonlinear version of parametric resonance, where the pumping wave is an anomaly-retarded electromagnetic wave $N_2 >> 1$. Experimentally, such a wave can be produced in a magnetized plasma waveguide (see Formula 11.37 and relevant comments). This field obviously is a proper field of the system. We pass to limits of the Form 11.120 in truncated Equations 11.108 to obtain

$$\frac{dE_{11}}{dz} = s_1\sigma B_{1m}(H_{21}E_3\delta_{\sigma,-1} + H_{22}E_3\delta_{\sigma,+1});$$

$$\frac{dE_{12}}{dz} = s_1\sigma B_{1m}^*(H_{22}E_3^*\delta_{\sigma,-1} + H_{21}E_3^*\delta_{\sigma,+1});$$

$$\frac{dH_{21}}{dz} = B_{2m}(E_{11}E_3^*\delta_{\sigma.-1} + E_{12}E_3\delta_{\sigma,+1});$$

$$\frac{dH_{22}}{dz} = B_{2m}^*(E_{12}E_3\delta_{\sigma,-1} + E_{11}E_3^*\delta_{\sigma,+1});$$

$$\frac{dE_3}{dz} + \Delta E_3 = \sigma r B_{3m}\Big[(H_{21}^*E_{11} + H_{22}E_{12}^*\delta_{\sigma,-1})$$

$$+(H_{21}E_{12}^* + H_{22}^*E_{11})\delta_{\sigma,+1}\Big], \tag{11.181}$$

where B_{jm}, B_{3m} are matrix elements given by

$$B_{1m} = \frac{ek_3\kappa\Lambda}{4\pi mc^2};\ B_{2m} = \frac{ek_3\kappa}{2\pi mc\omega_1};\ B_{3m} = \frac{ek_3\Omega_{30}\kappa\Lambda}{8\pi m v_0\omega_1};\ \Omega_{30} = \omega_3 - k_3 v_0, \tag{11.182}$$

Λ is the magnetic field undulation period ($\Lambda = 2\pi/k_2$), and the rest of the notation is standard for this chapter (see, for example, Equations 11.104 and 11.108).

Except the definition of matrix elements $B_{qm}(q = j,3)$, the mathematical structure of the system of Equation 11.181 reproduces Equation 11.108 for $s_2 = +1$. Hence, solutions of Equation 11.181 formally reproduce relevant solutions of Equation 11.108. In particular, we find that for $\sigma = s_s = r = +1$, the H-ubitron field grows explosively.

In discussing the obtained results, an important aspect should be noted. Regarding the field of wave $j = 2$, the H-Ubitron field is quite reasonable in the study of the equation of motion. However, it cannot be treated as such in the electrodynamic part of the problem described by the Maxwell equations. In that case, the mechanism of H-Ubitron field generation is due to the fact that the field of the wave $j = 2$ has a small electric component that is nevertheless essential. This observation is necessary for the comparison of this mechanism and the above-mentioned others.

11.10.3 Diamagnetic Mechanism

Now, we discuss the mechanism of generation of improper H-Ubitron fields [46,47]. Following the scheme discussed in Section 11.1.5, we can obtain the truncated equation system (corresponding to the self-consistent H-Ubitron model) similar to Equation 11.108 ($\Delta \cong 0$, $s_1 = \sigma = +1$)

$$
\begin{aligned}
H_{21}^{(g)} &= iB'_{2m}E_{11}^{*}E_3;\\
H_{22}^{(g)} &= iB'_{2m}E_{12}^{*}E_3;\\
\frac{dE_{11}}{dz} &= B'_{1m}H_{210}^{*}E_3;\\
\frac{dE_{12}}{dz} &= B'^{*}_{1m}H_{220}E_3^{*};\\
\frac{dE_3}{dz} &= rB'_{3m}(E_{11}H_{210} + E_{12}^{*}H_{220}^{*}),
\end{aligned}
\tag{11.183}
$$

where $B'_{1m} = \dfrac{ek_3\kappa}{2mc^2k_2}$, $B'_{2m} = \dfrac{ek_3ck_2\gamma\kappa}{2m\omega_1(c^2k_2^2\gamma + \omega_p^2)}$, and $B'_3 = \dfrac{ek_3\omega_p\kappa}{2m\omega_1\beta c^2k_2\gamma^3}$ are the matrix elements, $H_{2j}^{(g)}$ are complex amplitudes of the generated field, H_{2j0} are relevant amplitudes of the given part of pumping field, and $H_{2i}^{(g)}$ are the complex amplitudes of additional (nonproper) magnetic pumping field. We supplement Equations 11.183 with the boundary conditions

$$
z = 0, \quad E_{ji} = E_{ji0}, \quad E_3 = E_{30}. \tag{11.184}
$$

Besides that, we introduce the following notation:

$$
H_{20} = \sqrt{H_{210}^2 + H_{220}^2}, \; \mathcal{E}_{10} = \sqrt{|E_{110}|^2 + |E_{120}|^2}. \tag{11.185}
$$

Then the solutions for the electromagnetic wave and SCW E_3 can be written as

$$E_{11} = E_{110}\left(1 + 2r\frac{H_{210}^2}{H_{20}^2}sh^2\left(\frac{\alpha}{2}z\right)\right) + \sqrt{\frac{B'_{1m}}{B'_{3m}}}\frac{H_{210}}{H_{20}}E_{30}sh(\alpha z) + r\frac{H_{210}H_{220}}{H_{20}^2}E_{120}sh(\alpha z);$$

$$E_{12} = E_{120}\left(1 + 2r\frac{H_{220}^2}{H_{20}^2}sh^2\left(\frac{\alpha}{2}z\right)\right) + \sqrt{\frac{B'_1}{B'_3}}\frac{H_{220}}{H_{20}}E_{30}sh(\alpha z) + r\frac{H_{210}H_{220}}{H_{20}^2}E_{110}sh(\alpha z);$$

$$E_3 = E_{30}ch(\alpha z) + \sqrt{\frac{B'_2}{B'_1}}\left(\frac{H_{210}}{H_{20}}E_{110} + \frac{H_{220}}{H_{20}}E_{120}\right)sh(\alpha z), \quad (11.186)$$

where $\alpha = H_{20}\sqrt{B'_{1m}B'_{3m}}$. Substituting Equations 11.186 in the first two equations of Equation 11.183, we find additional magnetic field components $H_{2i}^{(g)}$. It can be verified that, in general, polarizations and initial phases of all the fields—the additional magnetic field among them—are slowly varying functions of coordinate z. Moreover, polarizations of the signal wave and additional magnetic field are similar and tend to external magnetic field polarization, whereas the phase of the additional magnetic field differs by $\pi/2$ from the external magnetic field phase.

If the interaction region is sufficiently large, the magnetic field amplitude can attain considerable values, i.e.,

$$\left|H_2^{(g)}\right|_{z=L} = \frac{e\omega_p \mathcal{E}_{10}^2}{mc^3k_2^2\beta^2\gamma^3}\left[\frac{ck_2\gamma^{1/2}}{\omega_p(1+\beta)}\right]^{\frac{1}{2}} \exp\left\{\frac{eH_{20}L}{2mc^2\beta}\left[\frac{\omega_p(1+\beta)}{ck_2\gamma^{1/2}}\right]^{\frac{1}{2}}\right\}, \quad (11.187)$$

where $H_2^{(g)} = \sqrt{\left(H_{21}^{(g)}\right)^2 + \left(H_{22}^{(g)}\right)^2}$ is the magnetic field amplitude, and L is the total system length. For example, for L ~ 1m linearly polarized waves, the electron beam energy ~1MeV, $\omega_p = 10^{10}$ s^{-1}, $\mathcal{E}_{10} = 10^2$ CGS units, $H_{20} = 1$ kGs, $\Lambda = 1$ cm, we have $H_2^{(g)} \sim 30$Gs, i.e., in this numerical example, the magnetic field generation is not significant at the quadratic-nonlinear interaction: $H_{2g}^{(g)}/H_{20} \sim 3\cdot 10^{-2}$. However, as is demonstrated in Chapter 12, this effect can exert remarkable influence at the cubic-nonlinear stage of the interaction.

References

1. Kulish, V.V. 1998. *Methods of averaging in non-linear problems of relativistic electrodynamics.* Atlanta: World Scientific Publishers.
2. Kulish, V.V. 2002. *Undulative electrodynamic systems,* Vol. 2 of *Hierarchic methods.* Dordrecht: Kluwer Academic Publishers.
3. Kulish, V.V. 2002. *Hierarchy and hierarchic asymptotic methods in electrodynamics,* Vol. 1 of *Hierarchic methods.* Dordrecht: Kluwer Academic Publishers.

4. Weiland, J., and Wilhelmsson, H. 1977. *Coherent nonlinear interactions of waves in plasmas*. Oxford: Pergamon Press.
5. Sukhorukov, A.P. 1988. *Nonlinear wave-interactions in optics and radiophysics*. Moscow: Nauka.
6. Bloembergen, N. 1965. *Nonlinear optics*. Singapore: World Scientific Publishing.
7. Berezhnoi, I.A., Kulish, V.V., and Zakharov, V.P. 1981. On the explosive instability of relativistic electron beams in the fields of transverse electromagnetic waves. *Zh.Tekh.Fiz.* [Sov.Phys.-Tech. Phys.] 51:660.
8. Zakharov, V.P., and Kulish, V.V. 1998. Parametric instability of a relativistic electron beam in the field of two arbitrary-polarization electromagnetic waves. Deposited manuscript. Deposited in VINITI, Moscow. 5804-82.
9. Butuzov, V.V., Zakharov, V.P., and Kulish, V.V. 1983. *Parametric instability of a flux of high current relativistic electron in the field of dispersed electromagnetic waves*. Deposited manuscript. Deposited in Ukrainian Scientific Research Institute of Technical Information, Kiev. 297 Uk-83:67.
10. Butuzov, V.V., Zakharov, V.P., and Kulish V.V. 1984. Stationary state establishment under parametric interaction of electron waves in the high-current relativistic electron beam plasma. *Radiotech.Electron.* [*Radio Eng. and Electron. (USSR)*] 29(11):2192–2198.
11. Zakharov, V.P., and Kulish, V.V. 1983. Explosive instability of a high-current relativistic electron beam in the field of two electromagnetic waves. *Zh. Tekh. Fiz.* [*Sov. Phys.-Tech. Phys.*] 53(6):1226–1228.
12. Kulish V.V. 1990. *The physics of free electron lasers. General principles*. Deposited in Ukrainian Scientific Research Institute of Technical Information, Kiev 1526, Uk-90:192.
13. Kalmykov A.M., Kotsarenko N.Y., Kulish V.V. 1978. Laser radiation frequency transformation in electron beams. *Pis'ma Zh.Tech.Fiz. [Sov.Tech.Phys.Lett.]*, 4(14):820–822.
14. Patent of USSR SU N705914. Electronic device. Authors: Silin, R.A., Kulish, V.V., Klimenko, Ju.I., priority of 18.05.72. Published in non-secret press of USSR after taking off of the relevant stamp of secrecy: 15.05.91, *Inventions Bulletin* N26, 1991.
15. Kalmykov, A.N., Kotsarenko, N.Y., and Kulish, V.V. 1977. Parametric generation and amplification of electromagnetic waves with frequencies higher than the pump wave frequency in electron beams. *Izv. Vyssh. Uchebn. Zaved. Radioelectron.* [*Soviet Radioelectronics*] 10:76.
16. Kalmykov, A.N., Kotsarenko, N.Y., and Kulish, V.V. 1979. On the theory of parametric frequency-increasing transformation in electron beams. *Radiotech. Electron.* [*Radio Eng. and Electron.* (USSR)] 24(10):2084–2088.
17. Kotsarenko, N.Ya., and Kulish V.V. 1980. On the effect of superheterodyne amplification of electromagnetic waves in a plasma-beam system. *Radiotekhnika I Electronika* 35(N11):2470–2471.
18. Zakharov, V.P., and Kulish, V.V. 1983. Discrimination effects of parametric wave interaction in an electron beam plasma. *Ukr. Fiz. Zh.* [*Ukrainian J. Phys.*] 28(3):406–411.
19. Zakharov, V.P., and Kulish, V.V. 1983. Polarization effects under parametric interaction of transverse electromagnetic waves with a high-current relativistic electron beam. *Ukr. Fiz. Zh.* [*Ukrainian J. Phys.*] 53(10):1904–1908.
20. Zakharov, V.P., and Kulish, V.V. 1984. Polarization effects under electromagnetic wave interaction in the high-current relativistic electron beam plasma. *Radiotech. Electron.* [*Radio Eng. and Electron.* (USSR)] 29(6):1162–1170.
21. Zakharov, V.P., and Kulish, V.V. 1985. Electron beam explosive instability in the field of disperse electromagnetic waves. *Ukr. Fiz. Zh.* [*Ukrainian J. Phys.*] 30(6):878–881.
22. Zakharov, V.P., Kochmanski, S.S., Kulish, V.V. 1983. Phase discrimination of electromagnetic signals in modulated relativistic electron beams. *Radiotech. Electron Radiotech. Electron.* [*Radio Eng. and Electron.* (USSR)] 28(11):2217–2224.
23. Kulish, V.V., and Kuleshov, S.A. 1993. Nonlinear self-consistent theory of free electron lasers. *Ukr. Fiz. Zh.* [*Ukrainian J. Phys.*] 38(2):198–205.
24. Kulish, V.V., Kuleshov, S.A., and Lysenko, A.V. 1993. Averaged kinetic equation method in the theory of three-wave parametric resonance in relativistic electron beam plasmas. *Fiz. Plazmy* [*Sov. J. Plasma Phys.*] 19(2):199–216.

25. Kulish, V.V., and Lysenko, A.V. 1993. Method of averaged kinetic equation and its use in the nonlinear problems of plasma electrodynamics. *Fizika Plazmy* (*Sov. Plasma Physics*) 19(2):216–227.
26. Kulish, V.V., Kuleshov, S.A., and Lysenko, A.V. 1993. Nonlinear self-consistent theory of superheterodyne and free electron lasers. *The International Journal of Infrared and Millimeter Waves* 14(3):451–568.
27. Kulish, V.V. 1997. Hierarchic oscillations and averaging methods in nonlinear problems of relativistic electronics. *The International Journal of Infrared and Millimeter Waves* 18(5):1053–1117.
28. Kulish, V.V., and Miliukov, V.V. 1984. Explosive instability of a high-current relativistic electron beam in H-ubitron fields. *Ukr. Fiz. Zh.* [*Ukrainian J. Phys.*] 2:389–390.
29. Kulish, V.V., and Kotsarenko, N.Y. 1977. Soviet Inventor's Certificate SU No 668491. Priority 17.05.77.
30. Kotsarenko, N.Ya., and Kulish V.V. 1980. On the effect of superheterodyne amplification of electromagnetic waves in a plasma-beam system. *Radiotekhnika I Electronika* 35(N11):2470–2471.
31. Kondratenko, A.N. 1976. *Plasma waveguides*. Moscow: Atomizdat.
32. Korn, G.A., and Korn, T.W. 1961. *Mathematical handbook for scientists and engineers*. NY: McGraw Hill.
33. Berezhnoi, I.A., Zakharov, V.P., Kulish, V.V. 1980. Soviet Inventor's Certificate SU No 1023949. Priority 22.12.80.
34. Silin, R.A., and Sazonov, V.P. 1966. *Retarding systems*. Moscow: Sovetskoe Radio.
35. Collin, A.E. 1960. *Field theory and guided waves*. New York: McGraw Hill.
36. Sokolov, A.A., and Ternov, I.M. 1974. *Relativistic electron*. Moscow: Nauka.
37. Kotsarenko, N.Y., and Fedorchenko, A.M. 1970. Absolute and convective instability criteria and absolute to convective instability transition. *Zh. Tekh. Fiz.* [*Sov. Phys.-Tech. Journ.*] 40(1):41–46.
38. Akhiezer, A.I., and Polovin, R.V. 1971. Wave increase criteria. *Usp. Fiz. Nauk* [*Sov. Phys.-Usp.*] 104(3):185–200.
39. Davis, G.R. 1976. Navy researchers develop new submillimeter wave power source. *Microwave* 12:12,17.
40. Granatstein, V.L., and Sprangle, P. 1977. Mechanisms for coherent scattering of electromagnetic wave from relativistic electron beams. *IEEE Trans. Microwave Theory and Techn.* 25(6):545–550.
41. Weiland, J., and Wilhelmsson, H. 1977. *Coherent nonlinear interactions of waves in plasmas*. Oxford: Pergamon Press.
42. Davydova, T.A., Zakharov, V.P., and Kulish, V.V. 1987. Three-wave parametric resonance in relativistic electron beam plasmas. *Zh. Tekh. Fiz.* [*Sov. Phys.-Tech. Phys.*] 57(4):687–694.
43. Davydova, T.A., Pavlenko, V.P., Taranov, V.B., and Shamrai, K.P. 1977. Self-similar dynamics of nonlinear wave interaction near the instability threshold. *Preprint of the Institute for Nuclear Research*, No 78-41, Kiev.
44. Kulish, V.V., and Miliukov, V.V. 1984. Explosive instability of a high-current relativistic electron beam in H-Ubitron fields. *Ukr. Fiz. Zh.* [*Ukrainian J. Phys.*] 2:389–390.
45. Kulish, V.V., Kuleshov, S.A., and Lysenko, A.V. 1993. Nonlinear self-consistent theory of superheterodyne and free electron lasers. *The International Journal of Infrared and Millimeter Waves* 14(3):451–568.
46. Zakharov, V.P., Kisletsov, A.V., and Kulish, V.V. 1986. Generation of static transverse periodic magnetic field in the relativistic electron beam plasma. *Fiz.Plazmy* [*Sov. J. Plasma Phys.*] 12(1):77–79.
47. Ginzburg, N.S. 1987. Diamagnetic and paramagnetic effects in free-electron laser. *IEEE Transactions in Plasma Science* 15(4):411–417.
48. Dzedolik, I.V., Zakharov, V.P., and Kulish, V.V. 1988. Nonlinear theory of parametric resonance interaction of electromagnetic waves in relativistic electron beam plasmas. *Radiotech. Electron.* [*Radio Eng. and Electron.* (*USSR*)] 33(6):1255–1264.
49. Dzedolik, I.V., Zakharov, V.P., and Kulish, V.V. 1988. Self-consistent nonlinear analysis of parametric resonance interaction of electromagnetic waves in relativistic electron beam plasmas. *Radiotech. Electron.* [*Radio Eng. and Electron.* (*USSR*)] 33(6):1264–1271.

12

Ordinary (Parametrical) Free Electron Lasers: Cubic-Nonlinear Theory

Considering the theme, this chapter is a continuation of Chapter 11. The physics of the ordinary (parametrical) FELs is studied, yet, as one can see, the given material below is essentially different from that of the previous chapter. Contrary to Chapter 11, where the weak-signal theory is the main object of attention, in the present chapter, we will concentrate on the cubic-nonlinear theory of the Dopplertrons and H-Ubitrons. This includes the study of the effect of the saturation of the amplification of the electromagnetic signal as the key physical mechanism for the nonlinear theory of any electronic device with long-time interaction. In addition to this, we describe a peculiar hierarchic calculation technology as well as the methodology of the nonlinear analysis of such kind, both of which deserve attention.

Somehow apart, such nonlinear effects like the generation of the longitudinal electric and the transverse H-Ubitron fields are described here. However, in this connection, the following should be noted. As it is shown by the magnitude, the action of both of these effects on the basic working mechanism is not essential. Therefore, one could think that they do not deserve such attention as they received. But the author has a different opinion. He is convinced that learning them is reasonable at least from the general-physics point of view, because by their issue they appear, by far, not banal or trivial. In reality, just the very fact of the generation of the static fields as a result of the nonlinear-resonance interaction of wave fields appears physically unusual, and it is interesting already because of this. In addition to this, the accomplished work might carry some applied value, because an unexpected generation of these fields might lead to malfunctions in the performance of some design elements of FEL.

For example, in a series of the design versions of FEL-klystrons, the significantly modulated electron beam in the transition section moves in the field of the intense signal electromagnetic wave. According to the theory, in such situations, an additional H-Ubitron pumping field might be simultaneously generated in the transportation channel. *A priori*, the dynamics of the interaction of these three wave fields, for which the parametric resonance condition holds, is far from obvious. Here, there might be conditions when the longitudinal velocity of a beam as a whole is changed, for example, by the beam's compression or depression or simply by acceleration (see Chapter 9). As a consequence, the nature of the resonance interaction can change essentially from a banal passive interaction to amplification or attenuation of the wave signal. Of course, in an experiment, an uncontrolled realization of amplification as well as of attenuation effects is undesirable.

Regrettably, even the large volume of this book does not allow the author to expand the analysis of these types of interaction mechanisms. However, this and the subsequently given material in this chapter are enough to remove doubts about the expediency of the study of both mentioned effects in the generation of the static fields.

12.1 Truncated Equations: Dopplertron Model

Compared with Chapter 11, the main peculiarities of the nonlinear FEL theory developed below are

1. The kinetic equation in canonical form (Equation 10.26) is used to describe electron beam motion in the fields (Equation 10.36).
2. The method of averaged characteristics (for the kinetic Equation 10.26—see Chapter 6) is used for solving the beam motion part of the problem.
3. The simplified version of the slowly varying amplitude method (see the previous chapter) is used for solving the field part of the considered problem.
4. All calculations are performed in third (cubic) order with respect to amplitude small parameters (Equations 11.1 and 11.11).

12.1.1 Formulation of the Problem

A general one-dimensional model of the FEL of Dopplertron type with arbitrary polarized signal and pumping waves is chosen as a suitable illustration of the main peculiarities of considered hierarchic calculation technology. This model is described in Chapters 10 and 11. Here, we assume only additionally that the case of main resonance $l, m_j, n_j = 1$ is realized (see Equation 10.41).

To solve the problem of electron beam motion, we employ the hierarchic method of averaged characteristics (for the kinetic equation) described in Section 6.5. We take into account the integral of motion for the transverse canonical momentum (Equation 7.21)

$$\vec{\mathcal{P}}_{\perp} = \vec{\mathcal{P}} - \left(\vec{n}\vec{\mathcal{P}}\right)\vec{n} = \vec{\mathcal{P}}_{0\perp} = \text{const.} \tag{12.1}$$

Following the main calculation scheme of the method, we consider the phases p_j and p_3 as parameters.

Then, we transform kinetic equation 10.27 into parametric form and pass from partial to total derivatives. Thus, we can rewrite kinetic equation 10.27 in the form

$$\frac{\mathrm{d}}{\mathrm{d}t} f\left(t, z, \mathcal{P}_z, \vec{\mathcal{P}}_{0\perp}, p_1, p_2, p_3\right) = \zeta^{-1} I_{st}(f); \tag{12.2}$$

$$\frac{\mathrm{d}p_j}{\mathrm{d}t} = \omega_j - s_j k_j \frac{\partial \mathcal{H}}{\partial \mathcal{P}_z}; \tag{12.3}$$

$$\frac{\mathrm{d}p_3}{\mathrm{d}t} = \omega_3 - k_3 \frac{\partial \mathcal{H}}{\partial \mathcal{P}_z}; \tag{12.4}$$

$$\frac{\mathrm{d}\mathcal{P}_z}{\mathrm{d}t} = -\frac{\mathrm{d}\mathcal{H}}{\mathrm{d}z}; \quad \frac{\mathrm{d}z}{\mathrm{d}t} = \frac{\partial H}{\partial \mathcal{P}_z}, \tag{12.5}$$

where $j = 1, 2$; $\mathcal{H} = \left[m^2c^4 + c^2\left(\vec{\mathcal{P}} + (e/c)\vec{A}\right)^2 \right]^{1/2} - e\varphi$, as before, is the Hamiltonian of electron in the fields (Equation 10.36), and $\vec{A}$ and ϕ are the vector and scalar potentials of the fields acting on electrons. They are related to the force field characteristics

$$\vec{B} = \left[\vec{\nabla}\vec{A}\right]; \vec{E} = -\frac{1}{c}\frac{\partial\vec{A}}{\partial t} - \left(\vec{\nabla}\varphi\right). \tag{12.6}$$

Here, we determine the scalar potential ϕ as

$$\varphi = \hat{\varphi}_3 + \varphi_0^{(n)} + \varphi_0^{(g)}, \tag{12.7}$$

where potential $\hat{\varphi}_3$ describes the electrostatic field of SCWs

$$\hat{\varphi}_3 = \frac{1}{2}\varphi_3 e^{i(\omega_3 t - k_3 z)} + c.c., \tag{12.8}$$

where φ_3 is the complex amplitude. The potential $\varphi_0^{(n)}$ describes the field of longitudinal external electrostatic support (see Chapter 7)

$$\vec{E}_0^{(n)} = \vec{n}E_0^{(n)} = -(\vec{\nabla}\varphi_0^{(n)}) = -\vec{n}\left(\partial\varphi_0^{(n)}/\partial z\right). \tag{12.9}$$

The last term in Equation 12.7 describes the longitudinal quasistationary electrostatic field generated as a result of nonlinear resonant-wave interactions in the system

$$\vec{E}_0^{(g)} = \vec{n}E_0^{(g)} = -\left(\vec{\nabla}\varphi_0^{(g)}\right) = -\vec{n}\left(\partial\varphi_0^{(g)}/\partial z\right). \tag{12.10}$$

In general, solutions of kinetic equation 10.26 describe the dynamics of rather complex processes occurring in the FEL interaction range. The nonlinear generation of longitudinal electric field is one of these processes [1–8]. Its essence is as follows. In the course of parametric resonance, a portion of the electron kinetic energy is converted into the energy of the signal wave (the same can happen with respect to pumping waves also in the Dopplertron models, for example, due to explosive instability). This results in a decrease in the whole beam longitudinal velocity along the system length. Therefore, as compared to undisturbed states, a certain excess of electrons is formed in the end region of the electron beam. This implies a local violation of the initial averaged quasineutrality of the model accepted: the longer the passed distance, the larger the surplus electron charge. As a result, an additional decelerating potential difference, that is, longitudinal decelerating electric field, arises between the input and output parts of the interaction region. We calculate it in the following.

We expand fields acting on beam electrons in Fourier series with respect to all mentioned phases. Therein, the zero electric field harmonic formally corresponds to the quasistatic scalar part of the electric field. The latter, in general, is a superposition of the external field (electric support) $E_0^{(n)}$ (Equation 12.9) and the proper field (generated in the interaction region by the above mechanism) $E_0^{(g)}$ (Equation 12.10). To calculate the field (Equation

12.10), we use Equations 10.10, 10.11, and 10.13 to derive the continuity equation for the zero harmonics of Fourier components of current density $J_z^{(0)}$ and space charge $\rho^{(0)}$, that is,

$$\frac{\partial J_z^{(0)}}{\partial z} + \frac{\partial \rho^{(0)}}{\partial t} = 0, \tag{12.11}$$

where

$$J_z^{(0)} = -en_0 \left\langle \int_{-\infty}^{+\infty} \frac{c^2 \mathcal{P}_z}{\mathcal{H}(\mathcal{P}_z)} f\left(t, \vec{r}, \vec{\mathcal{P}}\right) d^3\mathcal{P} \right\rangle_{p_{j,3}} ; \tag{12.12}$$

$$\rho^{(0)} = -en_0 \left\langle \int_{-\infty}^{+\infty} f\left(t, \vec{r}, \vec{\mathcal{P}}\right) d^3\mathcal{P} \right\rangle_{p_{j,3}} , \tag{12.13}$$

where $\langle \; \rangle_{p_{j,3}}$ implies the averaging overall fast and slow phases of electron oscillations. Therein, we take into consideration thin features of the discussed model, that is, we account that electron beam exists between only the planes $z = 0$ and $z = L$. It means that if $z < 0$ or $z < L$, beam density should really be equal to zero.

The potential of total electric field (Equation 12.7) enters in the Hamiltonian of electron H as a term in potential energy $-e\varphi$ (see comments to Equation 12.5). Therefore, the distribution function $f\left(t, \vec{r}, \vec{P}\right)$ depends on $\varphi_0^{(g)}$, and by virtue of Equations 12.12 and 12.13, this dependence is involved in Equation 12.11. We substitute the solution for distribution function into Equations 12.12 and 12.13, and then substitute the results into Equation 12.11. By this, we obtain an equation determining the generated electric field strength $E_0^{(g)}$. We will show later that the field strength $E_0^{(g)}$ is proportional to the gradient of squared amplitudes of interacting fields. Hence, a relevant equation for $E_0^{(g)}$ should be supplemented with the truncated equations for slowly varying wave amplitudes. On the other hand, for reasons above, the field strength $E_0^{(g)}$ enters in the truncated equations for amplitudes. The result of such mutual influence, however, can be calculated only in the cubic order of the considered nonlinear self-consistent theory. Thus, the effect of nonlinear generation of the longitudinal electric field is related to the cubic effects.

Furthermore, we turn to the problem of obtaining the cubic-nonlinear truncated equations. For simplicity, we neglect the electron collisions in Equation 12.2, that is,

$$\zeta >> \xi_1^m, \tag{12.14}$$

where ξ_1 is, as before, the scalar large parameter of the first hierarchy (see Chapters 6 and 7 for details)

$$\xi_1 \sim \frac{|d\psi/dt|}{|d\theta/dt|}; \left|\frac{d\theta}{dt}\right| \sim \left|\frac{dp_3}{dt}\right| << \left|\frac{d\psi}{dt}\right|; \tag{12.15}$$

$$\psi = p_1 - \sigma p_2; \quad \theta = p_1 + \sigma p_2, \tag{12.16}$$

$\sigma = \pm 1$; m is the highest order of approximation in the averaging method accepted.

12.1.2 Motion Problem

Following the calculation scheme set forth earlier in Chapter 6, we obtain from Equations 12.2 through 12.5 the averaged kinetic equation in the form

$$\left\{\frac{\partial}{\partial t}+\frac{c^2\bar{\mathcal{P}}_z}{\bar{\bar{\mathcal{H}}}}\frac{\partial}{\partial \bar{z}}-\left(\frac{e^2}{2\bar{\bar{\mathcal{H}}}}\frac{\partial \bar{A}^2}{\partial \bar{z}}-e\frac{\partial \bar{\varphi}}{\partial \bar{z}}\right)\frac{\partial}{\partial \bar{\mathcal{P}}_z}\right\}\bar{f}\left(t,\bar{z},\bar{\mathcal{P}}_z,\mathcal{P}_{0\perp}\right)=0; \tag{12.17}$$

where

$$\bar{\bar{\mathcal{H}}}=\sqrt{m^2c^4+c^2\bar{\mathcal{P}}^2+e^2\bar{A}^2}=\bar{\mathcal{H}}+\bar{\varphi};\,\bar{\varphi}=\bar{\varphi}_0+\tilde{\bar{\varphi}};$$

$$\tilde{\bar{\varphi}}=(1/2)\left[\varphi_3\exp\left(ip_3\right)+c.c.\right];\,\bar{\varphi}_0=\varphi_0^{(n)}+\varphi_0^{(g)};$$

$$A^2=(1/2)\left(\left|\vec{A}_1\right|^2+\left|\vec{A}_2\right|^2\right)+\mathrm{Re}\left\{\vec{A}_1\vec{A}_2\exp\left(i\bar{\theta}\right)\delta_{\sigma,+1}+\vec{A}_1\vec{A}_2^*\exp\left(i\bar{\theta}\right)\delta_{\sigma,-1}\right\};$$

$\varphi_3=-iE_3/k_3$; $\vec{A}_j=ic\vec{E}_j/\omega_j$; $\bar{p}_3=\omega_3 t-k_3\bar{z}$; ω_3, k_3 are the frequency and wave number of SCW; and $\delta_{\sigma,\pm1}$ are the Kronecker deltas. In the approximation accepted, the quasi-uniformity criterion for transverse field can be obtained from relevant single-particle equations

$$\left|-\frac{e^2}{2\bar{\bar{\mathcal{H}}}}\bar{\vec{\nabla}}_\perp\vec{A}^2+e\bar{\vec{\nabla}}_\perp\bar{\varphi}\right|\le\frac{1}{\xi_1^2}\frac{\bar{\bar{\mathcal{H}}}_0}{L}, \tag{12.18}$$

where $\bar{\vec{\nabla}}_\perp$ is the transverse part of averaged nabla operator; $\bar{\bar{\mathcal{H}}}_0=\bar{\bar{\mathcal{H}}}(\bar{z}=0)$; and L is the interaction region length. Then, we obtain analytical solutions of Equation 12.17 in the cubic approximation (with respect to small parameters $\varepsilon_j\sim 1/\xi_1$) using the iterative method.

We assume that the wave parametric resonance condition 10.41 along with one-particle resonance condition 12.15 is satisfied roughly ($n_j=m_j=l=1$; $\sigma_{12}\equiv\sigma$)

$$\omega_1+\sigma\omega_2=\omega_3+\Delta\omega;\ s_1k_1+\sigma s_2k_2=k_3+\Delta k, \tag{12.19}$$

where $\Delta\omega$ and Δk are relevant small mismatches ($|\Delta\omega|\ll\omega_j$, $|\Delta k|\ll k_j$). Thus, in view of Equations 12.15 and 12.19, the wave parametric resonant conditions can be written in the form

$$\bar{\theta}=\mathrm{Re}\left\{\bar{p}_3\right\}+\Delta\bar{\theta}, \tag{12.20}$$

where $\Delta\bar{\theta}=(\Delta\omega)t-(\Delta k)\bar{z}$ is a small phase mismatch.

In view of Equation 12.17, the distribution is biperiodic with respect to phases $\bar{\theta}$ and $\bar{p}_3$. It is seen that Equation 12.20 enables us to regard it as a one-period function with respect to one of the phases, either $\bar{\theta}$ or $\bar{p}_3$. Considering this, we expand the distribution function $\bar{f}$ in the Fourier series with respect to phase $\bar{\theta}$, that is,

$$\bar{f} = \bar{f}_0 + \frac{1}{2}\left(\bar{f}_1 e^{i\bar{\theta}} + c.c.\right) + \frac{1}{2}\left(\bar{f}_2 e^{i2\bar{\theta}} + c.c.\right) + \frac{1}{2}\left(\bar{f}_3 e^{i3\bar{\theta}} + c.c.\right) + \ldots, \tag{12.21}$$

where $\bar{f}_k$ $(k = 0,1,2,3,\ldots)$ are the slowly varying amplitudes of once-averaged distribution function. Therein, we accept that similarly to Equation 11.11, the following conditions are satisfied:

$$\frac{\partial \bar{f}_k}{\partial t} \sim \frac{\partial \bar{f}}{\partial \bar{z}} \le \varepsilon_j^2 \quad \left(\varepsilon_j = \frac{e\mathcal{E}_j}{mc\omega_j \gamma_0} << 1\right). \tag{12.22}$$

Moreover, for amplitudes $\bar{f}_k$, we additionally assume

$$\left|\bar{f}_k\right| \sim \varepsilon_j^k, \text{ i.e., } \left|\bar{f}_2/\bar{f}_1\right| \sim \left|\bar{f}_3/\bar{f}_2\right| \sim \ldots \le \varepsilon_j << 1. \tag{12.23}$$

We substitute Equation 12.21 into Equation 12.17 bearing in mind the above assumptions. Collecting coefficients of similar exponential functions, we put expressions obtained equal to zero. Therefore, we derive a system of related equations (we retain only amplitudes up to $\bar{f}_3$) of the form

$$\frac{d\bar{f}_0}{dt} + \frac{\bar{\mathcal{P}}_z}{m}\left[\frac{1}{\gamma_0}\frac{\partial \bar{f}_0}{\partial \bar{z}} + \frac{1}{4\gamma_1}\left(\frac{\partial \bar{f}_1^*}{\partial \bar{z}} + \frac{i\Phi}{c}\bar{f}_1^*\right) + \frac{1}{4\gamma_1}\left(\frac{\partial \bar{f}_1}{\partial \bar{z}} - \frac{i\Phi}{c}\bar{f}_1\right)\right]$$
$$- \frac{e^2}{2mc^2\gamma_0}\left[\tilde{A}_{02}\frac{\partial \bar{f}_0}{\bar{\mathrm{P}}\partial_z} + \frac{1}{4}\frac{\partial \bar{f}_1^*}{\partial \bar{\mathcal{P}}_z}\tilde{A}_{12} + \frac{1}{4}\frac{\partial \bar{f}_1}{\partial \bar{\mathcal{P}}_z}\bar{A}_{12}^*\right] + e\left[\frac{\partial \varphi_0}{\partial \bar{z}}\frac{\partial \bar{f}_0}{\partial \bar{\mathcal{P}}_z}\right.$$
$$\left. + \frac{1}{4}\left(\frac{\partial \varphi_3}{\partial \bar{z}} - ik_3\varphi_3\right)\frac{\partial \bar{f}_1^*}{\partial \bar{\mathcal{P}}_z}e^{i\Delta\bar{\theta}} + \frac{1}{4}\left(\frac{\partial \varphi_3^*}{\partial \bar{z}} + ik_3\varphi_3^*\right)\frac{\partial \bar{f}_1}{\partial \bar{\mathcal{P}}_z}e^{-i\Delta\bar{\theta}}\right] = 0;$$

$$\frac{\partial \bar{f}_1}{\partial t} + i\mu\bar{f}_1 + \frac{\bar{\mathcal{P}}_z}{m}\left[\frac{1}{\gamma_0}\left(\frac{\partial \bar{f}_1}{\partial \bar{z}} - \frac{i\Phi}{c}\bar{f}_1\right) + \frac{1}{\gamma_1}\frac{\partial \bar{f}_0}{\partial \bar{z}}\right] - \frac{e^2\tilde{A}_{12}}{2mc^2\gamma_0}\frac{\partial \bar{f}_0}{\partial \bar{\mathcal{P}}_z}$$
$$+ e\frac{\partial \varphi_0}{\partial \bar{z}}\frac{\partial \bar{f}_1}{\partial \bar{\mathcal{P}}_z} + e\left(\frac{\partial \varphi_3}{\partial \bar{z}} - ik_3\varphi_3\right)\frac{\partial \bar{f}_0}{\partial \bar{\mathcal{P}}_z}e^{i\Delta\bar{\theta}}$$
$$+ \frac{e}{2}\left(\frac{\partial \varphi_3^*}{\partial \bar{z}} + ik_3\varphi_3^*\right)\frac{\partial \bar{f}_1}{\partial \bar{\mathcal{P}}_z}e^{-i\Delta\bar{\theta}} = 0;$$

$$\frac{\partial \bar{f}_2}{\partial t} + 2i\mu\bar{f}_2 + \frac{\bar{\mathcal{P}}_z}{m}\left[\frac{1}{\gamma_0}\left(\frac{\partial \bar{f}_2}{\partial \bar{z}} - \frac{2i\Phi}{c}\bar{f}_2\right) + \frac{1}{2\gamma_1}\left(\frac{\partial \bar{f}_1}{\partial \bar{z}} - \frac{i\Phi}{c}\right)\bar{f}_1\right]$$
$$- \frac{e^2\tilde{A}_{12}}{4mc^2\gamma_0}\frac{\partial \bar{f}_1}{\partial \bar{\mathcal{P}}_z} + \frac{e}{2}\left(\frac{\partial \phi_3}{\partial \bar{z}} - ik_3\phi_3\right)\frac{\partial \bar{f}_1}{\partial \bar{\mathcal{P}}_z}e^{i\Delta\bar{\vartheta}} = 0;$$

$$\frac{\partial \bar{f}_3}{\partial \bar{z}} + 3i\mu \bar{f}_3 + \frac{\bar{\mathcal{P}}_3}{m\gamma_0}\left(\frac{\partial \bar{f}_3}{\partial \bar{z}} - \frac{3i\Phi}{c}\bar{f}_3\right) + \frac{e}{2}\left(\frac{\partial \varphi_3}{\partial \bar{z}} - ik_3\phi_3\right)\frac{\partial \bar{f}_2}{\partial \bar{\mathcal{P}}_z}e^{i\Delta\bar{\theta}} = 0, \tag{12.24}$$

where $\gamma = \widetilde{H}/mc^2$, as before, is the relativistic factor; $1/\gamma_{0,1}$ are the zero and first Fourier amplitudes of the function $1/\gamma(\theta)$;

$$\frac{1}{\gamma_0} = \frac{1}{\gamma_0^{(0)}} + \frac{1}{\gamma_0^2}; \frac{1}{\gamma_0^{(0)}} = \frac{1}{\sqrt{1+\left(\bar{\mathcal{P}}_z/mc\right)^2}};$$

$$\frac{1}{\gamma_0^{(2)}} = -\frac{\left(\mathcal{P}_{0\perp}/mc\right)^2}{2\left[1+\left(\bar{\mathcal{P}}_z/mc\right)^2\right]^{3/2}} - \frac{e^2\left(\left|\vec{A}_1\right|_2 + \left|\vec{A}_2\right|^2\right)}{4m^2c^4\sqrt{1+\left(\bar{\mathcal{P}}_z/mc\right)^2}};$$

$$\frac{1}{\gamma_1} = -\frac{e^2 B_\theta}{2m^2c^4\left[1+\left(\bar{\mathcal{P}}_z/mc\right)^2\right]^{3/2}}; \quad \mu = \omega_1 + \sigma\omega_2;$$

$$B_\theta = \vec{A}_1\vec{A}_2\delta_{\sigma,+1} + \vec{A}_1\vec{A}_2^*\delta_{\sigma.-1}; \quad \Phi = c\left(s_1k_1 + \sigma s_2k_2\right);$$

$$\tilde{A}_{02} = \frac{1}{2}\frac{\partial}{\partial \bar{z}}\left(\left|\vec{A}_1\right|^2 + \left|\vec{A}_2\right|^2\right); \quad \tilde{A}_{12} = \frac{\partial B_\theta}{\partial \bar{z}} - i\frac{\Phi}{c}B_\theta;$$

$$\Delta\theta = \mathrm{Re}\{p_3\} - \bar{\theta} = \left(\omega_3 - \mu\right)t - \left(k_3 - \Phi/c\right)\bar{z}.$$

The structure of Equation 12.24 allows solving it either numerically or by approximate analytical methods. To illustrate this assertion, we solve Equation 12.24 by successive approximations.

We write solutions for amplitudes $\bar{f}_0$, $\bar{f}_1$, $\bar{f}_2$, and $\bar{f}_3$ in terms of divergent series in powers of the wave amplitude smallness, that is,

$$\bar{f}_0 = \bar{f}_0^{(0)} + \bar{f}_0^{(1)} + \bar{f}_0^{(2)} + \bar{f}_0^{(3)} + \ldots;$$

$$\bar{f}_1 = \bar{f}_1^{(0)} + \bar{f}_1^{(1)} + \bar{f}_1^{(2)} + \ldots;$$

$$\bar{f}_2 = \bar{f}_2^{(0)} + \bar{f}_2^{(1)} + \ldots; \bar{f}_3 = \bar{f}_3^{(0)} + \ldots, \tag{12.25}$$

where superscripts in parentheses indicate approximation numbers.

In the zero approximation, solutions of Equation 12.24 are evident, that is,

$$\bar{f}_0^{(0)} = \bar{f}_{00}; \quad \bar{f}_1^{(0)} = \bar{f}_2^{(0)} = \bar{f}_3^{(0)} = 0, \tag{12.26}$$

where $\bar{f}_{00} \leq$ is the initial averaged distribution function. For convenience, we assume $\bar{f}_{00} = \bar{f}_{00}\left(\vec{\bar{\mathcal{P}}}\right)$.

In the first approximation, assuming

$$\frac{\partial \bar{f}_k^{(i)}}{\partial \bar{z}}, \frac{\partial \bar{f}_k^{(i)}}{\partial t} \sim \left|E_{j,3}\right|^{i+1}, \tag{12.27}$$

we find corrections to be $\bar{f}_0^{(1)} = \bar{f}_2^{(1)} = \bar{f}_3^{(1)} = 0$;

$$\bar{f}_1^{(1)} = \frac{ek_3\varphi_3}{\Omega_3} e^{i\Delta\bar{\theta}} \frac{\partial \bar{f}_{00}}{\partial \bar{\mathcal{P}}_z}. \tag{12.28}$$

In the second approximation, we have

$$\bar{f}_3^{(2)} = 0; \tag{12.29}$$

$$\frac{\partial \bar{f}_0^{(2)}}{\partial t} + \bar{v}_z \frac{\partial \bar{f}_0^{(2)}}{\partial \bar{z}} = eE_0 \frac{\partial \bar{f}_{00}}{\partial \bar{\mathcal{P}}_z}; \tag{12.30}$$

$$\bar{f}_1^{(2)} = -\frac{1}{\Omega_3}\left\{\frac{e^2\Phi B_\theta}{2mc^3\gamma_0^{(0)}} - ie\frac{\partial \varphi_3}{\partial \bar{z}} e^{i\Delta\bar{\theta}} - \frac{ek_3}{\Omega_3}\left(\frac{\partial \varphi_3}{\partial t} - \bar{v}_z \frac{\partial \varphi_3}{\partial \bar{z}}\right) e^{i\Delta\bar{\theta}}\right\} \bar{f}_{00}; \tag{12.31}$$

$$\bar{f}_2^{(2)} = \frac{e^2k_3^2\varphi_3^2}{4\Omega_3^2} e^{i2\Delta\bar{\theta}} \frac{\partial^2 \bar{f}_{00}}{\partial \bar{\mathcal{P}}_z^2}. \tag{12.32}$$

In the third approximation, we have

$$\begin{aligned} \frac{\partial \bar{f}_0^{(3)}}{\partial t} + \bar{v}_z \frac{\partial \bar{f}_0^{(3)}}{\partial \bar{z}} = &-i\frac{e^3\Phi k_3}{8mc^3\Omega_3}\left(\frac{1}{\gamma_0^{(0)}}\frac{\partial^2 \bar{f}_{00}}{\partial \bar{\mathcal{P}}_z^2} - \frac{\bar{\mathcal{P}}_z}{m^2c^2\left(\gamma_0^{(0)}\right)^3}\frac{\partial \bar{f}_{00}}{\partial \bar{\mathcal{P}}_z}\right) \times \left(B_\theta\varphi_3^* e^{-i\Delta\bar{\theta}} - B_\theta^*\varphi_3 e^{i\Delta\bar{\theta}}\right) \\ &- \frac{e^2k_3}{4\Omega_3}\left(\frac{\partial \varphi_3}{\partial \bar{z}}\varphi_3^* + \frac{\partial \varphi_3^*}{\partial \bar{z}}\varphi_3\right)\frac{\partial^2 \bar{f}_{00}}{\partial \bar{\mathcal{P}}_z^2} - e\frac{\partial \varphi_0}{\partial \bar{z}}\frac{\partial \bar{f}_{00}}{\partial \bar{\mathcal{P}}_z} + \frac{e^2\tilde{A}_{02}}{2mc^2\gamma_0^{(0)}}\frac{\partial \bar{f}_{00}}{\partial \bar{\mathcal{P}}_z}; \end{aligned} \tag{12.33}$$

$$\begin{aligned} \bar{f}_1^{(3)} = \frac{1}{\Omega_3}\Bigg\{ &i\left(\frac{\partial \bar{f}_1^{(2)}}{\partial t} + \bar{v}_z \frac{\partial \bar{f}_1^{(2)}}{\partial \bar{z}}\right) - \frac{ie^2}{2mc^2\gamma_0^{(0)}}\frac{\partial B_\theta}{\partial \bar{z}}\frac{\partial \bar{f}_{00}}{\partial \bar{\mathcal{P}}_z} \\ &+ ek_3\varphi_3 e^{i\Delta\bar{\theta}}\frac{\partial \bar{f}_0^{(2)}}{\partial \bar{\mathcal{P}}_z} + ie\frac{\partial \varphi_0}{\partial \bar{z}}\frac{\partial \bar{f}_1^{(1)}}{\partial \bar{\mathcal{P}}_z} + \frac{\Phi\bar{\mathcal{P}}_z\bar{f}_1^{(1)}}{mc\gamma_0^{(0)}}\Bigg\}; \end{aligned} \tag{12.34}$$

$$\bar{f}_2^{(3)} = \frac{1}{2\Omega_3}\left\{ i\left(\frac{\partial \bar{f}_1^{(2)}}{\partial t} + \bar{v}_z \frac{\partial \bar{f}_1^{(2)}}{\partial \bar{z}}\right) - \frac{e^2 \Phi B_\theta \bar{f}_1^{(1)}}{4m^3c^5\left(\gamma_0^{(0)}\right)^3} - \frac{e^2\Phi B_\theta}{4mc^3\gamma_0^{(0)}}\frac{\partial \bar{f}_1^{(1)}}{\partial \bar{\mathcal{P}}_z}e^{i\Delta\bar{\theta}} + \frac{ek_3\phi_3}{2}\frac{\partial \bar{f}_1^{(2)}}{\partial \bar{\mathcal{P}}_z}e^{i\Delta\bar{\theta}} \right\} \tag{12.35}$$

$$\bar{f}_3^{(3)} = \frac{ek_3\varphi_3}{6\Omega_3}\frac{\partial \bar{f}_2^{(2)}}{\partial \bar{\mathcal{P}}_z}e^{i\Delta\bar{\theta}}, \tag{12.36}$$

where $\Omega_3 = \mu - \bar{v}_z(\Phi/c); \bar{v}_z = \bar{P}_z/m\gamma_0^{(0)}$, as before, is the averaged longitudinal electron velocity.

To describe self-consistent resonance interaction of electromagnetic waves in the beam plasma, we calculate both the average as well as oscillating parts of the distribution function. According to the hierarchic methods (see Chapter 6), the solution for the distribution function f can be represented in the form of Bogolyubov substitution

$$f = \bar{f}(\bar{x}) + \sum_{n=1}^{\infty}\frac{1}{\xi_1^n}u_f^{(n)}(\bar{x},\bar{\psi}), \tag{12.37}$$

where x is the total vector of slow variables. As the uncomplicated analysis shows, it is sufficient to retain terms up to ~$1/\xi_1$ for obtaining the cubic-order terms. Therefore, we account for nonlinear terms up to the third order in wave amplitudes and write the solution for the nonaveraged distribution function as a multiple Taylor series in z, $\mathcal{P}_z$ in the vicinity of $z = \bar{z}, \mathcal{P}_z = \bar{\mathcal{P}}_z$ (see the calculative scheme of back transformation in Chapter 6). Thus,

$$f\left(t,z,\mathcal{P}_z\right) = \bar{f}\left(t,\bar{z},\bar{\mathcal{P}}_z\right)\Big|_{\substack{\bar{z}=z\\ \bar{\mathcal{P}}_z=\mathcal{P}_z}} + \frac{\partial \bar{f}\left(t,\bar{z},\bar{\mathcal{P}}_z\right)}{\partial \bar{z}}\Big|_{\substack{\bar{z}=z\\ \bar{\mathcal{P}}_z=\mathcal{P}_z}}(\bar{z}-z)$$
$$+ \frac{\partial \bar{f}\left(t,\bar{z},\bar{\mathcal{P}}_z\right)}{\partial \bar{\mathcal{P}}_z}\Big|_{\substack{\bar{z}=z\\ \bar{\mathcal{P}}_z=\mathcal{P}_z}}\left(\bar{\mathcal{P}}_z - \mathcal{P}_z\right) + \ldots. \tag{12.38}$$

Within the context of the above scheme (see Chapter 6), we find differences $(\bar{z}-z)$ and $\left(\bar{\mathcal{P}}_z - \mathcal{P}_z\right)$ to be

$$(\bar{z}-z) = \tilde{z}\left(t,z,\mathcal{P}_z\right) \equiv \frac{1}{\Omega_3 m^2c^2\left(\gamma_0^{(0)}\right)^4}\left\{\frac{e^2}{4}\left[is_1k_1\vec{A}_1^2e^{2ip_1} + is_2k_2\vec{A}_2^2e^{2ip_2}\right.\right.$$
$$\left.\left. + B_\psi\left(s_1k_1 - \sigma s_2k_2\right)e^{i(p_1-\sigma p_2)} + c.c.\right]\right\}$$
$$+ \frac{e^2\mathcal{P}_z}{\Omega_3 m^3c^4\left(\gamma_0^{(0)}\right)^3}\left[2iB_\psi e^{i(p_1-\sigma p_2)} + \frac{i}{8}\vec{A}_1^2e^{2ip_1} - \frac{i\sigma}{8}\vec{A}_2^2e^{2ip_2} + c.c.\right]; \tag{12.39}$$

$$(\bar{\mathcal{P}}_z - \mathcal{P}_z) = \tilde{\mathcal{P}}_z(t, z, \mathcal{P}_z)$$
$$\cong -\frac{2e^2}{8\Omega_3 mc^2\gamma_0^{(0)}}\Big[s_1k_1\vec{A}_1^2 e^{2ip_1} - \sigma s_2k_2\vec{A}_2^2 e^{2ip_2}$$
$$+B_\psi\left(s_1k_1 - \sigma s_2k_2\right)e^{i(p_1-\sigma p_2)} + c.c.\Big]. \tag{12.40}$$

Here, we neglect initial transverse emission [i.e., assume motion integral (Equation 12.1), $\vec{P}_\perp = 0$] and write the initial distribution function in the form

$$f_{00}(\vec{\mathcal{P}}) = \delta(\mathcal{P}_x)\delta(\mathcal{P}_y)g_{00}(\mathcal{P}_z), \tag{12.41}$$

where $g_{00}(\mathcal{P}_z)$ is a (normalized to 1) function given at the input and $\delta(\mathcal{P}_{x,y})$ are Dirac delta-functions.

12.1.3 Field Problem

The next step concerns Maxwell equations 10.10. Similar to the previous case, we transform equations with an assumption that the model is one-dimensional, that is, all relevant derivatives with respect to x and y vanish,

$$\frac{\partial}{\partial x}, \frac{\partial}{\partial y} \to 0. \tag{12.42}$$

We carry out the calculation procedure of the simplified slowly varying amplitude method (analogous to the previous section). As a result, the following expressions for dispersion functions of relevant linear waves can be found (see Chapter 9):

$$D_j = (-i)\left(\hat{\mu}_j^{-1}k_j^2 - \hat{\varepsilon}_j\frac{\omega_j^2}{c^2} + \frac{\omega_p^2}{c^2}\int_{-\infty}^{+\infty}\frac{\bar{f}_{00}}{\gamma_0^{(0)}}\mathrm{d}\mathcal{P}_z\right) = 0;$$

$$D_3 = (-1)\left(k_3 + \omega_p^2 m\int_{-\infty}^{+\infty}\frac{1}{\omega_3 - k_3v_z}\frac{\partial\bar{f}_{00}}{\partial\mathcal{P}_z}\mathrm{d}\mathcal{P}_z\right) = 0, \tag{12.43}$$

where $\hat{\mu}_j$ and $\hat{\varepsilon}_j$ ($j = 1.2$) are the spectral magnetic permeability and dielectric permittivity of simulating magnetodielectric for frequencies ω_j (see Chapter 11), respectively. Solutions 12.43) introduce corrections to generating (linear) equations derived earlier (i.e., similar to Equation 11.9).

We continue all the necessary calculation procedures of the simplified method of slowly varying amplitudes. The calculation technique does not imply either noticeable methodological details or difficulties not discussed in the previous example (see Chapter 11). The routine results in a system of cubic nonlinear provide truncated equations like the quadratic one (Equation 11.12). Here, the nonlinear current and space charge densities on the right-hand side are determined by amplitudes $\bar{f}_k$ written in terms of nonaveraged coordinates z and P_z (according to the procedure above). In this case, the result is rather involved. To simplify it, we assume

1. The electromagnetic signal and pumping wave polarizations are linear and oriented in the same plane.
2. The initial thermal spread of beam electrons is disregarded [$g_{00}(P_z) = \delta(P_{z0} - P_z)$; we accept the model of the cold beam].
3. Analogous to the previous example, the model is quasistationary ($\partial/\partial t \to 0$, $\partial/\partial z \to \mathrm{d}/\mathrm{d}t$).

Then, the required system of cubic equations reduces to the simplified form [1–7]

$$\left(\frac{\mathrm{d}^2}{\mathrm{d}z^2} + L_0\frac{\mathrm{d}}{\mathrm{d}z}\right)E_1 = \sigma L_1\left[E_2^*E_3\delta_{\sigma,+1} - E_2E_3\delta_{\sigma,-1}\right] + L_2E_1\left|E_3\right|^2 + L_3E_1\left|E_1\right|^2 + L_4E_1\left|E_2\right|^2 + \sigma L_5\frac{\mathrm{d}}{\mathrm{d}z}\left[E_2^*E_3\delta_{\sigma,+1} - E_2E_3\delta_{\sigma,-1}\right] + \sigma L_6\frac{\mathrm{d}E_3}{\mathrm{d}z}\left[E_2^*E_3\delta_{\sigma,+1} - E_2E_3\delta_{\sigma,-1}\right] + L_7E_1\int_0^z\left(E_0^{(g)} + E_0^{(n)}\right)dz; \quad (12.44)$$

$$\left(\frac{\mathrm{d}^2}{\mathrm{d}z^2} + \left[F_0 - a_2\left(\tilde{k}_2,\tilde{N}_2\right)\right]\frac{\mathrm{d}}{\mathrm{d}z}\right)E_2 = F_1b_2\left(\tilde{k}_2,\tilde{N}_2\right)\left[E_1^*E_3\delta_{\sigma,+1} + E_1E_3^*\delta_{\sigma,-1}\right] + \sigma F_2E_2\left|E_3\right|^2 + \sigma F_3E_2\left|E_2\right|^2 + \sigma F_4E_2\left|E_1\right|^2 + F_5\frac{\mathrm{d}}{\mathrm{d}z}\left[E_1^*E_3\delta_{\sigma,+1} + E_1E_3^*\delta_{\sigma,-1}\right] + F_6\left[E_1^*\frac{\mathrm{d}E_3}{\mathrm{d}z}\delta_{\sigma,+1} + E_1\frac{\mathrm{d}E_3^*}{\mathrm{d}z}\delta_{\sigma,-1}\right] + \sigma\left[F_7E_2\int_0^z\left(E_0^{(g)} + E_0^{(n)}\right)\mathrm{d}z + c_2\left(\tilde{k}_2,\tilde{N}_2\right)\right]; \quad (12.45)$$

$$\left(\frac{\mathrm{d}^2}{\mathrm{d}z^2} + G_0\frac{\mathrm{d}}{\mathrm{d}z}\right)E_3 = \sigma G_1E_1\left[E_2\delta_{\sigma,+1} - E_2^*\delta_{\sigma,-1}\right] + G_2E_3\left|E_3\right|^2 + G_3E_3\sum_{j=1}^{2}\frac{1}{\omega_j^2}\left|E_j\right|^2 + \sigma G_4\frac{\mathrm{d}}{\mathrm{d}z}\left[E_1E_2\delta_{\sigma,+1} - E_1E_2^*\delta_{\sigma,-1}\right] + G_5E_3\int_0^z\left(E_0^{(g)} + E_0^{(n)}\right)\mathrm{d}z; \quad (12.46)$$

$$E_0^{(g)} = \hat{D}_{01}\frac{\mathrm{d}}{\mathrm{d}z}\left|E_3\right|^2 + \hat{D}_{02}\frac{\mathrm{d}}{\mathrm{d}z}\sum_{j=1}^{2}\frac{1}{\omega_j^2}\left|E_j\right|^2. \quad (12.47)$$

Unfortunately, relevant expressions for nonlinear coefficients $L_k, F_l, G_m, \hat{D}_{0n}$ are unacceptably cumbrous. Therefore, we omit them.

The system equations 12.44 through 12.47 take into account both above-mentioned isochronization methods (see relevant examples in Chapter 7). The first one applies the electric field of support $E_0^{(n)}$ to the interaction region (see Equation 12.9). The second one is the varying spatial period of the pumping wave. The latter in framework of considered Dopplertron model is described by the slowly varying correction $\tilde{N}_2(z)$ to the pumping wave retardation factor modeled by the simulated magnetodielectric

$$N_2(z) = \sqrt{\hat{\varepsilon}_2(z)\hat{\mu}_2(z)} = N_2 + \tilde{N}_2(z); \left|\tilde{N}_2(z)\right| << N_2 , \tag{12.48}$$

where $\hat{\varepsilon}_2(z) = \hat{\varepsilon}_2 + \tilde{\hat{\varepsilon}}_2(z)$ and $\hat{\mu}_2(z) = \hat{\mu}_2 + \tilde{\hat{\mu}}_2(z)$ are the slowly varying dielectric permittivity and magnetic permeability of SM at the pumping frequency ω_2. We assume $\left|\tilde{N}_2(z)\right|/N_2 \sim \varepsilon_j << 1$. The notation for $\tilde{k}_2$ is obvious: $\tilde{k}_2 = \omega_2\tilde{N}/c$. We introduce the dependence 12.48 into Equations 12.44 through 12.47 in terms of functions

$$a_2\left(\tilde{k}_2, \tilde{N}_2\right) = \frac{2}{3} i k_2 \left(R_2^2 \frac{\tilde{N}_2}{N_2} - 3\frac{\tilde{k}_2}{k_2} \right); \quad b_2\left(\tilde{k}_2\right) = 1 + \frac{\tilde{k}_2}{k_2};$$

$$c_2\left(\tilde{k}_2, \tilde{N}_2\right) = \frac{2}{3} k_2^2 \left[\frac{\tilde{k}_2}{k_2} + \frac{3}{2}\frac{\tilde{k}_2^2}{k_2^2} - R_2^2 \left(\frac{\tilde{N}_2}{N_2} + \frac{\tilde{N}_2}{N_2}\frac{\tilde{k}_2}{k_2} + \frac{\tilde{N}_2^2}{N_2^2} \right) \right.$$

$$\left. + z\frac{\mathrm{d}}{\mathrm{d}z}\frac{\tilde{k}_2}{k_2} \right] + \frac{2}{3} i s_2 k_2^2 \left[\frac{3}{k_2}\frac{\mathrm{d}}{\mathrm{d}z}\left(\frac{\tilde{k}_2}{k_2}\right) - \frac{R_2^2}{k_2^2}\frac{\mathrm{d}}{\mathrm{d}z}\left(\frac{\tilde{N}_2}{N_2}\right) \right].$$

12.1.4 Raman and Compton Interaction Modes

The analysis of application range shows that Equations 12.44 through 12.47 are valid for the description of both Raman and Compton interaction modes (see Section 11.1.6). In the framework of cubic-nonlinear theory, the corresponding criteria (similar to Equation 11.25) can be formulated in the form

$$\left| G_0 \frac{\mathrm{d}E_3}{\mathrm{d}z} \right| \Big/ \left| \frac{\mathrm{d}^2 E_3}{\mathrm{d}z^2} \right| \geq \frac{1}{\varepsilon_j} \tag{12.49}$$

associated with Raman mode. The inverse inequality corresponds to Compton mode,

$$\left| G_0 \frac{\mathrm{d}E_3}{\mathrm{d}z} \right| \Big/ \left| \frac{\mathrm{d}^2 E_3}{\mathrm{d}z^2} \right| \leq \frac{1}{\varepsilon_j} \tag{12.50}$$

and for

$$\left| G_0 \frac{\mathrm{d}E_3}{\mathrm{d}z} \right| \Big/ \left| \frac{\mathrm{d}^2 E_3}{\mathrm{d}z^2} \right| \sim 1 \tag{12.51}$$

we have the combined Compton–Raman mode.

12.2 Truncated Equations: The H-Ubitron Model

The method of the averaged kinetic equation in the FEL theory can have a few calculation versions. One of them was discussed in the previous section. Its peculiarity is the use of the two-level hierarchic transformation scheme. Initial standard equations 12.2 through 12.5 (zero hierarchic level) transform here into relevant equations of the first hierarchic level. Averaged kinetic equation 12.17 of the first hierarchic level is constructed based on this equation set. Then we found approximation solutions 12.25 through 12.36 of averaged equation 12.17 and by inverse transformation has obtained nonaveraged approximate solutions 12.38 through 12.40. We classify this calculation scheme as the two-level hierarchic scheme.

The other version of the discussed hierarchic calculation scheme consists of the expanding hierarchic calculation scheme at the next (second) hierarchic level, that is, it can be classified as the three-level hierarchic calculation scheme. For its realization, we take the averaged kinetic equation 12.17 as an initial one and again perform all above-mentioned transformation procedures. As a result, a two-averaged kinetic equation of the second hierarchy can be obtained this way. After solving it and performing analogous (with the preceding case) two-fold inverse transformations, we can obtain approximate nonaveraged solutions of zero level such as Equations 12.38 through 12.40.

We illustrate the described three-level hierarchic scheme on the example of a one-dimension model of FEL-H-Ubitron. Contrary to the previous case, for solution of wave part of the problem, we use the rigorous version of the modified slowly varying amplitude method described in the preceding author's monograph [1].

12.2.1 Formulation of the Problem

Formulation of the problem of parametric-wave resonance in the H-Ubitron model is similar in many aspects to the preceding case. However, there exist some distinctions, and we would like to discuss briefly some of them.

Let us remember that the pumping in classical H-Ubitron is produced by an external electrodynamic system, that is, unlike the Dopplertron pumping field, it is not proper for the system. We show in what follows that this fact introduces specific features as well into the physics of processes considered and in the calculative part of the problem.

Let us assume that the magnetic pumping field is linearly polarized (in plane YZ) and harmonic

$$\hat{\vec{H}}_{20} = (1/2)\left(H_{20}e^{ip_2} + c.c.\right)\vec{e}_2, \tag{12.52}$$

where the phase p_2 is given by the expression

$$p_2 = -k_2 z, \tag{12.53}$$

$k_2 = 2\pi/\Lambda$, Λ is the magnetic field undulative period. We also bear in mind that nonlinear processes in the beam plasma give rise to an additional magnetostatic field [1–3,6,7,9–11]

$$\hat{\vec{H}}_{2g} = (1/2)\left(H_{2g}e^{ip_2} + c.c.\right)\vec{e}_y. \tag{12.54}$$

Thus, the resulting pumping field, influencing beam electrons, can be written as a sum of the external (given) and generated (self-consistent) fields

$$\hat{\vec{H}}_2 = \hat{\vec{H}}_{20} + \hat{\vec{H}}_{2g}. \tag{12.55}$$

We take the signal wave field linearly polarized in the YZ plane,

$$\hat{\vec{E}}_1 = (1/2)\left(E_1 e^{ip_1} + c.c.\right)\vec{e}_x, \tag{12.56}$$

with the phase

$$p_1 = \omega_1 t - s_1 k_1 z. \tag{12.57}$$

Analogous to the previous case way, we divide the general three-wave parametric resonance problem into two more particular ones:

1. The problem of electron beam motion in given pumping and signal fields (i.e., the problem of integration of initial kinetic equation 10.27)
2. The problem of electromagnetic field excitation for a given beam motion (i.e., the problem of integration of Maxwell equations 10.10 with given currents)

The general algorithm for solving the first problem has been described in Chapter 6. Besides that, except for configuration of the pumping field, all the results obtained by asymptotic integration of Equation 10.27 are also valid in this case (see Equations 12.21 through 12.51). However, as is discussed above, the problems of this type can be treated in terms of another version of calculation procedure, which we refer to as the three-hierarchic calculation scheme. As is mentioned above, the idea is to treat once-averaged kinetic equation 12.17 once more in terms of the same hierarchic averaged kinetic equation method. To do this, we introduce new combination phases of the second hierarchic level

$$\Theta = \bar{p}_3' - \bar{\theta} \equiv \Delta\bar{\theta}; \tag{12.58}$$

$$\Psi = \bar{p}_3' + \bar{\theta}. \tag{12.59}$$

In what follows, we put the new combination Θ phase as a slow one and the phase Ψ as the fast phase. Here, we used the notation $\bar{p}_3' = \mathrm{Re}\left\{\bar{p}_3\right\}$.

Within general definitions (see Chapter 6), a new large parameter for the second hierarchic level ξ_2

$$\xi_2 \sim \left|\frac{d\Psi}{dt}\right| \Big/ \left|\frac{d\Theta}{dt}\right| >> 1. \tag{12.60}$$

Thus, relevant *hierarchy series* (Equation 6.6; Chapter 6) in the considered case consists only of two terms, that is,

$$\xi_1 >> \xi_2 >> 1, \tag{12.61}$$

where $\xi_1 \equiv \xi$ is the large parameter of the first hierarchy level (see Equation 12.15). Now we discuss a rather peculiar methodological detail of the discussed double-averaging calculative technique. We have mentioned in Chapter 6 that asymptotic solutions obtained are valid for the system length $L_r \sim \xi_r^{n_r}$ (where r is the number of hierarchic level, and n_r is the approximation number for the rth hierarchic level). Therefore, the solution of Equation 12.17 is valid for the length

$$L_1 \sim \xi_1. \tag{12.62}$$

Thus, the averaging on the second hierarchy level requires $L_2 \sim \xi_2^{n_2}$. However, to adjust accuracies of the first- and second-level solutions, we require $L_2 \sim L_1$. This yields $n_1 < n_2$ (in view of Equation 12.61). Hence, in spite of that, the first approximation $n_1 = 1$ is sufficient for averaging on the first hierarchy level; the calculation for the second hierarchy level should be done at least in the second approximation ($n_2 \geq 2$). In what follows, we take $n_2 = 3$.

12.2.2 Motion Problem

At the next stage of calculation, we obtain the so-called double-averaged kinetic equation [1,2,12]. For this, we perform the substitutions following from Equations 12.58 and 12.59, that is,

$$\bar{p}_3' = (1/2)(\Theta + \Psi); \quad \bar{\theta} = (1/2)(\Theta - \Psi). \tag{12.63}$$

Then, we carry out standard (for the discussed method) calculations and obtain the double-averaged kinetic equation

$$\left\{ \frac{\partial}{\partial t} + \frac{\partial \bar{\bar{\mathcal{H}}}}{\partial \bar{\bar{\mathcal{P}}}_z} \frac{\partial}{\partial \bar{\bar{z}}} - \frac{\partial \bar{\bar{\mathcal{H}}}}{\partial \bar{\bar{z}}} \frac{\partial}{\partial \bar{\bar{\mathcal{P}}}_z} \right\} \bar{\bar{f}}\left(t, \bar{\bar{z}}, \bar{\bar{\mathcal{P}}}_z \right) = 0, \tag{12.64}$$

with the double-averaged electron Hamiltonian described by

$$\bar{\bar{\mathcal{H}}} = mc^2 \bar{\bar{\gamma}}_0^{(0)} + e \int_0^{\bar{\bar{z}}} \left(E_0^{(n)} + E_0^{(g)} \right) d\bar{\bar{z}} - \frac{1}{2} e \bar{\bar{l}}_2 \left(\left| \vec{A}_1 \right|^2 + \left| \vec{A}_2 \right|^2 \right)$$

$$+ e^2 \left(\frac{1}{\bar{\bar{\Omega}}_\Psi^2} \frac{\partial \bar{\bar{v}}_z}{\partial \bar{\bar{\mathcal{P}}}_z} + \frac{2 \bar{\bar{v}}_z}{\bar{\bar{\Omega}}_\Psi^3} \frac{\partial \bar{\bar{\Omega}}_\Psi}{\partial \bar{\bar{\mathcal{P}}}_z} \right) \left| E_3 \right|^2. \tag{12.65}$$

where $\bar{\bar{l}}_2 = -\left(e/mc^2 \bar{\bar{\gamma}}_0^{(0)} \right)$; $\bar{\bar{\Omega}}_\Psi = \omega_3 + \omega_1 - \left(k_3 + s_1 k_1 - s_2 \sigma k_2 \right) \bar{\bar{v}}_z$; $\bar{\bar{v}}_z = \bar{\bar{\mathcal{P}}}_z / m \bar{\bar{\gamma}}_0^{(0)}$ is the double-averaged z-component of electron velocity; $\bar{\bar{\gamma}}_0^{(0)} = \sqrt{1 + \left(\bar{\bar{\mathcal{P}}}_z / mc \right)^2}$ $\vec{A}_1$ is the signal wave vector potential amplitude; and $\vec{A}_2$ is the amplitude of the pumping field vector-potential. Comparing Equations 10.27, 12.17, and 12.64, we can see that the principle of hierarchic resemblance (self-modeling principle—see Chapter 2) holds. Indeed, the form of initial equation 10.27 is similar at each hierarchic level. In our case, it is the Boltzmann kinetic equation on each hierarchic level. Therein, the set of proper dynamic variables for each hierarchic level (nonaveraged, once-averaged, double-averaged ones, and so on) is changed only.

In view of Equation 12.65, Equation 12.64 does not contain any periodic coefficients before derivatives unlike Equations 10.27, 11.90, and 12.17. Moreover, it is much simpler (the principle of information compression—see Chapter 2) and can be solved by successive approximations. Solving we have

$$\bar{\bar{f}}\left(t,\bar{\bar{z}},\bar{\bar{\mathcal{P}}}_z\right)=\bar{\bar{f}}_{00}\left(\bar{\bar{\mathcal{P}}}_z\right)+\frac{e}{\bar{\bar{v}}_z}\frac{\partial\bar{\bar{f}}_{00}}{\partial\bar{\bar{P}}_z}\int_0^{\bar{\bar{z}}}\left(E_0^{(n)}+E_0^{(g)}\right)\mathrm{d}\bar{\bar{z}}-\frac{e\bar{\bar{l}}_2}{2\bar{\bar{v}}_z}\frac{\partial\bar{\bar{f}}_{00}}{\partial\bar{\bar{\mathcal{P}}}_z}\left(\left|\vec{A}_1\right|^2-\left|\vec{A}_{01}\right|^2+\left|\vec{A}_2\right|^2-\left|\vec{A}_{02}\right|^2\right)$$
$$+\frac{e^2}{\bar{\bar{v}}_z}\left(\frac{1}{\bar{\bar{\Omega}}_\Psi^2}\frac{\partial\bar{\bar{v}}_z}{\partial\bar{\bar{\mathcal{P}}}_z}+\frac{2\bar{\bar{v}}_z}{\bar{\bar{\Omega}}_\Psi^3}\frac{\partial\bar{\bar{\Omega}}_\Psi}{\partial\bar{\bar{\mathcal{P}}}_z}\right)\left(\left|E_3\right|^2-\left|E_{30}\right|^2\right). \tag{12.66}$$

Now we pass from the double-averaged distribution function to once-averaged one, that is,

$$\bar{f}\left(t,\bar{z},\bar{\mathcal{P}}_z\right)=\bar{\bar{f}}\left(t,\bar{\bar{z}},\bar{\bar{\mathcal{P}}}_z\right)\Big|_{\substack{\bar{\bar{z}}=\bar{z}\\ \bar{\bar{\mathcal{P}}}_z=\bar{\mathcal{P}}_z}}+\frac{\partial\bar{\bar{f}}\left(t,\bar{\bar{z}},\bar{\bar{\mathcal{P}}}_z\right)}{\partial\bar{\bar{z}}}\Big|_{\substack{\bar{\bar{z}}=\bar{z}\\ \bar{\bar{P}}_z=\bar{P}_z}}\left(\bar{\bar{z}}-\bar{z}\right)$$
$$+\frac{\partial\bar{\bar{f}}\left(t,\bar{\bar{z}},\bar{\bar{\mathcal{P}}}_z\right)}{\partial\bar{\bar{\mathcal{P}}}_z}\Big|_{\substack{\bar{\bar{z}}=\bar{z}\\ \bar{\bar{\mathcal{P}}}_z=\bar{\mathcal{P}}_z}}\left(\bar{\bar{\mathcal{P}}}_z-\bar{\mathcal{P}}_z\right)+\ldots, \tag{12.67}$$

where

$$\left(\bar{\bar{z}}-\bar{z}\right)=\left(\frac{\bar{\mathcal{P}}_z g_2}{m\bar{\gamma}_0^{(0)}\bar{\Omega}_\Psi}iB_\theta e^{i\bar{\theta}}+c.c.\right)-\left(\frac{2e}{\bar{\Omega}_\Psi^2}\frac{\partial\bar{v}_z}{\partial\bar{\mathcal{P}}_z}E_3e^{i\bar{p}_3}+c.c.\right)$$
$$-\left[\frac{2}{\bar{\Omega}_\Psi}\left(\frac{\bar{v}_z\bar{\mathcal{P}}_z g_2}{m\bar{\gamma}_0^{(0)}\bar{\Omega}_\Psi}-\frac{e\bar{l}_2}{\bar{\Omega}_\Psi}\frac{\partial\bar{v}_z}{\partial\bar{\mathcal{P}}_z}\right)\frac{\partial B_\theta}{\partial\bar{z}}e^{i\bar{\theta}}+c.c.\right]$$
$$-\left(\frac{2e\bar{l}_2k_\theta}{\bar{\Omega}_\Psi^2}\frac{\partial\bar{v}_z}{\partial\bar{\mathcal{P}}_z}iB_\theta e^{i\bar{\theta}}+c.c.\right)-\left[\frac{e}{\bar{\Omega}_\Psi^2}\frac{\partial}{\partial\bar{\mathcal{P}}_z}\left(\frac{\bar{\mathcal{P}}_z g_2}{m\bar{\gamma}_0^{(0)}}\right)E_3A_0e^{i\bar{p}_3}+c.c,\right]$$
$$-\left\{\frac{1}{\bar{\Omega}_\Psi}\left[\frac{e}{\bar{\Omega}_\Psi}\frac{\partial}{\partial\bar{\mathcal{P}}_z}\left(\frac{\bar{\mathcal{P}}_z g_2}{2m\bar{\gamma}_0^{(0)}}\right)+\frac{\bar{\mathcal{P}}_z g_2 e}{2m\bar{\gamma}_0^{(0)}\bar{\Omega}_\Psi^2}\frac{\partial\bar{\Omega}_\Psi}{\partial\bar{\mathcal{P}}_z}\right]E_3B_\theta e^{i\Psi}+c.c.\right\}$$
$$+\left\{\left[\frac{ek_\theta}{2\bar{\Omega}_\Psi}\frac{\partial\bar{l}_2}{\partial\bar{\mathcal{P}}_z}+\frac{e}{4}\left(\frac{k_\Psi\bar{\mathcal{P}}_z g_2}{m\bar{\gamma}_0^{(0)}\bar{\Omega}_\Psi}+\frac{2e\bar{l}_2k_\theta}{\bar{\Omega}_\Psi^2}\frac{\partial\bar{\Omega}_\Psi}{\partial\bar{\mathcal{P}}_z}\right)+\frac{e^2\bar{l}_2k_\theta}{2\bar{\Omega}_\Psi^2}\frac{\partial\bar{\Omega}_\Psi}{\partial\bar{\mathcal{P}}_z}\right]\times\frac{1}{\bar{\Omega}_\Psi^2}E_3B_\theta e^{i\Psi}+c.c.\right\}$$
$$-\left(\frac{e^2}{\bar{\Omega}_\Psi^4}\frac{\partial\bar{\Omega}_\Psi}{\partial\bar{\mathcal{P}}_z}\frac{\partial\bar{v}_z}{\partial\bar{\mathcal{P}}_z}iE_3^2e^{i2\bar{p}_3}+c.c.\right)-\left(\frac{4e\bar{v}_z}{\bar{\Omega}_\Psi^3}\frac{\partial\bar{v}_z}{\partial\bar{\mathcal{P}}_z}i\frac{\partial E_3}{\partial\bar{z}}e^{i\bar{p}_3}+c.c.\right)$$
$$-\left(\frac{8\bar{l}_2\bar{v}_zk_\theta}{\bar{\Omega}_\Psi^3}\frac{\partial\bar{v}_z}{\partial\bar{\mathcal{P}}_z}\frac{\partial B_\theta}{\partial\bar{z}}e^{i\bar{\theta}}+c.c.\right)+\left(\frac{4e\bar{\Omega}_\Theta}{\bar{\Omega}_\Psi^3}\frac{\partial\bar{v}_z}{\partial\bar{\mathcal{P}}_z}E_3e^{i\bar{p}_3}+c.c.\right)$$
$$-\left(\frac{2k_\Psi\bar{\mathcal{P}}_z g_2 e}{m\bar{\gamma}_0^{(0)}\bar{\Omega}_\Psi^3}\frac{\partial\bar{v}_z}{\partial\bar{\mathcal{P}}_z}E_3A_0e^{i\bar{p}_3}+c.c.\right);$$

$$\left(\bar{\bar{\mathcal{P}}}_z-\bar{\mathcal{P}}_z\right)=-\frac{1}{\bar{\Omega}_\Psi}\left[eiE_3e^{i\bar{p}_3}-e\bar{l}_2i\left(\frac{\partial B_\theta}{\partial \bar{z}}-ik_\theta B_\theta\right)e^{i\bar{\theta}}+c.c.\right]$$

$$-\left\{\left[\frac{e^2k_\theta}{2\bar{\Omega}_\Psi}\frac{\partial \bar{l}_2}{\partial \bar{\mathcal{P}}_z}+\frac{e}{4}\left(\frac{k_\Psi\bar{\mathcal{P}}_z g_2}{m\bar{\gamma}_0^{(0)}\bar{\Omega}_\Psi}+\frac{2e\bar{l}_2k_\theta}{\bar{\Omega}_\Psi^2}\frac{\partial\bar{\Omega}_\Psi}{\partial\bar{\mathcal{P}}_z}\right)+\frac{e^2\bar{l}_2k_\theta}{2\bar{\Omega}_\Psi^2}\frac{\partial\bar{\Omega}_\Psi}{\partial\bar{\mathcal{P}}_z}\right]\times\frac{1}{i\bar{\Omega}_\Psi}E_3B_\theta e^{i\Psi}+c.c.\right\}$$

$$+\left(\frac{e^2}{2\bar{\Omega}_\Psi}\frac{\partial\bar{\Omega}_\Psi}{\partial\bar{\mathcal{P}}_z}E_3^2e^{i2\bar{p}_3}+c.c.\right)+\left(\frac{k_\Psi\bar{\mathcal{P}}_z g_2 e}{i2m\gamma_0^{(0)}\bar{\Omega}_\Psi^2}E_3A_0e^{i\bar{p}_3}+c.c.\right)$$

$$-\left(\frac{2e\bar{l}_2\bar{v}_zk_\theta}{i\bar{\Omega}_\Psi^2}\frac{\partial B_\theta}{\partial\bar{z}}e^{i\bar{\theta}}+c.c.\right)-\left(\frac{e\bar{\Omega}_\Theta}{i\bar{\Omega}_\Psi^2}E_3e^{i\bar{p}_3}+c.c.\right)+\left(\frac{k_\Psi\bar{\mathcal{P}}_z g_2 e}{i2m\gamma_0^{(0)}\bar{\Omega}_\Psi^2}E_3A_0e^{i\bar{p}_3}+c.c.\right)$$

$$+\left(\frac{4e\bar{v}_z^2}{\bar{\Omega}_\Psi^3}i\frac{\partial^2E_3}{\partial\bar{z}^2}e^{i\bar{p}_3}+c.c.\right)-\left\{\frac{e^3}{4\bar{\Omega}_\Psi^5}\left[\left(\frac{\partial\bar{\Omega}_\Psi}{\partial\bar{\mathcal{P}}_z}\right)^2+\frac{1}{2}\frac{\partial}{\partial\bar{\mathcal{P}}_z}\left(\bar{\Omega}_\Psi\frac{\partial\bar{\Omega}_\Psi}{\partial\bar{\mathcal{P}}_z}\right)\right]\frac{2}{i3}E_3^3e^{i3\bar{p}_3}+c.c.\right\}$$

$$-\left\{\left[\frac{3e^3}{4\bar{\Omega}_\Psi^4}\left(\frac{\partial\bar{\Omega}_\Psi}{\partial\bar{\mathcal{P}}_z}\right)^2-\frac{e^3}{8\bar{\Omega}_\Psi^4}\frac{\partial}{\partial\bar{\mathcal{P}}_z}\left(\bar{\Omega}_\Psi\frac{\partial\bar{\Omega}_Z}{\partial\bar{\mathcal{P}}_z}\right)+\frac{e^3}{2\bar{\Omega}_\Psi^3}\frac{\partial^2\bar{\Omega}_\Psi}{\partial\bar{\mathcal{P}}_z^2}\right]\times\frac{2}{i}E_3\left|E_3\right|^2e^{i\bar{p}_3}+c.c.\right\}$$

$$+i\left\{\left[-\frac{e^2}{\bar{\Omega}_\Psi^2}\frac{\partial\bar{v}_z}{\partial\bar{\mathcal{P}}_z}+\frac{2e^2\bar{v}_z}{\bar{\Omega}_\Psi^3}\frac{\partial\bar{\Omega}_\Psi}{\partial\bar{\mathcal{P}}_z}+\frac{e^2\bar{v}_z}{\bar{\Omega}_\Psi^3}\frac{\partial\bar{v}_z}{\partial\bar{\mathcal{P}}_z}\right]\times E_3\frac{\partial E_3}{\partial\bar{z}}e^{i2p_3}+c.c.\right\},$$

(12.68)

where $g_2=-\frac{e^2}{2m^2c^4\left(\bar{\gamma}_0^{(0)}\right)^2}$; $\bar{\Omega}_\Theta=\omega_3+\omega_1-\left(k_3+s_1k_1+s_2\sigma k_2\right)\bar{v}_z$; $\bar{\Omega}_\Theta=\omega_3+\omega_1-\left(k_3+s_1k_1+s_2\sigma k_2\right)\bar{v}_z$; $k_\theta=s_1k_1+s_2\sigma k_2$; $k_\psi=k_3-s_1k_1-\sigma s_2k_2$. Similar to the previous case, we put $\mathcal{P}_x=\mathcal{P}_y=0$.

We retain in Equations 12.67 and 12.68 only three lowest-orders harmonics of the distribution function and nonlinear terms of the orders not higher than cubic with respect to amplitudes. Having done this, we come to results (Equations 12.25 through 12.36) obtained in the previous section by the first calculation procedure. It should be mentioned that, as will be shown later (in Chapters 13 and 14), generalization of the discussed calculation scheme for the case of many (hundreds, for instance) harmonics is possible and can be performed without especial calculation difficulties.

Thus, we have two methodological versions of application of the hierarchic method of averaged characteristics. Each version has its own advantages and disadvantages. In particular, the first calculation procedure is effective and sufficient in terms of the second-order nonlinear theory. If, however, more harmonics are taken into account (e.g., in the third-order theory), the second procedure is preferable. The natural conclusion is that both procedures are useful and the decision on which of these should be employed is made after the problem is specified.

Experience shows that both calculation procedures can be automated using ordinary personal computers. It means that all labor-intensive calculation can be entrusted with a computer. It is obvious that this prospect is rather attractive. However, taking in view the educational aspects of the book, we have given here the analytical version of discussed methods.

12.2.3 Truncated Equations

We will then obtain the cubic truncated equations for complex amplitudes. Using distribution functions 12.38 and 12.67, we calculate current density and space charge density (Equations 10.11 and 10.12). This enables us to regard these as given functions in future analysis. Then, we return to initial Maxwell equations 10.10 and asymptotically integrate them in terms of the algorithm of the rigorous slowly varying amplitude method. After all necessary calculations, we obtain a system of third-order equations describing the dynamics of processes in the H-Ubitron interaction region [1–3,7]

$$\left\{\frac{\partial}{\partial t}-s_1\frac{(\partial D_1/\partial k_1)}{(\partial D_1/\partial\omega_1)}\frac{\partial}{\partial z}-\frac{i}{2}\frac{d}{dk_1}\left[\frac{(\partial D_1/\partial k_1)}{(\partial D_1/\partial\omega_1)}\right]\frac{\partial^2}{\partial z^2}+M_1\right\}E_1=J_1^{(2)}$$

$$+J_1^{(3)}+\mathrm{K}_1^{(3)}+\frac{i}{2(\partial D_1/\partial\omega_1)}\left\{\left[\frac{\partial^2 D_1}{\partial\omega_1^2}-\frac{2}{\omega_1}\frac{\partial D_1}{\partial\omega_1}\right]\frac{\partial J_1^{(2)}}{\partial t}\right.$$

$$\left.-s_1\left[2\frac{\partial^2 D_1}{\partial\omega_1\partial k_1}-\frac{(\partial D_1/\partial k_1)}{(\partial D_1/\partial\omega_1)}\frac{\partial^2 D_1}{\partial\omega_1^2}\right]\frac{\partial J_1^{(2)}}{\partial z}\right\}; \tag{12.69}$$

$$\left\{\frac{\partial}{\partial t}-\frac{(\partial D_3/\partial k_3)}{(\partial D_3/\partial\omega_3)}\frac{\partial}{\partial z}-\frac{i}{2}\frac{d}{dk_3}\left[\frac{(\partial D_3/\partial k_3)}{(\partial D_3/\partial\omega_3)}\right]\frac{\partial^2}{\partial z^2}\right\}E_3=J_3^{(2)}$$

$$+J_3^{(3)}+\mathrm{K}_3^{(3)}+\frac{i}{2(\partial D_3/\partial\omega_3)}\left\{\frac{\partial^2 D_3}{\partial\omega_3^2}\frac{\partial J_3^{(2)}}{\partial t}\right.$$

$$\left.-\left[2\frac{\partial^2 D_3}{\partial\omega_3\partial k_3}-\frac{(\partial D_3/\partial k_3)}{(\partial D_3/\partial\omega_3)}\frac{\partial^2 D_3}{\partial\omega_3^2}\right]\frac{\partial J_3^{(2)}}{\partial z}\right\}. \tag{12.70}$$

Explicit expressions for $M_1, J_1^{(2)}, J_1^{(3)}\mathcal{K}_1, J_3^{(2)}, J_3^{(3)}, \mathcal{K}_3$ are unacceptably cumbrous. Therefore, we omit them.

Dispersion functions $D_{1,3}$ are determined by Equation 12.43. Performing relevant calculations, we find that contrary to Dopplertron, the dispersion function of H-Ubitron pumping field does not vanish (see Equation 10.58 and corresponding discussion), that is,

$$D_2=D_2(k_2)=\left(k_2^2+\frac{\omega_p^2}{c^2}\int_{-\infty}^{+\infty}\frac{f_{00}(\mathcal{P}_z)}{\gamma_0^{(0)}(\mathcal{P}_z)}d^3\mathcal{P}\right)\neq 0. \tag{12.71}$$

This observation reflects the evident fact that the H-Ubitron pumping field produced by external magnets and currents is not proper for the system under consideration (concerning the concepts of proper and nonproper waves; see Chapter 4 for more details).

Equations 12.69 and 12.70 are universal. In particular, they describe the evolution of processes of interest both in time and space, both in the Raman and Compton modes (see criteria 12.49 through 12.51). A disadvantageous feature of this system is that it is rather intricate and inconvenient for analytical and numerical study. That is why we simplify the initial model. We assume that the model is stationary ($\partial/\partial t \to 0$), the electron beam is cold, and longitudinal electric field generation can be neglected. Then the system (Equation 12.69 and 12.70) is transformed into

$$\left(A_1\frac{d}{dz}+A_2\frac{d^2}{dz^2}\right)E_1 = A_5E_3H_2^* + A_{6a}E_1\left|E_1\right|^2 + A_{6b}E_1\left(\left|E_1\right|^2-\left|E_{10}\right|^2\right)$$
$$+A_7E_1\left|H_2\right|^2 + A_{8b}E_1\left(\left|E_3\right|^2-\left|E_{30}\right|^2\right)+A_9H_2^*\frac{dE_3}{dz}; \qquad (12.72)$$

$$\left(C_1\frac{d}{dz}+C_2\frac{d^2}{dz^2}\right)E_3 = C_5E_1H_2 + C_{6a}E_3\left|E_1\right|^2 + C_{6b}E_3\left(\left|E_1\right|^2-\left|E_{10}\right|^2 + C_7E_3\left|H_2\right|^2\right)$$
$$+C_{8a}E_3\left|E_3\right|^2 + C_{8b}E_3\left(\left|E_3\right|^2-\left|E_{30}\right|^2\right)+C_9H_2\frac{dE_1}{dz}. \qquad (12.73)$$

Here, $E_{10} = E_1(z = 0)$, $E_{30} = E_3(z = 0)$. Expressions for the nonlinear coefficients A_k and C_k are unacceptably cumbrous. Therefore, we omit them.

12.3 Effect of Nonlinear Generation of the Longitudinal Electric Field

We discussed in Chapter 11 the quadratic-nonlinear models of FELs. Therein, main attention was paid to the solution of relevant truncated equations by analytical methods. Here, we illustrate the numerical technology of analysis using the cubic-nonlinear FEL models as examples.

We take truncated equations 12.44 through 12.47 as a basic system. We begin analysis of processes in the model with the nonlinear generation of longitudinal electric field (see Equation 12.47).

12.3.1 Physical Nature of the Generated Electric Field

As follows from the structure of Equation 12.47,

$$E_0^{(g)} = \hat{D}_{01}\frac{d}{dz}\left|E_3\right|^2 + \hat{D}_{02}\frac{d}{dz}\sum_{j=1}^{2}\frac{1}{\omega_j^2}\left|E_j\right|^2, \qquad (12.74)$$

the electric field is generated provided that at least one amplitude is a slowly varying function of coordinate z (expressions for the nonlinear coefficients $\hat{D}_{01}, \hat{D}_{02}$ are unacceptably cumbrous, and therefore, we omit them here). This results from causal relation between beam deceleration and wave amplification (wave energy increases at the expense of decreasing kinetic energy of the relativistic electron beam). It is manifested only in cubic approximation (with respect to wave amplitudes—see the mathematical structure of Equation 12.74), that is, inherently nonlinear. That is why its influence on the FEL operation is appreciable only for sufficient beam deceleration.

We mentioned in this chapter that the physical mechanism of this phenomenon is associated with signal wave amplification by trapping [1–8]. Qualitatively, this means that the trapping results in the whole beam deceleration caused by radiation energy losses of electrons. Hence, longitudinal electron velocities at interaction region input and output are different. Therefore, electron excess is formed in the output part of the working region, producing a decelerating electric field.

Thus, the nature of generation effect is self-consistent. The whole beam deceleration is, as is mentioned, determined by third-order nonlinear dynamics of wave amplitudes. In turn, the latter depends on deceleration dynamics and electric field generation as well.

12.3.2 Analysis

In view of further analysis, it is convenient to introduce wave efficiency η_w. Similar to one-particle efficiency (see Chapter 7), the wave (many-particle) efficiency is the ratio of useful energy density (signal wave energy density increase Δw_1) to total energy density. We determine the latter as the sum of energy densities of relativistic electron beam, w_{30}, pumping wave, w_{20}, and longitudinal electric support (see Equation 12.9, etc.), w_0, that is,

$$\eta_w = \frac{\Delta w_1}{w_{30} + w_{20} + w_0}. \tag{12.75}$$

Pumping and signal wave energy densities are known [1–3,13,14] to be

$$w_j = \frac{\left|\vec{E}_j \vec{E}_j^*\right| + \left|\vec{B}_j \vec{B}_j^*\right|}{8\pi}, \tag{12.76}$$

where all notions are self-evident. Making use of the relation between plane wave field strength $\vec{E}_j$ and induction, $\vec{B}_j$, in simulating magnetodielectric,

$$ck_j \vec{E}_j = \omega_j \vec{B}_j, \tag{12.77}$$

we find the expression for the increase in signal wave density

$$\Delta w_1 = w_1 - w_{10} = \frac{c^2 k_1^2}{8\pi\omega_1^2}\left(\left|E_1\right|^2 - \left|E_{10}\right|^2\right), \tag{12.78}$$

where $w_{10} = w_1(z = 0)$, E_1 is the complex amplitude of signal wave, and $E_{10} = E_1(z = 0)$. On the other hand, electron beam kinetic energy density can be written as

$$w_{30} = n_0 \int_{-\infty}^{\infty} \mathcal{H}_0 f_0^{(0)}\left(\vec{\mathcal{P}}_0\right) d^3\mathcal{P}_0, \tag{12.79}$$

where $\mathcal{H}_0$ is the electron energy (Hamiltonian) at system input; n_0 is the average electron concentration of the undisturbed electron beam; and $f_0^{(0)} = f^{(0)}(t = 0, z = 0)$ is the undisturbed initial canonical $\vec{\mathcal{P}}_0$ momentum distribution function. We substitute explicit expression for Hamiltonian $\mathcal{H}_0$ and take into account beam coldness [see Equation 12.41 in the case $g_{00}(\mathcal{P}_z) = \delta(\mathcal{P}_z - \mathcal{P}_{z0})$].

$$f_0^{(0)} = \delta\left(\vec{\mathcal{P}}_\perp\right)\delta\left(\mathcal{P}_z - \mathcal{P}_{z0}\right). \tag{12.80}$$

Here, $\delta\left(\vec{P}_\perp\right) = \delta\left(P_x\right)\delta\left(P_y\right); \delta\left(P_{x,y}\right), \delta\left(P_z - P_{z0}\right)$ are the relevant Dirac delta-functions. Thus, Equation 12.79 takes the form

$$w_{30} = n_0 mc^2(\gamma_0 - 1). \tag{12.81}$$

We again use notion γ_0—the initial electron relativistic factor. We remind that within the framework of the above quadratic approximation theory (see Chapter 11), $\gamma \cong \gamma_0$ (therefore we ignored subscript "0"), whereas in "cubic" theory, $\gamma \neq \gamma_0$.

Electric support energy density can be calculated obviously. Therefore, we have

$$w_0 = n_0 e\varphi_0^{(n)}, \tag{12.82}$$

where scalar potential $\varphi_0^{(n)}$ is determined by expression 12.9. Substituting Equations 12.78, 12.81, and 12.82 into Equation 12.75 yields wave electron efficiency for the FEL of Dopplertron type with longitudinal electrostatic support, that is,

$$\eta_w = \frac{k_1^2\left(\left|E_1\right|^2 - \left|E_{10}\right|^2\right)}{\left|E_{20}\right|^2 k_2^2\left(\omega_1/\omega_2\right)^2 + 2m^2\omega_p^2\left(\omega_1/e\right)^2\left(\gamma_0 - 1 + e\varphi_{00}/mc^2\right)}, \tag{12.83}$$

where $E_{20} = E_2(z = 0)$ (i.e., we suppose that $s_1 = +1$), $\varphi_{00} = \varphi_0^{(n)}(z = L)$, and L is the total length of the system.

Then we numerically integrate initial equations 12.44 through 12.47 to illustrate details of electric field generation. In particular, Figure 12.1 represents results of such calculations and it shows gain factor

$$K_p = \frac{\left|E_1(z = L)\right|}{\left|E_1(z = 0)\right|}, \tag{12.84}$$

and wave efficiency Equation 12.83 as functions of reduced coordinate $T = z/L$. The subscript p in Equation 12.84 indicates that we deal with the model of parametrical FEL (below we also discuss analogous superheterodyne FEL models).

Figure 12.2 gives the dynamics of efficiency $\eta_w(T)$ for this model, whereas Figure 12.1, as is mentioned, represents the analogous dependencies for the gain factor $K_p(T)$ (where $T = z/L$). In both figures, curve 1 corresponds to the case when electric field generation influence on signal wave amplification is disregarded. Analyzing the result of these figures, we draw at least two conclusions. The first one consists of the fact that in the framework of the cubic-nonlinear self-consistent model, the saturation effects appear. The general physical picture forms here on the background of two well-known independent particular mechanisms. The first is the deceleration of the electron beam as a result of radiation losses of its kinetic energy. The second is the nonlinear shift of frequency [1–3].

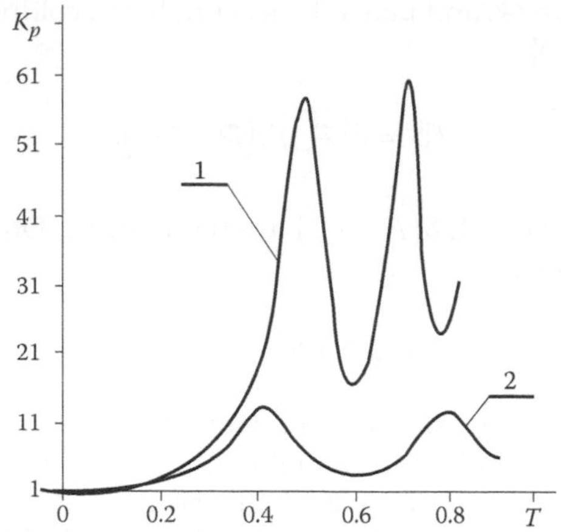

FIGURE 12.1
Influence of the nonlinear electric field generation on the dynamics of the signal gain factor $K_p(T)$. The Raman nonisochronous model of the Dopplertron FEL amplifier is considered. Curves 1 and 2 show the dependencies $K_p(T)$ without and with influence of the effect of generation of electric field, respectively; $T = z/L$ is the unidimensional normalized longitudinal coordinate. Here, $L = 7$ m, $\mathcal{E}_{10} = 0.5 \cdot 10^3$ V/cm, $\mathcal{E}_{02} = 10^3$ V/cm, $\mathcal{E}_{03} \cong 0$, $\omega_1 = 10^{14}$ s^{-1}, $\omega_2 = 10^{12}$ s^{-1}, $\omega_p = 5 \cdot 10^{10}$ s^{-1}, $\gamma_0 = 5.03$, $N_1 = 1$, $N_2 = 1.3$, $s_1 = s_2 = \sigma = +1$ (explosive instability mode). (From Kulish, V.V., et al., *Int. J. Infrared Millimeter Waves*, 14, 3, 117 pp., 1993. With permission.)

The second observation consists of the following. Comparing curves 1 and 2 in Figure 12.2, we see that, apart from the known, in the FEL, once more saturation mechanism exists. It is additional deceleration of the electron beam under longitudinal electric field $E_0^{(g)}$. It is clear that at least in Raman models of FELs, this effect can be appreciable.

However, electric field generation influence steeply decreases with decreasing plasma frequency. In the model, this can be interpreted as the shift along the ω_p-scale toward the Compton interaction mode. The latter is shown in Figure 12.3.

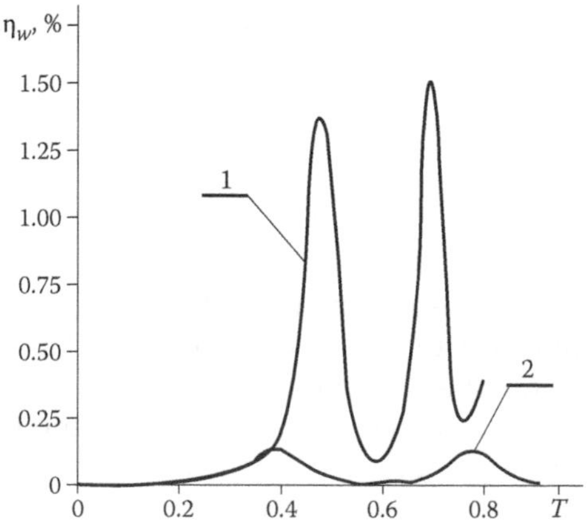

FIGURE 12.2
Dependency of efficiency η_w on the normalized longitudinal coordinate *T*. Here, curve 1 is calculated disregarding and curve 2 taking into account the effect of generation of longitudinal electric field (explosive instability). Parameters of the system are the same as in Figure 12.1. The details of generation dynamics for the nonisochronous model are shown in Figure 12.4. (From Kulish, V.V., et al., *Int. J. Infrared Millimeter Waves*, 14, 3, 117 pp., 1993. With permission.)

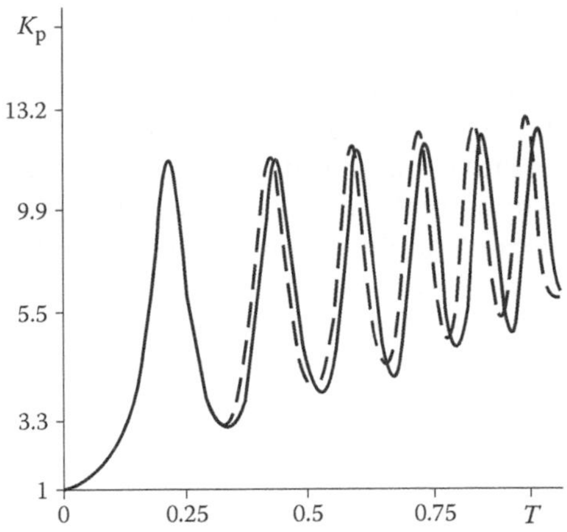

FIGURE 12.3
Influence of the effect of the electric field $E_0^{(g)}$ generation on the dynamics of the gain factor K_p in the Compton Dopplertron FEL amplifier (explosive instability). Here, the solid curve corresponds to the case of disregarding the effect of generation of longitudinal electric field and the dotted curve corresponds to the case of taking into account the latter. Here, the total length of the system $L = 20$ m, $\omega_p = 0.9 \cdot 10$ s^{-1}. The other parameters are the same as in Figure 12.1.

The solid line is calculated disregarding electric field generation; the dotted line is calculated considering this effect. This figure also illustrates one rather interesting calculation feature of nonisochronous FELs, that is, as is readily seen, dependency $K_p(T)$ has explicit oscillatory nature. It means that here once more (i.e., third) hierarchic-wave level can be formed in the given model. Hence, separating relevant hidden oscillation phase, we represent truncated Equations 12.44 through 12.47 in the form of some two-level hierarchic standard system. Respectively, in this way, we can repeat calculation technology of the hierarchic method; corresponding solutions of third hierarchy can be found, and so on.

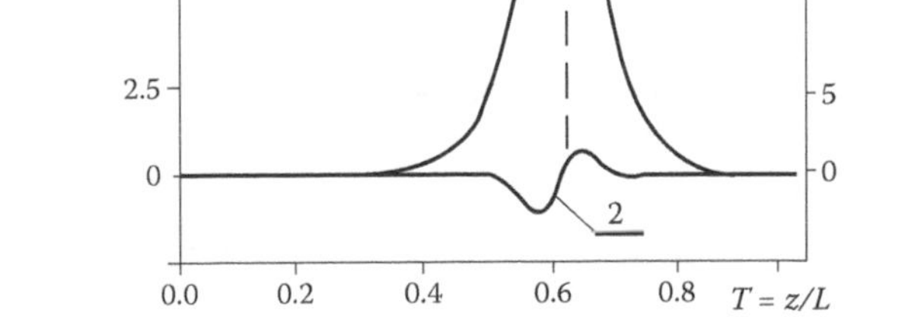

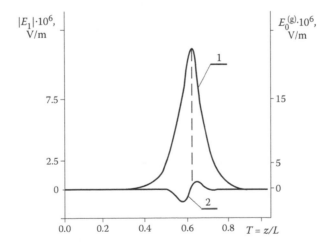

FIGURE 12.4
Dynamics of the generated electric field in the nonisochronous Raman Dopplertron FEL amplifier (explosive instability). Here, curves 1 and 2 are the dependencies of signal amplitude $E_1(T)$ and intensity of generated field $E_0^{(g)}(T)$, respectively. Here, $L = 7$ m, $\mathcal{E}_{01} = 500$ V/cm, $\mathcal{E}_{02} = 5 \cdot 10^4$ V/cm, $\mathcal{E}_{30} \cong 0$, $\omega_1 = 10^{12}$ s^{-1}, $\omega_2 = 5 \cdot 10^{11}$ s^{-1}, $\omega_p = 10^{11}$ s^{-1}, $N_1 = 1$, $N_2 = 1.3$, $s_1 = s_2 = \sigma = +1$.

However, in practice, it is not interesting really. This connects to that active region length of nonisochronous FELs usually chosen approximately equal to the length corresponding to the first maximum of curves $K_p(T)$, $\eta_v(T)$ (see Figures 12.1 through 12.3).

Figure 12.4 clearly reveals the correlation of dependencies $E_1(T)$ (curve 1) $-E_0^{(g)}$ (curve 2). The function $-E_0^{(g)}$ attains negative maximum for $T = 0.595$ (decelerating electric field). This is just the point at which exponential (quasilinear) law of signal wave amplitude increase gives place to a more complicated nonlinear law. At the point $T = 0.605$, signal wave amplification is saturated—beam deceleration stops. It is readily seen that at this very point $E_0^{(g)} = 0$ (see Equation 12.74). In the range $T > 0.605$, signal wave is absorbed (electron beam is accelerated). This process is also accompanied by electric field generation; this time the field is accelerating.

12.4 Isochronous Model of a Dopplertron Amplifier

Equations 12.44 through 12.47 provide two ways of isochronization of the model under consideration. The first way is to apply an optimum longitudinal electric field to the interaction region (longitudinal support). The second way is to optimally change the pumping wavelength (period or its retardation coefficient (see Equation 12.48)). The mathematical problem is to find the optimum distribution of support electric field $E_{0\text{opt}}^{(n)}(z)$ or retardation coefficient $N_1 = N_{1\text{opt}}(z)$ for which the gain factor and the efficiency are maximum. In general, the problem can be solved by optimum control theory. However, this way is rather complicated and involved. A simplifying factor is that another isochronization method can be employed equivalent to model optimization [1–4,7]. This method is based on some physical features of quadratic (weak-signal) and cubic-nonlinear models, respectively. Let us illustrate the main idea of the mentioned simplified optimization scheme.

At first, we return to the quadratic-nonlinear linearly polarized model discussed in the previous section. For this, we use truncated equations in scalar form (Equation 11.68)

$$\frac{d\mathcal{E}_1}{dz} = s_1 \sigma B_1 \mathcal{E}_1 \mathcal{E}_2 \cos\phi;$$

$$\frac{d\mathcal{E}_1}{dz} = s_1 \sigma B_1 \mathcal{E}_1 \mathcal{E}_2 \cos\phi;$$

$$\frac{d\mathcal{E}_1}{dz} = s_1 \sigma B_1 \mathcal{E}_1 \mathcal{E}_2 \cos\phi;$$

$$\frac{d\phi}{dz} = -\left(\sigma s_1 B_1 \frac{\mathcal{E}_2 \mathcal{E}_3}{\mathcal{E}_1} + s_2 B_2 \frac{\mathcal{E}_1 \mathcal{E}_3}{\mathcal{E}_2} + \sigma r B_3 \frac{\mathcal{E}_1 \mathcal{E}_2}{\mathcal{E}_3}\right) \sin\phi, \tag{12.85}$$

where all designations are given above in Chapter 11 (see Equation 11.68). We also remind the form of exact analytical solutions for this system, in particular, the case of explosive instability Equation 11.76

$$\mathcal{E}_1 = \mathcal{E}_{10} X_1 dc\left(\alpha_2' z + H_1\right);$$

$$\mathcal{E}_2 = \mathcal{E}_{20} X_2 nc\left(\alpha_2' z + H_1\right);$$

$$\mathcal{E}_3 = \mathcal{E}_{30} X_3 sc\left(\alpha_2' z + H_1\right), \tag{12.86}$$

where all designations are given in the comments to Equation 1.76. As is mentioned in the previous chapter (see the discussion concerning Equation 11.75), amplitudes $\mathcal{E}_q$ grow explosively if system length $L \to L_{cr}$, where L_{cr} is the critical (explosive) length determined by formula 11.76 (see Figure 11.2). The following question arises: What is an energy source for unlimited growth of electromagnetic and space charge waves in the model? The point is that in the quadratic nonlinear model, we neglect varying of the kinetic energy of the electron beam (i.e., deceleration of the electron beam; here we considered that $\beta \cong \beta_0 = \text{const}$). Strictly speaking, it means that the explosive instability in the quadratic model is accompanied by the violation of the energy conservation law. It can be treated that we suppose there is mystical unlimited energy stored in the working bulk of the electron beam. However, in reality, the energy of all electron beams is limited. Besides that, any mystical energy sources are unknown in physics. Thus, the above-mentioned quadratic model of explosive-unstable FEL should be considered as an inner-conflicting one.

This contradiction has its solution in the framework of the cubic model. Here, the main obvious (nonmystical) source of electron beam energy is the kinetic energy of longitudinal beam motion. Therefore, it is natural that exhaustion of the latter leads to realization of various mechanisms of saturation of signal amplification. Therein, as is mentioned already, there are two possible characteristic physical situations. The first is discussed in the previous section (see Figures 12.1 and 12.2)—the so-called nonisochronous version of interaction mechanism. Its main drawback is low efficiency of interaction (see Figure 12.2). The second is isochronous mechanism of interaction. Single-particle versions of the latter are discussed in Chapter 7. The main advantage is a possibility of effectively attaining high levels of efficiency. It can be realized using the above-mentioned isochronization methods, including methods of optimal longitudinal support $E_{0opt}^{(n)}(z)$ (see Chapter 7). Concerning the self-consistent cubic-nonlinear model, it is interesting to point out the following observation: dynamics of such isochronous optimal cubic-nonlinear systems is similar to that of quadratic-nonlinear models. We base this observation on peculiar optimization of the model with respect to parameters of isochronization. The essence of the method will be illustrated below at the simplest example of a Dopplertron isochronous spatially explosive-unstable model with longitudinal optimum support $E_{0opt}^{(n)}(z)$.

Thus, the problem is to find optimum dependence $E_{0opt}^{(n)}(z)$ that isochronizes the interaction. First, for convenience, we reformulate quadratic-nonlinear system 12.85 substituting for wave amplitudes

$$\mathcal{E}_q = a_q e_q, \tag{12.87}$$

where $a_1 = (B_2 B_3)^{-1/2}$; $a_2 = (B_1 B_3)^{-1/2}$; and $a_3 = (B_1 B_2)^{-1/2}$. Therefore, system 12.85 is rewritten as (it is assumed that $s_1 = s_2 = \sigma = r = +1$—explosive instability)

$$\frac{de_1}{dz} = e_2 e_3 \cos\phi;$$

$$\frac{de_2}{dz} = e_1 e_3 \cos\phi;$$

$$\frac{de_3}{dz} = e_1 e_2 \cos\phi;$$

$$\frac{d\phi}{dz} = -\left(\frac{e_2 e_3}{e_1} + \frac{e_1 e_3}{2} + \frac{e_1 e_2}{e_3}\right)\sin\phi. \tag{12.88}$$

Then we analogously transform cubic systems 12.44 through 12.47 considering Raman mode of interaction (see criteria 12.49) ($s_1 = s_2 = r = \sigma = +1$—explosive instability)

$$\frac{de_1}{dz} = e_2 e_3 \cos\phi;$$

$$\frac{de_2}{dz} = e_1 e_3 \cos\phi;$$

$$\frac{de_3}{dz} = e_1 e_2 \cos\phi;$$

$$\frac{d\phi}{dz} = -\left(\frac{e_2 e_3}{e_1} + \frac{e_1 e_3}{e_2} + \frac{e_1 e_2}{e_3}\right)\sin\phi + G^{(g)} + G^{(n)} + F;$$

$$e_0^{(g)} = M_1 \frac{de_3^2}{dz} + M_2\left(\frac{1}{\mathcal{K}_2\mathcal{K}_3}\frac{de_1^2}{dz} + \frac{1}{\mathcal{K}_1\mathcal{K}_3}\frac{de_2^2}{dz}\right), \tag{12.89}$$

where $e_0^{(g)} = \left(eE_0^{(g)}/mc\omega_p\right)$ is the normalized strength of the generated electric field; explicit forms of functions $G^{(g)}$, $G^{(n)}$, F are unacceptably cumbrous and therefore we omit them here. We must bear in mind that $G^{(g)}$ describes the saturation influence of above-mentioned generation of electric field; $G^{(n)}$ describes the isochronization possibilities of longitudinal electric support; and F takes into account the saturation influence of nonlinear electron beam deceleration and nonlinear shift of frequency.

Comparing systems 12.88 and 12.89, we see their unexpected resemblance, that is, cubic system 12.89 reduces to the form reproducing analogous result of quadratic theory (Equation 12.88) in special case

$$G^{(g)} + G^{(n)} + F = 0. \tag{12.90}$$

We remind that $G^{(n)}$ entering Equations 12.89 and 12.90 (i.e., the support) is considered as a known given function. Therefore, the dependence of optimal support field $E_0^{(n)}$ (Equation 12.9) on longitudinal coordinate z can be fitted in a way to satisfy condition 12.90. It is known that the quadratic theory does not describe nonlinear effects associated with the relativistic deceleration or nonlinear frequency shift. Therefore, formal coincidence of cubic equations 12.89 with equivalent quadratic system 12.88 (provided conditions 12.90 hold) can be interpreted as attaining the isochronous state. Then Equation 12.90 acquires the meaning of interaction isochronism condition. We employ this observation further. Namely, we assume that condition 12.90 holds and thus find analytic solutions to Equation 12.89 in weak-signal explosive form Equation 11.75. Taking into consideration the above-accepted normalization Equation 12.87, the relevant cubic-nonlinear isochronous solution can be written

$$e_1 = e_{10} X_1 dc\left(\alpha_2' z + H_1\right);$$
$$e_2 = e_{20} X_2 nc\left(\alpha_2' z + H_1\right);$$
$$e_3 = e_{30} X_3 sc\left(\alpha_2' z + H_1\right), \tag{12.91}$$

where all designations are determined in comments to Equation 11.75 with normalization Equation 12.87.

Substituting Equation 12.91 into Equation 12.90 and taking into account relevant expression for $G^{(g)}$, $G^{(n)}$, and F (whose evident expression we omit here) yield explicit expression of the normalized strength of support field $E_0^{(n)}$. Analogously, explicit expression for strength of generated field $E_0^{(g)}$ can be found substituting solutions Equation 12.91 into the last of Equation 12.89.

Numerical nonisochronous solutions and analytic isochronous solutions obtained are illustrated by curves 2, 4, and 1, 3 of Figure 12.5, respectively.

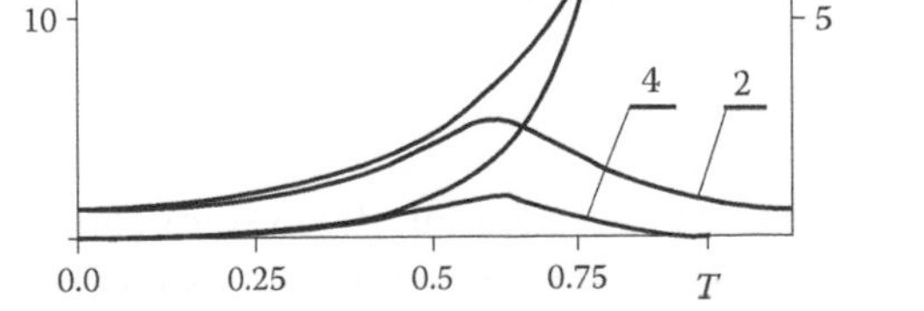

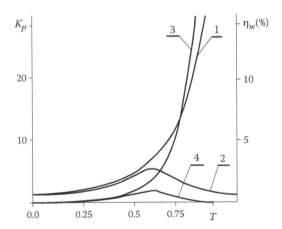

FIGURE 12.5
Signal wave gain factor K_p and the wave efficiency η_w as functions of the dimensionless coordinate $T = z/L$. Curve 1 corresponds to the dependence $K_{1p}(T)$; it is calculated under the assumption that the isochronization condition 12.90 is satisfied; curve 2 corresponds to the case of nonisochronous model. Curves 3 and 4 show analogous dependencies of the wave efficiency $\eta_w(T)$ (explosive instability). Here, $\mathcal{E}_{10} = 0.5 \cdot 10^3$ V/cm, $\mathcal{E}_{20} = 10^3$ V/cm, $\mathcal{E}_{30} \cong 0$, $\omega_1 = 10^{13}$ s^{-1}, $\omega_2 = 10^{11}$ s^{-1}, $\gamma_0 = 3.03$ (this corresponds to the electron beam energy about 1.55 MeV), $\omega_p = 5 \cdot 10^9$ s^{-1}, $L = 5$ m, $N_1 = 1$, $N_2 = 1.3$, $s_1 = s_2 = \sigma = +1$.

Having optimized support $E_0^{(n)}(z)$, we can obtain an explosion-like cubic dependence. Comparing Figures 11.2 and 12.5, we see that the dynamics of the cubic-optimized nonlinear model is analogous to the dynamics predicted by quadratic equations. However, contrary to the weak (quadratic) signal case, the efficiency of amplification is limited. The limit implicitly enters energy restrictions for the system producing support electric field $E_0^{(n)}$. If energy transfer from support field to electrons is much greater than initial electron beam kinetic energy, that is,

$$\frac{e\varphi_{00}}{mc^2} >> \gamma_0 - 1, \tag{12.92}$$

then the efficiency η_w due to Equation 12.83 tends to 1. Here, φ_{00} is the support field potential at plane $z = L$ (see Equation 12.9).

Thus, the longitudinal electric field energy of support is directly converted effectively into electromagnetic radiation energy in the considered isochronous model. The electron beam acts here as an intermediary in transforming the energy of electric support into the energy of electromagnetic pumping and signal waves. From an engineering point of view, such a system can be regarded as a peculiar hybrid of FEL and linear induction accelerator (see Chapter 9).

The model with optimum variation of pumping period can be studied in the same manner.

12.5 Generation of the Additional H-Ubitron Magnetic Field

We mentioned that FELs with undulation magnetic pumping (H-Ubitrons) belong to the most popular models for both theoreticians and experimentalists. Up to now, these models were studied most completely and described in details in numerous references (see, for instance, references [1–3,15–19] and many others). That is why it is of interest to show how too abstract, at the first sight, hierarchic ideology (see Part I) applies to obtain rather peculiar physical results in this very exhausting research field.

We consider a stationary amplifying linearly polarized H-Ubitron model with a cold electron beam. The consideration is restricted by Raman interaction mode analysis. Truncated Equations 12.72 and 12.73 are taken for the initial system.

12.5.1 Adapted Truncated Equations

With the above assumptions taken into account, the system could be considerably simplified—we can obtain the so-called adapted system of truncated equations. The equations of such type are widely known in the theory of parametrical interaction waves in plasmas and plasma-like systems. However, the original method obtained, unfortunately, is not rigorous from a mathematical point of view. So, let us illustrate the topic of accuracy of the adapted system of truncated equations.

Let us reduce Equations 12.72 and 12.73 to the form without second derivatives d^2/dz^2 and gradient terms of the type $\sim H_2^*(dE_3/dz), H_2(dE_1/dz)$. To estimate second derivatives, we use the quadratic approximation results. We put cubic terms equal to zero, that is,

$$A_1 \frac{dE_1}{dz} \cong A_5 E_3 H_2^*; \quad C_1 \frac{dE_3}{dz} \cong C_5 E_1 H_2. \tag{12.93}$$

We differentiate Equation 12.93 with respect to z and carry out some transformations obtaining estimates

$$A_2 \frac{d^2E_1}{dz^2} \cong \frac{A_2 A_5 C_5}{A_1 C_1} E_1 |H_2|^2; \quad C_2 \frac{d^2E}{dz^2} \cong \frac{C_2 C_5 A_5}{A_1 C_1} E_3 |H_2|^2. \tag{12.94}$$

Using Equations 12.93 and 12.94, we derive from Equations 12.72 and 12.73 the required adapted truncated equations [5–8],

$$A_1 \frac{dE_1}{dz} = A_5 E_3 H_2^* + A_{6a} E_1 |E_1|^2 + A_{6b} E_1 \left(|E_1|^2 - |E_{10}|^2\right)$$
$$+ \left(A_7 - \frac{A_2 A_5 C_5}{A_1 C_1} + A_9 \frac{C_5}{C_1}\right) E_1 |H_2|^2 + A_{8b} E_1 \left(|E_3|^2 - |E_{30}|^2\right); \tag{12.95}$$

$$C_1 \frac{dE_3}{dz} = C_5 E_1 H_2 + C_{6a} E_3 |E_1|^2 + C_{6b} E_3 \left(|E_1|^2 - |E_{10}|^2\right)$$
$$+ \left(C_7 - \frac{C_2 C_5 A_5}{C_1 A_1} + C_9 \frac{A_5}{A_1}\right) E_3 |H_2|^2 + C_{8a} E_3 |E_3|^2 + C_{8b} E_3 \left(|E_3|^2 - |E_{30}|^2\right). \tag{12.96}$$

Unfortunately, relevant definitions for coefficients A_l, C_k are unacceptably cumbrous. Therefore, we omit them. The mathematical structure of Equations 12.95 and 12.96 resembles the structure of equations obtained in plasma electrodynamics by simplified approaches [14]. We stress that our system is derived from complete systems 12.72 through 12.73 by omitting some cubic terms on the right-hand side. The latter procedure is controlled by expressions 12.93. Such criterial analysis is rather complicated in conventional approaches [1–3,13,14,20] because of specificity of numerical methods employed. We show practical consequences later. In our approach, we employ Equations 12.93 and 12.94 to formulate applicability criteria for adapted Equations 12.95 and 12.96. We have

$$\Xi_1 \sim \left|A_2 \frac{d^2E_1}{dz^2}\right| \Big/ \left|A_1 \frac{dE_1}{dz}\right| \cong \left|\frac{A_2 A_6 C_5}{A_1 C_1} E_3 |H_2|^2\right| \Big/ \left|A_1 \frac{dE_1}{dz}\right| << 1;$$

$$\Xi_2 \sim \Xi_1 \sim \Xi = \left|C_2 \frac{d^2E_3}{dz^2}\right| \Big/ \left|C_1 \frac{dE_3}{dz}\right| \cong \left|\frac{C_2 C_5 A_5}{A_1 C_1} E_3 |H_2|^2\right| \Big/ \left|C_1 \frac{dE_3}{dz}\right|. \tag{12.97}$$

We compare Equation 12.97 with criterion 12.49 and find that, unlike the systems 12.72 and 12.73, the applicability range of adapted systems 12.95 and 12.96 is restricted only by the Raman interaction mode. The sensitivity of the system to criteria 12.97 is illustrated by results of Figures 12.6 and 12.7.

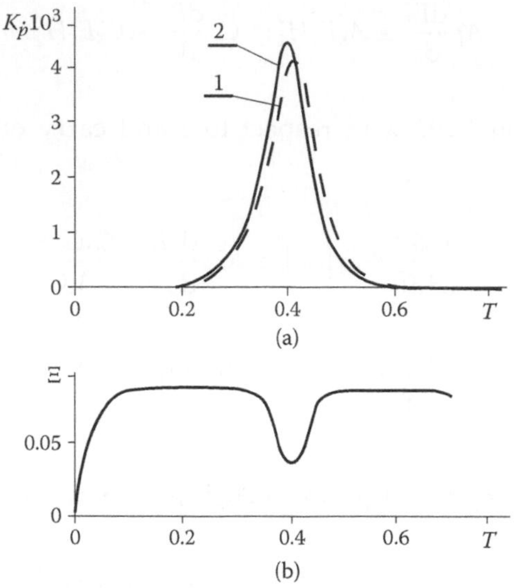

FIGURE 12.6
Dependencies of the signal wave gain factor K_p (upper figure a) and the criterion parameter Ξ (down figures b) as functions of the normalized longitudinal coordinate $T = z/L$. Curve 1 is calculated taking into account second derivatives and gradient terms; curve 2 disregards these terms (i.e., calculated in terms of Equations 12.95 and 12.96). Here, $\omega_p = 3 \cdot 10^{11}$ s^{-1}, $\omega_1 = 2 \cdot 10^{12}$ s^{-1}, the wiggler period $\Lambda = 1$ cm, $L = 2$ m, $\mathcal{E}_{10} = 500$ V/cm, $\mathcal{E}_{30} \cong 0$, $\mathcal{H}_2 = 1$ KGs.

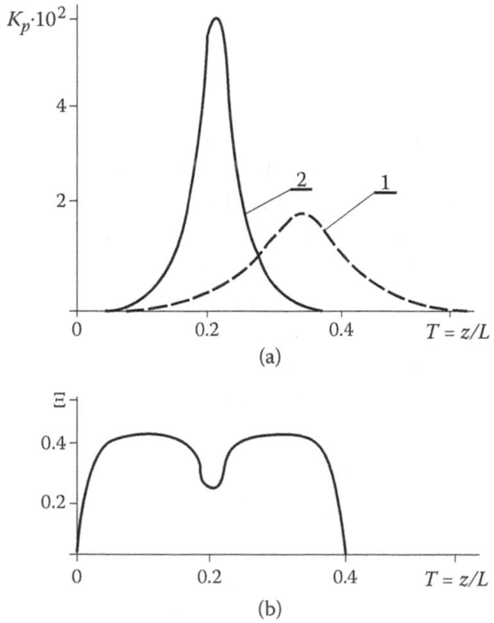

FIGURE 12.7
Dependencies of the signal wave gain factor K_p (upper figure a) and the criterion parameter Ξ (down figure b) on the normalized longitudinal coordinate $T = z/L$. Curve 1 is calculated taking into account the second derivatives and the gradient terms; curve 2 disregards these terms. Here, $\omega_p = 3 \cdot 10^{11}$ s^{-1}, $\omega_1 = 2 \cdot 10^{12}$ s^{-1}, $\Lambda = 1$ cm, $L = 2$ m, $\mathcal{E}_{10} = 500$ V/cm, $\mathcal{E}_{30} \cong 0$, $\mathcal{H}_2 = 3$ KGs. (From Kulish, V.V., et al., *Int. J. Infrared Millimeter Waves*, 14, 3, 117 pp., 1993. With permission.)

As is clear from the figures, ignoring criteria 12.97 (rather popular in papers, including that concerning the FEL theory) can lead to serious calculation errors.

12.5.2 Generation of the Improper H-Ubitron Fields

Then we turn to generation of additional improper H-Ubitron fields, in which weak-signal theory was discussed in the previous chapter. As distinct from the latter, this time we employ the cubic-nonlinear theory.

We have mentioned that two mechanisms of H-Ubitron field generation occur: diamagnetic effect and nonlinear generation. For diamagnetic effect to occur, only electron beam and external H-Ubitron field should be present in the system. Nonlinear generation needs the only presence of signal wave and SCW. Let us consider both physical mechanisms. We restrict the analysis to the case when H-Ubitron field is not a proper field of the system (see definitions similar to Equation 12.71 and others), that is, we assume that external permanent magnets or currents produce this field.

Let us study the diamagnetic effect in FELs, considering the electron beam is warm, that is,

$$f_{00}(\vec{\mathcal{P}}) = \delta(\mathcal{P}_x)\delta(\mathcal{P}_y)g_{00}(\mathcal{P}_z), \tag{12.98}$$

where all definitions are determined by comments to Equation 12.41. We write the additional magnetic field generated by the diamagnetic effect as

$$\vec{H}_2^d = \frac{1}{2}\left(H_{2g}^d \exp\{ip_2\} + c.c.\right)\vec{e}_y, \tag{12.99}$$

where H_{2g}^d is the complex amplitude of additional magnetic field generated by diamagnetic effect; and $\vec{e}_y$ is the unit vector along the y-axis. Then the amplitude of the effective pump magnetic field acting on beam electrons is

$$H_{2ef} = H_2 + H_{2g}^d. \tag{12.100}$$

First we assume $E_1 = 0$ and the beam is nonmodulated at the input. We apply standard calculation procedure to the Maxwell–Boltzmann Equations 10.10 through 10.12 and 10.26 to find the amplitude H_{2g}^d to be

$$H_{2g}^d = -\frac{\omega_p}{c^2k_2^2}\left(H_{2g}^d + H_2\right)\left[\int_{-\infty}^{+\infty}\left(\frac{g_{00}}{\gamma_0^{(0)}}\right)d\mathcal{P}_z + \frac{e^2\left|H_{2g}^d + H_2\right|^2}{8mc^2k_2^2}\int_{-\infty}^{+\infty}\left\{\frac{g'_{00}}{v_z\left(\gamma_0^{(0)}\right)^2} - \frac{3g_{00}}{mc^2\left(\gamma_0^{(0)}\right)^3}\right\}d\mathcal{P}_z\right], \tag{12.101}$$

where $g'_{00} = \left(\partial g_{00}/\partial\mathcal{P}_z\right)$; g_{00} is the initial distribution function of electrons with respect to longitudinal canonical momentum $\mathcal{P}_z$ determined by Equation 12.98; and $\gamma_0^{(0)} = \sqrt{1+\left(\mathcal{P}_z/mc\right)^2}$. Other designations are the same as earlier. Then we assume

$$\left(H_{2ef}\right)^2 = \left(H_2 + H_{2g}^d\right)^2 << D_2 \times \left(\left\| \frac{\omega_p^2 e^2}{8mc^4 k_2^2} \int_{-\infty}^{+\infty} \left\{ \frac{g'_{00}}{v_z \left(\gamma_0^{(0)}\right)^2} - \frac{3g_{00}}{mc^2 \left(\gamma_0^{(0)}\right)^2} \right\} \mathrm{d}\mathcal{P}_z \right\|\right)^{-1}, \quad (12.102)$$

where D_2 is determined by

$$D_2 = D_2\left(k_2\right) = \left(k_2^2 + \frac{\omega_p^2}{c^2} \int_{-\infty}^{+\infty} \frac{g_{00}\left(\mathcal{P}_z\right)}{\gamma_0^{(0)}\left(\mathcal{P}_z\right)} \mathrm{d}^3\mathcal{P}_z \right) \neq 0. \quad (12.103)$$

With Equation 12.102, Equation 12.101 yields [11]

$$H_{2ef} \cong k_2 H_2 / D_2. \quad (12.104)$$

As follows from Equation 12.104, diamagnetic effect can be considered provided that

$$\frac{\omega_p^2}{k_2^2 c^2} \int_{-\infty}^{+\infty} \left(\frac{g_{00}}{\gamma_0^{(0)}} \right) \mathrm{d}\mathcal{P}_z \sim 1. \quad (12.105)$$

The estimates show that the allowance for the difference between amplitudes H_2 and H_{2ef} becomes necessary in practical calculations of high-current H-Ubitron FELs. Unfortunately, authors of numerous papers concerning the theory of nonlinear high current FEL often ignore this observation instead of involving it into consideration. It should be mentioned that, as will be shown in Chapter 13, the considered effect can play a much more essential role in the case of superheterodyne FEL (two-stream and plasma-beam).

12.5.3 Generation of the Proper H-Ubitron Fields

Furthermore, we turn to another version of discussed effect—the nonlinear wave-resonant generation of H-Ubitron magnetic field. We proceed with the calculation procedure used in this chapter to obtain an adapted cubic truncated equation for generated field amplitude $H_2^{(g)}$ (see analogous quadratic results in the previous chapter). Thus, we find

$$\left(K_1 \frac{\mathrm{d}}{\mathrm{d}z} + K_0 \right) H_2^{(g)} = K_5 E_1^* E_3 + \left(K_{6a} + K_9 \frac{C_5}{C_1} \right) H_2 \left|E_1\right|^2$$
$$+ K_{6b} H_2 \left(\left|E_1\right|^2 - \left|E_{10}\right|^2 \right) + K_{8b} H_2 \left(\left|E_3\right|^2 - \left|E_{30}\right|^2 \right), \quad (12.106)$$

where

$$H_2 = H_{2ef} + H_2^{(g)}, \quad (12.107)$$

$$K_0 = -i D_2 / 2k_2^2, \quad K_1 = 1 - D_2 / 2k_2^2; \quad K_5 = ic^2 k_2 A_5 / 2\omega_1; \, K_{6b} = -\frac{ie^2 \omega_p^2}{8k_2 mc^2 \omega_1^2} \int_{-\infty}^{+\infty} \left[\frac{g'_{00}}{v_z \left(\gamma_0^{(0)}\right)^2} \right] \mathrm{d}\mathcal{P}_z;$$

$$K_{6a} = -\frac{ie^2\omega_p^2}{8k_2 mc^2\omega_1^2}\int_{-\infty}^{+\infty}\left[\frac{3g_{00}}{mc^2\left(\gamma_0^{(0)}\right)^3} + \frac{(\Phi/c)g'_{00}}{\Omega_3\left(\gamma_0^{(0)}\right)^2} + \frac{(k_1-k_2)g'_{00}}{\Omega\left(\gamma_0^{(0)}\right)^2}\right]d\mathcal{P}_z;\; K_{8b} = -\frac{i}{2k_2}A_{8b};$$

$$K_9 = -\frac{ic^2k_2}{2\omega_1}A_9;$$

$\Omega = \omega_1 - \frac{(k_1-k_2)P_z}{m\gamma_0^{(0)}}$; $\Omega_3 = \omega_3 - k_3 v_z$; $\omega_3 = \omega_1$; $\Phi = c(k_1 + k_2)$; $s_1 = \sigma = +1$, and dispersion function D_2 is determined by Equation 12.103.

Numerical integration results for Equations 12.95, 12.96, and 12.106 allow doing the following conclusions.

It is of interest to compare Equations 12.106 and 12.107 to the data of nonlinear generation of electric field (see Figure 12.1, etc.). A common feature of two effects is that, in the case of high-current beams, both cause a steep decrease in gain factor K_p. The distinctions, however, are more considerable than common features [1–3]. Recall that the main peculiarity of elec tric field generation is that the effect develops as an amplification saturation mechanism (it causes an appreciable decrease in the saturation level—see Figure 12.1). In the last case, the saturation level remains nearly unchanged. The influence of the magnetic field generation is manifested only in amplification maximum shift toward greater lengths of the system. This means that the phenomenon does not belong to saturation mechanisms (as distinct from electric field generation). It can be referred to as a reactive depression of the basic FEL mechanism [1–3]. The dynamics of the phase of complex amplitude $H_2^{(g)}$ is also interesting. As the analysis shows, the initial phase sign, $\varphi_2^{(g)} = \arg\{H_2^{(g)}\}$, is abruptly changed to the opposite value at saturation point ($+\pi/2$ is replaced by $-\pi/2$). The value $+\pi/2$ is then recovered approximately in the middle of the distance between neighboring amplification maxima.

Finally, it should be mentioned that the theory of the considered effect is now far from complete. Preliminary semiqualitative analysis shows that, basically, the models with the additional H-Ubitron field, where it increases the signal amplification, are also possible. But, this area of research waits for its investigators.

References

1. Kulish, V.V. 1998. *Methods of averaging in nonlinear problems of relativistic electrodynamics*. Atlanta: World Scientific Publishers.
2. Kulish, V.V. 2002. *Hierarchic methods. Undulative electrodynamic systems*, Vol. 2. Dordrecht: Kluwer Academic Publishers.
3. Kulish, V.V. 2002. *Hierarchic methods. Hierarchy and hierarchical asymptotic methods in electrodynamics*, Vol. 1. Dordrecht: Kluwer Academic Publishers.
4. Kulish, V.V., and Kuleshov, S.A. 1993. Nonlinear self-consistent theory of free electron lasers. *Ukr. Fiz. Zh. [Ukr. J. Phys.]* 38(2):198–205.
5. Kulish, V.V., Kuleshov, S.A., and Lysenko, A.V. 1993. Averaged kinetic equation method in the theory of three-wave parametric resonance in relativistic electron beam plasmas. *Fiz. Plazmy [Sov. J. Plasma Phys.]* 19(2):199–216.

6. Kulish, V.V., and Lysenko, A.V. 1993. Method of averaged kinetic equation and its use in the nonlinear problems of plasma electrodynamics. *Fiz. Plazmy [Sov. Plasma Phys.]* 19(2):216–227.
7. Kulish, V.V., Kuleshov, S.A., and Lysenko, A.V. 1993. Nonlinear self-consistent theory of superheterodyne and free electron lasers. *Int. J. Infrared Millimeter Waves* 14(3):451–568.
8. Kulish, V.V. 1997. Hierarchic oscillations and averaging methods in nonlinear problems of relativistic electronics. *Int. J. Infrared Millimeter Waves* 18(5):1053–1117.
9. Kulish, V.V., and Miliukov, V.V. 1984. Explosive instability of a high-current relativistic electron beam in H-Ubitron fields. *Ukr. Fiz. Zh. [Ukr. J. Phys.]* (2):389–390.
10. Zakharov, V.P., Kisletsov, A.V., and Kulish, V.V. 1986. Generation of static transverse periodic magnetic field in the relativistic electron beam plasma. *Fiz. Plazmy [Sov. J. Plasma Phys.]* 12(1):77–79.
11. Ginzburg, N.S. 1987. Diamagnetic and paramagnetic effects in free-electron laser. *IEEE Trans. Plasma Sci.* 15(4):411–417.
12. Kulish, V.V., and Krutko, O.B. 1995. Nonlinear self-consistent theory of free electron lasers pumped by crossed periodic magnetic and electric fields. In *Proceedings of the International Conference Physics and Technology of electron systems*, Sumy, May 18–20, 1995.
13. Aleksandrov, A.F., Bogdankevich, L.S., and Rukhadze, A.A. 1988. *Fundamentals of plasma electrodynamics*. Moscow: Vysshaya Shkola.
14. Weiland, J., and Wilhelmsson, H. 1977. *Coherent nonlinear interaction of waves in plasmas*. Oxford: Pergamon.
15. Marshall, T.C. 1985. *Free electron laser*. New York: MacMillan.
16. Brau, C. 1990. *Free electron laser.* Boston: Academic Press.
17. Luchini, P., and Motz, U. 1990. *Undulators and free electron lasers*. Oxford: Clarendon Press.
18. Saldin, E.L., Schneidmiller, E.V., and Yurkov, M.V. 2000. *The physics of free electron lasers (Advanced texts in physics)*. Springer.
19. Wu, J., Bolton, P.R., Murphy, J.B., and Zhong, X. 2007. Free electron laser seeded by IR laser driven high-order harmonic generation. *Appl. Phys. Lett.* 90(02): 1109.
20. Kuraev, A.A. 1982. *High-power microwave devices: Methods of analysis and parameter optimization*. Moscow: Radio i Sviaz'.

13

Two-Stream Superheterodyne Free Electron Lasers

The first experiments on FELs, performed in the United States in 1976 (Chapter 9), revealed that the realized FEL designs do not completely meet the demands of their customers, that is, the developers of the Star Wars systems (Chapter 1). This had something to do with key FEL properties such as large amplification coefficients for the one-signal passage through the interaction area, a moderate longitudinal size, and large mean and pulsed power. As is mentioned in Chapter 1, a known scientific–political excitement over FEL developed in full blast during these historical times. Hence, the reaction of the managers of the corresponding programs was the same in the USSR as well as in the United States, that is, to find a new, more perfect FEL design scheme. They planned to not only improve the basic properties of the already in-plan systems but also give them principally new directions for their further development. Femtosecond lasers are examples of such new FEL versions. Therefore, historically, the appearance of the first designs of the *superheterodyne* FEL (SFEL) in 1979 to 1980 [1,2] was a completely expected event that adequately fulfilled the aim of the highest management.

The collapse of the USSR became one of the basic work-related reasons that resulted in the interruption of the creation of the first experimental model of SFEL at the most interesting stage. However, further theoretical studies (references [3–39] and many others) revealed that the SFEL might become one of the key design elements of future generations of the Star Wars systems, as is discussed in Chapters 1 and 9 cluster point of view. Hence, taking into account this discussion, it should not be surprising that this and the following two chapters are dedicated exclusively to all kinds of SFEL and their different purposes.

General preliminary information about the history and the basic principles of SFEL is given in Section 9.3. As illustrated in Figures 9.17 and 9.18, this class of systems represents at present a separate and significantly branched family of FEL. Most of its representatives are discussed in various degrees in Chapters 9 and 14 as well as in this chapter. Here, attention is drawn mainly to the physics of the basic processes involved in the two-stream linear ("ordinary") SFEL (linear TSFEL) with monochromatic signal wave (see classification in Figures 9.17 and 9.18).

As is presented in Chapter 9, the so-called two-stream instability is the basic working mechanism of the TSFEL. In the considered class of devices, it appears doubly as the key submechanism of the so-called effect of the superheterodyne amplification (Chapters 4 and 9) as well as the basic working mechanism for the transit sections in the klystron TSFEL (Chapter 9). Therefore, when considering its importance for the complete understanding of the physics of TSFEL, it is logical to begin our analysis with the theory behind this unique physical phenomenon.

13.1 Two-Stream Instability*

13.1.1 Initial Model and the Problem Statement

Consider a transversely unlimited and uniform model of the two-stream relativistic electron beam. We suppose that the beams move along the positive z-axis with partial velocities u_1 and u_2. Assume that the averaging space charge of the two-velocity beam is taken as compensated by some ion background. It is assumed also that the electron beam moves, in general, in some external H-Ubitron or EH-Ubitron (pumping) fields

$$\vec{E}_2 = \frac{1}{2}(E_2 \exp\{ip_2\} + c.c.)\vec{e}_x \,;\; \vec{B}_2 = \frac{1}{2}(B_2 \exp\{ip_2\} + c.c.)\vec{e}_y, \tag{13.1}$$

where $p_2 = k_2 z$ is the phase of undulated fields; k_2 is the wave number; E_2 and B_2 are the complex amplitudes of corresponding undulated fields; and $\vec{e}_x$, $\vec{e}_y$ are the unit vectors along x- and y-axis. Accomplishing the passage $E_2, B_2 \to 0_2$, we obtain the considered particular case of pure transit TSFEL section.

The strength of SCWs in the considered two-stream system can be written as

$$\vec{E}_3 = \frac{1}{2}\sum_{m=1}^{N}(E_{3m} \exp\{\mathrm{Im}\, mp_3\} + c.c.)\vec{e}_z, \tag{13.2}$$

where $p_3 = \omega_3 - k_3 z$ is the phase, E_{3m} is the complex amplitude of the mth harmonic, $\vec{e}_z$ is the unite vector along z-axis, and N is the total number of harmonics taken into account.

It is considered that the motion of electron beam in Equations 13.1 and 13.2 is described by the current–density equation, which can be easily obtained combining definitions for current and charge densities (Equations 10.13 and 10.14), quasihydrodynamic equation (Equation 10.35), and continuity equation [3–5]

$$\frac{\partial \vec{j}_\alpha}{\partial t} + \frac{j_{\alpha z}}{\rho}\frac{\partial \vec{j}_\alpha}{\partial z} = -\nu \frac{\vec{j}_\alpha}{\gamma_\alpha^2(j_\alpha)} + \left(\frac{\vec{j}_\alpha j_{\alpha z}}{\rho_\alpha^2} - \frac{v_T^2 \vec{e}_z}{3\gamma_\alpha(j_\alpha)}\right)\frac{\partial \rho_\alpha}{\partial z}$$

$$+\frac{q_\alpha}{m_\alpha \gamma_\alpha(j_\alpha)}\left\{\rho_\alpha \vec{E} + \frac{1}{c}[\vec{j}_\alpha \vec{B}_2] - \frac{1}{c^2 \rho_\alpha}\vec{j}(\vec{j}\vec{E})\right\} + \frac{\vec{j}_\alpha}{\rho_\alpha}\frac{\partial \rho_\alpha}{\partial t}, \tag{13.3}$$

where $\alpha = 1,2$ is the number of the partial beams; quantities ρ_α, $\vec{E} = \vec{E}_2 + \vec{E}_3$, $\vec{B}_2$ are assumed to be predetermined (in the framework of motion part of the general self-consistent problem) functions of time t and coordinate z; relativistic factor γ_α, as before, is

$$\gamma_\alpha(j_\alpha) = \left(1 - \frac{\vec{j}_\alpha^{\,2}}{\rho_\alpha^2}\right)^{-\frac{1}{2}} \equiv \left(1 - \frac{\vec{u}_\alpha^{\,2}}{c^2}\right)^{-\frac{1}{2}}, \tag{13.4}$$

* From Kulish, V.V. et al., *Int. J. Infrared Millimetre Waves*, 15, 77–120, 1994. With permission.

ν is the effective particle collision frequency; $\vec{u}_\alpha$ is the velocity of the αth beam as a whole; q_α and m_α are the beam particle charge and mass; and υ_T is the root-mean-square particle thermal velocity. Densities ρ and $\vec{j}$ are coupled, as mentioned above, by the continuity equation

$$\frac{\partial\rho}{\partial t}+(\vec{\nabla}\vec{j})=0 \tag{13.5}$$

and are determined as is done earlier in Chapter 10 [Equations (10.13) and (10.14)]

$$\vec{j}_\alpha=q_\alpha n_\alpha \vec{u}_\alpha\,,\ \rho=q_\alpha n_\alpha. \tag{13.6}$$

Let us note that here, $n_{0\alpha}$ and n_α are the unperturbed and induced densities of respective particles. Taking this into account, the field part of the self-consistent problem, which is described by Maxwell equations, may be written in the following form:

$$[\vec{\nabla}\vec{E}]=-\frac{1}{c}\frac{\partial\vec{B}}{\partial t};$$

$$[\vec{\nabla}\vec{H}]=\frac{1}{c}\frac{\partial\vec{D}}{\partial t}+\frac{4\pi}{c}(\vec{j}_0+\vec{j});$$

$$(\vec{\nabla}\vec{D})=4\pi(\rho_0+\rho);\ (\vec{\nabla}\vec{B})=0, \tag{13.7}$$

where $\vec{E}$, $\vec{D}$, $\vec{B}$, and $\vec{H}$ are standard notations for the electric and magnetic field vectors; $\vec{j}_0$ and $\vec{j}$ are the current density vectors arising from external and intrinsic field sources; ρ_0 and ρ are the space charge densities caused by similar sources; and $\vec{\nabla}$ is the nabla operator. Herewith, we neglect the diamagnetic effect and the effect of nonlinear generation of longitudinal electric field discussed earlier in Chapters 11 and 12.

To solve the motion part of the problem, we use the hierarchic method of averaged characteristics (see Chapter 6). According to the general calculation scheme of this method, the first step in this direction is solving the so-called linear-approximation problem.

13.1.2 Linear (Weak Signal) Approximation

A specific calculation feature of the two-stream problem (in comparison to the parametric FEL problem studied in Chapters 11 and 12) is that increased solutions for the electron wave amplitudes are already found in the first approximation. Thus, the role of higher approximations here is reduced to finding relevant nonlinear additions to the already found increasing linear solutions. This means that the first approximation plays a much more important role in the theory of TSFEL than what it plays in the case of parametric FEL.

Let us put all nonlinear (with respect to amplitudes) terms in initial Maxwell equations (Equation 13.7) and current density equations (Equation 13.3) equal to zero. Performing relevant calculations (see details in Chapters 6, 11, and 12), we obtain the linear dispersion equation for SCWs in the considered system

$$1-\omega_p^2\sum_{\alpha=1}^{2}\frac{1}{\gamma_{0\alpha}^3(\omega_3-u_{\alpha 0}k_3)^2}=0, \tag{13.8}$$

where $u_{\alpha 0} = u_\alpha$ ($z = 0$), ω_p is the plasma frequency of each of partial (i.e., one-velocity) electron beams (we accept: $\omega_{p1} = \omega_{p2} = \omega_p$), $\gamma_{0\alpha} = (1-(u_{0\alpha}^2/c^2))^{-1/2}$, and $\alpha = 1,2$. We look for solutions of Equation 13.8 in the form

$$k_3=\frac{\omega_3}{v_0}+i\Gamma', \tag{13.9}$$

where addends Γ' after a series of noncomplex mathematical calculations can be found as

$$\Gamma'=\pm\frac{\omega_p}{v_0\gamma_0^{3/2}}\left\{\left(\frac{\omega_3\delta}{\omega_p}\gamma_0^{3/2}\right)^2+1\pm\left[4\left(\frac{\omega_3\delta}{\omega_p}\gamma_0^{3/2}\right)^2+1\right]^{\frac{1}{2}}\right\}^{\frac{1}{2}}. \tag{13.10}$$

Here the following condition $|v_0\Gamma'\delta/\omega_p| \ll 1$ has been used, $v_0 = (u_{01} + u_{02})/2$; $\delta = (u_{01} - u_{02})/v_0$ is the normalized beam velocity difference, $\gamma_0 = (1-(v_0^2/c^2))^{1/2}$ ($v_0 = u_0$ ($z = 0$)).

Thus, four different proper (inherent) SCWs can propagate in the considered two-velocity beam model. They are the increasing, damped, fast, and slow electron waves, respectively. For SCW frequency

$$\omega_3=\omega_{3\text{opt}}=\frac{\sqrt{3}\omega_p\gamma_0^{3/2}}{\Delta\gamma}(1-\gamma_0^{-2}), \tag{13.11}$$

the growth rate (increment) for the increasing wave and the decrement for the damped wave attain their maximums, that is,

$$\Gamma'=\pm\Gamma\cong\pm\frac{\omega_p}{2v_0\gamma_0^{3/2}}. \tag{13.12}$$

Here $\Delta\gamma = \gamma_{10} - \gamma_{20}$. When relationship 13.11 is satisfied, the corrections Γ' for slow and fast SCWs are purely imaginary

$$\Gamma'=\Gamma_{1,2}=\pm i\frac{\sqrt{15}}{2}\frac{\omega_p}{v_0\gamma_0^{3/2}}=\pm i\sqrt{15}\Gamma. \tag{13.13}$$

This automatically means that these waves are not increasing in principle. Substituting Equations 13.9 through 13.13 into the definition for the phase $p_3 = \omega t - k_3 z$, we can find that it is a complex value in the case of two-stream instability

$$p_3=p_3'+ip_3''=\text{Re}\{p_3\}+i\,\text{Im}\{p_3\}. \tag{13.14}$$

Hence, performing relevant mathematical transformations for the following orders, we should bear in mind the possible complexity of the phase p_3. Strictly saying, the analogous mathematical situation we met in the previous case of parametric FEL (see Chapters 11 and 12). The complexity of electron-beam wave (SCW) phase in that case was caused by the electron thermal spread and the collisions. However, we neglected these effects because they are not critical for the considered model of three-wave parametric instability. In contrast, the fact of complexity of the phase p_3 determines the basic working mechanism of TSFEL because it provides the required addition amplification for the SCW even in the first approximation.

Let us continue the analysis. Solutions of the dispersion equation (Equation 13.8) can be written in the form 13.9 through 13.13 only for the frequency range

$$\omega_3 \leq \omega_{cr} = \frac{\sqrt{2}\omega_p}{\gamma_0^{3/2}\delta}, \tag{13.15}$$

where ω_{cr} is the critical frequency. Therefore, the frequency range $\omega_3 < \omega_{cr}$ is called the "precritical" and the opposite case $\omega_3 > \omega_{cr}$ is referred to as the "overcritical." In this case ($\omega_3 > \omega_{cr}$), all four electron waves are described by the imaginary corrections Γ' given by Equation 13.9. In particular, for $\omega_3\gamma_0^{3/2}\delta >> \omega_p$, the dispersion dependencies k_3 (ω_3) can be represented as

$$k_3^{(1,2)} = \frac{\omega_3}{u_{01}} \pm \frac{\omega_p}{v_0\gamma_0^{3/2}};$$

$$k_3^{(3,4)} = \frac{\omega_3}{u_{02}} \pm \frac{\omega_p}{v_0\gamma_0^{3/2}}. \tag{13.16}$$

It is readily seen from the structure of expression 13.16 that two of four electron waves (fast and slow) can be regarded as belonging to the first partial beam, and the other two to the second partial beam. In other words, the first pair of waves (Equation 13.16) (which are proper for the first partial electron beam) can be conditionally considered as external (stimulated) waves for the second electron beam. On the contrary, the second pair of waves (Equation 13.16) can be treated as external for the first beam.

As it follows from Equation 13.16, the electron waves in the case $\omega_3 > \omega_{cr}$ are not increasing (opposite to Equations 13.9 through 13.12) in the first approximation on the wave amplitudes. However, they can increase in the second approximation, if required conditions for the three-wave parametric resonance are satisfied (see Chapter 14 for more details). This means that both classes (corresponded two different physical mechanisms) of the above-mentioned electron-beam waves can be used for the additional longitudinal amplification in SFEL. The first mechanism [which appears in the first approximation and has been treated above as linear (quasilinear)] is used as the physical basis of the ordinary (linear) TSFEL (see the classification in Figure 9.2 and the relevant discussion). Correspondingly, the second mechanism, which carries a purely quadratic nature (because it appears in the second approximation only), is used as the basis of quadratic (parametric) TSFEL (Figure 9.2).

Then let us perform some numerical estimates. We compare the expressions for optimum frequencies in relativistic (Equation 13.11) and traditional nonrelativistic cases [40–43]. The latter can also be obtained from Equation 13.11 assuming $\gamma_0 \sim 1$

$$\omega_3 = \omega_{3opt} \cong \frac{\sqrt{3}\omega_p}{\delta}. \tag{13.17}$$

Based on noncomplex estimations (Chapter 9), the nonrelativistic systems can work only in the microwave range. Comparing, we can find that the optimal frequency ω_{3opt} in the relativistic case for some fixed energy difference $\Delta\gamma \cdot mc^2$ grows as $\gamma_0^{3/2}$ in Equation 13.11. Assuming also $\omega_p \sim 5 \cdot 10^{11}\ s^{-1}$, $\Delta\gamma \sim 10^{-2}$, $\gamma_0 \sim 10$ (beam energy ~5MeV), we find $\omega_{3opt} \cong 2.7 \cdot 10^{15}\ s^{-1}$, that is, the working frequency of the two-stream instability, in principle, can lie in the visible wavelength range. This is a very important conclusion because of two reasons. First, the mechanism of a steep increase in working frequency has no direct association with the Doppler effect, unlike the analogous basic effect in parametric FEL (see details in Chapter 9). Second, this can have bright prospects for various practical applications.

The physical meaning of this phenomenon is as follows (Chapter 9). Both partial beam velocities u_{10} and u_{20}, in the case of an increase in the relativistic electron beam energy, asymptotically tend to the same limit value—light velocity in vacuum c. This means that for $\gamma_0 \to \infty$, the parameter δ in Equation 13.17 tends to zero and, consequently, the optimal frequency $\omega_{3opt} \to \infty$.

Unfortunately, the discussed progress related with the working frequency is not gratuitous. According to relationship 13.12, the gain factor Γ decreases (rather steeply) with growing γ_0. Besides that, we have limitations originating from the applicability of representation of an electron beam as a plasma flow. It is obvious that the length of electron wave $\lambda_e = 2\pi c/\omega_{3opt}$ cannot be lesser (in the reference frame connected with the electron beam) than the average distance between particles $l \sim 1/\sqrt[3]{n_0}$ (where n_0 is the average electron concentration of the beam). All this suggests the conclusion that the optimum range for practical application of a relativistic version of the two-stream instability seems to be in the range of medium relativistic electron beam energies (hundred of kiloelectron volts to megaelectron volts) only. Let us continue this interesting discussion.

13.1.3 Other Key Properties of the Two-Stream Instability

As revealed in elementary analyses, the two-stream instability in the high-current relativistic electron beams has a few key characteristics, which basically make them interesting to study. Some of them were mentioned shortly in the previous section. Let us continue to discuss these characteristics by the previously used method of the simple numerical evaluation, which are used for the more outstanding illustration of some of the described properties of the two-stream systems.

Let us assume, for example, $\omega_p \sim 2 \cdot 10^{11}\ c^{-1}$ (which might be real for the high-current electron beams) and $\gamma_0 \sim 10$ (which corresponds to the electron energy ~5 MeV; at average beam velocity approaching the light velocity: $v_0 \sim c$). Then, by using Equation 13.12 for the maximal increment increase Γ_{max}, it is easy to obtain $\Gamma_{max} \sim 10^{-1}$ cm. The last means that at the length of the interaction region L, for instance, 3 m (which, in principle, can be relatively simple to obtain for the strong-current FEL), the linear coefficient of amplification for some weak input monochromatic SCW $K_{lin} = e^{\Gamma L}$ might reach the fantastic value of $\sim e^{30}$. This is the reason why the two-stream instability is considered one of the strongest in electrodynamics. The larger amplification increments are able to secure perhaps only the acoustic-plasma instability in the semiconductors. Notice that, in this case, the length of the crystal cannot exceed a few centimeters (see references [44–46]).

Therewith, let us mention that, obviously, such linear amplifications cannot be realized in practice in the simple systems. First, various kinds of nonlinear effects, which appear at significantly shorter length L, are interfering (see the analogical case with parametrical FEL in Chapters 11 and 12). However, in the given case, this is essentially not important. More so, as the theory shows, in principle, these kinds of saturation mechanisms can be subdued in the by far not simple isochronous systems (examples are in Chapter 12). The mentioned numerical evaluations demonstrated the important issue of the potentially high amplification ability of the two-stream systems (and consequently, of its plasma-beam variety—Chapter 14). Historically, just this circumstance stimulated the appearance of the first studies on the two-stream and plasma-beam FEL [1–39].

Let us continue our numerical evaluations. This time, we turn again to the discussion of the problem of the working frequency range of the relativistic version of the two-stream instability. Like before, we take the equation of the optimal frequency ω_{opt} as the basis (Equation 13.11). Specifically, at the same value of the parameters ω_p, γ_0 as in the first case, and also assuming $\Delta\gamma_0 \sim 10^{-2} - 10^{-1}$ [which corresponds to the value of splitting of the partial beams in energies $(5 - 50)$ keV] for the optimal frequency ω_{opt}, we obtain $\omega_{opt} \sim 10^{14} - 10^{15}$ c^{-1}. In other words, we support the above obtained deduction that, in the case of the significantly strong-current and qualitative relativistic beams, the situation is possible when the optimal frequency of the SCWs of the two-stream system is within the optical range. One should stress the fact that the advance in the visible and higher frequency range, in turn, involves a complex technological problem of creating high-quality, high-current electron beams. In other words, we get again a confirmation of the old wisdom "you have to pay for everything that is good."

One should mention that, within the present technologic scope, a transformation of the longitudinal SCWs into the transverse electromagnetic waves is not an unsolvable problem. First, it happens thanks to the achievement of the contemporary technique of the high-current FEL. As is mentioned in Chapter 9, these lasers are built based on the physical mechanisms of the mutual transformation of the transverse electromagnetic waves and longitudinal space-charge waves. Therefore, we may state that this really unique peculiarity of the two-stream instability makes it a very promising physical–technological basis for the creation of a whole gamma of various TSFELs, including the cluster femtosecond TSFEL.

Let us mention one more deduction made in the previous section: the two-stream instability is accomplished only within the precritical frequency range $\omega_3 \leq \omega_{cr}$. It means that the weak incoming SCW's signal of frequency ω_{31} is always amplified when it is within the range of the two-stream instability. In the one-frequency TSFEL discussed here, the frequency of the first harmonic of the SCW ω_{31} is traditionally chosen near the optimal frequency ω_{opt}. First, this is therefore in agreement with Equation 13.12; especially in this case, one expects the maximum of the amplification. Therefore, one of the main aims in the creation of any of the discussed monochromatic amplification (e.g., designed to work in the cluster systems) is the achievement of the maximal amplification at the minimal length of the interaction region.

A completely different situation appears in the case of multiharmonic (cluster) TSFEL. Because only one of the SCW harmonics in one moment can be within the optimal amplification range (Equation 13.11), all the remaining harmonics are amplified in the two-stream system with smaller increments Γ_m (where m is the harmonic number). In other words, in general, all the SCW harmonics undergo amplification in various increments of the increase.

In the case of the cluster TSFEL, this circumstance performs the key role, because here, as is stressed in Chapters 1 and 9, the basic aim is to form narrow beams of the multiharmonic transverse clusters of the electromagnetic field. By virtue of the basic physics of the transformation of the longitudinal SCW into the transverse electromagnetic signal (details in Chapter 1), the technological achievement of the given aim is possible only in the case, when in the spectrum of SCW, a significantly long anomalous part is present, specifically, the part at which the spectral components have a large number of harmonics and larger amplitudes. The obligatory presence of the anomalous segment is explained by the spectra of the magnetic undulators, which are most often used for the pumping in cluster FEL, and as a rule, obviously appear to drop (Chapter 9).

Then, we take into account the basic contribution to the formation of each m spectral component of the electromagnetic signal E_{1m} (distant off the saturation), given by the quadratic components of the abbreviated equations, which are proportional to $\sim E_{3m}H_{2m}$, where E_{3m} and H_{2m} are the amplitudes of the mth harmonics of the electric field SCW and the magnetic field of pumping, respectively. In this connection, it becomes obvious that it is possible to obtain sufficiently long part of the spectrum with the approximate equal spectral amplitudes $E_{1\text{m}}$ only in the case of the anomalous dependence of $E_{3m}(m)$ at the dropping analogous dependence $H_{3m}(m)$. Let us mention that this type of spectrum is peculiar for the femtosecond cluster electromagnetic waves.

As is mentioned in Chapter 9, in some systems of the Dopplertron pumping, principally it is possible to create the cluster multistage SFEL, for which the demand for the anomaly of the SCW spectra is not that strict. Regrettably, at the present time, these models are insignificantly studied. What concerns the H-Ubitron systems, based on a project analysis of the majority of the most interesting, and from a practical point of view, special models reveal that here we can formulate the following rule: the longer the anomalous segment of the SCW spectrum is, the narrower the electromagnetic clusters appear in the final results.

However, in the light of the previous discussion, it is still not completely clear by which peculiarity of the discussed physical mechanisms the task of the formation of the long anomalous segment of the spectrum might be carried out. An adequate physical answer was presented in literature [25–29]. To be specific, for this, one has to turn to the models, in which the frequency of the first harmonic of SCW ω_{31} is significantly smaller than the optimal frequency ω_{opt}. In this case, each subsequent harmonic (including the harmonic $n_3 \approx \omega_{\text{opt}}/\omega_{31}$) is amplified with a larger increment of the amplification Γ than the previous. By taking as the basis, for example, the previously discussed variant of the numerical evaluation of Equation 13.11 and $\omega_{\text{opt}} \sim 10^{15}\,c^{-1}$, it is easy to realize that, at a choice of the frequency of the first harmonic within the submillimeter range ($\omega_{31} \sim 10^{13}\,c^{-1}$), the anomalous segment of the cluster SCW spectrum might contain ~100 harmonics. In this case, for the half-width of the cluster $\tau_{p2} \cong \pi/n_3\omega_{31} \approx \pi/\omega_{\text{opt}}$ (for details, see Ref. [47]), one can obtain the numerical value, specifically, in the *femtosecond* range: $\tau_{p2} \sim \pi \cdot 10^{-15}\,c^{-1}$ (at the cluster wave period $T_{31} \sim 2\pi \cdot 10^{-13}\,c$). In this case, the compression coefficient f_{com} (Chapter 1) reaches the value of $f_{\text{com}} \sim 100$.

Hence, the performed evaluation clearly illustrates three key characteristics of the two-stream instability in high-current relativistic electron beams:

1. A possibility to secure exclusively high level of amplification
2. A possibility to work within the visible range ($\omega_{\text{opt}} \sim 10^{14} - 10^{15}\,c^{-1}$)
3. A clearly expressed inclination to the generation of the wide spectra of harmonics: SCW containing long anomalous segments

On one hand, the derived conclusions to a great deal clarify the physical meaning of the basic principles of the construction of the cluster FEL, as discussed earlier in Chapter 9. On the other hand, they demonstrate the circumstance that, even within the framework of such simple (weak-signal) theory of the two-stream high-current systems, the physics of the basic processes appears by far not as simple and banal as one would like it to be in the case of a simple model. Therefore, further development of the theory, by taking into account the effect of the various types and the higher-order nonlinearities, should lead to a more complex physical picture of the interaction as well of the very theory. We will illustrate this further in the following sections, in a brief discussion of the effect of plural parametric resonances in the two-stream high-current relativistic beams. This phenomenon, as is illustrated in references [38,48], appears to be the key for understanding the principally multiharmonic nature of the two-stream instability.

13.1.4 Harmonic Parametric Resonances: Resonant Conditions

As is shown here, within the framework of the weak-signal theory, a significant quantity of the SCW harmonic can propagate in the precritical frequency range of a two-stream high-current electron beam. It is important that, in the weak-signal approximation, they do not interact with each other. Principally, the situation changes when a higher nonlinearity is taken into account; for example, the quadratic or the cubic nonlinearities, as is given below.

Therefore, an enormous number of harmonics, tens or even hundreds of them, have a possibility of interacting with each other by the three-wave parametric resonance mechanism (Chapter 4). Examples of the phenomenon of the three-wave resonance are discussed in detail in Chapters 11 and 12. It is done within the framework of the basic working mechanism theory of the parametrical FEL. However, in contrast to the described mechanism, the principal plurality appears as the principal difference of thus-analyzed models. Specifically, each of the SCW harmonics has a possibility to simultaneously get involved in many parametric-resonance interactions; this means that we get an enormous number of three-component parametric resonances mutually bound by plural harmonics.

As a result, we obtained, by a strange way, the self-consistent system when all the harmonics participate directly or indirectly in the formation of the amplitude dynamics of any other of the harmonics. Then we take into account the following circumstance: the higher number harmonics might have higher amplitudes than the lower number harmonics. This means automatically that we cannot use in such a situation any Fourier–Taylor series that is generally accepted in the traditional electrodynamics as solution for these kinds of problems. Ultimately, it becomes self-evident that, considering the mathematics, this problem can be considered as the most complicated in the whole electrodynamics of the plasma-like systems. Therefore, let us solve the problem.

First, let us formulate the conditions of the plural parametric resonances of SCW harmonics. Recall that, in agreement with the general theory of FEL [3–5], the kinematical analysis allows only to formulate the necessary conditions of the realization of this or another wave resonance when the sufficient conditions might be found only within the framework of the corresponding amplitude analysis, which we perform later.

We will use the previously obtained (Sections 13.1.1 and 13.1.2) linear laws of dispersion of the high-current two-stream relativistic system. This includes, for the increasing and decreasing waves

$$k_3(\omega_3)=\frac{\omega_3}{v_0}\pm\Gamma(\omega_3)\,,\ (\omega_3\le\omega_{cr}), \tag{13.18}$$

for fast and slow waves

$$k_3(\omega_3)=\frac{\omega_3}{v_0}+r_{s,f}\frac{\sqrt{15}}{2}\frac{\omega_p}{v_0\gamma_0^{3/2}}\,,\ (\omega_3\le\omega_{cr}), \tag{13.19}$$

and for the overcritical waves

$$\begin{aligned} k_3^{(1,2)}(\omega_3)&=\frac{\omega_3}{v_{01}}+r_{(1,2)}\frac{\omega_p}{v_0\gamma^{3/2}}\,; \\ & \qquad\qquad (\omega_3>\omega_{cr}) \\ k_3^{(3,4)}(\omega_3)&=\frac{\omega_3}{v_{02}}+r_{(3,4)}\frac{\omega_p}{v_0\gamma^{3/2}}\,, \end{aligned} \tag{13.20}$$

where $r_{s,f}=\pm1$ and $r_{1,2,3,4}=\pm1$ are the sign functions describing the slow (index *s*, corresponding to sign "+") and fast (index *f* corresponding to the sign "−"). All other notations are given above in Section 13.1.2.

As is mentioned before, the waves in Equations 13.18 and 13.19 correspond to the precritical region $\omega_3\le\omega_{cr}$, whereas the waves in Equation 13.20 correspond to the overcritical region $\omega_3>\omega_{cr}$. Then, to give our analysis a more universal character, we introduce the idea of the generalized law of dispersion

$$k_\chi=\frac{\omega_3}{v_0(1+\sigma\delta')}+r_\chi\frac{\omega_p}{v_0\gamma_0^{3/2}}\,, \tag{13.21}$$

where $\chi=1,2,3,4,5$ is the index that describes the type of SCW participating in the three-wave resonance. Here 1 corresponds to the growing waves, 2 and 3 are slow and fast precritical waves (all of them are within the $\omega_3\le\omega_{cr}$ region), and 4 and 5 are slow and fast overcritical waves ($\omega_3>\omega_{cr}$). The precritical region is described by the value of the sign function $\sigma=0$, where at $r_\chi=0$ by Equation 13.21, we obtain the increasing waves ($\chi=1$, see Equation 13.18), at $r_\chi=r_s=+\sqrt{15}/2$ the slow $\chi=2$, and at $r_\chi=r_f=-\sqrt{15}/2$ the fast SCW $\chi=3$ (see Equation 13.19). The remaining combinations of the parameters σ, r_χ correspond to the overcritical waves ($\omega_3>\omega_{cr}$). Particularly, at $\sigma=-1$, $r_\chi=r_{(1,2)}=\pm1$ [the slow ($\chi=4$) and the fast ($\chi=5$) waves of the first partial beam], and at $\sigma=+1$, $r_\chi=r_{(1,2)}=\pm1$ [the slow ($\chi=4$) and the fast ($\chi=5$) waves of the second partial beam]. Here we also used the renormalized shift of velocity of the partial beams $\delta'=(v_{01}-v_{02})/2v_0=\delta/2$.

Let us designate the frequencies and the wave numbers of the wave harmonics, which belong to those chosen for the analysis of the parametrical-resonant three waves, by indexes α, β, and γ. The peculiarity of the investigated variant of the parametric resonance is that here, in contrast to the standard approach to this type of problem (Chapters 11 and 12), the frequencies and the wave numbers of the interacting waves might have only certain discrete values. Therefore, each of the waves is a harmonic of one and the same lowest frequency ω_{31}.

$$\omega_\alpha = n_{3\alpha}\omega_{31};\quad k_{\chi\alpha} = n_{3\alpha}k_{\chi 31};$$

$$\omega_\beta = n_{3\beta}\omega_{31};\, k_{\chi\beta} = n_{3\beta}k_{\chi 31};$$

$$\omega_\gamma = n_{3\gamma}\omega_{31};\, k_{\chi\gamma} = n_{3\gamma}k_{\chi 31}, \tag{13.22}$$

where $n_{3\kappa} = n_{3\alpha}, n_{3\beta}, n_{3\gamma}$ are the numbers of harmonics, and the value $k_{\chi 31}$ is the generalized wave number of the first harmonic of SCW of the χth type. We rewrite generalized law of dispersion (Equation 13.22); this time, the frequency ω_3 acquires the value $\omega_\kappa = \{\omega_\alpha, \omega_\beta, \omega_\gamma\}$ (compare with Equation 13.21)

$$k_{\chi\kappa} = \frac{\omega_\kappa}{v_0(1+\sigma_\kappa\delta')} + r_{\chi\kappa}\frac{\omega_p}{v_0\gamma_0^{3/2}}, \tag{13.23}$$

where $\kappa = \alpha, \beta, \gamma$ (see Table 13.1). The remaining symbols were already described in the explanations to Equation 13.21 and in the Section 13.1.2.

By following the tradition [3–5], and by using the accepted symbols, the generalized conditions for the parametric three-wave resonance can be described by

$$\omega_\alpha + \nu\omega_\beta = \omega_\gamma;\; k_{\chi\alpha} + \nu k_{\chi'\beta} = k_{\chi''\lambda}, \tag{13.24}$$

where the indexes χ', χ'', χ''' reflect the different waves from the set κ; in general, they might belong to different waves χ. The introduction of the sign function $\nu = \pm 1$ is explained by the need to identify which of the frequencies ω_κ in the conditions (Equation 13.24) has the highest frequency in this condition. In particular, it is easy to see that, at $\nu = \pm 1$, the highest is the frequency ω_γ and at $\nu = -1$, the frequency ω_α.

By taking the designation of Equation 13.22, and the further conditions of Equation 13.24, we might rewrite it in a more specific form

$$n_{3\alpha}\omega_{31} + \nu n_{3\beta}\omega_{31} = n_{3\gamma}\omega_{31}; \tag{13.25}$$

$$n_{3\alpha}k_{\chi' 31} + \nu n_{3\beta}k_{\chi'' 31} = n_{3\gamma}k_{\chi''' 31}. \tag{13.26}$$

It is obvious that, in this case, the condition for the frequency (Equation 13.25) might be easily reduced to the form that we name the first condition of the resonance for a number of harmonics

$$n_{3\alpha} + \nu n_{3\beta} = n_{3\gamma}. \tag{13.27}$$

By using the definition of the generalized wave numbers from Equation 13.26, we obtain an analogous meaning of the second condition of resonance for harmonics

$$n_{3\alpha}\frac{\sigma_\gamma - \sigma_\alpha}{1+\sigma_\alpha\delta'} + \nu n_{3\beta}\frac{\sigma_\gamma - \sigma_\beta}{1+\sigma_\beta\delta'} = (r_{\chi\gamma} - r_{\chi\alpha} - \nu r_{\chi\beta})\frac{\omega_p(1+\sigma_\gamma\delta')}{\omega_{31}\gamma_0^{3/2}\delta'}. \tag{13.28}$$

The conditions 13.27 and 13.28 define all the possible types of the parametric-resonance interactions in the investigated model. Let us investigate briefly the most interesting of them.

13.1.5 Most Interesting Types of Resonances

Type 1: $\sigma_\kappa = 0$; $r_{\chi\kappa} = 0$: all waves belong to the precritical region ($\omega_3 \le \omega_{cr}$). In addition to this only the increasing waves participate in the three-wave interaction. The second condition (Equation 13.28) becomes degenerated and is reduced to the form that coincides with the first condition (Equation 13.27). By considering the physics, it means that all partial resonances for all harmonics are characterized by one and the same condition (Equation 13.27). Consequently, they have the possibility to participate simultaneously in all other parametric resonances. One should stress that this situation is really unique, because one cannot find another example of this type in the wave electrodynamics. Remember that the plasma beam system (Chapter 14) is not the second example because as is mentioned many times earlier, it represents a particular case of the two-stream system when one of the beams has average velocity equal to zero.

Type 2: $\sigma_\kappa = 0$, $r_{\chi\gamma} = 0$, $r_{\chi\alpha}, r_{\chi\beta} = \pm\sqrt{15}/2$, all waves, like in the previous case, belong to the precritical range ($\omega_3 \le \omega_{cr}$); however, in contrast to the previous case, here only one wave (e.g., $\kappa = \gamma$) among the waves, is the increasing wave. The other two waves correspond to the fast and the slow SCW, the dispersion law of which is presented in Equation 13.19. A simple analysis shows that apart from the first condition for the number of the harmonic (Equation 13.27), here we have one more (the second) condition: $r_{\chi\alpha} = -\nu r_{\chi\beta}$, which we obtained from Equation 13.28. In other words, in each set of three interacting waves, in the present case, there is only one slow and only one fast wave. The third of the waves must be an increasing wave. It is interesting that this type of plural wave resonances can develop simultaneously with the first-type resonances and, what is especially interesting, with the participation of the same waves.

Type 3: $\sigma_\gamma = 0$, $r_{\chi\gamma} = 0$, $r_{1,2}, r_{3,4} = \pm 1$. Here, one increasing precritical ($\omega_3 \le \omega_{cr}$) wave and two overcritical waves ($\omega_3 > \omega_{cr}$) participate in the resonance. Of course, because of the chosen scheme configuration for the interaction, the increasing wave of the frequency ω_γ cannot have the highest frequency. Therefore, we obtain the value for the sign function $\nu = -1$. It means that here, the wave of the frequency ω_α has the highest frequency (look for the condition of the parametric link in Equation 13.29). The first of the conditions of the parametric wave resonance for this model preserves the previous form (Equation 13.27), whereas the second condition (Equation 13.28) is transformed to

$$n_{3\alpha}\frac{\sigma_\alpha}{1+\sigma_\alpha\delta'} - n_{3\beta}\frac{\sigma_\beta}{1+\sigma_\beta\delta'} = (r_{\chi\alpha} - r_{\chi\beta})\frac{\omega_p}{\omega_{31}\gamma_0^{3/2}\delta'}. \tag{13.29}$$

It is obvious that at $r_{\chi\alpha} = r_{\chi\beta}$, when

$$n_{3\alpha} = n_{3\beta}\sigma_\alpha\sigma_\beta\frac{1+\sigma_\alpha\delta'}{1+\sigma_\beta\delta'}, \tag{13.30}$$

the resonance is impossible as long as this violates the first condition (Equation 13.27). The point is that, as is mentioned before, in the investigated system, the wave frequency might change only discreetly for the value no smaller than the frequency of the first harmonic. This means, in turn, that the numbers of these waves' harmonics can change for values not smaller than 1. However, in practice, as a rule, the normalized shift of the beam velocities δ' is a significantly small value as $\delta' \cong \Delta\gamma_0/2\beta_0\gamma_0 << 1$. Therefore, the second condition of the resonance of harmonics in Equation 13.30 cannot be fulfilled simultaneously with the first condition (Equation 13.27).

We can confirm that this contradiction is removed in the case when the condition $r_{\chi\alpha} = -r_{\chi\beta}$ is fulfilled

$$n_{3\alpha} = \sigma_\beta \sigma_\beta n_{3\beta} \frac{1+\sigma_\alpha \delta'}{1+\sigma_\beta \delta'} + 2r_{\chi\alpha} \frac{\omega_p(1+\sigma_\alpha)}{\omega_{31}\gamma_0^{3/2}\delta'} . \tag{13.31}$$

It is interesting to mention that this type of resonance interaction might take place in the process of the slow and fast wave of one and the same partial beam $\sigma_\alpha = \sigma_\beta$, and in a case when the increasing wave interacts with each individual overcritical wave of the beams ($\sigma_\alpha = -\sigma_\beta$). In the first case, the condition 13.29 acquires the shape of

$$n_{3\alpha} - n_{3\beta} = n_{3\gamma} = 2\sigma_\alpha r_{\chi\alpha} \frac{\omega_p(1+\sigma_\alpha\delta')}{\omega_{31}\gamma_0^{3/2}\delta'} , \tag{13.32}$$

and in the second case, we get the following result:

$$n_{3\alpha} + n_{3\beta} = 2\sigma_\alpha r_{\chi\alpha} \frac{\omega_p(1-\delta'^2)}{\omega_{31}\gamma_0^{3/2}\delta'} + \sigma_\alpha \delta' n_{3\gamma} . \tag{13.33}$$

One should mention that the given type of the resonance might be of interest only in some problems of the parametric amplification of electronic waves. However, for the investigated problems of the multiharmonic parametric interaction, it is of very limited interest. First, it might be found only in a case of a special tuning, when the value $2\omega_p/\omega_{31}\gamma_0^{3/2}\delta'$ is equal to an integer number. Such tunings are not discussed here because it is of no practical interest. Second, even in a most-impossible case, when such special tuning is used in an experiment, it will be revealed very soon that this is technologically not a simple problem as it appeared at first. For a better understanding of the essence of this case, we will perform some numerical evaluations.

Assuming for example that $\gamma_0 \sim 10$, $\omega_p \sim 1.65 \cdot 10^{11}\ s^{-1}$, $\Delta\gamma_0 \sim 2 \cdot 10^{-2}$ (what corresponds to $\delta' \sim 10^{-3}$), $\omega_{31} \sim 10^{13}\ s^{-1}$, from the conditions 13.31 through 13.33, it is easy to obtain: $n_{3\alpha} \sim n_{3\beta} \pm 1$, that is, it is the interaction of two neighboring harmonics. However, a real experiment reveals that the creation of this class of the wide high-current beams is a significantly complex and expensive problem. Consequently, the perspective to dedicate the significant attention and effort should require a strong motivation, what we regrettably do not have, considering the study of this class of systems.

Type 4: $\sigma_\kappa = \pm 1$, $r_{\chi\kappa} = \pm 1$: a parametric-resonance interaction of three overcritical waves of overcritical waves type (Equation 13.20). This type of resonance is significantly well known in the theory of FEL (Chapter 14) [38,48–52]. Its realization in the investigated model of harmonic resonance, in principle, does not bring anything new to its theory.

The fact that the given type of the resonances, in contrast to the one described above, does not have any correlation with the processes in the precritical region shows that it appears as if it is electrodynamically isolated from them. This means that the primary excitation of the overcritical waves cannot be a consequence of the other processes within the precritical region. In each resonance set of three waves, it needs a system input of at least two nonzero amplitudes of waves. For an illustration of this phenomenon, let us recall that a completely different situation is realized, for instance, in the case of a parametric interaction of three increasing harmonics (type 1).

In principle, it is enough to have only one first harmonic of SCW frequency ω_{31} for a further excitation of a cascade of all remaining harmonics, if possible, in the precritical region. In this case, the first harmonic interacts in a parametrically resonant manner with itself and excites the second harmonic; the second harmonic, in turn, resonantly interacts with the first and excites the third, etc. Consequently, a practical use of the overcritical resonances for the prolongation of the working part of the SCW spectra within the working volume of the cluster FEL, is possible only in the case when using the special sections of the modulation, which are able to provide the formation of the multiharmonic input spectra of the required configuration SCW.

13.1.6 Modes of Weak and Strong Interactions

In the theory of the traditional superheterodyne and parametric effects [3–5], it is known that the behavior of amplitudes of the interacting waves during the amplification process essentially determines the interacting mode, which is realized in the system. As is discussed in Chapters 11 and 12, in parametric FEL, such modes are predominantly called the Raman and Compton modes. In the case of superheterodyne FEL, they are called the Raman and Quasi-Compton modes. One meets terms like weak and strong wave interaction modes in the theory of other types of electrodynamics of plasma and plasma-like systems. Basically, in all these situations, we have to deal with the same physical concepts.

The physical essence of the discussed concepts is illustrated in Figure 13.1. In contrast to the material of the preceding Chapters 11 and 12, where processes in plasma-like objects like the parametrical FEL have been studied, in the case of theory of two-stream instability, we will use the traditional terminology of the electrodynamics of plasma. Instead of the above-mentioned Raman, Compton, and quasi-Compton modes, we will talk about the weak and strong interaction modes. Therefore, let us discuss them shortly in the here considered model of the two-stream instability.

Then we return to a further discussion of a semiqualitative physics of processes in the two-stream system. Considering the general physical nature of these class effects, we especially point to the following: As is already noted, there is a well-known fact in electrodynamics and optics that slow change of amplitude of a spectral component with time or spatial coordinates means that, in the real spectrum, it cannot be presented by the Dirac's delta function. In other words, it is in contrast to a case of true Fourier spectra, where any spectral line has infinitesimally small width; in real situations, we always have to do with spectral lines of the final (nonzero) width. Such lines are called the resonance curves. The key characteristic of each such curve is its half-width (see items 3 and 4 in Figure 13.1).

This circumstance makes us recall that, in the precritical region, besides the decreasing and the increasing wave, there exist also fast and slow waves. Their laws of dispersion are presented by Equation 13.19. By using the designation given in Figure 13.1, we rearrange it to the form

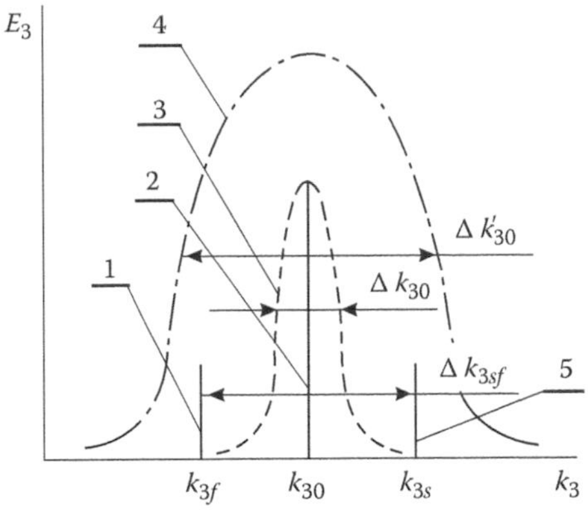

FIGURE 13.1
Illustration of the concepts strong and weak wave interactions in a plasma-like system. Here 1 is the spectral line of the slow SCW with the wave number k_{3f}; 2 is the spectral line of the increasing SCW with the wave number k_{30}; 3 is the real half-width of the spectral line for a case of a weak interaction; 4 is the real half-width of the spectral line for a case of a strong interaction; 5 is the spectral line of the fast SCW with the wave number k_{3s}; E_3 is the amplitude of the spectral line; $\Delta k_{3s,f} = \pm\left(\sqrt{15}/2\right)\left(\omega_p/v_0\gamma_0^{3/2}\right)$ is the dimension of the slit between the slow and fast SCW (Equation 13.24); Δk_{30} is the half-width of the resonance line of an increasing SCW for a case of a weak wave interaction; $\Delta k'_{30}$ is the half-width of an increasing SCW for a case of a strong wave interaction.

$$k_{3s,f} = \frac{\omega_3}{v_0} \pm \frac{\sqrt{15}}{2}\frac{\omega_p}{v_0\gamma_0^{3/2}} = k_{30} \pm \frac{\sqrt{15}}{2}\frac{\omega_p}{v_0\gamma_0^{3/2}}, \tag{13.34}$$

where k_0 is the wave number (real part) of the increasing and decreasing waves; k_{3s} and k_{3f} are, like before, the wave vectors of the slow and fast SCW; and the remaining values are defined above.

Graphically, this physical situation is presented in Figure 13.1 by the positions 1, 2, and 5. It is easy to see that the slow and the fast SCW are described by the split $\Delta k_{3s,f} = \pm\left(\sqrt{15}/2\right)\left(\omega_p/v_0\gamma_0^{3/2}\right)$, respectively, to the wave number of the increasing wave k_{30}. Correspondingly, a legitimate question arises: To what physical picture does the formal situation correspond, when the half-width of an increasing SCW is larger than the splitting between the fast and the slow SCW? The answer is well known: this is the so-called strong wave interaction system in a plasma-like system. Specifically, such physical picture is realized for the increasing SCW of a two-stream system in a case corresponding to the position 4 in Figure 13.1. For simplicity, we consider that for the very slow and fast SCW (see positions 1 and 5 in Figure 13.1), we can consider their spectral amplitudes, as before, as the lines of the zero width.

Let us clarify the above formulated deduction from the analysis of Figure 13.1 in detail. It is easy to deduct that, depending on the mutual correlation of the spectral half-widths of lines Δk_{30} and $\Delta k'_{30}$, we might have two principally different scenarios of the mutual interaction of waves in the system. The first case, described by the half-wave $\Delta k_{30} < \Delta k_{3sf}$, curve 3 in Figure 13.1, corresponds to the above-mentioned system of the weak amplification. As is shown above, in such case, the increasing wave with the wave number k_{30} might, in principle, interact by the parametric-resonance way with the other increasing spectral components (harmonics) separately by the present slow (k_{3s}) and fast (k_{3f}) waves. In contrast, in the second case ($\Delta k'_{30} = \Delta k_{30} > \Delta k_{3sf}$, curve 4 in Figure 13.1), the system stops to differentiate between the increasing wave and both of its satellite (slow and fast) SCWs. In this case, in the parametric interaction with other harmonics, the SCW does not exhibit the degenerate nature. The whole packet imaged in Figure 13.1 waves (the increasing 2 and both the satellite waves 1 and 5, respectively) interacts simultaneously with the rest of the two resonant waves. It is the main difference from the previous case, when only the single increasing wave 1 takes part in the interaction. Then, let us mention that all three waves in the packet have different energy signs (zero—the increasing, negative—the slow, and positive—the fast wave; see [3–5]). In correlation with this, it becomes obvious that the resulting dynamics of the amplitude of the discussed process of the strong interaction, in general, might be significantly different from the analogous dynamics in the case of a weak interaction. The last deduction confirms the performed calculation results (see below).

It should be mentioned that, in the purely precritical case, when the possibility of the harmonic amplification by the parametric resonance effect is not taken into account in calculations, the strong interaction cannot be realized in principle. It appears that in such a precritical model, as far as the spectral line widening Δk_{30} (owing to the amplification of the spectral harmonics only by the two-stream instability) is concerned, an equivalent amplification of the magnitude of the split between the slow and the fast SCWΔk_{3sf} (Figure 13.1) simultaneously takes place.

13.1.7 Truncated Equations

Thus, let us complete the above-composed semiqualitative view of the multiple resonances by corresponding mathematical description. For this, we will apply the method of the averaged characteristics (Chapter 6), that is, the calculation we perform in the cubic (or, more accurately, the so-called improved quadratic) approximation for wave amplitudes. The key methodic specificity of this type calculation is illustrated adequately in detail in the monograph [5]. However, notice that, as is clearly shown in the monograph, one specificity of the given calculation is that the small parameter of the problem is not proportional to the largest among the wave amplitudes as is traditionally accepted in the similar problems in electrodynamics. In the final calculation, this circumstance allows to calculate the full spectrum of SCW, including its anomalous parts.

As a result of the extensive analytical calculations, we obtain a system of the abbreviated equations for the amplitudes of harmonics $E_{\kappa,m}$ (where $m = n_{3\alpha}, n_{3\beta}, n_{3\gamma}$ is a generalized number of the harmonic), and each of them is considered slowly changing along the coordinate z

$$C_{2,\alpha,m}\frac{d^2E_{\alpha,m}}{dz^2}+C_{1,\alpha,m}\frac{dE_{\alpha,m}}{dz}+D_{\alpha,m}E_{\alpha,m}=C_{3,\alpha,m}E_{\beta,m}E_{\gamma,m}+F_{\alpha,m},$$

$$C_{2,\beta,m}\frac{d^2E_{\beta,m}}{dz^2}+C_{1,\beta,m}\frac{dE_{\beta,m}}{dz}+D_{\beta,m}E_{\beta,m}=C_{3,\beta,m}E_{\alpha,m}E^*_{\gamma,m}+F_{\beta,m},$$

$$C_{2,\gamma,m}\frac{d^2E_{\gamma,m}}{dz^2}+C_{1,\gamma,m}\frac{dE_{\gamma,m}}{dz}+D_{\gamma,m}E_{\gamma,m}=C_{3,\gamma,m}E_{\alpha,m}E^*_{\beta,m}+F_{\gamma,m}. \qquad (13.35)$$

The coefficients of the equations are determined by the parameters of the system, corresponding to the wave numbers and frequencies of the harmonics

$$D_{\chi,m} \equiv D(m\omega_\chi, k_{\chi,m}) =$$

$$= -ik_{\chi,m}\left(1 - \frac{\omega_p^2}{(m\omega_\chi - k_{\chi,m}\upsilon_1)^2\gamma_1^3} - \frac{\omega_p^2}{(m\omega_\chi - k_{\chi,m}\upsilon_2)^2\gamma_2^3}\right),$$

$$C_{1,\chi,m} = \partial D_{\chi,m}/\partial(-ik_{\chi,m}), C_{2,\chi,m} = \partial^2 D_{\chi,m}/\partial(-ik_{\chi,m})^2/2,$$

$$C_{3,\alpha,m} =$$

$$-k_{\alpha,m}\sum_{q=1}^{2}\left[\frac{\omega_p^2 e/m_e}{\Omega_{\alpha,q,m}\Omega_{\beta,q,m}\Omega_{\gamma,q,m}\gamma_q^6}\left(\frac{k_{\alpha,m}}{\Omega_{\alpha,q,m}} + \frac{k_{\beta,m}}{\Omega_{\beta,q,m}} + \frac{k_{\gamma,m}}{\Omega_{\gamma,q,m}} - 3\upsilon_q\gamma_q^2/c^2\right)\right],$$

$$C_{3,\beta,m} = -k_{\beta,m}C_{3,\alpha,m}/k_{\alpha,m}, C_{3,\gamma,m} = -k_{\gamma,m}C_{3,\alpha,m}/k_{\alpha,m}, \Omega_{\chi,q,m} = m\omega_\chi - k_{\chi,m}\upsilon_q,$$

$$\gamma_q = 1/\sqrt{1-(\upsilon_q/c)^2}, \; F_{\chi,m} = F_{\chi,m}(E_\alpha, E_\beta, E_\gamma) \qquad (13.36)$$

are functions, which take into account the nonlinear addenda and depend on harmonic amplitudes of the interacting waves (these functions in a manifested form are not given here because of its enormous awkwardness). Notice that Equation 13.36 determines the dispersion function of the mth harmonic of the type χ wave.

Let us return to the question of the previously discussed problem of the strong and weak interaction in the two-stream multiharmonic high-current system. Specifically, let us show how to create an adequate numerical criterion, by help of which it will be possible to classify the really occurring resonance-wave processes in a system into the strong and weak modes. In the theory of the FEL, a general idea of the construction of this type criterion is known (Chapters 4, 11, and 12, and monographs [3–5]). Applying it to the given case of the multiharmonic interactions in the two-stream system, this criterion can be obtained in the form

$$\left|\frac{C_{1\kappa m}}{C_{2\kappa m}}\right| \times \left|\frac{dE_{\kappa m}/dz}{d^2E_{\kappa m}/dz^2}\right| \sim \varsigma; \quad \kappa = \alpha, \beta, \gamma. \qquad (13.37)$$

At $\varsigma >> 1$, the weak interaction mode is realized; at $\varsigma << 1$, we have the strong interaction mode, and at $\varsigma \sim 1$, we have the so-called intermediate mode of interaction. Here all symbols coincide with those in Equations 13.34 and 11.35.

13.1.8 Dynamics of SCW Amplitudes: Resonances Only between the Increasing Waves

Let us assume that the plural parametrical resonances of the Type 1 (see Section 13.1.5) are realized. This means that we consider the resonant interactions only between the increasing waves. The results of the relevant numerical calculations are presented in Figures 13.2 through 13.5. In addition to this, the dependencies of the real amplitude of the first harmonic of the SCW $|E_{31}|$ in the dimensionless system length $T = z/L$ for the different two-stream models are presented in Figure 13.2. Here curve 1 describes the nonlinear amplification dynamics of the first harmonic $|E_{31}|$ in a case of the pure two-stream instability. Let us discuss shortly the presented results.

First, we turn to the circumstance, of which the numerical example, corresponding to the infrared frequency range (optimum frequency $\omega_{3opt} \sim 5.4 \cdot 10^{14}\ s^{-1}$), is presented. Hence, this result, as is mentioned above, is rather unusual for the conventional perception: of experts in the microwave vacuum electronics, who are accustomed to deal with the magnitudes of the SCW working frequency $\omega_{3\mathrm{opt}}$ in the microwave-millimeter range [53,54] only.

Second, we see a real possibility to realize the extremely high levels of the amplification of electron waves within the millimeter-infrared range. Hence, we can conclude that the two-stream instability is a rather promising means to solve the problem of the development of compact high-power relativistic electronic devices in the above-mentioned frequency range.

However, to use this mechanism in practice, at first we should solve the problem of the effective transformation of the amplified longitudinal electrostatic electron waves (the proper waves of the two-stream system) into the corresponding transverse electromagnetic waves. As is mentioned above, this problem can be solved by using specific FEL technologies. However, there arises a problem: how will the two-stream instability behave in the field of the FEL pumping, when the parametric resonant conditions are not satisfied. The curves 2 and 3 in Figure 13.2 are the answer to this question. It looks that, from the practical point of view, the effect of the traverse undulated pumping fields is not too important although it could have some physical interest. It is readily seen that the presence of the H-Ubitron pumping (curve 2 in Figure 13.2) suppresses the two-stream amplification process (compare the curve 2 with the curve 1 in Figure 13.2): and the saturation amplitude decreases from ~7.44 until 7.1 kV/m. At the same time, the saturation length increases from 1.29 until 1.36 m.

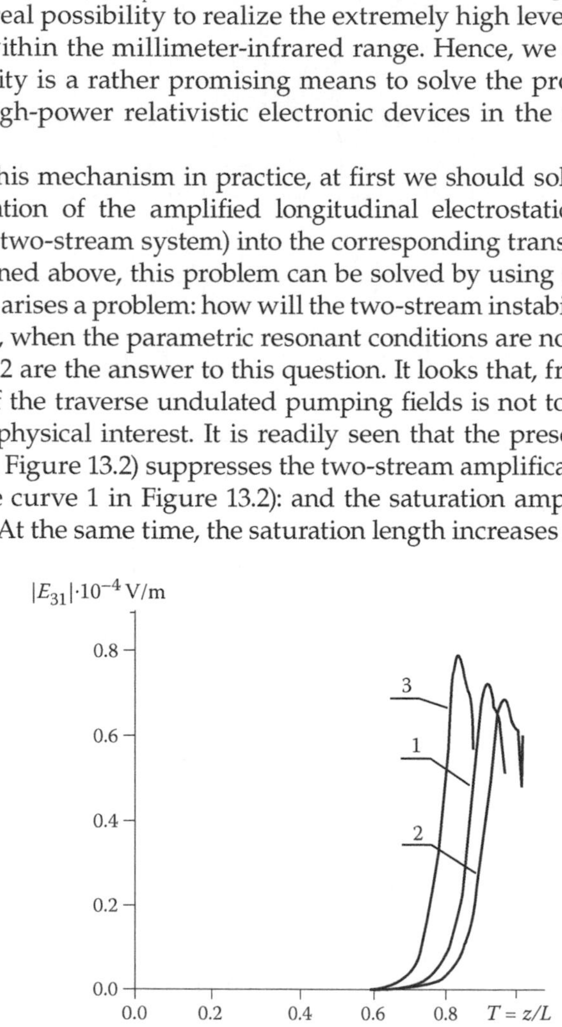

FIGURE 13.2
Dependencies of the amplitude of the first harmonic of SCW $|E_{31}|$ on the dimensionless system length $T = z/L$ for various models. Here curve 1 corresponds to the case of pure two-stream instability (external fields are absent); curve 2 is calculated for a two-stream beam moving within an H-Ubitron system; and curve 3 corresponds to the similar case of an EH-Ubitron model. Here $\omega_p = 10^{11}\ s^{-1}$; $L = 1.5$ m; $\gamma_0 = 10$; $|E_2| = 10^5$ V/m; $|B_2| = 200$ Gs; $|E_3| = 15$ V/m; $\Delta\gamma = 10^{-2}$; and $\omega_{3\mathrm{opt}} \cong 5.4 \cdot 10^{14}$ s^{-1} (it corresponds to the infrared range).

This suppression can be explained by the effect of the demodulation of the longitudinal electron bunches, which, in turn, is caused by the effect of the electron dispersion. An opposite situation is observed in the case of EH-pumping (curve 3 in Figure 13.2). Namely, there the saturation amplitude increases from 7.44 kV/m (curve 2) until 8.1 kV/m (curve 3) and the saturation decreases from 1.36 to 1.19 m.

As shown in the analysis, this phenomenon is related to the influence of the cooling effects, which is discussed in Chapter 7. The point is that, as is well known in microwave electronics, the saturation in vacuum devices with long-time interaction, as a rule, is accompanied by the appearance of an overgrouping effect. In Chapter 7, we clarified that a cooling effect is characterized by an opposite action; its influence leads to some degrouping of the electron bunches, that is, it can particularly compensate the above-mentioned overgrouping effect. The latter appears in Figure 13.2 as an increase in the saturation amplitude and, simultaneously, a decrease in the saturation length.

The dependencies of the amplitudes of the first ten harmonics on the dimensionless system length $T = z/L$ are shown in Figure 13.3. The characteristic feature of the considered physical process is that, as is readily seen, the waves with higher harmonic numbers can have larger amplitudes. Herewith, the amplitudes of the even-number harmonics turn out to be higher than the neighboring amplitudes of the odd-number harmonics. The same process is more clearly illustrated in Figure 13.4.

Another important observation is that the higher harmonics might exert an essential influence on the amplification process of the first electron-wave harmonic. Figure 13.5 illustrates the nature of this influence. The dependencies $|E_{31}(T)|$ are calculated for the case when such influence is neglected (curve 1) and for the case when this influence is taken into account (curve 2).

13.1.9 Dynamics of the SCW Amplitudes: Resonances of Different Types of Waves

We will examine the dependence of the amplitude of first harmonic of the increasing SCW (wave α, wave type 5) on the longitudinal coordinate. This dependence illustrates a

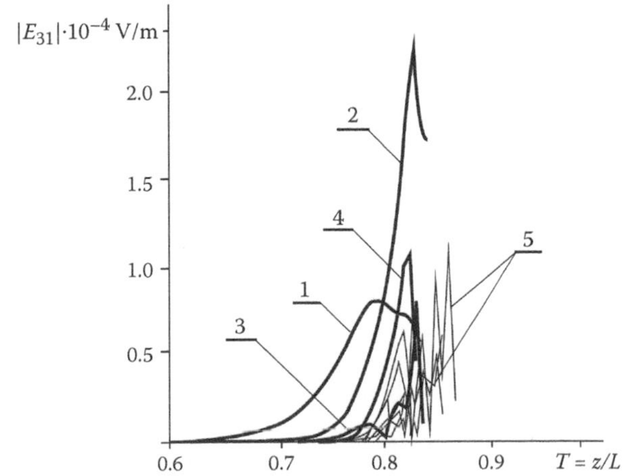

FIGURE 13.3
Dependencies of the amplitudes of the first ten harmonics on the dimensionless length $T = z/L$ for the case of the pure two-stream instability. Here curve 1 corresponds to the first harmonic; curve 2 to the second harmonic; curve 3 to the third harmonic; curve 4 to the fourth harmonic; and 5 are other higher harmonics. All system parameters are the same with Figure 13.2.

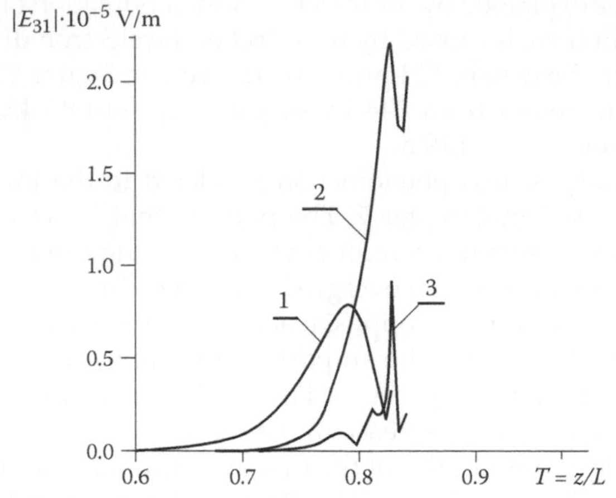

FIGURE 13.4
Dependencies of the amplitudes of the first three harmonics $|E_{3m}(T)|$ (m = 1,2,3) in a case of the presence of the EH-Ubitron pumping field. Curve 1 corresponds to the amplitude of the first harmonic; curve 2 is calculated for the second harmonic; and curve 3 corresponds to the third harmonic. Here $\Delta\gamma = 10^{-1}$ and $\omega_{3opt} \cong 5.4 \cdot 10^{13}\ s^{-1}$. Other parameters are the same as in Figure 13.2.

simultaneous effect of the first two types of resonances (see the relevant classification in Section 13.1.5) on the dynamics of development of the two-stream instability. The results of the calculation are given in Figure 13.6. To assure the possibility of a comparative analysis, the calculation was performed for the following cases:

1) when the effect only of the first harmonic of an increasing SCW wave was considered in the calculations
2) when the mutual interaction of the parametric resonance of 50 harmonics (see, for instance, Figures 13.3 and 13.4) of an increasing wave is considered, the various-

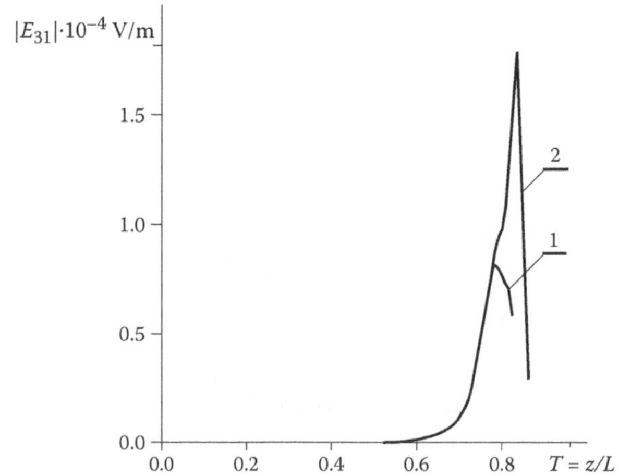

FIGURE 13.5
Illustration of the effect of the higher harmonic on the amplification process of the first harmonic. Curve 1 is calculated by taking into account the reciprocal effect of the first 20 harmonics, whereas curve 2 corresponds to the opposite situation. Here all parameters are the same as in Figure 13.2.

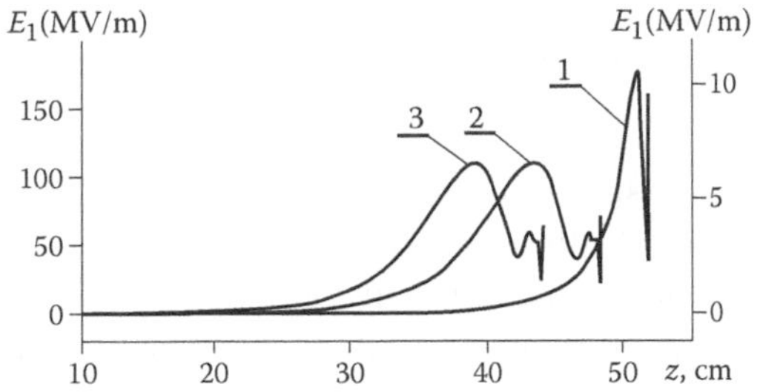

FIGURE 13.6
The dependence of the amplitude of the first harmonic of the increasing SCW on the longitudinal coordinate z. Here curve 1 corresponds to the case when only one harmonic of the increasing SCW wave is taken into account; curve 2 (the right-hand axis) corresponds to the case when 50 harmonics of the increasing wave are accounted for. The parametric interaction among them is also accounted, but the various-type parametric resonances are absent; curve 3 (left-hand axis) corresponds to the case when 50 harmonics participate in interactions. The action of parametric resonances of the waves of the same type as well as different types is considered in the calculation. The calculations were performed in parameters: $\omega_p = 1.5 \cdot 10^{11}\ s^{-1}$; $\gamma_0 = 4.5$; $\Delta\gamma = 0.5$; $E_{10} = 1$ V/cm.

type parametric wave resonance is absent, interaction of the Type 1 (see Section 13.1.5), curve 2

3) when the calculations consider also the 50 harmonics as the increasing waves as well as the harmonics of the precritical slow and fast waves, interaction of the Type 2 (see Section 13.1.5), curve 3

In the calculations, presented in Figure 13.6, the frequency of the first harmonic of the increasing wave SCW (wave α, wave type 1; see Table 13.1) was chosen such that the increment of its increasing Γ would be maximal. The frequency, which satisfies this condition, is equal to

$$\omega_{opt} = \sqrt{3}\omega_p / (2\delta\gamma_0^{3/2}) = \sqrt{3/8} \cdot \omega_{cr}. \tag{13.38}$$

Specifically such choice of the SCW frequency is used in the majority of the cases of the TSFEL.

Let us compare curve 1 with curve 2 in Figure 13.6. As is already mentioned, curve 1 was calculated without taking into account the effect of the higher harmonics; curve 2, on the

TABLE 13.1
Types of SCW in the Two-Stream Electron System

χ	σ_κ	$r_{\chi\chi}$	Wave Type
1	0	0	Increasing wave ($\omega < \omega_{cr}$)
2	0	$-\dfrac{\sqrt{15}}{2}$	Fast wave ($\omega < \omega_{cr}$)
3	0	$+\dfrac{\sqrt{15}}{2}$	Slow wave ($\omega < \omega_{cr}$)
4	+1	−1	Fast wave of the first beam ($\omega > \omega_{cr}$)
5	+1	+1	Slow wave of the first beam ($\omega > \omega_{cr}$)
6	−1	−1	Fast wave of the second beam ($\omega > \omega_{cr}$)
7	−1	+1	Slow wave of the second beam ($\omega > \omega_{cr}$)

other hand, took into account this effect. It is obvious that the difference between them is almost 30 times. By comparing this with the analogous result presented in Figure 13.5, we may convince ourselves that

1. An increase in the considered number of the harmonics from 20 (Figure 13.5) to 50 (Figure 13.6) parametrically interacting harmonics significantly decreases the magnitude of the first harmonic of the amplified SCW.
2. Included in the calculation are additional resonance interactions with the precritical slow and fast SCW, which leads to an unexpected decrease in the length of the amplification saturation, at approximately the same level of the saturation.

In the calculation of curve 3, we assume that on the system input, the two-velocity electron beam has also the nonzero first harmonics of the waves β and γ, which have the electric field potential equal to 500 V/cm, besides the first harmonic of the increasing wave α.

Once again, we stress that a calculation was performed of 50 harmonics, each of them of different types of waves, which participate in a parametric resonance. As it will be mentioned in the following section, in the given model, the amplitudes of the younger harmonics might appear (significantly) smaller than the amplitudes of the older harmonics (see, for instance, Figures 13.3 and 13.4). The investigation reveals that, in the electrodynamics of the plasma-like systems, the very fact of the successful performance of the nonlinear calculations with the participation of that quantity of harmonics, which are not described by the Fourier–Taylor divergence series, is a significantly serious achievement.

13.1.10 Dynamics of the SCW Spectra

For the formation of a wide multiharmonic spectrum of the two-stream wave system, the frequency of the first harmonic of an increasing SCW should be chosen significantly smaller than the optimal frequency ω_{opt}; this applies also to the critical frequency ω_{cr} [26–29,38,48]. Therewith, the effect of two-stream instability will be present for all harmonics, the frequency of which is smaller than the critical frequency ω_{cr}. In addition to this, the increments of the increase for the harmonics within the interval from ω_{31} to ω_{opt} are larger the larger is the harmonic number. Therefore, in the two-stream system, it is possible, to form an increasing wave with a wide multiharmonic spectrum, which has a significantly large anomalous segment, where the higher frequency harmonics exhibit a higher amplitude.

In Figure 13.7, an example of a spectrum of increasing harmonics SCW is presented. The spectrum represents a case, when in the system input ($z = 0$), the amplitude of the first harmonics of the increasing wave is equal to 10 V/cm and its frequency is 25 times smaller than the critical frequency ($\omega_{cr}/\omega_{\alpha,1} = 25$). In the calculation, 50 harmonics were taken into account. The remaining parameters are the same as those presented in Figure 13.6. One should mention that the spectrum presented in Figure 13.7 was obtained for the case of the Type 1 parametric interactions, that is, by taking into account the three-wave mutual parametrical-resonance interaction of harmonics of an increasing wave.

As is expected, the presented spectrum exhibits an anomalous segment from 1st to 15th harmonic, where the higher harmonic has a larger amplitude. Therewith, the maximal amplitude has a harmonic of optimal frequency ω_{opt}. As we see, the real width of this spectrum is determined by the frequency of the first harmonic $\omega_{\alpha,1}$ and the frequency ω_{min}, which corresponds to the harmonic with the minimal amplitude. Note that the frequency ω_{min}, as follows from the Figure 13.7, is higher than the critical frequency: $\omega_{min} > \omega_{cr}$.

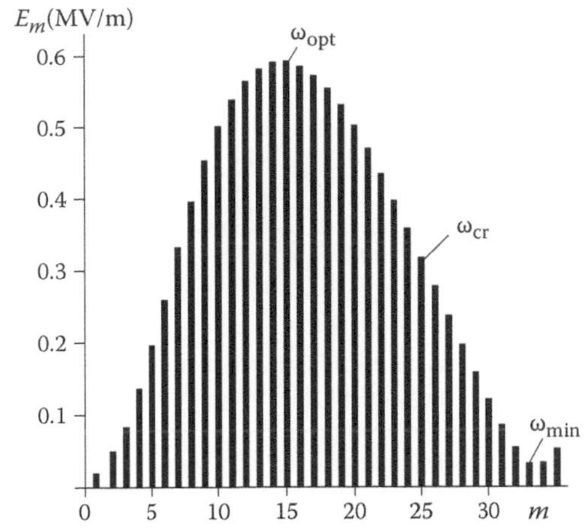

FIGURE 13.7
The distribution of the amplitudes of spectral harmonics E_m of an increasing SCW on its harmonic number m. On the system input ($z = 0$), the amplitude of the first harmonic is equal to 10 V/cm, and the amplitudes of the remaining harmonics are equal to zero. The frequency of the first harmonic is $\omega_1 = 3.1 \cdot 10^{11}$ s^{-1}. The spectrum is for the longitudinal coordinate $z = 110$ cm. The remaining parameters of the system are the same as in Figure 13.5.

Let us analyze how the shape of a spectrum of an increasing wave will change under the effect of the second-type parametric resonance. The dynamics of the spectrum of an increasing SCW wave is presented in Figures 13.8 and 13.9, which take into account such effect. The calculation was performed under the same conditions as in the case presented in Figure 13.6.

Figure 13.8 presents the spectrum of the increasing wave (wave α, wave type 1) at z = 50 cm, Figure 13.9 at $z = 90$ cm, and Figure 13.10 at $z = 109$ cm. The amplitudes of the wave β (wave type 2) and wave γ (wave type 3), which are in the parametric resonance with the 20th harmonic of the increasing wave α (wave type 1) are equal to 0.5 V/cm. In the calculation, we took into account the effect of 50 harmonics of each of the interacting waves.

As we can see, at the initial stage of the formation of the multiharmonic spectrum (Figure 13.8), plural parametric resonances excite the higher harmonics of the increasing wave α,

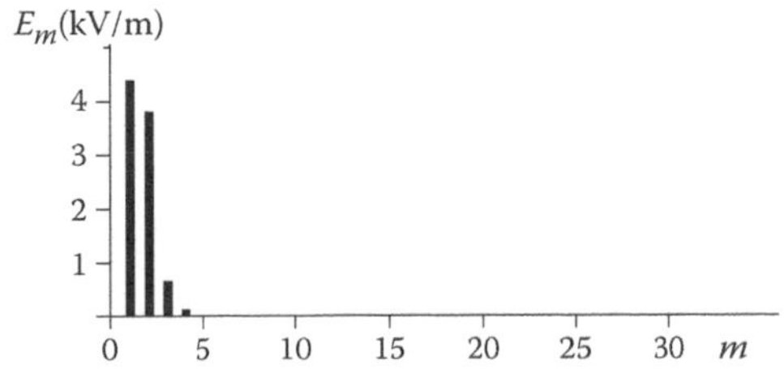

FIGURE 13.8
The distribution of the amplitudes of the harmonic E_m of an increasing wave α (wave type 1) on harmonic number m at $z = 50$ cm. On the system input ($z = 0$), the amplitudes of the first harmonic of the wave α is equal to 10 V/cm, those of the 20th harmonic of wave β and γ are equal to 0.5 V/cm, and the remaining harmonics are equal to zero. The frequency of the first harmonic of the wave α is $\omega_1 = 3.1 \cdot 10^{11}$ s^{-1}. The remaining parameters of the system are the same as those in Figure 13.5.

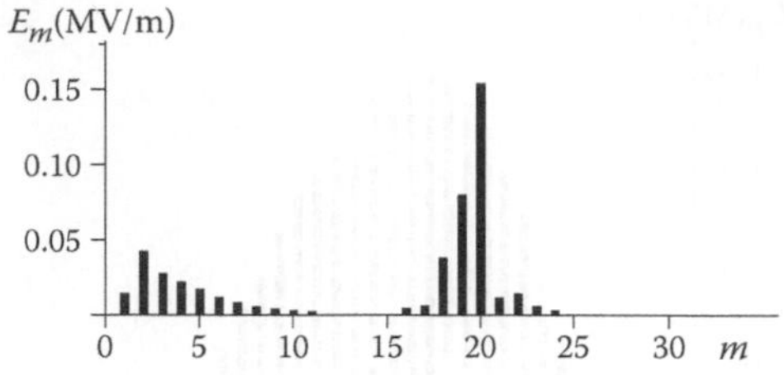

FIGURE 13.9
The distribution of the amplitudes of the harmonics E_m of the increasing wave α (wave type 1) by its harmonic number m at $z = 90$ cm. The remaining parameters of the system are the same as in Figure 13.5.

the amplitude of which decreases by an increasing number of harmonics. In the investigated system, the amplitudes of the waves β and γ, at the system input, are 20 times smaller than the amplitude of the first harmonic of the increasing wave α. Therefore, here (Figure 13.8), the remarkable effect of the parametrical resonance of the various-type longitudinal waves is not observed.

Furthermore (Figure 13.9), an anomalous spectrum forms owing to the two-stream instability. Therewith, the second harmonic has larger amplitude than the first harmonic, in contrast to the case presented in Figure 13.8. In addition to this, the 20th and the neighboring harmonics of the increasing wave α are generated owing to the parametrical resonance of the various types. In addition to this, their formation involved the parametric resonances of various types of wave as well as the two-stream instability. In this process, the role of the parametric resonance is reduced to the excitation of the corresponding harmonic of the increasing wave α, owing their subsequent growth to two-stream instability. As for the 20th harmonic of the wave α, the increment of the increase owing to the

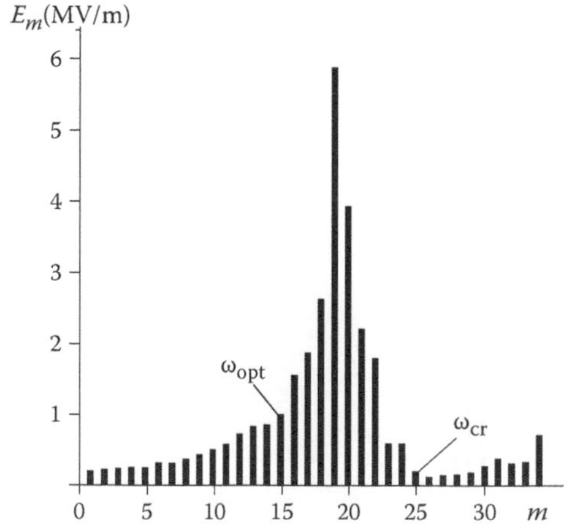

FIGURE 13.10
The distribution of the amplitudes of the harmonic E_m of the increasing wave α (wave type 1) as a function of its harmonic number m at $z = 109$ cm. At the system input ($z = 0$), the amplitudes of the first harmonic of the wave α are equal to 10 V/cm, those of the 20th harmonic of wave β and γ are equal to 0.5 V/cm, and the remaining harmonics are equal to zero. The frequency of the first harmonic of the wave α is $\omega_1 = 3.1 \cdot 10^{11}$ s^{-1}. The remaining parameters of the system are the same as in Figure 13.5.

two-stream instability is higher than that for the first harmonic so that at this stage of the process development, the amplitude of the 20th harmonic exceeds the amplitude of the first harmonic. It should be mentioned that, to secure this mechanism of the process development, it is enough, at the system input, to have the waves β and γ with relatively small amplitudes (0.5 V/cm).

Finally, the wide spectrum of the SCW wave, which is presented in Figure 13.10, is formed. It differs essentially from the case when the parametrical resonance of the different types of longitudinal waves is absent (Figure 13.7). The maximum in this spectrum is at the 19th harmonic. This is caused by the parametric resonance (of the various types of longitudinal waves), which is between the 20 harmonics corresponding to various types of waves and also the fact that the increase increment of the 19th harmonic is higher than that for the 20th harmonic. The form of the spectrum changed essentially, as seen by comparing Figures 13.7 and 13.10. The anomalous segment of the spectrum increased from 15 harmonics in Figure 13.7 to 19 harmonics in Figure 13.10. It should be mentioned also that the maximal amplitude increased significantly (10 times) from 0.6 V/m in Figure 13.7 to 6 MV/m in Figure 13.10.

Thus, the parametric resonance of the various types of longitudinal waves exhibits an important effect on the form of the multiharmonic spectrum of the SCW increasing longitudinal wave. Because of this, the given effect might be used for the formation of the cluster femtosecond waves by using the multiharmonic TSFELs (Section 13.3).

13.2 Ordinary Two-Stream Superheterodyne Free Electron Lasers

13.2.1 Model and Fields

Formulation and solving the problem of electromagnetic signal amplification in a TSFEL has much in common with the analogous parametric FEL problem (see Chapters 11 and 12). However, some peculiarities exist in the case of TSFEL. As is mentioned above, the main distinction is that even in linear approximation, the solutions for SCWs are increasing. This observation introduces some modifications in the integration of motion equations (Equation 10.27) as well as in the solution of the field part of the problem (Equation 10.10).

We consider the same, as before in Chapter 10, transversely unbounded FEL models with a charge-compensated relativistic electron beam. We assume that the beam, as a whole, drifts toward the positive z-axis.

We are oriented at using the hierarchic method of averaged characteristics (see Chapter 6).

The initial state of the system was determined by the electron distribution function $f_{00}\left(\vec{\mathcal{P}}\right) = \delta\left(\mathcal{P}_x\right)\delta\left(\mathcal{P}_y\right)g_{00}\left(\mathcal{P}_z\right)$ [where $\delta\left(\mathcal{P}_{x,y}\right)$ are the Dirac's delta-functions]. The function g_{00} describes the longitudinal kinetic of the beam. It is assumed that it has two bumps and depends evidently only on the longitudinal momentum $\mathcal{P}_z$, that is, we assume that the beam emittance in the transverse plane is neglected: $\mathcal{P}_\perp = \sqrt{\mathcal{P}_x^2 + \mathcal{P}_y^2} = const = 0$. It is considered that the external magnetic field is superweak (see classification in Chapter 7) and the model is stationary.

We assume also that the signal wave linearly polarized in plane XZ and propagating collinearly to the z-axis. The dispersion and impedance properties of the signal wave are described in terms of the method of simulating magnetodielectric (Chapter 10). We write the vector of strength of electric signal field in the traditional for this book form

$$\hat{\vec{E}}_1(z,t) = \frac{1}{2}(E_1 \exp(ip_1) + c.c.)\vec{e}_x, \quad (13.39)$$

where E_1 is the slowly varying amplitude,

$$p_1 = \omega_1 t - s_1 k_z \quad (13.40)$$

is the phase, $k_1 = \left|\vec{k}_1\right|$ is the wave number, $\vec{k}_1$ is the wave vector, $s_1 = \text{sign}\left\{\vec{k}_1\vec{n}\right\} = \pm 1$ is the sign function, and $\vec{e}_x, \vec{n}$ are the unit vectors along the x- and z-axes.

We consider the H-Ubitron pumping,

$$\hat{\vec{H}}(z,t) = \left[\hat{H}_2(z) + \hat{H}^{(g)}(z)\right]\vec{e}_y, \quad (13.41)$$

where

$$\hat{H}_2(z) = \frac{1}{2}[H_2 \exp\{ip_2\} + c.c.] \quad (13.42)$$

is the strength of the acting magnetic pump field,

$$p_2 = -k_2 z \quad (13.43)$$

is its phase, $k_2 = 1/\Lambda$, Λ is the pumping period,

$$\vec{H}^{(g)}(z) = \frac{1}{2}\left(H_2^{(g)} \exp\{ip_2\} + c.c.\right) \quad (13.44)$$

is the strength of additional magnetic field associated with the diamagnetic mechanism of generation (Chapters 11 and 12), and $H_2^{(g)}$ is the slowly varying complex amplitude. The diamagnetic effect is taken into account in the definition of the pumping field amplitude H_2.

13.2.2 Two-Stream Superheterodyne Free Electron Laser as a Hierarchic Wave-Oscillation System

Let us turn to the question about the hierarchic nature of the system under consideration. The hierarchic tree shown in Figure 13.11 evidently illustrates the discussed TSFEL model as a wave-oscillation hierarchic system.

As in the case of parametric FEL (Chapters 11 and 12), nonlinear interaction of electrons with the superposition of signal p_1 (Equation 13.38) and pumping waves p_2 (Equation 13.42) leads to excitation of the stimulated (induced) electron waves in the electron beam plasma. Some of these waves are characterized by the combination phases θ and ψ. As it has been noted earlier in Chapters 3, 4, and 7, the phases $p_{1,2}$ (and ψ, consequently) belong to the zero hierarchic level (see Figure 13.11), whereas the phase θ is related to the first hierarchic level. Apart from the stimulated waves, the proper electron waves also are excited in the electron beam. In contrast to the phases $p_{1,2}$, these waves are characterized by *initially*

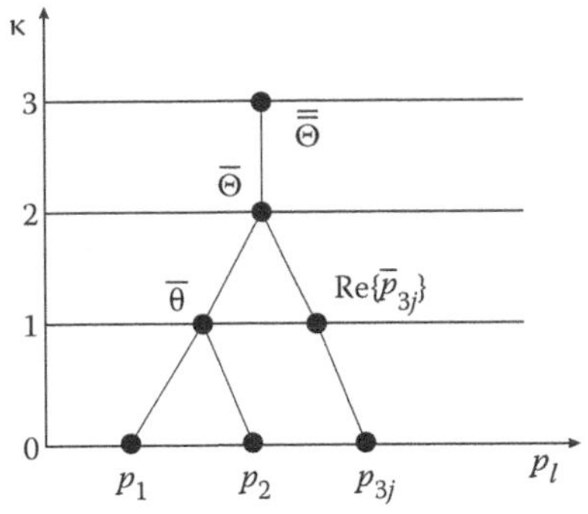

FIGURE 13.11
Two-stream SFEL as a hierarchic wave-oscillation system. Here $p_{1,2}$ are the phases of signal and pumping waves, p_{3j} (all zero hierarchic level) is the phase of the jth electron wave, κ is the number of hierarchic level, $\bar{\theta}$ (first hierarchic level) is the averaged slow combinative phase $\theta = p_1 - p_2$, $\bar{\Theta}$ (second hierarchic level) is the averaged combinative phase $\Theta = \theta - \mathrm{Re}\{\bar{p}_{3j}\}$ (where $\mathrm{Re}\{\bar{p}_{3j}\}$ is the real averaged part of the phase p_{3j}), and $\bar{\bar{\Theta}} = \mathrm{const}$ is the combinative phase of the third (highest) hierarchic level (see Chapters 3 and 4 for more details).

slowly wearying oscillation phases p_q. Herewith, one (or a few) of the proper waves (for instance, with the phase p_{3j}; see Figure 13.11) can satisfy the single particle resonant condition of the zero hierarchic level similar to $|\dot{\theta}|/|\dot{p}_{1.2}| \sim |\dot{\theta}|/|\dot{\psi}| << 1$ (Chapters 3, 4, and 7). As a result, two different phases $\bar{\theta}$ and p_{3j} exist on the first hierarchic level. Correspondingly, these phases also can be bounded by a resonant condition of the first hierarchic level similar to $|\dot{\Theta}|/|\dot{\Psi}| << 1$ (where $\Theta = \theta - \mathrm{Re}\{\bar{p}_{3j}\}$ and $\Psi = \theta + \mathrm{Re}\{\bar{p}_{3j}\}$ are the new superslow and quasifast combination phases), and so on. However, let us discus this topic in more detail.

Let us consider, for simplicity, the only case of main resonance (Chapter 3), when first harmonics of electron and electromagnetic waves take part in resonant conditions. According to Chapter 10, the strength of the electric field of the stimulated combination wave can be written as

$$\vec{E}_{st} = E_{st}\vec{n} = \frac{1}{2}[E_\theta \exp\{i\theta\} + E_\psi \exp\{i\psi\} + c.c.]\vec{n}, \tag{13.45}$$

where $E_{\theta,\psi}$ are the slowly varying amplitudes of combination waves with phases

$$\theta = p_1 + \sigma p_2; \ \psi = p_1 - \sigma p_2, \tag{13.46}$$

where $\sigma = \pm 1$, as before, is the sign function. The phase θ is considered as a slowly varying value that is determined by the condition of the single particle parametric resonance (Chapters 3 and 7). This resonance belongs to the zero hierarchic level.

Strength vector of the proper electron wave field is given by

$$\vec{E}_i = \vec{E}_i\vec{n} = \frac{1}{2}\sum_{q=1}^{M}(E_{iq}\exp\{ip_q\} + c.c.). \tag{13.47}$$

Here E_{iq} is the slowly varying complex amplitude,

$$p_q = \omega_q t - k_q z \tag{13.48}$$

is the phase, ω_q is the cyclic frequency, and k_q is the wave number of the qth proper wave.

Then we remind that relevant resonant condition for the wave parametric resonance can be formulated as the closeness of one of proper electron wave (Equation 13.46) and some partially stimulated electron wave (Equation 13.44) (see the first hierarchic level in Figure 13.11),

$$\bar{\theta} \approx \mathrm{Re}\{\bar{p}_{3j}\}, \tag{13.49}$$

where j is the number of the proper electron wave. The one that is similar to relationship 13.48, as is mentioned already, is the classical definition of wave resonance (Chapter 4). As a result, both the stimulated (Equation 13.44) as well as proper (Equation 13.46) waves merging form only SCWs of the two-stream electron beam.

13.2.3 Formulation of the Problem

Similar to the previous case (see Chapters 11 and 12), we employ the electric support to suppress the amplification saturation mechanisms. To isochronize the interaction, we apply to the interaction region the electric support field

$$\vec{E}_0^{(n)} = \vec{n}E_0^{(n)}. \tag{13.50}$$

Besides that, we take into account nonlinear generation of longitudinal electric field and corresponding comments

$$\vec{E}_0^{(g)} = \vec{n}E_0^{(g)}. \tag{13.51}$$

We take kinetic equation 10.27 and Maxwell's equations 10.10 for the initial ones. We supplement the Maxwell–Boltzmann system with definitions of current and space charge densities (Equations 10.11 and 10.12). Then, as before, we reduced the general superheterodyne amplification problem to solve two particular problems, namely, the problem of electron beam motion in a given field (i.e., solving, for instance, kinetic equation 10.27) and the problem of field excitation for given beam motion (i.e., solving Maxwell equations 10.10).

Let us discuss peculiarities of the motion problem in the TSFEL theory. We solve in this section the problem of two-velocity beam motion in a given signal, pumping, and SCW fields by the method of averaged characteristics. General calculation procedure is similar to that described in Chapter 7, and illustrated earlier in Section 13.1 and Chapter 11. Therefore, let us confine ourselves by discussing only some aspects of methodological importance. Details of calculation technique are analogous to the procedure described in Chapter 12.

In the case under consideration, one-particle resonance condition can be formulated in two ways. On one hand, for a parametric mechanism, we have a condition that is analogous to that discussed in Chapter 7, that is,

$$\left|\frac{d\psi}{dt}\right| \Big/ \left|\frac{d\theta}{dt}\right| \sim \xi_1 >> 1, \tag{13.52}$$

where all definitions are self-evident in view of that set forth in Chapter 7 and other chapters. On the other hand, the two-stream instability can occur in the system. Therefore, for this process, we also have to formulate conditions of the type similar to Equation 13.52. Such conditions must have distinctions for the Raman and Compton resonance mechanisms. As follows from the above comments to Figure 4.7 and definition 13.48, in the degenerate Raman case, only two SCWs (growing and damped) of the four proper electron waves get in the parametric resonance band. Below we show that these SCWs can have phases with equal real parts. We introduce the following notation for SCW phases:

p_{1d}—slow SCW
p_{2d}—fast SCW
p_{3d}—increasing SCW
p_4—damped (decreased) SCW

The latter classification will be substantiated in terms of dispersion analysis of the kinetic version of the two-stream system. Relevant conditions for the one-particle resonance under degenerate Raman mode can be written as

$$\left|\frac{dp_{3d}}{dt}\right| \Big/ \left|\frac{d\psi}{dt}\right| \sim \left|\frac{dp_{4d}}{dt}\right| \Big/ \left|\frac{d\psi}{dt}\right| \sim \frac{1}{\xi_1} << 1;$$

$$\left|\frac{dp_{1d}}{dt}\right| \Big/ \left|\frac{d\psi}{dt}\right| \sim \left|\frac{dp_{2d}}{dt}\right| \Big/ \left|\frac{d\psi}{dt}\right| \sim 1. \tag{13.53}$$

For the quasi-Compton mode, analogous conditions are given by

$$\left|\frac{dp_{1d}}{dt}\right| \Big/ \left|\frac{d\psi}{dt}\right| \sim \left|\frac{dp_{2d}}{dt}\right| \Big/ \left|\frac{d\psi}{dt}\right| \sim \left|\frac{dp_{3d}}{dt}\right| \Big/ \left|\frac{d\psi}{dt}\right| \sim \left|\frac{dp_{4d}}{dt}\right| \Big/ \left|\frac{d\psi}{dt}\right| \sim 1. \tag{13.54}$$

Thus, we see in the former case that only two of four phases are slow; in the latter case, all four phases are slow.

13.2.4 Concept of the Space Charge Waves in the TSFEL Nonlinear Theory

Then we turn to wave resonance condition 13.48. In the case of Raman mode, this condition can be reformulated as

$$p'_{3d} \approx p'_{4d} \approx \theta, \tag{13.55}$$

and for the quasi-Compton mode we have

$$p'_{1d} \approx p'_{2d} \approx p'_{3d} \approx p'_{4d} \approx \theta, \tag{13.56}$$

where $p'_{qd} = \mathrm{Re}\{p_{qd}\} \approx p'_3$. In what follows, we see that within the context of relationships 13.55 and 13.56, the expression for slowly varying part of the field of SCWs (Equation 13.49) can be written as

$$\hat{E}_3(z,t) = \frac{1}{2}\left(\tilde{E}_3 \exp\{ip'_3\} + c.c.\right), \tag{13.57}$$

where the imaginary parts of phases p_{qd} are taken into account in complex amplitude $\tilde{E}_3$. Thus, the latter turns out to be a slowly varying function of z and t even in the first approximation, and even for a cold electron beam (see solutions 13.9 through 13.14). This point makes an important methodological distinction of this superheterodyne problem as compared to the previous parametric one. To specify z- and t-dependencies of the amplitude, we have to solve the linear problem.

We put all nonlinear terms in initial Maxwell–Boltzmann equations 10.10 and 10.27 equal to zero. Let the initial distribution function be given by

$$f_0^{(0)} = f_0^{(0)}(\vec{\mathcal{P}}) = \sum_{\alpha=1}^{2} f_{0\alpha}^{(0)}(\vec{\mathcal{P}}) = \sum_{\alpha=1}^{2} \delta(\mathcal{P}_x)\delta(\mathcal{P}_y) g_\alpha(\mathcal{P}_z), \tag{13.58}$$

where $\alpha = 1{,}2$ is the number of a partial one-velocity electron beam, and the rest of the notations are traditional for this book. It is not difficult to find the linear dispersion law for electromagnetic waves. We have (compare with an equivalent expression for the case of parametric FEL given in Chapter 10)

$$D_1 = -i\left(\mu^{-1}k_1^2 - \varepsilon_1\frac{\omega_1^2}{c^2} + \frac{\omega_p^2}{c^2}\sum_{\alpha=1}^{2}\int_{-\infty}^{+\infty}\frac{g_\alpha(\mathcal{P}_z)}{\gamma_{z\alpha}}d\mathcal{P}_z\right) = 0;$$

$$k_1(\omega_1) = k_1 = \frac{N_1\omega_1}{c}\left(1 - \frac{\omega_p^2}{\varepsilon_1^2\omega_1^2}\sum_{\alpha=1}^{2}\int_{-\infty}^{+\infty}\frac{g_\alpha(\mathcal{P}_z)}{\gamma_{z\alpha}}d\mathcal{P}_z\right)^{\frac{1}{2}}, \tag{13.59}$$

where $\gamma_{z\alpha} = \left(1 - \beta_{z\alpha}^2\right)^{-1/2}$, $\beta_{z\alpha} = v_{z\alpha}/c$ is the dimensionless velocity of an electron belonging to the αth REB; ε_1 is the dielectric permittivity of simulating magnetodielectric; ω_p is the plasma frequency of a partial (each one-velocity) electron beam ($\omega_{p1} = \omega_{p2} = \omega_p$); $N_1 = \sqrt{\varepsilon_1\mu_2}$ is the retardation factor; and μ_1 is the magnetic permeability of simulating magnetodielectric (Chapter 10). The basic form of solutions reproduces Equation 13.38.

Other dispersion equations can be easily shown to be of the form

$$D_2 = \left(k_2^2 + \frac{\omega_p^2}{c^2}\sum_{\alpha=1}^{2}\int_{-\infty}^{+\infty}\frac{g_\alpha(\mathcal{P}_z)}{\gamma_{z\alpha}}d\mathcal{P}_z\right) \neq 0, \tag{13.60}$$

(i.e., we have the improper pumping field)

$$D_3 = \left(k_3 + m\omega_p^2 \sum_{\alpha=1}^{2} \int_{-\infty}^{+\infty} \frac{1}{\omega_3 - k_3 v_{z\alpha}} \frac{\partial g_\alpha(\mathcal{P}_z)}{\partial \mathcal{P}_z} d\mathcal{P}_z \right) = 0. \tag{13.61}$$

Then we assume relativistic electron beam cold

$$g_\alpha(\mathcal{P}_z) = \delta(\mathcal{P}_z - \mathcal{P}_{z\alpha}), \tag{13.62}$$

and thus dispersion equation for SCWs ($D_3 = 0$) can be reduced to the form Equation 13.8.

$$1 - \omega_p^2 \sum_{\alpha=1}^{2} \frac{1}{\gamma_{0l}^3(\omega_3 - v_{\alpha 0}k_3)^2} = 0, \tag{13.63}$$

where, as before, $v_{\alpha 0} = v_\alpha(z=0)$; $\gamma_{0\alpha} = (1 - (v_{0\alpha}^2/c^2))^{-1/2}$. Hence, using solutions 13.10 through 13.14, we can write the following explicit expressions for phases p_{qd} for the case $|v_0\Gamma'\delta/\omega_p|$ << 1:

$$p_{1d} = \omega_3 t - k_3' z + \sqrt{15}\Gamma z \text{ - slow SCW } (\mathrm{Im}\{p_{1d}\} = 0)$$

$$p_{2d} = \omega_3 t - k_3' z - \sqrt{15}\Gamma z \text{ —fast SCW } (\mathrm{Im}\{p_{2d}\} = 0)$$

$$p_{3d} = \omega_3 t - k_3' z - i\Gamma z = p_{3d}' + ip_{3d}'' \text{ —increasing SCW}$$

$$p_4 = \omega_3 t - k_3' z + i\Gamma z = p_{4d}' + ip_{4d}'' \text{ —damped (decreasing) SCW,} \tag{13.64}$$

where $p_{1d}' = p_{2d}' = p_{3d}' = p_{4d}' = p_3'$ (see definition 13.13); $k_3' = \omega_3/v_0$; $v_0 = (v_1 + v_2)/2$; $\delta = (v_1 - v_2)/v_0$. It is considered that optimum condition 13.11 is satisfied for increasing and damped SCWs

$$\omega_3 = \omega_{3\text{opt}} = \frac{\sqrt{3}\omega_p \gamma_0^{3/2}}{\Delta\gamma}(1 - \gamma_0^{-2}), \tag{13.65}$$

where $\Delta\gamma = \gamma_{10} - \gamma_{20}$, $\gamma_0 = (1 - (v_0^2/c^2))^{1/2}$. Then, according to Equation 13.12, increasing wave growth rate and the decrement of damped attain their maximums

$$\Gamma \cong \frac{\omega_p}{2v_0\gamma_0^{3/2}}. \tag{13.66}$$

Thus, the solution of linear system of equations for SCW can be expressed in terms of four integration constants C_q, that is,

$$\hat{E}_3(z,t) = \frac{1}{2}\sum_{q=1}^{4} C_q \exp\{ip_{qd}\} + c.c. \tag{13.67}$$

To apply the method of slowly varying amplitudes, we have to reduce Equation 13.67 to the form 13.57, that is, to eliminate three of four constants C_q. We consider the Raman interaction mode first and impose the boundary conditions

$$z = 0, \quad \hat{E}_3(z,t) = 0. \tag{13.68}$$

The expression for SCW field strength $\hat{E}_3(z,t)$ can be written as

$$\hat{E}_3(z,t) = \frac{1}{2}\sum_{q=1}^{2} C_q \exp\{ip_{qd}\} + \frac{1}{2}\sum_{q=3}^{4} C_q \exp\{ip_{qd}\} + c.c. \tag{13.69}$$

As follows from Equations 13.52 and 13.55, the Lagrange electron phases p_{3d} and p_{4d} (i.e., phases of electron oscillations under action of the field 13.67) are slow. We assume that condition 13.68 holds for the fast and slow (resonance) parts of the SCW field separately (i.e., quite reasonable). Therefore, we obtain

$$C_1 = -C_2 = C'/2\,;\ C_3 = -C_4 = C/2. \tag{13.70}$$

Correspondingly, we can rewrite Equation 13.69 as

$$\hat{E}_3 = \frac{1}{4}C(\exp\{ip_{3d}\} + \exp\{ip_{4d}\} + c.c.) + \hat{E}_{3\text{fast}} = \frac{1}{2}(\tilde{E}_3 \exp\{ip_3'\} + c.c.) + \hat{E}_{3\text{fast}}, \tag{13.71}$$

where $\hat{E}_{3\text{fast}}$ is fast oscillating on the first hierarchic level (with respect to electron oscillation phases) part of the SCW field. As calculations show, the latter does not exert any essential influence on the accuracy of results obtained in cubic-nonlinear approximation. Therefore, in what follows, we neglect the addition of $\hat{E}_{3\text{fast}}$ on the right-hand side of Equation 13.71. In such a case, we rewrite Equation 13.71 to a simpler form

$$\hat{E}_3 \cong \frac{1}{2}(\tilde{E}_3 \exp\{ip_3'\} + c.c.), \tag{13.72}$$

where

$$\tilde{E}_3 = E_3 sh(\Gamma z) \tag{13.73}$$

is the total SCW amplitude, and $E_3 \equiv C$ (i.e., the constant that at the following steps of the calculation is a slowly varying variable). It is clear that definition 13.73 reproduces Equation 13.57.

In the quasi-Compton mode case, the situation is somewhat more complicated. Within the context of Equations 13.54 and 13.56, all four proper waves (Equation 13.60) are in resonance with the stimulated (induced) oscillation waves (Equation 13.44). Hence, all four corresponding electron phases p_{qd} are slow. The only fast phase is combination phase ψ given by Equation 13.45. This infers that all four integration constants C_q must enter the solutions and influence the dynamics thereof. To eliminate three of them, additional boundary conditions must be imposed. For these, we take Equation 13.70 and the condition

$$C' = C, \tag{13.74}$$

where constants C' and C are determined by Equation 13.70. Above assumptions are made for qualitative reasons based on the analogy with linear wave properties in a one-stream system [3–5]. Then relation 13.67 can be written in the form

$$E_3(z,t) = \frac{1}{2}\{C[sh(\Gamma z) + i\sin(\Gamma' z)]\exp\{ip_3'\} + c.c.\} = \frac{1}{2}(\tilde{E}_3'\exp\{ip_3'\} + c.c.), \tag{13.75}$$

that reproduces Equation 13.69 except for complex amplitude, that is,

$$\tilde{E}_3' = E_3[sh(\Gamma z) + i\sin(\Gamma' z)]. \tag{13.76}$$

Here $\Gamma' = \sqrt{15}\Gamma$ (see solutions 13.13).

Thus, different expressions for amplitudes E_3 (Equations 13.73 and 13.75) introduce distinctions in general formulation of the problem for quasi-Compton and Raman interaction modes. In fact, the situation in one-stream parametric models is analogous. We disregard this aspect in the theory developed in Chapter 11. This does not mean, however, that under certain conditions (e.g., for combined Compton–Raman mode), these distinctions make no effect on the general character of processes in the parametric FEL. These points should be always borne in mind (and taken into account) in criteria analysis of the applicability of the theory developed.

13.2.5 Truncated Equations in the Cubic Nonlinear Approximation

Calculation technology of asymptotic integration in treating initial kinetic equation 10.27 is almost similar to the procedure described in Chapter 12. The only distinctive methodological point is as follows. Inasmuch as in two-stream model SCW amplitudes are slowly varying functions even in linear approximation and for cold electron beam (see expressions 13.72 and 13.66), the linear amplitude $\bar{f}_1^{(1)}$ is a slowly varying function, too. Thus, we have

$$\bar{f}_1^{(1)} = \bar{f}_{10}^{(1)}\chi_1(z,t), \tag{13.77}$$

where $\bar{f}_{10}^{(1)} = \bar{f}_1^{(1)}(z=0,t=0)$ is the initial averaged first-order approximation amplitude, and $\chi_1(z,t)$ is the slowly varying part of the latter. Therefore, in the second- and higher-order approximations, unlike the previous parametric case, we have to vary only the constant $\bar{f}_{10}^{(1)}$ rather than the whole amplitude $\bar{f}_1^{(1)}$.

Let us evaluate advantages and disadvantages of calculation procedures described in Chapter 12 (approaches that employed once- and twice-averaged kinetic equations). The first troublesome peculiarity of the problem under consideration is unsatisfactory convergence of amplitude hierarchy series for distribution function $\bar{f}$. This infers that attaining acceptable (for practical purposes) accuracy can require taking into account more than three harmonics. In such a case, twice-averaging procedure is more preferable because it makes it possible for a given length of the system $L \sim \xi_1$ to involve more harmonics (up to several tens). Herewith, the most effective way is to program this calculation scheme together with the method of slowly varying amplitudes (analogous to calculation strategy used in Section 13.1 for the nonlinear analysis of two-stream instability). Unfortunately, the volume of the book is too confined. Therefore, in this section, we discuss only the simplest and most obvious calculation schemes similar to those in Chapter 12 (some calculation examples with a larger number of the accounting harmonics is given in the next section). Apart from that, we simplify the problem assuming the model stationary, electron beam cold, and the mode Raman. As a result, we obtain the following system of cubic-nonlinear truncated equations for slowly varying amplitudes:

$$\left(\bar{A}_1\frac{d}{dz}+\bar{A}_2\frac{d^2}{dz^2}\right)E_1=\bar{A}_3\tilde{E}_3H_2^*+A_{4a}E_1\left|E_1\right|^2+\bar{A}_{4b}E_1\left(\left|E_1\right|^2-\left|E_{10}\right|^2\right)+\bar{A}_5E_1\left|H_2\right|^2+\bar{A}_6H_2^*$$

$$\left[\frac{d\tilde{E}_3}{dz}-\Gamma cth(\Gamma z)\tilde{E}_3\right]+\bar{A}_7E_1\int_0^z\left|\tilde{E}_3\right|^2dz'+\bar{A}_8E_1\int_0^z\left\{\tilde{E}_3^*\left[\frac{d\tilde{E}_3}{dz'}-\Gamma cth(\Gamma z')\tilde{E}_3\right]\right\}dz'$$

$$+\bar{A}_9E_1\int_0^z\left\{\tilde{E}_3\left[\frac{d\tilde{E}_3^*}{dz'}-\Gamma cth(\Gamma z')\tilde{E}_3^*\right]\right\}dz'+\bar{A}_{10}E_1\int_0^z\left\{E_1^*H_2^*\tilde{E}_3\right\}dz'+\bar{A}_{11}E_1\int_0^z\left\{E_1H_2\tilde{E}_3^*\right\}dz'$$

$$+\bar{A}_{12}\tilde{E}_3H_2^*\int_0^z\left\{\left|\tilde{E}_3\right|^2\right\}dz'+F_1\left(E_0^{(g)}\right);\qquad(13.78)$$

$$\left\{\bar{C}_2\frac{d^2}{dz^2}+\left[\bar{C}_1-2C_2\Gamma cth(\Gamma z)\right]\frac{d}{dz}+\bar{C}_2\Gamma^2\left[2cth(\Gamma z)-1\right]-\bar{C}_1\Gamma cth(\Gamma z)\right\}$$

$$\times\tilde{E}_3=\bar{C}_3E_1H_2+\bar{C}_{4a}\tilde{E}_3\left|E_1\right|^2+\bar{C}_{4b}\tilde{E}_3\left(\left|E_1\right|^2-\left|E_{10}\right|^2\right)+\bar{C}_5\tilde{E}_3\left|H_2\right|^2$$

$$+C_6\tilde{E}_3\left|\tilde{E}_3\right|^2+\bar{C}_7H_2\frac{dE_1}{dz}+\bar{C}_8\tilde{E}_3\int_0^z\left\{\left|\tilde{E}_3\right|^2\right\}dz'+\bar{C}_9\tilde{E}_3\int_0^z\left\{\tilde{E}_3^*\left[\frac{d\tilde{E}_3}{dz'}-\Gamma cth(\Gamma z')\tilde{E}_3\right]\right\}dz'$$

$$+\bar{C}_{10}\tilde{E}_3\int_0^z\left\{\tilde{E}_3\left[\frac{d\tilde{E}_3^*}{dz'}-\Gamma cth(\Gamma z')\tilde{E}_3^*\right]\right\}dz'+\bar{C}_{11}\tilde{E}_3\int_0^z\left\{E_1^*H_2^*\tilde{E}_3\right\}dz'+\bar{C}_{12}\tilde{E}_3\int_0^z\left\{E_1H_2\tilde{E}_3^*\right\}dz'$$

$$+\bar{C}_{13}E_1H_2^*\int_0^z\left\{\left|\tilde{E}_3\right|^2\right\}dz'+\bar{C}_{14}\left[\frac{d\tilde{E}_3}{dz}-\Gamma cth(\Gamma z)\right]\int_0^z\left\{\left|\tilde{E}_3\right|^2\right\}dz'+F_3(E_0^{(g)}),\qquad(13.79)$$

$$\left(\bar{K}_1\frac{d}{dz}+K_0\right)H_2^{(g)}=\bar{K}_3E_1^*E_3+\bar{K}_{4a}H_2\left|E_1\right|^2+\bar{K}_{4b}H_2\left(\left|E_1\right|^2-\left|E_{10}\right|^2\right)$$

$$+\bar{K}_5E_1^*\left[\frac{d\tilde{E}_3}{dz}-\Gamma cth(\Gamma z)\tilde{E}_3\right]+\bar{K}_7H_2\int_0^z\left\{\left|\tilde{E}_3\right|^2\right\}dz'+\bar{K}_8H_2\int_0^z\left\{\tilde{E}_3^*\left[\frac{d\tilde{E}_3}{dz}-\Gamma ctg(\Gamma z)\tilde{E}_3\right]\right\}dz'$$

$$+\bar{K}_9H_2\int_0^z\left\{\tilde{E}_3\left[\frac{d\tilde{E}_3^*}{dz}-\Gamma cth(\Gamma z)\tilde{E}_3^*\right]\right\}dz'\,\bar{K}_{10}H_2\int_0^z\left\{E_1^*H_2^*\tilde{E}_3\right\}dz'+\bar{K}_{11}H_2\int_0^z\left\{E_1H_2\tilde{E}_3^*\right\}dz'$$

$$+\bar{K}_{12}E_1^*\tilde{E}_3\int_0^z\left|\tilde{E}_3\right|^2dz'+F_2\left(E_0^{(g)}\right), \tag{13.80}$$

where $F_k(E_0^{(g)})$ $(k=1,2,3)$ are associated with the generated longitudinal electric field contribution, and z' is the dummy integration variable. Unfortunately, the evident form of the expressions for nonlinear coefficients $\bar{A}_l,\bar{C}_m,\bar{K}_n$ $(l,m,n=1,2,3,\ldots)$ is unacceptably cumbrous. Therefore, we omit them here.

Analogous, in principle, sets of equations can be also obtained for the case of Dopplertron pumping, or for the models with focusing longitudinal magnetic field. However, let us confine ourselves by representation equations (Equation 13.80) only because the mentioned systems are also too cumbersome. Nevertheless, furthermore, we will use such equations for numerical illustrations of some physical situations in these types of TSFEL.

The next step is to consider basic nonlinear effects in the model under consideration.

13.2.6 TSFEL Gain Factor

The analysis shows [1–30] that the physics of processes in TSFEL is much more diversified and involved than in the case of one-stream devices. Let us illustrate this assertion more evidently.

It is convenient to analyze the TSFEL in terms of the so-called *equivalent one-stream model.* This model is similar to the TSFEL model except that the electron beam is assumed one-velocity. This enables us to illustrate all typical distinctions of physical processes in SFEL comparing them with analogous effects in the one-stream system. We take the set of truncated equations (Equations 13.78 through 13.80) for basic system, and regard analogous parametric equations (see Chapter 12) as an equivalent system of equations.

The dynamics of gain factor as a function of the dimensionless longitudinal coordinate is shown in Figure 13.12. Figures 13.12a and b correspond to two-stream superheterodyne and equivalent one-stream models, correspondingly. It can be verified that the results of nonlinear theory satisfy the main conclusion of the weak signal theory: the gain factor of TSFEL can be high for a moderate length of the system. For example, for dimensionless saturation length $T=T_s^{sat}$ (see Figure 13.12a) of about 140 cm, the gain factor attains a value ~6.82 · 10^5, whereas an analogous parameter of an equivalent model at the same length is negligible (Figure 13.12). Gain factors of such magnitude can be attained for lengths more than six times larger: gain factor ~4 · 10^5 is attained for $T=T_p^{sat}\cong 888$ cm.

Now we consider the typical qualitative distinctions of signal wave amplification dynamics in two models to be compared. In Figure 13.12a, curve 1 shows the damping of oscillations of the gain factor for the segment $T>T_s^{sat}$. Both oscillation period and amplitude

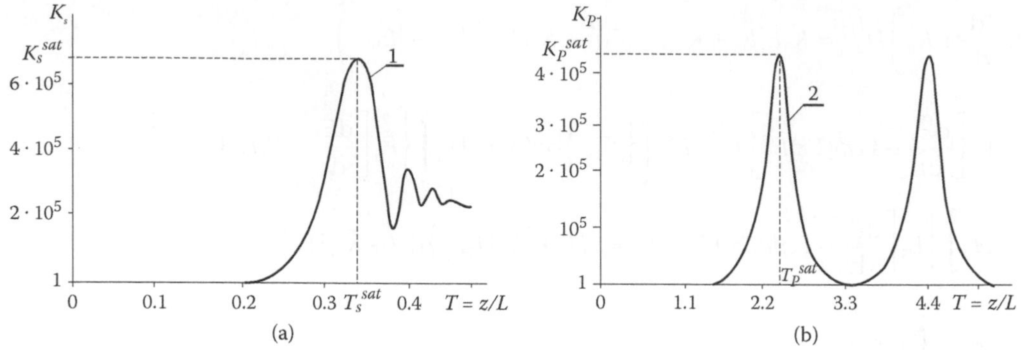

FIGURE 13.12
Dependence of the signal wave gain factors K_s (TSFEL model) and K_p (parametric model) as a function of the normalized coordinate $T = L/z$. Here $\omega_p = 9 \cdot 10^{10}$ s^{-1}; $\omega_1 = 2.9 \cdot 10^{12}$ s^{-1}; $\Lambda = 2.3$ cm; $E_{10} = 100$ V/m; $E_{30} = 10^{-2}$ V/m; $H_2 = 210$ Gs; $L = 400$ cm. (Both the magnetic and electric field generation is disregarded; the plasma frequency ω_p corresponds to each of the partial electron beams: $\omega_{p1} = \omega_{p2} = \omega_p$).

decrease as dimensionless length T increases. In an equivalent one-stream model (Figure 13.12b), these values remain unchanged for this segment. The explanation is rather simple. Unlike the one-stream model, amplifying effect on SCW in the TSFEL does not stop when the system attains saturated state (i.e., the point $T = T_s^{sat}$ in Figure 13.12a). Just this observation causes the characteristic form of the amplification curve in Figure 13.13.

Curve 2 of this Figure 13.13 shows the growth of SCW amplitude due to two-stream instability mechanism only ($E_1 = H_2 = 0$). When the parametric and two-stream instabilities are overlapped, the position of point $\tilde{\tilde{T}}_s^{sat}$ on curve 1 is determined by the competition of two SCW amplification mechanisms. Parametric mechanism weakens SCW for $T > T_s^{sat}$

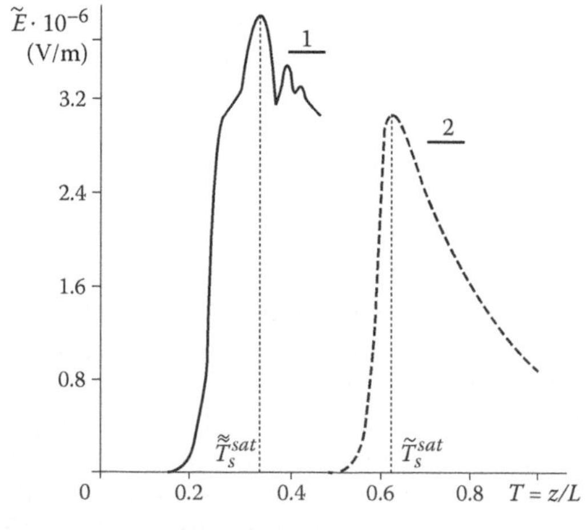

FIGURE 13.13
The SCW electric field amplitude $|\tilde{E}_3|$ as a function of the dimensionless length $T = z/L$ in the superheterodyne (curve 1) and pure two-stream (i.e., in the case $E_1 = H_2 = 0$). Here all parameters are similar to those of Figure 13.12.

(see Figure 13.12), whereas two-stream mechanism amplifies this wave (curve 2 in Figure 13.13). Therefore, saturation point $\tilde{\tilde{T}}_s^{sat}$ on curve 1 of Figure 13.13 is shifted to the right. Moreover, the bucket [3–5], which is formed due to the superposition of the electromagnetic ponderomotive and SCW field potentials, is much higher and, under other similar conditions, its height increases with T. This infers that both the amplitude and the period of electron bunch oscillations in such buckets (that just determine the character of curves in Figure 13.13) must gradually decrease.

13.2.7 TSFEL Efficiency

Efficiency dynamics for the two-stream and one-stream models is shown in Figures 13.14a and b.

First, we turn our attention to the fact of too low magnitudes of the maximal electron efficiency ($\eta_{smax} \approx 1.5\%$). Therefore, the question arises: Can this system be used effectively as a basis for the cluster systems (Chapter 1)? The correct answer contains the two key points. The first is that practically the same level of the maximal electron efficiency η_{pmax} can also be observed in the case of parametric equivalent FEL model (see Figure 13.14b). This means that the parametrical saturation mechanism determines the maximal electron efficiency in both these types of models.

The second of the mentioned key points is a direct sequence from the first one: similar to the parametrical FEL (Chapter 12), such saturation mechanism can be successfully suppressed by the use of one of the known isochronization methods (Chapters 7 and 12). It should be mentioned that even old experiments in the 1980s [54] on the isochronous first FELs have shown the possibility to attain the maximal efficiency $\eta_{smax} \sim 40\%$. The modern theory promises even higher efficiency levels (Chapters 7 and 12) [3,5,55–58]. All this means the following. On one hand, the discussed problem is not critical technologically and, on the other hand, theoretical ways for its solution are already illustrated in Chapters 7 and 12. Basically, a possibility of isochronization of the considered model is presented in the developed theory [see, for instance, the electric support (Equation 13.50)]. Nevertheless, accounting all that was said above, let us continue further discussion in the framework of the nonisochronous simplest TSFEL model.

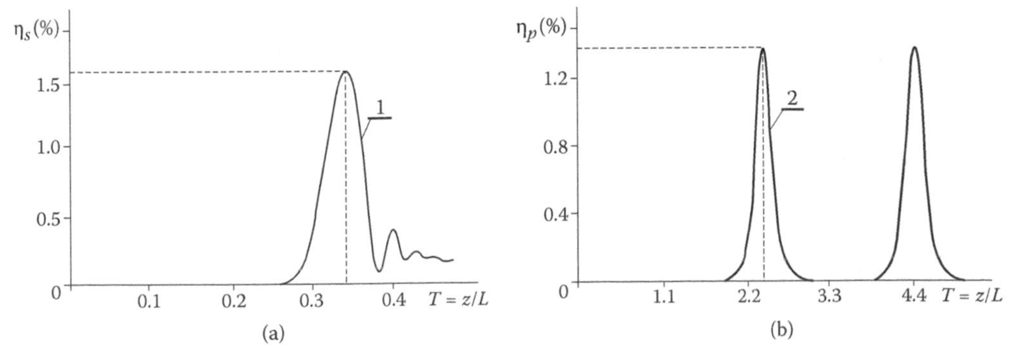

FIGURE 13.14
Dependencies of the efficiencies η_s, η_p of interaction in the two-stream (Fig. 13.14a) and equivalent parametric (one-stream) (Fig. 13.14b) models, correspondingly, on the dimensionless coordinate $T = z/L$. Here all parameters are similar to those in Figure 13.12.

Concerning the content of Figure 13.14, we can say that its material does not reveal any qualitative difference from the above-mentioned material on amplification dynamics (see Section 13.2.7). Let us remind once again of the discussion about the similarity of parametrical and superheterodyne saturation mechanisms. As is noted already, the efficiencies in both cases are nearly equal at saturation points of two-stream model $T_s^{sat} \sim 0.35$, and one-stream model, $T_p^{sat} = 2.4$. This implies that the effect produced by two-stream instability in SFEL reduces to effective additional bunching of the electron beam. Beam kinetic energy conversion into signal wave energy is completely governed by the parametric mechanism, like in parametric one-stream systems. The state of two-stream beam seems to make no appreciable effect on the mechanism occurrence. However, this impression is apparent. To verify this assertion, let us analyze the influence of longitudinal electric field nonlinear generation on the efficiency of signal wave amplification.

13.2.8 Generated Longitudinal Electric Field

The functions $F_k(E_0^{(g)})$, entering Equations 13.78 through 13.80, describe the influence of the effect of nonlinear generation of electric field. Inasmuch as for $\Gamma z > 1$ $sh(\Gamma z) \sim ch(\Gamma z) \sim (1/2)\exp\{\Gamma z\}$, we can write these functions in a simpler form

$$F_1\left(E_0^{(g)}\right) = \bar{A}_{15}E_1\left(\int_0^z E_0^{(g)}\,dz'\right)^2 + \bar{A}_{16}H_2\tilde{E}_3\left(\int_0^z E_0^{(g)}\,dz'\right)^2 + \bar{A}_{17}E_1E_0^{(g)}$$

$$+\bar{A}_{18}\frac{dE_1}{dz}\int_0^z E_0^{(g)}\,dz' + \bar{A}_{19}E_1\int_0^z E_0^{(g)}\int_0^{z'}\left|\tilde{E}_3\right|^2 dz'dz'' + \left(\bar{A}_{20}+\bar{A}_{21}\right)E_1\int_0^z\left|\tilde{E}_3\right|^2\int_0^{z'}E_0^{(g)}\,dz'dz';$$

$$F_2\left(E_0^{(g)}\right) = \bar{K}_{13}\left(H_2+H_2^{(g)}\right)\int_0^z E_0^{(g)}\,dz' + \bar{K}_{14}H_2\left(\int_0^z E_0^{(g)}\,dz'\right)^2$$

$$+\left(\bar{K}_{17}+\bar{K}_{18}\right)H_2\int_0^z\left|\tilde{E}_3\right|^2\int_0^z E_0^{(g)}\,dz'dz'' + \bar{K}_{19}H_2\int_0^z E_0^{(g)}\int_0^z\left|\tilde{E}_3\right|^2 dz'dz'';$$

$$F_3\left(E_0^{(g)}\right) = \bar{C}_{16}\frac{d\tilde{E}_3}{dz}\int_0^z E_0^{(g)}dz' + C_{17}\tilde{E}_3E_0^{(g)} + \bar{C}_{18}\tilde{E}_3\int_0^z E_0^{(g)}\,dz' + \bar{K}_{15}E_1^*\tilde{E}_3\int_0^z E_0^{(g)}\,dz' + \bar{K}_{16}H_2E_0^{(g)}$$

$$+\bar{C}_{19}E_1H_2\int_0^z E_0^{(g)}dz' + \bar{C}_{20}\tilde{E}_3\left(\int_0^z E_0^{(g)}\,dz'\right)^2 + \bar{C}_{21}\tilde{E}_3\int_0^z E_0^{(g)}\int_0^{z'}\left|\tilde{E}_3\right|dz'dz''$$

$$+\left(\bar{C}_{22}+\bar{C}_{23}\right)\tilde{E}_3\int_0^z\left|\tilde{E}_3\right|^2\int_0^{z'}E_0^{(g)}\,dz'dz'' \tag{13.81}$$

where $\bar{A}_l, \bar{K}_n, \bar{C}_m$ are the nonlinear coefficients. Unfortunately, their evident form turns out to be too cumbersome. Therefore, we omit them here. The following nonlinear equation can be obtained for generated longitudinal electric field $E_0^{(g)}$

$$E_0^{(g)}\left(\mathcal{D}_0-2\mathcal{D}_1\int_0^z E_0^{(g)}\,dz'+\mathcal{D}_2\int_0^z\left|\tilde{E}_3\right|^2 dz'\right)+\left|\tilde{E}_3\right|^2\left(\mathcal{D}_3\int_0^z E_0^{(g)}\,dz'+c.c.\right)=$$

$$+\mathcal{D}_6\left|\tilde{E}_3\right|^2\int_0^z\left|\tilde{E}_3\right|^2 dz'+\left(\mathcal{D}_7\tilde{E}_3^*\frac{\partial\tilde{E}_3}{\partial z}+\mathcal{D}_8E_1H_0\tilde{E}_3^*+c.c.\right), \qquad (13.82)$$

where evident expressions for the coefficients $\mathcal{D}_s$ ($s = 1,2,\ldots,8$) are unacceptably cumbrous. Therefore, we omit them here.

Let us discuss the results of numerical calculations.

The influence of the effect of nonlinear generation on the amplification is shown in Figures 13.15 and 13.16. Figure 13.15a shows the gain factor as a function of dimensionless length without (curve 1) and with (curve 2) nonlinear generation for the two-stream superheterodyne model. Figure 13.15b gives analogous curves for the equivalent one-beam model. It is not difficult to draw general conclusion that the generated electric field influences the amplification in TSFEL much stronger than in analogous one-stream devices. This observation is confirmed more clearly by wave efficiency dependencies on the length of the system. The gain factor (under saturation) in TSFEL becomes 5.2 times lower (Figure 13.15a), whereas the efficiency (Figure 13.16a) becomes 22 times lower. For equivalent one-stream FEL model, analogous values are about ~1.58 and ~2.2 times (Figures 13.15b and b). In the discussed cases, an appreciable portion of the kinetic beam energy is spent for the electric field generation.

Thus, the effect of nonlinear generation of longitudinal electric field belongs to one of the dominant mechanisms of amplification saturation in TSFEL. Therefore, though being physical exotics rather than real effect in one-stream systems, this phenomenon cannot be disregarded in calculations of TSFELs.

13.2.9 Generated Magnetic Field

In the previous discussion in Chapters 11 and 12, we have mentioned that two mechanisms of improper H-Ubitron magnetic field generation are known to occur in FEL. The

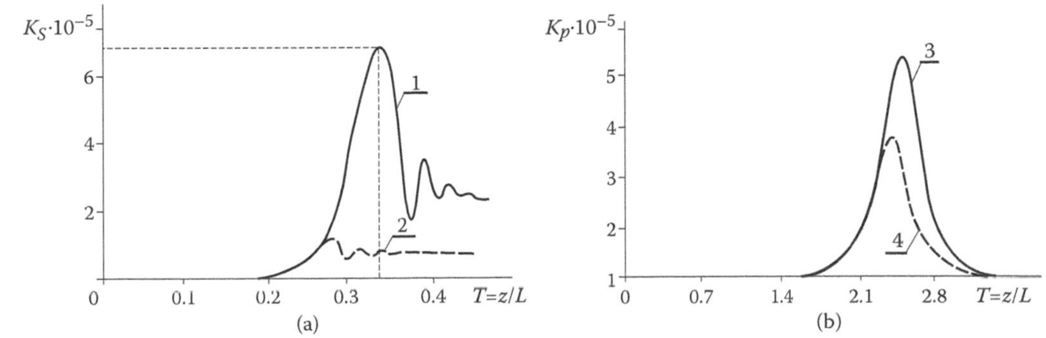

FIGURE 13.15
Influence of the effect of nonlinear generation of longitudinal electric field on the gain factors of the two-stream (Fig. 13.15a) and equivalent one-stream (Fig. 13.15b) models. Here curves 1 and 3 are calculated disregarding this effect; curves 2 and 4 take the effect into account. The parameters of the models are similar to those given in Figure 13.12.

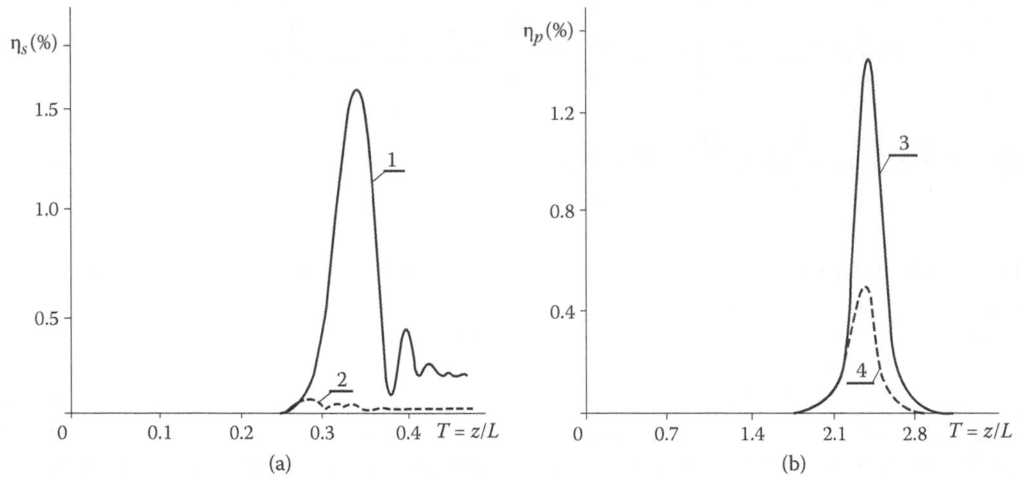

FIGURE 13.16
Dependencies of the efficiencies η_s, η_p of interaction in the two-stream (Fig. 13.16a) and equivalent parametric (one-stream) (Fig. 13.16b) models, correspondingly, on the dimensionless coordinate $T = z/L$. Here curves 1 and 3 are calculated ignoring the nonlinear electric field generation; curves 2 and 4 take this effect into account. All parameters are similar to those in Figure 13.12.

first mechanism is diamagnetic. The analysis does not reveal any distinctions between manifestations of this mechanism in one-stream and two-stream systems. The situation is quite different as far as the nonlinear wave-resonant mechanism is concerned. Let us discuss this aspect in more detail.

Comparing the results of calculations concerning the influence of effect of generation of magnetic field on processes in one-beam and two-stream systems suggests rather interesting conclusions. For example, fundamental qualitative distinctions exist between relevant dependencies. In particular, in Section 12.5, the result of generated field influence was classified as a reactive depression of the basic operation mechanism. Amplification saturation level remains practically unchanged and the effect actually reduces to increasing the saturation length T_p^{sat}. In the two-stream case, there occurs a pronounced decrease of the amplification saturation level and insignificant increase of the length T_s^{sat}. This justifies regarding magnetic field generation as one more mechanism of amplification saturation in TSFEL.

The physical picture of the behavior of the generated magnetic field in a two-stream model is a very interesting find. Comparing corresponding results, we can see that the generated magnetic field can be in antiphase with the pumping field in the saturation domain of a TSFEL. Thus, we can say that the phenomenon manifests pronounced diamagnetic features. The second distinction is that nonlinear evolution of the phase $\arg\left\{H_2^{(g)}\right\}$ starts long before quasilinear (exponential) amplification law $K_s(T)$ is violated. This illustrates the insufficiency of traditional definition of the nonlinear interaction stage associated with the violation of the exponential amplification law. Actually, everything is determined by specifying the effects under consideration. In the above case, conventional definition does not work.

Let us draw some conclusions of this section. The general conclusion is as follows: nonlinear superposition of two instabilities (two-stream and parametric) gives rise to the effect that does not resemble any of these two. This forms the ground for classifying this effect as a new phenomenon—superheterodyne amplification of electromagnetic waves [2–19]. FEL based on this effect has an enormously high gain factor (e.g., K_s ~10^5) for relatively moderate working lengths L~ (1 ÷ 1.5)m). The penalty for this advantage is strong manifestation of amplification saturation mechanisms, such as the nonlinear generation of electric and magnetic fields that are negligible in traditional one-beam (parametric) systems. This makes TSFEL an attractive research field in view of both applications and scientific knowledge.

13.2.10 Multiharmonic Processes

As is mentioned above, one of the characteristic features of the TSFEL mechanism is the presence of a large number of higher electron harmonics of approximately equal order (Section 13.1). Hence, the question arises about the action of this harmonics at the signal amplification process. Apart from that, the pumping wave as well as the initial signal wave also can have multiharmonic frequency specter. Therefore, let us further make clear some possible answers for the formulated question.

Some results of relevant calculations in this direction are given in Figures 13.17 through 13.22. Moreover, the results presented in Figures 13.17 through 13.20 illustrate the TSFEL-amplifier constructed on the basis of doubled 1.15-MeV induction injectors (see Chapter 9 for more details). Figures 13.20 and 13.22 are concerned with the TSFEL constructed on the basis of 20-MeV undulation accelerator (see also Chapter 9). Owing to this, the signal working frequency of devices of both types is essentially different.

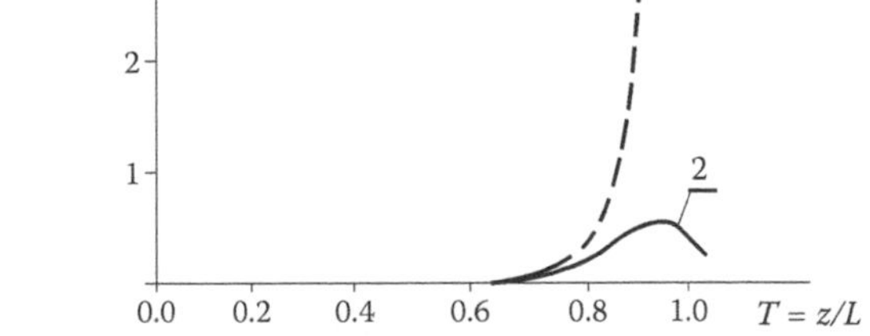

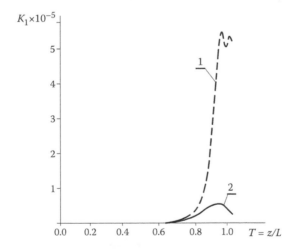

FIGURE 13.17
Dependency of the signal gain factor K_1 on the nondimensional longitudinal coordinate $T = z/L$ for the first signal harmonic. Here curve 1 corresponds to the case of one-harmonic superheterodyne amplification, whereas curve 2 describes analogous dependency for the case of degenerated superheterodyne amplification (here and everywhere below with a similar situation, we account ten harmonics SCW). Here the normalizing system length $L = 1$ m; the partial beam plasma frequency $\omega_p = 3.0 \cdot 10^{10}$ s^{-1}; the averaged (for both beams) relativistic factor $\gamma_0 = 3.3$; the difference of partial beam relativistic factors $\Delta\gamma = 0.3$; the signal wave frequency $\omega_1 = 9.43 \cdot 10^{11}$ s^{-1}; the input signal-wave-amplitude $E_{011} = 100$ V/m; the period of H-Ubitron pumping system $\Lambda_2 = 2\pi/k_2 = 4.0$ cm; the amplitude of the first pumping field harmonic $H_{21} = 200$ Gs. The relationship $H_{2m} = H_{21}/m$ is accepted for determining the amplitude in all other mth H-Ubitron pumping harmonics; m is the harmonic number.

The dependency of the signal gain factor K_1 on the nondimension longitudinal coordinate $T = z/L$ for the first signal harmonic is shown in Figure 13.17. The dotted curve corresponds to the case of one-harmonic superheterodyne amplification, whereas the solid curve describes analogous dependency for the case of degenerated superheterodyne amplification (here and in the succeeding discussions, we account ten harmonics SCW and pumping field). As is readily seen in Figure 13.17, the presence of signal high harmonics exerts essential action at the amplification process on the first signal harmonic. Indeed, this influence leads to reducing the gain factor K_1 ten times: from $5.5 \cdot 10^5$ until $0.53 \cdot 10^5$.

Thus, as the analysis shows, one characteristic feature of the discussed multiharmonic TSFEL that is evidently expressed is the presence of higher harmonics in the input and output signals. As before, the one-section H-Ubitron model is studied in this case. It is important to note that many first signal harmonics can have approximately the same order during the amplification process: the typical situation for the cluster waves (Chapter 1). This means that the proposed multiharmonic TSFEL can be effectively used as a generator (or amplifier) of electromagnetic wave signals with complex multiharmonic spectrum (including the cluster waves). The other variant is the generation of higher harmonics for the initially monochromatic signal. Just this situation is illustrated in Figure 13.18. The dependencies of signal gain factors K_j (for the first three signal harmonic ($j = 1,2,3$) on the nondimension longitudinal coordinate $T = z/L$) are shown there. We obtained that the first signal harmonic possesses maximal amplitude in the particular case discussed. It is important that initially, it is accepted that the amplitudes for all higher harmonics ($j \geq 2$) are zero. This means that all higher harmonics generated here begin from zero initial amplitude.

The influence of the longitudinal focusing magnetic field is demonstrated in Figure 13.18, too. It is well known that the focusing magnetic field is introduced in device working

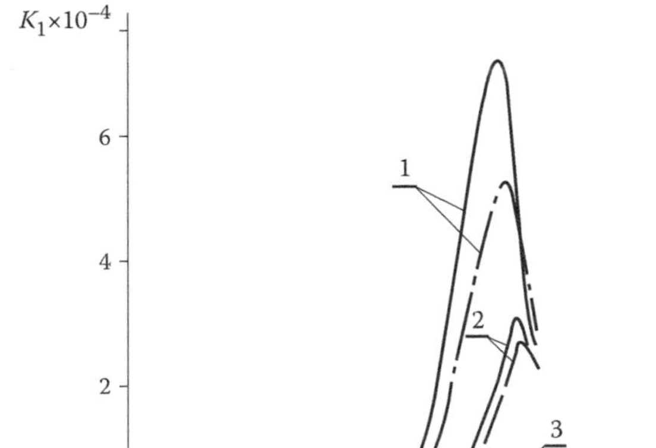

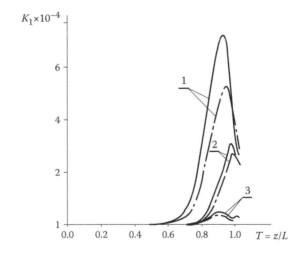

FIGURE 13.18
Dependency of the signal gain factors K_j on the nondimensional longitudinal coordinate $T = z/L$ (for the first three harmonic $j = 1,2,3$ for the one-section H-Ubitron-TSFEL). Here the solid lines correspond to the model without focusing magnetic field, whereas the dotted lines illustrate the case of a presence of the magnetic field. All calculations have been performed with accounting ten first harmonics SCW, signal, and pumping wave fields. Here 1 corresponds to the first harmonic, 2 corresponds to the second harmonic, and curve 3 corresponds to the three harmonics SCW, signal and pumping wave fields. It is assumed that induction of the focusing magnetic field $B_0 = 5k$ Gs. All other parameters are similar to those in Figure 13.17.

bulk for providing the beam transportation. We see in Figure 13.18 that the magnetic field some decreases the signal gain factors K_j, but this increase is not essential in the scale of amplitude of the first harmonic.

13.2.11 Klystron TSFEL on the Basis of Pumping Systems of Different Types

As is noted in Chapters 1 and 9, the klystron SFEL designs are most promising for construction of cluster systems. Therefore, let us discuss the physical features of the different types of this model in detail.

We will study TSFEL models of the three following types (see their description in Section 9.3):

1. The Dopplertron–klystrons
2. The H-Ubitron klystrons
3. The mixed (H-Ubitron plus Dopplertron) klystrons

The one-section Dopplertron model is used for comparison.

The dependencies of the signal gain factor K_1 on the nondimensional longitudinal coordinate $T = z/L$ (where L is, as before, the total length of the model) for these models are presented in Figure 13.19. Here the process of superheterodyne amplification in the one-

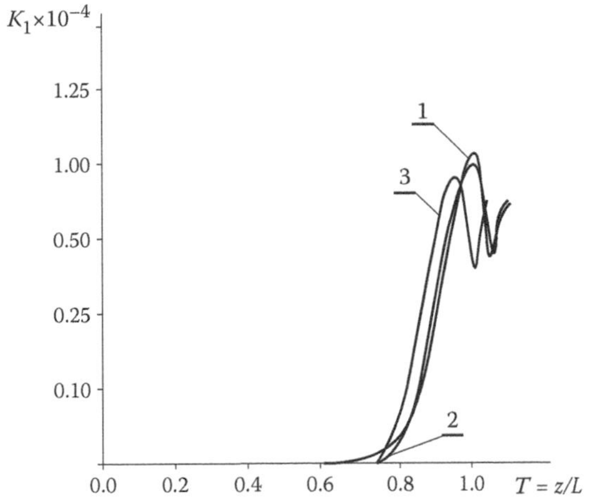

FIGURE 13.19
Dependencies of the gain factors K_1 on nondimensional coordinate $T = z/L$ for the one-section Dopplertron (curve 1), mixed klystron (curve 2), and Dopplertron–klystron models (curve 3), respectively. Here all calculations have been performed with accounting ten first harmonics SCW, signal, and pumping wave fields. Here the plasma frequency of one partial beam is $\omega_p = 3.0 \cdot 10^{10}$ s^{-1}; the averaged relativistic factor in the system input is $\gamma_0 = 3.3$; the difference of partial beam relativistic factors $\Delta\gamma = 0.3$; the signal wave frequency $\omega_1 = 8.43 \cdot 10^{11}$ s^{-1}; the frequency of the retarded pumping electromagnetic field $\omega_2 = 1.0 \cdot 10^{11}$ s^{-1}; the intensity of input signal amplitude for the first harmonic $E_{11} = 100$ V/m; the intensity of input pumping amplitude for the first harmonic $E_{21} = 2 \cdot 10^6$ V/m; the intensity of pumping amplitudes for the mth harmonic can be found from the ratio $E_{2m} = E_{21}/m$; the retardation coefficient for the pumping wave $N_2 = (ck_2)/\omega_2 \cong 1.47$; the undulation period for the H-Ubitron pumping system $\Lambda_2 = 2\pi/k_2 = 4.0$ cm; the amplitude of the first pumping harmonic (for the H-Ubitron pumping) $H_{21} = 200$ Gs; the intensity of magnetic pumping amplitudes for the mth harmonics can be found from the ratio $H_{2m} = H_{21}/m$; induction of the magnetic focusing field $B_0 = 5k$ Gs; the total system length $L = 1.0$ m, including the first section length 30 cm, the second section length 45 cm, and the third section length 35 cm.

section Dopplertron is described by curve 1. Curve 2 corresponds to the mixed klystron model, and, last, dynamics of the Dopplertron–klystron model are described by curve 3.

Let us once more make the reader pay more attention to the klystron TSFEL, which is similar to the one-sectional one; on one hand, it is also characterized by extremely high magnitudes of the gain (K_1 ~10^4), and, on the other hand, they have moderate (in comparison with traditional parametric FEL) requirements for the model parameters. Indeed, the power density for the Dopplertron-pumping wave here is only ~294 kWt/cm^2, the amplitude of magnetostatic field of H-Ubitron pumping is 200 Gs, the work length of the system is ~ 1.1 m (for curves 1 and 2) and ~0.95 m (for curve 3), etc. In traditional parametric FEL, the analogous characteristic parameters are essentially higher (see relevant examples in Chapter 12). For instance, typical density of power in the parametric Dopplertron for the pumping wave is a few tens MWt/cm^2, the amplitude of H-Ubitron pumping usually is a few kGs, and so on.

Comparison of the gain properties of the systems, which are presented in Figure 13.19, shows that the mixed klystron SFEL turns out to be the most promising for practical use.

Dependencies for the signal gain factors for the 1.15-MeV TSFEL, as a function of the normalized longitudinal coordinate $T = z/L$, are presented in Figure 13.20. They have been calculated for the one-section (curve 1) and klystron-type H-Ubitron (curve 2) amplifiers. As is readily seen, both systems have approximately equal gain factors; in spite of that, the total length of pumping systems is essentially shorter for the klystron system. Besides that, the klystron system appears to be essentially simpler than the one-section one, owing to shorter total pumping length. The klystron system, as is mentioned earlier, also allows very reliable decoupling of input and output of the amplifier.

The analogous dependencies, which illustrate dynamics of the 20-MeV TSFEL–Dopplertrons, are represented in Figures 13.20, 13.21, and 13.22. Moreover, the dependencies of the gain factors K_1 on the nondimensional coordinate $T = z/L$ for the one-section Dopplertron (curve 1), mixed klystron (curve 2), and Dopplertron-klystron models (curve 3)

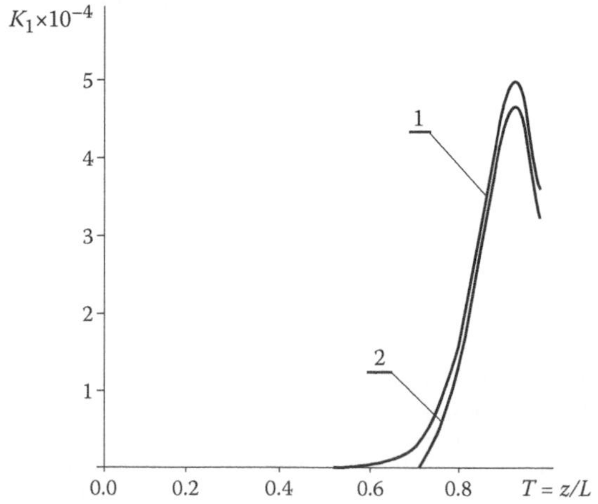

FIGURE 13.20
Dependencies of the gain factors K_1 on nondimensional coordinate $T = z/L$ for the one-section H-Ubitron (curve 1) and H-Ubitron klystron (curve 2). Here all calculations have been performed with accounting ten first harmonics SCW, signal, and pumping wave fields. Here the total system length in both cases is $L = 1.0$ m, including the first section length 30 cm, the second section length 40 cm, and the third section length 30 cm. All other parameters are similar to those in Figure 13.17.

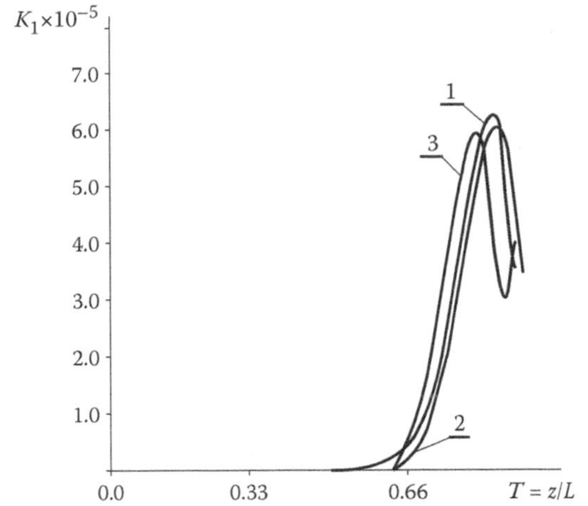

FIGURE 13.21
Dependencies of the gain factors K_1 on nondimensional coordinate $T = z/L$ for the one-section 20 MeV-Dopplertron (curve 1), mixed klystron (curve 2), and Dopplertron–klystron models (curve 3), respectively. Here all calculations have been performed with accounting ten first harmonics SCW, signal, and pumping wave fields. Here the plasma frequency of one partial beam $\omega_p = 3.0 \cdot 10^{11}$ s^{-1}; the averaged relativistic factor in the system input $\gamma_0 = 40$; the difference of partial beam relativistic factors $\Delta\gamma = 1.16$; the signal wave frequency $\omega_1 = 1.89 \cdot 10^{14}$ s^{-1}; the frequency of the retarded pumping electromagnetic field $\omega_2 = 8.43 \cdot 10^{11}$ s^{-1}; the input signal intensity amplitude for the first harmonic $E_{11} = 100$ V/m; the input pumping intensity amplitude for the first harmonic $E_{21} = 4 \cdot 10^6$ V/m; the intensity pumping amplitudes for the *m*th harmonics can be found from the ratio $E_{2m} = E_{21/m}$; the retardation coefficient for the pumping wave $N_2 = (ck_2)/\omega_2 \cong 1.30$; the undulation period for the H-Ubitron pumping system $\Lambda_2 = 2\pi/k_2 = 3.2$ cm; the amplitude of the first pumping harmonic (for the H-Ubitron pumping) $H_{21} = 100$ Gs; the intensity magnetic pumping amplitudes for the *m*th harmonics can be found from the ratio $H_{2m} = H_{21}/m$; the induction of the magnetic focusing field $B_0 = 5k$ Gs, the total system length $L = 1.1$ m, including the first section length 90 cm, the second section length 190 cm, and the third section length 80 cm.

are represented in Figure 13.20. As before, all calculations have been performed with accounting ten first harmonics SCW, signal, and pumping wave fields.

It is readily seen that in comparison with traditional (parametric) FEL, the discussed TSFELs are characterized by a unique parameter set.

First, the TSFELs have a possibility to work effectively in the mode with a multiharmonic (*including cluster*) electromagnetic signal. This can be attained owing to specific multiharmonic properties of the two-stream instability (Section 13.1). This means that analogous designs (when the initial signal is monochromatic) cannot be realized in the traditional FEL, in principle.

Second, the TSFEL can work as an amplifier with very high levels of input–output decoupling. This result can be reached when the main amplification is attained in the second section, which, in turn, works on the basis of longitudinal electrostatic beam waves. Such change of the physical nature of the amplified signal gives the above-mentioned unique decoupling TSFEL characteristics. Apart from that, in the case of mixed klystron TSFEL, the decoupling improves additionally owing to the difference of input and output signal frequencies.

Third, the TSFELs are characterized extremely by high levels of the gain factor. As is easily seen, the gain factors at the level 10^4–10^7 can be typical for the system of the considered class in mm-IR ranges. Some specific designs are known (in particular those, where

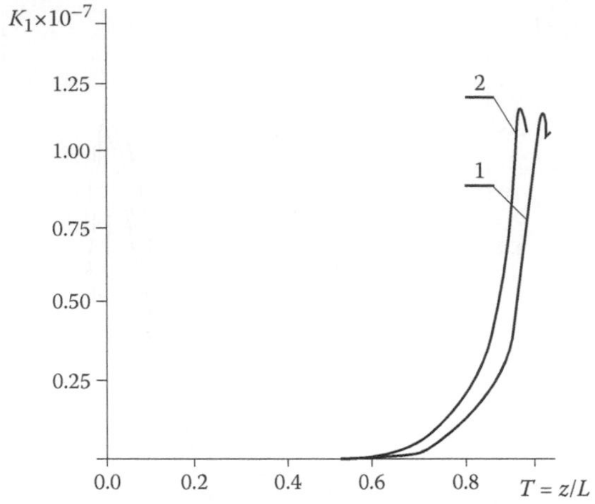

FIGURE 13.22
Dependencies of the gain factors K_1 on the nondimensional coordinate $T = z/L$ for the one-section 20-MeV H-Ubitron (curve 1) and H-Ubitron klystron (curve 2). Here all calculations have been performed with accounting ten first harmonics SCW, signal, and pumping wave fields. Here the plasma frequency of one partial beam is $\omega_p = 3.0 \cdot 10^{11}$ s^{-1}; the averaged relativistic factor in the system input $\gamma_0 = 40$; the difference of partial beam relativistic factors $\Delta\gamma = 1.16$; the signal wave frequency $\omega_1 = 1.89 \cdot 10^{14}$ s^{-1}; the input signal intensity amplitude for the first harmonic $E_{11} = 100$ V/m; the undulation period for the H-Ubitron pumping system $\Lambda_2 = 2\pi/k_2 = 4.0$ cm; the amplitude of the first pumping harmonic (for the H-Ubitron pumping) $H_{21} = 200$ Gs; the intensity magnetic pumping amplitudes for the mth harmonics can be found from the ratio $H_{2m} = H_{21}/m$; the induction of the magnetic focusing field $B_0 = 5k$ Gs, the total system length $L = 2.5$ m, including the first section length 70 cm, the second section length 70 cm, and the third section length 110 cm.

the terminal section works on the highest SCW harmonics), which allows to continue the signal working frequency range until the UV waves and more.

Fourth, the TSFELs are characterized relatively small (for their class of devices) total sizes. This result, as is mentioned already, can be attained by virtue of the two following peculiarities of the proposed designs. The first is the use of the multichannel induction undulation accelerators (see Section 9.4) as an electron beam source. As is shown in Chapter 9, these accelerators can be rather compact. The second is compactness of the TSFEL section itself. The combination of both these design solutions allows us to attain the total system compactness.

Thus, it looks like that just the TSFELs are the most promising candidate for the role of a key technological elements for a future New Star Wars System (Chapter 1). However, this conclusion we can draw only now, on the basis of only the existing level of our knowledge, when many important physical factors and mechanisms are not known elsewhere. For instance, it is well known that the traditional two-stream systems are characterized by an increased level in electromagnetic noises [59]. What influence does this have on the process of synthesis of electromagnetic clusters? It is because that, as is noted above, the synthesis itself has an explicitly expressed interferential nature (Chapter 1). This means, in turn, that only the *coherent* part of the field of electromagnetic harmonics can be involved in this forming process, and so on. Therefore, now, in such a situation, we are stressed to study other possible candidates, too. There are the ordinary parametric FEL (Chapters 11 and 12), the superheterodyne plasma-beam and electron-wave parametric FEL (Chapter 14), etc. This means, basically, that long and hard work awaits us on the way to the eventual acceptance of a final decision concerning the above-mentioned best candidate.

And finally, let us discuss some problems of the nonlinear theory of TSFEL, as a device class.

The first is the developing model with high efficiency. It should be mentioned that efficiency of the above-studied models, as a rule, does not exceed a few percentages. As is mentioned above, the designs with 70%–80% can be realized, basically, by using the various isochronous schemes. This field of theory of multiharmonic TSFEL awaits now its investigators.

The second class of problems concerns the mathematical methods, which are suitable for calculations of such high-efficiency models. Unfortunately, existing today are *traditional* methods, which were elaborated in the framework of traditional (parametric) FEL, which cannot be used in the multiharmonic nonlinear TSFEL theory in view of specific mathematical problems. As it has been shown in this book, the new elaborated hierarchic calculation technologies (Part I) give a real possibility to overcome this difficulty. Today in this area, we have only one problem: the autoimmunization of the proposed hierarchic calculation algorithms. Our preliminary analysis shows that it is absolutely possible and the only problem is the researchers who are ready to do it.

Last, the third problem is the limited frequency range, within which the TSFEL physical mechanisms can work effectively. These limitations, from a physical point of view, are determined by the plasma nature of the main working TSFEL mechanism (two-stream instability). This means that the working signal wavelength cannot be smaller than the average distance between beam electrons (in the beam reference frame). As a result, we have the highest frequency boundary for the TSFEL (1–10 μm). However, based on the analysis shown, the discussed limitations concern only the simplest designs, some of which are described above. Apart from that, a few more perfect designs, which do not have such limitations, can be realized, too.

13.3 Project of the Simplest Femtosecond TSFEL Former

13.3.1 Design Scheme

Basing on the obtained systems of truncated equations (Equations 13.78 through 13.80) (written for harmonic wave amplitudes only), we accomplish relevant numerical simulation of the femtosecond multiharmonic klystron TSFEL. Let us discuss shortly the project of a computer experiment with this device. We take the design scheme shown in Figure 13.1 as a basic technological solution. It is assumed that the device discussed is destined for use as an experimental prototype of the TSFEL for a cluster system of the second order (see relevant discussion to Figure 1.14).

The characteristic feature of the illustrated design version is the use of the one-harmonic (monochromatic) pumping in the first pumping section and the multiharmonic pumping in the second pumping sections. Therefore, the proposed system can be regarded as a transformer of the sine-like electromagnetic signal (see curve 1 in Figure 1.12) into a periodical sequence of the femtosecond clusters (see curve 2 in Figure 1.11). However, it should be mentioned that today, a few tens other, more perfect design versions of the sources of the femtosecond clusters are proposed (see Section 9.3). The design scheme shown in Figure 13.23 is the simplest and, therefore, the most convenient for the explanation of the basic operation principles of the femtosecond TSFEL.

The device presented in Figure 13.23 works in the following manner. One-harmonic electromagnetic signal 1 enters into the input system 2 and then is directed in the work

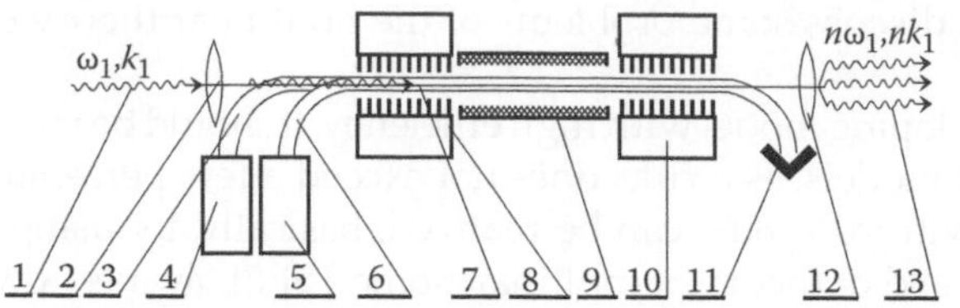

FIGURE 13.23
Simplest variant of the design scheme of the multiharmonic femtosecond TSFEL. Here 1 is the input harmonic wave signal (where ω_1, k_1 are its frequency and wave vector), 2 is the input of the system, 3 is the first one-velocity high-current relativistic electron beam, 4 is the first high-current electron accelerator, 5 is the second high-current electron accelerator, 6 is the second high-current electron beam, 7 is the two-velocity (doubled) high-current electron beam, 8 is the first pumping system, 9 is the transit system, 10 is the second (multiharmonic) pumping system, 11 is the electron collector (system for recuperation of energy of the electron beam 7), 12 is the output signal system, and 13 is the output multiharmonic signal in the form of a sequence of femtosecond clusters (where $n\omega_1$, nk_1 are the harmonic frequencies and the wave vectors, n is the number of signal Fourier harmonic).

bulk of the TSFEL. High-current relativistic electron beams 3 and 6, which have different initial energies, are formed by means of high-current accelerators 4 and 5. Furthermore, both beams 3 and 6 are merged into one two-velocity (two-energy) high-current electron beam 7. This beam, in turn, is directed in the operational part of the first one-harmonic pumping system 8.

The main destination of the pumping system 8 is the multiharmonic modulation of initial two-stream beam 7 with respect to the electron density. The excitation (generation) of many of *electron-beam wave harmonics* occurs due to the realization of the mechanism of plural parametric resonances (Section 13.1). As a result, the initially monochromatic input transverse electromagnetic signal 1 transforms in pumping system 8 into the form of the longitudinal multiharmonic electron-beam wave. The worked up electromagnetic signal absorbs further in the walls of transit system 9.

Thus, the initial transverse electromagnetic one-harmonic signal 1 transforms in the first pumping system 8 in the form of the longitudinal multiharmonic electron-beam wave.

Then this electron-beam wave is amplified strongly in the transit section 9 owing to the mechanism of the two-stream instability. Strongly modulated and essentially multiharmonic electron beam 7 enters in the input of second multiharmonic pumping system 10. The generation of output femtosecond transverse electromagnetic signal 13 occurs within the work volume of second pumping system 10. It happens due to the realization of the mechanism of multiharmonic three-wave superheterodyne resonant interaction.

As is mentioned above, the most important physical feature of this mechanism is its explicitly expressed degenerated nature. Namely, each three harmonic waves (harmonics of signal, pumping, and electron-beam waves) with the same harmonic number interact with another in the superheterodyne-resonant manner. If we have a total of, for example, N harmonics for each of three waves, then N parallel three-wave superheterodyne resonances are realized simultaneously. As a result, the longitudinal multiharmonic electron-beam wave transforms (with amplification) into the transverse multiharmonic output electromagnetic signal 13. In other words, in this case, we have a unique resonant interaction of three multiharmonic clusters of different origins: the longitudinal electron-beam wave, the transverse pumping wave, and the transverse output signal clusters, respectively.

In accordance with the basic principles, which were set forth earlier in Chapter 1, the Fourier spectrum of output signal 13 should be close to the Delta-function Fourier spectrum (see Figure 1.13 and corresponding discussion). However, as is shown in Section 13.1, the specific characteristic of the two-stream cluster SFEL is that the SCW spectrum actually is not similar to the Delta-function Fourier spectrum. However, in spite of this, as it will be shown below, the formation of the femtosecond clusters is also possible in such systems.

13.3.2 General Arrangement

Let us then discuss the general arrangement of the experimental system, given in Figures 13.24 and 13.25, where the system consists of two main parts: the acceleration block 1 and MTSFEL section 4. The turning system 2 and the compression system 3 serve for their junction. The main project parameters of the system are given in Table 13.2. The acceleration block 1 is made on the basis of two parallel linear acceleration blocks with attached high-current electron injectors (Chapter 9) [21,22]. The beam diameter on the block output is ~0.8 cm.

Then both beams move within the worked bulk of the turning system 2. It executes the three following functions at the same time. The first is turning the direction of both beams at an angle of 180°. The second is the compression of the beams and the third is preparation of the beams for further compression and merging together in the next section 3.

The turning systems 2 are made on the basis of toroidal magnetic sections (Section 9.3) [21] like those used in the TOKAMAKs (for keeping magnetized plasma in the process of plasma fusion). Its main design feature is that the vector of the magnetic field along the circular-like axis of tore (in the electron-beam channel within the toroidal sections) periodically changes its direction [21]. The longitudinal (along the axis) component of the magnetic field could be described mathematically in this case by a periodically reversion (oscillation) function. Owing to this, we can suppress the effect of radial electron shifting, which is one of the main technological difficulties of this class of turning systems [21]. Besides that, the amplitude of these oscillations increases with the increasing angle of electron-beam turn. Due to this, the preliminary compression of both beams happened during the beams' motion in the turning system 2.

Both beams, generated by acceleration block 1 should be merging further in the compression system 3. For the sake of fulfilling this procedure, both turning systems 2 are oriented in such a manner that between their turning planes, some oblique angle appears (see Figure 13.25 for more details). Owing to the above-mentioned design properties of

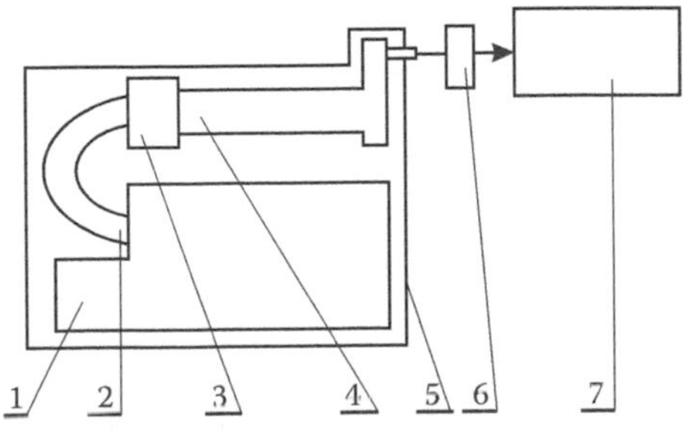

FIGURE 13.24
General design scheme of the TSFEL for technological purposes (for instance, the EMI-test system). Here 1 is the two-beam and two-channel undulation induction accelerator, 2 is the beam-turning system, which includes also the curvilinear part of the compression and system for merging together the electron beams, 3 is the linear part of the compression system, 4 is the TSFEL block, 5 is the screen, 6 is the operation system, and 7 is the subject of treatment.

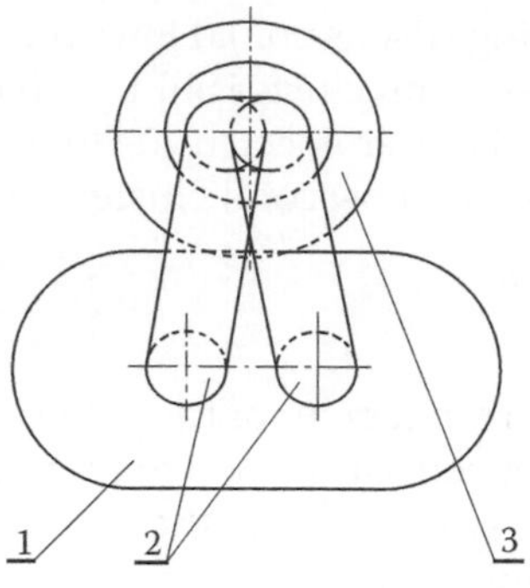

FIGURE 13.25
General design scheme of the TSFEL (the view from left of the design in Fig. 13.24). Here the designation of items coincides with Figure 13.24.

the turning systems 2, both beams come to the input of compression system 3 particularly merged and compressed already. The further merging and compression occur in the work bulk of the system 3. After accomplishing all of the above described procedures, the two-velocity and strongly compressed (from the initial diameter ~8 cm into the terminal beam diameter ~2.48 mm) is directed on the TSFEL-block 4 input.

TABLE 13.2
Device Parameters

Parameters	Magnitudes
Accelerator block	
Pulsed current strength of the electron beam (A)	250
Duration of two-velocity electron pulse (ns)	100
Averaged beam energy of the two-velocity electron beam (MeV)	1.2
Difference of energy of one-velocity beams (MeV)	0.051
Energy thermal spread of the beam (%)	1
Beam diameter (mm)	2.48
Plasma frequency of the beam (s^{-1})	$6 \cdot 10^{10}$
Electromagnetic signal	
Ratio of the signal and optimal frequencies	15
Signal wavelength (mm)	4.9
Power of the input signal (W)	1.67
Pulsed power of the femtosecond pulse (W)	$1.73 \cdot 10^7$
Duration of the output femtosecond wave packages (fs)	340
Other design characteristics	
Period of the undulator (cm)	8.85
Induction of the undulator magnetic field (first harmonic) (Gs)	600
Focusing magnetic field (Gs)	443
Length of the first pumping section (m)	1
Length of the transit section (m)	0.5
Length of the second pumping section (m)	1
Total length of the MTSFEL block (m)	2.5

13.3.3 Computer Simulation

The main device parameters are given in Table 13.2 and Figure 13.26.

As is readily seen, the device discussed allows testing a possibility of computer experimental realization of the compression effect (see Figure 1.13 and relevant discussion). On the other hand, it is obvious that the compression factor (1.4) f_{com} ~12, which is planned to be attained in the discussed experiment, is not enough for most interesting practical applications. This means automatically that we should look for more perfect design solutions of the considered femtosecond TSFEL former. Relevant estimations show that the compression factor f_{com} could be obtained on the level ~100 and more times. The analysis allows also clearing up so that the improved design solutions of such a kind can really be found in the existing level of FEL technologies. Analysis of these improved femtosecond designs (like those shown in Chapter 9) gives the following key parameters for the future femtosecond TSFEL: the compression factor f_{com} ~120, the efficiency ~30%, the pulsed (instantaneous) power ~$5 \cdot 10^{13}$ W, the beam energy ~ 5 MeV, and the averaged power ~50–100 kW or higher. This looks rather promising for the modern state of technologies of cluster FEL. Analysis shows that the best chance to approach this level of technological parameters gives the constructing of the second order cluster systems (see discussion to Figure 1.14). The considered here FEL-design is dedicated for the use (as a key design element) just in such kind of cluster systems.

It should be mentioned especially that appearance of first real experimental femtosecond TSFEL would be a great event for electrodynamics, in particular, as well as for physics as a whole, in general. First, this will give a powerful push for development of a number of new scientific directions. For instance, at present we do not have any satisfactory nonlinear theory of propagation of such signals in different mediums. The commercial technology of forming such kind of signals today is not developed yet. Its military significance is discussed in Chapter 1 already. We also do not have any suitable experiments in this area. Therefore, we can say that a project of this kind, apart from the obvious purely practical objectives, would play an important role as a powerful stimulator for further development of various fundamental aspects of modern physics.

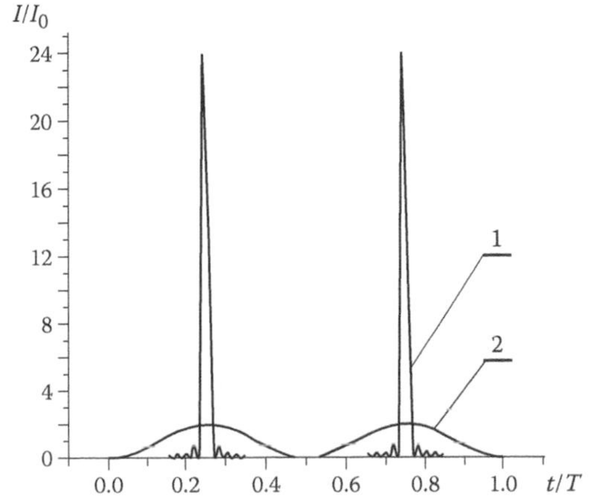

FIGURE 13.26
Dependency of the normalized signal intensity I/I_0 on the normalized duration of the femtosecond wave clusters t/T. Here I_0 is the averaged intensity of the equivalent sine-like signal. All calculations have been accomplished for the project parameters given in Table 13.1.

References

1. Kotsarenko, N.Y., and Kulish, V.V. 1980. Superheterodyne amplification of electromagnetic waves in a beam-plasma system. *Radiotech. Electron [Radio Eng. and Electron. (USSR)]*, 25(11):2470–2471.
2. Kotsarenko, N.Y., and Kulish, V.V. 1980. On the possibility of superheterodyne amplification of electromagnetic waves. *Zh. Tekh. Fiz. [Sov. Phys. Tech. Phys.]*, 50: 220–222.
3. Kulish, V.V. 1998. *Methods of Averaging in Non-Linear Problems of Relativistic Electrodynamics*. Atlanta: World Scientific Publishers.
4. Kulish, V.V. 2002. *Hierarchic Methods. Undulative Electrodynamic Systems. Vol. 2*. Boston: Kluwer Academic Publishers.
5. Kulish, V.V. 2002. *Hierarchic Methods. Hierarchy and Hierarchical Asymptotic Methods in Electrodynamics. Vol. 1*. Boston: Kluwer Academic Publishers.
6. Kulish, V.V., Bolonin, O.N., Kochmanski, S.S., and Kulish, V.V. 1989. Coupled parametric resonances. *Acta Phys. Polonica*, A76(3):455–472.
7. Kulish, V.V., Storizhko, V.E. Patent of USSR No 1809934 (cl. H 01 J 25/00). Free electron laser. Priority 18.07.90.
8. Perekupko, V.A., Silivra, A.A., Kotsarenko, N.Y., Kulish, V.V. Patent of USSR No 835259 (cl. H 01 J 25/00). Electronic device. Priority 28.01.80.
9. Bekefi, G., and Jacobs, K.D. 1982. Two-stream FELs. *Jour. Appl. Phys.*, 53:4113–4121.
10. Botton, M., and Ron, A. 1990. Two-stream instability in FELs. *IEEE Trans. Plasma Science*, 18(3):416–423.
11. Bolonin, O.N., Kulish, V.V., and Pugachev, V.P. 1988. Superheterodyne amplification of electromagnetic waves in a relativistic electron two-stream system. *Ukr. Fiz. Zh. [Ukrainian Journal of Physics]*, 33(10):1465–1468.
12. Kulish, V.V., and Pugachev, V.P. 1991. On the theory of superheterodyne wave amplification in a two-stream plasma. *Fiz. Plazmy [Sov. J. Plasma Phys.]*, 17(6):696–705.
13. Kulish, V.V. 1992. Physics of two-stream free electron lasers. *Moscow Univ. Bull., Physics, Astronomy*, 33(3):64–78.
14. Kulish, V.V. 1991. On the theory of klystron-type superheterodyne free electron lasers. *Ukr. Fiz. Zh. [Ukrainian Journal of Physics]*, 36(1):28–33.
15. Kulish, V.V. 1991. On the theory of relativistic electron-wave free electron lasers. *Ukr. Fiz. Zh. [Ukrainian Journal of Physics]*, 36(5):682–694.
16. Kulish, V.V., and Storizhko, V.E. 1991. Patent of USSR No 1837722 (cl. H 01 J 25/00). Free electron laser. Date of Patent 15.02.91.
17. Kulish, V.V. 1993. Superheterodyne electron-wave free-electron laser. *Int. Jour. Infrared & Millimeter Waves*, 14(3).
18. Kulish, V.V., Kuleshov, S.A., and Lysenko, A.V. 1994. Nonlinear self-consistent theory of two stream superheterodyne free electron laser. *Int. Jour. Infrared & Millimeter Waves*, 15(1).
19. Wilhelmsson, H. 1991. Double beam free electron laser. *Phys.Scripta*, 44:603–605.
20. Davydova, T.A., and Wilhelmsson, H. 1992. Resonant and nonresonant wave excitation in a double beam free electron laser. *Phys. Scripta*, 45:607–607.
21. Bekefi, G. 1992. Double-stream cyclotron maser. *Jour. Appl. Phys*, 71(9):4128–4131.
22. Kulish, V.V., Lysenko, A.V., and Savchenko, V.I. 2003. Two-stream free electron lasers. General properties. *Int. Jour. Infrared & Millimeter Waves*, 24(2):129–172.
23. Kulish, V.V., Lysenko, A.V., and Savchenko, V.I. 2003. Two-stream free electron lasers. Analysis of the system with monochromatic pumping. *Int. Jour. Infrared & Millimeter Waves*, 24(3):285–309.
24. Kulish, V.V., Lysenko, A.V., and Savchenko, V.I. 2003. Two-stream free electron lasers. Physical and project analysis of the multiharmonic models. *Int. Jour. Infrared & Millimeter Waves*, 24(4):501–524.

25. Kulish, V.V., Lysenko, A.V., Savchenko, V.I., and Majornikov, I.G. 2005. The two-stream free electron laser as a source of electromagnetic femto-second wave packages. *Materials of the International Workshop "Microwave, Radar and Remote Sensing" MRRS 2005, September 19-21, Kyiv, Ukraine*: 304–309.
26. Kulish, V.V., Lysenko, A.V., and Majornikov, I.G. 2004. A two-stream free electron laser for generation of the electromagnetic femto-second wave packages. *Abstracts of the "International Conference on Infrared and Millimeter Waves & International Conference on Terahertz Electronics (THz) (IRMMW2004/THz2004), Karlsruhe, Germany, September 27 - October 01, 2004"*.
27. Kulish, V.V., Lysenko, O.V., and Majornikov, I.G. 2004. The two-stream multi-harmonic FEL as a powerful source of femto-second wave packages. *Proceedings for the European Radar Conference, Amsterdam, Netherlands, 14–15 October 2004*.
28. Kulish, V.V., Lysenko, A.V., Savchenko, V.I., and Majornikov, I.G. 2005. The two-stream free electron laser as a source of electromagnetic femto-second wave packages. *Laser Physics*, (12):1629–1633.
29. Kulish, V.V., Lysenko, A.V., and Mayornikov, I.G. 2005. The two-stream multi-harmonic FEL as a powerful source of femto-second wave package. *Proceedings of the National Aviation University*, (Kyiv), (2):126–130.
30. Kulish, V.V., and Savchenko, V.I. 2002. The method of averaged characteristics in nonlinear theory of the two-stream instability. *Visnyk Sumskoho Universytetu (Ukraine), ser. Physics and Mathematics*, (2):5–12.
31. Kulish, V.V. 1985. Physical processes in the parametrical electronic lasers (free electron lasers). *Thesis for the Doctor of Sciences Degree*. Institute of Physics of Ukrainian Academy of Sciences, Kiev, 405 pp.
32. Kulish, V.V., Lysenko, A.V., and Koval, V.V. 2007. On the theory of the plasma–beam superheterodyne free electron laser. *Visnyk Sumskoho Universytetu (Ukraine), ser. Physics and Mathematics*, (2):112–119.
33. Kulish, V.V., Lysenko, A.V., and Koval, V.V. 2009. Multi-harmonic cubic-nonlinear theory of the superheterodyne plasma-beam free electron lasers with dopplertron pumping. *Prikladnaya Fizika (Russia)*, (5):76–82.
34. Kulish, V.V., Lysenko, A.V., and Koval, V.V. 2009. On the theory of the plasma–beam superheterodyne free electron laser with H-Ubitron pumping. *Technical Physics Letters (Russia)*, 35(8):696–699.
35. Kulish, V.V., Lysenko, A.V., Gubanov, I.V., and Brusnik, A.Ju. 2009. Superheterodyne free electron laser with longitudinal electric undulator. Patent UA No 87750. *Published in Bulletin of Invention (Ukraine)* 08.10., No 15.
36. Kulish, V.V., Lysenko, A.V., and Koval, V.V. 2009. Multi-harmonic cubic-nonlinear theory of the superheterodyne plasma-beam free electron lasers with H-Ubitron pumping. *Radiotechnika I Elektronika (Ukraine)*, 14(3):383–389.
37. Kulish, V.V., Lysenko, A.V., and Brusnik, A.Ju. 2010. Active FEL-klystrons as formers of femto-second clusters of electromagnetic field. *Journal of Nano- and Electron. Physics (Ukraine)*, 2(2):50–78.
38. Kulish, V.V., Lysenko, A.V., and Rombovsky, M. Yu. 2010. Effect of parametric resonance on the formation of waves with a broad multi-harmonic spectrum during the development of two-stream instability. *Plasma Physics Reports*, 36(7):594–600.
39. Kulish, V.V., Lysenko, A.V., and Koval, V.V. 2010. Multi-harmonic cubic-nonlinear theory of the superheterodyne plasma-beam free electron lasers with dopplertron pumping. *Plasma Phys. Reports*, 36(13):1185–1190.
40. Lebedev, I.V. 1964. *Microwave Technology and Devices. vol. II.*, Moscow: Energiya.
41. Haiduk, V.I., Palatov, K.I., and Petrov, D.N. 1971. *Physical principles of microwave electronics.* Moscow: Sovetskoye Radio.

42. Lopoukhin, V.M. 1953. *Excitation of Electromagnetic Oscillations and Waves by Electron Beams.* Moscow: Gostekhizdat.
43. Haeff, A.V. 1948. Space charge wave amplification effect. *Phys. Rev.* 74(1):1532–1532.
44. Jermolajev, Ju. M., Kalmykov, A.M., Kotsarenko, N.Ja., and Kulish, V.V. 1978. On the theory of separation of and simultaneous amplification of a different frequency in a semiconductor. *Radio Eng. and Electron* (USSR), (10):2133–2137.
45. Kulish, V.V., and Jermolajev, Ju. M. Demodulator, Patent No. USSR 563882, Date of the Patent March 9, 1977, Application number No2061065, Filed September 20, 1974.
46. Kulish, V.V., and Kotsarenko, Ju.Ya. Optical transformer of frequency, Patent No. USSR 563882, Date of the Patent March 21, 1978, Application number No 2368844, Filed June 4, 1976.
47. Kulish, V.V., Lysenko, A.V., and Brusnik, A.Ju. 2010. Active FEL-klystrons as formers of femtosecond cluster electromagnetic waves. General description. J. *Nano- Electron. Phys. (Ukraine),* 2(2):50–78.
48. Kulish, V.V., Lysenko, A.V., and Rombovsky, M.Yu. 2010. Effect of parametric resonance on the formation of waves with a broad multi-harmonic spectrum during the development of two-stream instability. *Plasma Physics (Russ.),* 36(7):637–643.
49. Kulish, V.V., Lysenko, A.V., and Rombovsky, M.Yu. 2005. Theory of the two-stream electron-wave free electron lasers with H-Ubitron pumping *Visnyk Sumskoho Universytetu (Ukraine), Ser. Physics and Mathematics,* (4):58–70.
50. Kulish, V.V., Lysenko, A.V., and Rombovsky, M.Yu. 2009. Parametrical resonance for the two-velocity beam waves. *Applied Physics (Moscow),* (1):71–78.
51. Kulish, V.V., Lysenko, A.V., and Rombovsky, M.Yu. 2010, The cubic-nonlinear of electron-wave two-stream free electron lasers with the H-Ubitron pumping. *VANT (Ukraine), ser. Nuclear-Physics Res.,* 54(3):78–82.
52. Kulish, V.V. 1991. On the theory of relativistic electron-wave free electron lasers. *Ukrainiam Phys. Journ.,* 36(5):686–693.
53. Trubetskov, D.I., and Khramov, A.E. 2003. *Lections on the Microwave Electronics for Physicists. In two volumes.* Moscow: Fizmatlit.
54. Marshall, T.C. 1985. *Free Electron Lasers.* New York, London: MacMillan.
55. Freund, H.P., and Antonsen, T.M. 1996. *Principles of Free Electron Lasers.* Springer: New York.
56. Saldin, E.L., Scheidmiller, E.V., and Yurkov, M.V. 2000. *The Physics of Free Electron Lasers.* Springer: New York.
57. Toshiyuki Shozawa. 2004. *Classical Relativistic Electrodynamics: Theory of Light Emission and Application to Free Electron Lasers.* Springer: New York.
58. Schmuser, P., Ohlus, M., and Rossbach, J. 2008. *Ultraviolet and Soft X-Ray Free Electron Lasers: Introduction to Physical Principles, Experimental Results, Technological Challenges.* Springer: New York.
59. Haiduk, V.I., Palatov, K.I., and Petrov, D.N. 1971. *Physical Principles of Microwave Electronics.* Moscow: Sovetskoye Radio.

14

Plasma-Beam and Parametrical Electron-Wave Superheterodyne FEL

As we already mentioned in Chapter 1, at the time of the unforgettable Star Wars program, it became clear that, besides the usual H-Ubitrons, a whole series of other types of the FELs might be constructed. The plasma-beam superheterodyne FEL (PBSFEL) and the parametrical electron-wave superheterodyne FEL (PEWSFEL) have a special place in this series.

Historically, the first PBSFELs were proposed in 1979 to 1980 [1,2]. Subsequently, this type of system was studied in references [3–7] and some other sources. Already the primary analysis of Dopplertron variants showed very promising perspectives of their practical application. The high estimations were supported by the exceptionally high level of the linear amplification, relatively small for their class dimensions, high potential effectiveness, etc. This was explained by the simple circumstance that, considering the physics, the plasma-beam instability is a particular case of the two-stream instability, when the velocity of one of the partial beams is equal to zero. As was mentioned many times before, the last corresponds to the strongest instabilities, known in electrodynamics. The undoubted merit of the PBSFEL is also the possibility to make the Dopplertrons with the anomalously retarded pumping: the retardation coefficient of pumping wave can reach the value ~100 (see Chapter 11 for details). This, in turn, opens a favorable possibility of a sharp increase in the transformation coefficient of frequency without a stepped increase in the beam relativism (Chapter 11) and, related with it, a train of technological problems.

The shortcoming of the PBSFEL is in its limited working wavelength range; the optimal range is in the millimeter wave range. Therefore, originally it was proposed that the basic application of this device is in various types of EMI systems. Soon, it became clear that PBSFEL (the Dopplertrons as well as the H-Ubitrons) might be promising as the technological basis of the cluster systems of millimeter range. In such a case, this range cluster signal spectrum might contain tens or more harmonics. An example of this design scheme is given in Figure 14.1.

The system's performance involves the following processes. The monochromatic partial plasma-beam FEL amplifiers 5 each generate its own harmonics 7 of the future cluster wave 11. By the help of the quasioptical system 8, all this emission is collected into one multiharmonic beam 9 and is directed into the multiharmonic synthesis system 10. The given system might, in principle, have two variants of performance: the passive and the active. In the first case, the entering signal 9 is optimized along the parameter of the distribution of the spectral amplitudes by harmonic numbers to make the emitting signal 11 maximally resembling the Dirac's Delta-function spectrum (Chapter 1). In the case of the active variant, in addition to the above, it provides also amplification of the multiharmonic signal 9 as a whole. Such active synthesis might be fulfilled for instance on the basis of the ordinary multiharmonic parametric FEL.

What concerns the PEWSFEL, its history in many aspects resembles the one described above. Historically, the first PEWSFEL was proposed in 1991 [8–10] and subsequently this idea was developed in references [11–16]. The stimulus for their appearance was the

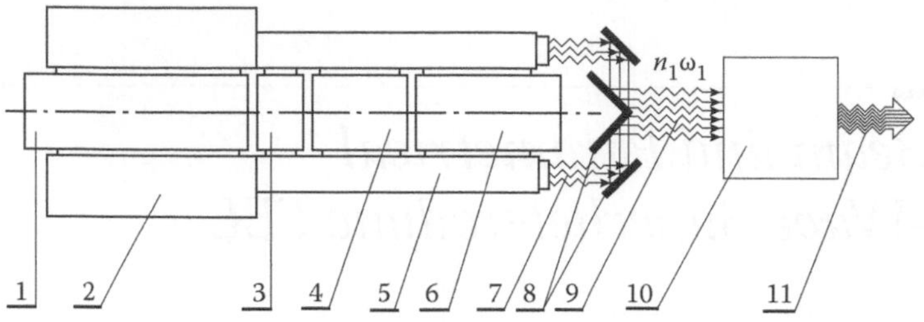

FIGURE 14.1
An example of arrangement of the cluster FEL on the basis of many parallel harmonic plasma-beam superheterodyne FEL. Here, 1 is the driving and operation systems of the multichannel and multibeam accelerator 2; 3 is the block of harmonic input signals; 4 is the microwave pumping oscillator for FEL; 5 are the monochromatic partial plasma-beam FEL amplifiers; 6 are the driving and operation systems of the FEL 5; 7 are the beams of output FEL radiation with the frequencies $n_1\omega_1$; 8 is the quasioptical system; 9 is the input beam of multiharmonic radiation for synthesis system 10; 11 is the output femtosecond cluster signal directed to an antenna system; $n_1 = 1,2,\ldots,$ are the numbers of harmonics of the master oscillator frequency ω_1 (which is not shown in the figure).

impressive success obtained at the time of the Star Wars Program in the field of the radiofrequency accelerators (RFAs) or, what is the same, the linear accelerators (LINACs).

The characteristic peculiarity of the given type of system is that the electron beam, which is emitted from the RFA, is always strongly modulated on its density. There appears an essential question: Can one use this circumstance for an increase in the linear coefficient of amplification of the FEL? One should admit that this is a very actual question because the current of an RFA-formed beam is always significantly lower than that formed by the induction accelerators (Chapter 9). This, in turn, in a better case, leads to a significant increase in the general length of FEL, and, in a worse case, it makes the given problem technologically unsolvable. Let us be reminded that it is the problem of producing compact FEL amplifiers with a high linear amplification (see Chapter 1).

However, there appears another circumstance: the RFA has a serious advantage over the induction devices, considering the weight–dimension characteristics (at the condition that the inductors of the past contain magnetic cores; Chapter 9). This advantage acquires a special weight in the case of the space boarded systems, when lifting any kilogram into the orbit costs a lot of money. However, money, as it is widely known, has a magical power over the activity of the inventive minds. As a result of an organized brainstorm, the problem of the insufficient beam current produced by the RFA was proposed to be solved by making the developed multibeam systems. In this case, the shortcoming of the emission power was compensated with interest by introducing the multibeam and multichannel arrangement for the acceleration system as a whole. Simultaneously, the problem of the small linear amplification of the FEL signal was solved by a transition from the ordinary parametrical (Chapters 11 and 12) to the electron-wave parametrical mechanism of the superheterodyne amplification. As an additional electron-wave pumping, it was proposed to use the previously mentioned strong beam modulation of RFA.

An example of designs of this type is given in Figure 14.2. The system works in the following way: PEWSFEL amplifiers 3 generate electromagnetic signals (each on its harmonic), which by the help of the optic system 1 are collected into the multiharmonic signal 2. Then this signal is directed to the active cluster synthesizer, which in this case is presented by the structural elements 4, 6, and 10. This includes in 4 a passive mixing of spatially inhomogeneous (with respect to harmonic number) multiharmonic signal 2 into a uniform signal. The system 6 is designed to assure the above-mentioned optimization of

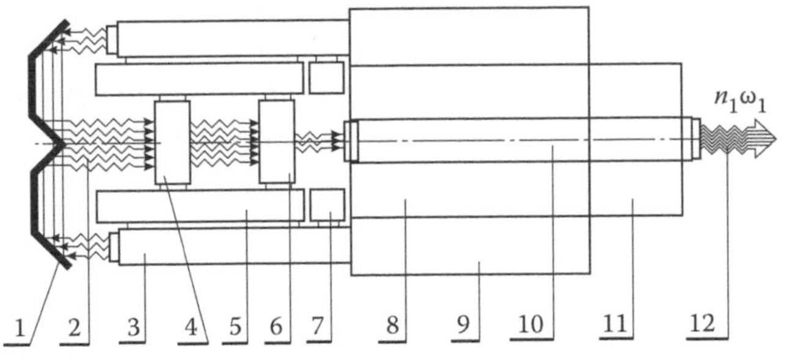

FIGURE 14.2
Example of the arrangement of the cluster FEL on the basis of many parallel harmonic FEL and active system for synthesis of cluster waves. Here, 1 is the optical system; 2 are the monochromatic radiation signal beams from the partial FEL 3; 3 are the PEWSFEL amplifiers; 4 is the system for blend of many monochromatic radiation beams into one multiharmonic beam; 5 are the driving and operation systems of the partial FEL 3; 6 is the system for optimization of spectrum of the multiharmonic radiation beam 2; 7 is the block of harmonic input signals for the partial FEL 3; 8 is the multichannel acceleration undulation block for the FEL synthesizer 10; 9 is the multichannel and multibeam acceleration block for the partial FEL 3; 10 is the multiharmonic active FEL synthesizer of electromagnetic clusters; 11 is the driving and operation systems of the FEL synthesizer 10; 12 is the output cluster signal.

the spectral amplitude distribution by harmonic numbers. Notice that the latter does not at all mean a formation of any kind of the cluster signal. The point is that the multiharmonic signal further undergoes amplification in the FEL synthesizer 10 and the very process of the amplification has a disperse character. Specifically, there are smaller amplitudes of the signal multiharmonic spectrum with higher numbers than the harmonics with lower numbers (Chapters 11 and 12). On the other hand, as was mentioned a few times (Chapter 1), the cluster signal 12 at the system output should have a spectrum maximally resembling the Dirac's Delta-function. It means to resemble the uniform harmonic amplitude distribution by their numbers. This, in turn, means that the multiharmonic signal 2, after passing the system for optimization of the spectrum of the multiharmonic radiation 6, should, figuratively speaking, exhibit as if the dislocated distribution of the spectral amplitudes. Specifically, the amplitudes with higher numbers should be larger than the amplitudes with smaller numbers. In such a case, in the process of its amplification in FEL synthesizer 10, this dislocation is compensated by the mentioned dispersion effect. As a result, the emitted signal 12 obtains the wanted spectrum, which in its form appears maximally to resemble the Dirac's Delta-function's spectrum.

One of the most characteristic technological peculiarities of the design solution, illustrated in Figure 14.2, is the technological idea of the performance of the acceleration blocks 8 and 9. It was suggested to make them in one shape common for all FEL (8 and 9) multichannel and multibeam accelerator systems. It is supposed that all acceleration channels have the same common source of the microwave driving system and relevant control system. Therewith, a part of this system, which is designed to form a high-current beam for the FEL synthesizer 10, was intended to perform like an undulator-type multichannel multibeam RFA. The single one- or two-velocity (depending on the FEL type) high-current beam performs the key role. Therefore, it was proposed to be used like the conceptual base of the design schemes, which was developed earlier for the multichannel undulation induction accelerators (Chapter 9).

In contrast to the previously illustrated example, presented in Figure 14.1, the discussed system, given in Figure 14.2, is basically designed to work within the optical wavelength range.

Thus, the performed discussion allows us to make the deduction that like the PBSFEL, the PEWSFELs, owing to their applied possibilities, are very promising devices. This, however, does not decrease their attraction as the purely fundamental object for investigation. Furthermore, we will show that the physics of these systems appears quite interesting and certainly far from banal.

14.1 Plasma-Beam Superheterodyne Free Electron Lasers: H-Ubitron Model*

14.1.1 Theoretical Model

We start the discussion with a theoretical analysis of a simplest H-Ubitron-type PBSFEL, the model of which is presented in Figure 14.3. Here, the incoming electromagnetic signal 1 moves with the relativistic electron beam 2, having the Langmuir frequency ω_b, into the system's input. Within the system's interaction region, they move in the medium of the magnetized plasma 3, which has the Langmuir frequency ω_p. Here, the magnetoundulator system 5 creates a spiral H-Ubitron pumping field. The amplified electromagnetic signal 7 evolves from the system.

We consider beam 2 to be sufficiently wide, to be transverse uniform, and to satisfy all the conditions of its applicability in a transverse nonlimited model (Chapter 10). We assume that the pumping and the signal fields are circularly polarized, uniform within the transverse plane XY, and periodic along the longitudinal z-axis.

The parameters of the system are chosen to secure the three-wave parametric-resonance effect. Specifically, the frequencies and the wave numbers of the signal waves $\{\omega_1, k_1\}$ and SCW $\{\omega_3, k_3\}$, on one side, and the wave number of the undulator field k_2, on the other side, appear to be interrelated by the conditions of the parametric wave resonance (Chapters 4, 11, and 12).

$$p_3 = p_1 + p_2, \tag{14.1}$$

where $p_1 = \omega_1 t - k_1 z$ is the phase of the electromagnetic wave signal; $p_2 = -k_2 z$ is the phase of the pumping H-Ubitron field; and $p_3 = \omega_3 t - k_3 z$ is the phase of the SCW electron wave. We assume that, in analogy with two-stream heterodyne FEL (TSFEL; Chapter 13), the SCW wave $\{\omega_3, k_3\}$ has the maximal amplification, which it obtains on account of the plasma-beam instability. Then, condition 14.1 can acquire a more traditional form for H-Ubitrons (Chapters 11 and 12)

$$\omega_3 = \omega_1, k_3 = k_1 + k_2. \tag{14.2}$$

The relativistic quasihydrodynamic equation (Chapter 10) is used as the initial equation

* From Kulish, V.V., and Lysenko, A.V., On the Theory of a Plasma-Superheterodyne Beam Free Electron Laser with H-Ubitron Pumping, *Plasma Physics Reports*, Pleiades Publishing, Ltd., vol. 35, No. 8, 696-699, 2009. With permission.

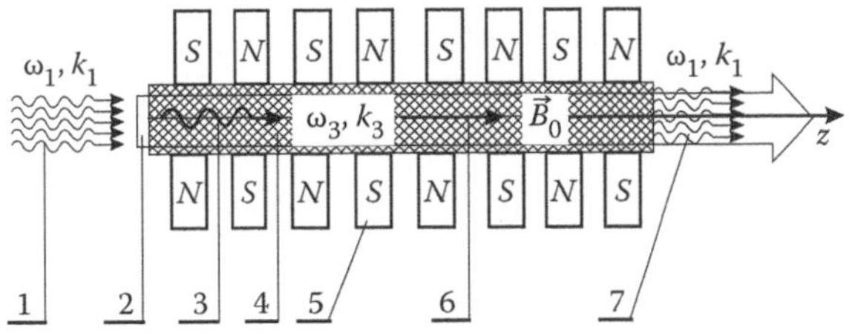

FIGURE 14.3
A theoretical model of the plasma-beam superheterodyne free electron laser (PBSFEL). Here, 1 is the input electromagnetic signal; 2 is the relativistic electron beam; 3 is the SCW; 4 are the plasmas; 5 is the spiral H-Ubitron pumping; 6 is the longitudinal magnetic field; 7 is the output electromagnetic signal; 8 is the worked-up relativistic electron beam; ω_1 and ω_3 are the cyclic frequencies of the electromagnetic signal; SCW, k_1, and k_3 are their wave numbers; $\vec{B}_0$ is the vector of magnetic induction of the longitudinal focusing magnetic field.

$$\left(\frac{\partial}{\partial t}+\vec{v}_\alpha\frac{\partial}{\partial \vec{\mathbf{r}}}+\frac{\nu}{\gamma_\alpha^2}\right)\vec{v}_\alpha=\frac{e}{m_e\gamma_\alpha}\times\left\{\vec{E}+\frac{1}{c}\left[\vec{v}_\alpha\vec{B}\right]-\frac{\vec{v}_\alpha}{c^2}\left(\vec{v}_\alpha\vec{E}\right)\right\}-\frac{\upsilon_T^2}{n_\alpha\gamma_\alpha}\left[\frac{\partial n_\alpha}{\partial \vec{\mathbf{r}}}-\frac{\vec{v}_\alpha}{c^2}\left(\vec{v}_\alpha\frac{\partial}{\partial \vec{\mathbf{r}}}\right)n_\alpha\right], \quad (14.3)$$

the continuity equation

$$\frac{\partial n_\alpha}{\partial t}+\frac{\partial(n_\alpha v_{z,\alpha})}{\partial z}=0, \quad (14.4)$$

and the Maxwell equation

$$\mathrm{rot}\vec{B}=\frac{1}{c}\frac{\partial \vec{\mathrm{E}}}{\partial t}+\frac{4\pi}{c}\sum_{q=1}^{2}(en_\alpha\vec{v}_\alpha), \quad \mathrm{rot}\vec{E}=-\frac{1}{c}\frac{\partial \vec{B}}{\partial t}, \quad (14.5)$$

where $\vec{v}_\alpha$ is the velocity vector of the αth component of the electron plasma ($\alpha = 1,2$); ν is the particle collision frequency; $\vec{E}$ is the electric field strength; $\vec{B}$ is the magnetic field induction; $\gamma_\alpha = 1/\sqrt{1-v_\alpha^2/c^2}$ is the relativistic factor of the partial α beam; n_α is the concentration of the particles of the α-sort; υ_T is the mean root square velocity of the particle thermal motion; $\vec{\mathbf{r}}$ is the spatial coordinate of the observation point; c is the light velocity in vacuum; and $e = -|e|$, m_e are the electron charge and rest mass, respectively. Furthermore, we will assume that one can disregard the interparticle collisions and the thermal dispersion of electrons; it means that $\upsilon_T = 0$ and $\nu = 0$.

14.1.2 Truncated Equations

The methods of averaging characteristics (Chapter 6) and of slowly changing amplitudes (Chapters 4 and 11) were used to solve the problems of the motion and electron concentration determination in a beam and plasma. The problem of the electromagnetic field excitation was solved by the method of slowly changing amplitudes.

Recall, what is a widely known fact, that analogous to the two-stream instability (Chapter 13), the law of dispersion for the electron wave SCW $\{\omega_3, k_3\}$ is quasilinear. Therefore, in the investigated system, in principle, the effect of plural parametric resonances in respect to the wave harmonic SCW (Chapter 13) can also be realized. Because of it, in general, we chose the fields in the form of the multiharmonic spectra.

$$\vec{E}_1 = \sum_{m=1}^{N} \left[\left(E_{1x,m} \vec{\mathbf{e}}_x + E_{1y,m} \vec{\mathbf{e}}_y \right) e^{imp_1} + c.c. \right],$$

$$\vec{B}_1 = \sum_{m=1}^{N} \left[\left(B_{1x,m} \vec{\mathbf{e}}_x + B_{1y,m} \vec{\mathbf{e}}_y \right) e^{imp_1} + c.c. \right],$$

$$\vec{E}_3 = \sum_{m=1}^{N} \left[E_{3z,m} e^{imp_3} + c.c. \right] \vec{\mathbf{e}}_z,$$

$$\vec{B}_2 = \sum_{m=1}^{N} \left[\left(B_{2x,m} \vec{\mathbf{e}}_x + B_{2y,m} \vec{\mathbf{e}}_y \right) e^{imp_2} + c.c. \right], \tag{14.6}$$

where m is the number of the corresponding harmonic, and N is the quantity of harmonics, which are taken into account in the calculation of the problem.

As a result of the standard procedure in the asymptotic integration (analogous with Chapters 4, 6, 11, and 12) for the harmonics of the electric field strength of the electromagnetic wave $\{\omega_1, k_1\}$ and the electron wave of SCW $\{\omega_3, k_3\}$, we obtain a system of differential equations in the cubic approximation

$$K_{2,m} \frac{\partial^2 E_{1x,m}}{\partial t^2} + K_{1,m} \frac{\partial E_{1x,m}}{\partial t} + D_{1,m} E_{1x,m} = K_{3,m} E_{3z,m} B_{2y,m}^{*} + FX_{1,m},$$

$$K_{2,m} \frac{\partial^2 E_{1y,m}}{\partial t^2} + K_{1,m} \frac{\partial E_{1y,m}}{\partial t} + D_{1,m} E_{1y,m} = K_{3,m} E_{3z,m} B_{2x,m}^{*} + FY_{1,m},$$

$$C_{2,m} \frac{\partial^2 E_{3z,m}}{\partial t^2} + C_{1,m} \frac{\partial E_{3z,m}}{\partial t} + D_{3,m} E_{3z,m} = C_{3,m} E_{1x,m} B_{2y,m} + FZ_{3,m}. \tag{14.7}$$

Here,

$D_{1,m} = \dfrac{im}{c\omega_1} \left(k_1^2 c^2 - \omega_1^2 + \sum_{\alpha}^{b,p} \dfrac{\omega_\alpha^2 \Omega_{1,\alpha} \left(m\Omega_{1,\alpha} + \eta_1 \omega_{H,\alpha} \right)}{m\bar{\gamma}_\alpha \left(\left(m\Omega_{1,\alpha} \right)^2 - \left(\omega_{H,\alpha} \right)^2 \right)} \right)$ is the dispersion function of the signal electromagnetic wave

$$K_{1,m} = \frac{\partial D_{1,m}}{\partial \left(im\omega_1 \right)} = -\frac{c^2 k_1^2 + \omega_1^2}{\omega_1^2 c} - \sum_{\alpha}^{b,p} \frac{\omega_\alpha^2 \left(\Omega_{1,\alpha} \omega_1 + (\omega_{H,\alpha} \eta_1 - m\Omega_{1,\alpha}) \bar{v}_{z,\alpha} k_1 / m \right)}{\omega_1^2 c \bar{\gamma}_\alpha \left(\omega_{H,\alpha} \eta_1 - m\Omega_{1,\alpha} \right)^2};$$

$$K_{2,m} = \frac{1}{2}\frac{\partial^2 D_{1,m}}{\partial(im\omega_1)^2} = \frac{ck_1^2}{im\omega_1^3} + \sum_{\alpha}^{b,p} \frac{\omega_\alpha^2\left(\Omega_{1,\alpha}\omega_1^2 - \left(\Omega_{1,\alpha} - \frac{\omega_{H,\alpha}\eta_1}{m}\right)\bar{v}_{z,\alpha}k_1\left(\omega_1 + \left(\Omega_{1,\alpha} - \frac{\omega_{H,\alpha}\eta_1}{m}\right)\right)\right)}{i\omega_1^3 c\bar{\gamma}_\alpha\left(m\Omega_{1,\alpha} - \omega_{H,\alpha}\eta_1\right)^3};$$

$$K_{3,m} = -\frac{1}{c}\sum_{\alpha}^{b,p}\frac{\omega_\alpha^2}{cm}\frac{e}{m_e}\frac{1}{\bar{\gamma}_\alpha^2\Omega_{3\alpha}} \times \left(\frac{\left(m\Omega_{1\alpha} - \eta_2\omega_{H,\alpha}\right)}{\left(\omega_{H\alpha}\right)^2 - \left(m\Omega_{1\alpha}\right)^2}\left(\left(1 - 2\frac{\bar{v}_{z,\alpha}^2}{c^2}\right) + \frac{\bar{v}_{z,\alpha}^2}{c^2}\frac{\left(m\Omega_{3\alpha} - \eta_2\omega_{H,\alpha}\right)}{\left(m\Omega_{2\alpha} - \eta_2\omega_{H,\alpha}\right)}\right)\right)$$

$$+\frac{1}{c}\sum_{\alpha=1}^{2}\frac{\omega_\alpha^2 k_3}{\bar{\gamma}_\alpha^2\Omega_{3\alpha}^2}\left(1 - \frac{\bar{v}_{z\alpha}^2}{c^2}\right)\frac{e}{m_e}\frac{\bar{v}_{z\alpha}}{cm}\frac{\left(m\Omega_{2\alpha} + \eta_2\omega_{H,\alpha}\right)}{\left(\omega_{H,\alpha}\right)^2 - \left(m\Omega_{2\alpha}\right)^2};$$

$$D_{3,m} = \frac{-im\omega_3}{c}\left(1 + \sum_{\alpha}^{b,p}\frac{\omega_\alpha^2}{\Omega_{3,\alpha}^2(im)^2\bar{\gamma}_\alpha^3}\right)$$

is the disperse function of the electronic SCW

$$C_{1,m} = \frac{\partial D_{3,m}}{\partial(im\omega_3)} = -\frac{1}{c} + \sum_{\alpha}^{b,p}\frac{\omega_\alpha^2\left(1 + 2k_3\bar{v}_{z,\alpha}/\Omega_{3,\alpha}\right)}{c\bar{\gamma}_\alpha^3\left(im\Omega_{3,\alpha}\right)^2};$$

$$C_{2,m} = \frac{1}{2}\frac{\partial^2 D_{3,m}}{\partial(im\omega_3)^2} = -\sum_{\alpha}^{b,p}\frac{\omega_\alpha^2\left(1 + 3k_3 \upsilon\bar{v}_{z,\alpha}/\Omega_{3,\alpha}\right)}{c\bar{\gamma}_\alpha^3\left(im\Omega_{3,\alpha}\right)^3};$$

$$C_{3,m} = \sum_{\alpha}^{b,p}\frac{e}{m_e}\frac{\omega_\alpha^2\omega_3\left(1 - \eta_1\eta_2\right)}{c^2\bar{\gamma}_\alpha^2 im\Omega_{3\alpha}^2}\times$$

$$\left[\frac{\Omega_{1\alpha}}{\omega_{1\alpha}}\frac{im\Omega_{1\alpha} + i\eta_1\omega_{H,\alpha}}{\left(im\Omega_{1\alpha}\right)^2 + \left(\omega_{H,\alpha}\right)^2} - \left(\frac{ck_1}{\omega_1} - \frac{\bar{v}_{z,\alpha}}{c}\right)\frac{\bar{v}_{z,\alpha}}{c}\frac{im\Omega_{2\alpha} + i\eta_2\omega_{H,\alpha}}{\left(im\Omega_{2\alpha}\right)^2 + \left(\omega_{H,\alpha}\right)^2}\right];$$

$FX_{1,m} = FX_{1,m}(\vec{E}_1, \vec{B}_2, \vec{E}_3)$, $FY_{1,m} = FY_{1,m}(\vec{E}_1, \vec{B}_2, \vec{E}_3)$, $FZ_{1,m} = FZ_{1,m}(\vec{E}_1, \vec{B}_2, \vec{E}_3)$ are functions that take into account the cubic-nonlinear additions to the corresponding equations and depend on the harmonics of the interacting waves (these functions are not written here because of their awkwardness); $\eta_1 = E_{1y,m}/(iE_{1x,m}) = \pm 1$ is the sign function that describes the rotation direction of the vector of electric strength of the circularly polarized signal wave; $\eta_2 = B_{2y,m}/(iB_{2x,m}) = \pm 1$ is the sign function that describes the direction of the rotation of the spiral H-Ubitron field of pumping; c is the light velocity; e is the electron charge; m_e is the electron rest mass; by index "b" we describe the beam parameters; by index "p" we describe the plasma parameters; $\bar{v}_{z,b}$ is the average velocity of the electron beam; $\bar{v}_{z,p} = 0$ is the average electron velocity of plasmas; $\bar{\gamma}_\alpha = 1/\sqrt{1-(\bar{v}_{z,\alpha}/c)^2}$; $\Omega_{1,\alpha} = \omega_1 - k_1\vec{v}_{z,\alpha}$; $\Omega_{2,\alpha} = -k_2\bar{v}_{z,\alpha}$; $\Omega_{3,\alpha} = \omega_3 - k_3\bar{v}_{z,\alpha}$; $\omega_{H,\alpha} = eH_0/(\bar{\gamma}_\alpha m_e c)$ is the cyclotron frequency of the α-sort particle (it means that the electron belongs to the beam or plasma).

It follows from the structure of the coefficient $C_{3,m}$ that, for realization of the parametric resonance, it is necessary that the direction of the rotation vector of the electric field strength signal $\vec{E}_1$ and of the pumping magnetic field induction $\vec{B}_2$ were alike. The rotation of the vectors of both the waves should be clockwise ($\eta_1 = +1$, $\eta_2 = +1$) or counterclockwise ($\eta_1 = -1$, $\eta_2 = -1$). Otherwise, the coefficient $C_{3,m}$ is equal to zero. Recall the similar results we obtained in Chapter 11 in the analysis of the parametric-resonance interaction of the circularly polarized waves in the ordinary FEL.

14.1.3 Quadratic-Nonlinear Approximation

Before we get down with the quadratic-nonlinear analysis, let us recall one important characteristic of the quasilinear superheterodyne FEL (Chapter 13). It concerns the determination of the concepts of the nonlinear and the linear (more accurately quasilinear) interaction. In a simplest model of the ordinary parametric FEL (Chapter 11), the quadratic approximation describes linear (quasilinear) interaction. The explanation of this paradox consists of the fact that the nonzero result for the amplification of the signal wave appears here only in the quadratic (formally nonlinear) approximation. In this sense, the quadratic mechanism appears to be the basic mechanism because, in essence, it behaves like linear (weak signal, quasilinear). Specifically, therefore, in the FEL theory, the linear (quasilinear) mechanisms in reality appear to be quadratic (Chapter 9).

The situation in the superheterodyne systems appears different (Chapter 13). Here, we have the simultaneous imposition of two different-hierarchy mechanisms of amplification. One of them is really linear (two-stream or plasma-beam instability) and the second is quadratic (usual parametric interaction). This all induces a large confusion in the terminology and, what is worse, it often leads to serious mistakes and saturates them with a series of the known works in quadratic-nonlinear theory of the superheterodyne systems.

The problem in this is that, traditionally, the calculation is performed in the framework of one or another version of the successive approximation scheme. For example, in the case of the superheterodyne FEL, the lowest nonzero approximation for the really linear mechanism appears linear. Then, when we move to the quadratic approximation, we obtain a completely complex calculation subtlety. Specifically, for the parametric cases, as we already mentioned, it appears linear (weak-signal) and for the plasma-beam (or two-stream), the quadratic approximation appears already really nonlinear. Consequently, it is necessary to perform this part of the calculation within the framework of the quadratic-nonlinear (and not in the weak-signal) theory. Regrettably, the majority of the investigators forget about this and in the framework of the quadratic theory of PBSFEL, they lose the quadratic-nonlinear effects. This significantly distorts the general physical picture of the investigated process.

Thus, by using the mentioned precaution, we turn further to the quadratic-nonlinear analysis of the investigated model of PBSFEL. Thus, in the model presented in Figure 14.3, the plasma-beam instability effect is realized. From the mathematical point of view, this means that some of the solutions of the dispersion equation for the mth harmonic of SCW

$$D_{3,m}(\omega'_{3,m}, mk_3) = 0 \tag{14.8}$$

are complex. It means that, by supposing that the wave number mk_3 is real and known, we obtain from Equation 14.8 the solution for the frequency $\omega'_{3,m}$ (and not of the wave number like we did before) which in this case appears to be complex

$$\omega'_{3,m} = \omega_{3,m} - i\Gamma_m, \tag{14.9}$$

where $\omega_{3,m}$ is the real part of the complex frequency $\omega'_{3,m}$, and Γ_m is the growth increment of the mth wave harmonic SCW in time. Here, only the real part of frequency is included in the dispersion equation

$$\omega_{3,m} \approx m\frac{k_3}{v_{z0}} = m\omega_3. \tag{14.10}$$

It means, in analogy with the two-stream instability case, that the dispersion law for the real part of the increasing wave SCW, as before, is linear. In the conditions of the parametric resonance (Condition 14.1) and in the obtained system of the field amplitude equations (Equation 14.7), real parts of the frequency SCW ω_3 are used. Formally, the dispersion function in Equation 14.7, which depends on the real ω_3 and k_3, is not equal to zero, $D_{3,m}(m\omega_3, mk_3) \neq 0$. Therefore, for the elimination of the methodical ambiguity, let us transform the dispersion function $D_{3,m}(m\omega_3, mk_3) \neq 0$ by expanding Equation 14.8 in Taylor series near the real values of $m\omega_3$, mk_3.

$$0 = D_{3,m}(\omega'_{3,m}, mk_3) = D_{3,m}(m\omega_{3,m}, mk_3) + \frac{\partial D_{3,m}}{\partial(m\omega_3)}(-i\Gamma_m) + \frac{\partial^2 D_{3,m}}{2\partial(m\omega_3)}(-i\Gamma_m) + \dots .$$

From this we find

$$D_{3,m}(m\omega_{3,m}, mk_3) = -\frac{\partial D_{3,m}}{\partial(im\omega_3)}\Gamma_m - \frac{1}{2}\frac{\partial^2 D_{3,m}}{\partial(im\omega_3)^2}(\Gamma_m)^2 + \dots = -C_{1,m}\Gamma_m + C_{2,m}\Gamma_m^2 + \dots .$$

When the system of equations (Equation 14.7) acquires (in quadratic approximation) the methodologically more correct form

$$K_{1,m}\frac{\partial E_{1x,m}}{\partial t} + D_{1,m}E_{1x,m} = K_{3,m}E_{3,m}B_{2y,m},$$

$$K_{1,m}\frac{\partial E_{1y,m}}{\partial t} + D_{1,m}E_{1y,m} = K_{3,m}E_{3,m}B_{2x,m},$$

$$C_{2,m}\frac{\partial^2 E_{3,m}}{\partial t^2} + C_{1,m}\frac{\partial E_{3,m}}{\partial t} - (C_{1,m}\Gamma_m + C_{2,m}\Gamma_m^2)E_{3,m}$$

$$= C_{3,m}E_{1x,m}B^*_{2y,m} + C_{4,m}\left\langle E_3\left(\int E_3\,dp_3\right)\right\rangle_{mp3} + C_{5,m}\left\langle\left(\int E_3\,dp_3\right)^2\right\rangle_{mp3}. \tag{14.11}$$

Furthermore, we will perform an analysis of the system (Equation 14.11). Let us analyze a simple case with the monochromatic work mode (for each wave the harmonic number of harmonics $N = 1$). We look for a solution in the form $E_{1x,1}$, $E_{3z,1}$ ~exp(αt), substitute it in Equation 14.11 and, for the determination of the increment of superheterodyne amplification α, we obtain the following equation

$$\left(\frac{C_{2,1}}{C_{1,1}}(\alpha^2-\Gamma_1^2)+(\alpha-\Gamma_1)\right)\alpha=\frac{C_{3,1}K_{3,1}}{C_{1,1}K_{1,1}}\cdot|B_{2y,1}|^2\equiv\alpha_{par}^2. \tag{14.12}$$

We assume that according to the conditions of the superheterodyne amplification, the growth increment of plasma-beam instability Γ_1 should always be significantly larger than the parametric increment α_{par} ($\Gamma_1 >> \alpha_{par}$) (Chapter 4). Then, for the superheterodyne increment α, we obtain

$$\alpha=\Gamma_1\left(1+\frac{\alpha_{par}^2}{\Gamma_1^2(1+2C_{2,1}\Gamma_1/C_{1,1})}\right). \tag{14.13}$$

It follows from Equation 14.13 that according to the mathematical structure, the increment of the superheterodyne amplification α coincides with the increment, which is obtained within the framework of the model problem (Chapter 4) without the realization of a concrete mechanism of the additional amplification of SCW. Also we observe that in the condition $\Gamma_1 >> \alpha_{par}$, the increment α of the wave in the system practically coincides with the increment of the plasma-beam instability Γ_1. By using the known correlation for Γ_1, in the case when $k_3\bar{v}_{z,b}=\omega_p$, we obtain

$$\alpha\approx\Gamma_1=\frac{\sqrt{3}\cdot\omega_b^{2/3}\cdot\omega_p^{1/3}}{\sqrt[3]{16}\cdot\bar{\gamma}_b}. \tag{14.14}$$

A closest physical analog of the investigated phenomenon appears to be the effect of the superheterodyne amplification in the TSFEL (Chapter 13). To illustrate the advantage of the plasma-beam SFEL above the TSFEL, we will perform a comparison of the obtained increment α with the maximal increase increment of an equivalent two-stream system Γ_2 (Chapter 13)

$$\Gamma_2=\frac{\omega_b}{2\cdot\bar{\gamma}_b^{3/2}}. \tag{14.15}$$

In this case, under the equivalent two-stream system, we understand the system in which the partial electron beam has the same parameters as in the case of the plasma-beam SFEL.

From the analysis of the correlation of Equations 14.14 and 14.15, one can derive the conclusion that the increment of the plasma-beam system Γ_1 is significantly larger than that of the equivalent two-stream system Γ_2. For an illustration of the conclusion, we will perform numerical estimation. Considering that $\omega_p \sim 10^{12}\ s^{-1}$, $\omega_b \sim 10^9\ s^{-1}$, $\bar{\gamma}_b = 2$ and using correlations (Equations 14.14 and 14.15), we obtain that $\alpha = 3.44\cdot10^9\ s^{-1}$ and $\Gamma_2 = 1.77\cdot10^8\ s^{-1}$.

Thus, the increment of the investigated plasma-beam system might be significantly larger (e.g., almost 20 times larger, as is illustrated in this numerical example) than in the case of the equivalent TSFEL. One should not forget, however, that the working range of PBSFEL is limited, as a rule, to the millimeter range, whereas in the case of TSFEL, the range expands up to the visible range.

14.1.4 Cubic-Nonlinear Approximation

Let us recall that, in the above-performed analysis, we limited ourselves in Equation 14.7 to take into account only the quasilinear components for the signal wave and to partially consider quadratic for SCW. Therefore, to retain the methodic integrity, we will further study the influence of those quadratic and cubic nonlinear effects, which we previously disregarded.

For the integration of the system (Equation 14.7), we will use the numerical methods.

The tuning of the model is chosen in such a way that the SCW in the case of the plasma-beam instability would have the optimal frequency $\omega_3 = \omega_{opt}$. It means, in analogy with two-stream instability, that it has the maximal growth increment. To achieve this, the wave number of the first harmonic of wave SCW is assumed to be equal to $k_3 = \omega_p/\bar{v}_b = 38.5\ \text{cm}^{-1}$. In this case, the wavelength of the signal in vacuum will be equal to 1.6 mm at the undulation period of the magnetic field of pumping equal to 5.8 cm. At this stage of the calculation, we assume that the circularly polarized undulator is monochromatic. It means that we assume that the higher harmonics of the H-Ubitron field is zero. It follows that the questions of the formation of the multiharmonic electromagnetic clusters, in this case, are outside the field of our attention. Thus, in this case, the role of the higher harmonics of SCW is reduced to their effect on the dynamics of the amplification of the first harmonic of the signal. This effect is illustrated in Figures 14.4 and 14.5.

Figure 14.4 presents the dependence of the resulting amplitude of the first harmonic of the signal $E_{1,1} = \sqrt{|E_{1x,1}|^2 + |E_{1y,1}|^2}$ on normalized time $\tau = \Gamma_1 t$, whereas before (see Equation 14.14),

$$\Gamma_1 = \frac{\sqrt{3}\cdot\omega_b^{2/3}\cdot\omega_p^{1/3}}{\sqrt[3]{16}\cdot\bar{\gamma}_b} \tag{14.16}$$

is the growth increment of the SCW amplitude at the expense of the plasma-beam instability.

Two variants of numerical calculation were performed. In the first case (curve 1), only the first harmonics of the interacting waves of the signal of pumping and SCW were considered and, in the second case (curve 2), ten harmonics of SCW were considered. By analyzing Figure 14.4, we see that the saturation level, in the case when taking into account the higher harmonics (curve 2), is five times smaller as compared with the calculation variant, when the very fact of the higher harmonic generation was not taken into account (curve 1). This result was expected because analogous deductions were made before in the case of the simple parametrical FEL (Chapter 12) as well as in the case of the TSFEL (Chapter 13). However, even in this unfavorable case, the amplification coefficient in comparison with the simple parametrical FEL remains fantastically high. On the other hand, one should remember (Chapter 12) that in real systems, the saturation effect for the interacting waves might be suppressed by isochronization of the system. The electron efficiency in this case might be significantly high (to 70% and, theoretically, even higher).

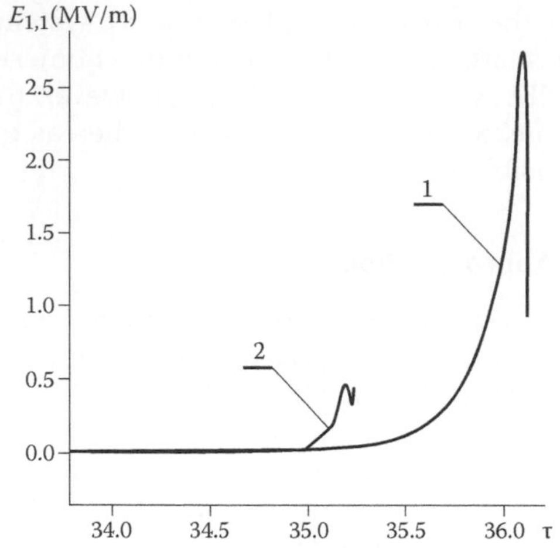

FIGURE 14.4
Dependence of the amplitude of the first harmonic's wave signal $E_{1,1}$ on normalized time $\tau = \Gamma_1 t$. Here, curve 1 represents the case when only the first harmonics are taken into account; curve 2 represents the case when the first ten harmonics of the interacting waves are taken into account. The calculations were performed at the following parameters: plasma Langmuir frequency is $\omega_p = 10^{12}$ s^{-1}; the beam Langmuir frequency is $\omega_b = 2.10^9$ s^{-1}; the relativistic beam factor is $\Gamma = 2$; the tension of the longitudinal magnetic field is $H_0 = 2.23\ 10^5$ A/m; the tension of the longitudinal magnetic field of pumping is $H_2 = 7.96\ 10^4$ A/m; the amplitude of signal wave at input is $E_{1,10} = 10$ V/m.

It is interesting to investigate synchronously the dynamics of the first harmonic SCW. Figure 14.5 presents such dependencies. There, the dependence of the amplitudes of the first harmonics of SCW on the normalized time coordinate $\tau = \Gamma_1 t$, curve 2, corresponds to the case when taking into account the first ten harmonics SCW and corresponds to curve 1 when only the first harmonics of all interacting waves are considered. A comparison of

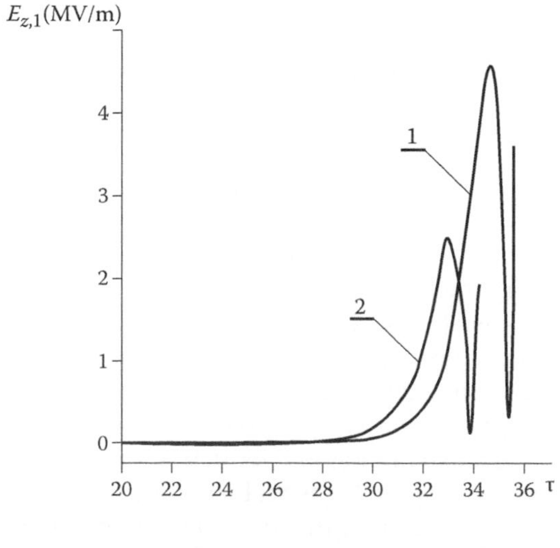

FIGURE 14.5
Dependence of the amplitude of the first harmonic of the electric field tension of SCW E_{z3} on the normalized time $\tau = \Gamma_1 t$. Here, curve 1 represents the case when only the first harmonic is considered in the calculation; curve 2 represents the case when the first ten harmonics of the interactive waves are taken into account. The calculations were performed in the same parameters as in Figure 14.4.

curves 1 and 2 reveals that, when taking into account the effect of the higher harmonics, the saturation level of the first harmonic SCW decreases for more than two times; however, what is quite interesting is that it also simultaneously decreases the time of the saturation. Also, like in the case of the two-stream systems (Chapter 13), the amplitudes of the higher harmonics of wave SCW appear to be commensurable with the amplitude of the first harmonic SCW.

Such behavior of the SCW harmonics is explained by the fact that their dispersion curve is close to linear. Therefore, in analogy with TSFEL, the effect of the plural parametrical resonances between the SCW harmonics (Chapter 13) might take place also between the wave SCW harmonics. As we see, the result of such interaction leads, on one hand, to an intensification of the generation process of higher SCW harmonics and, on the other hand, to a decrease in the amplitude of the first harmonic SCW (as compared with the case when the higher harmonics were disregarded).

This observation opens many interesting variants for a practical application of the investigated systems. A simple technological analysis reveals that the given-type SFELs are significantly complicated for a practical realization. This situation appears to be most striking when considering a reliable isolation of plasmas in the interaction region from the remaining vacuum part of the system. The point is that a realization of the most simple and simultaneously reliable variants of design solutions becomes complicated by a series of the technological contradictions, for instance, between the demands of the electron beam input and output windows into a plasma, on one hand, and analogous signal windows, on the other hand (see model in Figure 14.3). These and others of this type of technological peculiarities make the use of the PBSFEL sections (not complete system) as the most acceptable variant of the highly effective modulators of electron beams in SFEL-klystron (Chapter 9). In this case, the following section of the energy transformation (from kinetic to electromagnetic signal energy; Chapter 9) can be performed on the basis of the more ordinary parametrical FEL. To be specific, this type of design scheme appears in the example of the cluster system, presented in Figure 14.1. Not less but, in reality, the multiharmonic versions of such klystrons appear even more promising: the first PBSFEL section is used as a multiharmonic modulator (recall that the generated SCW harmonics might have commensurate amplitudes within a fairly wide frequency range), and the second PBSFEL section is an active cluster synthesizer.

We will continue the analysis of the obtained results. From Figure 14.4, it follows that, as it should be, in the initial stage, the process of the development of the superheterodyne amplification is described by the exponential law with the increment (Equation 14.13)

$$\alpha = \Gamma_1 \left(1 + \frac{\alpha_{\text{par}}^2}{(\Gamma_1)^2 (1 + 2C_{2,1}\delta\omega / C_{1,1})} \right) \approx \Gamma_1, \tag{14.17}$$

where

$$\alpha_{\text{par}} = \sqrt{\frac{C_{3,1}K_{3,1}}{2C_{1,1}K_{1,1}} \cdot (|B_{2x,1}|^2 + |B_{2y,1}|^2)}$$

is the growth increment corresponding to the parametric wave interaction in the system. In this regard, we may say that the cubic-nonlinear theory confirms the conclusion, which was obtained within the framework of the quadratic theory. It concerns the possibility to achieve the exceptionally high increments of wave increase in the investigated system. In the given situation, the role of the cubic theory consists, basically, only in determining the moment when the exponential law of the amplification begins to break. It means that when the saturation of the amplification process occurs, as we see, even at moderate requirements to the electron beam ($\omega_b = 2 \cdot 10^9$ s^{-1}, $\bar{\gamma}_b = 2$), before the saturation of the amplification, one can obtain an intense electromagnetic emission $E_{1,1}$ = 0.75 MeV/m at the output when the input signal was ~100 V/m within the millimeter wavelength range. Therewith, let us turn our attention to what is possible to achieve at the interaction region length of only ~ 1.5 m.

Historically, we had a circumstance [1] that served as the principal argument for determining the EMI systems as the basic field of practical application of the PBSFEL. In time, however, it became clear that not a lesser perspective has its application as the technological basis for the cluster system (see Figures 14.1 and 14.2, and the corresponding material in Chapter 1). In particular, as was already mentioned, the most interesting design solutions were obtained in cases when the sections of PBSFEL were used as the multiharmonic modulators in the cluster klystrons.

Simultaneously, we mention that at the evaluation of the plasma-beam superheterodyne systems, one should not excessively overestimate the applied side of the discussed problems. First, therefore, as is revealed from the discussions, the physics of the PBSFEL and of TSFEL are exclusively interesting physical objects, independent of the practical application that they have.

14.2 Plasma-Beam Superheterodyne Free Electron Lasers: Dopplertron Model

14.2.1 Model and the Problem Formulation

Let us investigate the model PBSFEL with the Dopplertron pumping [5,22], which is illustrated in Figure 14.6. Here, the relativistic electron beam 1, which is moving along the z-axis, passes through the plasma 2. The plasma-beam system, as a whole, is located in the longitudinal focusing magnetic field of tension H_0 5, also oriented along the z-axis. The pumping waves 3 and 7 were chosen in the form of unusual electromagnetic wave of frequency ω_2 and wave number k_2. In the case when the wave propagates into the positive direction of z-axis, we describe it by the function $s_2 = +1$ (see position 3) and, in the case of the opposite direction, by the function $s_2 = -1$ (position 7). Also, for the input of the system we convey the electromagnetic signal 3 of frequency ω_1 and wave number k_1 moving along the z-axis. At the output, we have output amplified signal 8, which has the same frequency and wave number.

As a consequence of the parametric resonance between the signal wave $\{\omega_1, k_1\}$ and pumping wave $\{\omega_2, k_2\}$ in the plasma-beam system, the SCW 6 is excited. It has frequency ω_3 and wave number k_3. The SCW is further amplified simultaneously by two mechanisms: the parametric and the plasma beam. The frequency and the wave numbers of these SCW are interrelated by the traditional condition of the parametric resonance.

$$p_3 = p_1 + \nu p_2$$

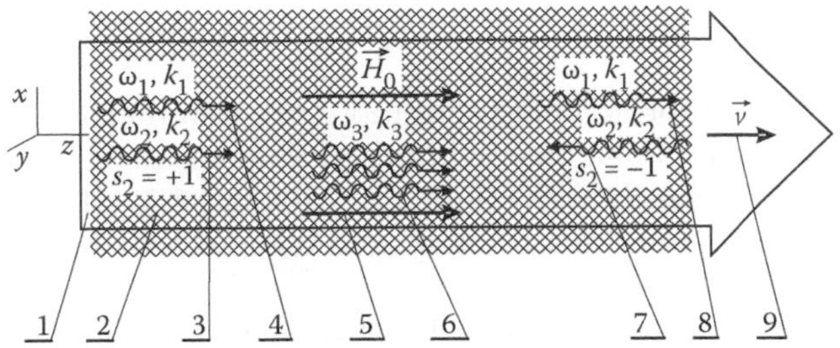

FIGURE 14.6
Model of the plasma-beam superheterodyne FEL of the Dopplertron type. Here, 1 is the relativistic electron beam; 2 are the magnetized plasmas; 3 is the pumping wave in the case of its propagation along the positive direction of z axis ($s_2 = +1$); 4 is the input signal wave; 5 is the vector of intensity of the magnetic focusing field $\vec{H}_0$; 6 is the SCW; 7 is the pumping wave in the case of its propagation along the negative direction of z axis ($s_2 = -1$); 8 is the output signal wave; 9 is the averaged velocity vector of electron beam 1.

or

$$\omega_3 = \omega_1 + \nu\omega_2,\ k_3 = k_1 + \nu s_2 k_2, \tag{14.18}$$

where $p_1 = \omega_1 t - k_1 z$ is the phase of the signal of the electromagnetic wave; $p_2 = \omega_2 t - s_2 k_2 z$ is the phase of the electromagnetic pumping; and $p_3 = \omega_3 t - k_3 z$ is the phase of the SCW. By comparing Equations 14.1 and 14.2, it becomes apparent that here we used the generalized variant of the parametric resonance conditions, which was studied in detail in Chapter 11. Obviously, as in the previous case, the system's parameters were chosen in the way that the SCW $\{\omega_3, k_3\}$ became within the band of the amplification due to plasma-beam instability.

The peculiarity of any kind of Dopplertron models, which makes them unique among the other models of FEL, is the use of an intense electromagnetic wave for pumping. In the studied model (positions 3 and 7 in Figure 14.6), the Dopplertron pumping performs an unusual retarded electromagnetic wave in the magnetized plasma, the dispersion law of which is described as (Chapters 4 and 11)

$$k_2 = \frac{\omega_2}{c}\sqrt{1 - \frac{\omega_p^2}{\omega_2(\omega_2 - \omega_H)}} \approx \frac{\omega_2}{c} \cdot N_2, \tag{14.19}$$

where c is the light velocity; ω_H is the cyclotron frequency; and $N_2 = \omega_p / \sqrt{\omega_2 \omega_H}$ is the retardation coefficient. The parameters of the system are chosen in such a way to make $N_2 >> 1$. Because of this there is a possibility for an amplification of the electromagnetic signal with increased frequency. It is important that this result in this case is achieved without a significant increase of energy of the electron beam

$$\omega_1 = \nu\omega_2 \frac{s_2 N_2 \beta + 1}{1 - \beta}, \tag{14.20}$$

where $\beta = v_z / c$ is the standard dimensionless velocity of the electric beam. In the relativistic case $1 - \beta << 1$ and $N_2 >> 1$, correlation 14.20 might be more conveniently rewritten into the more informative form

$$\omega_1 \approx 2\nu s_2 N_2 \gamma^2 \omega_2. \tag{14.21}$$

Relation 14.21 shows that, at the accepted conditions, the parametric resonance might happen only in the case when $\nu s_2 = +1$. Exactly this correlation is used in the case of explosive instability, which was sufficiently well studied in Chapters 11 and 12. The quasihydrodynamic equation, the equation of the continuity, and the Maxwell equation (Chapter 10) are used as the initial point.

We used the method of averaged characteristics (Chapter 6) to solve the problem on motion and determination of concentration. To solve the problem on the excitation of the electromagnetic fields, we used the method of the slowly varying amplitudes (Chapters 4 and 11).

By taking into account that the transverse waves of electromagnetic signal and pumping and, in general, the longitudinal electron wave might have the multiharmonic nature

$$\vec{E}_1 = \sum_{m=1}^{N} \left[\left(E_{1x,m} \vec{e}_x + E_{1y,m} \vec{e}_y \right) e^{imp_1} + c.c. \right],$$

$$\vec{B}_1 = \sum_{m=1}^{N} \left[\left(B_{1x,m} \vec{e}_x + B_{1y,m} \vec{e}_y \right) e^{imp_1} + c.c. \right],$$

$$\vec{E}_2 = \sum_{m=1}^{N} \left[\left(E_{2x,m} \vec{e}_x + E_{2y,m} \vec{e}_y \right) e^{imp_2} + c.c. \right],$$

$$\vec{B}_2 = \sum_{m=1}^{N} \left[\left(B_{2x,m} \vec{e}_x + B_{2y,m} \vec{e}_y \right) e^{imp_2} + c.c. \right],$$

$$\vec{E}_3 = \sum_{m=1}^{N} \left[E_{3,m} e^{imp_3} + c.c. \right] \vec{e}_z, \tag{14.22}$$

where N is the number of harmonics, which are taken into account when solving the problem, and m here and further is the number of the corresponding harmonic. Therewith, we consider both the electromagnetic waves to be circularly polarized.

As we mentioned before, the dispersion law of the electron wave SCW $\{\omega_3, k_3\}$, which increases owing to the plasma-beam instability, is linear (Section 14.1) Therefore, like in the case of the two-stream instability (Chapter 13), in the given system, the plural multiharmonic resonance effect for the SCW harmonics might also be realized.

14.2.2 Truncated Equations for the Complex Amplitudes

By taking into account the above-mentioned problem for the complex amplitudes of the interacting waves, we obtain the following system of truncated equations:

$$K_{2,m}^{(1)}\frac{\partial^2 E_{1x,m}}{\partial t^2}+K_{1,m}^{(1)}\frac{\partial E_{1x,m}}{\partial t}+D_{1,m}E_{1x,m}=-\nu K_{3,m}^{(1)}\left(E_{2x,m}E_{3z,m}\delta_{\nu,-1}+E_{2x,m}^{*}E_{3z,m}\delta_{\nu,+1}\right)+F_1X_{1,m};$$

$$K_{2,m}^{(1)}\frac{\partial^2 E_{1y,m}}{\partial t^2}+K_{1,m}^{(1)}\frac{\partial E_{1y,m}}{\partial t}+D_{1,m}E_{1y,m}=-\nu K_{3,m}^{(1)}\left(E_{2y,m}E_{3z,m}\delta_{\nu,-1}+E_{2y,m}^{*}E_{3z,m}\delta_{\nu,+1}\right)+F_1Y_{1,m};$$

$$K_{2,m}^{(2)}\frac{\partial^2 E_{2x,m}}{\partial t^2}+K_{1,m}^{(2)}\frac{\partial E_{2x,m}}{\partial t}+D_{2,m}E_{2x,m}=-s_2K_{3,m}^{(2)}\left(E_{1x,m}E_{3z,m}^{*}\delta_{\nu,-1}+E_{1x,m}^{*}E_{3z,m}\delta_{\nu,+1}\right)+F_2X_{2,m};$$

$$K_{2,m}^{(2)}\frac{\partial^2 E_{2y,m}}{\partial t^2}+K_{2,m}^{(1)}\frac{\partial E_{2y,m}}{\partial t}+D_{2,m}E_{2y,m}=-s_2K_{3,m}^{(2)}\left(E_{1y,m}E_{3z,m}^{*}\delta_{\nu,-1}+E_{1y,m}^{*}E_{3z,m}\delta_{\nu,+1}\right)+F_2Y_{2,m};$$

$$C_{2,m}\frac{\partial^2 E_{3z,m}}{\partial t^2}+C_{1,m}\frac{\partial E_{3z,m}}{\partial t}+D_{3,m}E_{3z,m}=\nu C_{3,m}\left[\left(E_{1x,m}E_{2x,m}^{*}+E_{1y,m}E_{2y,m}^{*}\right)\delta_{\nu,-1}\right.$$
$$\left.+\left[\left(E_{1x,m}E_{2x,m}+E_{1y,m}E_{2y,m}\right)+F_3Z_{3,m}\right.,\qquad(14.23)$$

where

$$D_{j,m}=a_j\frac{im}{c\omega_j}\left(k_j^2c^2-\omega_j^2+\sum_{\alpha}^{b,p}\frac{\omega_\alpha^2\Omega_{j,\alpha}\left(m\Omega_{j,\alpha}+\eta_1\omega_{H,\alpha}\right)}{m\bar{\gamma}_\alpha\left(\left(m\Omega_{j,\alpha}\right)^2-\left(\omega_{H,\alpha}\right)^2\right)}\right)$$

is the dispersion function of the electromagnetic waves of signal ($j = 1$) and pumping ($j = 2$);

$$K_{1,m}^{(j)}=\frac{\partial D_{j,m}}{\alpha\left(im\omega_j\right)}=-\frac{c^2k^2+\omega_j^2}{\omega_j^2c}+\sum_{\alpha}^{b,p}\frac{\omega_\alpha^2\left(a_j\Omega_{j,\alpha}\omega_j+b_j\sigma_j(\Omega_{j,\alpha}-\omega_{H,\alpha}\eta_j/m)\bar{v}_{z,\alpha}k_j\right)}{\omega_j^2c\bar{\gamma}_\alpha\left(m\Omega_{j,\alpha}-\omega_{H,\alpha}\eta_j\right)^2};$$

$$K_{2,m}^{(j)}=\frac{1}{2}\frac{\partial^2 D_{j,m}}{\partial\left(ia_jm\omega_j\right)^2}=\frac{ck_j^2}{ia_jm\omega_j^3}$$

$$+\sum_{\alpha}^{b,p}\frac{\omega_\alpha^2\left(\Omega_{j,\alpha}\omega_j^2-\sigma_j(\Omega_{j,\alpha}-b_j\omega_{H,\alpha}\eta_j/m)\bar{v}_{z,\alpha}k_j(a_j\omega_j+\Omega_{j,\alpha}-\omega_{H,\alpha}\eta_j/m)\right)}{ia_j\omega_j^3c\bar{\gamma}_\alpha\left(m\Omega_{j,\alpha}-\omega_{H,\alpha}\eta_j\right)^3};$$

$\alpha_j = \delta_{j,1} - \nu\delta_{j,2}$, $b_j = \delta_{j,1} + \nu s_2\delta_{j,2}$ are the sign-operators;

$$K_{3,m}^{(j)} = \sum_{\alpha}^{b,p} \frac{\omega_\alpha^2}{c\bar{\gamma}_\alpha^2} \frac{e}{m_e} \left[\frac{d_{ij}\Omega_{j,\alpha}k_3}{\bar{\gamma}_\alpha^2 m\Omega_{3,\alpha}^2\omega_2\left(\omega_{H,\alpha}\eta_2 - md_{ij}\Omega_{j,\alpha}\right)} + \frac{\left(\omega_{H,\alpha}\eta_2 + m\Omega_{j\alpha}\right)}{\left(\omega_{H\alpha}\right)^2 - \left(m\Omega_{j\alpha}\right)^2} \times \right.$$

$$\left. \times \left(\frac{(-1)}{m\Omega_{3,\alpha}} \left(\frac{\bar{\upsilon}_{z,\alpha}}{c^2} \frac{d_{ij}\Omega_{j,\alpha}}{d_{ij}\omega_j} - \frac{d_{ij}k_j}{\bar{\gamma}_\alpha^2 d_{ij}\omega_j} \right) \times + \frac{\Omega_{2,\alpha}\left(\bar{\upsilon}_{z,\alpha}\left(m\Omega_{3,\alpha} + \omega_{H,\alpha}\eta_2\right) + md_{ij}k_j c^2 / \bar{\gamma}_\alpha^2\right)}{\omega_2 c^2 m\Omega_{3,\alpha}\left(\omega_{H,\alpha}\eta_2 - md_{ij}\Omega_{2,\alpha}\right)} \right) \right];$$

d_{ij} are the operators of the subscripts exchanging ($i = 1,2$, $j = 1,2$ $I \neq j$): $d_{ij}\omega_j = a_i\omega_i$, $d_{ij}k_j = b_ik_i$, where a_i and b_i are the above-determined sign operators in the case of obvious indexes

exchanging $I \to j$; $D_{3,m} = \dfrac{-im\omega_3}{c}\left(1 + \sum_{\alpha}^{b,p} \dfrac{\omega_\alpha^2}{\Omega_{3,\alpha}^2 (im)^2 \bar{\gamma}_\alpha^3}\right)$ is the SCW dispersion function;

$$C_{1,m} = \frac{\partial D_{3,m}}{\partial(im\omega_3)} = -\frac{1}{c} + \sum_{\alpha}^{b,p} \frac{\omega_\alpha^2\left(1 + 2k_3\bar{\upsilon}_{z,\alpha}/\Omega_{3,\alpha}\right)}{c\bar{\gamma}_\alpha^3\left(im\Omega_{3,\alpha}\right)^2};$$

$$C_{2,m} = \frac{1}{2}\frac{\partial^2 D_{3,m}}{\partial(im\omega_3)^2} = -\sum_{\alpha}^{b,p} \frac{\omega_\alpha^2\left(1 + 3k_3\bar{\upsilon}_{z,\alpha}/\Omega_{3,\alpha}\right)}{c\bar{\gamma}_\alpha^3\left(im\Omega_{3,\alpha}\right)^3};$$

$$C_{3,m} = \sum_{\alpha}^{b,p} \frac{-\nu e}{m_e c^3} \frac{\omega_\alpha^2\omega_3\left(1 + \eta_1\eta_2\right)}{m\Omega_{3,\alpha}^2\bar{\gamma}_\alpha^2\omega_2\omega_1} \times \left[\frac{\Omega_{1,\alpha}\left(-\nu s_2 c^2 k_2 - \omega_2\bar{\upsilon}_{z,\alpha}\right)}{\left(\omega_{H,\alpha}\eta_1 - m\Omega_{1,\alpha}\right)} - \frac{\Omega_{2,\alpha}\left(c^2k_1 - \omega_1\bar{\upsilon}_{z,\alpha}\right)}{\left(\omega_{H,\alpha}\eta_2 - m\Omega_{2,\alpha}\right)} \right];$$

$FX_{j,m} = F_jX_{j,m}(\vec{E}_1, \vec{E}_2, \vec{E}_3)$, $FY_{j,m} = F_jY_{j,m}(\vec{E}_1, \vec{E}_2, \vec{E}_3)$, and $F_3Z_{1,m} = FZ_{1,m}(\vec{E}_1, \vec{E}_2, \vec{E}_3)$ are the functions that take into account the cubic-nonlinear additions to corresponding equations and depend on harmonic wave interactions (these functions are obviously not written because of their enormous bulk); $\eta_j = E_{jy,m}/(iE_{jx,m}) = \pm 1$ are the sign functions that describe the rotational direction of the electric field tension of circularly polarized waves of the signal ($j = 1$) and pumping; c is the light velocity; e is the charge of electron; m_e is the rest mass of electron; $\alpha = \{b, p\}$, that is, b index describes beam parameters and p index describes plasma parameters, respectively; $\bar{\upsilon}_{z,b}$ is the average velocity of beam electrons; $\bar{\upsilon}_{z,p} = 0$ is the average velocity of plasma electrons; $\bar{\gamma}_\alpha = 1/\sqrt{1 - (\bar{\upsilon}_{z,\alpha}/c)^2}$; $\Omega_{1,\alpha} = \omega_1 - k_1\bar{\upsilon}_{z,\alpha}$; $\Omega_{2,\alpha} = -\nu\omega_2 + \nu s_2 k_2\bar{\upsilon}_{z,\alpha}$; $\Omega_{3,\alpha} = \omega_3 - k_3\bar{\upsilon}_{z,\alpha}$; $\omega_{H,\alpha} = eH_0/(\bar{\gamma}_\alpha m_e c)$; and $<\ldots>_{mp3} = \dfrac{1}{2\pi}\displaystyle\int_0^{2\pi}(\ldots \cdot \exp(imp_3))dp_3$.

It follows from the above-presented formulation of the problem that the circularly polarized electromagnetic waves (Equation 14.22) are proper for the given system, which is included in Equation 14.23. The coefficients $K_{3,m}$ and $C_{3,m}$ describe the parametric-resonance interaction of waves in the investigated system in quadratic approximation. It follows from the analysis of $C_{3,m}$ that for the description of the parametric resonance in the quadratic approximation, it is necessary for the electromagnetic waves of the signal and pumping to have the same rotational direction of the vector of electric field tension. It means that the rotation of the vector of the electric field tension should be either clockwise along the vector of the magnetic field tension ($\eta_2 = +1$, $\eta_1 = +1$) or counterclockwise ($\eta_1 = -1$), ($\eta_2 = -1$). In an opposite case, $C_{3,m}$ will be equal to zero. This deduction completely coincides with the deductions made in the framework of the theory of the ordinary parametric FEL (Chapters 11 and 12).

Equation 14.23 describes completely the cubic-nonlinear dynamics of the multiharmonic waves of pumping, signal, and SCW within the framework of the above-described model (Figure 14.6). Notice, however, that principally all the results of the analysis, which can be obtained by integration, can hardly give us something really new, as compared with the results described in Chapters 11 through 13 and in Section 14.1. Therefore, considering the overload of the present book, further in this section, we will limit ourselves to a short discussion of the most interesting results concerning the physics of our topics.

14.2.3 Analysis

First, let us confirm the above-formulated observation by a study of the effect of the excitation of a larger number of the SCW harmonics on the process of the amplification of the signal first harmonic. We stress that, in contrast to the above-analyzed case of the H-Ubitron FEL, here the pumping wave is proper for the system and its dynamics has a principally self-consistent nature. Therefore, for a proper comparison of the two models, let us analyze first the discussed model of Dopplertron SFEL in the given approximation of pumping field. Therewith, we take into consideration the effect of all the remaining saturation mechanisms. The results of the analysis of this model are presented in Figures 14.7 through 14.9. There, we limit ourselves to the study of the model with the monochromatic pumping wave. The calculations were performed for two cases. In the first case, only the first harmonics of the interacting waves were considered, whereas in the second case, each of the ten harmonics was considered.

In Figure 14.7, the dependence was performed on the amplitude of the first harmonic of the signal wave $E_{1,1} = \sqrt{|E_{1x,1}|^2 + |E_{1y,1}|^2}$ on the normalized time $\tau = \Gamma_1 t$, where (see also Equation 14.16)

$$\Gamma_1 = \frac{\sqrt{3}\cdot\omega_b^{2/3}\cdot\omega_p^{1/3}}{\sqrt[3]{16}\cdot\bar{\gamma}_b} \qquad (14.24)$$

is the growth increment for the plasma-beam instability (Section 14.1). In this drawing, curve 1 is obtained by calculations that considered the effect of the first ten harmonics of waves that participated in the nonlinear interaction process, and curve 2 presents the calculation result that took into account the interaction of only first SCW harmonics and pumping and signal waves.

The analysis of the signal wave saturation level within the framework of the cubic-nonlinear theory (Figure 14.7, curve 1) reveals that, as was mentioned before, in principle, the obtained results differ very little from those discussed in the previous section (Figure 14.4). Therefore, we again come to the conclusion that in the investigated model, at a moderate demand on the electric beam ($\omega_b = 2\cdot 10^9\ c^{-1}$, $\bar{\gamma}_b = 2$), a significantly intense electromagnetic emission in the millimeter-range wavelength might form.

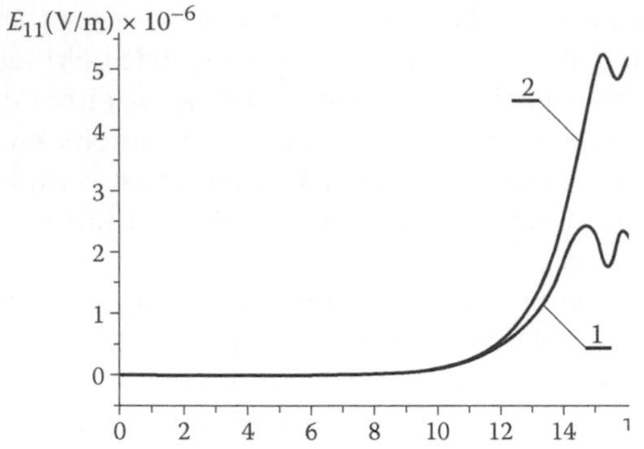

FIGURE 14.7
Dependence of the amplitude of the first signal harmonic on the normalized time $\tau = \Gamma_1 t$. Here, curve 1 corresponds to the case when only one harmonic of the signal wave and SCW are taken into account, whereas curve 2 corresponds to the case accounting for their first ten harmonics. The value Γ_1 is the increment growth for the plasma-beam system (see also the definition in Equation 14.16). All calculations were performed for the following parameters: $\omega_p = 10^{12}$ s^{-1}; $\omega_b = 2.10^9$ s^{-1}; $\bar{\gamma}_b = 2$; $H_0 = 2.8\ 10^3$ Gs; the resulting pumping amplitude $E_{2,1} = \sqrt{|E_{2x,1}|^2 + |E_{2y,1}|^2} \approx 5$ MV/m; the input signal amplitude $E_{01,1} = 10$ V/m; $k_3 = \omega_p/\bar{v}_b = 38.5$ cm^{-1}; the signal wavelength in vacuum $\lambda_{1,1} = 1.8$ mm; the pumping wavelength in vacuum $\lambda_{2,1} = 6.61$ cm.

By a comparison of curves 1 and 2 of Figure 14.7, obtained without and with taking into account the effect of the higher wave harmonics, respectively, like in the case of the H-Ubitron pumping (Figure 14.4), we see that the effect of the higher wave harmonics leads to a decrease in the saturation level, etc. The fact of the self-consistency of the pumping field brings some variety into the general physical picture. This includes, at $s_2 = -1$, $\nu = -1$; it is in the case when the asymptotic mode of interaction is realized (Chapters 11 and 12) that the pumping wave rapidly becomes exhausted and the saturation advances still in the stage of the quadratic amplification. In contrast, in the case of $s_2 = +1$, $\nu = +1$, the explosive instability mode is realized, when in the quadratic stage of interaction, all three waves increase by explosion (Chapters 11 and 12) and the length of the cubic-nonlinear saturation is essentially determined by the initial pumping wave amplitude. What concerns the saturation level of the signal, in this situation, is that it changes insignificantly.

As was expected, in the case of the initially monochromatic pumping and signal waves, their interaction process with SCW proceeds in practically monochromatic nature. A significantly intense excitation of the multiharmonic SCW spectrum appears as a peculiar secondary result in this process. As we mentioned in the previous section, this circumstance specifically opens a favorable possibility to use the plasma-beam SFEL as a modulation section in the cluster klystron-SFEL (Chapter 9). In this regard, we will dedicate more attention to the multiharmonic SCW dynamics in the investigated Dopplertron model. The results of this consideration are illustrated in Figures 14.8 and 14.9.

Figure 14.8 presents the dependence of the first harmonic's amplitudes of the electric field tension of SCW $E_{3,1} = |E_{3z,1}|$ on normalized time τ. Curve 1 corresponds to the case when nine higher harmonics of SCW are considered, and curve 2 corresponds to the case when only the first harmonic of SCW is considered in the calculation. It is obvious that the saturation level of the first harmonic of the SCW decreases for ~25% when the effect of the higher harmonics is taken into account. As we mentioned in the previous section, in this case, the excitation of the large number of the higher harmonics of SCW is explained by the dispersion curve of the SCW waves approaching linear law (curve 1 in Figure 14.5). Therefore,

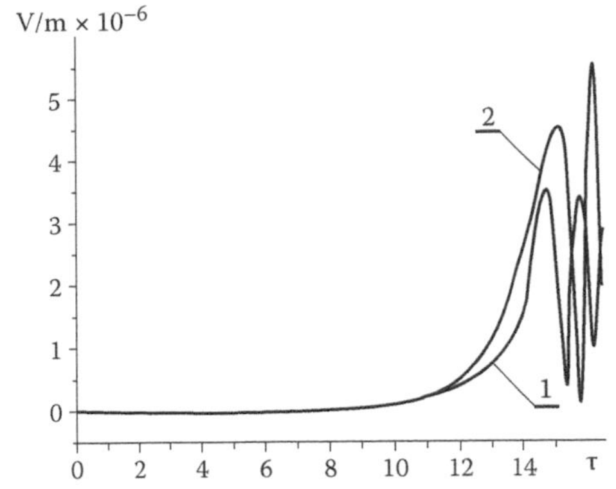

FIGURE 14.8
Dependence of amplitude of the first SCW harmonic on the normalized time $\tau = \Gamma_1 t$. Here, curve 1 corresponds to the case when one only harmonic of the signal wave and SCW are taken into account, whereas curve 2 corresponds to the case accounting for their first ten harmonics. All system parameters are the same with Figure 14.7.

similar to the two-stream system (Chapter 13), an intense parametrical-resonance interaction takes place between the harmonics of the SCW waves. The specificity of the chosen variant of the numerical calculation is that, at a chosen collection of parameters, the amplitudes of higher harmonics have a significantly lower level than the amplitude of the first harmonic. Nevertheless, the joint effect of the higher harmonics leads to significant reduction of the first harmonic's amplitude.

It should be stressed that by changing the correlation of the parameters of a system, a significant increase in the level of the higher harmonics can be achieved. For instance, by an increase in the plasma frequency of a beam for ten times, the higher harmonics of SCW become more intense and commensurable with the first harmonic. More so, there are situations when the amplitudes of the higher harmonics become higher than those of the first harmonic. There is a striking case in models, in which the frequency of the first harmonic

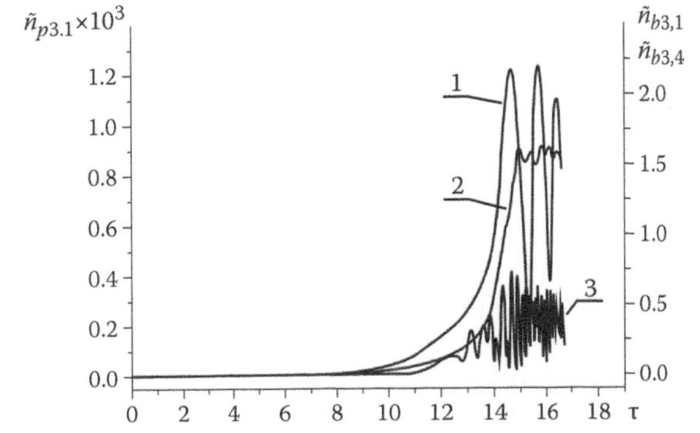

FIGURE 14.9
Dependencies of first harmonic of the plasma electron concentration $\tilde{n}_{p3,1}$ (curve 1, left-hand axis), the same value for the electron beam $\tilde{n}_{b3,1}$ (curve 2, the right-hand axis), and the fourth harmonic of the electron beam concentration $\tilde{n}_{b3,4}$ (curve 3, right-hand axis) on the normalized time $\tau = \Gamma_1 t$. Here, the concentration $\tilde{n}_{p3,1}$ is normalized at the averaged electron concentration of plasmas n_{p0}; the concentrations $\tilde{n}_{b3,1}$ and $\tilde{n}_{b3,4}$ are normalized at the averaged electron concentration of the electron beam n_{b0}. All system parameters are the same with Figure 14.7.

of SCW is significantly lower than the optimal frequency. However, these and some other related problems (e.g., the multiharmonicity of the pumping and the signal) are not the subjects of this chapter. They are adequately discussed already in the previous chapters. In addition to this, we mentioned several times before that physical differences between the two-stream and plasma-beam models are not principal differences: the latter are the particular cases of the first when the velocity of one of the partial beams is equal to zero.

Let us analyze further some peculiarity of the saturation mechanisms in a given system. For this, we investigate the behavior of the amplitude of the first harmonic of the plasma concentration $\tilde{n}_{p3,1}$ (Figure 14.9, curve 1) and the beam wave concentration $\tilde{n}_{b3,1}$ (Figure 14.9, curve 2) on the normalized time $\tau = \Gamma_1 t$. Here, the value $\tilde{n}_{p3,1}$ is normalized by the plasma-electron concentration constant n_{p0}, and the value $\tilde{n}_{b3,1}$ is normalized by the constant of the beam-electron concentration constant n_{b0}. It appears obvious that, at the initial stage, both values undergo an exponential increase. Subsequently they undergo a nonlinear stabilization. Simultaneously, the amplitude of the first harmonic of the increasing plasma performs regular large-dimension vibrations. Like it is known, these vibrations are explained by a capture of the beam electrons by the potential field of the plasma wave. In this case, it might be imagined as a chain of the potential wells.

The capture becomes possible, owing to a simultaneous action of two mechanisms. The first is related with the beam breaking within the process of the interaction. A decrease in the beam velocity to the velocity comparable to the plasma wave phase velocity results in the electron kinetic energy in the beam reference frame becoming lower than the height of each of the mentioned potential wells. In the second mechanism, the height of the same potential wells (SCW amplitudes) increases in the process of the plasma wave increase. Clusters form as a result of the electron capture and they vibrate within the range of each potential well. In the vibration process, the clusters periodically give and take away the energy from the plasma waves causing the previously mentioned large-dimension vibrations of SCW amplitude within the range of saturation (Figure 14.9).

The specificity of the discussed process has been studied such that for each one vibration of plasma wave, the clusters collide twice with the edges of a potential well. This results in a twice higher oscillation of the beam amplitude $\tilde{n}_{b3,1}$ than that of the plasma wave amplitude $\tilde{n}_{p3,1}$ (Figure 14.9, curves 1 and 2). In this case, the level of the electron beam modulation is approximately equal to 1. This way, in the given case, practically total modulation of the beam occurs.

Curve 3 in Figure 14.9 represents the dependence of the amplitude of the fourth harmonic of the $\tilde{n}_{b3,4}$ (normalized by the constant of the electron beam concentration n_{b0}) on the normalized time τ. We can see here that the saturation processes begin before the electron capture time for the first harmonics. It is explained by the smaller oscillation amplitude of the fourth harmonic.

An analysis of the higher harmonic's behavior of the beam wave concentration reveals that, in the given system, besides the capture-type processes, the processes of a nonlinear frequency shift also occur (Chapters 11 and 12). As a matter of fact, in the considered model, and similar with the situation in ordinary FEL, we observed the results of superimposition of both these saturation mechanisms.

Then, let us note that, apart from the above-mentioned plasma potential wells, we have here also one more potential (quasipotential) relief. In contrast to the above-mentioned plasma-wave (i.e., longitudinal) variant, it, being also longitudinal, is related with the transverse signal and pumping waves. This physical phenomenon is known in the theory of the ordinary parametric FEL (Chapters 11 and 12) as the ponderomotor potential or the Miller–Gaponov's potential [17]. From the above-described analysis, this physical mechanism, basically, also takes part in the nonlinear processes in the considered PBSFEL. However, contrary to the case of ordinary FEL, it does not play any significant role here. There the plasma-beam saturation mechanisms essentially prevail under their FEL analogs. In other words, the saturation of the signal amplification in PBSFEL depends mainly on the saturation of the plasma-beam instability. However, in the case of the isochronous models, those typical for the ordinary FEL saturation mechanism should not be rejected completely.

An analysis of the higher harmonic's behavior of the beam wave concentration reveals that, in the given system, besides the capture-type processes, the processes of a nonlinear frequency shift also occur (Chapters 11 and 12).

Finally, we present a quick discussion of the isochronous models, which are models of high efficiency. The dynamics of the processes in the isochronous mode, in general, differs only insignificantly from that described in Chapters 11 and 12, related with ordinary parametric FEL. The basic difference consists of the two following characteristics. The first of them appears quite obvious. Specifically, owing to the additional, essentially superheterodyne amplification, related with the effect of the plasma-beam instability, all processes of the energy exchange in the system become sharply activated. This includes a remarkable shortening of the length of the system, which is necessary for obtaining a high efficiency.

The second consists of a simultaneous isochronization of the two saturation mechanisms of different physical nature: the plasma-beam and ordinary parametric mechanisms. As a result, the isochronizing function (the longitudinal electric field or the retardation coefficient of the pumping wave) becomes nonmonotonous, contrary to the case of the ordinary parametric FEL. To be specific, here appear the large-scale oscillations, which are related with the above-mentioned electron capture mechanism by the two physical types of the potential wells: the plasma and ponderomotor Miller–Gaponov's potential wells, respectively. Therefore, the tuning of PBSFEL in isochronous mode appears technologically significantly more complex than that of the ordinary parametric FEL. However, as the analysis reveals, even in this complicated case, the achievement of 50% of the electronic efficiency is completely realistic.

14.3 Parametrical Three-Wave Instability in Two-Velocity High-Current Beams

In what follows, we turn to the discussion of the mentioned above type of SFEL, the PEWSFEL. The effect of the three-wave parametrical resonant interaction of longitudinal SCW is the basis of the performance of this class of devices. Therefore, we will start our study with the theory of this physical phenomenon.

14.3.1 Model and the Problem Formulation

The considered model is presented in Figure 14.10. Here, the mean spatial charge of the two-velocity electron beams 2 and 4 are considered to be compensated by the stationary ionic background. We assume the beam to be uniform in the transverse plane and transverse unlimited. The partial mutually penetrating beams 2 and 4 are described by the partial plasma frequencies $\omega_{p1} = \omega_{p2} = \omega_p$ and velocities v_1, v_2, which are directed along the z-axis. We suppose that $v_1 - v_2 << v_1, v_2$.

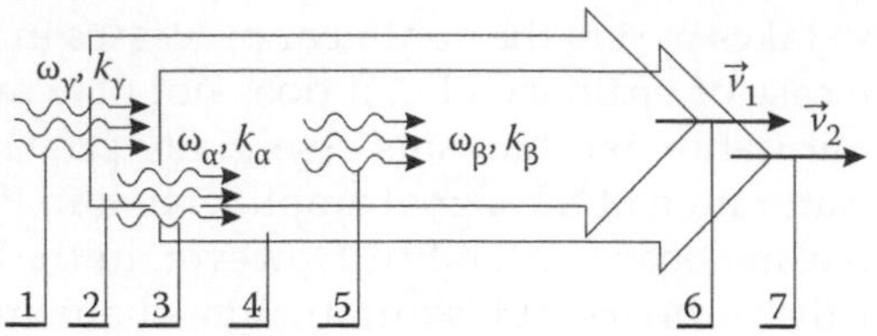

FIGURE 14.10
Theoretical model of the three-wave resonant parametrical system. Here, 1 is the γth electron-beam wave (pumping, for instance); 2 is the first partial electron beam; 3 is the αth electron-beam wave; 4 is the second partial electron beam; 5 is the βth electron-beam wave; 6 is the velocity vector of the beam 2; 7 is the velocity vector of the beam 4.

In the general case, we consider the electric fields of the beam waves (SCW) to be multiharmonic

$$\vec{E}_\chi = \sum_{m=1}^{N} \left[E_{\chi,m} \exp(imp_\chi) + k.c. \right] \vec{e}_z. \tag{14.25}$$

Here, the type of SCW is marked by the index χ (this index acquires the value from 1 to 5; Table 14.1). It can be excited in the plasma of the two-velocity electron beam; $p_\chi = \omega_\chi t - k_\chi z$ is the phase of the wave χ; ω_χ and k_χ are its frequency and wave number, respectively; N is the total number of harmonics, which are considered in the calculation; and m is the number of the corresponding harmonic.

We consider that in a two-velocity electron beam, the three-wave parametric resonance is fulfilled for SCW (Equation 14.18)

$$p_\alpha + \nu p_\beta = p_\gamma,$$

or, considering the above-presented phase definitions,

$$\omega_\alpha + \nu\omega_\beta = \omega_\gamma, \quad k_\alpha + \nu k_\beta = k_\gamma. \tag{14.26}$$

TABLE 14.1
Types of Electronic Waves in the Two-Stream System

χ	σ_κ	$r_{\chi\kappa}$	Wave Type
1	0	0	Increasing wave ($\omega < \omega_{cr}$)
2	0	$-\frac{\sqrt{15}}{2}$	Fast wave ($\omega < \omega_{cr}$)
3	0	$+\frac{\sqrt{15}}{2}$	Slow wave ($\omega < \omega_{cr}$)
4	+1	−1	Fast wave of the first beam ($\omega > \omega_{cr}$)
5	+1	+1	Slow wave of the first beam ($\omega > \omega_{cr}$)
6	−1	−1	Fast wave of the second beam ($\omega > \omega_{cr}$)
7	−1	+1	Slow wave of the second beam ($\omega > \omega_{cr}$)

Here, the indexes α, β, $\gamma = \kappa$ describe the specific SCW, which participates in the parametric resonances. The sign function $\nu = \pm 1$ is presented for the convenience in the performance of the analysis. Specifically, in the situation when, for example, the wave of the frequency ω_γ is assumed to be the pumping, it is important to know which among the frequencies ω_κ is the maximal wave. Then, at $\nu = +1$, the maximal appears $\omega_\gamma = \omega_{max}$ and at $\nu = -1$, the maximal is the wave $\omega_\alpha = \omega_{max}$.

14.3.2 Truncated Equations in the Quadratic Approximation

We use the relativistic quasihydrodynamic equation, the equation of continuity, the Maxwell equation (Chapter 10), and the method of averaged characteristics (Chapter 6). In this stage of the analysis, we limit ourselves with the analysis of the given model within the framework of the quadratic-nonlinear approximation in amplitudes of waves. Then, for the complex amplitudes of SCW, α, β, and γ, we obtain the system of the truncated equations

$$C_{2,\alpha,m}\frac{d^2E_{\alpha,m}}{dz^2}+C_{1,\alpha,m}\frac{dE_{\alpha,m}}{dz}+D_{\alpha,m}E_{\alpha,m}=C_{3,\alpha,m}E_{\beta,m}E_{\gamma,m},$$

$$C_{2,\beta,m}\frac{d^2E_{\beta,m}}{dz^2}+C_{1,\beta,m}\frac{dE_{\beta,m}}{dz}+D_{\beta,m}E_{\beta,m}=C_{3,\beta,m}E_{\alpha,m}E^*_{\gamma,m},$$

$$C_{2,\gamma,m}\frac{d^2E_{\gamma,m}}{dz^2}+C_{1,\gamma,m}\frac{dE_{\gamma,m}}{dz}+D_{\gamma,m}E_{\gamma,m}=C_{3,\gamma,m}E_{\alpha,m}E^*_{\beta,m}. \quad (14.27)$$

The beam parameters, corresponding wave numbers, and frequencies of the mth harmonics determine the coefficients of Equation 14.27

$$D_{\chi,m}=-imk_\chi\left(1-\frac{\omega_p^2}{m^2(\omega_\chi-k_\chi v_1)^2\gamma_1^3}-\frac{\omega_p^2}{m^2(\omega_\chi-k_\chi v_2)^2\gamma_2^3}\right), \quad (14.28)$$

$$C_{1,\chi,m}=\partial D_{\chi,m}/\partial(-imk_\chi),\; C_{2,\chi,m}=\partial^2 D_{\chi,m}/\partial(-imk_\chi)^2/2,$$

$$C_{3,\alpha,m}=-k_\alpha\sum_{q=1}^{2}\left[\frac{\omega_p^2 e/m_e}{\Omega_{\alpha,q}\Omega_{\beta,q}\Omega_{\gamma,q}\gamma_q^6 m^2}\left(\frac{k_\alpha}{\Omega_{\alpha,q}}+\frac{k_\beta}{\Omega_{\beta,q}}+\frac{k_\gamma}{\Omega_{\gamma,q}}-3v_q\gamma_q^2/c^2\right)\right],$$

$$C_{3,\beta,m}=-k_\beta C_{3,\alpha,m}/k_\alpha,\; C_{3,\gamma,m}=-k_\gamma C_{3,\alpha,m}/k_\alpha,\; \Omega_{\chi,q}=\omega_\chi-k_\chi v_q,\; \gamma_q=1/\sqrt{1-(v_q/c)^2}.$$

In Equation 14.27, $D_{\chi,m}$ is the dispersion function of the mth harmonic of the type-χ wave. In this case, these functions for the first harmonic are equal to zero (because all waves are proper) and determine the dispersion law for the given type of waves. Therewith, the solution of the dispersion equations $D_{\chi,m}=0$ for k_χ at $\omega_{1,\ldots,4} > \omega_{cr}$ is real while in the case when $\omega_5 < \omega_{cr}$, part of the solution of the equation is complex. Here, ω_{cr} is the critical frequency, determination of which is given in Chapter 13 (see also Equation 14.30 below). As shown in Chapter 13, at $\omega_{1,\ldots,4} > \omega_{cr}$, the two-stream instability effect is realized.

14.3.3 Types of the Parametrical Resonances: Interactions of the Precritical Waves

When considering mathematics, the developed theory of the parametric three-wave resonances represents a kind of generalization of the previously discussed theory of the plural multiharmonic resonances (Section 13.1). In this case, the generalization is that the harmonic numbers $n_{3\kappa}$, considered there as especially discrete values, perform here a certain formal role, normalized by the first harmonic ω_{31} wave frequencies ω_κ (Equation 13.22), the spectrum of which is believed to be continuous. In this regard, the performed analysis of all the possible resonance states of the system (kinematic analysis; Chapter 11), including the discussion, retains its validity also for the examined case. Therefore, further we will limit ourselves to short remarks only about the most interesting physical peculiarity of the discussed phenomenon.

As was mentioned in Section 13.1, seven types of the longitudinal SCW (Table 14.1) can exist in a two-stream electron system. The frequencies and the wave numbers of these waves satisfy the dispersion equation 14.28, the linear solutions of which one can uniformly write as the generalized law of linear dispersion 13.21.

$$k_{\chi\kappa} = \frac{\omega_\kappa}{v_0\left(1+\sigma_\kappa\delta'\right)} + r_{\chi\kappa}\frac{\omega_p}{v_0\gamma_0^{3/2}}, \tag{14.29}$$

where, like before, $\kappa = \alpha, \beta, \gamma$; $v_0 = (v_{01} + v_{02})/2$, $\delta' = (v_{01} - v_{02})/(v_{01} + v_{02})$, $\gamma_0 = 1/(1-(v_0/c)^2)^{1/2}$, and $\sigma_\kappa = \pm 1, 0$ are the sign functions. The index $\chi = 1$ to 7 and corresponds to the longitudinal wave types, classification of which is given in Table 14.1 (see also Table 13.1).

As was mentioned many times, in the studied two-stream system, there exists frequency ω_{cr}

$$\omega_{cr} = \sqrt{2}\omega_p/(\delta'\gamma_0^{3/2}), \tag{14.30}$$

which conventionally divides all totality of the possible types of waves (Table 14.1) into two large groups. The waves of the first group are called precritical and those of the second group are called overcritical. Following the standard procedure (Condition 14.26), one can obtain the so-called generalized resonance condition, which considers as possible the mutual interaction of waves within their group as well as with waves of different groups

$$\omega_\alpha\frac{\sigma_\gamma - \sigma_\alpha}{1+\sigma_\alpha\delta'} + \nu\omega_\beta\frac{\sigma_\gamma - \sigma_\beta}{1+\sigma_\beta\delta'} = \frac{\omega_p\left(1+\sigma_\gamma\delta'\right)\left(r_{\chi\gamma} - r_{\chi\alpha} - \nu r_{\chi\beta}\right)}{\delta'\gamma_0^{3/2}}. \tag{14.31}$$

Furthermore, we will comment briefly on condition 14.31, concerning some types of the possible resonances.

Let us start with the resonances group, which exists between the precritical group waves ($\omega_\kappa \le \omega_{cr}$). In this case, $\sigma_\kappa = 0$. Besides this, in this model, there are $r_{\chi\kappa} = 0$ for increasing waves and $r_{\chi\kappa} = \pm\sqrt{15}/2$ for the slow and fast SCW (Table 14.1). The generalized resonance condition 14.31 is reduced to the simplified form

$$r_{\chi\gamma} - \nu r_{\chi\beta} - r_{\chi\alpha} = 0. \tag{14.32}$$

Considering the data in Table 14.1, it is easy to assure oneself that condition 14.32 might be fulfilled in two cases:

1. when all three waves are increasing $r_{\chi\gamma} = r_{\chi\beta} = r_{\chi\alpha} = 0$
2. when one of the waves is considered increasing and the two others are presented as slow and fast precritical SCW, respectively

The second peculiarity of the interaction of this group of waves is that the resonance condition 14.32 is practically independent of the system's parameters such as the partial v_j (j = 1,2) and average v_0 beam velocities, the plasma frequency ω_p, etc. One should mention that the situation is quite uncommon for the general theory of the three-wave parametric resonances. This type of model is called wideband. By considering everything that has been mentioned, let us recall once again that within the framework of the accepted definitions (condition 14.26) at $\nu = +1$, the wave γ possesses the maximal frequency ω_{max}. In analogy, at $\nu = -1$, the frequency ω_α appears to be maximal. Consequently, in the given model, the wave β cannot possess the maximal frequency in any of the possible individual configurations of the system.

The value for a practical application of the first of the mentioned resonance configurations of the PEWSFEL is obvious, for example, in situations when for some reason a wave of the lowest frequency, for instance, wave α at $\nu = +1$, is the pumping wave. At the same time, a realization of the superheterodyne mechanism of the amplification requires the highest frequency wave ω_γ. Then, giving at the input the initial amplitudes of the waves α and β, at the output of the system, we obtain the transformed to a higher frequency and simultaneously amplified wave γ. Obviously, the maximal amplification will happen at $\omega_\gamma = \omega_{opt}$. Notice that the explosive instability cannot appear here because all three interactive waves possess zero energy, although formally, when considering the fact of the simultaneously increasing amplitudes of all three waves, the given type of the parametric resonance strongly resembles the explosive instability (Chapter 11).

Let us consider the second of the mentioned cases. Here, in turn, we can also separate two peculiar variants of interactions. Specifically

1. the increase ($r_{\chi\kappa} = 0$; Table 14.1) is one wave, having the highest frequency (the wave γ or α, respectively);
2. the increasing is the wave β.

In the first case, $r_{\chi\gamma} = 0$; from condition 14.32, we easily obtain

$$r_{\chi\alpha} = -\nu r_{\chi\beta}; \tag{14.33}$$

it means at $\nu = +1$, $\omega_\gamma = \omega_{max}$, the waves with different sign functions $r_{\chi\alpha,\beta}$ (Table 14.1) are the resonance waves. When one of the waves α,β is slow, then the other one is certainly fast, and vice versa.

14.3.4 Types of the Parametrical Resonances: Interactions of the Overcritical Waves

Let us turn to the discussion of the resonance condition 14.31 in a case when there is only the overcritical wave ($\omega_\kappa > \omega_{cr}$) interactions. Its mathematical structure shows clearly that the parametric resonance of the overcritical waves cannot take place in a case when all three waves belong to one and the same beam ($\sigma_\alpha = \sigma_\beta = \sigma_\gamma$). Therefore, for the sign functions σ_κ, we might have only three variants of their combinations. In the first variant,

$$\sigma_\alpha = -\sigma_\beta = \sigma_\gamma, \tag{14.34}$$

the wave β belongs to one beam and the waves α and γ to another beam. The specificity of such combination is that each of the frequencies of waves α or γ might be the maximum in the investigated three-wave $\omega_{\alpha,\beta,\gamma}$. In such a case, the condition of the resonance acquires the form

$$\omega_\beta = \nu\sigma_\gamma \frac{\omega_p\left(1-\delta'^2\right)}{2\delta'\gamma_0^{3/2}}\left(r_{\chi\gamma} - r_{\chi\alpha} - \nu r_{\chi\gamma}\right), \quad \nu\sigma_\gamma\left(r_{\chi\gamma} - r_{\chi\alpha} - \nu r_{\chi\beta}\right) = +3. \tag{14.35}$$

In the case of the combination

$$\sigma_\alpha = -\sigma_\beta = -\sigma_\gamma, \tag{14.36}$$

the wave α belongs to one beam, whereas the remaining two belong to another. Here, the waves α and γ correspond to different beams. For resonance condition 14.31, we obtain

$$\omega_\alpha = \sigma_\gamma \frac{\omega_p\left(1-\delta'^2\right)}{2\delta'\gamma_0^{3/2}}\left(r_{\chi\gamma} - r_{\chi\alpha} - \nu r_{\chi\gamma}\right), \quad \sigma_\gamma\left(r_{\chi\gamma} - r_{\chi\alpha} - \nu r_{\chi\beta}\right) = +3. \tag{14.37}$$

And, finally, the third variant of the possible combinations

$$\sigma_\alpha = \sigma_\beta = -\sigma_\gamma, \tag{14.38}$$

when the wave γ belongs to one beam and the waves β and γ belong to another beam. Here, similarly to the previous case, the waves α and γ each correspond to a different beam. In such a case, the condition of the resonance can acquire the form

$$\omega_\gamma = \sigma_\gamma \frac{\omega_p\left(1-\delta'^2\right)}{2\delta'\gamma_0^{3/2}}\left(r_{\chi\gamma} - r_{\chi\alpha} - \nu r_{\chi\gamma}\right), \quad \sigma_\gamma\left(r_{\chi\gamma} - r_{\chi\alpha} - \nu r_{\chi\beta}\right) = +3. \tag{14.39}$$

Let us give the necessary explanation to the criterion of the sign function $r_{\chi\alpha,\beta,\gamma}$ in conditions 14.35, 14.37, and 14.39

$$\sigma_\gamma\left(r_{\chi\gamma} - r_{\chi\alpha} - \nu r_{\chi\beta}\right) = +3.$$

At first glance, the formulated request might appear illogical because on its right-hand side, instead of +3, in principle, one can also write +1 (Table 14.1). However, when we turn to the determination of the critical frequency (Equation 14.30)

$$\omega_{cr} = \sqrt{2}\omega_p/(\delta'\gamma_0^{3/2}),$$

it is easy to be convinced that, in this case, we inevitably obtain $\omega_\gamma < \omega_{cr}$. The latter, in turn, leads to a contradiction with the earlier accepted supposition that all interacting waves are overcritical.

ω_κ appears to be close to the critical (Equation 14.30). This circumstance by far not always appears acceptable in practice.

One should turn attention to the fact that here and further in this section, a full symmetry among the indexes α, β, and γ does not exist. This is explained by the same lack of the symmetry in the numeration of beam numbers, because earlier, we accepted a simplification condition that the average velocity of the first beam is higher than that of the second beam (Figure 14.10). Or, the normalized difference of the velocity δ' always appears as a positive determined value. The symmetry is established when an additional sign function is introduced for the difference δ'. However, in practice, this is not necessary because the complication of the mathematical expressions is not compensated in the course of obtaining any advantage in physical analysis.

A comparison of conditions 14.35, 14.37, and 14.39 reveals that one of the three overcritical wave frequencies is described by the system's parameters ω_p, γ_0, δ'. Here, for the remaining two waves, there exists a definite arbitrariness, which is limited only by the combination condition for the frequencies

$$\omega_\alpha + \nu\omega_\beta = \omega_\gamma \tag{14.40}$$

(see resonance condition 14.26). As we mentioned in the previous discussion, we consider the cases of such variants of interaction as the wideband interactions.

Let us fix the frequency of the pumping wave by one of the conditions 14.35, 14.37, and 14.39. Then we recall that one of the remaining waves functionally appears as the SCW signal and the second performs the role of the intermediate (it is also called the idle wave). The frequency of this idle wave is determined as a sum or a difference of the pumping and the signal waves (condition 14.40). This way, as fixed by the pointed method of choice of the pumping frequency, formally we have an arbitrary way to choose the resonance frequency of the signal. By gratifying the condition 14.40, the idle wave is always generated with the needed frequency. By considering the above discussion, it is beneficial to choose the frequencies 14.35, 14.37, and 14.39 for the waves by considering their functional designation and use as the electron-wave pumping.

Furthermore, we shall mention especially that in fulfilling conditions 14.35, 14.37, and 14.39, the input signal does not have to be presented by the monochromatic electron wave. In other words, in the case when it is in the form of a multiharmonic spectrum, then its entire spectral harmonics will be amplified simultaneously. It follows that all schemes of the resonance interacting type (conditions 14.35, 14.37, and 14.39) are fit to be used in the cluster systems, for instance, in multiharmonic transit sections of cluster klystrons (Chapter 9), or, in another variant: as a supplementary multiharmonic mechanism of the superheterodyne amplification in cluster PEWSFEL-monotrons.

14.3.5 Types of the Parametrical Resonances: Interactions of the Precritical and Overcritical Waves

Finally, we conclude the presented kinematical analysis of the discussion of the third type of the mentioned interactions, specifically the case when the parametric-resonance interactions participate as the precritical as well as overcritical waves.

By analyzing the general resonance condition 14.31, we conclude that all possible resonances in the given model might be divided into two characteristic groups. The resonances with participating two precritical and one overcritical waves form the first group. The resonances of the one precritical and two overcritical waves we ascribe to the second group. We will begin the discussion with the resonances of the first group.

Hence, we consider that the condition $\sigma_\alpha = \sigma_\beta = 0$ is fulfilled; it means the waves α and β are precritical when the wave γ is overcritical. It is obvious that such a case of interactions can exist only in the case when $\sigma_\gamma = +1$ because the wave γ possesses the highest frequency. Then the general condition 14.31 can be presented in the form of two more particular conditions

$$\omega_\gamma = \frac{\omega_{cr}(1+\delta')(r_{\chi\gamma} - r_{\chi\alpha} - r_{\chi\beta})}{\sqrt{2}}; \quad \frac{r_{\chi\gamma} - r_{\chi\beta} - r_{\chi\alpha}}{\sqrt{2}} > 1+\delta'. \tag{14.41}$$

It is easy to see that in the analyzed resonance condition 14.41, we can see three more distinct variants of resonances:

1. $r_{\chi\alpha} = 0$ or $r_{\chi\beta} = 0$, where one of the precritical waves is increasing
2. $r_{\chi\alpha} = r_{\chi\beta} = 0$, where both precritical waves are increasing
3. $r_{\chi\alpha} = \pm\sqrt{15}/2$, $r_{\chi\beta} = \pm\sqrt{15}/2$ (remember that $r_{\chi\gamma} = \pm 1$, Table 14.1), where both the precritical waves are not increasing

It is easy to see that, in the analyzed resonance condition 14.41, we can see three more distinct variants of resonances. It is obvious that only the two first variants among the three presented ones have a specific practical value. We can consider the third one as some kind of physical exotics. Such variant is hardly useful and can be considered in an experiment as a parasite phenomenon.

Let us turn to the discussion of the first two variants.

In the first case, the ($r_\alpha = 0$) condition of the resonance condition 14.41 may be rewritten as

$$\omega_\gamma = \frac{\omega_{cr}(1+\delta')(r_{\chi\gamma} - r_{\chi\beta})}{\sqrt{2}}; \quad \frac{r_{\chi\gamma} - r_{\chi\beta}}{\sqrt{2}} > 1+\delta'. \tag{14.42}$$

Here, the resonance is possible only in the case of different signs $r_{\chi\gamma}$ and $r_{\chi\beta}$, for example, at $r_{\chi\gamma} = +1$ (the slow overcritical SCW) and $r_{\chi\beta} = -\sqrt{15}/2$ (the fast precritical SCW). One can be easily assured that, in this case, the explosive instability condition is possible. Remember that the observed asymmetry is related with the earlier accepted assumption that the velocity of the first partial beam is higher than the velocity of the second ($\delta' > 0$). The symmetry is established by a change of δ' for $-\delta'$.

What concerns the second of the mentioned variants ($r_{\chi\gamma} = r_{\chi\beta} = 0$) is that it is clear that it contradicts the criterion of sign functions in condition 14.41; therefore, it cannot be realized physically.

Even so, the discussed variant (condition 14.42) presents an interest for practice; however, this interest is significantly limited. First, by the applied condition, both waves α and β are precritical; therefore, we are limited with the possibility of obtaining higher frequencies for the third, the overcritical wave γ. As was illustrated previously, the value of the last appears to be close to the critical frequency ω_{cr}.

Consequently, it is easy to be convinced that the highest practical interest evokes the interaction type of one precritical + two overcritical waves [e.g., $\sigma_{\alpha,\beta}$, $r_{\chi\alpha,\beta} = \pm 1$, $\nu = -1$ ($\omega_\alpha = \omega_\beta + \omega_\beta$), $\sigma_\gamma = r_{\chi\gamma} = 0$; Table 14.1]. From the applied consideration, we consider the overcritical wave α as having the highest frequency as the wave of the electron-wave signal. Accordingly, also the overcritical wave β, in such situation, performs the role of an idle wave. A precritical increasing wave χ performs the role of the electron-wave pumping. In this case, we can rewrite condition 14.41 in the form

$$\omega_\alpha \frac{\sigma_\alpha}{1+\sigma_\alpha\delta'} - \omega_\beta \frac{\sigma_\beta}{1+\sigma_\beta\delta'} = \frac{\omega_p\left(r_{\chi\alpha} - r_{\chi\beta}\right)}{\delta'\gamma_0^{3/2}}. \quad (14.43)$$

It is clear that the given type of resonance interactions is possible even in a case when both the overcritical waves belong to one and the same beam: $\sigma_\alpha = \sigma_\beta$. In this case, condition 14.41 acquires somehow a simpler form

$$\omega_\gamma = \sigma_\alpha \frac{\omega_p\left(1+\sigma_\alpha\delta'\right)\left(r_{\chi\alpha} - r_{\chi\beta}\right)}{\delta'\gamma_0^{3/2}}, \quad \sigma_\alpha\left(r_{\chi\alpha} - r_{\chi\beta}\right) = +2. \quad (14.44)$$

The explosive instability condition is possible in a case when $\sigma_\alpha = +1$, $r_{\chi\alpha} = +1$, $r_{\chi\beta} = -1$. One can be convinced that the given variant of the resonance interaction can also be classified as the wideband interaction. Therefore, it might also be used successfully for amplification of the multiharmonic electron-wave signals.

As it could be expected, the resonance might be found in a case when the overcritical waves belong to different beams: $\sigma_\alpha = -\sigma_\beta$. In such a case, condition 14.43 acquires the form

$$\omega_\alpha + \omega_\beta \approx \sigma_\alpha \frac{\omega_p\left(1-\delta'^2\right)\left(r_{\chi\alpha} - r_{\chi\beta}\right)}{\delta'\gamma_0^{3/2}}, \quad \sigma_\alpha(r_{\chi\alpha} - r_{\chi\beta}) = +2. \quad (14.45)$$

For simplification, the assumption $\sigma_\alpha\delta'\omega_\gamma \ll \omega_{\alpha,\beta}$ was used in derivation of Equation 14.44.

The applied attractiveness of the given variants is determined as follows. Let us assume that at least one of the partial beams is modulated at the input into the system. As was mentioned before, in the introduction to this chapter, such case takes place in the case when the RFAs are used as the technological basis for PEWSFEL (Chapter 9). Let us assume that the transit section of such SFEL-klystron [8,9] is performed on the basis of the here-studied two-beam parametrical electron-wave system. The peculiarity of such class of designs is that the input SCW, for example, a wave of the frequency ω_γ, which is excited in the beam on expense of its initial modulation, appears to be relatively of low frequency. Simultaneously, in a next section of such PEWSFEL [8,9], the presence of a strongly amplified SCW with an essentially higher frequency is needed. It is obvious that both of these interaction schemes are well suited for the solution of the given technological problem. Indeed, we consider the wave χ, related with the initial beam modulation, as the increasing pumping wave. As was mentioned before, the wave of the highest frequency (i.e., wave ω_α) is considered the wave of the electron-wave signal. The idle wave becomes excited by a summed up or by differenced frequency.

14.3.6 Gain Factors

We will use the approximation of a given field of pumping wave. By integration of Equation 14.27, we obtain the growth increment for the signal wave. In addition to this, we will use for each of the resonant combinations (Section 14.3.4) the mentioned classification rule for the functional wave destination. Specifically, we will always interpret a wave of the highest frequency (ω_α or ω_γ) as the electron-wave signal, whereas the wave with the smallest frequency is considered as pumping. The remaining third wave is considered as the idle wave.

By following the standard methodic procedures, we obtain a corresponding dispersion equation for each of the interesting schemes of interaction. For illustration, we analyze, for instance, the case of three overcritical interaction waves (Section 14.3.4). The wave γ is considered as the pumping; it means that the given dispersion equation acquires the form

$$\Gamma^4 \cdot C_{2,\alpha,m}C_{2,\beta,m} + \Gamma^3 \cdot (C_{2,\alpha,m}C_{1,\beta,m} + C_{1,\alpha,m}C_{2,\beta,m}) + \Gamma^2 \left(C_{2,\alpha,m}D_{\beta,m} + C_{1,\alpha,m}C_{1,\beta,m} + D_{\alpha,m}D_{\beta,m} \right) +$$

$$+ \Gamma \cdot (C_{1,\alpha,m}D_{\beta,m} + D_{\alpha,m}C_{1,\beta,m}) + (D_{\alpha,m}D_{\beta,m} - |E_{\gamma,m}|^2 C_{3,\alpha,m}C_{3,\beta,m}) = 0. \quad (14.46)$$

Equation 14.46 is the fourth degree algebraic equation with respect to the unknown Γ value. Basically, corresponding solutions in an analytical form can be obtained for it. However, owing to their enormous size, they are unsuitable for our analysis. Therefore, we shall find numerically the solution of Equation 14.46. Let us determine the dependence of the growth increment as a function of the frequency ω_α for some types of the parametric-resonance interaction, which was studied above. Three types of such interactions shall be chosen for this analysis. Therewith, two of them were chosen by assuming that only the overcritical waves interact in resonance conditions 14.38 and 14.39, and one of the waves, increasing precritical wave, is characteristic for the third resonance type (see condition 14.45).

In Figure 14.11, the dependence of the growth increment on the frequency ω_α for the parametric-resonance interaction determined by condition 14.39 is presented. It corresponds to

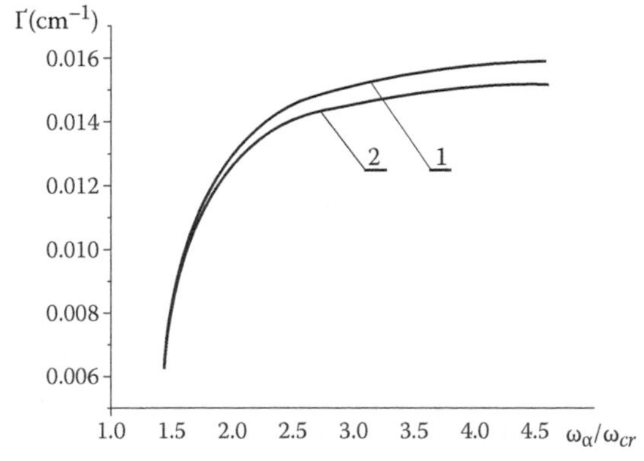

FIGURE 14.11
Dependencies of the gain factors on the normalized coordinate $\omega_\alpha/\omega_{cr}$ for the interaction types 1 and 3. Here, curve 1 corresponds to the interaction with the set of parameters: $\sigma_\alpha = \sigma_\sigma = -\sigma_\gamma = +1$; $r_{\chi\alpha} = -r_{\chi\beta} = -r_{\chi\gamma} = +1$; $\nu = -1$; $\omega_\beta = 5\omega_{cr}$. Curve 2 corresponds to the interaction with parameters: $\sigma_\alpha = \sigma_\sigma = -\sigma_\gamma = -1$; $r_{\chi\alpha} = -r_{\chi\beta} = -r_{\chi\gamma} = -1$; $\nu = -1$; $\omega_\beta = 5\omega_{cr}$. The plasma frequency of each partial beam is $\omega_p = 6.0\ 10^{10}\ s^{-1}$; the averaged relativistic factor of the two-beam system $\gamma_0 = 3.348$; the difference of relativistic factors $\Delta\gamma = 0.1$; the critical frequency $\omega_{cr} = 9.5\ 10^{12}\ s^{-1}$; the amplitude of the pumping wave $E_\gamma = 10^5$ V/m.

a case when the explosive instability condition is satisfied in the considered system. As is readily seen, the increment Γ increases with the increase in the frequency ω_α. One should agree that it is not a common phenomenon for electrodynamics (compared with similar results in Chapters 11 through 13).

Another type of the investigated dependence for the case one increasing + two overcritical is presented in Figure 14.12 (see resonance condition 14.45). In this case, the peculiar superheterodyne version of the explosive instability is realized. It is characterized by an increase in the precritical wave γ (pumping) which takes part actively in the three-wave parametrical resonance. What is very interesting in this case is that the growth increment is much higher than in a case of pure explosive instability (compare the values of the increments in Figures 14.11 and 14.12). The most interesting is the maximum, which occurs at the point $\omega_\alpha = 0.6\omega_{cr}$.

The other discussed types of the resonance interaction might be analyzed analogously.

14.3.7 Short Conclusions to the Section

Thus, the performed analysis revealed that, from the applied point of view, the effect of the three-wave parametric resonance in a two-stream system might be evaluated in two ways. On one side, there appears clearly a potential inclination of the system for a simultaneous realization of many resonance interaction schemes, which cannot be welcomed in an experiment. On the other side, disregarding this, it gives a perspective of an alternate way for using the two-stream instability for creating the SFEL amplifiers for the cluster systems. Therefore, those other possible ways appear often even more complex and difficult for practical solution.

Considering this, let us recall that in the case of the ordinary TSFEL (Chapter 13), the possibility for moving up in frequency is related with two technological possibilities. The first of them consists of using the very dense high-current relativistic beams and the second consists of a decrease in the difference δ′ for the partial beam velocities. A simple project analysis shows that technologically, both of these possibilities appear far from unapproachable. Let us analyze some of the appearing problems in detail.

First, notice that the primary requirements (high current of the beam and its density) limit significantly the application framework of the ordinary TSFEL (Chapter 13). Basically,

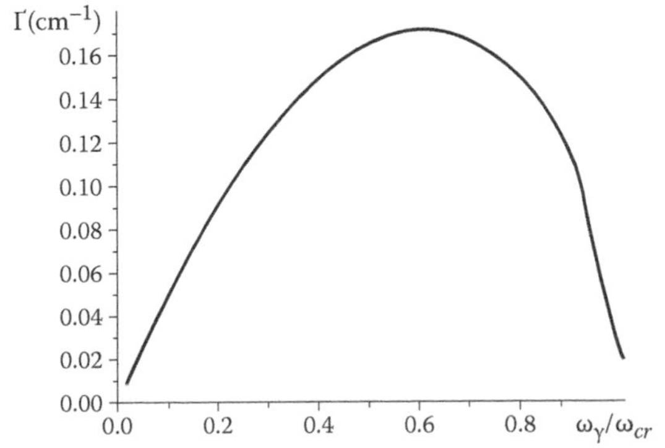

FIGURE 14.12
Dependencies of the gain factors on the normalized coordinate $\omega_\gamma/\omega_{cr}$. Here, the wave γ is the electron-wave pumping; $\nu = +1$ (the signal wave α has the maximal frequency); $\sigma_\alpha = -\sigma_\beta = +1$; $r_\alpha = -r_\beta = +1$. All other parameters are the same with Figure 14.11.

in this case, the TSFEL should be the high and superhigh powerful. However, by far, it is not always the main of the project's aims. For instance, at the present time, in the tactics of the creation of the cluster systems, it is not clear which of the design schemes, described in Chapter 1, is preferred. Either the basic source of the cluster signals should be performed as an only superpowerful former of clusters (Figures 1.16 and 1.17) or it should be performed as an essentially multichannel system for generation of tens of the parallel simple powerful harmonic signal waves (Figure 1.15). Today, construction-wise, the second method appears clearer. Simultaneously, owing to a series of reasons, it becomes obvious that the future belongs to the combination variants including those structural schemes simultaneously (see, for example, Figure 14.2 and the corresponding discussion).

What concerns the use of the two-velocity beams with a small difference δ' is that, today, we have more questions about it than the answers. Therefore, there is a predominant problem related with the creation of the very high-quality, high-current beams with a very stable velocities' difference δ', which should be exceptionally stable at possible fluctuations of the system. Disregarding the presence of a high number of the proposed projects (Chapter 9), no real trial was performed in this direction. In addition to this, considering the physics, the behavior of the two-stream systems in the working TSFEL in many aspects is still not clear. Therefore, at the very low δ' values, a saturation of the two-stream instability appears very fast. In such a case, the key amplification mechanism is in the form where the superheterodyne effect completely cannot exist. In reality, in this case, the role of the two-stream instability is reduced to the excitation of a significantly rich SCW harmonic's spectra. In this situation, a subsequent amplification of the wave of the electromagnetic signal occurs only by the traditional parametric FEL mechanism. Theoretically, the effect of the swift saturation of the two-stream instability can be overcome, for example, by applying a longitudinal isochronous electric field (Chapters 7 and 12). However, until today, this type of model is not studied. Therefore, we still do not have an intelligible answer to the question on what the lower technological limitations are for the minimal possible velocities' difference δ'.

Thus, from the project as well as purely physical points of view, appearance of a model of some kind of simple powerful amplifier, which contains a source of a moderately high-current beam and also is significantly compact and effective, looks very attractive. In this amplifier, an increase in the frequency of the signal wave should not be directly related with the necessity of the formation of the fantastically high-quality dense and high-current beams or any other problems of this kind. It should be noted that, in many important practical situations, the parametrical electron-wave TSFEL (PEWTSFEL), on the background of the above-mentioned ordinary TSFEL, looks just as such simple powerful amplifiers.

By considering everything discussed above, we will swiftly analyze this type of system in the following section.

14.4 Parametrical Electron-Wave Two-Stream Superheterodyne Free Electron Lasers

14.4.1 Theoretical Model

In this section, we will analyze PEWTSFEL (Figure 14.13). We consider that a two-velocity relativistic electron beam, which consists of two partial mutually penetrating electron beams 3 and 4, moves along the *z*-axis of the system. Each of the beams exhibits close value

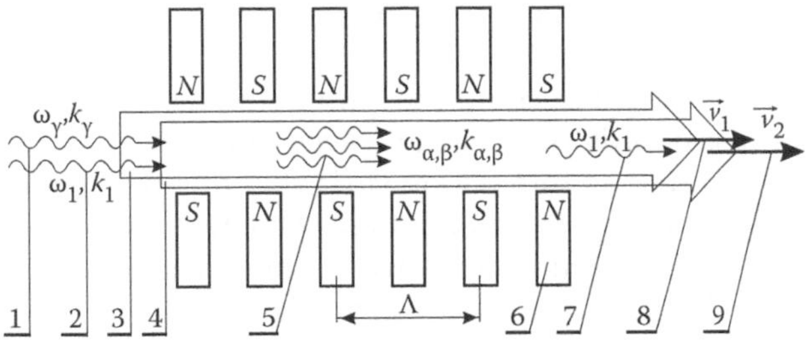

FIGURE 14.13
Theoretical model of the studied parametrical electron-wave superheterodyne free electron lasers (PEWSFEL). Here, 1 is the electron-wave pumping (ω_γ, k_γ); 2 is the input electromagnetic signal (ω_1, k_1); 3 is the first partial electron beam; 4 is the second partial electron beam; 5 are the idle (ω_β, k_β) and signal (ω_α, k_α) electronic waves; 6 is the magnetic undulator; 7 is the output electromagnetic signal (ω_1, k_1); 8 is the velocity vector of the first partial electron beam 3; 9 is the velocity vector of the second partial electron beam 4; $\Lambda = 2\pi/k_2$ is period of the magnetic undulator 6; k_2 is its wave number.

partial relativistic velocities v_1, v_2 so we consider their difference to be very small ($v_1 - v_2 << v_1, v_2$). We consider also the plasma frequencies of the electric beams to be equal ($\omega_{p1} = \omega_{p2} = \omega_p$) and the space charge of the two-velocity beam compensated by the ionic background. In the transverse plane, we consider the beam to be unlimited and uniform.

The two-velocity relativistic electron beam passes through the H-Ubitron magnetic field, which is created by the magnetic undulator 6 with the undulation period Λ (wave number $k_2 = 2\pi/\Lambda$). The magnetic field of the undulator 2 performs the role of the first (magnetic) pumping. The electron beam 1, which is entering the system at the input, is preliminarily modulated in order to excite it in a longitudinal wave of the spatial charge (SCW), with the frequency ω_γ and the wave number k_γ. We will consider this electron wave to be the second (electron wave) pumping. A transverse amplifying electromagnetic signal wave 2 with frequency ω_1 and the wave number k_1 enters the system at the input.

The operation principles of the PEWTSFEL, which is illustrated in Figure 14.13, consist of the following. The device's parameters are chosen in a way that will satisfy the resonance conditions simultaneously for the two three-wave systems. The first one is the resonance three-wave system of the transverse electromagnetic pumping and signal + longitudinal SCW beam, which is traditional for the ordinary parametric FEL (Chapters 11 and 12). The second is one of the resonances of longitudinal SCW, which was studied in the previous section. The key peculiarity of the analyzed system is that one of the longitudinal SCWs is common for both these three-wave systems.

As a result of the parametric-resonance wave interaction in the first three-wave system, the longitudinal SCW, having frequency ω_α and wave number k_α, becomes excited in the electron beam. We will consider this wave as a common for both three wave systems. Functionally, it performs the role of an idle wave in the first one. In contrast, in the second, it becomes the electron-wave signal.

Thus, both of the discussed three-wave systems appear to be resonant. The resonance conditions of the first one may be written in the standard form (Chapters 11 and 12)

$$\omega_\alpha = \omega_1, \quad k_\alpha = k_1 + k_2. \tag{14.47}$$

We assume that the frequencies and the wave numbers of the longitudinal SCW of the second three-wave satisfy the conditions of the three-wave parametric resonance (Equation 14.26)

$$\omega_\alpha + \nu\omega_\beta = \omega_\gamma, \quad k_\alpha + \nu k_\beta = k_\gamma. \tag{14.48}$$

We limit ourselves to the study of a case of the Raman interaction and to the analysis of the resonance interaction case of only overcritical waves (conditions 14.38 and 14.39).

We start with a discussion of an interaction variant with participation of the overcritical SCW only. At $\nu = -1$, as it follows from Equation 14.48, the wave α possesses the highest frequency. We assume that the pumping wave frequency ω_γ is the smallest such that it satisfies condition 14.39. For specificity, we also assume that $\sigma_\gamma = -\sigma_\alpha = -\sigma_\beta = -1$. At $r_\alpha = +1$, $r_\beta = r_\gamma = -1$, the explosive instability is realized for SCWs of the second subsystem (Equation 14.48). In this case the waves α, β, and γ simultaneously increase (Section 14.3). This particularly means that, in such situation, we can expect a maximum of the superheterodyne amplification also for the electromagnetic wave of the signal.

However, in the second system, the given interaction system can also be used differently. For instance, we assume that what is common for both three-wave systems is not the slow SCW α, as we assumed before, but the fast SCW β wave. At a first glance, such system configuration appears somehow paradoxical. Indeed, in the ordinary purely parametrical FEL, as shown in Chapter 11, the fast SCW always appears damping for the three-wave electromagnetic pumping and signal + SCW. However, in the considered superheterodyne model, we can observe its superheterodyne amplification. Hence, in the PEWTSFEL, contrary to the ordinary FEL, the electromagnetic signal can be amplified due to the explosive instability in the second three-wave system. But, at that time, it occurs with a somehow smaller amplification coefficient than in the above-mentioned case of common slow SCW. The rational core in this case consists of something different. In such situation we lose a little considering the magnitude of the superheterodyne amplification, but it is compensated by something different and by far not less interesting for practice.

The point is that, as is well known in the traditional microwave electronics [18–21], the slow SCW always is a source of electromagnetic noises. It is also known that, on the contrary, the fast SCWs are not noisy [18–21]. And finally, as we mentioned in Chapter 1, one of the important requirements for the generators of the cluster signals is the minimization of their level of electromagnetic noises. By considering all this information, it is easy to conclude that the generator should be made in a way that will allow the amplified electromagnetic signal to interact directly with the fast increasing SCW. Specifically such situation is realized and is discussed here in a case of the explosive instability for two SCW α and γ + common for both three-wave fast SCW β.

Basically, a case might be of significant interest also when one of the waves of the second three-wave system is the increasing precritical SCW γ ($\sigma_\gamma = r_{\chi\gamma} = 0$). We assume that in this case, condition 14.45 is fulfilled. For the sake of definiteness, we accept $\sigma_\alpha = -\sigma_\beta = +1$, $\nu = -1$ (the wave α possesses the maximal frequency), $r_\alpha = +1$, $r_\beta = r_\gamma = -1$. An analysis reveals that a peculiar version of a superheterodyne explosive instability is realized in this case. Its basic peculiarity is a parametric-resonance superposition of the two-stream and explosive instabilities. However, further in the analysis, we will limit ourselves with the discussion of the system variant when the second three-wave system is presented only by the overcritical waves (resonance condition 14.39).

14.4.2 Formulation of the Problem

We consider that the wave of the electromagnetic signal, the longitudinal electron waves, and the magnetic field of pumping have a multiharmonic nature

$$\vec{E}_1 = E_1 \vec{e}_x = \sum_{m=1}^{N} \left[E_{1,m} \exp(imp_1) + c.c. \right] \vec{e}_x,$$

$$\vec{B}_1 = B_1 \vec{e}_y = \sum_{m=1}^{N} \left[B_{1,m} \exp(imp_1) + c.c. \right] \vec{e}_y,$$

$$\vec{B}_2 = B_2 \vec{e}_y = \sum_{m=1}^{N} \left[B_{2,m} \exp(imp_2) + c.c. \right] \vec{e}_y,$$

$$\vec{E}_\chi = E_\chi \vec{e}_z = \sum_{m=1}^{N} \left[E_{\chi,m} \exp(imp_\chi) + c.c. \right] \vec{e}_z, \tag{14.49}$$

where the index χ, like before, has the meaning α,β,γ; N is the number of harmonics used in the solution of problems; m is, here and further, the number of the corresponding harmonic; and $p_2 = k_2 z$, $p_{\chi 1} = \omega_{\chi 1} t - k_{\chi 1} z$ ($\chi_1 = \alpha,\beta,\gamma,1$). Thus, the electric and magnetic fields in the working volume of the investigated FEL (Figure 14.13) might be presented as

$$\vec{E} = \vec{E}_1 + \vec{E}_\alpha + \vec{E}_\beta + \vec{E}_\gamma; \quad \vec{B} = \vec{B}_1 + \vec{B}_2. \tag{14.50}$$

The relativistic quasihydrodynamic equation, the equation of continuity, and the Maxwell equation are used as initial system. Apart from that, we chose the quasihydrodynamics equation (Chapter 10) for description of the two-velocity beam motion

$$\left(\frac{\partial}{\partial t} + \vec{v}_q \frac{\partial}{\partial \vec{r}} + \frac{\nu}{\gamma_q^2} \right) \vec{v}_q = \frac{e}{m_e \gamma_q} \left\{ \vec{E} + \frac{1}{c} \left[\vec{v}_q \vec{B} \right] - \frac{\vec{v}_q}{c^2} \left(\vec{v}_q \vec{E} \right) \right\}$$

$$- \frac{\upsilon_T^2}{n_q \gamma_q} \left[\frac{\partial n_q}{\partial \vec{r}} - \frac{\vec{v}_q}{c^2} \left(\vec{v}_q \frac{\partial}{\partial \vec{r}} \right) n_q \right], \tag{14.51}$$

where $\vec{v}_q$ is the vector of velocity of the qth beam component ($q = 1,2$); n_q is the concentration of the particles of the qth beam component; υ_T is the mean-square velocity of the particle thermal motion; $\vec{r}$ is the radius-vector; c is the light velocity in vacuum; and $e = -|e|$ and m_e are the charge and the rest mass of electron, respectively. In this section, we consider that we can disregard the interparticle collision and the thermal electron scattering; it means $\nu = 0$ and $\upsilon_T = 0$.

To solve the problems of the motion and concentration determination (for the given fields), we will use the method of the averaging characteristics (Chapter 6). To solve the problem of the electromagnetic field excitation (for the given motion), we will use the method of slowly changing amplitudes (Chapters 4, 11, and 12).

As the result of the performed calculations, we obtain a system of the truncated differential equations in the cubic-nonlinear approximation for the complex wave amplitudes

$$C_{1,1,m}\frac{dE_{1,m}}{dz}+D_{1,m}E_{1,m}=C^{I}_{1,3,m}E_{\alpha,m}B_{2,m}+F_{1,m},$$

$$C_{\alpha,1,m}\frac{dE_{\alpha,m}}{dz}+D_{\alpha,m}E_{\alpha,m}=C^{I}_{\alpha,3,m}E_{1,m}B^{*}_{2,m}$$

$$+C^{II}_{\alpha,3,m}E_{\beta,m}E_{\gamma,m}+C_{\alpha,4,m}\left\langle E_{\alpha}E^{\text{int}}_{\alpha}\right\rangle_{mp_{\alpha}}+F_{\alpha,m},$$

$$C_{\beta,1,m}\frac{dE_{\beta,m}}{dz}+D_{\beta,m}E_{\beta,m}=C^{II}_{\beta,3,m}E_{\alpha,m}E^{*}_{\gamma,m}+C_{\beta,4,m}\left\langle E_{\beta}E^{\text{int}}_{\beta}\right\rangle_{mp_{\beta}}+F_{\beta,m},$$

$$C_{\gamma,1,m}\frac{dE_{\gamma,m}}{dz}+D_{\gamma,m}E_{\gamma,m}=C^{II}_{\gamma,3,m}E_{\alpha,m}E^{*}_{\beta,m}+C_{\gamma,4,m}\left\langle E_{\gamma}E^{\text{int}}_{\gamma}\right\rangle_{mp_{\gamma}}+F_{\gamma,m}. \tag{14.52}$$

Here,

$$D_{1,m}=\left(m^{2}/c^{2}\right)\left[k_{1}^{2}c^{2}-\omega_{1}^{2}-\sum_{q}\left(\omega^{2}_{p,q}/m^{2}\bar{\gamma}_{q}\right)\right],$$

$$D_{\chi,m}\equiv D(m\omega_{\chi},mk_{\chi})=-i(mk_{\chi})\cdot\left(1-\sum_{q=1,2}\frac{\omega^{2}_{p,q}}{\Omega^{2}_{\chi,q}m^{2}\bar{\gamma}^{3}_{q}}\right),$$

$$C_{1,1,m}=\partial D_{1,m}/\partial\left(-imk_{1}\right),\; C_{\chi,1,m}=\partial D_{\chi,m}/\partial(-imk_{\chi}),$$

$$C^{I}_{1,3,m}=\sum_{q=1,2}\frac{\omega^{2}_{p,q}\omega_{1}e}{2m^{2}\Omega_{\alpha,q}m_{e}\bar{\gamma}^{4}_{q}k_{1}k_{2}c^{3}}\left(\frac{\bar{v}_{q}\bar{\gamma}^{2}_{q}}{c^{2}}-\frac{k_{\alpha}}{\Omega_{\alpha,q}}\right),$$

$$C^{I}_{\alpha,3,m}=\sum_{q=1,2}\left(\frac{\omega^{2}_{p,q}ek_{\alpha}}{m^{2}\Omega^{2}_{\alpha,q}m_{e}\bar{\gamma}^{2}_{q}c^{2}k_{2}}\left(\frac{\bar{v}_{q}}{c}-\frac{k_{\alpha}c}{\omega_{\alpha}}\right)\right),$$

$$C^{II}_{\alpha,3,m}=-k_{\alpha}\sum_{q=1,2}\frac{\omega^{2}_{p,q}e/m_{e}}{\Omega_{\alpha,q}\Omega_{\beta,q}\Omega_{\gamma,q}\bar{\gamma}^{6}_{q}m^{2}}\times\left(\frac{k_{\alpha}}{\Omega_{\alpha,q}}+\frac{k_{\beta}}{\Omega_{\beta,q}}+\frac{k_{\gamma}}{\Omega_{\gamma,q}}-3\bar{v}_{q}\bar{\gamma}^{2}_{q}/c^{2}\right),$$

$$C^{II}_{\beta,3,m}=-k_{\beta}C^{II}_{\alpha,3,m}/k_{\alpha},\; C^{II}_{\gamma,3,m}=-k_{\gamma}C^{II}_{\alpha,3,m}/k_{\alpha},$$

$$C_{\chi,4,m}=\sum_{q=1,2}\frac{3\omega_{p,q}^2 ek_\chi}{im\Omega_{\chi,q}^3 m_e\bar{\gamma}_q^6}\cdot\left(\frac{k_\chi}{\Omega_{\chi,q}}-\frac{\bar{v}\gamma_q^2}{c^2}\right),$$

$$\Omega_{\chi,q}=\omega_\chi-k_\chi\bar{v}_q,\ \bar{\gamma}_q=\frac{1}{\sqrt{1-\left(\bar{v}_q/c\right)^2}},\ \omega_{p,q}^2=\frac{4\pi\bar{n}_q e^2}{m_e}.$$

$$\langle\ldots\rangle_{mp_{\chi 2}}=\frac{1}{(2\pi)^5}\int_0^{2\pi}\left(\ldots\ \exp(-imp_{\chi_2})\mathrm{d}p_1\mathrm{d}p_2\mathrm{d}p_\alpha\mathrm{d}p_\beta\mathrm{d}p_\gamma\right),$$

$$E_{\chi_2}^{\text{int}}=\sum_{m=1}^{N}\left[E_{\chi_2,m}\frac{\exp(imp_{\chi_2})}{im}+c.c.\right],\ (\chi_2=\alpha,\beta,\gamma,1,2).$$

Notice that $D_{\chi,m}$ is a dispersion function of the mth harmonic of the SCW wave. In Equation 14.52, $F_{1,m}$ and $F_{\chi,m}$ are the functions containing cubic-nonlinear terms. This type of function was obtained many times previously in Chapters 11 through 13, where, however, their manifested appearance was not written out, and each time, this was motivated by their enormous volume. Therefore, in relevant cubic-nonlinear calculations, we always used the numerical form for analogous functions. However, this time, for the sake of evident illustration of the above motivation, we make an exception

$$F_{1,m}=-C_{1,2,m}\left(\mathrm{d}^2E_{1,m}/\mathrm{d}z^2\right)+C_{1,5,m}E_{\alpha,m}'^{\text{int}}B_{2,m}+C_{1,6,m}B_{2,m}\left\langle E_\alpha E_\alpha^{\text{int}}\right\rangle_{mp_\alpha}+C_{1,7,m}B_{2,m}\left\langle E_\alpha^{\text{int}}E_\alpha^{\text{int}}\right\rangle_{mp_\alpha}$$
$$+C_{8,1,m}\left\langle E_1E_1^{\text{int}}E_1^{\text{int}}\right\rangle_{mp_1}+C_{1,9,m}B_{2,m}E_{\beta,m}E_{\gamma,m}+\left\langle E_1\sum_{l=1}^{N}\left(C_{1,10,m,l}E_{\alpha,l}B_{2,l}e^{ilp_1}+c.c.\right)\right\rangle_{mp_1}$$
$$+E_{1,m}\sum_{l=1}^{N}\left(C_{1,11,m,l}\left|E_{\alpha,l}\right|^2+C_{1,12,m,l}\left|E_{\beta,l}\right|^2+C_{1,13,m,l}\left|E_{\gamma,l}\right|^2+C_{1,14,m,l}\left|B_{2,l}\right|^2\right),\qquad(14.53)$$

$$F_{\alpha,m}=K_{\alpha,m}+C_{\alpha,8,m}E_{\beta,m}'E_{\gamma,m}+C_{\alpha,9,m}E_{\gamma,m}'E_{\beta,m}+E_{\beta,m}\left(C_{\alpha,10,m}\left\langle E_\gamma^{\text{int}}E_\gamma^{\text{int}}\right\rangle_{mp_\gamma}+C_{\alpha,11,m}\left\langle E_\gamma E_\gamma^{\text{int}}\right\rangle_{mp_\gamma}\right)$$
$$+E_{\gamma,m}\left(C_{\alpha,12,m}\left\langle E_\beta^{\text{int}}E_\beta^{\text{int}}\right\rangle_{mp_\beta}+C_{\alpha,13,m}\left\langle E_\beta E_\beta^{\text{int}}\right\rangle_{mp_\beta}\right)+C_{\alpha,14,m}\left\langle E_\alpha E_\alpha^{\text{int}}E_\alpha^{\text{int}}\right\rangle_{mp_\alpha}+C_{\alpha,15,m}E_{1,m}'B_{2,m}^*$$
$$+\left\langle E_\alpha\sum_{l=1}^{N}\left(C_{\alpha,16,m,l}E_{\beta,l}E_{\gamma,l}e^{ilp_\alpha}+c.c.\right)\right\rangle_{mp_\alpha}+E_{\alpha,m}\sum_{l=1}^{N}\left(C_{\alpha,17,m,l}\left|E_{1,l}\right|^2+C_{\alpha,18,m,l}\left|E_{\beta,l}\right|^2\right.$$
$$\left.+C_{\alpha,19,m,l}\left|E_{\gamma,l}\right|^2+C_{\alpha,20,m,l}\left|B_{2,l}\right|^2\right)+\left\langle E_\alpha\sum_{l=1}^{N}\left(C_{\alpha,21,m,l}E_{1,l}B_{2,l}^*e^{ilp_\alpha}+c.c.\right)\right\rangle_{mp_\alpha},\qquad(14.54)$$

$$F_{\beta,m} = K_{\beta,m} + C_{\beta,8,m}E'_{\alpha,m}E^*_{\gamma,m} + C_{\beta,9,m}E'^*_{\gamma,m}E_{\alpha,m} + E_{\alpha,m}\left(C_{\beta,10,m}\left\langle E^{\text{int}}_{\gamma}E^{\text{int}}_{\gamma}\right\rangle_{-mp_\gamma} + C_{\beta,11,m}\left\langle E_{\gamma}E^{\text{int}}_{\gamma}\right\rangle_{-mp_\gamma}\right)$$

$$+ E^*_{\gamma,m}\left(C_{\beta,12,m}\left\langle E^{\text{int}}_{\alpha}E^{\text{int}}_{\alpha}\right\rangle_{mp_\alpha} + C_{\beta,13,m}\left\langle E_{\alpha}E^{\text{int}}_{\alpha}\right\rangle_{mp_\alpha}\right) + C_{\beta,14,m}\left\langle E_{\beta}E^{\text{int}}_{\beta}E^{\text{int}}_{\beta}\right\rangle_{mp_\beta} + C_{\beta,15,m}E^*_{\gamma,m}E_{1,m}B^*_{2,m}$$

$$+\left\langle E_{\beta}\sum_{l=1}^{N}\left(C_{\beta,16,m,l}E_{\alpha,l}E^*_{\gamma,l}e^{ilp_\beta} + c.c.\right)\right\rangle_{mp_\beta} + E_{\beta,m}\sum_{l=1}^{N}\left(C_{\beta,17,m,l}\left|E_{1,l}\right|^2 + C_{\beta,18,m,l}\left|E_{\alpha,l}\right|^2\right.$$

$$\left. + C_{\beta,19,m,l}\left|E_{\gamma,l}\right|^2 + C_{\beta,20,m,l}\left|B_{2,l}\right|^2\right), \tag{14.55}$$

$$F_{\gamma,m} = K_{\gamma,m} + C_{\gamma,8,m}E'_{\alpha,m}E^*_{\beta,m} + C_{\gamma,9,m}E'^*_{\beta,m}E_{\alpha,m} + E_{\alpha,m}\left(C_{\gamma,10,m}\left\langle E^{\text{int}}_{\beta}E^{\text{int}}_{\beta}\right\rangle_{-mp_\beta} + C_{\gamma,11,m}\left\langle E_{\beta}E^{\text{int}}_{\beta}\right\rangle_{-mp_\beta}\right)$$

$$+ E^*_{\beta,m}\left(C_{\gamma,12,m}\left\langle E^{\text{int}}_{\alpha}E^{\text{int}}_{\alpha}\right\rangle_{mp_\alpha} + C_{\gamma,13,m}\left\langle E_{\alpha}E^{\text{int}}_{\alpha}\right\rangle_{mp_\alpha}\right) + C_{\gamma,14,m}\left\langle E_{\gamma}E^{\text{int}}_{\gamma}E^{\text{int}}_{\gamma}\right\rangle_{mp_\gamma} + C_{\gamma,15,m}E^*_{\beta,m}E_{1,m}B^*_{2,m}$$

$$+\left\langle E_{\gamma}\sum_{l=1}^{N}\left(C_{\gamma,16,m,l}E_{\alpha,l}E^*_{\beta,l}e^{ilp_\gamma} + c.c.\right)\right\rangle_{mp_\gamma} + E_{\gamma,m}\sum_{l=1}^{N}\left(C_{\gamma,17,m,l}\left|E_{1,l}\right|^2 + C_{\gamma,18,m,l}\left|E_{\alpha,l}\right|^2\right.$$

$$\left. + C_{\gamma,18,m,l}\left|E_{\beta,l}\right|^2 + C_{\gamma,19,m,l}\left|B_{2,l}\right|^2\right). \tag{14.56}$$

In Equations 14.53 through 14.56, we use the notation

$$K_{\chi,m} = -C_{\chi,2,m}\left(d^2E_{\chi,m}/dz^2\right) + C_{\chi,5,m}\left\langle E'_{\chi}E^{\text{int}}_{\chi}\right\rangle_{mp_\chi} + C_{\chi,6,m}\left\langle E_{\chi}E'^{\,\text{int}}_{\chi}\right\rangle_{mp_\chi} + C_{\chi,7,m}\left\langle E^{\text{int}}_{\chi}E'^{\,\text{int}}_{\chi}\right\rangle_{mp_\chi};$$

$$E'_{\chi_1} = \sum_{m=1}^{N}\left[\frac{dE_{\chi_1,m}}{dz}\exp(imp_{\chi 1}) + c.c.\right], (\chi_1 = \alpha,\beta,\gamma,1).$$

The coefficient C depends on the wave numbers, frequencies, constant components of beam velocity $\bar{v}_q$, and beam concentration $\bar{n}_q$. We supplement the system (Equation 14.52) by equations for their slowly varying components

$$\frac{d\bar{v}_q}{dz} = V_{1,q}\left\langle E_1' \, E_1 \right\rangle_0 + \sum_{l=1}^{N}\left(V_{2,q,l}E_{\alpha,l}E_{\beta,l}^{*}E_{\gamma,l}^{*} + c.c.\right)$$

$$+\sum_{l=1}^{N}\left(V_{3,q,l}E_{\alpha,l}E_{1,l}^{*}B_{2,l}^{*} + c.c.\right) + \sum_{\chi}^{\alpha,\beta,\gamma}\left(V_{4,q,\chi}\left\langle E_\chi' \, E_\chi \right\rangle_0\right.$$

$$\left. + V_{5,q,\chi}\left\langle E_\chi E_\chi^{\mathrm{int}} E_\chi^{\mathrm{int}} \right\rangle_0 + V_{6,q,\chi}\left\langle E_\chi E_\chi E_\chi^{\mathrm{int}} \right\rangle_0\right). \tag{14.57}$$

$$\frac{d\bar{n}_q}{dz} = N_{1,q}\left\langle E_1' \, E_1 \right\rangle_0 + \sum_{l=1}^{N}\left(N_{2,q,l}E_{\alpha,l}E_{\beta,l}^{*}E_{\gamma,l}^{*} + c.c.\right)$$

$$+\sum_{l=1}^{N}\left(N_{3,q,l}E_{\alpha,l}E_{1,l}^{*}B_{2,l} + c.c.\right) + \sum_{\chi}^{\alpha,\beta,\gamma}\left(N_{4,q,\chi}\left\langle E_\chi' \, E_\chi \right\rangle_0\right.$$

$$\left. + N_{5,q,\chi}\left\langle E_\chi E_\chi^{\mathrm{int}} E_\chi^{\mathrm{int}} \right\rangle_0 + N_{6,q,\chi}\left\langle E_\chi E_\chi E_\chi^{\mathrm{int}} \right\rangle_0\right). \tag{14.58}$$

The coefficients V and N depend on the wave numbers, frequencies, slowly varying components of the velocity $\bar{v}_q$, and the partial beam concentrations $\bar{n}_q$. Unfortunately, as before in other similar situations, we are stressed to omit writing out their evident analytical forms.

14.4.3 Analysis

It should be mentioned that in the theory of the FELs, the most interesting (applied) results were obtained, as a rule, in the cubic-nonlinear approximation. By using the same scheme, we will also build our cubic-nonlinear analysis. The only exclusion we will make is for the growth increments, which describe the wave dynamics in a direct vicinity of the system's input. For instance, for a case when all three waves of the second three-wave systems are overcritical, it is easy to obtain

$$\Gamma' = \sqrt{-\frac{C_{\alpha,3,1}^{II}C_{\beta,3,1}^{II}}{C_{\alpha,1,1}C_{\beta,1,1}}\cdot\left|E_{\gamma,1}\right|^2 + \frac{C_{\alpha,3,1}^{I}C_{1,3,1}^{I}}{C_{1,1,1}C_{\alpha,1,1}}\cdot\left|B_{2,1}\right|^2}\,. \tag{14.59}$$

It was assumed that the wave β, which previously was considered as the idle wave, appears to be common for both the three-wave systems. As we mentioned before in the description of the model, we assume that for the second three-wave system, the explosive instability is realized. In the other case, when the slow SCW α is the common wave, the equation for the increment like Equation 14.59 has a similar mathematical structure. In such a situation, the only important difference appears: in this case, we should change the sign of the first term under the square root in Equation 14.59 from "−" into "+."

The parameters of the system were chosen in such a way that the second term under the square root in Equation 14.59 is always significantly higher than the first one.

Furthermore, we perform an analysis of the nonlinear dynamics of the wave processes in the investigated PEWTSFEL. We consider the system of the nonlinear equations 14.52 through 14.58 as the basic equations and use the standard numerical methods.

Figure 14.14 shows the dependence of the normalized amplitude of the first harmonic of the electromagnetic wave of the signal $E_{1,1}$ on the normalized length $T = z/L$ ($L = 300$ cm) in PEWTSFEL (curve 1) in comparison with the analogous dependence for the equivalent traditional parametrical FEL (curve 2). In this case, the model of ordinary parametrical FEL, with one-velocity electron beam of the same current and energy (Chapters 11 and 12), is used for the comparison. We also chose the same values of the parameters of the pumping magnetic field and the input electromagnetic signal in both models.

As we see it, the saturation levels of the signal electromagnetic wave in PEWTSFEL and the ordinary equivalent FEL appear to be approximately equal. However, the saturation length of the electron-wave FEL T_{s1} is almost four times shorter than in the case of the equivalent parametric FEL T_{s2} (Figure 14.14). Thus, as was expected, the electron-wave FEL at the normalized length $T \leq 0.82$ possesses incommensurably higher amplification than the equivalent ordinary FEL.

Figure 14.15 illustrates the dynamics of the longitudinal three SCWs with the parametric-resonance interaction (explosive instability). It illustrates the dependence of the first harmonics of the idle wave SCW $E_{\beta,1}$ (curve 1), the signal SCW $E_{\alpha,1}$ (curve 2), and electron-wave pumping $E_{\gamma,1}$ (curve 3) on the normalized system's length T. In this case, the idle wave β is the common wave for both three-wave systems. It is apparent that, because of the presence of the explosive instability, the amplitudes of all three interacting SCWs simultaneously increase. Therefore, by using significantly powerful low-frequency electron-wave pumping γ (curve 3), it is possible to form the strong idle signal wave β (curve 1), through which

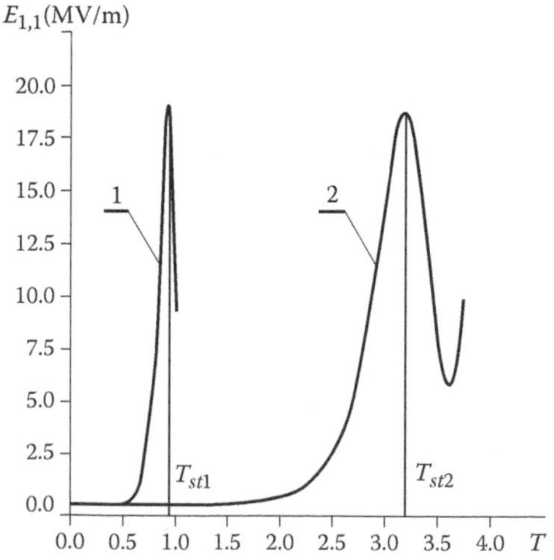

FIGURE 14.14
Dependency of the first harmonic of the electromagnetic signal $E_{1,1}$ on the normalized system length $T = z/L$. Here, curve 1 corresponds to the PEWSFEL and curve 2 describes the equivalent ordinary parametrical FEL. The system parameters: the signal wavelength $\lambda_1 = 0.08$ mm; the pumping undulation period $\Lambda = 5$ mm; the induction of magnetic pumping field $B_{2,1} = 200$ Gs; the partial beam Langmuir frequency $\omega_p = 3.10^{10}\ s^{-1}$; the averaged relativistic factor $\gamma_0 = 6$; the difference of relativistic factors $\Delta\gamma = 0.7$; the system length $L = 3$ m. It is considered the case, when the common wave β is the fast SCW.

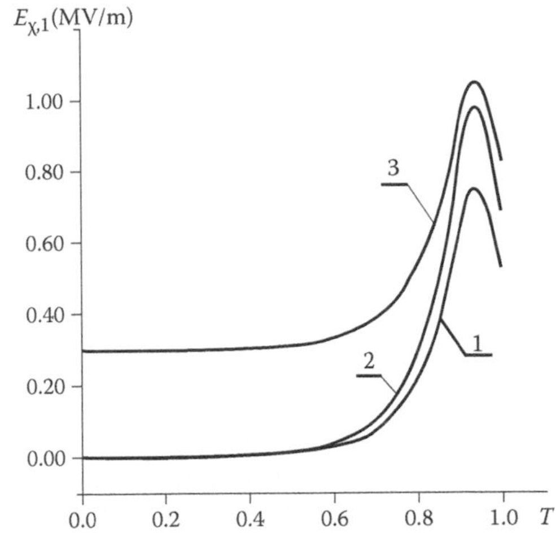

FIGURE 14.15
Dependence of the amplitudes of first SCW harmonics of the second three-wave systems on the normalized system length $T = z/L$ in the PEWTSFEL model. Here, curve 1 corresponds to the amplitude of the common idle SCW $E_{\beta,1}$; curve 2 is related to the amplitude of the signal SCW $E_{\alpha,1}$; and curve 3 describes the electron-wave pumping dynamics $E_{\gamma,1}$. All other parameters are the same with Figure 14.14.

the amplification is transferred from the electron waves to the electromagnetic signal wave $\{\omega_1,k_1\}$. In this process, the level of the amplitude of the wave β appears lower than the level of the amplitude of the signal and pumping waves. This is related with the fact that in the considered model, the amplification of the wave β is determined by two simultaneously competing mechanisms that are characterized by reciprocally opposing dynamics. Specifically, the resonance of the longitudinal SCW in the second three-wave system causes its increase, whereas the action of resonance in the first three-wave system leads to its damping. However, the choice of the parameters of the system is such that the contribution of the second three-wave system to the superheterodyne amplification of the electromagnetic signal wave is significantly higher than that of the first three-wave system.

References

1. Kotsarenko, N.Y., and Kulish, V.V. 1980. Superheterodyne amplification of electromagnetic waves in a beam–plasma system. *Radiotech. Electron. [Radio Eng. Electron. (USSR)]* 25(11):2470–2471.
2. Perekupko, V.A., Silivra, A.A., Kotsarenko, N.Y., and Kulish, V.V. Patent of USSR No. 835259 (cl. H 01 J 25/00). Electronic device. Priority 28.01.80.
3. Kulish, V.V., Lysenko, A.V., and Koval, V.V. 2009. On the theory of the plasma-beam superheterodyne free electron laser with H-ubitron pumping. *Tech. Phys. Lett.* (Russia) 35(8):696–699.
4. Kulish, V.V., Lysenko, A.V., and Koval, V.V. 2007. On the theory of the plasma-beam superheterodyne free electron laser. *Visnyk Sumskoho Universytetu* (Ukraine). *Seria Phys. Math. Mech.* 2:112–119.
5. Kulish, V.V., Lysenko, A.V., and Koval, V.V. 2009. Multi-harmonic cubic-nonlinear theory of the superheterodyne plasma-beam free electron lasers with Dopplertron pumping. *Prikladnaya Fizika* (Russia) 5:76–82.

6. Kulish, V.V., Lysenko, A.V., and Koval, V.V. 2009. On the theory of the plasma-beam superheterodyne free electron laser with H-ubitron pumping. *Tech. Phys. Lett.* (Russia) 35(8):696–699.
7. Kulish, V.V., Lysenko, A.V., and Koval, V.V. 2009. Multi-harmonic cubic-nonlinear theory of the superheterodyne plasma-beam free electron lasers with H-ubitron pumping. *Radiotechnika I Elektronika* (Ukraine) 14(3):383–389.
8. Kulish, V.V. 1991. On the theory of relativistic electron-wave free electron lasers. *Ukr. Fiz. Zh. [Ukr. J. Phys.]* 36(5):682–694.
9. Kulish, V.V., and Storizhko, V.E. Patent of USSR No. 1837722 (cl. H 01 J 25/00). Free electron laser. Priority 15.02.91.
10. Kulish, V.V. 1993. Superheterodyne electron-wave free-electron laser. *Int. J. Infrared Millimeter Waves* 14(3):415–450.
11. Kulish, V.V., Lysenko, A.V., and Rombovsky, M.Yu. 2010. Effect of parametric resonance on the formation of waves with a broad multi-harmonic spectrum during the development of two-stream instability. *Plasma Phys. Rep.* 36(7):594–600.
12. Kulish, V.V., Lysenko, A.V., and Koval, V.V. 2010. Multi-harmonic cubic-nonlinear theory of the superheterodyne plasma-beam free electron lasers with Dopplertron pumping. *Plasma Phys. Rep.* 36(13):1185–1190.
13. Kulish, V.V., Lysenko, A.V., and Rombovsky, M.Yu. 2010. Effect of parametric resonance on the formation of waves with a broad multi-harmonic spectrum during the development of two-stream instability. *Plasma Phys.* (Russia) 36(7):637–643.
14. Kulish, V.V., Lysenko, A.V., and Rombovsky, M.Yu. 2005. Theory of the two-stream electron-wave free electron lasers with H-Ubitron pumping. *Visnyk Sums'kogo Unoversytetu* (Ukraine). *Seria Phys. Math. Mech.* 4:58–70.
15. Kulish, V.V., Lysenko, A.V., and Rombovsky, M.Yu. 2009. Parametrical resonance for the two-velocity beam waves. *Appl. Phys.* (Russia) 1:71–78.
16. Kulish, V.V., Lysenko, A.V., and Rombovsky, M.Yu. 2010, The cubic-nonlinear of electron-wave two-stream free electron lasers with the H-ubitron pumping. *VANT* (Ukraine). *Nucl. Phys. Res.* 54(3):C78–82.
17. Kulish, V.V. 2002. *Hierarchic methods. Undulative electrodynamic systems*, Vol. 2. Dordrecht: Kluwer Academic Publishers.
18. Lebedev, I.V. 1964. *Microwave technology and devices*, Vol. II. Moscow: Energiya.
19. Haiduk, V.I., Palatov, K.I., and Petrov, D.N. 1971. *Physical principles of microwave electronics*. Moscow: Sovetskoye Radio.
20. Lopoukhin, V.M. 1953. *Excitation of electromagnetic oscillations and waves by electron beams*. Moscow: Gostekhizdat.
21. Trubetskov, D.I., and Khramov, A.E. 2003. *Lectures on the microwave electronics for physicists*, In two volumes. Moscow: Fizmatlit.
22. Kulish, V.V., Lysenko, A.V., and Koval, V.V. 2010. Multi-harmonic cubic-nonlinear theory of the superheterodyne plasma-beam free electron lasers with Dopplertron pumping. *Plasma Phys. Rep.* 36(13):1185–1190.

Index

F

G

H

M

N

O

P

U

V

W

X

Z